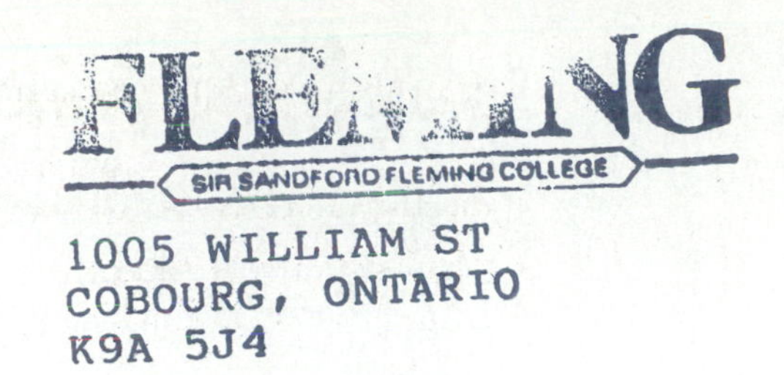

ESSENTIAL MATHEMATICS

A WORKTEXT

THIRD EDITION

D. Franklin Wright
Cerritos College

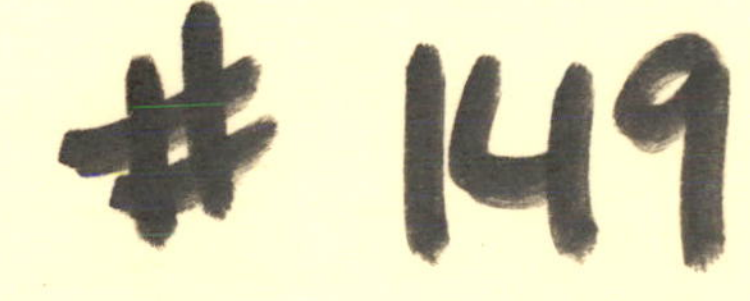

L

Address editorial correspondence to:

D. C. Heath and Company
125 Spring Street
Lexington, MA 02173

Acquisitions Editor: Charles W. Hartford
Developmental Editor: Philip Charles Lanza
Production Editor: Melissa Ray
Designer: Judith Miller
Photo Researcher: Lauretta Surprenant
Art Editor: Diane B. Grossman
Production Coordinator: Lisa Merrill

Photograph credits; p. 2, The Image Works/M. Greenlar; p. 88, NASA; p. 150, PhotoEdit/Richard Hutchings; p. 224, The Image Works/H. Gans; p. 288, Tony Stone Images/John Garrett; p. 376, PhotoEdit/Tony Freeman; p. 422, PhotoEdit/Robert Brenner; p. 486, Tony Stone Images; p. 524, The Image Works; p. M2, Tony Stone Images; p. G2, Photo provided by Art Resource; p. T2, Tony Stone Images/David Hijer

Published simultaneously in Canada.

Printed in the United States of America.

International Standard Book Number: 0–669–35287-X

10 9 8 7 6 5 4 3 2

Preface

Purpose and Style

The purpose of *Essential Mathematics: A Worktext,* Third Edition, is to provide students with a learning tool that will help them develop basic arithmetic skills, develop reasoning and problem-solving skills, and achieve satisfaction in learning so that they will be encouraged to continue their education in mathematics. The style of writing gives carefully worded, thorough explanations that are direct, easy to understand, and mathematically accurate. The use of color, subheadings, and shaded boxes helps students understand and reference important topics.

Students are expected to participate actively by working margin exercises and completion examples as they progress through each section. The answers to the margin exercises are in the back of the text and the answers to the completion examples are at the end of the section. This approach provides the students with immediate reinforcement and the instructor with immediate feedback as to how well the students are understanding the material. Space is provided in the text for students to calculate and write answers on the text pages themselves.

The text's flexible style allows for teaching based on the standard lecture method and for programs based on cooperative learning (small group studies) or independent study in a mathematics laboratory setting.

The NCTM curriculum standards have been taken into consideration in the development of the topics throughout the text.

New Special Features

In each Chapter:

- *Mathematics at Work!* presents a brief discussion related to an upcoming concept from the chapter ahead, and an example of mathematics used in daily life situations.
- *What to Expect in this Chapter* opens each chapter and offers an overview of the topics that will be covered.
- *Learning Objectives* are listed at the beginning of each section.
- *Cumulative Review* exercises appear in each chapter beyond Chapter 1 to provide continuous, cumulative review.

In the Exercise sets:

- *Writing and Thinking about Mathematics* exercises encourage students to express their ideas, interpretations, and understanding in writing.
- *Check Your Number Sense* exercises are designed to help students develop confidence in their judgment, estimation, and mental calculation skills.
- *Collaborative Learning Exercises* are designed to be done in interactive groups.
- *The Recycle Bin* is a skills refresher that provides maintenance exercises on important topics from previous chapters.

Customizing Options

Any of four chapters, **Measurement, Statistics and Reading Graphs, Geometry,** and **Additional Topics from Algebra,** can be included as part of the text. The choice of using any or all of these chapters will allow instructors to better meet the diverse requirements of their courses. Contact your local D. C. Heath Representative for more information about these custom-publishing options.

Pedagogical Features

In each section, the presentation and development are based on the following format for learning and teaching:

1. Each section begins with a list of learning objectives for that section.
2. Subsections are introduced by color subheadings for easy reference.
3. Thorough, mathematically accurate discussions feature several detailed examples completely worked out with step-by-step explanations.
4. Completion examples give students the opportunity to reinforce thinking patterns developed in previous examples. (Answers are provided at the end of the section.)
5. Students are prompted to complete margin exercises as they read the lessons, which provide them with immediate reinforcement of the basic ideas presented. (Answers are provided in the back of the text.)
6. Graded exercises offer variety in their style and level of difficulty including drill, multiple choice, matching, applications, written responses, estimating, and group discussion exercises.
7. Real-life applications are presented in most sections with several sections devoted solely to applications.
8. A new two-color design provides visual support to the text's pedagogy.

Each chapter contains

1. What to Expect in this Chapter
2. Chapter-opening Mathematics at Work!
3. An Index of Key Ideas and Terms with page references
4. A Chapter Test
5. A Cumulative Review of topics discussed in that chapter and previous chapters (beginning with Chapter 2)

Important Topics and Ideas

- In Chapter 1, vertical lines have been placed in long division exercises to help students get in the habit of keeping numbers aligned properly.
- Emphasis has been placed on the development of reading and writing skills as they relate to mathematics.
- The exercises are designed to be motivating and interesting. Applications are varied and practical and contain many facts of interest.
- The use of calculators is encouraged beginning with Chapter 5. A new section on scientific notation has been added to help students understand this type of display on their calculators.

- Estimating is an integral part of many discussions throughout the text.
- Geometric concepts such as finding perimeter and area and recognizing geometric figures are integrated throughout the text.

Content

There is sufficient material for a three- or four-semester-hour course. The topics in Chapters 1–7 form the core material for a three-hour course. The topics in Chapters 8 and 9 provide for additional flexibility in the course depending on students' background and the goals of the course.

Chapter 1, Whole Numbers, reviews the fundamental operations of addition, subtraction, multiplication, and division with whole numbers. Estimation is used to develop a better understanding of whole number concepts, and word problems help to reinforce the need for these ideas and skills in common situations such as finding averages and making purchases.

Chapter 2, Prime Numbers, introduces exponents, shows how to use the rules for order of operations, and defines prime numbers. The concepts of divisibility and factors are emphasized and related to finding prime factorizations, which are, in turn, used to develop skills needed for finding the least common multiple (LCM) of a set of numbers. All of these ideas form the foundation of the development of fractions in Chapter 3.

Chapter 3, Fractions, discusses the operations of multiplication, division, addition, and subtraction with fractions. A special effort is made to demonstrate the validity of the use of improper fractions, and exponents and the rules for order of operations are applied to fractions. Knowledge of prime numbers (and prime factorizations) underlies all of the discussions about fractions.

Chapter 4, Mixed Numbers and Denominate Numbers, shows how mixed numbers are related to whole numbers and fractions. Included are the basic operations, complex fractions, and the rules for order of operations. The last section discusses the difference between abstract numbers and denominate numbers and operations with denominate numbers.

Chapter 5, Decimal Numbers, covers the basic operations with decimal numbers, estimating, and the use of calculators (including scientific notation and finding square roots). To emphasize number concepts, one section presents operating with decimal numbers, fractions, and mixed numbers in a single expression. As an application of decimal numbers and the use of calculators, the Pythagorean Theorem is presented in the final section, and problems are given in finding the length of the hypotenuse.

Chapter 6, Ratios and Proportions, develops an understanding of ratios and introduces the idea of a variable. Techniques for solving equations are developed through finding the unknown term in a proportion.

Chapter 7, Percent (Calculators Recommended), approaches percent as hundredths and uses this idea to find equivalent numbers in the form of percents, decimals, and fractions. Applications are developed around proportions, the formula $R \times B = A$, and the skills of solving equations. A special section on estimating with percent is included to reinforce basic understanding.

Chapter 8, Consumer Applications, addresses the topics of simple interest, compound interest, buying and owning a home and a car. The use of calculators is encouraged and necessary, as in the case of using the formula for compound interest: $A = P(1 + \frac{r}{n})^{nt}$. Calculating with this formula emphasizes the importance of following the rules for order of operations.

Chapter 9, Introduction to Algebra, provides a head start for those students planning to continue in their mathematics studies, either in a prealgebra course or a beginning course in algebra. Integers are introduced with the aid of number lines, and integers are graphed on number lines. Topics included are operations with integers, combining like terms, translating phrases, solving equations, and problem solving.

Customized Chapters

The following chapters are designed to complement the basic text material (Chapters 1–9) to fit any additional curriculum features at your school or any particular syllabus topics desired by instructors. Any one, two, three, or all four of these chapters may be requested by the instructor from the publisher to be included as part of the text in the order shown.

Measurement: This chapter discusses the metric system of measurement, the U.S. customary system of measurement, and how to change units of measure within each system and between the systems.

Statistics and Reading Graphs: This chapter includes the concepts of mean, median, and range and how to read various types of graphs: circle graphs, bar graphs, line graphs, pictographs, histograms, and frequency polygons.

Geometry: This chapter provides coverage of geometry beyond that which is integrated in the previous chapters. Included are length, perimeter, area, volume, and formulas related to specific figures. Also included is a discussion of types of angles and types of triangles (along with similar triangles).

Additional Topics from Algebra: This chapter introduces some topics that will be found in more detail in a beginning algebra course. Included are solving inequalities, graphing in two dimensions (ordered pairs and straight lines), and operating with polynomials.

Some Possible Course Offerings

Short Course (Chapters 1–7)	Longer Course (Chapters 1–9)	Optional Course (Chapters 2–9 with Customized Chapters.)
Whole Numbers	Whole Numbers	Prime Numbers
Prime Numbers	Prime Numbers	Fractions
Fractions	Fractions	Mixed Numbers and Denominate Numbers
Mixed Numbers and Denominate Numbers	Mixed Numbers and Denominate Numbers	Decimal Numbers
Decimal Numbers	Decimal Numbers	Ratios and Proportions
Ratios and Proportions	Ratios and Proportions	Percent (Calculators Recommended)
Percent (Calculators Recommended)	Percent (Calculators Recommended)	Consumer Applications
	Consumer Applications	Introduction to Algebra
	Introduction to Algebra	Selected topics from Measurement, Statistics and Reading Graphs, Geometry, and Additional Topics from Algebra

Practice and Review

There are more than 4100 exercises, carefully chosen and graded, proceeding from easy exercises to more difficult ones, plus more than 400 margin exercises and two or three completion examples in most sections. Many sections contain a feature entitled The Recycle Bin which generally has six to ten review exercises from previous chapters. Each chapter includes a Chapter Test and a Cumulative Review (after Chapter 1). The tests are similar in length and content to the tests provided in the *Instructor's Resource Manual.*

Many sections have exercises entitled Writing and Thinking about Mathematics, Check Your Number Sense, and Collaborative Learning Exercises. These exercises are an important part of the text and provide a chance for each student to improve communication skills, develop an understanding of general concepts, and communicate his or her ideas to the instructor. Written responses can be a great help to the instructor in understanding just what students do and do not understand.

Many of these questions are designed for the student to investigate ideas other than those presented in the text, with responses that are to be based on each student's own experiences and perceptions. In most cases there is no one right answer.

Answers to the odd-numbered exercises, all Recycle Bin questions, all Chapter Test questions, all Cumulative Review questions, and all margin exercises are provided in the back of the book. The answers to the completion examples are given at the end of each section.

Supplements

Essential Mathematics: A Worktext, Third Edition, is accompanied by a comprehensive supplement support package, with each item designed and created to provide maximum benefit to the students and instructors who use them.

Instructor's Annotated Edition: This is a special version of the text that provides in-place answers to all exercises, tests, and review items. It also includes extensive margin annotations to the instructor from the author.

Instructor's Resource Manual: This manual includes several prepared forms of chapter quizzes and tests, prepared forms of cumulative tests, and answers for all quizzes and tests.

Computerized Testing: The *Quisitor* computerized test item file is available in both IBM-PC and Macintosh versions. Instructors may edit or add questions and compose customized tests in multiple choice, free response, or mixed formats. Questions can be selected randomly or according to other specific criteria chosen by the insructor.

Student Study Guide: This supplement has been prepared to provide students with additional practice and review questions. Answers to all items are included in the guide, along with solutions to selected items that illustrate the working of all key problem types.

Student Solutions Manual: Worked-out solutions to odd section exercises and all Chapter Test and Cumulative Review items are included in this manual.

Videotapes: A series of videotapes, covering all major topics, provide concept review and additional examples to reinforce the text presentation.

Tutorial Software: A new student tutorial program is available in both IBM-PC and Macintosh platforms. The package parallels the text development and offers hundreds of problems for additional drill and practice. The tutorial portion provides interactive feedback as students proceed through concept review and examples.

Acknowledgments

I would like to thank Philip Lanza, developmental editor, and Melissa Ray, production editor, for their hard work and invaluable assistance in the development of this text. Again, Phil has contributed so much and taken such personal interest in the project that he has become my hidden "coauthor." Melissa really does understand that mathematics textbooks have special requirements in style and format and did not hesitate to follow up on many of my "suggestions." I feel that D. C. Heath provides its authors with the best editorial assistance possible.

Many thanks go to the following manuscript reviewers who offered their constructive and critical comments: Lenore Frank at SUNY—Stony Brook; Lionel Geller at Dawson College in Montreal; Sylvia Kennedy at Manatee Junior College; David Longshore at Victor Valley College; Gael Mericle at Mankato State University; William Naegele at South Suburban College; Lois Norris at Northern Virginia Community College; Susan Novelli at Kellogg Community College; and Ellen Van Driel at Iowa Lakes Community College. Additional thanks go to Sylvia Kennedy for her accuracy review of the complete answer key.

Finally, special thanks go to Charles Hartford, mathematics editor, for his faith in this edition and his willingness to commit so many resources throughout the development and production process to guarantee a top-quality product for students and teachers.

D. Franklin Wright

Contents

1

Whole Numbers

What to Expect in Chapter 1

Chapter 1 provides a review of the basic operations (addition, subtraction, multiplication, and division) with whole numbers. Section 1.1 discusses number systems in general, with emphasis on reading and writing whole numbers in the decimal system. The skills related to rounding off and estimating with whole numbers are developed in Section 1.3. These and other general number concepts are an integral part of the text and the exercises and are emphasized in a feature entitled **Check Your Number Sense.**

Section 1.6 (Problem Solving with Whole Numbers) concludes Chapter 1. A four-step process for solving word problems, developed by George Pōlya, a famous educator and mathematician from Stanford University, is outlined in this section and is reinforced throughout the text, since it is basic to problem solving of all types.

Mathematics at Work!

View of the main quad at the College of Arts and Sciences, Cornell University.

Whole Numbers are everywhere. Businesspeople, carpenters, auto mechanics, teachers, and bus drivers all deal with whole numbers on a daily basis, as do many other people.

As a student, you may be interested in information about college costs. The following list shows tuition, room, and board expenses for 15 well-known colleges and universities. Costs are given as annual figures (rounded to the nearest ten and subject to change) for the year 1992. (**Note:** If you want this type of information for any college or university, write to the registrar of that school.)

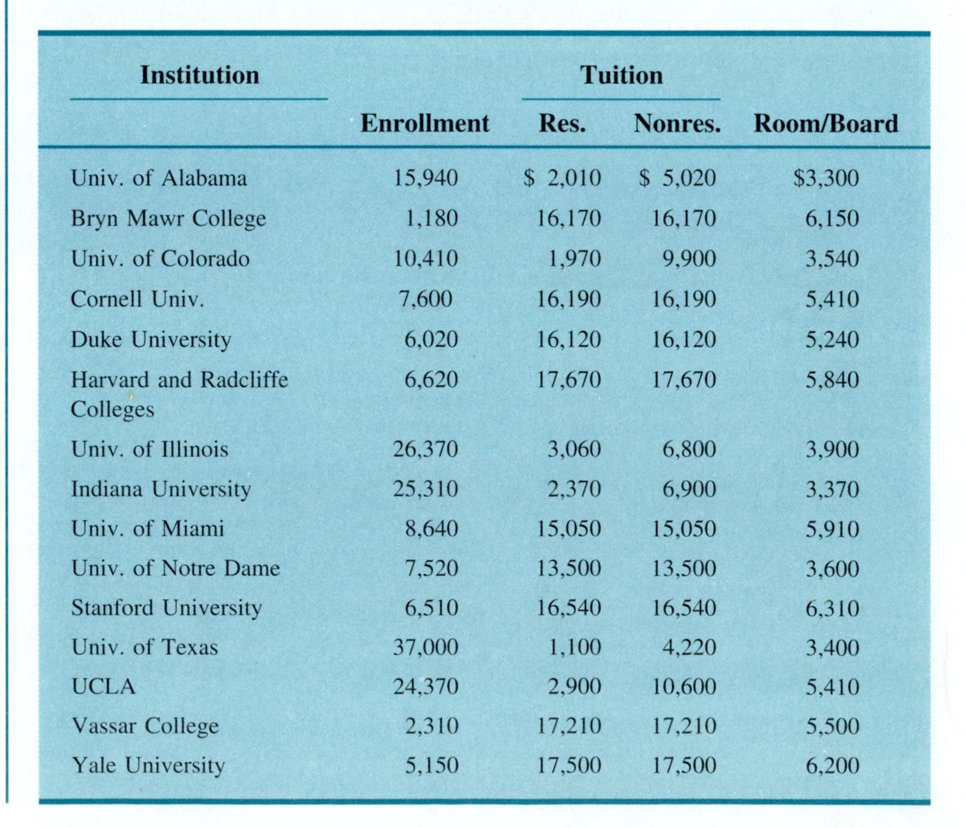

Institution		Tuition		
	Enrollment	Res.	Nonres.	Room/Board
Univ. of Alabama	15,940	$ 2,010	$ 5,020	$3,300
Bryn Mawr College	1,180	16,170	16,170	6,150
Univ. of Colorado	10,410	1,970	9,900	3,540
Cornell Univ.	7,600	16,190	16,190	5,410
Duke University	6,020	16,120	16,120	5,240
Harvard and Radcliffe Colleges	6,620	17,670	17,670	5,840
Univ. of Illinois	26,370	3,060	6,800	3,900
Indiana University	25,310	2,370	6,900	3,370
Univ. of Miami	8,640	15,050	15,050	5,910
Univ. of Notre Dame	7,520	13,500	13,500	3,600
Stanford University	6,510	16,540	16,540	6,310
Univ. of Texas	37,000	1,100	4,220	3,400
UCLA	24,370	2,900	10,600	5,410
Vassar College	2,310	17,210	17,210	5,500
Yale University	5,150	17,500	17,500	6,200

For each of these universities and colleges, (a) find the yearly expenses that a state resident might expect to pay for tuition and room and board, (b) find the yearly expenses that an out-of-state student (nonresident) might expect to pay, (c) determine which of these universities and colleges is the least expensive and which is the most expensive for tuition and room and board for residents, and (d) determine which of these universities and colleges is the least expensive and which is the most expensive for tuition and room and board for nonresidents. (See also Section 1.6, Exercise 33.)

1.1 The Decimal System

OBJECTIVES

1. *Understand the concept of a place value number system.*
2. *Know the powers of ten.*
3. *Know the values of the places for whole numbers in the decimal system.*
4. *Read and write whole numbers in the decimal system.*

The Decimal System

The abstract concept of numbers and the numeration systems used to represent numbers have been important, indeed indispensable, parts of intelligent human development. The Hindu-Arabic system of numeration that we use today was invented about 800 A.D. This system, called the **decimal system** (**deci** means **ten** in Latin), allows us to add, subtract, multiply, and divide faster and more easily than did any of the ancient number systems, such as Egyptian hieroglyphics and the Roman numeral system. The decimal system is more sophisticated than these earlier systems in that it uses a symbol for zero (0), the operation of multiplication, and the concept of place value. The symbols used are called **digits,** and for the **whole numbers** the value of a digit depends on its position to the left of a beginning point, called a **decimal point.** (See Figure 1.1.)

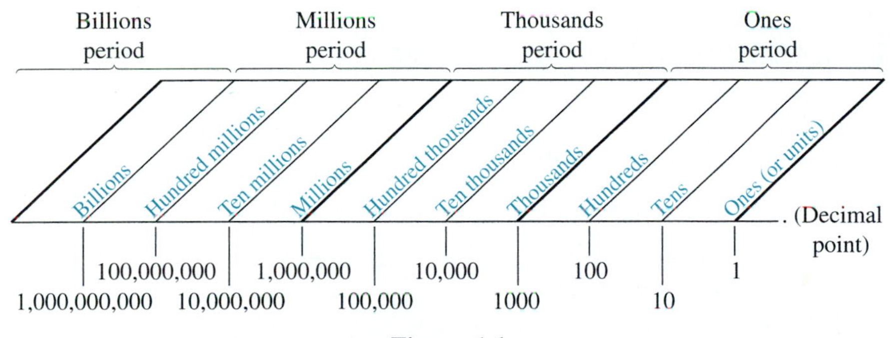

Figure 1.1

The decimal system (or base ten system) is a **place value system** that depends on three things:

1. The **ten digits:** 0, 1, 2, 3, 4, 5, 6, 7, 8, 9;
2. The **placement** of each digit; and
3. The **value** of each place.

1. Write 4635 in expanded notation.

2. Write 73,428 in expanded notation.

3. Write $500 + 20 + 1$ in standard notation.

Each place to the left of the decimal point has a **power of ten** as its place value. The powers of ten are:

1 10 100 1000 10,000 100,000 1,000,000 and so on.

You should **memorize** the material in Figure 1.1.

In **standard notation,** digits are written in places, and the value of each digit is to be multiplied by the value of the place. The value of the number is found by adding the results of multiplying the digits and place values. In **expanded notation,** the values represented by each digit in standard notation are written as a sum. The English word equivalents can then be easily read (or written) from these indicated sums. Examples 1–4 illustrate these ideas.

EXAMPLE 1

954 (standard notation)

9 5 4 ← digits (standard notation)
100 10 1 ← place values

$954 = 9(100) + 5(10) + 4(1) = 900 + 50 + 4$ (expanded notation)

EXAMPLE 2

6507 (standard notation)

6 5 0 7 ← digits (standard notation)
1000 100 10 1 ← place values

$6507 = 6(1000) + 5(100) + 0(10) + 7(1)$

$= 6000 + 500 + 0 + 7$ (expanded notation)

✓ COMPLETION EXAMPLE 3

Complete the expanded notation form of 532,081.

$532,081 = 500,000 + 30,000 +$ ________ $+$

________ $+$ ________ $+ 1$

✓ COMPLETION EXAMPLE 4

Complete the expanded notation form of 497,500.

$497,500 = 400,000 +$ ________ $+$ ________ $+$ ________ $+$

$0 + 0$

Now Work Exercises 1–3 in the Margin.

Reading and Writing Whole Numbers

Whole numbers are those numbers used for counting and the number 0. They are the decimal numbers with digits written to the left of the decimal point and with only 0's (or no digits at all) written to the right of the decimal point. (Decimal numbers with digits to the right of the decimal point will be discussed in Chapter 5.) We use the capital letter ***W*** to represent the set of all whole numbers.

Definition

The **whole numbers** are the counting numbers and the number 0.

$W = \{0, 1, 2, 3, 4, 5, 6, 7, 8, 9, 10, 11, 12, 13, 14, 15, \ldots\}$

The three dots in the definition indicate that the pattern continues without end.

Figure 1.2 shows how large whole numbers are written and read with the digits in groups of three. **Note that in writing decimal numbers in standard form the decimal point itself is optional.**

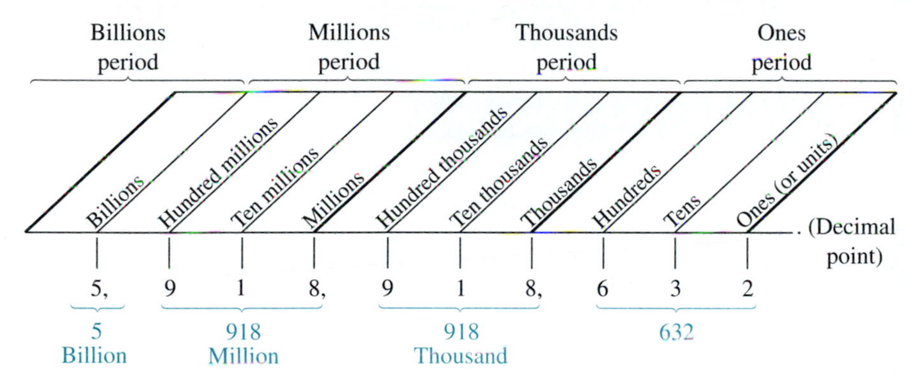

Figure 1.2

We can write whole numbers in standard form, using digits and the place value system, in expanded notation, and lastly in their English word equivalents. For example,

268 standard form

200 + 60 + 8 expanded notation

two hundred sixty-eight English word equivalent

Examples 5–8 illustrate whole numbers written in both expanded notation and English word equivalents.

EXAMPLE 5

653

653 = 600 + 50 + 3 expanded notation

six hundred fifty-three English word equivalent

4. Write 9567 in expanded notation and in its English word equivalent.

5. Write 25,400 in expanded notation and in its English word equivalent.

6. Write 6000 + 700 + 90 + 2 in standard notation and in its English word equivalent.

Example 6

75,900

75,900 = 70,000 + 5000 + 900 + 0 + 0

seventy-five thousand, nine hundred

☑ *Completion Example 7*

8407

8407 = 8000 + _____ + _____ + _____

eight thousand, _____________________

☑ *Completion Example 8*

15,352

15,352 = 10,000 + _____ + _____ + _____ + _____

fifteen thousand, _____________________________

Now Work Exercises 4–6 in the Margin.

You should note the following four things when reading or writing whole numbers:

1. The word **and** does not appear in English word equivalents. **And** is said only when reading a decimal point. (See Chapter 5.)
2. Digits are read in **groups of three.** (See Figure 1.2.)
3. Commas are used to separate groups of three digits **if a number has more than four digits.**
4. Hyphens (-) are used to write English words for the two-digit numbers from 21 to 99 except for those that end in 0.

Example 9

Write four hundred eighty thousand, five hundred thirty-three in standard notation.

480,533

Example 10

Write five hundred seventy-two thousand, six hundred in standard notation.

572,600

✔ COMPLETION EXAMPLE 11

Write two million, eight hundred thousand, thirty-five in standard notation.

2,800, ______

✔ COMPLETION EXAMPLE 12

Write five million, three hundred fifty thousand in standard notation.

5, ______, ______

Now Work Exercises 7–10 in the Margin.

Write each of the following numbers in standard notation.

7. Twenty-three thousand, six hundred forty-two

8. Three hundred sixty-three thousand, nine hundred seventy-five

Write each of the following numbers in its English word equivalent.

9. 6,300,500

10. 4,875,000

Completion Example Answers

3. 532,081 = 500,000 + 30,000 + **2000** + **0** + **80** + 1
4. 497,500 = 400,000 + **90,000** + **7000** + **500** + 0 + 0
7. 8407 = 8000 + **400** + **0** + **7**
   eight thousand, **four hundred seven**
8. 15,352 = 10,000 + **5000** + **300** + **50** + **2**
   fifteen thousand, **three hundred fifty-two**
11. 2,800,**035**
12. 5,**350,000**

NAME ____________ SECTION ____________ DATE ____________

Exercises 1.1

1. List the ten digits used to write whole numbers.

2. The beginning point for writing digits in the decimal system is called a ____________.

3. The powers of ten are ____________.

4. The word ____________ does not appear in English word equivalents for whole numbers.

5. Use a comma to separate groups of three digits if a number has more than ____________ digits.

6. Hyphens are used to write English words for numbers that have ____________.

Write the following whole numbers in expanded notation.

7. 37
8. 84
9. 56
10. 821
11. 1892
12. 2059
13. 25,658
14. 32,341

Write the following whole numbers in their English word equivalents.

15. 83
16. 122
17. 10,500
18. 683,100

ANSWERS

1. ____________
2. ____________
3. ____________
4. ____________
5. ____________
6. ____________
7. ____________
8. ____________
9. ____________
10. ____________
11. ____________
12. ____________
13. ____________
14. ____________
15. [Respond below exercise.]
16. [Respond below exercise.]
17. [Respond below exercise.]
18. [Respond below exercise.]

19. [Respond below exercise.] ______

20. [Respond below exercise.] ______

21. [Respond below exercise.] ______

22. ______

23. ______

24. ______

25. ______

26. ______

27. ______

28. ______

29. ______

30. ______

31. [Respond below exercise.] ______

19. 592,300

20. 16,302,590

21. 71,500,000

Write the following whole numbers in standard notation.

22. seventy-five

23. ninety-eight

24. one hundred forty-two

25. five hundred seventy-three

26. three thousand, eight hundred thirty-four

27. ten thousand, eleven

28. four hundred thousand, seven hundred thirty-six

29. five hundred thirty-seven thousand, eighty-two

30. sixty-three million, two hundred fifty-one thousand, sixty-five

31. Name the position of each nonzero digit in the following number: 2,403,189,500.

NAME ______________ SECTION ______________ DATE ______________

ANSWERS

32. [Respond below exercise.]

33. [Respond below exercise.]

34. [Respond below exercise.]

35. [Respond below exercise.]

Write the English word equivalent for the number(s) in each sentence.

32. The population of Los Angeles is 3,485,398.

33. The distance from the earth to the sun is about 93,000,000 miles, or 149,730,000 kilometers.

34. The country of Chile averages about 100 miles in width and is about 2700 miles long.

35. The Republic of Venezuela covers an area of 352,143 square miles, or about 912,050 square kilometers.

36. [Respond below exercise.]

37. [Respond below exercise.]

Writing and Thinking about Mathematics

36. A *googol* is the name of a very large power of ten. Look up the meaning of this term in a dictionary or in a book that discusses the history of mathematics. (**Note:** Ten to the power of a googol is called a *googolplex*.)

37. According to the 1990 census, Illinois is the sixth most populous state in the United States with 11,430,602 people. Name the value of each nonzero digit in this number. In an almanac or other source, find the populations of the five most populous states.

1.2 Addition and Subtraction

OBJECTIVES

1. *Know the basic facts of addition with whole numbers.*
2. *Be able to add whole numbers in both a vertical format and a horizontal format.*
3. *Know the following properties of addition: commutative, associative, additive identity (0).*
4. *Be able to subtract whole numbers with no borrowing.*
5. *Be able to subtract whole numbers with borrowing.*

The basic operations of addition, subtraction, multiplication, and division with whole numbers are part of our everyday lives. We add the scores at sporting events, add the number of cars in a parking lot, subtract to find the number of minutes left in a class period, add and subtract to balance a checking account, multiply to find the cost of constructing a new building or the cost of groceries, and divide to calculate grade point averages and the average gas mileage of a car. In this section we will discuss the terms and properties related to the operations of addition and subtraction. Multiplication and division will be discussed in Sections 1.4 and 1.5, respectively.

Addition with Whole Numbers

The operation of addition with whole numbers is indicated either by writing the numbers horizontally with a plus (+) sign between them or by writing the numbers vertically in columns with instructions to add. In the vertical format, the digits are aligned by place value.

Definition

Addition is the operation of counting that results in the sum, or total, of two or more whole numbers.

The **sum** is the result of addition.

Addends are the numbers being added.

EXAMPLE 1

$$\underset{\text{addend}}{\underset{\uparrow}{5}} + \underset{\text{addend}}{\underset{\uparrow}{3}} = \underset{\text{sum}}{\underset{\uparrow}{8}}$$

Or, writing the numbers vertically,

$$\begin{array}{rl} 5 & \leftarrow \text{addend} \\ +\ 3 & \leftarrow \text{addend} \\ \hline 8 & \leftarrow \text{sum} \end{array}$$

You should **memorize** the basic addition facts in Table 1.1 to be able to add larger numbers with speed and accuracy.

By looking at Table 1.1, we can see that reversing the **order** of any two addends does not change their sum. This fact is called the **commutative property of addition.** We state this property using letters to represent whole numbers.

Commutative Property of Addition

For whole numbers a and b, $\quad a + b = b + a$

For Exercises 1 and 2, find both sums and tell which property of addition is illustrated.

1. $\begin{array}{r} 8 \\ +\,5 \\ \hline \end{array}$ $\quad$ $\begin{array}{r} 5 \\ +\,8 \\ \hline \end{array}$

2. $9 + 6 =$ ______

$6 + 9 =$ ______

Table 1.1 Basic Addition Facts

+	0	1	2	3	4	5	6	7	8	9
0	**0**	1	2	3	4	5	6	7	8	9
1	1	**2**	3	4	5	6	7	8	9	10
2	2	3	**4**	5	6	7	8	9	10	11
3	3	4	5	**6**	7	8	9	10	11	(12)
4	4	5	6	7	**8**	9	10	11	12	13
5	5	6	7	8	9	**10**	11	12	13	14
6	6	7	8	9	10	11	**12**	13	14	15
7	7	8	9	10	11	12	13	**14**	15	16
8	8	9	10	11	12	(13)	14	15	**16**	17
9	9	10	11	12	13	14	15	16	17	**18**

$3 + 9 = 12$

$8 + 5 = 13$

***Example* 2**

$5 + 4 = 9$ and $4 + 5 = 9$

Thus, $5 + 4 = 4 + 5$.

***Example* 3**

$\begin{array}{r} 6 \\ +\,5 \\ \hline 11 \end{array}$ $\quad$ $\begin{array}{r} 5 \\ +\,6 \\ \hline 11 \end{array}$

Now Work Exercises 1 and 2 in the Margin.

Another property, called the **associative property of addition,** states that numbers may be **grouped** (or associated) differently and still have the same sum.

Associative Property of Addition

For whole numbers a, b, and c,

$$a + b + c = (a + b) + c = a + (b + c)$$

EXAMPLE 4

$$6 + 3 + 5 \qquad 6 + 3 + 5$$
$$= 9 + 5 \qquad = 6 + 8$$
$$= 14 \qquad = 14$$

Thus, $(6 + 3) + 5 = 6 + (3 + 5)$.

EXAMPLE 5

$7 + 1 + 4 = (7 + 1) + 4 = 8 + 4 = 12$ and

$7 + 1 + 4 = 7 + (1 + 4) = 7 + 5 = 12$

Now Work Exercises 3 and 4 in the Margin.

Another property of addition with whole numbers is addition with 0. Whenever 0 is added to a number, the result is the original number. Zero (0) is called the **additive identity** or the **identity element for addition.**

0 is the Additive Identity

For any whole number a, $\quad a + 0 = a$

EXAMPLE 6

$7 + 0 = 7$

EXAMPLE 7

$$\begin{array}{r} 8 \\ +\,0 \\ \hline 8 \end{array}$$

In Examples 8–10, find the sums and tell what property of addition is illustrated.

COMPLETION EXAMPLE 8

$1 + (6 + 5) = (1 + 6) + 5 =$ ______________

COMPLETION EXAMPLE 9

$8 + 3 = 3 + 8 =$ ______________

COMPLETION EXAMPLE 10

$5 + 0 =$ ______________

For Exercises 3 and 4, find both sums and tell which property of addition is illustrated.

3. $6 + (5 + 4) =$ ______

$(6 + 5) + 4 =$ ______

4. $(9 + 1) + 7 =$ ______

$9 + (1 + 7) =$ ______

Show that each of the following statements is true by performing the addition both ways in a vertical format.

5. $9 + 1 = 1 + 9$

$$\begin{array}{r} 9 \\ +\ 1 \\ \hline \end{array} \qquad \begin{array}{r} 1 \\ +\ 9 \\ \hline \end{array}$$

6. $(8 + 4) + 3 = 8 + (4 + 3)$

$$\begin{array}{r} 8 \\ 4 \\ +\ 3 \\ \hline \end{array} \qquad \begin{array}{r} 8 \\ 4 \\ +\ 3 \\ \hline \end{array}$$

7. $3 + 0 = 0 + 3$

$$\begin{array}{r} 3 \\ +\ 0 \\ \hline \end{array} \qquad \begin{array}{r} 0 \\ +\ 3 \\ \hline \end{array}$$

8. $5 + 7 = 7 + 5$

$$\begin{array}{r} 5 \\ +\ 7 \\ \hline \end{array} \qquad \begin{array}{r} 7 \\ +\ 5 \\ \hline \end{array}$$

Find each sum.

9. $\begin{array}{r} 4361 \\ +\ 2528 \\ \hline \end{array}$

10. $\begin{array}{r} 3590 \\ +\ 4207 \\ \hline \end{array}$

11. $\begin{array}{r} 6781 \\ +\ 213 \\ \hline \end{array}$

12. $\begin{array}{r} 5307 \\ +\ 401 \\ \hline \end{array}$

You should understand the following distinction between the associative and commutative properties of addition:

In the associative property, the order of the numbers is unchanged. (The grouping is changed.)

In the commutative property, the order of the numbers is changed.

Now Work Exercises 5–8 in the Margin.

To Add Whole Numbers with More Than One Digit:

1. Write the numbers vertically so that the place values are **lined up** in columns.
2. Add only the digits with the same place value.

Example 11

Add 5623 + 3172.

$$\begin{array}{r} 5\,6\,2\,3 \\ +\ 3\,1\,7\,2 \\ \hline 5 \end{array}$$
← addend
← addend
Add ones.

$$\begin{array}{r} 5\,6\,2\,3 \\ +\ 3\,1\,7\,2 \\ \hline 9\,5 \end{array}$$
Add tens.

$$\begin{array}{r} 5\,6\,2\,3 \\ +\ 3\,1\,7\,2 \\ \hline 7\,9\,5 \end{array}$$
Add hundreds.

$$\begin{array}{r} 5\,6\,2\,3 \\ +\ 3\,1\,7\,2 \\ \hline 8\,7\,9\,5 \end{array}$$
Add thousands.

$$\begin{array}{r} 5\,6\,2\,3 \\ +\ 3\,1\,7\,2 \\ \hline 8\,7\,9\,5 \end{array}$$
← You do not write all the steps. You write only the addends and the sum.
← sum

Now Work Exercises 9–12 in the Margin.

Carrying

If the sum of the digits in one place-value column is more than 9, then:

1. Write the ones digit in that column.
2. **Carry** the other digit to the next column to the left.

Example 12

Add 28 + 66.

$$\begin{array}{r} 28 \\ +\ 66 \\ \hline \end{array} \qquad \begin{array}{r} 8 \\ +\ 6 \\ \hline 14 \end{array}$$ You think: 8 + 6 = 14.

$$\begin{array}{r} {}^{1} \\ 28 \\ +\ 66 \\ \hline 4 \end{array}$$ Write only the digit 4 and carry 1 to the next column.

$$\begin{array}{r} {}^{1} \\ 28 \\ +\ 66 \\ \hline 94 \end{array}$$ Now add all the digits in the tens column, including the 1 that was carried.

Example 13

Example 12 can be explained using expanded notation.

$$\begin{array}{rcl} 28 = 20 + 8 & = & 2 \text{ tens} + 8 \text{ ones} \\ +\ 66 = 60 + 6 & = & 6 \text{ tens} + 6 \text{ ones} \\ \hline & & 8 \text{ tens} \quad 14 \text{ ones} \\ & & 1 \text{ ten} \quad 4 \text{ ones} \\ \hline & & 9 \text{ tens} \quad 4 \text{ ones} = 90 + 4 = 94 \end{array}$$

Now Work Exercises 13–15 in the Margin.

Addition with More Than Two Numbers

1. You can add the digits in one column in any order or from the top down or from the bottom up because of the commutative and associative properties of addition.
2. With several digits in a column, look for combinations of digits that total 10. This will increase your speed.
3. **Carry** digits just as before.

Find the following sums.

13. $\begin{array}{r} 359 \\ +\ 647 \\ \hline \end{array}$

14. $\begin{array}{r} 8793 \\ +\ 4595 \\ \hline \end{array}$

15. $\begin{array}{r} 53{,}895 \\ +\ 69{,}146 \\ \hline \end{array}$

Find the following sums.

16.
$$\begin{array}{r} 64 \\ 58 \\ +\ 38 \\ \hline \end{array}$$

17.
$$\begin{array}{r} 196 \\ 357 \\ 492 \\ 804 \\ +\ 621 \\ \hline \end{array}$$

18.
$$\begin{array}{r} 21{,}452 \\ 32{,}551 \\ 364{,}625 \\ +\ 75{,}807 \\ \hline \end{array}$$

EXAMPLE 14

$$\begin{array}{r} {}^{2} \\ 217 \\ 389 \\ 634 \\ +\ 536 \\ \hline 6 \end{array}$$

Add 7 + 9 + 4 + 6
= 7 + 9 + 10
= 26

(634, 536 → 10)

Carry the 2.

$$\begin{array}{r} {}^{12} \\ 217 \\ 389 \\ 634 \\ +\ 536 \\ \hline 76 \end{array}$$

(2, 8 → 10)

Add ② + 1 + ⑧ + 3 + 3
= 10 + 1 + 3 + 3
= 17

$$\begin{array}{r} {}^{12} \\ 217 \\ 389 \\ 634 \\ +\ 536 \\ \hline 1776 \end{array}$$

(1, 3, 6 → 10)

Add ① + 2 + ③ + ⑥ + 5
= 10 + 2 + 5
= 17

Now Work Exercises 16–18 in the Margin.

Subtraction with Whole Numbers

A driver who is 29 years old had driven 18 miles of his 23-mile delivery route when his van broke down and forced him to quit for the day. How many miles did he have left to drive on his route?

What *thinking* did you do to answer the question? You may have reasoned something like this: "Well, I don't need to know how old the driver is to answer the question, so his age is just extra information. Since he had already driven 18 miles, I need to know what to add to 18 to get 23. Since 18 + 5 = 23, he had 5 miles left to drive."

In this problem, the sum of two addends was given, and only one of the addends was given. The other addend was the unknown quantity.

18	+	____	=	23
↑		↑		↑
addend		missing addend		sum

This kind of addition problem is called **subtraction** and can be indicated by writing the sum minus the given addend in the following format:

23	−	18	=	____	(Read: "23 minus 18 equals blank.")
↑		↑		↑	
sum		addend		missing addend (or difference)	

We generally think of subtraction as the reverse of addition, or the process of "taking away" a number from the total to find the **difference.**

Definition

Subtraction is the operation of taking one amount, or number, away from another. (This is the opposite of adding the two amounts.)

The **difference** is the result of subtracting one addend (also called the **subtrahend**) from the sum (also called the **minuend**).

EXAMPLE 15

$$\underset{\text{sum}}{10} - \underset{\text{addend}}{7} = \underset{\substack{\text{missing addend}\\ \text{(or difference)}}}{\boxed{?}}$$

Think: What number added to 7 will give 10?
Since $7 + 3 = 10$, we have $10 - 7 = 3$; or

$$\begin{array}{rl} 10 & \leftarrow \text{minuend} \\ -\ 7 & \leftarrow \text{subtrahend} \\ \hline \boxed{3} & \leftarrow \text{difference or missing addend} \end{array}$$

To Subtract Whole Numbers With More Than One Digit:

1. Write the numbers vertically so that the place values are **lined up** in columns.
2. Subtract only the digits with the same place value.

EXAMPLE 16

Subtract 496 − 342.

$$\begin{array}{r} 4\,9\,6 \\ -\ 3\,4\,2 \\ \hline 4 \end{array}$$ Subtract ones. $6 - 2 = 4$

$$\begin{array}{r} 4\,9\,6 \\ -\ 3\,4\,2 \\ \hline 5\,4 \end{array}$$ Subtract tens. $9 - 4 = 5$

$$\begin{array}{r} 4\,9\,6 \\ -\ 3\,4\,2 \\ \hline 1\,5\,4 \end{array}$$ Subtract hundreds. $4 - 3 = 1$ ← difference

COMPLETION EXAMPLE 17

Subtract 897 − 364 by using expanded notation.

$$\begin{array}{rcl} 897 &=& 800 + 90 + \underline{\quad\quad} \\ -364 &=& 300 + \underline{\quad\quad} + \underline{\quad\quad} \\ \hline && \underline{\quad\quad} + \underline{\quad\quad} + \underline{\quad\quad} = \underline{\quad\quad\quad} \end{array}$$

Find the following differences. Use expanded notation only if you find it helpful.

19. $\begin{array}{r} 654 \\ -\ 421 \\ \hline \end{array}$

20. $\begin{array}{r} 1857 \\ -\ 346 \\ \hline \end{array}$

21. $\begin{array}{r} 2469 \\ -\ 1125 \\ \hline \end{array}$

Find the following differences. Use expanded notation to show any borrowing.

22. $\begin{array}{r} 867 \\ -\ 328 \\ \hline \end{array}$

23. $\begin{array}{r} 426 \\ -\ 388 \\ \hline \end{array}$

Now Work Exercises 19–21 in the Margin.

Borrowing

1. **Borrowing** is necessary when a digit is smaller than the digit being subtracted. We cannot add two whole numbers and get a smaller number.
2. The process starts from the rightmost digit. **Borrow** from the digit to the left.

Borrowing can be done more than once, as illustrated in Example 18.

EXAMPLE 18

$$\begin{array}{r} 736 = 700 + 30 + 6 \\ -258 = 200 + 50 + 8 \\ \hline \end{array}$$

Since 6 is smaller than 8, borrow 10 from 30.

Since 20 is smaller than 50, borrow 100 from 700.

$$\begin{array}{rcl} 736 = 700 + 30 + 6 = 700 + 20 + 16 = & 600 + 120 + 16 \\ -258 = 200 + 50 + 8 = 200 + 50 + \ \ 8 = & 200 + \ \ 50 + \ \ 8 \\ \hline & 400 + \ \ 70 + \ \ 8 = 478 \end{array}$$

Now Work Exercises 22 and 23 in the Margin.

A common practice is to indicate borrowing by crossing out digits and writing new digits instead of expanded notation.

EXAMPLE 19

This example uses the same numbers as in Example 18 to illustrate the different techniques.

$$\begin{array}{r} 736 \\ -\ 258 \\ \hline \end{array}$$

STEP 1: Since 6 is smaller than 8, borrow 10 from 30. This leaves 20, so cross out 3 and write 2.

$$\begin{array}{r} \ \ {}^{2}\ {}^{1} \\ 7\ \not{3}\ 6 \\ -\ 2\ 5\ 8 \\ \hline \end{array}$$

STEP 2: Since 2 is smaller than 5, borrow 100 from 700. This leaves 600, so cross out 7 and write 6.

$$\begin{array}{r} {}^{6}\ {}^{1}{}^{2}\ {}^{1} \\ \not{7}\ \not{3}\ 6 \\ -\ 2\ 5\ 8 \\ \hline \end{array}$$

STEP 3: Now subtract.

$$\begin{array}{r} {}^{6}\ {}^{1}{}^{2}\ {}^{1} \\ \not{7}\ \not{3}\ 6 \\ -\ 2\ 5\ 8 \\ \hline 4\ 7\ 8 \end{array}$$

Example 20

$$\begin{array}{r} 8000 \\ -\ 657 \\ \hline \end{array}$$

STEP 1: Trying to borrow from 0 each time, we end up borrowing 1000 from 8000. Cross out 8 and write 7.

$$\begin{array}{rrrr} \overset{7}{\cancel{8}} & \overset{1}{0} & 0 & 0 \\ - & 6 & 5 & 7 \\ \hline \end{array}$$

STEP 2: Now borrow 100 from 1000. Cross out 10 and write 9.

$$\begin{array}{rrrr} \overset{7}{\cancel{8}} & \overset{\overset{9}{\cancel{1}}}{\cancel{0}} & 0 & 0 \\ - & 6 & 5 & 7 \\ \hline \end{array}$$

STEP 3: Borrow 10 from 100. Cross out 10 and write 9.

$$\begin{array}{rrrr} \overset{7}{\cancel{8}} & \overset{\overset{9}{\cancel{1}}}{\cancel{0}} & \overset{\overset{9}{\cancel{1}}}{\cancel{0}} & \overset{1}{0} \\ - & 6 & 5 & 7 \\ \hline \end{array}$$

STEP 4: Now subtract.

$$\begin{array}{rrrr} \overset{7}{\cancel{8}} & \overset{\overset{9}{\cancel{1}}}{\cancel{0}} & \overset{\overset{9}{\cancel{1}}}{\cancel{0}} & \overset{1}{0} \\ - & 6 & 5 & 7 \\ \hline 7 & 3 & 4 & 3 \end{array}$$

Now Work Exercises 24–26 in the Margin.

Completion Example 21

Two painters bid on painting the same house. The first painter bid $2738 and the second painter bid $2950. What was the difference between the two bids?

Solution

This is a subtraction problem because we are asked for a difference.

$$\begin{array}{r} \$2950 \\ -\ 2738 \\ \hline \_\_\_\_\_ \end{array} \leftarrow \text{difference in bids}$$

Find the following differences without using expanded notation.

24. $\begin{array}{r} 537 \\ -\ 249 \\ \hline \end{array}$

25. $\begin{array}{r} 6545 \\ -\ 2687 \\ \hline \end{array}$

26. $\begin{array}{r} 4000 \\ -\ 3946 \\ \hline \end{array}$

☑ COMPLETION EXAMPLE 22

After selling their house for $132,000, the owners paid the realtor $7920, back taxes of $450, and $350 in other fees. If they also paid off the bank loan of $57,000, how much cash did the owners receive from the sale?

Solution

In this problem experience tells us that we must add and subtract even though there are no specific directions to do so. We add the expenses and then subtract this sum from the selling price to find the cash that the owners received.

```
  $ 7,920
      450
      350
 + 57,000
 ________
 ________  ← total expenses
```

```
$132,000   selling price
− _______  total expenses
  _______  cash to owners
```

Completion Example Answers

8. $1 + (6 + 5) = (1 + 6) + 5 = \mathbf{12}$ **Associative Property**

9. $8 + 3 = 3 + 8 = \mathbf{11}$ **Commutative Property**

10. $5 + 0 = \mathbf{5}$ **Additive Identity**

17.
$$\begin{array}{rcl} 897 & = & 800 + 90 + \mathbf{7} \\ -\ 364 & = & 300 + \mathbf{60} + \mathbf{4} \\ \hline & & \mathbf{500} + \mathbf{30} + \mathbf{3} = \mathbf{533} \end{array}$$

21.
```
  $2950
 − 2738
 ______
  $ 214  ← difference in bids
```

22.
```
  $ 7,920
      450
      350
  +57,000
  _______
  $65,720  ← total expenses
```

```
  $132,000   selling price
 − 65,720    total expenses
 _________
 $ 66,280    cash to owners
```

NAME ______________ SECTION ______________ DATE ______________

ANSWERS

1. ______________
2. ______________
3. ______________
4. ______________
5. ______________
6. ______________
7. ______________
8. ______________
9. ______________
10. ______________
11. ______________
12. ______________
13. ______________
14. ______________
15. ______________
16. ______________
17. ______________
18. ______________
19. ______________
20. ______________

Exercises 1.2

Show that the following statements are true by performing the addition mentally. State which property of addition is illustrated.

1. $6 + 3 = 3 + 6$

2. $9 + 7 = 7 + 9$

3. $9 + (8 + 3) = (9 + 8) + 3$

4. $(2 + 3) + 4 = 2 + (3 + 4)$

5. $11 + 0 = 11$

6. $0 + 17 = 17$

7. $4 + 1 + 5 = 4 + 6$

8. $12 + 3 + 2 = 12 + 5$

9. $6 + (2 + 7) = 6 + (7 + 2)$

10. $(3 + 4) + 5 = (4 + 3) + 5$

Find the following differences mentally.

11. $14 - 8$

12. $16 - 9$

13. $15 - 6$

14. $12 - 3$

15. $7 - 0$

16. $28 - 0$

17. $\begin{array}{r} 10 \\ -\ 9 \\ \hline \end{array}$

18. $\begin{array}{r} 15 \\ -\ 7 \\ \hline \end{array}$

19. $\begin{array}{r} 12 \\ -\ 6 \\ \hline \end{array}$

20. $\begin{array}{r} 14 \\ -\ 9 \\ \hline \end{array}$

21. [Respond in exercise.] ________

22. ________

23. ________

24. ________

25. ________

26. ________

27. ________

28. ________

29. ________

30. ________

31. ________

32. ________

33. ________

34. ________

35. ________

36. ________

37. ________

38. ________

39. ________

40. ________

41. ________

21. Complete the following table of sums.

+	5	8	7	9
3				
6				
5				
2				

Find the following sums.

22. 56 + 95

23. 37 + 88

24. 156 + 285

25. 816 + 736

26. 1076 + 3095

27. 7328 + 5996

28. 65, 43, + 54

29. 24, 78, + 95

30. 73, 68, + 98

31. 165, 276, + 394

32. 876, 279, + 143

33. 268, 93, + 192

34. 981, 146, 92, + 17

35. 2112, 147, 904, + 1005

36. 114, 5402, 710, + 643

37. 1403, 7010, 622, + 29

38. 213,116, 116,018, 722,988, 24,336, + 526,968

39. 21,442, 32,462, 564,792, 801,801, + 43,433

40. 438,966, 1,572,486, 327,462, 181,753, + 90,000

41. 123,456, 456,123, 879,282, 617,500, + 740,765

NAME ____________ SECTION ____________ DATE ____________

ANSWERS

Subtract to find each indicated difference.

42. $\begin{array}{r} 17 \\ -17 \\ \hline \end{array}$ **43.** $\begin{array}{r} 42 \\ -31 \\ \hline \end{array}$ **44.** $\begin{array}{r} 89 \\ -76 \\ \hline \end{array}$ **45.** $\begin{array}{r} 53 \\ -33 \\ \hline \end{array}$

46. $\begin{array}{r} 96 \\ -27 \\ \hline \end{array}$ **47.** $\begin{array}{r} 23 \\ -18 \\ \hline \end{array}$ **48.** $\begin{array}{r} 126 \\ -\ 32 \\ \hline \end{array}$ **49.** $\begin{array}{r} 174 \\ -\ 48 \\ \hline \end{array}$

50. $\begin{array}{r} 347 \\ -129 \\ \hline \end{array}$ **51.** $\begin{array}{r} 543 \\ -167 \\ \hline \end{array}$ **52.** $\begin{array}{r} 900 \\ -307 \\ \hline \end{array}$ **53.** $\begin{array}{r} 603 \\ -208 \\ \hline \end{array}$

54. $\begin{array}{r} 7843 \\ -6274 \\ \hline \end{array}$ **55.** $\begin{array}{r} 6793 \\ -5827 \\ \hline \end{array}$ **56.** $\begin{array}{r} 4376 \\ -2808 \\ \hline \end{array}$

57. $\begin{array}{r} 4900 \\ -3476 \\ \hline \end{array}$ **58.** $\begin{array}{r} 5070 \\ -4376 \\ \hline \end{array}$ **59.** $\begin{array}{r} 8007 \\ -2136 \\ \hline \end{array}$

60. $\begin{array}{r} 7,085,076 \\ -4,278,432 \\ \hline \end{array}$ **61.** $\begin{array}{r} 6,543,222 \\ -2,742,663 \\ \hline \end{array}$ **62.** $\begin{array}{r} 8,000,000 \\ -\ \ 647,561 \\ \hline \end{array}$

63. $\begin{array}{r} 6,000,000 \\ -\ \ 328,989 \\ \hline \end{array}$

42. ____________
43. ____________
44. ____________
45. ____________
46. ____________
47. ____________
48. ____________
49. ____________
50. ____________
51. ____________
52. ____________
53. ____________
54. ____________
55. ____________
56. ____________
57. ____________
58. ____________
59. ____________
60. ____________
61. ____________
62. ____________
63. ____________

64. ______________

65. ______________

66. ______________

67. ______________

64. Mr. Jones kept the mileage records indicated in the table shown here. How many miles did he drive during the six months?

Month	Mileage
January	546
February	378
March	496
April	357
May	503
June	482

65. The Modern Products Corporation showed profits as indicated in the table for the years 1990–1993. What were the company's total profits for the years 1990–1993?

Year	Profits
1990	$1,078,416
1991	1,270,842
1992	2,000,593
1993	1,963,472

66. Fred estimated his yearly expenses for six years of college, including two years of graduate school, as: $2035, $2786, $3300, $4000, $3500, and $4500. What were his total expenses for the six years of schooling? (**Note:** He had some financial aid.)

67. Apple County has the following items budgeted: highways, $270,455; salaries, $95,479; maintenance, $127,220. What is the county's total budget for these three items?

NAME ______________________ SECTION ______________ DATE __________

ANSWERS

68. ______________

69. ______________

70. ______________

71. ______________

72. ______________

68. The following numbers of students at South Junior College are enrolled in mathematics courses: 303 in arithmetic, 476 in algebra, 293 in trigonometry, 257 in college algebra, and 189 in calculus. Find the total number of students taking mathematics.

69. If seven hundred eighty-one is subtracted from the sum of one thousand seventy-five and four hundred ninety-six, what is the difference?

70. If the difference between two thousand five hundred twenty and one thousand six hundred forty-two is added to seven hundred thirty-three, what is the sum?

71. What number should be added to 860 to get a sum of 1000?

72. If the sum of two numbers is 537 and one of the numbers is 139, what is the other number?

73. ______________

74. ______________

75. ______________

76. [Respond below exercise.]

73. In June, Ms. Garcia opened a checking account and deposited \$1342, \$238, \$157, and \$486. She also wrote checks for \$132, \$76, \$480, and \$90. What was her balance at the end of June?

74. A house was sold for \$135,000. If the owners paid the realtor \$8100, paid the bank \$87,000 for the mortgage, and paid \$800 for other expenses of the sale, what were the net proceeds to the owners?

75. A woman bought a condominium for a price of \$150,000. She also had to pay other expenses of \$750. If the local savings and loan association agreed to loan her \$105,500 as a first trust deed on the house, how much cash did she need in order to buy the condominium?

Writing and Thinking about Mathematics

76. Nothing was said in the text about a commutative property for subtraction. Do you think that this omission was intentional? Is there a commutative property of subtraction? Give several examples that justify your answer and discuss this idea in class.

NAME ______________________ SECTION ______________________ DATE ______________

ANSWERS

77. [Respond below exercise.]

78. [Respond below exercise.]

79. [Respond in exercise.]

77. Nothing was said in the text about an associative property for subtraction. Do you think that this omission was intentional? Is there an associative property of subtraction? Give several examples that justify your answer and discuss this idea in class.

78. You may have heard someone say, "You cannot subtract a larger number from a smaller number." Do you think that this statement is true? Does a subtraction problem such as $6 - 10$ make sense to you? Can you think of any situation in which such a difference might seem reasonable?

Collaborative Learning Exercise

Separate the class into teams of two to four students. Each team is to read and fill out the partial 1040 Forms in Exercises 79–81. After these are completed, the team leader is to read the team's results and discuss any difficulties they had in reading and following directions while filling out the forms.

79. The Internal Revenue Service allows us to use whole numbers in calculating our income tax returns. Calculate the taxable income on the portion of each part of the Form 1040 shown here, assuming that you have three exemptions.

Line	Description		Line	Amount
32	Amount from line 31 (adjusted gross income)		32	56,458
33a	Check if: ☐ **You** were 65 or older, ☐ Blind; ☐ **Spouse** was 65 or older; ☐ Blind. Add the number of boxes checked above and enter the total here . . . . ▶ **33a**	☐		
b	If your parent (or someone else) can claim you as a dependent, check here . ▶ **33b**	☐		
c	If you are married filing separately and your spouse itemizes deductions or you are a dual-status alien, see page 22 and check here. . . . . . . ▶ **33c**	☐		
34	Enter the **larger** of your: **Itemized deductions** from Schedule A, line 26, **OR** **Standard deduction** shown below for your filing status. **But if you checked any box on line 33a or b,** go to page 22 to find your standard deduction. If you checked **box 33c,** your standard deduction is zero. • Single–\$3,600 • Head of household–\$5250 • Married filing jointly or Qualifying widow(er)–\$6,000 • Married filing separately–\$3,000		34	6,000
35	Subtract line 34 from line 32		35	
36	If line 32 is \$78,950 or less, multiply \$2,300 by the total number of exemptions claimed on line 6e. If line 32 is over \$78,950, see the worksheet on page 23 for the amount to enter .		36	
37	**Taxable income.** Subtract line 36 from line 35. If line 36 is more than line 35, enter "0" .		37	

[Respond in exercise.]

80. ________

80. Calculate the taxable income on the portion of the Form 1040 shown here, assuming that you have four exemptions.

Line	Description	Line no.	Amount
32	Amount from line 31 (adjusted gross income)	32	43,552
33a	Check if: ☐ **You** were 65 or older, ☐ Blind; ☐ **Spouse** was 65 or older; ☐ Blind. Add the number of boxes checked above and enter the total here . . . . ▶ 33a ☐		
b	If your parent (or someone else) can claim you as a dependent, check here . ▶ 33b ☐		
c	If you are married filing separately and your spouse itemizes deductions or you are a dual-status alien, see page 22 and check here. . . . . . . ▶ 33c ☐		
34	Enter the **larger** of your: **Itemized deductions** from Schedule A, line 26, **OR** **Standard deduction** shown below for your filing status. **But if you checked any box on line 33a or b**, go to page 22 to find your standard deduction. If you checked **box 33c,** your standard deduction is zero. • Single–$3,600 • Head of household–$5250 • Married filing jointly or Qualifying widow(er)–$6,000 • Married filing separately–$3,000	34	6,000
35	Subtract line 34 from line 32	35	
36	If line 32 is $78,950 or less, multiply $2,300 by the total number of exemptions claimed on line 6e. If line 32 is over $78,950, see the worksheet on page 23 for the amount to enter .	36	
37	**Taxable income.** Subtract line 36 from line 35. If line 36 is more than line 35, enter "0" .	37	

[Respond in exercise.]

81. ________

81. Calculate the taxable income on the portion of the Form 1040 shown here, assuming that you have two exemptions.

Line	Description	Line no.	Amount
32	Amount from line 31 (adjusted gross income)	32	75,950
33a	Check if: ☐ **You** were 65 or older, ☐ Blind; ☐ **Spouse** was 65 or older; ☐ Blind. Add the number of boxes checked above and enter the total here . . . . ▶ 33a ☐		
b	If your parent (or someone else) can claim you as a dependent, check here . ▶ 33b ☐		
c	If you are married filing separately and your spouse itemizes deductions or you are a dual-status alien, see page 22 and check here. . . . . . . ▶ 33c ☐		
34	Enter the **larger** of your: **Itemized deductions** from Schedule A, line 26, **OR** **Standard deduction** shown below for your filing status. **But if you checked any box on line 33a or b**, go to page 22 to find your standard deduction. If you checked **box 33c,** your standard deduction is zero. • Single–$3,600 • Head of household–$5250 • Married filing jointly or Qualifying widow(er)–$6,000 • Married filing separately–$3,000	34	6,000
35	Subtract line 34 from line 32	35	
36	If line 32 is $78,950 or less, multiply $2,300 by the total number of exemptions claimed on line 6e. If line 32 is over $78,950, see the worksheet on page 23 for the amount to enter .	36	
37	**Taxable income.** Subtract line 36 from line 35. If line 36 is more than line 35, enter "0" .	37	

1.3 Rounding Off and Estimating

OBJECTIVES

1. *Know the rule for rounding off whole numbers to any given place.*
2. *Use the rule of rounding off to estimate sums and differences.*
3. *Develop a sense of the expected size of a sum or difference so that major errors can be easily spotted.*

Rounding Off Whole Numbers

To **round off** a given number means to find another number close to the given number. The desired place of accuracy must be stated. For example, if you were asked to round off 872, you would not know what to do unless you were told the desired place of accuracy. The number lines in Figure 1.3 serve as visual aids in understanding the rounding-off process.

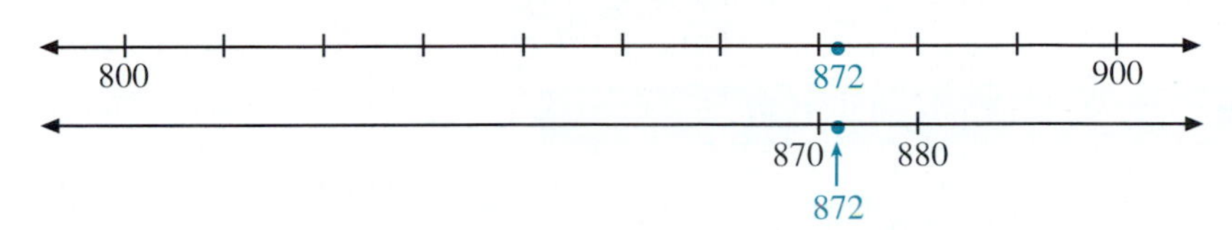

Figure 1.3

We can see that 872 is closer to 900 than to 800. So, **to the nearest hundred,** 872 rounds off to 900. Also, 872 is closer to 870 than to 880. So, **to the nearest ten,** 872 rounds off to 870.

In many situations rounded-off (or approximate) answers are quite acceptable. For example, we might say that the distance across the United States from east coast to west coast is approximately 3000 miles. Such an approximation (or rounded-off value) is common with the use of numbers that result from any type of measurement (not counting). These numbers are only approximate because the measuring devices are themselves somewhat inaccurate and, therefore, involve some form of rounding off.

EXAMPLE 1

Round 37 to the nearest ten.

Solution

A number line gives a visual reference.

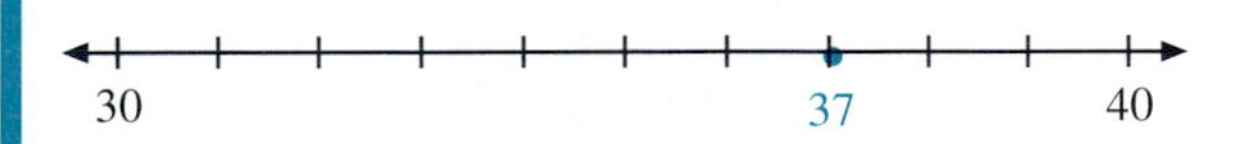

37 is closer to 40 than to 30. So, 37 rounds off to 40 (to the nearest ten).

EXAMPLE 2

Round 253 to the nearest ten.

Solution

Use a number line.

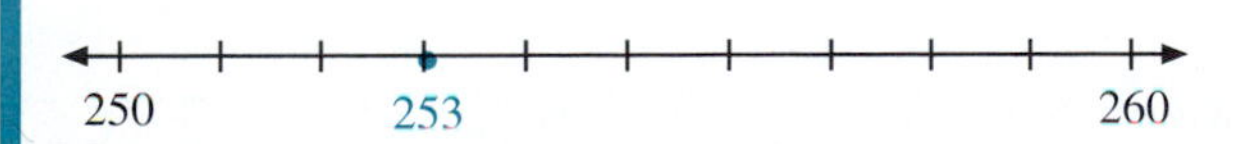

253 is closer to 250 than to 260. So, 253 rounds off to 250 (to the nearest ten).

Round off each number to the place indicated.

1. 57 (nearest ten)

2. 345 (nearest hundred)

3. 2345 (nearest thousand)

EXAMPLE 3

Round 278 to the nearest hundred.

Solution

Again, a number line is helpful.

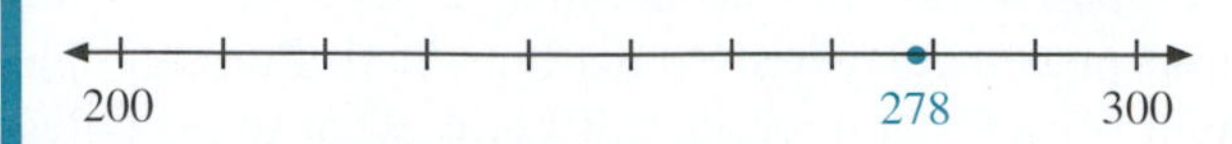

278 is closer to 300 than to 200. So, 278 rounds off to 300 (to the nearest hundred).

Now Work Exercises 1–3 in the Margin.

The following rule can be used in place of a number line.

Rounding-Off Rule for Whole Numbers

1. Look at the single digit just to the right of the place of desired accuracy.
2. If this digit is 5 or greater, make the digit in the desired place of accuracy one larger and replace all digits to the right with zeros. All digits to the left remain unchanged.
3. If this digit is less than 5, leave the digit in the desired place of accuracy as it is and replace all digits to the right with zeros. All digits to the left remain unchanged.

EXAMPLE 4

Round 5749 to the nearest hundred.

Solution

5749	5749	5700
↑	↑	↑
place of desired accuracy	One digit to the right; 4 is less than 5.	Leave 7 and fill in zeros.

So, 5749 rounds off to 5700 (to the nearest hundred).

EXAMPLE 5

Round 6500 to the nearest thousand.

Solution

6500	6500	7000
↑	↑	↑
place of desired accuracy	Look at 5; 5 is 5 or greater.	Increase 6 to 7 (one larger) and fill in zeros.

So, 6500 rounds off to 7000 (to the nearest thousand).

EXAMPLE ***6***

Round 397 to the nearest ten.

Solution

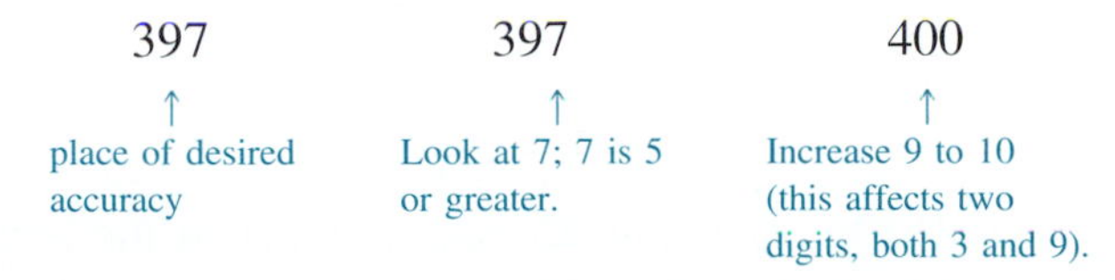

So, 397 rounds off to 400 (to the nearest ten).

Now Work Exercises 4–10 in the Margin.

Use the **Rounding-Off Rule for Whole Numbers** to round off the following numbers as indicated.

4. 576 (nearest ten)

5. 839 (nearest hundred)

6. 1500 (nearest thousand)

7. 2589 (nearest hundred)

8. 43,610 (nearest thousand)

9. 1983 (nearest hundred)

10. 1938 (nearest hundred)

Estimating Sums and Differences with Whole Numbers

One very good use for rounded-off numbers is to **estimate** (or **approximate**) an answer before any calculations are made with the given numbers. Thus, answers that are not reasonable (for example, because someone pushed a wrong key on a calculator or because some other large calculation error was made) can be spotted. Usually, we simply repeat the calculations and find the error.

To **estimate an answer** means to use rounded-off numbers in a calculation to get some idea of what the size of the actual answer should be. This is a form of checking your work before you do it. There are some situations when an estimated answer itself is sufficient. For example, a shopper may only need to estimate the total cost of purchases to be sure that he or she has enough cash to cover the cost.

To Estimate a Sum or Difference with Whole Numbers:

1. Round off each number to the place of the leftmost digit. (**Note:** This may not be the same place in all the numbers.)
2. Perform the indicated operation with these rounded-off numbers.

EXAMPLE 7

Add: $\begin{array}{r} 568 \\ 934 \\ +712 \\ \hline \end{array}$

Solution

(a) Estimate the sum first by rounding off each number to the leftmost digit, then adding. In actual practice, many of these steps can be done mentally.

$$\begin{array}{rcr} 568 & \rightarrow & 600 \\ 934 & \rightarrow & 900 \\ +712 & \rightarrow & +\ 700 \\ \hline & & 2200 \end{array}$$

(b) Now we find the sum with the knowledge that the answer should be close to 2200.

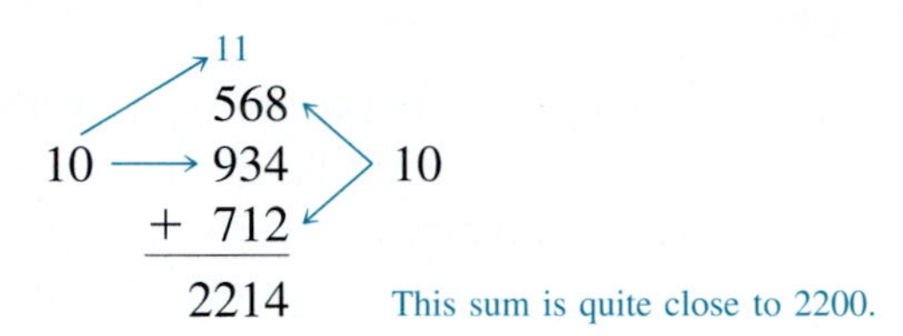

This sum is quite close to 2200.

EXAMPLE 8

Subtract: $\begin{array}{r} 658 \\ -189 \\ \hline \end{array}$

Solution

(a) Estimate the difference first by rounding off each number to the leftmost digit, then subtracting.

$$\begin{array}{rcr} 658 & \rightarrow & 700 \\ -189 & \rightarrow & -200 \\ \hline & & 500 \end{array}$$

(b) Now we find the difference with the knowledge that the answer should be approximately 500.

$$\begin{array}{r} \overset{5}{\cancel{6}}\ \overset{\overset{1}{4}}{\cancel{5}}\ \overset{1}{8} \\ -1\ 8\ 9 \\ \hline 4\ 6\ 9 \end{array}$$

This difference rounds off to 500, the **estimated** value.

✓ *Completion Example 9*

Add:

```
  5483
   232
+  657
```

Solution

(a) First estimate the sum by rounding off each number to the place of its leftmost digit and adding these rounded-off numbers.

```
  5483  →    5000
   232  →     200
+  657  →  + _____

           _______  ← estimated sum
```

(b) Find the sum and compare your answer with the estimated sum. They should be "close."

```
  5483
   232
+  657

_______  ← actual sum
```

Note: In the case where the leftmost digit is 9 and it is to be rounded up, it is rounded up to 10. For example, 983 rounds up to 1000 for approximation purposes.

Now Work Exercises 11 and 12 in the Margin.

Caution: The use of rounded-off numbers to approximate answers demands some understanding by the user of numbers in general and some judgment as to how "close" an estimate must be to be acceptable. In particular, when one number is rounded up and another is rounded down, the estimate might not be "close enough" to detect a large error. Still, the process is worthwhile and helpful in most cases.

11. First estimate the sum, then find the sum.

```
   785
  3467
+  624
```

12. First estimate the difference, then find the difference.

```
  5780
-  975
```

Completion Example Answers

9. (a)

$$\begin{array}{rcr} 5483 & \rightarrow & 5000 \\ 232 & \rightarrow & 200 \\ +\ 657 & \rightarrow & \mathbf{700} \\ \hline & & \mathbf{5900} \end{array} \leftarrow \text{estimated sum}$$

(b)

$$\begin{array}{r} 5483 \\ 232 \\ +\ 657 \\ \hline \mathbf{6372} \end{array} \leftarrow \text{actual sum}$$

NAME ____________ SECTION ____________ DATE ____________

ANSWERS

Exercises 1.3

Round off as indicated.

To the nearest ten:

1. 763 **2.** 31 **3.** 82 **4.** 503

5. 296 **6.** 722 **7.** 987 **8.** 347

To the nearest hundred:

9. 4163 **10.** 4475 **11.** 495 **12.** 572

13. 637 **14.** 3789 **15.** 76,523 **16.** 7007

To the nearest thousand:

17. 6912 **18.** 5500 **19.** 7500 **20.** 7499

21. 13,499 **22.** 13,501 **23.** 62,265 **24.** 47,800

To the nearest ten thousand:

25. 78,419 **26.** 125,000 **27.** 256,000 **28.** 62,200

1. ____________
2. ____________
3. ____________
4. ____________
5. ____________
6. ____________
7. ____________
8. ____________
9. ____________
10. ____________
11. ____________
12. ____________
13. ____________
14. ____________
15. ____________
16. ____________
17. ____________
18. ____________
19. ____________
20. ____________
21. ____________
22. ____________
23. ____________
24. ____________
25. ____________
26. ____________
27. ____________
28. ____________

29. ______________

30. ______________

31. ______________

32. ______________

33. ______________

34. ______________

35. ______________

36. ______________

37. ______________

38. ______________

39. ______________

40. ______________

41. ______________

42. ______________

43. ______________

44. ______________

29. 118,200 **30.** 312,500 **31.** 184,900 **32.** 615,000

33. 87 to the nearest hundred

34. 46 to the nearest hundred

35. 532 to the nearest thousand

First estimate the answers using rounded-off numbers (rounded to the leftmost nonzero digit); then find the following sums and differences.

36.
```
  83
  62
 +78
```

37.
```
  97
  46
 +25
```

38.
```
  146
  259
 +384
```

39.
```
  475
  126
 +572
```

40.
```
  600
  542
 +483
```

41.
```
  851
  736
 +294
```

42.
```
  5742
  6271
  8156
+  972
```

43.
```
   483
  1681
  3054
 +4006
```

44.
```
  22,506
  38,700
 +10,465
```

NAME ____________ SECTION ____________ DATE ____________

ANSWERS

45. 8742
−3275

46. 6421
−1652

47. 10,531
− 4,600

48. 275,600
− 94,300

49. 63,504
−42,700

50. 74,305
−33,082

51. If the population of the People's Republic of China is 1,165,800,000 and the population of the United States is 255,600,000, about how many more people live in mainland China than live in the United States?

52. In 1990, the University of California campuses had the following enrollments:

UC Berkeley	22,262	UC Riverside	7,310
UC Davis	17,877	UC San Diego	14,392
UC Irvine	13,811	UC Santa Barbara	15,975
UC Los Angeles	24,368	UC Santa Cruz	8,883

Approximately how many students were enrolled in the University of California system in 1990?

45. ____________

46. ____________

47. ____________

48. ____________

49. ____________

50. ____________

51. ____________

52. ____________

53. ______________

54. ______________

55. ______________

53. The perimeter of (distance around) a triangle (a three-sided plane figure) is found by adding the lengths of its sides. Suppose a triangle has sides of length 12 centimeters, 16 centimeters, and 25 centimeters. Without actually adding the numbers, which of the following numbers do you think is closest to the actual perimeter of the triangle: 20 centimeters, 30 centimeters, 60 centimeters, or 100 centimeters?

54. A quadrilateral is a four-sided plane figure. If a quadrilateral has sides of length 8 meters, 9 meters, 10 meters, and 11 meters, what is your estimate of the perimeter of (distance around) the quadrilateral: 20 meters, 40 meters, or 55 meters?

55. If you looked at the book list at school and found that the prices of the textbooks that you need to buy this semester are \$45, \$56, \$32, and \$17, about how much cash should you take with you, assuming you want to pay in cash: \$50, \$100, \$150, \$200, or \$250? (Don't forget that you will need to pay some sales tax.)

1.4 Multiplication

OBJECTIVES

1. *Know the basic multiplication facts (the products of the numbers from 0 to 9).*
2. *Develop the skill of mentally multiplying numbers involving powers of ten.*
3. *Be able to multiply whole numbers by using partial products.*
4. *Be able to multiply whole numbers by using the standard short method.*
5. *Estimate products.*

Basic Multiplication and Powers of Ten

Multiplication is a process that shortens repeated addition with the same number. For example, we can write

$$8 + 8 + 8 + 8 + 8 = 5 \cdot 8 = 40$$

The repeated addend (8) and the number of times it is used (5) are both called **factors** of 40, and the result (40) is called the **product** of the two factors.

$$8 + 8 + 8 + 8 + 8 = \underset{\text{factor}}{\underset{\uparrow}{5}} \cdot \underset{\text{factor}}{\underset{\uparrow}{8}} = \underset{\text{product}}{\underset{\uparrow}{40}}$$

Several notations can be used to indicate multiplication. In this text, we will use the raised dot (as with $5 \cdot 8$) and parentheses much of the time. The cross sign (×) will also be used; however, we must be careful not to confuse it with the letter x used in some of the later chapters and in algebra.

Symbols for Multiplication

Symbol	Example
· raised dot	$5 \cdot 8$
( ) numbers inside or next to parentheses	5(8) or (5)8 or (5)(8)
× cross sign	5×8 or $\begin{array}{r} 8 \\ \times\ 5 \\ \hline \end{array}$

To change a multiplication problem to a repeated addition problem every time we multiply two numbers would be ridiculous. For example, 24(36) would mean using 36 as an addend 24 times (or 24 as an addend 36 times). The first step in learning the multiplication process is to **memorize the basic multiplication facts** shown in Table 1.2.

EXAMPLE 1

$$6 + 6 + 6 + 6 = \underset{\text{factor}}{\underset{\uparrow}{4}} \cdot \underset{\text{factor}}{\underset{\uparrow}{6}} = \underset{\text{product}}{\underset{\uparrow}{24}}$$

4 and 6 are factors of 24.

EXAMPLE 2

$$9 + 9 + 9 + 9 + 9 + 9 + 9 = 7(9) = 63$$

factor factor product

Or, the numbers can be written vertically as

$$\begin{array}{r} 9 \\ \times\ 7 \\ \hline 63 \end{array}$$

← factor
← factor
← product

Table 1.2 Basic Multiplication Facts

·	**0**	**1**	**2**	**3**	**4**	**5**	**6**	**7**	**8**	**9**
0	**0**	0	0	0	0	0	0	0	0	0
1	0	**1**	2	3	4	5	6	7	8	9
2	0	2	**4**	6	8	10	12	14	16	18
3	0	3	6	**9**	12	15	18	21	24	27
4	0	4	8	12	**16**	20	24	28	32	36
5	0	5	10	15	20	**25**	30	35	40	45
6	0	6	12	18	24	30	**36**	42	48	54
7	0	7	14	21	28	35	42	**49**	56	63
8	0	8	16	24	32	40	48	56	**64**	72
9	0	9	18	27	36	45	54	63	72	**81**

Inspection of Table 1.2 indicates that multiplication is **commutative** (the table is a mirror image of itself on either side of the main diagonal). Also, multiplication by 0 gives a product of 0, and this result is called the **zero factor law.** Multiplication of any number by 1 leaves that number unchanged, and 1 is called the **multiplicative identity.** Also, multiplication can be shown to be **associative.** These properties are summarized in Table 1.3.

In Example 3, find the resulting products and tell what property of multiplication is illustrated.

COMPLETION EXAMPLE 3

(a) $2 \cdot 1 =$ ____________________

(b) $0 \cdot 8 =$ ____________________

(c) $2 \cdot (4 \cdot 6) = (2 \cdot 4) \cdot 6 =$ ____________________

(d) $5 \cdot 9 = 9 \cdot 5 =$ ____________________

Table 1.3 Properties of Multiplication with Whole Numbers

If a, b, c are whole numbers, then the following properties of multiplication are true.

Name	General Form	Example
Commutative Property (The order of multiplication can be reversed.)	$a \cdot b = b \cdot a$	$6 \cdot 7 = 7 \cdot 6$
Associative Property (The grouping can be changed.)	$(a \cdot b) \cdot c = a \cdot (b \cdot c)$	$(4 \cdot 5) \cdot 3 = 4 \cdot (5 \cdot 3)$
Multiplicative Identity (The product of a number and 1 is that same number. 1 is called the **multiplicative identity.**)	$a \cdot 1 = 1 \cdot a = a$	$85 \cdot 1 = 1 \cdot 85 = 85$
Zero Factor Law (Multiplication by 0 gives a product of 0.)	$a \cdot 0 = 0$	$29 \cdot 0 = 0$

Now Work Exercises 1–5 in the Margin.

Before we discuss the general technique of multiplication with whole numbers we discuss multiplication by **powers of ten.** The powers of ten are 1, 10, and all products of 10 multiplied by itself one or more times. Some of the powers of ten are

$$1$$
$$10$$
$$10 \cdot 10 = 100$$
$$10 \cdot 10 \cdot 10 = 1000$$
$$10 \cdot 10 \cdot 10 \cdot 10 = 10{,}000$$
$$10 \cdot 10 \cdot 10 \cdot 10 \cdot 10 = 100{,}000$$

and so on.

Find each of the indicated products and tell which property of multiplication is illustrated.

1. $7 \cdot 8 = 8 \cdot 7$

2. $5 \cdot 1 =$ ______

3. $2(3 \cdot 7) = (2 \cdot 3)7$

4. $(4 \cdot 4)2 = 4(4 \cdot 2)$

5. $16 \cdot 0 =$ ______

Find the following products mentally.

6. $8 \cdot 1000 =$ ________

7. $15 \cdot 100 =$ ________

8. $100 \cdot 25 =$ ________

9. $10 \cdot 300 =$ ________

10. $27 \cdot 10{,}000 =$ ________

As the following pattern of products illustrates, multiplication by powers of ten can be done mentally and quickly by writing the correct number of zeros:

$$5 \cdot 1 = 5 \quad \text{no zeros}$$
$$5 \cdot 10 = 50 \quad \text{one zero}$$
$$5 \cdot 100 = 500 \quad \text{two zeros}$$
$$5 \cdot 1000 = 5000 \quad \text{three zeros}$$
$$5 \cdot 10{,}000 = 50{,}000 \quad \text{four zeros}$$

and so on.

Example 4 illustrates how the commutative and associative properties of multiplication are related to the method of multiplication of numbers that end with 0's.

EXAMPLE 4

(a) $6 \cdot 90 = 6(9 \cdot 10) = (6 \cdot 9)10 = 54 \cdot 10 = 540$

(b) $3 \cdot 4000 = 3(4 \cdot 1000) = (3 \cdot 4)1000 = 12 \cdot 1000 = 12{,}000$

(c) $60 \cdot 900 = (6 \cdot 10)(9 \cdot 100) = (6 \cdot 9)(10 \cdot 100)$
$= 54 \cdot 1000 = 54{,}000$

(d) $500 \cdot 700 = (5 \cdot 7)(100 \cdot 100) = 35 \cdot 10{,}000 = 350{,}000$

Now Work Exercises 6–10 in the Margin.

Multiplication with Whole Numbers

We first illustrate multiplication by using expanded notation and multiplication by powers of ten. This technique will help you develop a more complete understanding of the more familiar multiplication process, which we will call the **short method.**

Multiplying Whole Numbers by Using Partial Products

1. Write the numbers vertically, one under the other.
2. Write each number (or think of each number) in expanded notation.
3. Find each partial product by using your knowledge of multiplication by powers of ten.
4. Add the partial products.

Note: The vertical lines inserted in the multiplication problems are there as a reminder and an aid to keep the digits neatly aligned to avoid mistakes.

EXAMPLE 5

Multiply $4 \cdot 68$.

$$\begin{array}{r} 68 \\ \times\ 4 \\ \hline \end{array} \qquad \begin{array}{r} 60 + 8 \\ 4 \\ \hline 240 + 32 = 272 \end{array}$$

240 + 32: partial products; 272: product

or

	6	8	
		4	
	3	2	$\leftarrow 4 \cdot 8$ (partial products)
2	4	0	$\leftarrow 4 \cdot 60$ (partial products)
2	7	2	← product

EXAMPLE 6

Multiply $37 \cdot 42$.

$$\begin{array}{r} 42 \\ \times 37 \\ \hline \end{array} \qquad \begin{array}{r} 40 + 2 \\ 30 + 7 \\ \hline 280 + 14 \\ 1200 + 60 \quad\ \ \\ \hline 1200 + 340 + 14 = 1554 \end{array}$$

or

		4	2	factor
		3	7	factor
		1	4	$(7 \cdot 2 = 14)$
	2	8	0	$(7 \cdot 40 = 280)$
		6	0	$(30 \cdot 2 = 60)$
1	2	0	0	$(30 \cdot 40 = 1200)$
1	5	5	4	product

✓ COMPLETION EXAMPLE 7

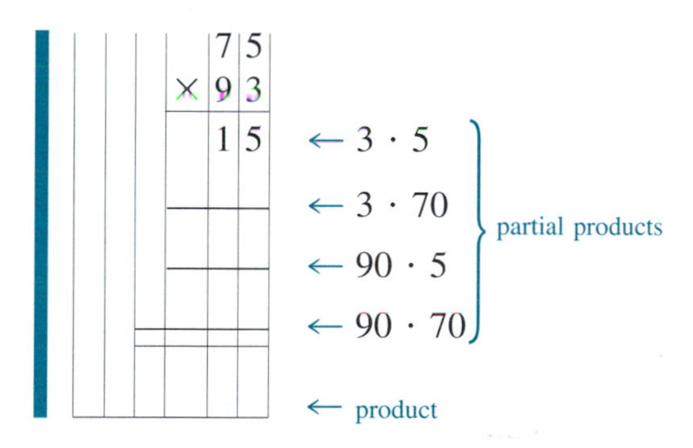

Now Work Exercises 11–13 in the Margin.

The method of writing all the partial products is recommended only as an aid in understanding the faster **short method,** which we use most of the time. The short method is discussed in detail in Example 8.

Find each product by using partial products.

11. $\begin{array}{r} 256 \\ \times\ \ 7 \\ \hline \end{array}$

12. $\begin{array}{r} 83 \\ \times\ 49 \\ \hline \end{array}$

13. $\begin{array}{r} 372 \\ \times\ 64 \\ \hline \end{array}$

EXAMPLE 8

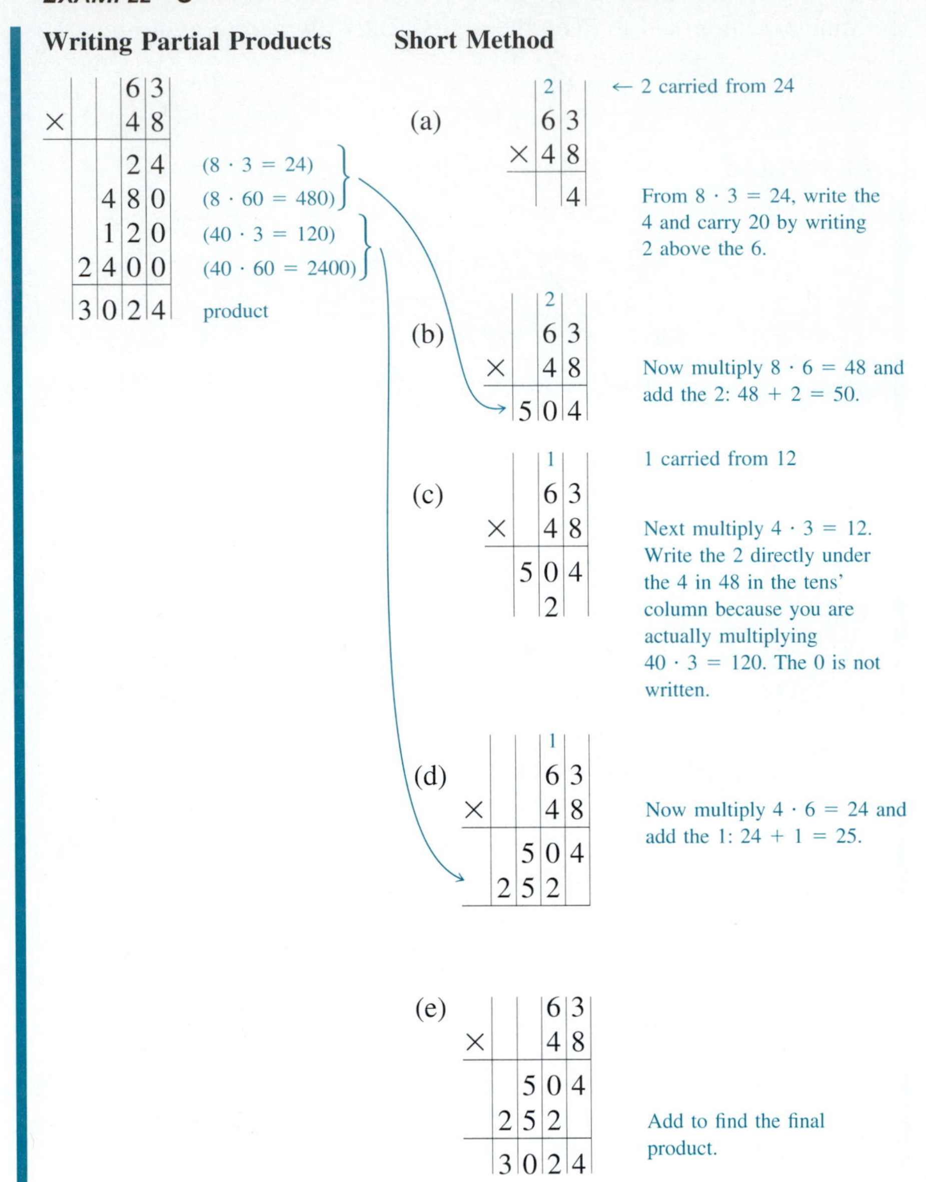

Steps (a), (b), (c), and (d) are shown for clarification of the Short Method. You will write only step (e).

In Examples 9 and 10, tell what product or sum gives the circled digits.

☑ COMPLETION EXAMPLE 9

$$
\begin{array}{r}
{}^{3}\\
{}^{1}\quad\\
45\\
\times\ 27\\
\hline
(31)5\\
(9)0\\
\hline
1215
\end{array}
$$

31 comes from 7 · 4 = 28 and 28 + 3 = 31

9 comes from ______________________

✓ COMPLETION EXAMPLE 10

```
      4
      1
      5 6
×     3 8
   (44)(8)
  (16) 8
  2 1 2 8
```

8 is from the product $8 \cdot 6 = 48$

44 comes from ____________________

16 comes from ____________________

Now Work Exercises 14–16 in the Margin.

Many times when large numbers are being multiplied, one or both of the numbers ends in 0. Multiplication by numbers ending in 0's can be simplified by using the following three steps:

1. Align the numbers so that the 0's are written to the right.
2. Multiply the numbers as if there were no 0's.
3. Rewrite the 0's in the product.

This technique is shown in Examples 11 and 12.

EXAMPLE 11

Multiply $423 \cdot 2000$.

```
    4 2 3
×       2 0 0 0
  8 4 6,0 0 0
```

Note that $2000 = 2 \cdot 1000$. So, multiply $423 \cdot 2$, and then multiply the result by 1000 by writing three 0's.

EXAMPLE 12

Multiply $265 \cdot 1500$.

```
      2 6 5
×       1 5 0 0
  1 3 2 5 0 0
  2 6 5
  3 9 7,5 0 0
```

Note that $1500 = 15 \cdot 100$. So, multiply $265 \cdot 15$, and then multiply the result by 100 by writing two 0's.

Now Work Exercises 17–19 in the Margin.

Find each of the following products by using the **Short Method.**

14. $\begin{array}{r} 142 \\ \times \quad 6 \\ \hline \end{array}$

15. $\begin{array}{r} 37 \\ \times 85 \\ \hline \end{array}$

16. $\begin{array}{r} 273 \\ \times \quad 24 \\ \hline \end{array}$

Find the following products.

17. $\begin{array}{r} 62 \\ \times \quad 800 \\ \hline \end{array}$

18. $\begin{array}{r} 47 \\ \times \quad 1300 \\ \hline \end{array}$

19. $\begin{array}{r} 354 \\ \times \quad 24{,}000 \\ \hline \end{array}$

Estimating Products with Whole Numbers

Estimating the product of two whole numbers before actually performing the multiplication can be of help in detecting errors. Just as with addition and subtraction, estimations of products are done with rounded-off numbers.

To Estimate a Product with Whole Numbers:

1. Round off each number to the place of the leftmost digit.
2. Using your knowledge of multiplying with powers of 10, multiply the rounded-off numbers.

Example 13

First (a) estimate the product $52 \cdot 38$, and then (b) find the product.

Solution

(a)

$$\begin{array}{r} 52 \\ \times\, 38 \\ \hline \end{array} \quad \begin{array}{c} \rightarrow \\ \rightarrow \end{array} \quad \begin{array}{r} 50 \\ \times\quad 40 \\ \hline 2000 \end{array}$$

50 rounded-off value of 52

40 rounded-off value of 38

2000 The actual product should be close to 2000.

(b)

$$\begin{array}{r} {\scriptstyle 1} \\ 52 \\ \times\, 38 \\ \hline 416 \\ 156 \\ \hline 1976 \end{array}$$

1976 The actual product is close to 2000.

The Concept of Area

Area is the measure of the interior, or enclosed region, of a plane surface and is measured in square units. The concept of area is illustrated in Figure 1.4 in terms of square inches.

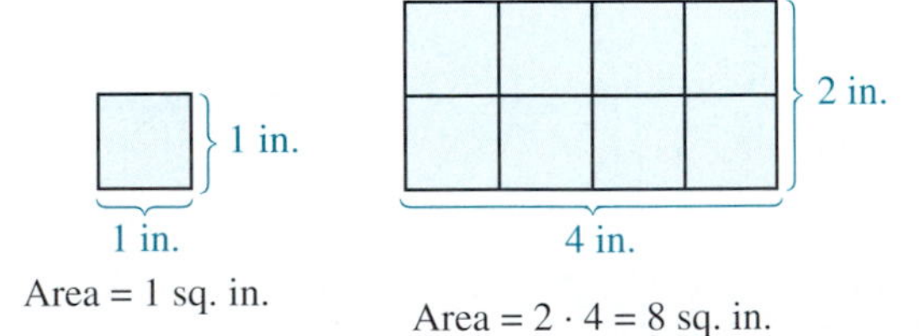

Area = 1 sq. in.

Area = 2 · 4 = 8 sq. in.

There are 8 squares that are each 1 sq. in. for a total of 8 sq. in.

Figure 1.4

Some of the units of area in the metric system are square meters, square decimeters, square centimeters, and square millimeters. In the U.S. customary system, some of the units of area are square feet, square inches, and square yards.

EXAMPLE 14

The area of a rectangle (measured in square units) is found by multiplying its length by its width. (a) Estimate the area of a rectangular lot for a house if the lot has a length of 123 feet and a width of 86 feet. (b) Find the exact area.

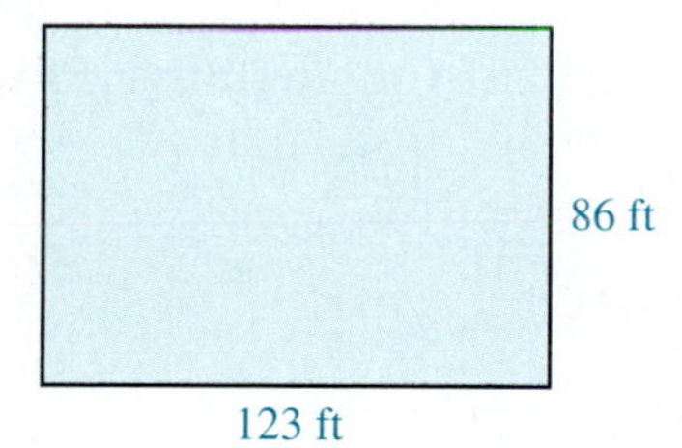

Solution

(a) We round off 123 to 100, round off 86 to 90, and then multiply $100 \cdot 90$.

$$\begin{array}{r} 100 \\ \times \quad 90 \\ \hline 9000 \end{array}$$ square feet (estimated area of lot size)

(b) To find the exact area, we multiply $123 \cdot 86$.

$$\begin{array}{r} 123 \\ \times \quad 86 \\ \hline 738 \\ 984 \\ \hline 10,578 \end{array}$$ square feet (exact area)

Completion Example Answers

3. (a) $2 \cdot 1 = \mathbf{2}$ **Multiplicative Identity**
   (b) $0 \cdot 8 = \mathbf{0}$ **Zero Factor Law**
   (c) $2 \cdot (4 \cdot 6) = (2 \cdot 4) \cdot 6 = \mathbf{48}$ **Associative Property of Multiplication**
   (d) $5 \cdot 9 = 9 \cdot 5 = \mathbf{45}$ **Commutative Property of Multiplication**

7.
$$\begin{array}{r} 75 \\ \times 93 \\ \hline 15 \\ \mathbf{210} \\ \mathbf{450} \\ \mathbf{6300} \\ \hline \mathbf{6975} \end{array} \begin{array}{l} \\ \\ \\ \leftarrow 3 \cdot 5 \\ \leftarrow 3 \cdot 70 \\ \leftarrow 90 \cdot 5 \\ \leftarrow 90 \cdot 70 \\ \\ \leftarrow \text{product} \end{array}$$
($3 \cdot 5$, $3 \cdot 70$, $90 \cdot 5$, $90 \cdot 70$: partial products)

9.
$$\begin{array}{r} {}^{3} \\ {}^{1} \\ 45 \\ \times 27 \\ \hline 315 \\ 90 \\ \hline 1215 \end{array}$$
31 comes from $7 \cdot 4 = 28$ and $28 + 3 = 31$
9 comes from $\mathbf{2 \cdot 4 = 8}$ **and** $\mathbf{8 + 1 = 9}$

10.
$$\begin{array}{r} {}_{1}\,{}^{4} \\ 56 \\ \times 38 \\ \hline 448 \\ 168 \\ \hline 2128 \end{array}$$
8 comes from the product $8 \cdot 6 = 48$
44 comes from $\mathbf{8 \cdot 5 = 40}$ **and** $\mathbf{40 + 4 = 44}$
16 comes from $\mathbf{3 \cdot 5 = 15}$ **and** $\mathbf{15 + 1 = 16}$

NAME ________________ SECTION ________________ DATE ________

ANSWERS

1. ______
2. ______
3. ______
4. ______
5. ______
6. ______
7. ______
8. ______
9. ______
10. ______
11. ______
12. ______
13. ______
14. ______
15. ______
16. ______
17. ______
18. ______
19. ______
20. ______
21. ______
22. ______
23. ______
24. ______
25. ______
26. ______
27. ______
28. ______

Exercises 1.4

Do the following problems mentally and write only the answers.

1. $8 \cdot 9$ **2.** $7 \cdot 6$ **3.** $6(4)$ **4.** $5(0)$

5. $1(7)$ **6.** $9 \cdot 5$ **7.** $0(3)$ **8.** $(7)(7)$

Find each product and tell which property of multiplication is illustrated.

9. $4 \cdot 7 = 7 \cdot 4$ **10.** $2(1 \cdot 6) = (2 \cdot 1)6$

11. $5 \cdot 1$ **12.** $8 \cdot 0$

Use the technique of multiplying by powers of ten to find the following products mentally.

13. $25 \cdot 10$ **14.** $47 \cdot 1000$ **15.** $20 \cdot 200$

16. $500 \cdot 700$ **17.** $200 \cdot 80$ **18.** $40 \cdot 2000$

19. $30 \cdot 30$ **20.** $300 \cdot 600$ **21.** $4000 \cdot 4000$

22. $900 \cdot 3000$ **23.** $10 \cdot 500$ **24.** $120 \cdot 300$

25. $\begin{array}{r} 6000 \\ \times \quad 6 \\ \hline \end{array}$ **26.** $\begin{array}{r} 90 \\ \times\ 90 \\ \hline \end{array}$ **27.** $\begin{array}{r} 300 \\ \times\ 500 \\ \hline \end{array}$ **28.** $\begin{array}{r} 700 \\ \times\ 80 \\ \hline \end{array}$

29. ____________
30. ____________
31. ____________
32. ____________
33. ____________
34. ____________
35. ____________
36. ____________
37. ____________
38. ____________
39. ____________
40. ____________
41. ____________
42. ____________
43. ____________
44. ____________
45. ____________
46. ____________
47. ____________
48. ____________
49. ____________
50. ____________
51. ____________
52. ____________

In the following problems, first estimate each product, and then find the product.

29. 56×4

30. 27×6

31. 48×9

32. 65×5

33. 84×3

34. 95×8

35. 42×56

36. 25×33

37. 48×20

38. 93×30

39. 83×85

40. 96×62

41. 17×32

42. 28×91

43. 20×44

44. 16×26

45. 25×15

46. 93×47

47. 24×86

48. 72×65

49. 12×13

50. 81×36

51. 126×41

52. 232×76

NAME ______________ SECTION ______________ DATE ______________

ANSWERS

53. $\begin{array}{r} 114 \\ \times\ 25 \\ \hline \end{array}$

54. $\begin{array}{r} 72 \\ \times 106 \\ \hline \end{array}$

55. $\begin{array}{r} 207 \\ \times 143 \\ \hline \end{array}$

56. $\begin{array}{r} 420 \\ \times 104 \\ \hline \end{array}$

57. $\begin{array}{r} 200 \\ \times\ 49 \\ \hline \end{array}$

58. $\begin{array}{r} 849 \\ \times 205 \\ \hline \end{array}$

59. $\begin{array}{r} 673 \\ \times 186 \\ \hline \end{array}$

60. $\begin{array}{r} 192 \\ \times 467 \\ \hline \end{array}$

61. Find the sum of eighty-four and one hundred forty-seven. Find the difference between ninety-six and thirty-eight. Then find the product of the sum and the difference.

62. Find the sum of thirty-four and fifty-five. Find the sum of one hundred seventeen and two hundred twenty. Find the product of the two sums.

63. If your beginning salary is $2300 per month and you are to get a raise once a year of $230 per month, by how much will your pay increase from year to year? How much money will you make over a five-year period?

53. ______________

54. ______________

55. ______________

56. ______________

57. ______________

58. ______________

59. ______________

60. ______________

61. ______________

62. ______________

63. ______________

64. ______________

65. ______________

66. [Respond below exercise.]

67. [Respond below exercise.]

64. If you rent an apartment with three bedrooms for $720 per month and you know that the rent will increase $40 per month every 12 months, what will you pay in rent during the first three years of living in this apartment? By how much will the rent increase each year?

65. Your company bought 18 new cars at a price of $12,800 per car. If each car had air-conditioning and antilock brakes, how much did your company pay for these cars?

In Exercises 66–70, first make an estimate of the answers. Then, after you have performed the actual calculation, check to see that your estimate and your answer are reasonably close.

66. A rectangular lot for a house measures 208 feet long by 175 feet wide. Find the area of the lot in square feet. (The area of a rectangle is found by multiplying the length times the width.)

67. Find the number of square meters of area needed for a rectangular swimming pool and deck if the plans call for a length of 215 meters and a width of 82 meters. (The area of a rectangle is found by multiplying the length times the width.)

NAME ______________________ SECTION ______________ DATE __________

ANSWERS

68. [Respond below exercise.] __________

69. [Respond below exercise.] __________

70. [Respond below exercise.] __________

71. __________

68. Network television has 12 minutes of commercial time in each hour. How many minutes of commercial time does a network have in one day's programming schedule of 20 hours? in one week?

69. According to the U.S. Fish and Wildlife Service, about 800,000 live birds (including about 250,000 parrots) are imported each year at a value of about $19 per bird. What is the total value of these imported birds?

70. According to the Department of Transportation, U.S. citizens drove some sort of vehicle an average of almost 6000 miles per person in 1990. If the 1990 census population is 248,709,873, how many miles did U.S. citizens drive?

Check Your Number Sense

71. You want to pay cash for your textbooks at the bookstore because you know that the cash line is shorter and moves faster than the other lines. If you are going to buy four books and their prices are between $35 and $45 each, about how much cash do you need to take with you: $100, $200, $300, or $400? (Don't forget the sales tax.)

72. [Respond in exercise.]

73. [Respond in exercise.]

74. [Respond below exercise.]

75. [Respond below exercise.]

72. Match each indicated product with the closest estimate of that product. Perform any calculations mentally.

Product	Estimate
___ (a) $16 \cdot 18$	A. 210
___ (b) 6(78)	B. 300
___ (c) $11 \cdot 32$	C. 400
___ (d) (8)(69)	D. 480
___ (e) 25(7)	E. 560

73. Match each indicated product with the closest estimate of that product. Perform any calculations mentally.

Product	Estimate
___ (a) 37(500)	A. 200
___ (b) 37(50)	B. 2000
___ (c) 37(5)	C. 20,000
___ (d) 37(5000)	D. 200,000

Writing and Thinking about Mathematics

74. Write, in your own words, the meaning of the Zero Factor Law.

75. Explain, in your own words, why 1 is called the Multiplicative Identity.

1.5 Division

OBJECTIVES

1. *Know the terms related to division: dividend, divisor, quotient, remainder.*
2. *Be able to divide with whole numbers.*
3. *Understand that when the remainder is 0, both the divisor and quotient are factors of the dividend.*
4. *Estimate quotients.*
5. *Know that division by 0 is not possible.*

Division with Whole Numbers

We know that $5 \cdot 12 = 60$ and that 5 and 12 are **factors** of 60. They are also called **divisors** of 60. In **division,** we want to find how many times one number is contained in another. How many 12's are in 60? There are five 12's in 60, and we say that 60 **divided by** 12 is 5 (or, using the division sign (÷), $60 \div 12 = 5$).

The process of division can be thought of as the reverse of multiplication. For example, if we know that the product of two numbers is 70 and one of the numbers is 10, then we know that the other number is 7 because $7 \cdot 10 = 70$, and this same idea can be written as $70 \div 10 = 7$. This relationship between multiplication and division can be seen by studying the following table format.

Division							**Multiplication**		
DIVIDEND		*DIVISOR*		*QUOTIENT*		*FACTORS*		*PRODUCT*
21	÷	7	=	3	since	$7 \cdot 3$	=	21
24	÷	6	=	4	since	$6 \cdot 4$	=	24
36	÷	4	=	9	since	$4 \cdot 9$	=	36
12	÷	2	=	6	since	$2 \cdot 6$	=	12

This table indicates that the number being divided is called the **dividend,** the number doing the dividing is called the **divisor,** and the result of division is called the **quotient.**

EXAMPLE 1

If one factor of 72 is 8, what is the corresponding factor?

Solution

We are looking for the quotient: $72 \div 8 =$ ______.
Since $72 \div 8 = 9$, the corresponding factor is 9.

Division does not always deal with **factors** (or **exact divisors**). Suppose we want to divide 23 by 4. By using repeated subtraction, we can find how many 4's are in 23. We continuously subtract 4 until the number left (called the **remainder**) is less than 4.

$$\begin{array}{r} 23 \\ -\ 4 \\ \hline 19 \end{array} \quad \begin{array}{r} 19 \\ -\ 4 \\ \hline 15 \end{array} \quad \begin{array}{r} 15 \\ -\ 4 \\ \hline 11 \end{array} \quad \begin{array}{r} 11 \\ -\ 4 \\ \hline 7 \end{array} \quad \begin{array}{r} 7 \\ -\ 4 \\ \hline 3 \end{array} \leftarrow \text{remainder}$$

(subtraction 5 times)

Another form used to indicate division as repeated subtraction is shown in Examples 2 and 3.

EXAMPLE 2

How many 7's are in 185?

```
 7)185
 − 140   ← subtract 20 sevens     (20 · 7 = 140)
                                  (30 · 7 = 210; too much since
    45                            210 is greater than 185)
  − 42   ← subtract  6 sevens     (6 · 7 = 42)
     3              26 sevens total
     ↑               ↑
 remainder        quotient
```

CHECK: Division is checked by multiplying the quotient and the divisor and then adding the remainder. The result should be the dividend.

```
  26  quotient        182
× 7   divisor       +   3  remainder
 182                  185  dividend
```

EXAMPLE 3

Find 275 ÷ 6 by using repeated subtraction, and check your work.

```
(a) 6)275
    −180   Subtract 30 sixes.    (30 · 6 = 180)
      95
    − 60   Subtract 10 sixes.    (10 · 6 = 60)
      35
    − 18   Subtract  3 sixes.    (3 · 6 = 18)
      17
    − 12   Subtract  2 sixes.    (2 · 6 = 12)
       5            45 sixes total
       ↑             ↑
   remainder      quotient
```

Note: You can subtract any number of sixes less than the quotient. But this will not lead to a good explanation of the shorter division algorithm. You should subtract the larger number of thousands, hundreds, tens, or units that you can at each step.

```
(b) 6)275
    −240   Subtract 40 sixes.    (40 · 6 = 240)
                                 (50 · 6 = 300; too much since
      35                         300 is greater than 275)
    − 30   Subtract  5 sixes.
       5            45 sixes total
       ↑             ↑
   remainder      quotient
```

CHECK: Division is checked by multiplying the quotient and the divisor and then adding the remainder. The result should be the dividend.

```
  45  quotient        270
× 6   divisor       +   5  remainder
 270                  275  dividend
```

Now Work Exercises 1 and 2 in the Margin.

The repeated subtraction technique for division provides a basis for understanding the **division algorithm,** a much shorter method of division with which we are generally familiar. (**Note:** An algorithm is a process or pattern of steps to be followed in working with numbers or solving a related problem.) Examples 4 and 5 illustrate the division algorithm in detail. Study them carefully before you work Completion Examples 6 and 7.

EXAMPLE 4

Find 2076 ÷ 8.

STEP 1:

```
      2      ← Write 2 in the hundreds position.
 8 ) 2076      200 eights (200 · 8 = 1600)
    −1600
      476
```

STEP 2:

```
     25      ← Write 5 in the tens position.
 8 ) 2076
    −1600
      476
     −400    ← 50 eights (50 · 8 = 400)
       76
```

STEP 3:

```
    259      ← Write 9 in the units position.
 8 ) 2076
    −1600
      476
     −400
       76
      −72    ← 9 eights (9 · 8 = 72)
        4
```

SUMMARY: The process can be shortened by not writing all the 0's and **bringing down** only one digit at a time.

```
   259 R4
8 ) 2076
    16
     47    ← Bring down the 7 only; then divide 8 into 47.
     40
      76   ← Bring down the 6; then divide 8 into 76.
      72
       4
```

Divide by using repeated subtraction, then check your answer.

1. 9) 8 4 9

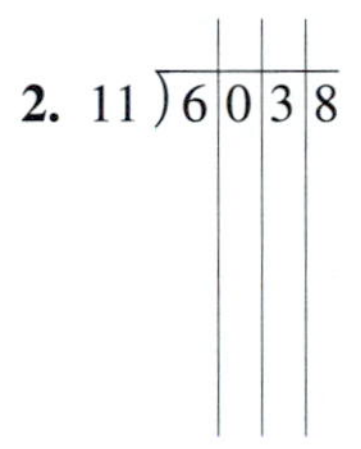

2. 11) 6 0 3 8

EXAMPLE 5

Find 9325 ÷ 45.

STEP 1:

```
       2
45 ) 9325
     90
      3
```

Trial divide 40 into 90 or 4 into 9, giving 2 in the hundreds position.

STEP 2:

```
       20
45 ) 9325
     90
      32
       0
```

45 will not divide into 32, so write 0 in the tens column and multiply $0 \cdot 45 = 0$.

STEP 3:

```
      208
45 ) 9325
     90
      32
       0
      325
      340
```

Trial divide 45 into 325 or 4 into 32. The trial quotient is too large, since $8 \cdot 45 = 340$ and 340 is larger than 325.

```
      207 R10
45 ) 9325
     90
      32
       0
      325
      315
       10
```

Now the trial divisor is 7. Since $7 \cdot 45 = 315$ and 315 is smaller than 325, 7 is the desired number.

CHECK:

```
   207          9315
 ×  45        +   10
  1035          9325
  828
  9315
```

Special Note about 0 in the Quotient: In Step 2 of Example 5 we wrote 0 in the quotient because 45 did not divide into 32. Be sure to write 0 in the quotient whenever the divisor does not divide into one of the partial remainders.

✓ COMPLETION EXAMPLE 6

Find the quotient and remainder.

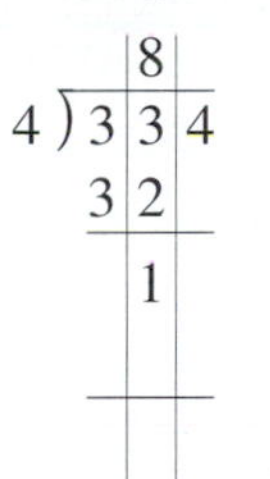

remainder

✓ COMPLETION EXAMPLE 7

Find the quotient and remainder.

$$\begin{array}{r} 2 \\ 12\overline{)2451} \\ \underline{24} \\ 05 \end{array}$$

remainder

Now Work Exercises 3–5 in the Margin.

If the remainder is 0, then the following statements are true:

1. Both the divisor and quotient are **factors** of the dividend.
2. We say that both factors **divide exactly** into the dividend.
3. Both factors are called **divisors** of the dividend.

Estimating Quotients with Whole Numbers

By rounding off both divisor and dividend, then dividing, we can estimate the quotient. As with any estimation, one purpose is to ensure that the actual value calculated is reasonable and does not contain a major error.

To Estimate a Quotient:

1. Round off both the divisor and the dividend to the place of the leftmost digit.
2. Divide using the rounded-off numbers.

This process is very similar to the trial dividing step in the division algorithm.

Divide by using the division algorithm, then check your answer.

3. $325 \div 7$

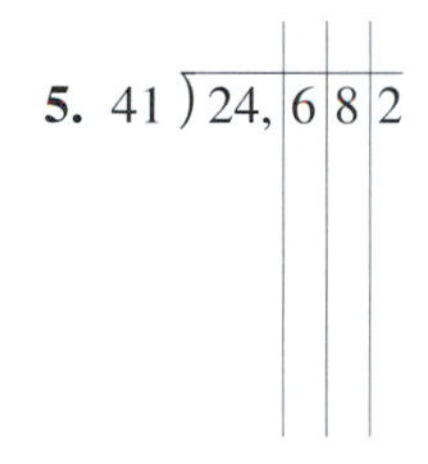

4. $16\overline{)324}$

5. $41\overline{)24,682}$

EXAMPLE 8

Estimate the quotient $325 \div 42$ by using rounded-off numbers; then find the quotient.

Solution

(a) Estimation: $325 \div 42 \rightarrow 300 \div 40$

$$\begin{array}{r} 7 \\ 40\overline{)300} \\ \underline{280} \\ 20 \end{array} \quad \text{estimated quotient}$$

(b) The quotient should be near 7.

$$\begin{array}{rl} 7 & \text{quotient} \\ 42\overline{)325} & \\ \underline{294} & \\ 31 & \text{remainder} \end{array}$$

In this case, the quotient is the same as the estimated value. The true remainder is different.

An Adjustment to the Process of Estimating Answers

Even though we have given a rule for rounding off to the leftmost digit for estimating an answer, this "rule" is flexible, and there are times when using two digits gives simpler calculations and more accurate estimates. Such a case is illustrated in Example 9, where 3 is easily seen to divide into 15. Thus, estimating answers does involve some basic understanding and intuitive judgment, and there is no one best way to estimate answers. However, this "adjustment" to the leftmost digit rule is more applicable to division than it is to addition, subtraction, or multiplication.

EXAMPLE 9

Estimate the quotient $148{,}062 \div 26$; then find the quotient.

Solution

(a) Estimation:

$$148{,}062 \div 26 \rightarrow 150{,}000 \div 30$$

In this case, we rounded off using the two leftmost digits for 148,062 because 150,000 can be divided evenly by 30.

```
         5 000   ← approximate quotient
   30 ) 150,000
        150
         0 0
         0 0
           00
           00
            00
            00
             0   remainder
```

(b) The quotient should be near 5000.

```
         5 694   ← quotient
   26 ) 148,062
        130
         18 0
         15 6
           2 46
           2 34
            122
            104
             18   remainder
```

☑ COMPLETION EXAMPLE 10

Find the quotient and remainder.

```
         3 0
   21 ) 6 4 6 1
        6 3
          1 6


                  remainder
```

Now estimate the quotient. Do this by mentally dividing rounded-off numbers:

```
               ← estimate
   20 ) 6000
```

Is your estimate close to the actual quotient? __________

What is the difference? __________

Now Work Exercises 6 and 7 in the Margin.

For completeness, we close this section with two rules about division with 0. These rules will be discussed again in detail in Chapter 3.

Estimate each of the following quotients, but do not find the actual quotients.

6. $12\overline{)1869}$

7. $39\overline{)83,700}$

Division with 0

1. If a is any nonzero whole number, then

$$0 \div a = 0.$$

2. If a is any whole number, then

$a \div 0$ is **undefined.**

Completion Example Answers

6.
```
     82
4 ) 334
    32
     14
     12
      2   remainder
```

7.
```
      204
12 ) 2451
     24
      05
       0
      51
      48
       3   remainder
```

10.
```
      307
21 ) 6461
     63
      16
       0
     161
     147
      14   remainder
```

Now estimate the quotient by mentally dividing rounded-off numbers:

```
      300   ← estimate
20 ) 6000
```

Is your estimate close to the actual quotient? **Yes**
What is the difference? **7**

NAME ______________ SECTION ______________ DATE ______________

Exercises 1.5

Find the following quotients. (You should be able to do these mentally since they are related to the multiplication tables.)

1. $12 \div 3$ **2.** $6 \div 2$ **3.** $18 \div 2$ **4.** $49 \div 7$

5. $64 \div 8$ **6.** $20 \div 5$ **7.** $20 \div 4$ **8.** $25 \div 5$

9. $30 \div 6$ **10.** $35 \div 7$ **11.** $40 \div 8$ **12.** $42 \div 7$

13. $56 \div 7$ **14.** $30 \div 5$ **15.** $72 \div 8$ **16.** $63 \div 9$

17. $63 \div 7$ **18.** $15 \div 3$ **19.** $24 \div 3$ **20.** $54 \div 6$

21. $0 \div 8$ **22.** $0 \div 5$ **23.** $6 \div 6$ **24.** $2 \div 2$

ANSWERS

1. ______
2. ______
3. ______
4. ______
5. ______
6. ______
7. ______
8. ______
9. ______
10. ______
11. ______
12. ______
13. ______
14. ______
15. ______
16. ______
17. ______
18. ______
19. ______
20. ______
21. ______
22. ______
23. ______
24. ______

25. [Respond below exercise.] ______________

26. [Respond below exercise.] ______________

27. [Respond below exercise.] ______________

28. [Respond below exercise.] ______________

29. [Respond below exercise.] ______________

30. [Respond below exercise.] ______________

31. [Respond below exercise.] ______________

32. [Respond below exercise.] ______________

33. [Respond below exercise.] ______________

34. [Respond below exercise.] ______________

35. [Respond below exercise.] ______________

36. [Respond below exercise.] ______________

37. [Respond below exercise.] ______________

38. ______________

39. ______________

40. ______________

In Exercises 25–37, find the quotient and remainder using the method of repeated subtraction. State whether or not the divisor and quotient are factors of the dividend.

25. $210 \div 7$

26. $168 \div 8$

27. $132 \div 11$

28. $75 \div 15$

29. $51 \div 3$

30. $600 \div 25$

31. $413 \div 20$

32. $161 \div 15$

33. $182 \div 13$

34. $3\overline{)98}$

35.

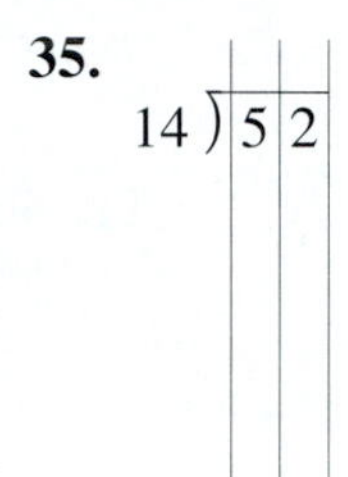

$14\overline{)52}$

36.

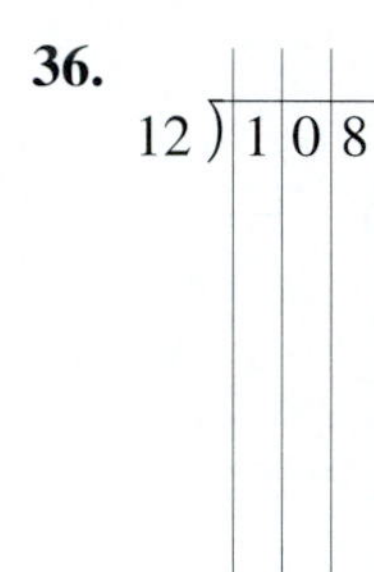

$12\overline{)108}$

37.

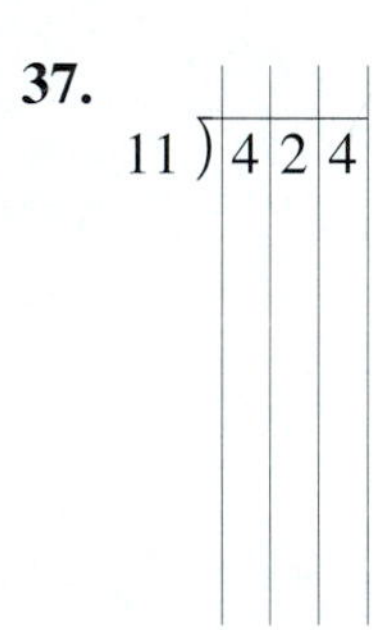

$11\overline{)424}$

In Exercises 38–55, divide using the division algorithm after first estimating the quotient using rounded-off numbers.

38. $16\overline{)128}$

39.

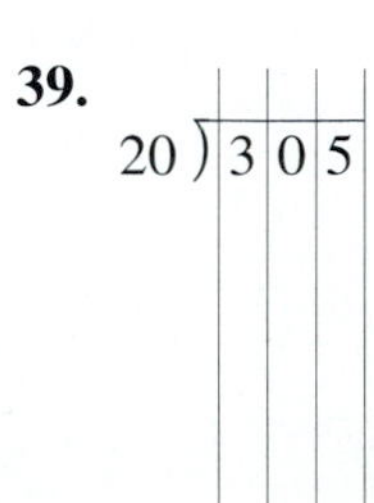

$20\overline{)305}$

40.

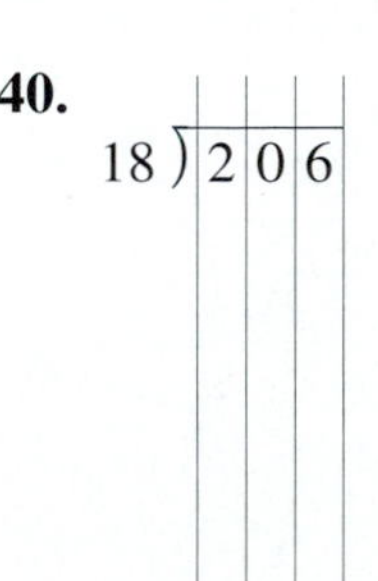

$18\overline{)206}$

NAME ______________ SECTION ______________ DATE ______________

41. 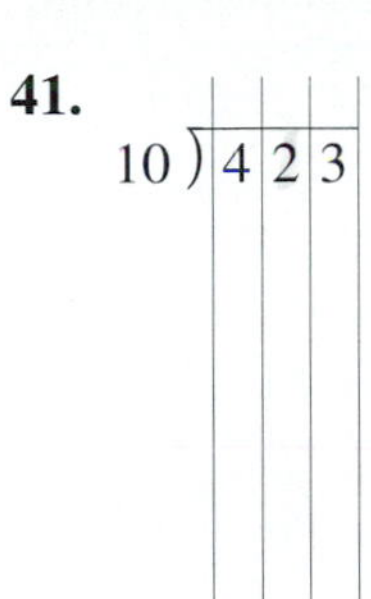 $10\overline{)423}$

42. $15\overline{)750}$

43. 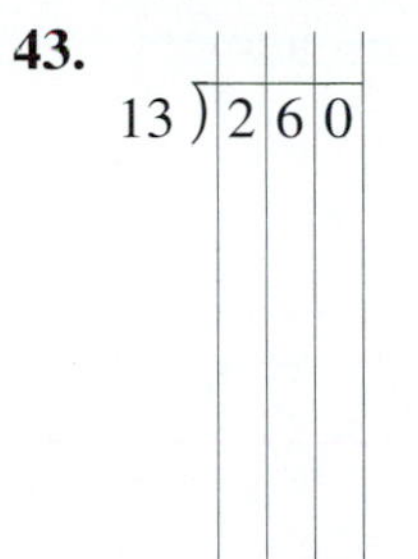 $13\overline{)260}$

44. $12\overline{)360}$

45. $19\overline{)7603}$

46. $16\overline{)4813}$

47. $13\overline{)3917}$

48. $73\overline{)148}$

49. $68\overline{)207}$

50. $50\overline{)3065}$

51. $40\overline{)2163}$

52. $105\overline{)210}$

53. $213\overline{)4760}$

54.   $716\overline{)3056}$

55. $630\overline{)4768}$

ANSWERS

41. ______________

42. ______________

43. ______________

44. ______________

45. ______________

46. ______________

47. ______________

48. ______________

49. ______________

50. ______________

51. ______________

52. ______________

53. ______________

54. ______________

55. ______________

56a. ______________

b. ______________

57a. ______________

b. ______________

58. [Respond below exercise.] ______________

59. [Respond below exercise.] ______________

56. Suppose your income for one year was $30,576 and your income was the same each month. (a) Estimate your monthly income. (b) What was your exact monthly income?

57. A high school bought 3075 new textbooks for a total price of $116,850. (a) What was the approximate price of each text? (b) What was the exact price of each text?

58. Show that 28 and 36 are both factors of 1008 by using division.

59. Show that 45 and 702 are both factors of 31,590 by using division.

NAME ________________ SECTION ________________ DATE ________

ANSWERS

60. ________

61. [Respond in exercise.]

62. [Respond in exercise.]

63. [Respond below exercise.]

60. The United States has a population of about 248,400,000 people and a land area of about 3,600,000 square miles. About how many people are there for each square mile? (**Note:** This may not be true in large cities such as New York and Chicago or in the state of New Mexico. In other words, the way raw numbers are used does not always give a true-to-life picture. In many applications, some judgment must be used as well as numerical skills.)

Check Your Number Sense

61. Match each indicated quotient with the closest estimate of that quotient. Perform as many calculations mentally as you can.

	Quotient	Estimate
____	(a) $9\overline{)910}$	A. 6
____	(b) $34 \div 5$	B. 10
____	(c) $34\overline{)12{,}000}$	C. 100
____	(d) $18\overline{)3900}$	D. 200
____	(e) $216 \div 18$	E. 300

62. Match each indicated quotient with the closest estimate of that quotient.

	Quotient	Estimate
____	(a) $3\overline{)870}$	A. 30
____	(b) $3\overline{)87{,}000}$	B. 300
____	(c) $3\overline{)87}$	C. 3000
____	(d) $3\overline{)8700}$	D. 30,000

Writing and Thinking about Mathematics

63. Nothing is said in this text about division being a commutative operation. Do you think that there might be a commutative property for division? Give several examples that help justify your answer.

64. [Respond below exercise.]

65. [Respond below exercise.]

64. There is no mention in this text of an associative property for division. Do you think that division is associative? Give several examples that help justify your answer.

Collaborative Learning Project

65. Separate the class into teams of two to four students. Each team is to list 10 jobs or work situations in which arithmetic skills are a necessary part of the work. After the lists are complete, each team leader is to read the team's list with classroom discussion to follow.

1.6 Problem Solving with Whole Numbers

OBJECTIVES

1. *Develop the reading skills necessary for understanding word problems.*
2. *Be able to analyze a word problem with confidence.*
3. *Realize that problem solving takes time and that the learning process also involves learning from making mistakes.*

In this section, the problem solving can involve various combinations of the four operations of addition, subtraction, multiplication, and division. Decisions on which operations to use are generally based on experience and practice as well as certain key words. **If you are unsure of which operations to use, at least try something. If you make an error in technique or judgment, at least you are learning what does not work. If you do nothing, then you learn nothing. Do not be embarrassed if you make mistakes.**

Each of the problems discussed here will come under one of the following headings: Number Problems, Consumer Items, Checking Accounts, Geometry, and Average. The steps in the basic strategy listed here will help give an organized approach to problem solving regardless of the type of problem. These steps were developed by George Pōlya, a famous educator and mathematician from Stanford University.

Basic Strategy for Solving Word Problems

1. Read each problem carefully until you understand the problem and know what is being asked for.
2. Draw any type of figure or diagram that might be helpful and decide what operations are needed.
3. Perform these operations.
4. Mentally check to see if your answer is reasonable and see if you can think of another more efficient or more interesting way to do the same problem.

Number problems usually contain key words or phrases that tell which operations are to be performed with the numbers. Learn to look for these key words.

Key Words that Indicate Operations

Addition	Subtraction	Multiplication	Division
add	subtract (from)	multiply	divide
sum	difference	product	quotient
plus	minus	times	
more than	less than	twice	
increased by	decreased by		

Problem Solving Examples

EXAMPLE 1 Number Problems

Find the **difference** between the **product** of 75 and 28 and the **sum** of 213 and 426.

Solution

The strategy is to identify the key words (in boldface print) and to perform the operations indicated by these words. [**Note:** The word **and** does **not** indicate any type of operation. It is used three times in the statement of the problem, each time only as a grammatical connector (a conjunction).]

Before we can find the difference, we need to find the numbers that are to be used in the subtraction—namely, the product and the sum.

Product:

$$\begin{array}{r} {}^{1\ 4} \\ 75 \\ \times 28 \\ \hline 600 \\ 150 \\ \hline 2100 \end{array} \quad \text{product}$$

Sum:

$$\begin{array}{r} 213 \\ +426 \\ \hline 639 \end{array} \quad \text{sum}$$

Now we can subtract:

$$\begin{array}{r} 2100 \\ -\ 639 \\ \hline 1461 \end{array} \quad \text{difference}$$

Thus, the requested difference is 1461. (As a quick mental check, calculate $80 \cdot 30 = 2400$ and $200 + 400 = 600$, and find the difference between these two estimates: $2400 - 600 = 1800$.)

EXAMPLE 2 Consumer Items

Carol bought a car for \$15,000. The salesperson added \$1200 for taxes and \$450 for license fees. If Carol made a down payment of \$4650 and financed the rest through her credit union, how much did she finance?

Solution

Find the total cost of the car by adding the expenses, and then subtract the down payment. (**Note:** The key word "adding" does help us here. However, only real-life experience tells us to perform both operations—addition and subtraction.)

$$\begin{array}{r} \$15,000 \\ 1,200 \\ +\quad 450 \\ \hline \$16,650 \end{array} \quad \text{total cost}$$

$$\begin{array}{rl} \$16,650 & \text{total cost} \\ -\quad 4,650 & \text{down payment} \\ \hline \$12,000 & \text{to be financed} \end{array}$$

Carol financed \$12,000.

MENTAL CHECK: The car cost about \$17,000 and she put down about \$5000. So, an answer of \$12,000 is very reasonable.

EXAMPLE 3 Checking Account

In July Mr. Martinez opened a checking account and deposited \$6850. During the month, he made another deposit of \$1500 and wrote checks for \$775, \$86, \$450, \$174, and \$1320. What was the balance in his account at the end of the month?

Solution

To find the balance, we find the difference between the sum of the deposit amounts and the sum of the check amounts.

$$\begin{array}{r} \$6850 \\ +\ 1500 \\ \hline \$8350 \end{array} \text{ sum of deposits} \qquad \begin{array}{r} \$775 \\ 86 \\ 450 \\ 174 \\ +1200 \\ \hline \$2685 \end{array} \text{ sum of checks} \qquad \begin{array}{r} \$8350 \\ -\ 2685 \\ \hline \$5665 \end{array} \text{ balance}$$

The balance in the account was \$5665 at the end of the month.

MENTAL CHECK: Using rounded-off numbers, \$8000 − \$3000 = \$5000.

EXAMPLE 4 Geometry

A triangle has three sides: a base of 3 feet, a height of 4 feet, and a third side of 5 feet. (See diagram below.) This triangle is called a right triangle because one angle is 90°. Find the perimeter of (distance around) the triangle. Find the area of the triangle in square feet. (To find area, multiply base times height, then divide by 2.)

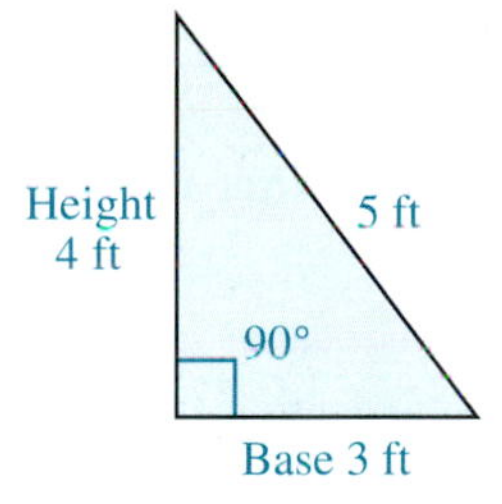

Solution

Perimeter (in feet)

$$\begin{array}{r} 3 \\ 4 \\ +\ 5 \\ \hline 12 \end{array} \text{ feet (perimeter)}$$

Area (in square feet)

$$\begin{array}{r} 3 \\ \times\ 4 \\ \hline 12 \end{array} \qquad \begin{array}{r} 6 \\ 2\,\overline{)\,12} \\ \underline{12} \\ 0 \end{array} \text{ square feet (area)}$$

The perimeter is 12 feet and the area is 6 square feet.

Average

A topic closely related to addition and division is **average.** Your grade in this course may be based on the average of your exam scores. Newspapers and magazines have information about the Dow-Jones stock averages, the average income of American families, the average life expectancy of dogs

1. Find the average of the four numbers 78, 93, 105, and 128.

2. The three sides of a triangle measure 25 centimeters, 60 centimeters, and 65 centimeters in length. This triangle is a right triangle (see Example 4) and the longest side is called the **hypotenuse.** (a) Find the perimeter of (distance around) the triangle. (b) Find the area of the triangle in square centimeters. (To find the area, multiply the base times the height and divide the product by 2.) As an aid, draw a sketch of the triangle and label the lengths of the sides.

and cats, and so on. The average of a set of numbers is a sort of "middle number" of the set.

The average of a set of numbers is found by dividing the sum of the numbers by the number of numbers in the set. The average is also called the **arithmetic average,** or **mean.**

The average of a set of whole numbers is not always a whole number. However, in this section, the problems are set up so that the averages *are* whole numbers. Other cases involving fractions and decimals will be discussed later.

EXAMPLE 5 Average

Find the average of the three numbers 32, 47, and 23.

Solution

$$\begin{array}{r} 32 \\ 47 \\ +\ 23 \\ \hline 102 \end{array} \qquad \begin{array}{r} 34 \\ 3\overline{)102} \\ \underline{9} \\ 12 \\ \underline{12} \\ 0 \end{array} \quad \text{average}$$

(The sum, 102, is divided by 3 because three numbers were added together to get 102.)

EXAMPLE 6 Average

On an English exam, two students score 95 points, five students score 86 points, one student scores 82 points, one student scores 78 points, and six students score 75 points. What is the mean score of the class?

Solution

$$\begin{array}{r} 95 \\ \times\ 2 \\ \hline 190 \end{array} \qquad \begin{array}{r} 86 \\ \times\ 5 \\ \hline 430 \end{array} \qquad \begin{array}{r} 82 \\ \times\ 1 \\ \hline 82 \end{array} \qquad \begin{array}{r} 78 \\ \times\ 1 \\ \hline 78 \end{array} \qquad \begin{array}{r} 75 \\ \times\ 6 \\ \hline 450 \end{array}$$

We multiply rather than write down all 15 scores. However, when the five products are added together, we divide the sum by 15 because the sum represents 15 scores.

$$\begin{array}{r} 190 \\ 430 \\ 82 \\ 78 \\ +\ 450 \\ \hline 1230 \end{array} \qquad \begin{array}{r} 82 \\ 15\overline{)1230} \\ \underline{120} \\ 30 \\ \underline{30} \end{array} \quad \text{mean score}$$

The class mean is 82 points.

Now Work Exercises 1 and 2 in the Margin.

NAME ______________ SECTION ______________ DATE ______________

ANSWERS

Exercises 1.6

Number Problems

1. Find the sum of the three numbers 846, 950, and 783. Then subtract 579. What is the quotient if the difference is divided by 125?

2. The difference between 8000 and 1895 is added to 1296. If this sum is then multiplied by 900, what is the product?

3. If the product of 607 and 93 is divided by 3, what is the quotient? Is 3 a factor of the product? Explain briefly.

4. If the difference between 347 and 196 is multiplied by 15, what is the product? Is 5 a factor of this product? Explain briefly.

1. ______________

2. ______________

3. [Respond below exercise.]

4. [Respond below exercise.]

5. ______________

6. ______________

7. ______________

8. ______________

Consumer Items

5. To purchase a new refrigerator for $1200 including tax, Mr. Kline paid $240 down and the remainder in six equal monthly payments. What were his monthly payments?

6. Mike decided to go shopping for school clothes before college started in the fall. How much did he spend if he bought four pairs of pants for $21 each, five shirts for $18 each, three pairs of socks for $4 a pair, and two pairs of shoes for $38 a pair?

7. To purchase a new dining room set for $1200, Mrs. Steel had to pay an additional $72 in sales tax. If she made a deposit of $486, how much did she still owe?

8. Alan wants to buy a new car. He could buy a red one for $8500 plus $510 in sales tax and $135 in fees; or he could buy a blue one for $8700 plus $522 in sales tax and $140 in fees. If the manufacturer is giving a $250 rebate on the blue model, which car would be cheaper for Alan? How much cheaper?

NAME ______________________ SECTION ______________ DATE __________

ANSWERS

9. ______________

10. ______________

11. ______________

12. ______________

9. Lynn decided to take up surfing. She bought a new surfboard for \$675, a wet suit for \$130, a beach towel for \$12, and a new swimsuit for \$57. How much money did she spend? (Sales tax was included in the prices.)

10. Pat needed art supplies for a new course at the local community college. She bought a portfolio for \$32, a zinc plate for \$44, etching ink for \$12, and three sheets of rag paper for a total of \$6. She received a student discount of \$9. How much did she spend on art supplies?

Checking Account

11. If you opened a checking account with \$875, then wrote checks for \$20, \$35, \$115, \$8, and \$212, what would be your balance?

12. Your friend had a checking account balance of \$1250 and wrote checks for \$375, \$52, \$83, and \$246. What was her new balance?

13. ______________

14. ______________

15a. ______________

b. ______________

16. ______________

13. On August 1, Matt had a balance of $250 in his checking account. During August, he made deposits of $200, $350, and $236. He wrote checks for $487, $25, $33, and $175. What was his balance on September 1?

14. Steve deposited $500, $2470, $800, $3562, and $2875 in his checking account over a five-month period. He wrote checks totaling $6742. If his beginning balance was $1400, what was his balance at the end of the five months?

Geometry

15. A rectangle is a four-sided figure with opposite sides equal and all four angles equal. (Each angle is 90°.) (a) Find the perimeter of a rectangle that has a width of 15 meters and a length of 37 meters. (b) Find its area (multiply length times width) in square meters.

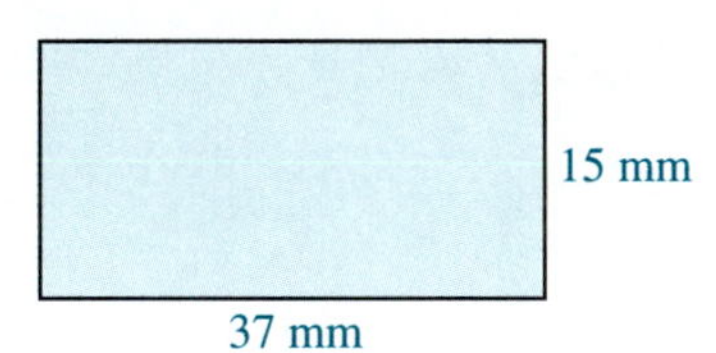

16. A regular hexagon is a six-sided figure with all six sides equal and all six angles equal. Find the perimeter of a regular hexagon if one side measures 19 centimeters.

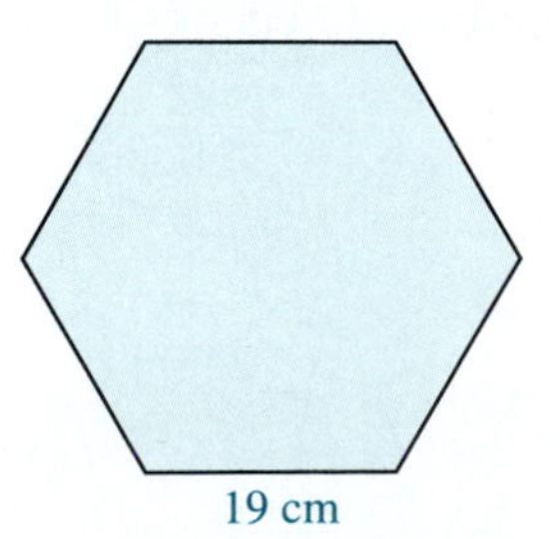

NAME ______________ SECTION ______________ DATE ______________

ANSWERS

17. ______________

18. ______________

19. ______________

20. ______________

21. ______________

22. ______________

23. ______________

24. ______________

25. ______________

17. An isosceles triangle (two sides equal) is placed on top of a square to form a window, as shown in the figure below. If each of the two equal sides of the triangle is 18 inches long and the square is 28 inches on a side, what is the perimeter of the window?

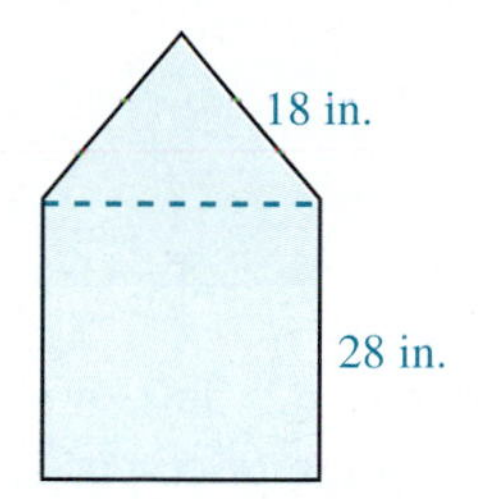

18. A rectangular picture is mounted in a rectangular frame with a border (called a mat). If the picture is 12 inches by 18 inches and the frame is 16 inches by 24 inches, what is the area of the mat? (Area of a rectangle is length times width.)

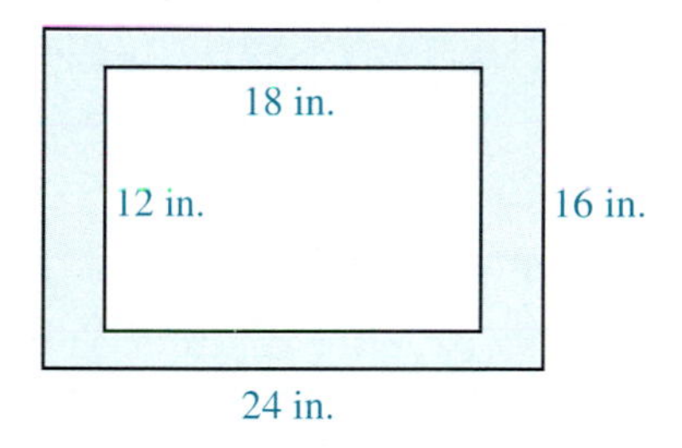

Average

In Exercises 19–24, find the average (or mean) of each set of numbers.

19. 102, 113, 97, 100

20. 56, 64, 38, 58

21. 6, 8, 7, 4, 4, 5, 6, 8

22. 5, 4, 5, 6, 5, 8, 9, 6

23. 512, 618, 332, 478

24. 436, 520, 630, 422

25. On a history exam, two students scored 95, six scored 90, three scored 80, and one scored 50. What was the class average?

26. ______________

27. ______________

28. ______________

29a. ______________

b. ______________

26. A salesman sold items from his sales list for $972, $834, $1005, $1050, and $799. What was the average price per item?

27. Ms. Lee bought 100 shares of stock in Microsoft at $14 per share. Two months later she bought another 200 shares at $20 per share. What average price per share did she pay? If she sold all 300 shares at $22 per share, what was her profit?

28. Three families, each with two children, had incomes of $15,942. Two families each with four children, had incomes of $18,512. Four families, each with two children, had incomes of $31,111. One family had no children and an income of $20,016. What was the mean income per family?

29. During July, Mr. Rodriguez made deposits in his checking account of $400 and $750 and wrote checks totaling $625. During August, his deposits were $632, $322, and $798, and his checks totaled $978. In September, his deposits were $520, $436, $200, and $376, and his checks totaled $836. (a) What was the average monthly difference between his deposits and his withdrawals? (b) What was his bank balance at the end of September if he had a balance of $500 on July 1?

NAME ______________ SECTION ______________ DATE ______________

ANSWERS

30. ______________

31. ______________

32. ______________

30. In order to fly 12 trips in one month (30 days), an airline pilot spent the following number of hours in preparation and flight time: 6, 8, 9, 6, 7, 7, 7, 5, 6, 6, 6, and 11 hours. What was the mean amount of time he spent per flight?

31. The 10 largest cities in South Carolina have the following approximate populations:

Columbia	98,000	Sumter	41,900
Charleston	80,400	Rock Hill	41,600
North Charleston	70,200	Mount Pleasant Town	30,100
Greenville	58,300	Florence	29,800
Spartanburg	44,000	Anderson	26,200

What is the average population of these cities?

32. The five longest rivers in the world are the:

Nile (Africa)	4180 miles
Amazon (South America)	3900 miles
Mississippi-Missouri-Red Rock (North America)	3880 miles
Yangtze (China)	3600 miles
Ob (Russia)	3460 miles

What is the mean length of these rivers?

33a. ______________

b. ______________

c. ______________

d. ______________

33. The following list is repeated from **Mathematics at Work!** at the beginning of this chapter. Tuition, room, and board are given as annual figures (rounded to the nearerst ten and subject to change) for the year 1992.

Institution		Tuition		
	Enrollment	Res.	Nonres.	Room/Board
Univ. of Alabama	15,940	$ 2,010	$ 5,020	$3,300
Bryn Mawr College	1,180	16,170	16,170	6,150
Univ. of Colorado	10,410	1,970	9,900	3,540
Cornell Univ.	7,600	16,190	16,190	5,410
Duke University	6,020	16,120	16,120	5,240
Harvard and Radcliffe Colleges	6,620	17,670	17,670	5,840
Univ. of Illinois	26,370	3,060	6,800	3,900
Indiana University	25,310	2,370	6,900	3,370
Univ. of Miami	8,640	15,050	15,050	5,910
Univ. of Notre Dame	7,520	13,500	13,500	3,600
Stanford University	6,510	16,540	16,540	6,310
Univ. of Texas	37,000	1,100	4,220	3,400
UCLA	24,370	2,900	10,600	5,410
Vassar College	2,310	17,210	17,210	5,500
Yale University	5,150	17,500	17,500	6,200

For these universities and colleges, find (a) the average enrollment, (b) the average tuition for residents, (c) the average tuition for nonresidents, and (d) the average cost of room and board.

1 Index of Key Ideas and Terms

NAME ______________ SECTION ______________ DATE ______________

Chapter 1 Test

Write the following whole numbers in their expanded notation and in their English word equivalents.

1. 653

2. 8952

3. Give an example that illustrates the commutative property of addition.

4. Give an illustration of the use of the multiplicative identity.

Round off as indicated.

5. 1342 (nearest ten)

6. 15,840 (nearest thousand)

7. 249,600 (nearest ten thousand)

First estimate each sum and then find the sum.

8.
$$\begin{array}{r} 9586 \\ 345 \\ +\ 2978 \\ \hline \end{array}$$

9.
$$\begin{array}{r} 1{,}480{,}900 \\ 2{,}576{,}850 \\ 340{,}200 \\ +\ \ 725{,}300 \\ \hline \end{array}$$

First estimate each difference and then find the difference.

10.
$$\begin{array}{r} 850 \\ -\ 138 \\ \hline \end{array}$$

11.
$$\begin{array}{r} 8000 \\ -\ 2783 \\ \hline \end{array}$$

First estimate each product and then find the product.

12.
$$\begin{array}{r} 34 \\ \times\ 76 \\ \hline \end{array}$$

13.
$$\begin{array}{r} 2593 \\ \times\ \ \ 85 \\ \hline \end{array}$$

Find the quotient and remainder.

14. $25\overline{)10{,}075}$

15. $462\overline{)79{,}952}$

ANSWERS

1. [Respond below exercise.] ______________
2. [Respond below exercise.] ______________
3. ______________
4. ______________
5. ______________
6. ______________
7. ______________
8. ______________
9. ______________
10. ______________
11. ______________
12. ______________
13. ______________
14. ______________
15. ______________

16. ______________

17. ______________

18. ______________

19a. ______________

b. ______________

c. ______________

20a. ______________

b. ______________

16. If the quotient of 51 and 17 is subtracted from the product of 19 and 3, what is the difference?

17. Find the average of the numbers 82, 96, 49, and 69.

18. On an English exam, two students scored 98 points, five scored 87 points, one scored 81 points, and six scored 75 points. What was the average score for the class?

19. Robert and his brother saved money to buy a new TV set for their parents. (a) If Robert saved $54 a week and his brother saved $38 a week, about how much had they saved in six weeks? (b) Exactly how much had they saved in six weeks? (c) If the set they wanted to buy cost $830 including tax, how much did they still need after the six weeks?

20. An elementary school playground is in the shape of a rectangle that is 50 meters wide and 70 meters long. (a) What is the perimeter of the playground? (b) What is the area of the playground?

2

Prime Numbers

What to Expect in Chapter 2

The topics in Chapter 2 form the foundation for all of the work in Chapter 3 (Fractions) and Chapter 4 (Mixed Numbers). Exponents are introduced in Section 2.1 and are used throughout Chapter 2 and in appropriate places throughout the rest of the text to simplify expressions and represent powers. The rules for order of operations, also presented in Section 2.1, are necessary so that everyone will arrive at the same answer when evaluating an expression involving more than one operation. These rules are used throughout all levels of mathematics and science and in technical fields, such as electronics and mechanics. Calculators and computers also use the rules for order of operations when performing calculations.

Section 2.2 provides a few tests for divisibility by 2, 3, 4, 5, 9, and 10, which indicate divisibility without actually performing the divisions. The tests are valuable tools for working with prime factorizations and simplifying fractions. The ideas discussed in the first two sections are carried over and used throughout Sections 2.3, 2.4, and 2.5 in discussions of prime numbers, prime factorizations, and the least common multiple (LCM). The concepts of prime numbers and prime factorizations will be particularly important in your understanding of fractions and mixed numbers in Chapters 3 and 4. Also, since much of algebra involves these topics and the related techniques, they should be studied carefully and thoroughly.

A tracking and data relay satellite is launched by space shuttle *Discovery*.

Mathematics at Work!

The concepts of multiples and least common multiples are an important part of this chapter. These ideas are used throughout our work with fractions and mixed numbers in Chapters 3 and 4. The following problem is an example of working with the least common multiple (or LCM). (See Example 7(a) on page 134.)

Suppose three weather satellites—A, B, and C—are orbiting the earth. Satellite A takes 24 hours, B takes 18 hours, and C takes 12 hours. If they are directly above each other now, as shown in part (a) of the figure below, in how many hours will they again be directly above each other in the positions shown in part (a)?

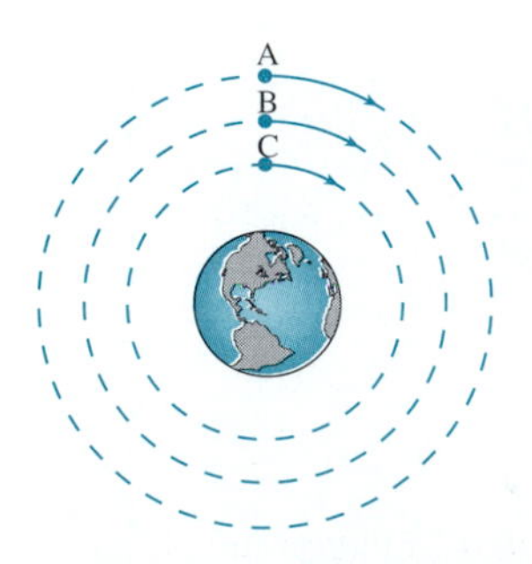

(a) Beginning positions

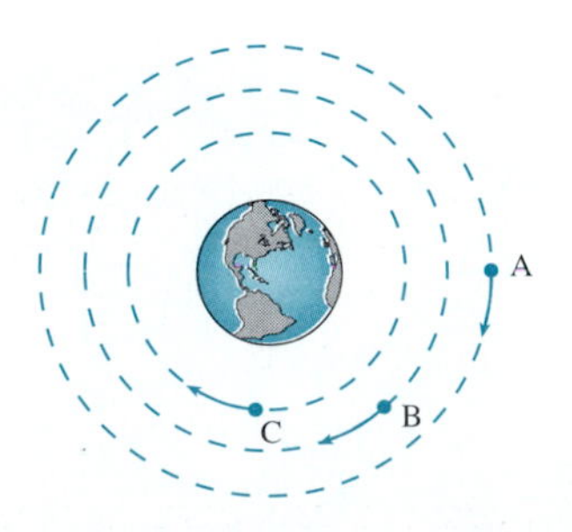

(b) Positions after 6 hours

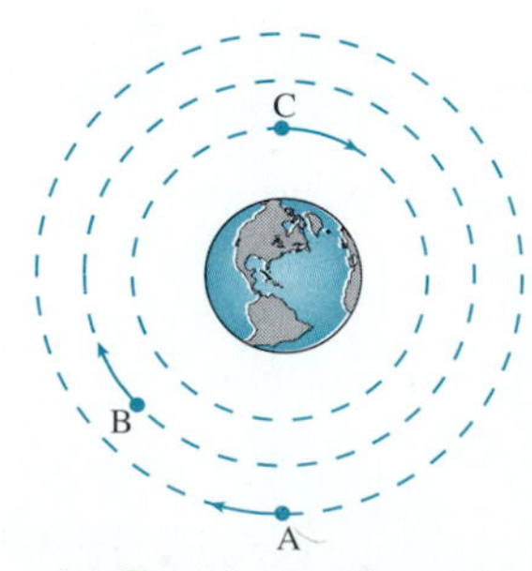

(c) Positions after 12 hours

2.1 Exponents and Order of Operations

OBJECTIVES

1. Know that an exponent is used to indicate repeated multiplication.
2. Know the rules for order of operations.
3. Be able to use the rules for order of operations to simplify numerical expressions.

Exponents

We know that repeated addition of the same number can be shortened by using multiplication:

$$5 + 5 + 5 = \underbrace{3 \cdot 5}_{\text{factors}} = \underbrace{15}_{\text{product}}$$

The result is called the **product,** and the whole numbers that are multiplied are called **factors** of the product.

In a similar manner, repeated multiplication by the same whole number can be shortened by using **exponents.** Thus, if 5 is used as a factor three times, we can write

$$5 \cdot 5 \cdot 5 = 5^3 = 125$$

Note that $3 \cdot 5 \neq 5^3$. (The symbol $\neq$ is read "is not equal to.")

In an expression such as $5^3 = 125$, 5 is called the **base,** 3 is called the **exponent,** and 125 is called the **power.** (Exponents are written slightly to the right of, and above, the base.)

exponent

$$5^3 = 125$$

base power

In this case, we would say that 125 is the third power of 5.

Example 1

With Repeated Multiplication	With Exponents
(a) $7 \cdot 7 = 49$	$7^2 = 49$
(b) $3 \cdot 3 = 9$	$3^2 = 9$
(c) $2 \cdot 2 \cdot 2 = 8$	$2^3 = 8$
(d) $10 \cdot 10 \cdot 10 = 1000$	$10^3 = 1000$
(e) $2 \cdot 2 \cdot 2 \cdot 2 \cdot 2 = 32$	$2^5 = 32$

Definition

An **exponent** is a number that tells how many times its base is to be used as a factor.

The two common terms **squared** and **cubed** are illustrated with exponential expressions in Examples 2 and 3.

In each expression, (a) name the exponent; (b) name the base; (c) find the value of the expression (find the power).

1. 8^2

2. 6^3

3. 9^2

4. 2^6

5. 1^4

EXAMPLE 2

In expressions with the exponent 2, the base is said to be **squared.**

$5^2 = 25$ is read "five squared is equal to twenty-five."

EXAMPLE 3

In expressions with the exponent 3, the base is said to be **cubed.**

$4^3 = 64$ is read "four cubed is equal to sixty-four."

Expressions with other exponents are read as the base **to the ____ power.** For example, $2^5 = 32$ is read "two to the fifth power is equal to thirty-two."

Special Note: You should understand the proper use of the word **power.** A power is **not** an exponent. A power is the product indicated by the exponent. Thus, for the equation $2^5 = 32$, we must think of the phrase "two to the fifth power" in its entirety, and the corresponding power is the product, 32.

If there is no exponent written with a number, then the exponent is understood to be 1; that is, any number is equal to itself raised to the first power. Thus,

$$8 = 8^1 \qquad 6 = 6^1 \qquad 93 = 93^1$$

Definition

For any whole number a, $a^1 = a$.

Now Work Exercises 1–5 in the Margin.

Property of Exponents

For any nonzero whole number a, $a^0 = 1$.

The Property of Exponents given here states that when the exponent 0 is used for any nonzero base, the value of the power is 1. To help in understanding the use of 0 as an exponent, consider the following rule for exponents, which will be studied in algebra.

When dividing numbers with the same base, subtract the exponents.

For example,

$$\frac{2^6}{2^2} = \frac{\not{2} \cdot \not{2} \cdot 2 \cdot 2 \cdot 2 \cdot 2}{\not{2} \cdot \not{2}} = 2 \cdot 2 \cdot 2 \cdot 2 = 2^4 \quad \text{or} \quad \frac{2^6}{2^2} = 2^{6-2} = 2^4$$

$$\frac{5^4}{5^3} = \frac{\not{5} \cdot \not{5} \cdot \not{5} \cdot 5}{\not{5} \cdot \not{5} \cdot \not{5}} = 5 \qquad \text{or} \qquad \frac{5^4}{5^3} = 5^{4-3} = 5^1$$

So,

$$\frac{3^4}{3^4} = 3^{4-4} = 3^0, \qquad \text{but} \qquad \frac{3^4}{3^4} = \frac{81}{81} = 1$$

$$\frac{5^2}{5^2} = 5^{2-2} = 5^0, \qquad \text{but} \qquad \frac{5^2}{5^2} = \frac{25}{25} = 1$$

Thus,

$$3^0 = 1 \qquad \text{and} \qquad 5^0 = 1.$$

Note: The expression 0^0 is not defined.

Example 4

(a) $6^0 = 1$ (b) $56^0 = 1$ (c) $10^0 = 1$

To improve your speed in factoring and in working with fractions and simplifying expressions, you should memorize all the squares of the whole numbers from 1 to 20. These numbers are called **perfect squares.** The following tables list these squares.

Number (n)	1	2	3	4	5	6	7	8	9	10
Square (n^2)	1	4	9	16	25	36	49	64	81	100

Number (n)	11	12	13	14	15	16	17	18	19	20
Square (n^2)	121	144	169	196	225	256	289	324	361	400

Now Work Exercises 6–10 in the Margin.

Rewrite each product by using exponents.

6. $8 \cdot 8 \cdot 8$

7. $2 \cdot 2 \cdot 3 \cdot 5 \cdot 5$

8. $3 \cdot 3 \cdot 4 \cdot 4 \cdot 4$

Find the following squares.

9. 16^2

10. 13^2

Rules for Order of Operations

Mathematicians have agreed on a set of rules for the order of operations in evaluating any numerical expression involving addition, subtraction, multiplication, division, and exponents. These rules are used in all branches of mathematics and computer science to ensure that there is one correct value for an expression regardless of how complicated it might be. For example, evaluate the following expression in whatever way you think is correct:

$$6(10 - 2^3) + 12 \div 2 \cdot 3$$

The correct value of this expression can be found by using the rules for order of operations, as follows:

$$\begin{aligned} 6(10 - 2^3) + 12 \div 2 \cdot 3 &= 6(10 - 8) + 12 \div 2 \cdot 3 \\ &= 6(2) + 12 \div 2 \cdot 3 \\ &= 12 + 12 \div 2 \cdot 3 \\ &= 12 + 6 \cdot 3 \\ &= 12 + 18 \\ &= 30 \end{aligned}$$

See if you can understand how the following rules for order of operations were applied at each step in the example above.

Rules for Order of Operations

1. First, simplify within grouping symbols, such as parentheses ( ), brackets [ ], and braces { }. Start with the innermost grouping.
2. Second, find any powers indicated by exponents.
3. Third, moving from **left to right,** perform any multiplications or divisions in the order in which they appear.
4. Fourth, moving from **left to right,** perform any additions or subtractions in the order in which they appear.

These rules are very explicit and should be studied carefully. Note that in Rule 3, neither multiplication nor division has priority over the other. Whichever operation occurs first, moving from **left to right,** is performed first. In Rule 4, addition and subtraction are handled in the same way. Unless they occur within grouping symbols, **addition and subtraction are the last operations to be performed.**

The following examples show how to apply the rules for order of operations. In some cases, more than one step can be performed at the same time. This is possible when parts are separated by + or − signs or are within separate symbols of inclusion. These parts are called **terms,** and they can be evaluated at the same time because addition and subtraction are performed last. The highlighted + and − signs in the examples separate terms.

EXAMPLE 5

Evaluate the expression $48 \div 6 \cdot 2 - 5 \cdot 2 + 1$.

Solution

$$\begin{aligned}
& 48 \div 6 \cdot 2 - 5 \cdot 2 + 1 && \text{Divide before multiplying in this case.} \\
&= 8 \cdot 2 - 5 \cdot 2 + 1 && \text{Multiply before adding or subtracting.} \\
&= 16 - 5 \cdot 2 + 1 && \text{Again, multiply before adding or subtracting.} \\
&= 16 - 10 + 1 && \text{Moving from left to right, subtract before adding.} \\
&= 6 + 1 && \text{Add.} \\
&= 7
\end{aligned}$$

EXAMPLE 6

Evaluate the expression $(6 + 4) + (9 + 1) \div 5$.

Solution

$$\begin{aligned}
& (6 + 4) + (9 + 1) \div 5 && \text{Operate within each pair of parentheses. This is okay since they are in separate terms.} \\
&= 10 + 10 \div 5 && \text{Divide before adding.} \\
&= 10 + 2 && \text{Add.} \\
&= 12
\end{aligned}$$

EXAMPLE 7

Evaluate the expression $3 \cdot 5^2 - 2(14 \div 2 - 7) + 10^3$.

Solution

$$\begin{aligned}
& 3 \cdot 5^2 - 2(14 \div 2 - 7) + 10^3 && \text{Operate within parentheses.} \\
&= 3 \cdot 5^2 - 2(7 - 7) + 10^3 && \text{Within parentheses, divide first, then subtract.} \\
&= 3 \cdot 5^2 - 2(0) + 10^3 && \text{In different terms, find the powers.} \\
&= 3 \cdot 25 - 2(0) + 1000 && \text{Multiply in different terms.} \\
&= 75 - 0 + 1000 && \text{Add and subtract, from left to right.} \\
&= 1075
\end{aligned}$$

Find the value of each expression by using the rules for order of operations.

11. $15 \div 5 + 10 \cdot 2$

12. $4 \div 2^2 + 3 \cdot 2^2$

13. $(5 + 7) \div 3 + 1$

14. $19 - 5(3 - 1)$

15. $3 \cdot 2^4 - 18 - 2 \cdot 3^2$

✔ COMPLETION EXAMPLE 8

Evaluate the expression $(15 - 10)[(9 + 3^2) \div 2 + 6]$.

Solution

$(15 - 10)[(9 + 3^2) \div 2 + 6]$	Operate within parentheses and find the power.
$= \_\_\_\_ [(9 + \_\_\_\_) \div 2 + 6]$	Operate within parentheses within the brackets.
$= \_\_\_\_ [(\_\_\_\_) \div 2 + 6]$	Divide within the brackets.
$= \_\_\_\_ [\_\_\_\_ + 6]$	Add within the brackets.
$= \_\_\_\_ [\_\_\_\_]$	Multiply.
$= \_\_\_\_$	

✔ COMPLETION EXAMPLE 9

Evaluate the expression $3(2 + 2^2) - 6 - 3 \cdot 2^2$.

Solution

$$3(2 + 2^2) - 6 - 3 \cdot 2^2$$
$$= 3(2 + \_\_\_\_) - 6 - 3 \cdot \_\_\_\_$$
$$= 3(\_\_\_\_) - 6 - \_\_\_\_$$
$$= \_\_\_\_ - 6 - \_\_\_\_$$
$$= \_\_\_\_ - \_\_\_\_$$
$$= \_\_\_\_$$

Now Work Exercises 11–15 in the Margin.

Completion Example Answers

8. $(15 - 10)[(9 + 3^2) \div 2 + 6]$	Operate within parentheses and find the power.
$= \mathbf{(5)}[(9 + \mathbf{9}) \div 2 + 6]$	Operate within parentheses within the brackets.
$= \mathbf{(5)}[(\mathbf{18}) \div 2 + 6]$	Divide within the brackets.
$= \mathbf{(5)}[\mathbf{9} + 6]$	Add within the brackets.
$= \mathbf{(5)}[\mathbf{15}]$	Multiply.
$= \mathbf{75}$	

9. $3(2 + 2^2) - 6 - 3 \cdot 2^2$
$= 3(2 + \mathbf{4}) - 6 - 3 \cdot \mathbf{4}$
$= 3(\mathbf{6}) - 6 - \mathbf{12}$
$= \mathbf{18} - 6 - \mathbf{12}$
$= \mathbf{12} - \mathbf{12}$
$= \mathbf{0}$

NAME ______ SECTION ______ DATE ______

Exercises 2.1

In each of the following expressions, (a) name the base, (b) name the exponent, and (c) find the power.

1. 2^3
(a) ______
(b) ______
(c) ______

2. 5^3
(a) ______
(b) ______
(c) ______

3. 4^2
(a) ______
(b) ______
(c) ______

4. 7^0
(a) ______
(b) ______
(c) ______

5. 9^2
(a) ______
(b) ______
(c) ______

6. 1^{13}
(a) ______
(b) ______
(c) ______

7. 10^4
(a) ______
(b) ______
(c) ______

8. 23^1
(a) ______
(b) ______
(c) ______

9. 3^6
(a) ______
(b) ______
(c) ______

10. 2^4
(a) ______
(b) ______
(c) ______

Find a base and exponent form for each of the following powers without using the exponent 1.

11. 25 **12.** 16 **13.** 27 **14.** 121

15. 8 **16.** 36 **17.** 100 **18.** 10,000

ANSWERS

1. [Respond in exercise.]
2. [Respond in exercise.]
3. [Respond in exercise.]
4. [Respond in exercise.]
5. [Respond in exercise.]
6. [Respond in exercise.]
7. [Respond in exercise.]
8. [Respond in exercise.]
9. [Respond in exercise.]
10. [Respond in exercise.]
11. ______
12. ______
13. ______
14. ______
15. ______
16. ______
17. ______
18. ______

19. ________
20. ________
21. ________
22. ________
23. ________
24. ________
25. ________
26. ________
27. ________
28. ________
29. ________
30. ________
31. ________
32. ________
33. [Respond in exercise.]
34. [Respond in exercise.]
35. [Respond in exercise.]
36. [Respond in exercise.]

Rewrite the following products by using exponents.

19. $6 \cdot 6 \cdot 6 \cdot 6 \cdot 6$ **20.** $7 \cdot 7 \cdot 7 \cdot 7$ **21.** $2 \cdot 2 \cdot 2 \cdot 3 \cdot 3$

22. $2 \cdot 2 \cdot 5 \cdot 5 \cdot 5$ **23.** $2 \cdot 3 \cdot 3 \cdot 11 \cdot 11$ **24.** $4 \cdot 4 \cdot 13 \cdot 13$

Find the value of each of the following squares. Write as many as you can from memory.

25. 8^2 **26.** 20^2 **27.** 14^2 **28.** 9^2

29. 12^2 **30.** 15^2 **31.** 40^2 **32.** 19^2

Fill in each blank with the word (or phrase) that indicates which operation (or operations) is (or are) performed at each step in evaluating the expression.

33. $15 + 6 \cdot 8 \div 2$ (a) ________
$= 15 + 48 \div 2$ (b) ________
$= 15 + 24$ (c) ________
$= 39$

34. $(9 - 3) - (5 + 2) \div 7$ (a) ________
$= 6 - (7) \div 7$ (b) ________
$= 6 - 1$ (c) ________
$= 5$

35. $(33 \div 3 + 3) \div 7$ (a) ________
$= (11 + 3) \div 7$ (b) ________
$= 14 \div 7$ (c) ________
$= 2$

36. $3 \cdot 2^2 - 8 \div 2^2 + 5(6)$ (a) ________
$= 3 \cdot 4 - 8 \div 4 + 5(6)$ (b) ________
$= 12 - 2 + 30$ (c) ________
$= 10 + 30$ (d) ________
$= 40$

NAME ______________ SECTION ______________ DATE ______________

Find the value of each of the following expressions by using the rules for order of operations.

37. $4 \div 2 + 7 - 3 \cdot 2$

38. $8 \cdot 3 \div 12 + 13$

39. $6 + 3 \cdot 2 - 10 \div 2$

40. $14 \cdot 3 \div 7 \div 2 + 6$

41. $6 \div 2 \cdot 3 - 1 + 2 \cdot 7$

42. $5 \cdot 1 \cdot 3 - 4 \div 2 + 6 \cdot 3$

43. $72 \div 4 \div 9 - 2 + 3$

44. $14 + 63 \div 3 - 35$

45. $(2 + 3 \cdot 4) \div 7 + 3$

46. $(2 + 3) \cdot 4 \div 5 + 3 \cdot 2$

47. $(7 - 3) + (2 + 5) \div 7$

48. $16(2 + 4) - 90 - 3 \cdot 2$

49. $35 \div (6 - 1) - 5 + 6 \div 2$

50. $22 - 11 \cdot 2 + 15 - 5 \cdot 3$

ANSWERS

37. ______________

38. ______________

39. ______________

40. ______________

41. ______________

42. ______________

43. ______________

44. ______________

45. ______________

46. ______________

47. ______________

48. ______________

49. ______________

50. ______________

51. ______
52. ______
53. ______
54. ______
55. ______
56. ______
57. ______
58. ______
59. ______
60. ______
61. ______
62. ______
63. ______
64. ______

51. $(42 - 2 \div 2 \cdot 3) \div 13$

52. $18 + 18 \div 2 \div 3 - 3 \cdot 1$

53. $4(7 - 2) \div 10 + 5$

54. $(33 - 2 \cdot 6) \div 7 + 3 - 6$

55. $72 \div 8 + 3 \cdot 4 - 105 \div 5$

56. $6(14 - 6 \div 2 - 11)$

57. $48 \div 12 \div 4 - 1 + 6$

58. $5 - 1 \cdot 2 + 4(6 - 18 \div 3)$

59. $8 - 1 \cdot 5 + 6(13 - 39 \div 3)$

60. $(21 \div 7 - 3)42 + 6$

61. $16 - 16 \div 2 - 2 + 7 \cdot 3$

62. $(135 \div 3 + 21 \div 7) \div 12 - 4$

63. $(13 - 5) \div 4 + 12 \cdot 4 \div 3 - 72 \div 18 \cdot 2 + 16$

64. $15 \div 3 + 2 - 6 + (3)(2)(18)(0)(5)$

NAME ______________ SECTION ______________ DATE ______________

65. $100 \div 10 \div 10 + 1000 \div 10 \div 10 \div 10 - 2$

66. $[(85 + 5) \div 3 \cdot 2 + 15] \div 15$

67. $2 \cdot 5^2 - 4 \div 2 + 3 \cdot 7$

68. $16 \div 2^4 - 9 \div 3^2$

69. $(4^2 - 7) \cdot 2^3 - 8 \cdot 5 \div 10$

70. $4^2 - 2^4 + 5 \cdot 6^2 - 10^2$

71. $(2^5 + 1) \div 11 - 3 + 7(3^3 - 7)$

72. $(6 + 8^2 - 10 \div 2) \div 5$

73. $(5 + 7) \div 4 + 2$

74. $(2^3 + 2) \div 5 + 5$

75. $(5^2 + 7) \div 8 - (14 \div 7 \cdot 2)$

76. $(3 \cdot 2^2 - 5 \cdot 2 + 2) - (1 \cdot 2^2 + 5 \cdot 2 - 10)$

ANSWERS

65. ______________

66. ______________

67. ______________

68. ______________

69. ______________

70. ______________

71. ______________

72. ______________

73. ______________

74. ______________

75. ______________

76. ______________

77. ______________

78. ______________

79. ______________

80. ______________

81. ______________

82. ______________

83. ______________

84. ______________

85. ______________

86. ______________

77. $2^3 \cdot 3^2 \div 24 - 3 + 6^2 \div 4$

78. $2 \cdot 3^2 + 5 \cdot 9 + 15^2 - (21 \cdot 3^2 + 6)$

79. $3 \cdot 8 - 2^2 + 4 \cdot 2 - 2^4$

80. $2 \cdot 5^2 - 4(21 \div 3 - 7) + 10^3 - 1000$

81. $(4 + 3)(4 + 3) - (2 + 3)(2 + 3)$

82. $(5 + 1)(5 + 1) + (3 + 1)(3 + 1)$

83. $40 \div 2 \cdot 5 + 1 \cdot 9 \cdot 2$

84. $20 - 2(3 - 1) + 6^2 \div 2 \cdot 3$

85. $(2^4 - 16)[13 - (5^2 - 20)]$

86. $100 + 2[(7^2 - 9)(5 + 1)^2]$

2.2 Tests for Divisibility (2, 3, 4, 5, 9, and 10)

OBJECTIVES

1. *Understand the terms* **exactly divisible, divisible,** *and* **divides.**
2. *Know the tests for easily checking divisibility by 2, 3, 4, 5, 9, and 10.*

Rules for Tests of Divisibility

In our work with factoring (Section 2.4) and fractions (Chapter 3), we will need to be able to divide quickly by small numbers. Knowing that a number is **exactly divisible** (remainder 0) by some number even **before** we actually divide, will save time and build confidence. There are simple tests we can use to determine whether a number is divisible by 2, 3, 4, 5, 9, or 10 without actually dividing.

Definition

If a number can be divided by another number so that the remainder is 0, then we say:

1. The first number is **exactly divisible by** (or **divisible by**) the second number; or
2. The second number **divides** the first.

Example 1

Is 360 divisible by 4?

```
    90
4 ) 360
    36
     00
      0
      0  remainder
```

Since the remainder is 0, 360 is divisible by 4.

Example 2

Is 360 divisible by 10?

```
     36
10 ) 360
     30
      60
      60
       0  remainder
```

Since the remainder is 0, 360 is divisible by 10.

EXAMPLE 3

Is 104 divisible by 5?

```
    20
5 ) 104
    10
    04
     0
     4   remainder
```

Since the remainder is not 0, 104 is **not** divisible by 5.

EXAMPLE 4

Does 9 divide 1,809,333?

We could divide by 9 to find the answer. Remember, we do not want to find the quotient. We just want to know if the remainder is 0. We can determine this without dividing. The answer is yes. [See the following test for 9 and Example 5(g).]

Tests for Divisibility by 2, 3, 4, 5, 9, and 10

For 2: If the last digit (units digit) of a whole number is 0, 2, 4, 6, or 8, then the whole number is divisible by 2.

For 3: If the sum of the digits of a whole number is divisible by 3, then the number is divisible by 3.

For 4: If the last two digits of a whole number form a number that is divisible by 4, then the number is divisible by 4. (00 is considered to be divisible by 4.)

For 5: If the last digit of a whole number is 0 or 5, then the number is divisible by 5.

For 9: If the sum of the digits of a whole number is divisible by 9, then the number is divisible by 9.

For 10: If the last digit (units digit) of a whole number is 0, then the number is divisible by 10.

Even and Odd Whole Numbers

Even whole numbers are divisible by 2.
(If a whole number is divided by 2 and the remainder is 0, then the whole number is even.)

Odd whole numbers are not divisible by 2.
(If a whole number is divided by 2 and the remainder is 1, then the whole number is odd.)

Example 5 illustrates the application of each of the tests of divisibility listed above.

EXAMPLE 5

(a) 356 is divisible by 2 since the last digit is 6, an even digit.
(b) 6801 is divisible by 3 since $6 + 8 + 0 + 1 = 15$ and 15 is divisible by 3.
(c) 9036 is divisible by 4 since 36 (last 2 digits) is divisible by 4. (9036 is also divisible by 2.)
(d) 1365 is divisible by 5 since the last digit is 5.
(e) 9657 is divisible by 9 since $9 + 6 + 5 + 7 = 27$ and 27 is divisible by 9.
(f) 3590 is divisible by 10 since the last digit is 0.
(g) 9 divides 1,809,333 because $1 + 8 + 0 + 9 + 3 + 3 + 3 = 27$ and 27 is divisible by 9.

Now Work Exercises 1–5 in the Margin.

1. Does 4 divide 9044? Explain why or why not.
2. Does 3 divide 106? Explain why or why not.
3. Is 306 divisible by 3? by 9? Explain your reasoning.
4. Is 3165 divisible by 5? by 10? Explain.
5. Is 463 divisible by 2? by 4? by 5? Explain.

Use all six tests to determine which of the numbers 2, 3, 4, 5, 9, and 10 will divide into the numbers in Examples 6 and 7.

EXAMPLE 6

2530
(a) divisible by 2 (last digit is 0, an even digit)
(b) not divisible by 3
($2 + 5 + 3 + 0 = 10$ and 10 is not divisible by 3)
(c) not divisible by 4 (30 is not divisible by 4)
(d) divisible by 5 (last digit is 0)
(e) not divisible by 9
($2 + 5 + 3 + 0 = 10$ and 10 is not divisible by 9)
(f) divisible by 10 (last digit is 0)

EXAMPLE 7

5712
(a) divisible by 2 (last digit is 2, an even digit)
(b) divisible by 3 ($5 + 7 + 1 + 2 = 15$ and 15 is divisible by 3)
(c) divisible by 4 (12 is divisible by 4)
(d) not divisible by 5 (last digit is not 0 or 5)
(e) not divisible by 9
($5 + 7 + 1 + 2 = 15$ and 15 is not divisible by 9)
(f) not divisible by 10 (last digit is not 0)

Determine which of the numbers 2, 3, 4, 5, 9, and 10 divides into each of the following numbers.

6. 842

7. 9030

8. 4031

COMPLETION EXAMPLE 8

(a) 250 is divisible by 10 because

______________________________.

(b) 250 is not divisible by 3 because

______________________________.

(c) 250 is not divisible by 4 because

______________________________.

COMPLETION EXAMPLE 9

(a) 512 is divisible by 4 because

______________________________.

(b) 512 is not divisible by 9 because

______________________________.

(c) 512 is not divisible by 5 because

______________________________.

Now Work Exercises 6–8 in the Margin.

Checking Divisibility of Products

To emphasize the relationships among the concepts of multiplication, factors, and divisibility, we now discuss these relationships in terms of given products. This discussion will form the basis of our work with least common multiple (LCM) in Section 2.5 and common denominators of fractions in Chapter 3.

Consider the fact that $3 \cdot 4 \cdot 5 \cdot 5 = 300$. This means that if any one factor (or the product of two or more factors) is divided into 300, the quotient will be the product of the remaining factors. For example, if 300 is divided by 3, then the product $4 \cdot 5 \cdot 5 = 100$ will be the quotient. Similarly, various groupings give

$$3 \cdot 4 \cdot 5 \cdot 5 = (3 \cdot 4)(5 \cdot 5) = 12 \cdot 25 = 300,$$

$$3 \cdot 4 \cdot 5 \cdot 5 = (4 \cdot 5)(3 \cdot 5) = 20 \cdot 15 = 300, \quad \text{and}$$

$$3 \cdot 4 \cdot 5 \cdot 5 = (3 \cdot 4 \cdot 5)(5) = 60 \cdot 5 = 300.$$

Thus, we can make the following statements:

12 divides 300 25 times and 25 divides 300 12 times
(or, $300 \div 12 = 25$ and $300 \div 25 = 12$);

20 divides 300 15 times and 15 divides 300 20 times
(or, $300 \div 20 = 15$ and $300 \div 15 = 20$);

60 divides 300 5 times and 5 divides 300 60 times
(or, $300 \div 60 = 5$ and $300 \div 5 = 60$).

EXAMPLE 10

Does 36 divide the product $4 \cdot 5 \cdot 9 \cdot 3 \cdot 7$? If so, how many times?

Solution

Since $36 = 4 \cdot 9$, we have

$$4 \cdot 5 \cdot 9 \cdot 3 \cdot 7 = (4 \cdot 9)(5 \cdot 3 \cdot 7) = 36 \cdot 105$$

Thus, 36 does divide the product, and it divides the product 105 times.

COMPLETION EXAMPLE 11

Does 15 divide the product $5 \cdot 7 \cdot 2 \cdot 3 \cdot 2$? If so, how many times?

Solution

Since $15 = \_\_\_\_ \cdot \_\_\_\_$, we have

$$5 \cdot 7 \cdot 2 \cdot 3 \cdot 2 = (\_\_\_\_ \cdot \_\_\_\_)(\_\_\_\_ \cdot \_\_\_\_ \cdot \_\_\_\_)$$

$$= (\_\_\_\_)(\_\_\_\_)$$

Thus, 15 does divide the product, and it divides the product ______ times.

EXAMPLE 12

Does 35 divide the product $3 \cdot 4 \cdot 5 \cdot 11$?

Solution

We know that $35 = 5 \cdot 7$, and while 5 is a factor of the product, 7 is not. Therefore, 35 does not divide the product $3 \cdot 4 \cdot 5 \cdot 11$. In other words, since $3 \cdot 4 \cdot 5 \cdot 11 = 660$, 660 is not divisible by 35.

Completion Example Answers

8. (a) 250 is divisible by 10 because **the units digit is 0.**
   (b) 250 is not divisible by 3 because **the sum of the digits is 7 and 7 is not divisible by 3.**
   (c) 250 is not divisible by 4 because **the number 50 (formed by the last two digits) is not divisible by 4.**

9. (a) 512 is divisible by 4 because **the number 12 (formed by the last two digits) is divisible by 4.**
   (b) 512 is not divisible by 9 because **the sum of the digits is 8 and 8 is not divisible by 9.**
   (c) 512 is not divisible by 5 because **the units digit is not 0 and not 5.**

11. Since 15 = $\mathbf{3 \cdot 5}$, we have

$$5 \cdot 7 \cdot 2 \cdot 3 \cdot 2 = (\mathbf{3} \cdot \mathbf{5})(\mathbf{7} \cdot \mathbf{2} \cdot \mathbf{2})$$
$$= (\mathbf{15})(\mathbf{28})$$

Thus, 15 does divide the product, and it divides the product **28** times.

NAME ______________ SECTION ______________ DATE ______________

ANSWERS

Exercises 2.2

Using the tests for divisibility, determine which of the numbers 2, 3, 4, 5, 9, and 10 will divide exactly into each of the following numbers.

1. 72 **2.** 81 **3.** 105 **4.** 333

5. 150 **6.** 471 **7.** 664 **8.** 154

9. 372 **10.** 375 **11.** 443 **12.** 173

13. 567 **14.** 480 **15.** 331 **16.** 370

17. 571 **18.** 466 **19.** 897 **20.** 695

21. 795 **22.** 777 **23.** 45,000 **24.** 885

25. 4422 **26.** 1234 **27.** 4321 **28.** 8765

1. ______________
2. ______________
3. ______________
4. ______________
5. ______________
6. ______________
7. ______________
8. ______________
9. ______________
10. ______________
11. ______________
12. ______________
13. ______________
14. ______________
15. ______________
16. ______________
17. ______________
18. ______________
19. ______________
20. ______________
21. ______________
22. ______________
23. ______________
24. ______________
25. ______________
26. ______________
27. ______________
28. ______________

29. ______
30. ______
31. ______
32. ______
33. ______
34. ______
35. ______
36. ______
37. ______
38. ______
39. ______
40. ______
41. ______
42. ______
43. ______
44. ______
45. ______
46. ______
47. ______
48. ______
49. ______
50. ______
51. ______
52. ______
53. ______
54. ______
55. ______
56. ______
57. ______
58. ______
59. ______
60. ______
61. [Respond below exercise.] ______
62. [Respond below exercise.] ______
63. [Respond below exercise.] ______
64. [Respond below exercise.] ______

29. 5678 **30.** 402 **31.** 705 **32.** 732

33. 441 **34.** 555 **35.** 666 **36.** 9000

37. 10,000 **38.** 576 **39.** 549 **40.** 792

41. 5700 **42.** 4391 **43.** 5476 **44.** 6930

45. 4380 **46.** 510 **47.** 8805 **48.** 7155

49. 8377 **50.** 2222 **51.** 35,622 **52.** 75,495

53. 12,324 **54.** 55,555 **55.** 632,448 **56.** 578,400

57. 9,737,001 **58.** 17,158,514 **59.** 36,762,252 **60.** 20,498,105

Determine whether each of the given numbers divides (or is a factor of) the given product. If it does divide the product, tell how many times. Find each product and make a written statement concerning the divisibility of the product by the given number.

61. 6; $2 \cdot 3 \cdot 3 \cdot 5$ **62.** 10; $2 \cdot 3 \cdot 3 \cdot 5$

63. 14; $2 \cdot 3 \cdot 5 \cdot 7$ **64.** 20; $3 \cdot 4 \cdot 5 \cdot 11$

NAME ______________ SECTION ______________ DATE ________

65. 10; $3 \cdot 3 \cdot 5 \cdot 7$

66. 25; $2 \cdot 3 \cdot 5 \cdot 7 \cdot 11$

67. 25; $2 \cdot 2 \cdot 3 \cdot 5 \cdot 5$

68. 35; $3 \cdot 4 \cdot 5 \cdot 7 \cdot 10$

69. 21; $3 \cdot 3 \cdot 5 \cdot 7 \cdot 11$

70. 30; $2 \cdot 3 \cdot 4 \cdot 5 \cdot 13$

Writing and Thinking about Mathematics

71. If a number is divisible by both 2 and 9, must it be divisible by 18? Explain your reasoning and give several examples to support your answer.

72. If a number is divisible by both 3 and 9, must it be divisible by 27? Explain your reasoning and give several examples to support your answer.

ANSWERS

65. [Respond below exercise.] ________

66. [Respond below exercise.] ________

67. [Respond below exercise.] ________

68. [Respond below exercise.] ________

69. [Respond below exercise.] ________

70. [Respond below exercise.] ________

71. [Respond below exercise.] ________

72. [Respond below exercise.] ________

73. [Respond below exercise.] ______________

73. With your understanding of factors and divisibility, make up a rule for divisibility by 6 and a rule for divisibility by 15.

RECYCLE BIN

1. ______________
2. ______________
3. ______________
4. ______________
5. ______________
6. ______________
7. ______________
8. ______________
9. ______________
10. ______________

The Recycle Bin (from Section 1.3)

Round off as indicated.

To the nearest ten: **1.** 847 **2.** 1931

To the nearest hundred: **3.** 439 **4.** 2563

To the nearest thousand: **5.** 13,612 **6.** 20,500

First estimate the answers using rounded-off numbers; then find the following sums and differences.

7. 485 + 93 + 115

$$\begin{array}{r} 485 \\ 93 \\ +\ 115 \\ \hline \end{array}$$

8.

$$\begin{array}{r} 661 \\ 1730 \\ +\ 2059 \\ \hline \end{array}$$

9.

$$\begin{array}{r} 8752 \\ -\ 3527 \\ \hline \end{array}$$

10.

$$\begin{array}{r} 74{,}605 \\ -\ 46{,}083 \\ \hline \end{array}$$

Prime Numbers and Composite Numbers

OBJECTIVES

1. *Know the definition of a prime number.*
2. *Know the definition of a composite number.*
3. *Be able to list all the prime numbers less than 50.*
4. *Be able to determine whether a number is prime or composite.*

Prime Numbers and Composite Numbers

The **counting numbers** (or **natural numbers**) are the numbers

1, 2, 3, 4, 5, 6, 7, 8, 9, 10, 11, 12, 13, 14, 15, . . .

Every counting number, except 1, has at least two factors, as illustrated in the following list. Note that, in this list, every number has **at least** two factors, but 5 and 17 have **only** two factors.

Counting Numbers		Factors
5	⟶	1, 5
12	⟶	1, 2, 3, 4, 6, 12
17	⟶	1, 17
20	⟶	1, 2, 4, 5, 10, 20
49	⟶	1, 7, 49

In Chapter 3, our work with fractions will be based on the use and understanding of counting numbers that have exactly two different factors. Such numbers (for example, 5 and 17 in the list above) are called **prime numbers.**

Definition

A **prime number** is a counting number greater than 1 that has only 1 and itself as factors.

OR

A **prime number** is a counting number with exactly two different factors (or divisors).

Definition

A **composite number** is a counting number with more than two different factors (or divisors).

Thus, in the previous list, 12, 20, and 49 are composite numbers.

Note: 1 is neither a prime nor a composite number. We know that $1 = 1 \cdot 1$, and that 1 is the only factor of 1. Thus, 1 does not have exactly two different factors, and it does not have more than two different factors.

Determine whether each of the following numbers is prime or composite. Explain your reasoning.

1. 17

2. 28

3. 25

4. 2

5. 31

Example 1

Some prime numbers:

2 2 has exactly two different factors: 1 and 2.

7 7 has exactly two different factors: 1 and 7.

13 13 has exactly two different factors: 1 and 13.

Example 2

Some composite numbers:

18 18 has the factors 1, 2, 3, 6, 9, and 18.

24 24 has the factors 1, 2, 3, 4, 6, 8, 12, and 24.

35 35 has the factors 1, 5, 7, and 35.

Now Work Exercises 1–5 in the Margin.

The Sieve of Eratosthenes

There is no formula for finding all the prime numbers. However, a famous Greek mathematician, Eratosthenes (c. 300 B.C.), developed a technique that helps in finding prime numbers. This technique is based on the concept of **multiples** and the fact that a multiple of a number is divisible by that number.

To find the multiples of a counting number larger than 1, multiply each of the counting numbers by that number, as shown in the following table.

Counting Numbers:	**1,**	**2,**	**3,**	**4,**	**5,**	**6,**	**7,**	**8,**	**. . .**
Multiples of 2:	2,	4,	6,	8,	10,	12,	14,	16,	. . .
Multiples of 5:	5,	10,	15,	20,	25,	30,	35,	40,	. . .
Multiples of 8:	8,	16,	24,	32,	40,	48,	56,	64,	. . .

None of the multiples of a number, except possibly the number itself, can be prime since each has that number as a factor. That is, all multiples of a number except possibly the number itself, must be composite. To sift out the prime numbers using the **Sieve of Eratosthenes,** we proceed by eliminating multiples as described in the following steps.

STEP 1: To find the prime numbers from 1 to 50, list all the counting numbers from 1 to 50 in rows of ten.

1	2	3	4	5	6	7	8	9	10
11	12	13	14	15	16	17	18	19	20
21	22	23	24	25	26	27	28	29	30
31	32	33	34	35	36	37	38	39	40
41	42	43	44	45	46	47	48	49	50

STEP 2: Start by crossing out 1 (since 1 is not a prime number). Next, circle 2 and cross out all the other multiples of 2; that is, cross out every second number.

~~1~~	②	3	~~4~~	5	~~6~~	7	~~8~~	9	~~10~~
11	~~12~~	13	~~14~~	15	~~16~~	17	~~18~~	19	~~20~~
21	~~22~~	23	~~24~~	25	~~26~~	27	~~28~~	29	~~30~~
31	~~32~~	33	~~34~~	35	~~36~~	37	~~38~~	39	~~40~~
41	~~42~~	43	~~44~~	45	~~46~~	47	~~48~~	49	~~50~~

STEP 3: The first number after 2 that is not crossed out is 3. Circle 3 and cross out all multiples of 3 that are not already crossed out; that is, after 3, every third number should be crossed out.

~~1~~	②	③	~~4~~	5	~~6~~	7	~~8~~	~~9~~	10
11	~~12~~	13	~~14~~	~~15~~	~~16~~	17	~~18~~	19	~~20~~
~~21~~	~~22~~	23	~~24~~	25	~~26~~	~~27~~	~~28~~	29	~~30~~
31	~~32~~	~~33~~	~~34~~	35	~~36~~	37	~~38~~	~~39~~	~~40~~
41	~~42~~	43	~~44~~	~~45~~	~~46~~	47	~~48~~	49	~~50~~

STEP 4: The next number that is not crossed out is 5. Circle 5 and cross out all multiples of 5 that are not already crossed out. If we proceed this way, we will have the prime numbers circled and the composite numbers crossed out. The final table is as follows.

~~1~~	②	③	~~4~~	⑤	~~6~~	⑦	~~8~~	~~9~~	~~10~~
⑪	~~12~~	⑬	~~14~~	~~15~~	~~16~~	⑰	~~18~~	⑲	~~20~~
~~21~~	~~22~~	㉓	~~24~~	~~25~~	~~26~~	~~27~~	~~28~~	㉙	~~30~~
㉛	~~32~~	~~33~~	~~34~~	~~35~~	~~36~~	㊲	~~38~~	~~39~~	~~40~~
㊶	~~42~~	㊸	~~44~~	~~45~~	~~46~~	㊼	~~48~~	~~49~~	~~50~~

The final table shows that the prime numbers less than 50 are

2, 3, 5, 7, 11, 13, 17, 19, 23, 29, 31, 37, 41, 43, and 47

You should memorize these numbers for use in factoring and dealing with fractions. You should also observe the following two facts about prime numbers:

1. The number 2 is the only even prime number; and

2. All other prime numbers are odd, but not all odd numbers are prime.

Determining Prime Numbers

Computers can be used to determine whether or not very large numbers are prime. The following procedure of dividing by prime numbers can be used to determine whether or not relatively small numbers are prime. If a prime number smaller than the given number is found to be a factor (or divisor), then the given number is composite.

To Determine Whether a Number is Prime:

Divide the number by progressively larger **prime numbers** (2, 3, 5, 7, 11, and so forth) until:

1. You find a remainder of 0 (meaning that the prime number is a factor and the given number is composite); or
2. You find a quotient smaller than the prime divisor (meaning that the given number has no smaller prime factors and is therefore prime itself).

Note: Reasoning that if a composite number were a factor, then one of its prime factors would have been found to be a factor in an earlier division, we divide only by prime numbers—that is, there is no need to divide by a composite number.

Example 3

Is 605 a prime number?

Solution

Since the units digit is 5, the number 605 is divisible by 5 (using the divisibility test in Section 2.2) and is not prime. The number 605 is a composite number. In fact, $605 = 5 \cdot 121 = 5 \cdot 11 \cdot 11$, and 605 has the factors 1, 5, 11, 55, 121, and 605.

Example 4

Is 103 prime?

Solution

Tests for 2, 3, and 5 fail. (The number 103 is not even; $1 + 0 + 3 = 4$ and 4 is not divisible by 3; and the last digit is not 0 or 5.)

Divide by 7:

$$\begin{array}{r} 14 \\ 7\overline{)103} \\ \underline{7} \\ 33 \\ \underline{28} \\ 5 \end{array}$$

quotient greater than divisor

remainder not 0

Divide by 11:

$$\begin{array}{r} 9 \\ 11\overline{)103} \\ \underline{99} \\ 4 \end{array}$$

quotient is less than divisor

The number 103 is prime.

Example 5

Is 221 prime or composite?

Solution

Tests for 2, 3, and 5 fail.

Divide by 7:

```
    31    quotient greater
 7)221    than divisor
   21
   11
    7
    4     remainder not 0
```

Divide by 11:

```
     20    quotient greater
 11)221    than divisor
    22
     01
      0
      1    remainder not 0
```

Divide by 13:

```
     17
 13)221
    13
     91
     91
      0    remainder is 0
```

The number 221 is composite and not prime.

> **Note:** $221 = 13 \cdot 17$; that is, 13 and 17 are factors of 221.

Completion Example 6

Is 211 prime or composite?

Solution

Tests for 2, 3, and 5 all fail.

Divide by 7: $7\overline{)211}$

Divide by 11: $11\overline{)211}$

Divide by 13: $13\overline{)211}$

Divide by ______: $\overline{)211}$

211 is ______________.

6. Determine whether 187 is prime or composite. Explain each step.

7. Determine whether 233 is prime or composite. Explain each step.

Now Work Exercises 6 and 7 in the Margin.

EXAMPLE 7

One interesting application of factors of counting numbers (very useful in beginning algebra) involves finding two factors whose sum is some specified number. For example, find two factors of 70 such that their product is 70 and their sum is 19.

Solution

The factors of 70 are 1, 2, 5, 7, 10, 14, 35, and 70, and the pairs whose products are 70 are

$$1 \cdot 70 = 70, \quad 2 \cdot 35 = 70, \quad 5 \cdot 14 = 70, \quad 7 \cdot 10 = 70$$

Thus, the numbers we are looking for are 5 and 14 because

$$5 \cdot 14 = 70 \quad \text{and} \quad 5 + 14 = 19.$$

Completion Example Answers

6. Tests for 2, 3, and 5 all fail.

Divide by 7:
$$\begin{array}{r} \mathbf{30} \\ 7\,\overline{)\,211} \\ \underline{\mathbf{21}} \\ \mathbf{01} \end{array}$$

Divide by 11:
$$\begin{array}{r} \mathbf{19} \\ 11\,\overline{)\,211} \\ \underline{\mathbf{11}} \\ \mathbf{101} \\ \underline{\mathbf{99}} \\ \mathbf{2} \end{array}$$

Divide by 13:
$$\begin{array}{r} \mathbf{16} \\ 13\,\overline{)\,211} \\ \underline{\mathbf{13}} \\ \mathbf{81} \\ \underline{\mathbf{78}} \\ \mathbf{3} \end{array}$$

Divide by 17:
$$\begin{array}{r} \mathbf{12} \\ 17\,\overline{)\,211} \\ \underline{\mathbf{17}} \\ \mathbf{41} \\ \underline{\mathbf{34}} \\ \mathbf{7} \end{array}$$

The number 211 is **prime.**

NAME ______________________ SECTION ______________ DATE __________

ANSWERS

Exercises 2.3

List the multiples of each of the following numbers.

1. 5 **2.** 7 **3.** 11 **4.** 13

5. 12 **6.** 20 **7.** 16 **8.** 25

9. Construct a Sieve of Eratosthenes for the numbers from 1 to 100. List the prime numbers that are less than 100.

Decide whether each of the following numbers is prime or composite. If the number is composite, find at least two pairs of factors for the number where the product of each pair is the number.

10. 31 **11.** 39 **12.** 45 **13.** 51

14. 73 **15.** 89 **16.** 150 **17.** 105

1. ________
2. ________
3. ________
4. ________
5. ________
6. ________
7. ________
8. ________
9. [Respond below exercise.]
10. ________
11. ________
12. ________
13. ________
14. ________
15. ________
16. ________
17. ________

18. ______________
19. ______________
20. ______________
21. ______________
22. ______________
23. ______________
24. ______________
25. ______________
26. ______________
27. ______________
28. ______________
29. ______________
30. ______________
31. ______________
32. ______________
33. ______________
34. ______________
35. ______________
36. ______________
37. ______________
38. ______________
39. ______________
40. ______________
41. ______________
42. ______________
43. ______________
44. ______________
45. ______________

18. 113 **19.** 317 **20.** 377 **21.** 619

22. 289 **23.** 713 **24.** 1147 **25.** 527

26. 839

Two numbers are given. Find two factors of the first number such that their product is the first number and their sum is the second number.

Example: 12, 8 Two factors of 12 whose product is 12 and whose sum is 8 are 6 and 2 since $6 \cdot 2 = 12$ and $6 + 2 = 8$.

27. 24, 10 **28.** 12, 7 **29.** 16, 10 **30.** 12, 13

31. 14, 9 **32.** 50, 27 **33.** 20, 9 **34.** 24, 11

35. 48, 19 **36.** 36, 15 **37.** 7, 8 **38.** 63, 24

39. 51, 20 **40.** 25, 10 **41.** 16, 8 **42.** 60, 17

43. 52, 17 **44.** 27, 12 **45.** 72, 22

NAME ______________ SECTION ______________ DATE ______________

ANSWERS

46. [Respond below exercise.]

47. [Respond below exercise.]

48. [Respond below exercise.]

49. [Respond below exercise.]

Writing and Thinking about Mathematics

46. Find the set of all prime numbers less than 1000 that are not odd.

47. Are all odd numbers prime? Explain your answer.

48. Explain why the number 1 is neither prime nor composite.

49. The number 1001 is composite. Find all the factors of 1001.

50. [Respond below exercise.] ______

51. [Respond below exercise.] ______

Collaborative Learning Exercises

50. With the class divided into teams of 2 to 4 students, each team is to try to find the largest prime number it can within 15 minutes. The team leader is to explain the team's reasoning to the class.

51. Mathematicians have been interested since ancient times in a search for **perfect numbers.** A **perfect number** is a counting number that is equal to the sum of its proper divisors (divisors not including itself). For example, the first perfect number is 6. The proper divisors of 6 are 1, 2, and 3, and $1 + 2 + 3 = 6$. The teams formed for Exercise 49 are to try to find the second and third perfect numbers. (*HINT:* The second perfect number is between 20 and 30, and the third perfect number is between 450 and 500.)

RECYCLE BIN

1. ______
2. ______
3. ______
4. ______
5. ______
6. ______
7. ______
8. ______

The Recycle Bin (from Section 1.4)

Tell what property of multiplication is illustrated.

1. $8 \cdot 7 = 7 \cdot 8$

2. $5\,(2 \cdot 6) = (5 \cdot 2)\,6$

Use the technique of multiplying by powers of 10 to find the following products mentally.

3. $40 \cdot 400$

4. $300 \cdot 500$

5. $120 \cdot 7000$

Find the following products.

6. $\begin{array}{r} 314 \\ \times\ 73 \\ \hline \end{array}$

7. $\begin{array}{r} 182 \\ \times\ 466 \\ \hline \end{array}$

8. $\begin{array}{r} 105 \\ \times\ 1700 \\ \hline \end{array}$

2.4 Prime Factorizations

OBJECTIVES

1. *Know the Fundamental Theorem of Arithmetic.*
2. *Know the meaning of the term* ***prime factorization.***
3. *Be able to find the prime factorization of a composite number.*

Finding a Prime Factorization

To add and subtract fractions (Chapter 3), we first need to find a common denominator for the fractions. Finding **all** the prime factors of a composite number will help us to accomplish this. For example, factoring 28 gives $28 = 4 \cdot 7$. While 7 is a prime number, 4 is not prime. So, continuing to factor 4 as $2 \cdot 2$, we have all prime factors, and the **prime factorization** of 28: $28 = 2 \cdot 2 \cdot 7$.

We could have started factoring 28 as $28 = 2 \cdot 14$; however, since $14 = 2 \cdot 7$, we still would have had the same end result: $28 = 2 \cdot 2 \cdot 7$. Regardless of the method used or the factors used in the beginning, **there is only one prime factorization for any composite number.** This fact is so important that it is called the **Fundamental Theorem of Arithmetic.**

The Fundamental Theorem of Arithmetic

Every composite number has exactly one prime factorization.

Remember that, because multiplication is a commutative operation, the order in which the prime factors are written is not important. That is, a different ordering of the factors is not a different prime factorization. Thus, $2 \cdot 2 \cdot 7$ and $2 \cdot 7 \cdot 2$ are the same prime factorizations of 28. What is important is that **all of the factors must be prime numbers.** One procedure for finding the prime factorization of a composite number is outlined in the following box.

To Find the Prime Factorization of a Composite Number.

1. Factor the composite number into any two factors.
2. Factor each factor that is not prime.
3. Continue this process until all factors are prime.

The **prime factorization** is the product of all the prime factors.

Many times the beginning factors needed to start the process of finding a prime factorization can be found by using the tests for divisibility by 2, 3, 4, 5, 9, and 10 discussed in Section 2.2. This was one reason for developing these tests, and you should review them or write them down for easy reference. They are used in the following examples.

EXAMPLE 1 Find the prime factorization of 60.

Solution

$60 = 6 \cdot 10$ Since the last digit is 0, we know 10 is a factor.

$= 2 \cdot 3 \cdot 2 \cdot 5$ 6 and 10 can both be factored so that each factor is a prime number. This is the prime factorization of 60.

or,

$60 = 3 \cdot 20$ 3 is prime, but 20 is not.

$= 3 \cdot 4 \cdot 5$ 4 is not prime.

$= 3 \cdot 2 \cdot 2 \cdot 5$ All factors are prime.

Since multiplication is commutative, the order of the factors is not important. What is important is that **all of the factors must be prime numbers.**

Writing the factors in order, we see the prime factorization of 60 is $2 \cdot 2 \cdot 3 \cdot 5$ or, using exponents, $2^2 \cdot 3 \cdot 5$.

EXAMPLE 2 Find the prime factorization of 70.

Solution

$70 = 7 \cdot 10$ 10 is a factor since the last digit is 0.

$= 7 \cdot 2 \cdot 5$

$= 2 \cdot 5 \cdot 7$ Writing the factors in order is not necessary, but it is convenient for comparing answers.

EXAMPLE 3 Find the prime factorization of each number.

Solution

(a) $85 = 5 \cdot 17$ 5 is a factor since the last digit is 5. Since both 5 and 17 are prime, $5 \cdot 17$ is the prime factorization.

(b) $72 = 8 \cdot 9$
$= 2 \cdot 4 \cdot 3 \cdot 3$
$= 2 \cdot 2 \cdot 2 \cdot 3 \cdot 3$
$= 2^3 \cdot 3^2$

or $72 = 2 \cdot 36$
$= 2 \cdot 6 \cdot 6$
$= 2 \cdot 2 \cdot 3 \cdot 2 \cdot 3$
$= 2^3 \cdot 3^2$ using exponents

(c) $245 = 5 \cdot 49$
$= 5 \cdot 7 \cdot 7$
$= 5 \cdot 7^2$

(d) $264 = 2 \cdot 132$
$= 2 \cdot 2 \cdot 66$
$= 2 \cdot 2 \cdot 2 \cdot 33$
$= 2 \cdot 2 \cdot 2 \cdot 3 \cdot 11$
$= 2^3 \cdot 3 \cdot 11$

or $264 = 4 \cdot 66$
$= 2 \cdot 2 \cdot 6 \cdot 11$
$= 2 \cdot 2 \cdot 2 \cdot 3 \cdot 11$
$= 2^3 \cdot 3 \cdot 11$

Regardless of your choices for the first two factors, there is only one prime factorization for any composite number.

✔ Completion Example 4

Find the prime factorization of 90.

Solution

$90 = 9 \cdot \_\_\_\_$

$= 3 \cdot 3 \cdot \_\_\_\_ \cdot \_\_\_\_$

$= \_\_\_\_\_\_\_\_$ using exponents

✔ Completion Example 5

Find the prime factorization of 925.

Solution

$925 = 5 \cdot \_\_\_\_$

$= 5 \cdot \_\_\_\_ \cdot \_\_\_\_$

$= \_\_\_\_\_\_\_\_$ using exponents

✔ Completion Example 6

Find the prime factorization of 196.

Solution

$196 = 2 \cdot 98$

$= 2 \cdot \_\_\_\_ \cdot \_\_\_\_$

$= 2 \cdot \_\_\_\_ \cdot \_\_\_\_ \cdot \_\_\_\_$

$= \_\_\_\_\_\_\_\_$ using exponents

Now Work Exercises 1–4 in the Margin.

Find the prime factorization of each of the following numbers.

1. 42

2. 56

3. 230

4. 165

Factors of Composite Numbers

Once the prime factorization of a composite number is known, all the factors (or divisors) of that number can be found. For a number to be a factor of a composite number, it must be either 1, the number itself, one of the prime factors, or the product of two or more of the prime factors.

The only factors (or divisors) of a composite number are:

1. 1 and the number itself;

2. Each prime factor; and

3. Products formed by all combinations of the prime factors (including repeated factors).

5. Find all the factors of 18.

6. Find all the factors of 63.

EXAMPLE 7

Find all the factors of 30.

Solution

Since $30 = 2 \cdot 3 \cdot 5$, the factors are
(a) 1 and the number itself: 1 and 30.
(b) Each prime factor: 2, 3, 5.
(c) Products of all combinations of the prime factors:

$$2 \cdot 3 = 6, \qquad 2 \cdot 5 = 10, \qquad 3 \cdot 5 = 15$$

The factors are 1, 30, 2, 3, 5, 6, 10, and 15. These are the only factors of 30.

EXAMPLE 8

Find all factors of 140.

Solution

$$\begin{aligned} 140 &= 14 \cdot 10 \\ &= 2 \cdot 7 \cdot 2 \cdot 5 \\ &= 2 \cdot 2 \cdot 5 \cdot 7 \end{aligned}$$

The factors are
(a) 1 and the number itself: 1 and 140.
(b) Each prime factor: 2, 5, 7.
(c) Products of all combinations of the prime factors:

$$2 \cdot 2 = 4 \qquad 2 \cdot 5 = 10, \qquad 2 \cdot 7 = 14, \qquad 5 \cdot 7 = 35,$$

$$2 \cdot 2 \cdot 5 = 20, \qquad 2 \cdot 2 \cdot 7 = 28, \qquad 2 \cdot 5 \cdot 7 = 70$$

The factors are

1, 140, 2, 5, 7, 4, 10, 14, 35, 20, 28, and 70.

There are no other factors (or divisors) of 140.

Now Work Exercises 5 and 6 in the Margin.

Completion Example Answers

4. $$\begin{aligned} 90 &= 9 \cdot \mathbf{10} \\ &= 3 \cdot 3 \cdot \mathbf{2} \cdot \mathbf{5} \\ &= \mathbf{2} \cdot \mathbf{3^2} \cdot \mathbf{5} \quad \text{using exponents} \end{aligned}$$

5. $$\begin{aligned} 925 &= 5 \cdot \mathbf{185} \\ &= 5 \cdot \mathbf{5} \cdot \mathbf{37} \\ &= \mathbf{5^2} \cdot \mathbf{37} \quad \text{using exponents} \end{aligned}$$

6. $$\begin{aligned} 196 &= 2 \cdot 98 \\ &= 2 \cdot \mathbf{2} \cdot \mathbf{49} \\ &= 2 \cdot \mathbf{2} \cdot \mathbf{7} \cdot \mathbf{7} \\ &= \mathbf{2^2} \cdot \mathbf{7^2} \quad \text{using exponents} \end{aligned}$$

NAME ______________________ SECTION ______________ DATE ________

Exercises 2.4

Find the prime factorization for each of the following numbers. Use the tests for divisibility for 2, 3, 4, 5, 9, and 10 whenever they help find beginning factors.

1. 24	**2.** 28	**3.** 27	**4.** 16
5. 36	**6.** 60	**7.** 72	**8.** 90
9. 81	**10.** 105	**11.** 125	**12.** 160
13. 75	**14.** 150	**15.** 210	**16.** 40
17. 250	**18.** 93	**19.** 168	**20.** 360
21. 126	**22.** 48	**23.** 17	**24.** 47
25. 51	**26.** 144	**27.** 121	**28.** 169
29. 225	**30.** 52	**31.** 32	**32.** 98

ANSWERS

1. ________
2. ________
3. ________
4. ________
5. ________
6. ________
7. ________
8. ________
9. ________
10. ________
11. ________
12. ________
13. ________
14. ________
15. ________
16. ________
17. ________
18. ________
19. ________
20. ________
21. ________
22. ________
23. ________
24. ________
25. ________
26. ________
27. ________
28. ________
29. ________
30. ________
31. ________
32. ________

33. ____________
34. ____________
35. ____________
36. ____________
37. ____________
38. ____________
39. ____________
40. ____________
41. ____________
42. ____________
43. ____________
44. ____________
45. ____________
46. ____________
47. ____________
48. ____________
49. ____________
50. ____________
51. [Respond below exercise.] ____________

33. 108 **34.** 103 **35.** 101 **36.** 202

37. 78 **38.** 500 **39.** 10,000 **40.** 100,000

Using the prime factorization of each number, find all the factors (or divisors) of each number.

41. 12 **42.** 18 **43.** 28 **44.** 98

45. 121 **46.** 45 **47.** 105 **48.** 54

49. 97 **50.** 144

Collaborative Learning Exercises

51. The product of the counting numbers from 1 to a given number is called the **factorial** of that number. For example, the symbol 5! is read "five factorial," and the value is $5! = 1 \cdot 2 \cdot 3 \cdot 4 \cdot 5$.

In teams of 2 to 4 students, analyze the following questions and statements and discuss your conclusions in class.

(a) What are the meanings of the symbols 6!, 7!, 8!, and 10!?
(b) Write the prime factorization of each of these factorials.
(c) Determine whether each of the following statements is true or false:

(1) 24 divides 6!. (2) 28 divides 6!.

(3) 75 divides 8!. (4) 75 divides 10!.

NAME ____________ SECTION ____________ DATE ____________

ANSWERS

52. [Respond below exercise.]

52. The following expressions are part of formulas used in probability and statistics. For example, the answer to part (a) indicates the number of committees that can be formed with two people from a group of six people.

(a) $\dfrac{6!}{2!(4!)}$ (b) $\dfrac{10!}{2!(8!)}$ (c) $\dfrac{20!}{6!(14!)}$

In teams of 2 to 4 students, evaluate each of these expressions and discuss how you arrived at these results in class. Did you find any shortcuts in the calculations? What are possible "committee" problems related to parts (b) and (c)?

RECYCLE BIN

1. ______________

2. ______________

3. ______________

4a. ______________

b. ______________

The Recycle Bin (from Section 1.6)

1. Find the difference between the product of 42 and 38 and the quotient of 1420 and 20.

2. Chris bought a car for $16,200. The salesperson added $972 for taxes and $520 for license fees. If Chris made a down payment of $4750 and financed the rest with his bank, how much did he finance?

3. Find the average (or mean) of the numbers 103, 114, 98, and 101.

4. A right triangle (one angle is 90°) has three sides: a base of 12 meters, a height of 5 meters, and a third side of 13 meters. (See diagram.)
   (a) Find the perimeter of (distance around) the triangle.
   (b) Find the area of the triangle in square meters. (To find the area, multiply base times height, then divide by 2.)

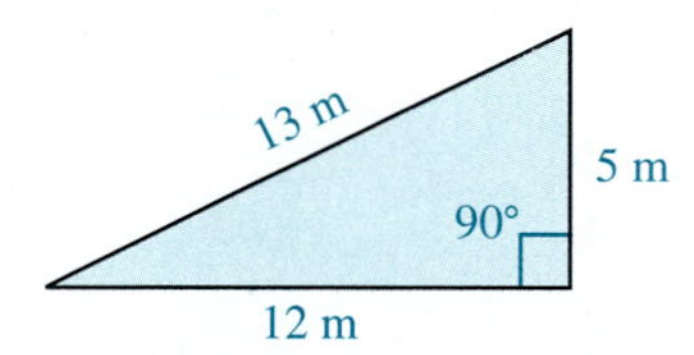

2.5 Least Common Multiple (LCM) with Applications

OBJECTIVES

1. *Understand the meaning of the term* ***multiple.***
2. *Use prime factorizations to find the least common multiple of a set of numbers.*
3. *Recognize the application of the LCM concept in a word problem.*

Finding the LCM of a Set of Counting Numbers

The ideas discussed in this section related to factors, prime factors, and **least common multiples** are used throughout Chapter 3 (Fractions). Study these ideas and the techniques involved carefully because they will make your work with fractions much easier.

Recall that the multiples of a counting number are the products of that number with all the counting numbers. The first multiple of any counting number is the number itself, and all other multiples are larger than that number.

Counting Numbers:	**1,**	**2,**	**3,**	**4,**	**5,**	**6,**	**7,**	**8,**	**. . .**
Multiples of 8:	8,	16,	(24,)	32,	40,	(48,)	56,	64,	. . .
Multiples of 12:	12,	(24,)	36,	(48,)	60,	(72,)	84,	96,	. . .

The common multiples of 8 and 12 are 24, 48, 72, 96, and so on. The **least common multiple (LCM)** is 24.

Listing all of the multiples (and we just did for 8 and 12) and then choosing the least common multiple (LCM) is not very efficient. The following technique involving prime factorizations is generally much easier to use.

To Find the LCM of a Set of Counting Numbers:

1. Find the prime factorization of each number.
2. Find the prime factors that appear in **any one** of the prime factorizations.
3. Form the product of these primes using each prime the most number of times it appears in **any one** of the prime factorizations.

Ease in using the LCM method depends on your ability to find prime factorizations quickly and accurately, and that comes with practice. STAY WITH IT!

EXAMPLE 1 Find the LCM of 6 and 10.

Solution

$6 \cdot 10 = 60$ and 60 is divisible by both 6 and 10, but 60 is not the smallest number divisible by both 6 and 10.

(a) Prime factorizations: $6 = 2 \cdot 3$
$10 = 2 \cdot 5$

(b) The prime factors: 2, 3, 5

(c) The factor 2 appears once in 6 and once in 10, so the most number of times 2 appears in any one prime factorization is once. The same is true for both 3 and 5.

$$\text{LCM} = 2 \cdot 3 \cdot 5 = 30$$

Each factor appears only once in the LCM, and 30 is the smallest number divisible by both 6 and 10.

EXAMPLE 2 Find the LCM of 18, 30, and 45.

Solution

(a) Prime factorizations:

$18 = 2 \cdot 9 = 2 \cdot 3 \cdot 3$ one 2, two 3's
$30 = 6 \cdot 5 = 2 \cdot 3 \cdot 5$ one 2, one 3, one 5
$45 = 9 \cdot 5 = 3 \cdot 3 \cdot 5$ two 3's, one 5

(b) 2, 3, and 5 are the only prime factors.

(c) Most of each factor in any one factorization:

one 2 (in 18 and in 30)
two 3's (in 18 and in 45)
one 5 (in 30 and in 45)

$$\text{LCM} = 2 \cdot 3 \cdot 3 \cdot 5 = 2 \cdot 3^2 \cdot 5 = 90$$

90 is the smallest number divisible by all the numbers 18, 30, and 45.

COMPLETION EXAMPLE 3 Find the LCM of 36, 24, and 48.

Solution

(a) Prime factorizations:

36 = ____________

24 = ____________

48 = ____________

(b) ______ and ______ are the only prime factors.

(c) Most of each factor in any one factorization:

__________ (in 48)

__________ (in 36)

LCM = ____________ = __________ = 144

__________ is the smallest number divisible by all the numbers 36, 24, and 48.

Now Work Exercise 1 in the Margin.

1. Find the LCM of 28 and 70.

Alternate Approach to Finding the LCM

Another, more intuitive, approach to finding the least common multiple (LCM) of a set of numbers involves a step-by-step building process that relies on the concept of divisibility. In this approach, we proceed as follows:

STEP 1: Find the prime factorization of each number in the set.

STEP 2: Write the prime factorization of **any one** of the numbers in the set.

STEP 3: Look at a second number and multiply what you have in Step 2 by any prime factors needed to insure divisibility by this second number.

STEP 4: Continue as in Step 3 with each of the remaining numbers in the set.

STEP 5: The resulting product will be the LCM of the original set of numbers.

Example 4

Find the LCM of the set of numbers 15, 18, and 24.

Solution

STEP 1: Find the prime factorization of each number:

$$15 = 3 \cdot 5$$
$$18 = 2 \cdot 3 \cdot 3$$
$$24 = 2 \cdot 2 \cdot 2 \cdot 3$$

STEP 2: Write the prime factorization of any one of the numbers:

$$\text{LCM} = 3 \cdot 5 \cdot \underline{\quad ? \quad}$$

Having all the factors of 15 insures divisibility by 15.

STEP 3: Multiply by whatever new factors you need to insure divisibility by 18:

$$\text{LCM} = 3 \cdot 5 \cdot 2 \cdot 3 \cdot \underline{\quad ? \quad}$$ One 2 and one more 3 are needed.

Having all the factors of 18 insures divisibility by 18.

STEP 4: Multiply by whatever new factors you need to insure divisibility by 24:

$$\text{LCM} = 3 \cdot 5 \cdot 2 \cdot 3 \cdot 2 \cdot 2$$ Two more 2's are needed.

Having all the factors of 24 insures divisibility by 24.

STEP 5: The LCM is now found, and it is divisible by each number in the original set of numbers:

$$\text{LCM} = 3 \cdot 5 \cdot 2 \cdot 3 \cdot 2 \cdot 2 = 360$$

Finding How Many Times Each Number Divides into the LCM

In our work with fractions (Chapter 3), we will want to know how many times each number in a set divides into the LCM. We could perform long division:

$$\begin{array}{r} 6 \\ 24\overline{)144} \\ \underline{144} \\ 0 \end{array}$$

Thus, $24 \cdot 6 = 144$. However, once we have the prime factorization of 144, as we do from the Completion Example 3, we can group the factors as follows to find how many times 24 divides into 144 without performing long division.

$$\begin{aligned} 144 &= 2 \cdot 2 \cdot 2 \cdot 2 \cdot 3 \cdot 3 \\ &= \underbrace{2 \cdot 2 \cdot 2 \cdot 3}_{} \cdot \underbrace{2 \cdot 3}_{} \\ &= \quad 24 \quad\quad \cdot \; 6 \end{aligned}$$

To Find How Many Times a Number in a Set of Counting Numbers Divides into the LCM of that Set of Numbers:

1. Find the LCM.
2. In the LCM, group together all the prime factors that make up the number.

The product of the remaining factors in the LCM tells how many times that number divides into the LCM.

Example 5

(a) Find the LCM of 12, 18, and 20. (b) State the number of times each number divides into the LCM.

Solution

(a)
$$\left.\begin{aligned} 12 &= 2 \cdot 2 \cdot 3 \\ 18 &= 2 \cdot 3 \cdot 3 \\ 20 &= 2 \cdot 2 \cdot 5 \end{aligned}\right\} \quad \begin{aligned} \text{LCM} &= 2 \cdot 2 \cdot 3 \cdot 3 \cdot 5 \\ &= 2^2 \cdot 3^2 \cdot 5 = 180 \end{aligned}$$

(b) Group the factors of 12:

$$\begin{aligned} 180 = 2 \cdot 2 \cdot 3 \cdot 3 \cdot 5 &= \underbrace{(2 \cdot 2 \cdot 3)}_{} \cdot \underbrace{(3 \cdot 5)}_{} \\ &= \quad 12 \quad\quad \cdot \; 15 \end{aligned}$$

Group the factors of 18:

$$180 = 2 \cdot 2 \cdot 3 \cdot 3 \cdot 5 = \underbrace{(2 \cdot 3 \cdot 3)}_{} \cdot \underbrace{(2 \cdot 5)}_{}$$
$$= 18 \cdot 10$$

Group the factors of 20:

$$180 = 2 \cdot 2 \cdot 3 \cdot 3 \cdot 5 = \underbrace{(2 \cdot 2 \cdot 5)}_{} \cdot \underbrace{(3 \cdot 3)}_{}$$
$$= 20 \cdot 9$$

Thus,

12 divides 15 times into 180.

18 divides 10 times into 180.

20 divides 9 times into 180.

☑ COMPLETION EXAMPLE 6

(a) Find the LCM of 27, 30, and 42. (b) State the number of times each number divides into the LCM.

Solution

(a) 27 = __________
30 = __________ } LCM = __________
42 = __________ = __________ = 1890

(b) 1890 = __________ = $(3 \cdot 3 \cdot 3) \cdot$ (______) = $27 \cdot$ ______
1890 = __________ = $(2 \cdot 3 \cdot 5) \cdot$ (______) = $30 \cdot$ ______
1890 = __________ = $(2 \cdot 3 \cdot 7) \cdot$ (______) = $42 \cdot$ ______

Now Work Exercises 2 and 3 in the Margin.

In the following exercises, (a) find the LCM of each set of numbers and (b) state the number of times each number in the set divides into the LCM.

2. 30, 50

3. 18, 20, 25

An Application

EXAMPLE 7(a)

Suppose three weather satellites—A, B, and C—are orbiting the earth. Satellite A takes 18 hours, B takes 14 hours, and C takes 10 hours. If they are directly above each other now, as shown in part (a) of the figure below, in how many hours will they again be directly above each other in the positions shown in part (a) of the figure?

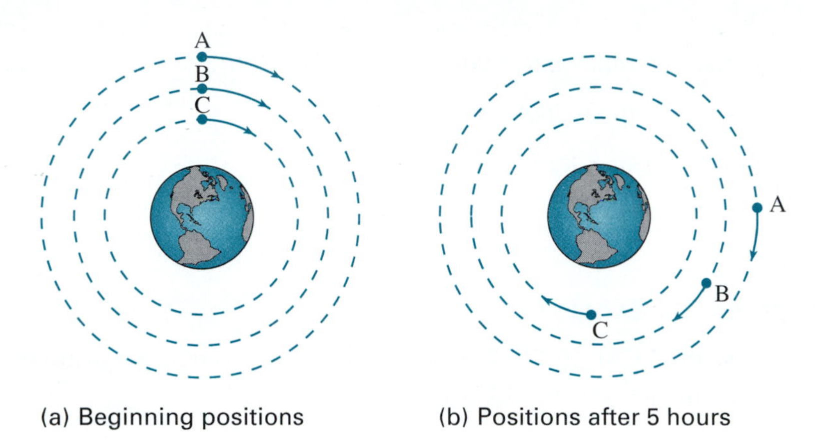

(a) Beginning positions (b) Positions after 5 hours

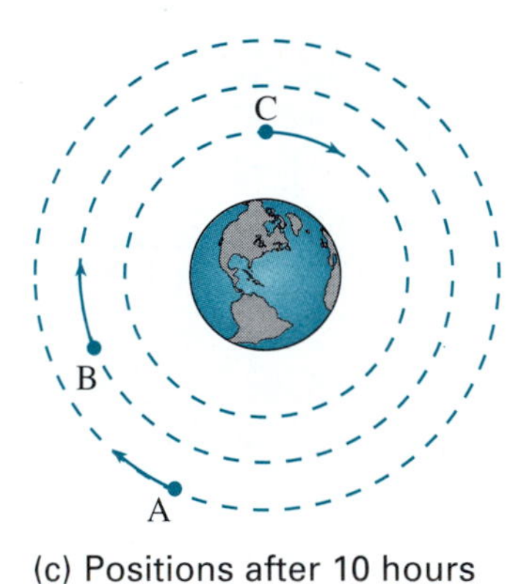

(c) Positions after 10 hours

Solution

Note that, when the three satellites are in this position again, each one will have made some number of complete orbits. Since satellite A takes 18 hours to make one complete orbit of the earth, our solution must be a multiple of 18. Similarly, our solution must also be a multiple of 14 and a multiple of 10 to account for the complete orbits of satellites B and C.

The LCM of 18, 14, and 10 will tell us the next time satellites A, B, and C will all complete an orbit at the same time.

$$\left.\begin{aligned} 18 &= 2 \cdot 3^2 \\ 14 &= 2 \cdot 7 \\ 10 &= 2 \cdot 5 \end{aligned}\right\} \quad \text{LCM} = 2 \cdot 3^2 \cdot 5 \cdot 7 = 630$$

Thus, the satellites will not align again for 630 hours (or 26 days and 6 hours).

Note: Satellite A will have made 35 orbits, since $630 = 18 \cdot 35$.
Satellite B will have made 45 orbits, since $630 = 14 \cdot 45$.
Satellite C will have made 63 orbits, since $630 = 10 \cdot 63$.

☑ *Completion Example 7(b)*

The problem stated in **Mathematics at Work!** at the beginning of this chapter is similar to Example 7(a). Only the orbiting times of the satellites are different. Redraw the figures from page 88 here. In how many hours will they again be directly above each other in the positions shown in the first figure? How many orbits will each satellite have completed at that time?

Solution

Figures:

They will be in the original position in ______ hours.

Satellite A will have made ______ orbits.

Satellite B will have made ______ orbits.

Satellite C will have made ______ orbits.

Completion Example Answers

3. (a) Prime factorizations:

   $36 = \mathbf{2 \cdot 2 \cdot 3 \cdot 3}$

   $24 = \mathbf{2 \cdot 2 \cdot 2 \cdot 3}$

   $48 = \mathbf{2 \cdot 2 \cdot 2 \cdot 2 \cdot 3}$

   (b) **2** and **3** are the only prime factors.

   (c) Most of each factor in any one factorization:

   4 2's (in 48)

   2 3's (in 36)

   $\text{LCM} = \mathbf{2 \cdot 2 \cdot 2 \cdot 2 \cdot 3 \cdot 3} = \mathbf{2^4 \cdot 3^2} = 144$

   144 is the smallest number divisible by all the numbers 36, 24, and 48.

6. (a)

$$\left.\begin{aligned} 27 &= \mathbf{3 \cdot 3 \cdot 3} \\ 30 &= \mathbf{2 \cdot 3 \cdot 5} \\ 42 &= \mathbf{2 \cdot 3 \cdot 7} \end{aligned}\right\} \quad \begin{aligned} \text{LCM} &= \mathbf{2 \cdot 3 \cdot 3 \cdot 3 \cdot 5 \cdot 7} \\ &= \mathbf{2 \cdot 3^3 \cdot 5 \cdot 7} = 1890 \end{aligned}$$

   (b) $1890 = \mathbf{2 \cdot 3 \cdot 3 \cdot 3 \cdot 5 \cdot 7} = (3 \cdot 3 \cdot 3) \cdot (\mathbf{2 \cdot 5 \cdot 7}) = 27 \cdot \mathbf{70}$

   $1890 = \mathbf{2 \cdot 3 \cdot 3 \cdot 3 \cdot 5 \cdot 7} = (2 \cdot 3 \cdot 5) \cdot (\mathbf{3 \cdot 3 \cdot 7}) = 30 \cdot \mathbf{63}$

   $1890 = \mathbf{2 \cdot 3 \cdot 3 \cdot 3 \cdot 5 \cdot 7} = (2 \cdot 3 \cdot 7) \cdot (\mathbf{3 \cdot 3 \cdot 5}) = 42 \cdot \mathbf{45}$

7. (b) They will be in the original position in **72** hours.
   Satellite A will have made **3** orbits.
   Satellite B will have made **4** orbits.
   Satellite C will have made **6** orbits.

NAME ______________ SECTION ______________ DATE ______________

ANSWERS

Respond below exercises 1–24.

Exercises 2.5

Find the LCM of each of the following sets of numbers.

1. 8, 12

2. 3, 5, 7

3. 4, 6, 9

4. 3, 5, 9

5. 2, 5, 11

6. 4, 14, 18

7. 6, 15, 12

8. 6, 8, 27

9. 25, 40

10. 40, 75

11. 28, 98

12. 30, 75

13. 30, 80

14. 16, 28

15. 25, 100

16. 20, 50

17. 35, 100

18. 144, 216

19. 36, 42

20. 40, 100

21. 2, 4, 8

22. 10, 15, 35

23. 8, 13, 15

24. 25, 35, 49

Respond below exercises 25–50.

25. 6, 12, 15

26. 8, 10, 120

27. 6, 15, 80

28. 13, 26, 169

29. 45, 125, 150

30. 34, 51, 54

31. 33, 66, 121

32. 36, 54, 72

33. 45, 145, 290

34. 54, 81, 108

35. 45, 75, 135

36. 35, 40, 72

37. 10, 20, 30, 40

38. 15, 25, 30, 40

39. 24, 40, 48, 56

40. 169, 637, 845

In Exercises 41–50, (a) find the LCM and (b) state the number of times each number divides into the LCM.

41. 8, 10, 15

42. 6, 15, 30

43. 10, 15, 24

44. 8, 10, 120

45. 6, 18, 27, 45

46. 12, 95, 228

47. 45, 63, 98

48. 40, 56, 196

49. 99, 143, 363

50. 125, 135, 225

NAME ______________________ SECTION ______________ DATE __________

ANSWERS

51a. ____________

b. ____________

52a. ____________

b. ____________

53a. ____________

b. ____________

54a. ____________

b. ____________

51. Two long-distance joggers are running on the same course in the same direction. They meet at the water fountain and say "Hi." One jogger goes around the course in 10 minutes and the other goes around the course in 14 minutes. They continue to jog until they meet again at the water fountain. (a) How many minutes elapsed between meetings at the fountain? (b) How many times will each have jogged around the course?

52. Three night watchmen walk around inspecting buildings at a shopping center. The watchmen take 9, 12, and 14 minutes, respectively, for the inspection trip. (a) If they start at the same time, in how many minutes will they meet? (b) How many inspection trips will each watchman have made?

53. Two astronauts miss connections at their first rendezvous in space. (a) If one astronaut circles the earth every 12 hours and the other every 16 hours, in how many hours will they rendezvous again? (b) How many orbits will each astronaut have made between the first and second rendezvous?

54. Three truck drivers eat lunch together whenever all three are at the routing station at the same time. The first driver's route takes 5 days, the second driver's takes 15 days, and the third driver's takes 6 days. (a) How often do the three drivers eat lunch together? (b) If the first driver's route is changed to take 6 days, how often would they eat lunch together?

55a. ______________

b. ______________

56. [Respond below exercise.]

57. [Respond below exercise.]

55. Four book salespersons leave the home office the same day. They take 10 days, 12 days, 15 days, and 18 days, respectively, to travel their own sales territories. (a) In how many days will they all meet again at the home office? (b) How many sales trips will each have made?

Writing and Thinking about Mathematics

56. From studying Example 7(a) and working Exercises 51–55, write a general statement about the types of word problems (or characteristics of these problems) in which the least common multiple is involved.

Collaborative Learning Project

57. Separate the class into teams of two to four students. Each team is to write one or two paragraphs explaining which topic in this chapter they found to be the most interesting and why. The team leader is to read the paragraphs(s) with classroom discussion to follow.

2 Index of Key Ideas and Terms

NAME ______________ SECTION ______________ DATE ______________

ANSWERS

1. ______________
2a. ______________
b. ______________
3. ______________
4. ______________
5. ______________
6. ______________
7. ______________
8. ______________
9. ______________
10. ______________
11. ______________
12. [Respond below exercise.]

Chapter 2 Test

1. In the equation $7^3 = 343$, ______ is the exponent, ______ is the base, and ______ is the power.

2. (a) List the prime numbers between 6 and 20.
(b) List the squares of these prime numbers.

In Exercises 3–6, find the value of each expression by using the rules for order of operations.

3. $12 + 9 \div 3 - 2$

4. $60 \div 4(4 - 1) + 3$

5. $2 \cdot 3^2 - (2^2 \cdot 3 \div 2) - 3(5 - 1)$

6. $12 \cdot 2^3 \div 4 \cdot 3$

In Exercises 7–10, use the tests for divisibility to determine if the given number can be divided exactly by 2, 3, 4, 5, 9, or 10.

7. 90

8. 221

9. 324

10. 1700

11. List the multiples of 13 that are less than 100.

12. Show, by division, that 81 is a factor of 7452.

13. [Respond below exercise.]

14. ______

15. ______

16. ______

17. ______

18. [Respond below exercise.]

19. [Respond below exercise.]

20. [Respond below exercise.]

21a. ______

b. ______

13. Does 42 divide the product $2 \cdot 5 \cdot 6 \cdot 7 \cdot 9$? If so, how many times? If not, explain briefly why not.

14. List all the factors of 60.

Find the prime factorization of each number.

15. 124

16. 165

17. Determine whether or not 107 is a prime number. Show each step in your reasoning.

In Exercises 18–20, find the LCM of each set of numbers and tell how many times each number divides into the LCM.

18. 4, 14, 21

19. 6, 15, 60

20. 8, 10, 15, 28

21. Three salesmen have lunch together each time they are in the home office on the same day. (a) If it takes Salesman A 6 days to cover his territory, Salesman B 9 days, and Salesman C 12 days, how often do they have lunch together? (b) If Salesman A's route is changed so that he takes only 5 days to cover his territory, how often do they have lunch together?

NAME ______________ SECTION ______________ DATE ______________

ANSWERS

Cumulative Review

1. Write the number 50,732 in (a) expanded notation and (b) its English word equivalent.

In Exercises 2–5, name the property illustrated.

2. $45 \cdot 2 = 2 \cdot 45$

3. $7 \cdot (3 \cdot 4) = (7 \cdot 3) \cdot 4$

4. $19 + 0 = 19$

5. $25 + 3 = 3 + 25$

6. Round off 41,624 to the nearest thousand.

In Exercises 7–10, (a) first estimate each answer and (b) then find the answer by performing the indicated operation.

7.
$$\begin{array}{r} 83 \\ 947 \\ +\ 1035 \\ \hline \end{array}$$

8.
$$\begin{array}{r} 6003 \\ -\ 759 \\ \hline \end{array}$$

9.
$$\begin{array}{r} 74 \\ \times\ 86 \\ \hline \end{array}$$

10. $17\overline{)2210}$

11. Multiply mentally: 90×300.

1. [Respond below exercise.] ______________

2. ______________

3. ______________

4. ______________

5. ______________

6. ______________

7a. ______________

b. ______________

8a. ______________

b. ______________

9a. ______________

b. ______________

10a. ______________

b. ______________

11. ______________

12. ______________

13. ______________

14. ______________

15. ______________

16. ______________

17. ______________

Use the rules for order of operations to evaluate the expressions in Exercises 12 and 13.

12. $15 + 2(8 - 2^3) - 3 \cdot 2^2$

13. $75 \div 5 \cdot 3 + 4(7 - 2^2)$

14. Without actually dividing, determine which of the numbers 2, 3, 4, 5, 9, and 10 will divide into 10,840. Give a brief reason for each decision.

15. Determine whether the number 307 is prime or not. Show all the steps you used.

16. List all the prime numbers less than 20,000 that are even.

17. List all the factors of 65.

NAME ______________________ SECTION ________________ DATE __________

ANSWERS

18. ______________

19. [Respond below exercise.]

20. [Respond below exercise.]

21. [Respond below exercise.]

22. ______________

18. Find the prime factorization of 475.

19. Does 56 divide the product $2 \cdot 3 \cdot 4 \cdot 6 \cdot 7 \cdot 9$? If so, how many times? If not, explain briefly why not.

In Exercises 20 and 21, (a) find the LCM of each set of numbers and (b) tell how many times each number divides into the LCM.

20. 15, 27, 35

21. 33, 44, 55, 121

22. The Danube River is 2842 kilometers long and flows into the Black Sea. It is 509 kilometers longer than the Colorado River which flows into the Gulf of California. How long is the Colorado River?

23. ______________

24. ______________

23. A painting is in a rectangular frame that is 36 inches wide and 48 inches high. How many square inches of wall space is covered when the painting is hung?

24. On a psychology exam, three students scored 75, four students scored 82, two students scored 85, one student scored 87, five students scored 91, and one student scored 95. What was their average score on the exam?

3

Fractions

What to Expect in Chapter 3

Chapter 3 will help you understand fractions (called rational numbers) and operations with fractions. Each section has some word problems to reinforce the various concepts as they are developed. The term **improper fraction** is introduced in Section 3.1, and improper fractions are used throughout the chapter. The use of the term *improper* is quite unfortunate because it implies that there is something wrong with such fractions. This is not the case, and you should understand that improper fractions are perfectly good and as useful as any other type of number.

As you will see, the commutative and associative properties of multiplication (Section 3.2) and addition (Section 3.4) apply to fractions just as they do to whole numbers. The prime factorization techniques that were developed in Chapter 2 are used to reduce fractions to lower terms and to raise fractions to higher terms. For many students, this prime factoring method of dealing with fractions provides valuable insight into the nature of fractions and helps build confidence in working with fractions.

The rules for order of operations (Section 3.5) are the same for evaluating expressions that contain fractions as they are for evaluating expressions with whole numbers. At each step, you must be careful to remember what you have learned in the previous sections about how to perform the appropriate operations with fractions.

Mathematics at Work!

Young voters meet at a local polling place.

Politics and the political process affect everyone in some way. From local to state and national elections, registered voters make decisions about who will represent them and about various ballot measures.

For major issues at the state and national level, pollsters use mathematics (in particular, statistics and statistical methods) to indicate attitudes and to predict, within certain percentages, how voters will vote. The next time there is an important election in your area, read the papers and magazines and listen to the television reports for mathematics-related statements predicting the outcome.

There are 8000 registered voters in Brownsville, $\frac{3}{8}$ of whom live in neighborhoods on the north side of town. A survey indicates that $\frac{4}{5}$ of these north-side voters are in favor of a bond measure for constructing a new recreation facility that would largely benefit their neighborhoods. Also, $1\frac{7}{10}$ of the registered voters from all other parts of town are in favor of the measure.

(a) How many north-side voters favor the bond measure?
(b) How many voters in all favor the bond measure?

(See Example 18, Section 3.1.)

3.1 Basic Multiplication and Building to Higher Terms

OBJECTIVES

1. *Know the terms* **numerator** *and* **denominator.**
2. *Learn two meanings of fractions: to indicate equal parts of a whole and to indicate division.*
3. *Learn how to multiply fractions.*
4. *Determine what to multiply by in order to build a fraction to higher terms.*
5. *Understand why no denominator can be 0.*

Introduction to Fractions (or Rational Numbers)

In this chapter we will deal with fractions in which the numerator (top number) and denominator (bottom number) are whole numbers, and the denominator is not 0. Such fractions are called **rational numbers.** (Note that there are other fractions, called **irrational numbers,** that are studied in algebra and other mathematics courses.)

Definition

A **rational number** is a number that can be written in the fraction form $\frac{a}{b}$, where a is a whole number and b is a nonzero whole number.

$$\frac{a}{b} \quad \begin{matrix} \leftarrow \text{numerator} \\ \leftarrow \text{denominator} \end{matrix}$$

Note: The numerator a can be 0, but the denominator b cannot be 0.

Examples of fractions (or rational numbers) are $\frac{1}{2}, \frac{3}{4}, \frac{9}{10}$, and $\frac{17}{3}$. Unless otherwise stated, we will use the terms **fraction** and **rational number** to mean the same thing.

Figure 3.1 shows how a whole may be separated into equal parts. We see that the fractions $\frac{1}{2}, \frac{2}{4}, \frac{4}{8}$, and $\frac{6}{12}$ all represent the same amount of the whole. These numbers are said to be **equivalent** (or **equal**).

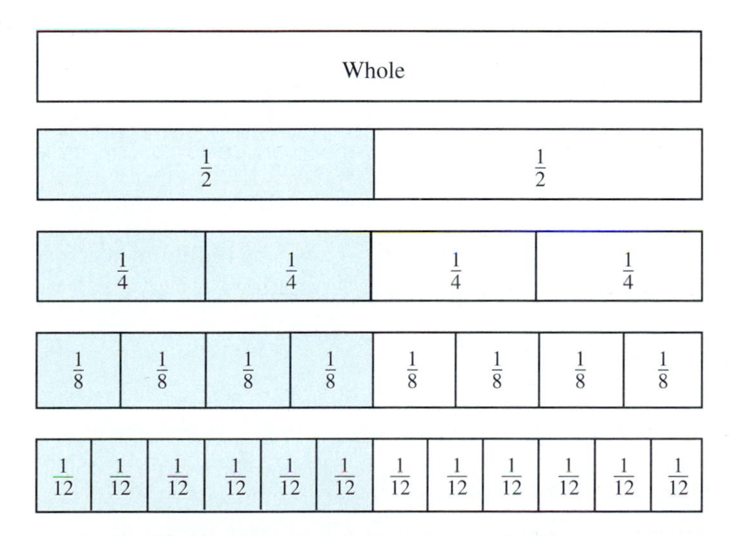

The shaded regions indicate equivalent (or equal) parts of the whole.

Figure 3.1

Fractions can be used to indicate:

1. Equal parts of a whole.

2. Division.

Example 1

$\frac{1}{2}$ can indicate 1 of 2 equal parts.

Example 2

$\frac{3}{4}$ can indicate 3 of 4 equal parts.

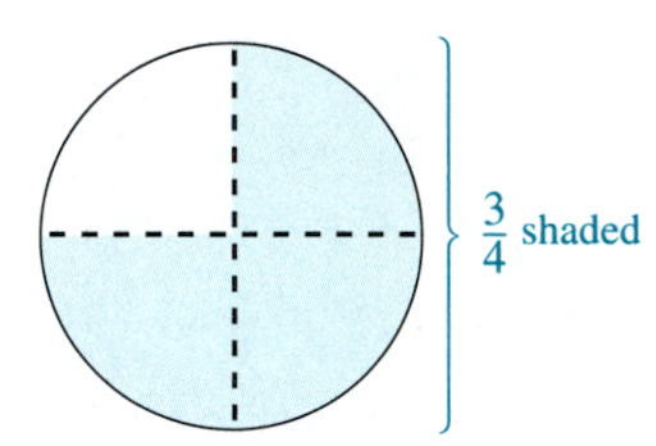

Example 3

$\frac{2}{3}$ can indicate 2 of 3 equal parts.

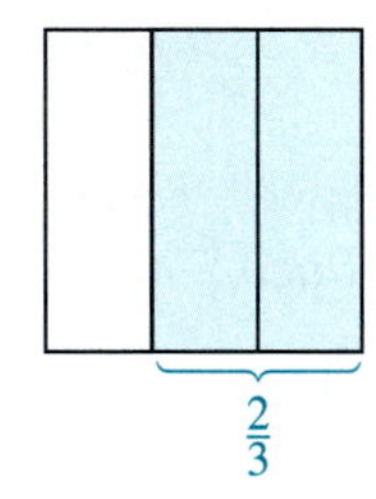

Improper fractions, such as $\frac{5}{3}$, $\frac{13}{6}$, and $\frac{18}{11}$, are fractions in which the numerator is larger than the denominator. The word **improper** is misleading because there is nothing improper about such fractions. In fact, in algebra and other mathematics courses, improper fractions are sometimes preferred over mixed numbers, such as $2\frac{3}{4}$ and $5\frac{1}{8}$. Mixed numbers will be discussed in Chapter 4.

Example 4

$\frac{5}{3}$ can indicate 5 of 3 equal parts (more than a whole).

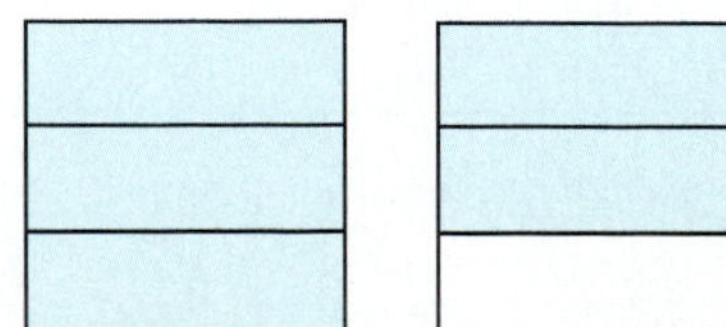

Whole numbers can be thought of as fractions with denominator 1. We say that each whole number is **equivalent** to (or **equal** to) its corresponding fraction. Thus,

$$0 = \frac{0}{1}, \quad 1 = \frac{1}{1}, \quad 2 = \frac{2}{1}, \quad 3 = \frac{3}{1}, \quad 4 = \frac{4}{1}, \quad \text{and so on.}$$

So, in working with fractions and whole numbers, we may rewrite a whole number in fraction form with denominator 1 if we so choose. This is particularly convenient when multiplying and dividing with fractions.

Now we want to clarify the use of 0 in the numerator and denominator. Since fractions may be used to indicate division, $\frac{24}{3}$ is the same as $24 \div 3$. And, because division is related to multiplication, we can make the following statement:

$$\frac{24}{3} = 8 \quad \text{because} \quad 24 = 3 \cdot 8$$

Similarly,

$$\frac{0}{5} = 0 \quad \text{because} \quad 0 = 5 \cdot 0$$

In general, if $b \neq 0$,

$$\frac{0}{b} = 0 \quad \text{because} \quad 0 = b \cdot 0.$$

However, if the denominator is 0, then the fraction has no meaning. We say that **division by 0 is undefined.** The following discussion explains why.

Division by 0 is Undefined.
No Denominator Can Be 0.

1. Consider $\frac{5}{0} = \boxed{?}$. Then we must have $5 = 0 \cdot \boxed{?}$. But this is impossible because $0 \cdot \boxed{?} = 0$ and $5 \neq 0$. Therefore, $\frac{5}{0}$ is undefined.

2. Next consider $\frac{0}{0} = \boxed{?}$. Then we must have $0 = 0 \cdot \boxed{?}$. But note that $0 = 0 \cdot \boxed{?}$ is true for any value of $\boxed{?}$, and in arithmetic, an operation such as division cannot give more than one answer. Thus, we agree that $\frac{0}{0}$ is undefined.

Thus, $\frac{a}{0}$ is undefined for any value of a.

Example 5

(a) $\frac{0}{36} = 0$ (b) $\frac{0}{124} = 0$

Example 6

(a) $\frac{17}{0}$ is undefined. (b) $\frac{233}{0}$ is undefined.

Multiplying Fractions

We now state the rule for multiplying fractions and discuss the use of the word **of** to indicate multiplication by fractions. Remember that any whole number can be written in fraction form with denominator 1.

To Multiply Fractions:

1. Multiply the numerators.
2. Multiply the denominators.

$$\frac{a}{b} \cdot \frac{c}{d} = \frac{a \cdot c}{b \cdot d}$$

Special Note about Fractions: We know that any whole number ***a*** can be considered as a fraction with denominator 1; that is, $a = \frac{a}{1}$. Similarly, any fraction can be considered to be the product of a whole number and a fraction; that is,

$$\frac{a}{b} = \frac{a}{1} \cdot \frac{1}{b} = a \cdot \frac{1}{b}$$

This idea will be helpful throughout the text in understanding how various forms of whole numbers, fractions, mixed numbers, and decimal numbers are related. For example,

$$\frac{2}{3} = 2 \cdot \frac{1}{3} \quad \text{and} \quad \frac{5}{8} = 5 \cdot \frac{1}{8}$$

Finding the product of two fractions can be thought of as finding one fractional part **of** of another fraction. For example, when we multiply

$$\frac{1}{2} \cdot \frac{3}{10}$$

we are finding

$$\frac{1}{2} \textbf{ of } \frac{3}{10}$$

Thus, $\frac{1}{2}$ **of** $\frac{3}{10}$ is $\frac{3}{20}$, because $\frac{1}{2} \cdot \frac{3}{10} = \frac{3}{20}$.

Example 7

Find the product of $\frac{2}{3}$ and $\frac{4}{5}$.

Solution

$$\frac{2}{3} \cdot \frac{4}{5} = \frac{2 \cdot 4}{3 \cdot 5} = \frac{8}{15}$$

This product can be illustrated graphically. We can think of the multiplication as finding $\frac{2}{3}$ **of** $\frac{4}{5}$.

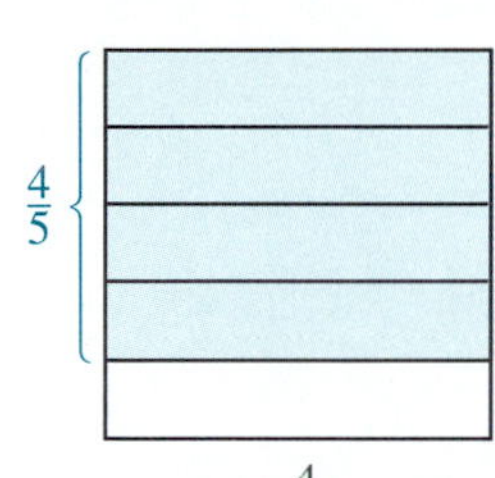

(a) Shade $\frac{4}{5}$.

(b) Shade $\frac{2}{3}$ of $\frac{4}{5}$.

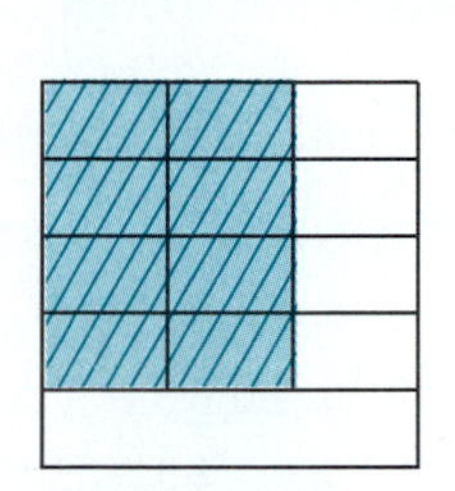

(c) Shaded region is $\frac{2}{3}$ of $\frac{4}{5}$ or $\frac{2}{3} \cdot \frac{4}{5} = \frac{8}{15}$.

Example 8

Find the product of $\frac{1}{3}$ and $\frac{2}{5}$ and illustrate the product as in Example 7 by shading parts of a square.

Solution

$$\frac{1}{3} \cdot \frac{2}{5} = \frac{1 \cdot 2}{3 \cdot 5} = \frac{2}{15}$$

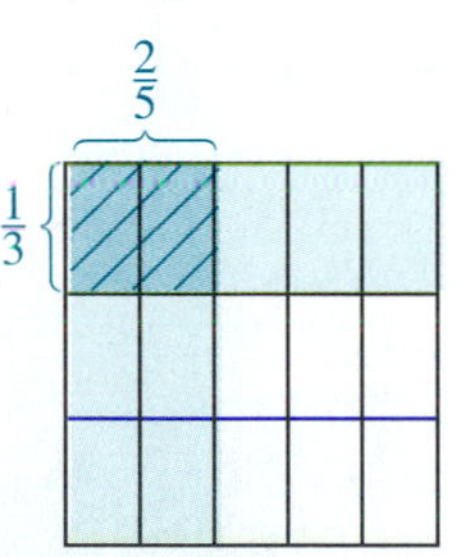

Now Work Exercise 1 in the Margin.

Completion Example 9

Find the following products.

(a) $\frac{2}{3} \cdot \frac{5}{7} = \frac{2 \cdot 5}{3 \cdot 7} =$ ______________

(b) $\frac{1}{4} \cdot \frac{3}{5} = \frac{1 \cdot 3}{4 \cdot 5} =$ ______________

(c) $\frac{7}{5} \cdot 2 = \frac{7}{5} \cdot \frac{2}{1} =$ ______________

(d) $\frac{1}{4} \cdot \frac{3}{5} \cdot \frac{7}{2} = \frac{1 \cdot 3 \cdot 7}{\_\_\_\_\_\_} =$ ______________

(e) $\frac{2}{3} \cdot \frac{11}{15} \cdot \frac{1}{7} =$ ______________

1. Find the product of $\frac{3}{4}$ and $\frac{3}{5}$ and illustrate the product with appropriate shading in a square

Find the following products.

2. $\frac{1}{7} \cdot \frac{3}{5} =$

3. $\frac{0}{4} \cdot \frac{5}{8} =$

4. $\frac{3}{14} \cdot 5 =$

5. $\frac{1}{2} \cdot \frac{3}{4} \cdot \frac{7}{5} =$

6. $\frac{7}{8} \cdot \frac{3}{4} \cdot 1 =$

Now Work Exercises 2–6 in the Margin.

Both the commutative property and the associative property of multiplication apply to fractions.

Commutative Property of Multiplication

If $\frac{a}{b}$ and $\frac{c}{d}$ are fractions, then

$$\frac{a}{b} \cdot \frac{c}{d} = \frac{c}{d} \cdot \frac{a}{b}$$

Associative Property of Multiplication

If $\frac{a}{b}$, $\frac{c}{d}$, and $\frac{e}{f}$ are fractions, then

$$\frac{a}{b} \cdot \frac{c}{d} \cdot \frac{e}{f} = \left(\frac{a}{b} \cdot \frac{c}{d}\right) \cdot \frac{e}{f} = \frac{a}{b} \cdot \left(\frac{c}{d} \cdot \frac{e}{f}\right)$$

EXAMPLE 10

$$\frac{3}{4} \cdot \frac{1}{7} = \frac{1}{7} \cdot \frac{3}{4}$$

Illustrates the commutative property of multiplication.

$$\frac{3}{4} \cdot \frac{1}{7} = \frac{3 \cdot 1}{4 \cdot 7} = \frac{3}{28} \quad \text{and} \quad \frac{1}{7} \cdot \frac{3}{4} = \frac{1 \cdot 3}{7 \cdot 4} = \frac{3}{28}$$

EXAMPLE 11

$$\left(\frac{5}{8} \cdot \frac{3}{2}\right) \cdot \frac{3}{11} = \frac{5}{8} \cdot \left(\frac{3}{2} \cdot \frac{3}{11}\right)$$

Illustrates the associative property of multiplication.

$$\left(\frac{5}{8} \cdot \frac{3}{2}\right) \cdot \frac{3}{11} = \left(\frac{5 \cdot 3}{8 \cdot 2}\right) \cdot \frac{3}{11} = \frac{15}{16} \cdot \frac{3}{11} = \frac{15 \cdot 3}{16 \cdot 11} = \frac{45}{176} \quad \text{and}$$

$$\frac{5}{8} \cdot \left(\frac{3}{2} \cdot \frac{3}{11}\right) = \frac{5}{8} \cdot \left(\frac{3 \cdot 3}{2 \cdot 11}\right) = \frac{5}{8} \cdot \frac{9}{22} = \frac{5 \cdot 9}{8 \cdot 22} = \frac{45}{176}$$

Now Work Exercises 7–10 in the Margin.

State which property of multiplication is illustrated in each problem.

7. $\frac{1}{2} \cdot \frac{5}{8} = \frac{5}{8} \cdot \frac{1}{2}$

8. $\frac{3}{4} \cdot 6 = 6 \cdot \frac{3}{4}$

9. $\frac{1}{3} \cdot \left(\frac{1}{2} \cdot \frac{1}{4}\right) = \left(\frac{1}{3} \cdot \frac{1}{2}\right) \cdot \frac{1}{4}$

10. $\frac{5}{6} \cdot \left(\frac{7}{8} \cdot \frac{11}{13}\right) = \left(\frac{5}{6} \cdot \frac{7}{8}\right) \cdot \frac{11}{13}$

Raising Fractions to Higher Terms

We know that 1 is the multiplicative identity for whole numbers; that is, for any whole number a, $a \cdot 1 = a$. The number 1 is also the multiplicative identity for fractions (or rational numbers), since

$$\frac{a}{b} \cdot 1 = \frac{a}{b} \cdot \frac{1}{1} = \frac{a \cdot 1}{b \cdot 1} = \frac{a}{b}$$

Multiplicative Identity

The number 1 is called the **multiplicative identity,** and for any fraction $\frac{a}{b}$,

$$\frac{a}{b} \cdot 1 = \frac{a}{b}.$$

This fact states that a fraction has the same value if it is multiplied by 1. It allows us to find a fraction equal to (or equivalent to) a given fraction with a larger denominator. In this case, the given fraction is said to be **raised to higher terms.**

To Raise a Fraction to Higher Terms:

Multiply the numerator and denominator by the same nonzero whole number.

$$\frac{a}{b} = \frac{a}{b} \cdot 1 = \frac{a}{b} \cdot \frac{k}{k} = \frac{a \cdot k}{b \cdot k}, \quad \text{where } k \neq 0 \quad \left(1 = \frac{k}{k}\right)$$

The following examples illustrate the use of this technique of raising a fraction to higher terms. Note carefully the importance of the choice of the form of $\frac{k}{k}$.

Example 12

$\frac{3}{4} = \frac{?}{28}$

We know $4 \cdot 7 = 28$, so use $1 = \frac{k}{k} = \frac{7}{7}$.

$$\frac{3}{4} = \frac{3}{4} \cdot 1 = \frac{3}{4} \cdot \frac{7}{7} = \frac{21}{28}$$

Find the missing numerator or denominator that will make the fractions equal.

11. $\frac{3}{5} = \frac{3}{5} \cdot \frac{?}{?} = \frac{?}{20}$

12. $\frac{1}{9} = \frac{1}{9} \cdot \frac{?}{?} = \frac{?}{72}$

13. $\frac{5}{8} = \frac{5}{8} \cdot \frac{?}{?} = \frac{25}{?}$

14. $\frac{3}{11} = \frac{3}{11} \cdot \frac{?}{?} = \frac{12}{?}$

EXAMPLE 13

$\frac{9}{10} = \frac{?}{30}$

Since $10 \cdot 3 = 30$, use $1 = \frac{k}{k} = \frac{3}{3}$.

$$\frac{9}{10} = \frac{9}{10} \cdot 1 = \frac{9}{10} \cdot \frac{3}{3} = \frac{27}{30}$$

EXAMPLE 14

$\frac{9}{10} = \frac{?}{40}$

Since $10 \cdot 4 = 40$, use $1 = \frac{k}{k} = \frac{4}{4}$.

$$\frac{9}{10} = \frac{9}{10} \cdot 1 = \frac{9}{10} \cdot \frac{4}{4} = \frac{36}{40}$$

EXAMPLE 15

$\frac{11}{8} = \frac{?}{40}$

Since $8 \cdot 5 = 40$, use $1 = \frac{5}{5}$.

$$\frac{11}{8} = \frac{11}{8} \cdot \frac{5}{5} = \frac{55}{40}$$

✓ COMPLETION EXAMPLE 16

Find a fraction with denominator 35 equal to $\frac{3}{5}$.

$$\frac{3}{5} = \frac{3}{5} \cdot 1 = \frac{3}{5} \cdot \underline{\qquad} = \frac{\quad}{35}$$

✓ COMPLETION EXAMPLE 17

Find a fraction with numerator 28 equal to $\frac{2}{3}$.

$$\frac{2}{3} = \frac{2}{3} \cdot 1 = \frac{2}{3} \cdot \underline{\qquad} = \frac{28}{\quad}$$

Now Work Exercises 11–14 in the Margin.

EXAMPLE 18

In a certain voting district, $\frac{3}{5}$ of the eligible voters are actually registered to vote. Of these registered voters, $\frac{2}{7}$ are independents (have no party affiliation). What fraction of the eligible voters are registered independents?

Solution

Since the independents are a fraction **of** the eligible voters, we multiply:

$$\frac{2}{7} \cdot \frac{3}{5} = \frac{2 \cdot 3}{7 \cdot 5} = \frac{6}{35}$$

Thus, $\frac{6}{35}$ of the eligible voters are registered as independents.

Completion Example Answers

9. (a) $\frac{2}{3} \cdot \frac{5}{7} = \frac{2 \cdot 5}{3 \cdot 7} = \mathbf{\frac{10}{21}}$

(b) $\frac{1}{4} \cdot \frac{3}{5} = \frac{1 \cdot 3}{4 \cdot 5} = \mathbf{\frac{3}{20}}$

(c) $\frac{7}{5} \cdot 2 = \frac{7}{5} \cdot \frac{2}{1} = \mathbf{\frac{7 \cdot 2}{5 \cdot 1}} = \mathbf{\frac{14}{5}}$

(d) $\frac{1}{4} \cdot \frac{3}{5} \cdot \frac{7}{2} = \frac{1 \cdot 3 \cdot 7}{\mathbf{4 \cdot 5 \cdot 2}} = \mathbf{\frac{21}{40}}$

(e) $\frac{2}{3} \cdot \frac{11}{15} \cdot \frac{1}{7} = \mathbf{\frac{2 \cdot 11 \cdot 1}{3 \cdot 15 \cdot 7}} = \mathbf{\frac{22}{315}}$

16. $\frac{3}{5} = \frac{3}{5} \cdot 1 = \frac{3}{5} \cdot \mathbf{\frac{7}{7}} = \frac{\mathbf{21}}{35}$

17. $\frac{2}{3} = \frac{2}{3} \cdot 1 = \frac{2}{3} \cdot \mathbf{\frac{14}{14}} = \frac{28}{\mathbf{42}}$

NAME ______________________ SECTION ______________ DATE ________

Exercises 3.1

What is the value, if any, of each of the following numbers or products?

1. $\frac{0}{6}$ **2.** $\frac{0}{35}$ **3.** $\frac{0}{6} \cdot \frac{1}{10}$ **4.** $\frac{0}{5} \cdot \frac{0}{11}$

5. $\frac{13}{0}$ **6.** $\frac{2}{0}$ **7.** $\frac{3}{2} \cdot \frac{1}{0}$ **8.** $\frac{76}{0} \cdot \frac{21}{0}$

9. Explain, in your own words, why no denominator can be 0.

10. What is the meaning of the term **rational number?**

Write a fraction that indicates the shaded parts in each of the following diagrams.

11.

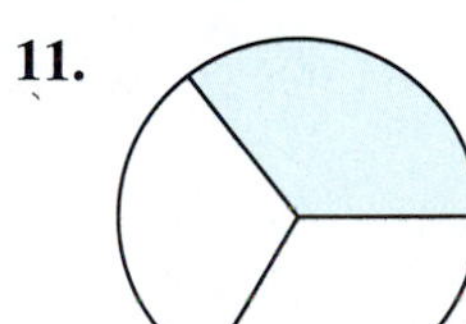

12.

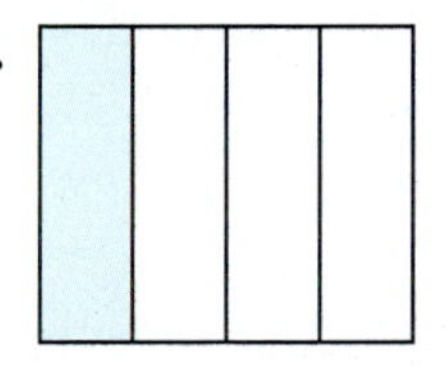

13.

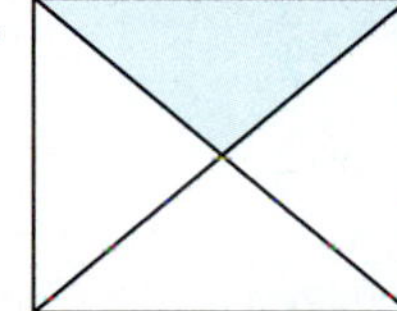

ANSWERS

1. ____________

2. ____________

3. ____________

4. ____________

5. ____________

6. ____________

7. ____________

8. ____________

9. [Respond below exercise.]

10. [Respond below exercise.]

11. ____________

12. ____________

13. ____________

14. ______

15. ______

16. ______

17. [Respond below exercise.]

18. [Respond below exercise.]

19. [Respond below exercise.]

20. [Respond below exercise.]

21. ______

22. ______

23. ______

24. ______

25. ______

26. ______

27. ______

28. ______

29. ______

30. ______

31. ______

32. ______

33. ______

34. ______

35. ______

36. ______

37. ______

38. ______

39. ______

40. ______

14. 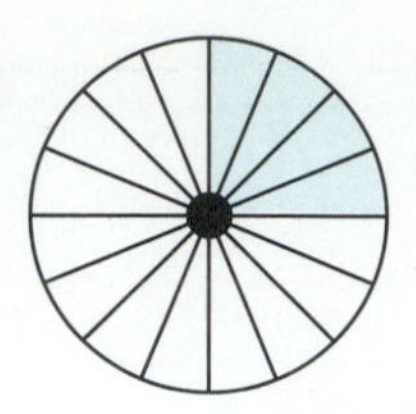

15. 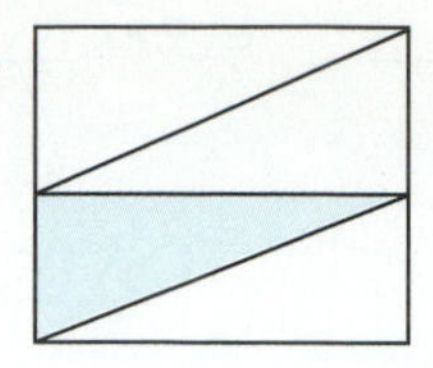

16. 

In Exercises 17–20, determine each product and then illustrate each product with appropriate shading in a square.

17. $\frac{1}{5} \cdot \frac{3}{4}$ 

18. $\frac{5}{6} \cdot \frac{5}{6}$ 

19. $\frac{2}{3} \cdot \frac{4}{7}$ 

20. $\frac{3}{4} \cdot \frac{3}{4}$ 

Find the following products.

21. $\frac{3}{16} \cdot \frac{1}{2}$

22. $\frac{2}{5} \cdot \frac{2}{5}$

23. $\frac{3}{7} \cdot \frac{3}{7}$

24. $\frac{1}{2} \cdot \frac{3}{4}$

25. $\frac{5}{8} \cdot \frac{3}{4}$

26. $\frac{1}{9} \cdot \frac{4}{9}$

27. $\frac{0}{3} \cdot \frac{5}{7}$

28. $\frac{0}{4} \cdot \frac{7}{6}$

29. $\frac{7}{6} \cdot \frac{5}{2}$

30. $\frac{4}{1} \cdot \frac{3}{1}$

31. $\frac{2}{1} \cdot \frac{5}{1}$

32. $\frac{14}{1} \cdot \frac{0}{2}$

33. $\frac{15}{1} \cdot \frac{3}{2}$

34. $\frac{6}{5} \cdot \frac{7}{1}$

35. $\frac{8}{5} \cdot \frac{4}{3}$

36. $\frac{5}{6} \cdot \frac{11}{3}$

37. $\frac{9}{4} \cdot \frac{11}{5}$

38. $\frac{1}{5} \cdot \frac{2}{7} \cdot \frac{3}{11}$

39. $\frac{4}{13} \cdot \frac{2}{5} \cdot \frac{6}{7}$

40. $\frac{7}{8} \cdot \frac{7}{9} \cdot \frac{7}{3}$

NAME ________________ SECTION ________________ DATE ________

ANSWERS

In Exercises 41–46, find the fractional parts as indicated.

41. Find $\frac{1}{5}$ of $\frac{1}{2}$

42. Find $\frac{1}{3}$ of $\frac{1}{2}$.

43. Find $\frac{1}{4}$ of $\frac{1}{2}$

44. Find $\frac{2}{3}$ of $\frac{2}{15}$.

45. Find $\frac{4}{7}$ of $\frac{3}{5}$

46. Find $\frac{1}{3}$ of $\frac{2}{3}$.

Tell which property of multiplication is illustrated in each exercise.

47. $\frac{5}{9} \cdot \frac{11}{14} = \frac{11}{14} \cdot \frac{5}{9}$

48. $\frac{3}{5} \cdot \left(\frac{7}{8} \cdot \frac{1}{2}\right) = \left(\frac{3}{5} \cdot \frac{7}{8}\right) \cdot \frac{1}{2}$

49. $\left(\frac{6}{11} \cdot \frac{13}{5}\right) \cdot \frac{7}{5} = \frac{6}{11} \cdot \left(\frac{13}{5} \cdot \frac{7}{5}\right)$

50. $\frac{2}{3} \cdot 4 = 4 \cdot \frac{2}{3}$

51. If you had \$20 and you spent \$7 for food and a soft drink, what fraction of your money did you spend? What fraction would you still have?

52. In a class of 35 students, 6 students received A's on a mathematics exam. What fraction of the class did not receive an A?

41. ________

42. ________

43. ________

44. ________

45. ________

46. ________

47. ________

48. ________

49. ________

50. ________

51. ________

52. ________

53. ______________

54. ______________

55. ______________

56. ______________

57. ______________

58. ______________

59. ______________

60. ______________

53. A recipe calls for $\frac{3}{4}$ cup of flour. How much flour should be used if only half the recipe is to be made?

54. One of Maria's birthday presents was a box of candy. Half of the candy was chocolate-covered and one-fourth of the chocolate-covered candy had cherries inside. What fraction of the candy was chocolate-covered cherries?

Find the missing numerators and denominators in building each fraction to higher terms.

55. $\frac{5}{8} = \frac{5}{8} \cdot \frac{?}{?} = \frac{?}{24}$

56. $\frac{1}{16} = \frac{1}{16} \cdot \frac{?}{?} = \frac{?}{64}$

57. $\frac{2}{5} = \frac{2}{5} \cdot \frac{?}{?} = \frac{?}{25}$

58. $\frac{6}{7} = \frac{6}{7} \cdot \frac{?}{?} = \frac{?}{49}$

59. $\frac{1}{9} = \frac{1}{9} \cdot \frac{?}{?} = \frac{5}{?}$

60. $\frac{3}{4} = \frac{3}{4} \cdot \frac{?}{?} = \frac{15}{?}$

NAME ______________ SECTION ______________ DATE ______________

61. $\frac{5}{8} = \frac{5}{8} \cdot \frac{?}{?} = \frac{10}{?}$

62. $\frac{6}{5} = \frac{6}{5} \cdot \frac{?}{?} = \frac{?}{45}$

63. $\frac{14}{3} = \frac{14}{3} \cdot \frac{?}{?} = \frac{?}{9}$

64. $\frac{5}{8} = \frac{5}{8} \cdot \frac{?}{?} = \frac{?}{96}$

65. $\frac{9}{16} = \frac{9}{16} \cdot \frac{?}{?} = \frac{?}{96}$

66. $\frac{7}{2} = \frac{7}{2} \cdot \frac{?}{?} = \frac{?}{20}$

67. $\frac{10}{11} = \frac{10}{11} \cdot \frac{?}{?} = \frac{?}{44}$

68. $\frac{3}{16} = \frac{3}{16} \cdot \frac{?}{?} = \frac{?}{80}$

69. $\frac{11}{12} = \frac{11}{12} \cdot \frac{?}{?} = \frac{?}{48}$

70. $\frac{5}{21} = \frac{5}{21} \cdot \frac{?}{?} = \frac{?}{42}$

71. $\frac{2}{3} = \frac{2}{3} \cdot \frac{?}{?} = \frac{?}{48}$

72. $\frac{1}{13} = \frac{1}{13} \cdot \frac{?}{?} = \frac{?}{39}$

73. $\frac{9}{10} = \frac{9}{10} \cdot \frac{?}{?} = \frac{?}{100}$

74. $\frac{3}{10} = \frac{3}{10} \cdot \frac{?}{?} = \frac{?}{100}$

75. $\frac{7}{10} = \frac{7}{10} \cdot \frac{?}{?} = \frac{?}{70}$

76. $\frac{9}{10} = \frac{9}{10} \cdot \frac{?}{?} = \frac{?}{90}$

ANSWERS

61. ______________
62. ______________
63. ______________
64. ______________
65. ______________
66. ______________
67. ______________
68. ______________
69. ______________
70. ______________
71. ______________
72. ______________
73. ______________
74. ______________
75. ______________
76. ______________

77. ______________

78. ______________

79. ______________

80. ______________

77. $\frac{5}{12} = \frac{5}{12} \cdot \frac{?}{?} = \frac{?}{108}$

78. $\frac{8}{5} = \frac{8}{5} \cdot \frac{?}{?} = \frac{?}{30}$

79. $\frac{7}{6} = \frac{7}{6} \cdot \frac{?}{?} = \frac{?}{36}$

80. $\frac{10}{3} = \frac{10}{3} \cdot \frac{?}{?} = \frac{?}{33}$

3.2 Multiplication and Reducing

Reducing Fractions

In Section 3.1, we used the multiplicative identity in the form $\frac{k}{k}$ to raise a fraction to higher terms. We can use this same idea to **change a fraction to lower terms** (to **reduce a fraction**). **A fraction is reduced to lowest terms if the numerator and the denominator have no common factors other than 1.**

To Reduce a Fraction to Lowest Terms:

1. Factor the numerator and denominator into prime factorizations.
2. Use the fact that $\frac{k}{k} = 1$ and **divide out** common factors.

Note that reduced fractions might be improper fractions. Changing a fraction to a mixed number is not the same as reducing it. Mixed numbers will be discussed in Chapter 4.

Example 1 illustrates how to reduce fractions to lowest terms.

OBJECTIVES

1. *Learn how to reduce a fraction to lowest terms.*
2. *Recognize when a fraction is in lowest terms.*
3. *Be able to multiply fractions and reduce at the same time.*

Example 1

(a) $\frac{14}{21} = \frac{2 \cdot 7}{3 \cdot 7} = \frac{2}{3} \cdot \frac{7}{7} = \frac{2}{3} \cdot 1 = \frac{2}{3}$

(b) $\frac{15}{20} = \frac{3 \cdot 5}{2 \cdot 2 \cdot 5} = \frac{3}{4} \cdot \frac{5}{5} = \frac{3}{4} \cdot 1 = \frac{3}{4}$

Example 2

$\frac{44}{20}$

We can just divide out common factors (prime or not) with the understanding that $\frac{k}{k} = 1$.

(a) $\frac{44}{20} = \frac{\cancel{4} \cdot 11}{\cancel{4} \cdot 5} = \frac{11}{5}$

(b) Or, using prime factors,

$$\frac{44}{20} = \frac{\cancel{2} \cdot \cancel{2} \cdot 11}{\cancel{2} \cdot \cancel{2} \cdot 5} = \frac{11}{5}$$

EXAMPLE 3

$\frac{8}{72}$

Remember that 1 is a factor of any whole number. So, if all of the factors in the numerator or denominator are divided out, 1 must be used as a factor.

(a) Using prime factors,

$$\frac{8}{72} = \frac{\cancel{2} \cdot \cancel{2} \cdot \cancel{2} \cdot 1}{\cancel{2} \cdot \cancel{2} \cdot \cancel{2} \cdot 3 \cdot 3} = \frac{1}{9}$$ 1 is used as a factor in the numerator.

(b) Or, if you see that 8 is a common factor, divide it out. But remember that 1 is a factor.

$$\frac{8}{72} = \frac{\cancel{8} \cdot 1}{\cancel{8} \cdot 9} = \frac{1}{9}$$

EXAMPLE 4

$\frac{36}{27}$

(a) Using prime factors,

$$\frac{36}{27} = \frac{2 \cdot 2 \cdot \cancel{3} \cdot \cancel{3}}{3 \cdot \cancel{3} \cdot \cancel{3}} = \frac{4}{3}$$

(b) Or, using 3 as a factor, we get an incomplete answer.

$$\frac{36}{27} = \frac{\cancel{3} \cdot 12}{\cancel{3} \cdot 9} = \frac{12}{9}$$ not in lowest terms

By using prime factors, you can be certain that the fraction is reduced to lowest terms. You may use larger numbers, but be sure you have the largest common factor.

$$\frac{36}{27} = \frac{4 \cdot \cancel{9}}{\cancel{9} \cdot 3} = \frac{4}{3}$$

☑ COMPLETION EXAMPLE 5

Reduce $\frac{12}{18}$ to lowest terms.

(a) $\frac{12}{18} = \frac{2 \cdot 2 \cdot 3}{2 \cdot 3 \cdot 3} =$ ________

(b) Or, $\frac{12}{18} = \frac{2 \cdot ?}{3 \cdot ?} =$ ________

☑ COMPLETION EXAMPLE 6

Reduce $\frac{52}{65}$ to lowest terms.

Finding a common factor could be difficult here. Prime factoring helps.

$\frac{52}{65} = \frac{2 \cdot 2 \cdot ?}{5 \cdot ?} =$ ________

Now Work Exercises 1–6 in the Margin.

Reduce each fraction to lowest terms.

1. $\frac{8}{36} =$

2. $\frac{35}{40} =$

3. $\frac{66}{44} =$

4. $\frac{12}{25} =$

5. $\frac{10}{60} =$

6. $\frac{78}{104} =$

Multiplying and Reducing Fractions at the Same Time

Now we can multiply fractions and reduce all in one step by using prime factors (or other common factors). If you have any difficulty understanding how to multiply and reduce, use prime factors. This will help you gain confidence that your answers are correct.

Examples 7–10 illustrate how to multiply and reduce at the same time by factoring the numerators and denominators. Note that **if all of the factors in the numerator or denominator divide out, then 1 must be used as a factor.** (See Completion Example 10.)

Example 7

$$\frac{15}{28} \cdot \frac{7}{9}$$

Poor Method (Multiplying first is a waste of time.)

$$\frac{15}{28} \cdot \frac{7}{9} = \frac{15 \cdot 7}{28 \cdot 9} = \frac{105}{252}$$

Now factor and reduce.

$$\frac{105}{252} = \frac{5 \cdot 21}{4 \cdot 63} = \frac{5 \cdot \cancel{3} \cdot \cancel{7}}{2 \cdot 2 \cdot 3 \cdot \cancel{3} \cdot \cancel{7}} = \frac{5}{12}$$

Good Method (Use prime factors.)

$$\frac{15}{28} \cdot \frac{7}{9} = \frac{15 \cdot 7}{28 \cdot 9} = \frac{\cancel{3} \cdot 5 \cdot \cancel{7}}{2 \cdot 2 \cdot \cancel{7} \cdot 3 \cdot \cancel{3}} = \frac{5}{2 \cdot 2 \cdot 3} = \frac{5}{12}$$

Only the good method is used to find the products in Examples 8 and 9.

Example 8

$$\frac{9}{10} \cdot \frac{25}{32} \cdot \frac{44}{33}$$

$$\frac{9}{10} \cdot \frac{25}{32} \cdot \frac{44}{33} = \frac{9 \cdot 25 \cdot 44}{10 \cdot 32 \cdot 33}$$

$$= \frac{\cancel{3} \cdot 3 \cdot \cancel{5} \cdot 5 \cdot \cancel{2} \cdot \cancel{2} \cdot \cancel{11}}{2 \cdot \cancel{5} \cdot 2 \cdot 2 \cdot 2 \cdot \cancel{2} \cdot \cancel{2} \cdot \cancel{3} \cdot \cancel{11}}$$

$$= \frac{3 \cdot 5}{2 \cdot 2 \cdot 2 \cdot 2}$$

$$= \frac{15}{16}$$

Find each product in lowest terms by using prime factors.

7. $\frac{2}{5} \cdot \frac{8}{5}$

8. $\frac{10}{9} \cdot \frac{3}{5}$

9. $\frac{5}{6} \cdot \frac{8}{7} \cdot \frac{14}{10}$

10. $6 \cdot \frac{7}{12} \cdot \frac{3}{28} \cdot \frac{1}{9}$

$\left(\text{Remember that } 6 = \frac{6}{1}.\right)$

EXAMPLE 9

$\frac{36}{49} \cdot \frac{14}{75} \cdot \frac{15}{18}$

$$\frac{36}{49} \cdot \frac{14}{75} \cdot \frac{15}{18} = \frac{36 \cdot 14 \cdot 15}{49 \cdot 75 \cdot 18}$$

$$= \frac{\cancel{2} \cdot 2 \cdot \cancel{3} \cdot \cancel{3} \cdot 2 \cdot \cancel{7} \cdot \cancel{3} \cdot \cancel{5}}{\cancel{7} \cdot 7 \cdot \cancel{3} \cdot 5 \cdot \cancel{5} \cdot \cancel{2} \cdot \cancel{3} \cdot \cancel{3}}$$

$$= \frac{2 \cdot 2}{7 \cdot 5}$$

$$= \frac{4}{35}$$

☑ *COMPLETION EXAMPLE 10*

$\frac{55}{26} \cdot \frac{8}{44} \cdot \frac{91}{35}$

$$\frac{55}{26} \cdot \frac{8}{44} \cdot \frac{91}{35} = \frac{55 \cdot 8 \cdot 91}{26 \cdot 44 \cdot 35}$$

$$= \frac{?}{?}$$

$$= \frac{?}{?}$$

$$= \underline{\quad ? \quad}$$

Now Work Exercises 7–10 in the Margin.

Another method frequently used to multiply and reduce at the same time is to divide numerators and denominators by common factors whether they are prime or not. If these factors are easily determined, then this method is probably faster. But common factors are sometimes missed with this method, whereas they are not missed with the prime factorization method. In either case, be careful and organized.

Examples 7–10 are shown again as Examples 11–14, this time using the division method.

EXAMPLE 11

$$\frac{\overset{5}{\cancel{15}}}{\underset{4}{\cancel{28}}} \cdot \frac{\overset{1}{\cancel{7}}}{\underset{3}{\cancel{9}}} = \frac{5}{12}$$

3 is divided into both 15 and 9.
7 is divided into both 7 and 28.

EXAMPLE 12

$$\frac{\overset{3}{\cancel{9}}}{\underset{2}{\cancel{10}}} \cdot \frac{\overset{5}{\cancel{25}}}{\underset{8}{\cancel{32}}} \cdot \frac{\overset{1}{\overset{\cancel{4}}{\cancel{44}}}}{\underset{1}{\underset{\cancel{3}}{\cancel{33}}}} = \frac{15}{16}$$

11 is divided into both 44 and 33.
5 is divided into both 25 and 10.
4 is divided into both 4 and 32.
3 is divided into both 3 and 9.

EXAMPLE 13

$$\frac{\overset{2}{\cancel{36}}}{\underset{7}{\cancel{49}}} \cdot \frac{\overset{2}{\cancel{14}}}{\underset{5}{\cancel{75}}} \cdot \frac{\overset{1}{\cancel{15}}}{\underset{1}{\cancel{18}}} = \frac{4}{35}$$

18 is divided into both 18 and 36.
7 is divided into both 14 and 49.
15 is divided into both 15 and 75.

EXAMPLE 14

$$\frac{\overset{1}{\overset{\cancel{5}}{\cancel{55}}}}{\underset{1}{\underset{\cancel{2}}{\cancel{26}}}} \cdot \frac{\overset{1}{\overset{\cancel{4}}{\cancel{8}}}}{\underset{1}{\underset{\cancel{4}}{\cancel{44}}}} \cdot \frac{\overset{1}{\overset{\cancel{7}}{\cancel{91}}}}{\underset{1}{\underset{\cancel{7}}{\cancel{35}}}} = \frac{1}{1}$$

11 is divided into both 55 and 44.
13 is divided into both 26 and 91.
5 is divided into 5 and 35.
7 is divided into both 7 and 7.
2 is divided into both 2 and 8.
4 is divided into both 4 and 4.

Now Work Exercises 11–13 in the Margin.

EXAMPLE 15

A study showed that $\frac{5}{8}$ of the members of a public service organization were in favor of a new set of bylaws. If the organization had a membership of 200 people, how many were in favor of the changes in the bylaws?

Solution

We want to find $\frac{5}{8}$ of 200 so we multiply:

$$\frac{5}{8} \cdot 200 = \frac{5}{8} \cdot \frac{200}{1} = \frac{5 \cdot 2 \cdot 10 \cdot 10}{2 \cdot 2 \cdot 2} = \frac{5 \cdot \cancel{2} \cdot \cancel{2} \cdot 5 \cdot \cancel{2} \cdot 5}{\cancel{2} \cdot \cancel{2} \cdot \cancel{2}}$$

$$= \frac{5 \cdot 5 \cdot 5}{1} = 125$$

Thus, there are 125 members in favor of the bylaw changes.

Multiply and reduce at the same time.

11. $\frac{16}{27} \cdot \frac{9}{4}$

12. $\frac{14}{15} \cdot \frac{25}{21} \cdot \frac{3}{10}$

13. $\frac{17}{100} \cdot \frac{27}{34} \cdot \frac{25}{9} \cdot 6$

Completion Example Answers

5. (a) $\frac{12}{18} = \frac{\cancel{2} \cdot 2 \cdot \cancel{3}}{\cancel{2} \cdot 3 \cdot \cancel{3}} = \mathbf{\frac{2}{3}}$

   (b) Or, $\frac{12}{18} = \frac{2 \cdot \cancel{\mathbf{6}}}{3 \cdot \cancel{\mathbf{6}}} = \mathbf{\frac{2}{3}}$

6. $\frac{52}{65} = \frac{2 \cdot 2 \cdot \cancel{\mathbf{13}}}{5 \cdot \cancel{\mathbf{13}}} = \mathbf{\frac{4}{5}}$

10. $\frac{55}{26} \cdot \frac{8}{44} \cdot \frac{91}{35} = \frac{55 \cdot 8 \cdot 91}{26 \cdot 44 \cdot 35} = \frac{\cancel{\mathbf{5}} \cdot \cancel{\mathbf{11}} \cdot \cancel{\mathbf{2}} \cdot \cancel{\mathbf{2}} \cdot \cancel{\mathbf{2}} \cdot \cancel{\mathbf{7}} \cdot \cancel{\mathbf{13}}}{\cancel{\mathbf{2}} \cdot \cancel{\mathbf{13}} \cdot \cancel{\mathbf{2}} \cdot \cancel{\mathbf{2}} \cdot \cancel{\mathbf{11}} \cdot \cancel{\mathbf{7}} \cdot \cancel{\mathbf{5}}} = \mathbf{\frac{1}{1}} = \mathbf{1}$

NAME ______________________ SECTION ______________________ DATE ______________

ANSWERS

1. ______
2. ______
3. ______
4. ______
5. ______
6. ______
7. ______
8. ______
9. ______
10. ______
11. ______
12. ______
13. ______
14. ______
15. ______
16. ______
17. ______
18. ______
19. ______
20. ______
21. ______
22. ______
23. ______
24. ______

Exercises 3.2

Reduce each fraction to lowest terms. If it is already in lowest terms, simply rewrite the fraction.

1. $\frac{3}{9}$ **2.** $\frac{16}{24}$ **3.** $\frac{9}{12}$ **4.** $\frac{6}{20}$

5. $\frac{16}{40}$ **6.** $\frac{24}{30}$ **7.** $\frac{14}{36}$ **8.** $\frac{5}{11}$

9. $\frac{0}{25}$ **10.** $\frac{75}{100}$ **11.** $\frac{22}{55}$ **12.** $\frac{60}{75}$

13. $\frac{30}{36}$ **14.** $\frac{7}{28}$ **15.** $\frac{26}{39}$ **16.** $\frac{27}{56}$

17. $\frac{34}{51}$ **18.** $\frac{36}{48}$ **19.** $\frac{24}{100}$ **20.** $\frac{16}{32}$

21. $\frac{30}{45}$ **22.** $\frac{28}{42}$ **23.** $\frac{12}{35}$ **24.** $\frac{66}{84}$

25. ______
26. ______
27. ______
28. ______
29. ______
30. ______
31. ______
32. ______
33. ______
34. ______
35. ______
36. ______
37. ______
38. ______
39. ______
40. ______
41. ______
42. ______
43. ______
44. ______
45. ______
46. ______
47. ______
48. ______
49. ______
50. ______
51. ______
52. ______

25. $\frac{14}{63}$ **26.** $\frac{30}{70}$ **27.** $\frac{25}{76}$ **28.** $\frac{70}{84}$

29. $\frac{50}{100}$ **30.** $\frac{48}{12}$ **31.** $\frac{27}{72}$ **32.** $\frac{18}{40}$

33. $\frac{144}{156}$ **34.** $\frac{150}{135}$ **35.** $\frac{121}{165}$ **36.** $\frac{140}{112}$

37. $\frac{96}{108}$ **38.** $\frac{72}{36}$ **39.** $\frac{84}{42}$ **40.** $\frac{51}{85}$

Find each product in lowest terms.

41. $\frac{2}{3} \cdot \frac{4}{3}$ **42.** $\frac{1}{5} \cdot \frac{4}{7}$ **43.** $\frac{3}{7} \cdot \frac{5}{3}$

44. $\frac{2}{11} \cdot \frac{3}{2}$ **45.** $\frac{5}{16} \cdot \frac{16}{15}$ **46.** $\frac{7}{18} \cdot \frac{9}{14}$

47. $\frac{10}{18} \cdot \frac{9}{5}$ **48.** $\frac{11}{22} \cdot \frac{6}{8}$ **49.** $\frac{15}{27} \cdot \frac{9}{30}$

50. $\frac{35}{20} \cdot \frac{36}{14}$ **51.** $\frac{25}{9} \cdot \frac{3}{100}$ **52.** $\frac{30}{42} \cdot \frac{7}{100}$

NAME ______________ SECTION ______________ DATE ______________

ANSWERS

53. $\frac{18}{42} \cdot \frac{14}{75}$

54. $\frac{42}{70} \cdot \frac{20}{12}$

55. $8 \cdot \frac{5}{12}$

56. $9 \cdot \frac{7}{24}$

57. $\frac{6}{85} \cdot \frac{34}{9}$

58. $\frac{13}{91} \cdot \frac{34}{65}$

Find each product. (**Hint:** Factor before multiplying.)

59. $\frac{23}{36} \cdot \frac{20}{46}$

60. $\frac{7}{8} \cdot \frac{4}{21}$

61. $\frac{5}{15} \cdot \frac{18}{24}$

62. $\frac{20}{32} \cdot \frac{9}{13} \cdot \frac{26}{7}$

63. $\frac{69}{15} \cdot \frac{30}{8} \cdot \frac{14}{46}$

64. $\frac{42}{52} \cdot \frac{27}{22} \cdot \frac{33}{9}$

65. $\frac{3}{4} \cdot 18 \cdot \frac{7}{2} \cdot \frac{22}{54}$

66. $\frac{9}{10} \cdot \frac{35}{40} \cdot \frac{65}{15}$

67. $\frac{66}{84} \cdot \frac{12}{5} \cdot \frac{28}{33}$

68. $\frac{24}{100} \cdot \frac{36}{48} \cdot \frac{15}{9}$

69. $\frac{17}{10} \cdot \frac{5}{42} \cdot \frac{18}{51} \cdot 4$

70. $\frac{75}{8} \cdot \frac{16}{36} \cdot 9 \cdot \frac{7}{25}$

53. ______________
54. ______________
55. ______________
56. ______________
57. ______________
58. ______________
59. ______________
60. ______________
61. ______________
62. ______________
63. ______________
64. ______________
65. ______________
66. ______________
67. ______________
68. ______________
69. ______________
70. ______________

71. ____________

72. ____________

73. ____________

74. ____________

71. Suppose that a ball is dropped from a height of 20 feet. If the ball bounces back to five-eighths the height from which it was dropped, how high will it bounce on its third bounce?

72. A study showed that one-fifth of the students in an elementary school were left-handed. If the school had an enrollment of 550 students, how many were left-handed?

73. For her sprinkler system, Patricia needs to cut a 20-foot piece of plastic pipe into four pieces. If each new piece of pipe is to be half the length of the previous piece, how long will each of the four pieces be?

74. If you go on a 100-mile bicycle trip in the mountains and one-fourth of the trip is downhill, how many miles are uphill?

NAME ____________________ SECTION ______________ DATE ________

ANSWERS

75. ____________

76. ____________

77. ____________

75. A pizza pie is to be cut into fourths. Each of these fourths is to be cut into thirds. What fraction of the pie is each of the final pieces?

76. Major league baseball teams play 162 games each season. If a team has played $\frac{4}{9}$ of its games by the All-Star break (around mid-season), how many games has it played by that time?

77. Venus orbits the sun in $\frac{45}{73}$ the time that the earth takes to orbit the sun. Assuming that one "earth-year" is 365 days long, how long is a "Venus-year" in terms of "earth-days"?

78. [Respond below exercise.] ____________

Check Your Number Sense

78. Consider the following statement:

"If a fraction is less than 1 and greater than 0, the product of this fraction with another number (fraction or whole number) must be less than this other number."

Do you agree with this statement? To help you understand this concept, answer the following questions involving the fraction $\frac{3}{4}$.

Is $\frac{3}{4}$ of 12 less than 12? Is $\frac{3}{4}$ of $\frac{2}{3}$ less than $\frac{2}{3}$?

Is $\frac{3}{4}$ of $\frac{1}{2}$ less than $\frac{1}{2}$? Is $\frac{3}{4}$ of $\frac{4}{5}$ less than $\frac{4}{5}$?

Answer these same questions using $\frac{2}{3}$ and $\frac{1}{2}$ in place of $\frac{3}{4}$.

RECYCLE BIN

1. ____________
2. ____________
3. ____________
4. ____________
5. ____________
6. ____________
7. ____________
8. ____________

The Recycle Bin (from Section 2.4)

Find the prime factorization for each of the following numbers. Use the tests for divisibility by 2, 3, 4, 5, 9, and 10 whenever they help in finding the beginning factors.

1. 45 **2.** 78 **3.** 130

4. 460 **5.** 1,000,000

Find all of the factors (or divisors) of each number.

6. 30 **7.** 48 **8.** 63

3.3 Division

Reciprocals

If the product of two fractions is 1, then the fractions are called reciprocals of each other. (Remember that whole numbers can also be written in fraction form.) For example,

$\frac{5}{8}$ and $\frac{8}{5}$ are reciprocals because $\frac{5}{8} \cdot \frac{8}{5} = \frac{40}{40} = 1$

Definition

The **reciprocal** of $\frac{a}{b}$ is $\frac{b}{a}$ ($a \neq 0$ and $b \neq 0$). The product of a nonzero number and its reciprocal is always 1.

$$\frac{a}{b} \cdot \frac{b}{a} = 1$$

Note: $0 = \frac{0}{1}$, but $\frac{1}{0}$ is undefined. That is, **the number 0 has no reciprocal.**

Example 1

The reciprocal of $\frac{2}{3}$ is $\frac{3}{2}$.

$$\frac{2}{3} \cdot \frac{3}{2} = \frac{2 \cdot 3}{3 \cdot 2} = 1$$

Example 2

The reciprocal of $\frac{5}{8}$ is $\frac{8}{5}$.

$$\frac{5}{8} \cdot \frac{8}{5} = \frac{5 \cdot 8}{8 \cdot 5} = 1$$

Example 3

The reciprocal of 10 is $\frac{1}{10}$.

$$10 \cdot \frac{1}{10} = \frac{10}{1} \cdot \frac{1}{10} = \frac{10 \cdot 1}{1 \cdot 10} = 1$$

Now Work Exercises 1–3 in the Margin.

OBJECTIVES

1. *Recognize and find the reciprocal of a number.*
2. *Know that division is accomplished by multiplication by the reciprocal of the divisor.*

State the reciprocal of each number.

1. $\frac{7}{8}$

2. $\frac{1}{10}$

3. 16

Division with Fractions

To develop an understanding of division, we first write a division problem in fraction form with fractions in the numerator and denominator. For example,

$$\frac{2}{3} \div \frac{5}{11} = \frac{\frac{2}{3}}{\frac{5}{11}}$$

Now we multiply the numerator and the denominator (both fractions) by the reciprocal of the denominator. This is the same as multiplying by 1 and does not change the value of the expression. The reciprocal of $\frac{5}{11}$ is $\frac{11}{5}$, so we multiply both the numerator and the denominator by $\frac{11}{5}$.

$$\frac{2}{3} \div \frac{5}{11} = \frac{\frac{2}{3}}{\frac{5}{11}} \cdot \frac{\frac{11}{5}}{\frac{11}{5}} = \frac{\frac{2}{3} \cdot \frac{11}{5}}{\frac{5}{11} \cdot \frac{11}{5}} = \frac{\frac{2}{3} \cdot \frac{11}{5}}{1} = \frac{2}{3} \cdot \frac{11}{5}$$

Thus, a division problem has been changed into a multiplication problem:

to divide — multiply by the reciprocal

$$\frac{2}{3} \div \frac{5}{11} = \frac{2}{3} \cdot \frac{11}{5} = \frac{22}{15}$$

> To divide by any nonzero number, multiply by its reciprocal. In general,
>
> $$\frac{a}{b} \div \frac{c}{d} = \frac{a}{b} \cdot \frac{d}{c} \quad \text{where } b, c, d \neq 0.$$

As illustrated in the following examples, once a division problem is in the form of a product, we can reduce by factoring.

EXAMPLE 4

$\frac{5}{6} \div \frac{1}{6}$ How many $\frac{1}{6}$'s are there in $\frac{5}{6}$?

$$\frac{5}{6} \div \frac{1}{6} = \frac{5}{\cancel{6}} \cdot \frac{\cancel{6}}{1} \quad \frac{6}{1} \text{ is the reciprocal of } \frac{1}{6}.$$

$$= 5$$

Thus, there are five $\frac{1}{6}$'s in $\frac{5}{6}$.

EXAMPLE 5

$\frac{2}{3} \div \frac{3}{4}$ The divisor is $\frac{3}{4}$. Its reciprocal is $\frac{4}{3}$.

$$\frac{2}{3} \div \frac{3}{4} = \frac{2}{3} \cdot \frac{4}{3} = \frac{8}{9}$$

Example 6

$\frac{7}{16} \div 7$ The divisor is 7. Its reciprocal is $\frac{1}{7}$.

$$\frac{7}{16} \div 7 = \frac{7}{16} \cdot \frac{1}{7} = \frac{\cancel{7} \cdot 1}{16 \cdot \cancel{7}} = \frac{1}{16}$$

Example 7

$\frac{16}{27} \div \frac{4}{9}$ The divisor is $\frac{4}{9}$. Its reciprocal is $\frac{9}{4}$.

$$\frac{16}{27} \div \frac{4}{9} = \frac{16}{27} \cdot \frac{9}{4} = \frac{4 \cdot \cancel{4} \cdot \cancel{9}}{\cancel{9} \cdot 3 \cdot \cancel{4}} = \frac{4}{3}$$

Example 8

$\dfrac{\frac{9}{4}}{\frac{9}{2}}$ The divisor is $\frac{9}{2}$. Its reciprocal is $\frac{2}{9}$.

$$\dfrac{\frac{9}{4}}{\frac{9}{2}} = \frac{9}{4} \cdot \frac{2}{9} = \frac{\cancel{9} \cdot \cancel{2} \cdot 1}{\cancel{2} \cdot 2 \cdot \cancel{9}} = \frac{1}{2}$$

☑ Completion Example 9

$\frac{13}{4} \div \frac{39}{5}$

$$\frac{13}{4} \div \frac{39}{5} = \frac{13}{4} \cdot \_\_\_\_\_\_ = \frac{13 \cdot \_\_}{4 \cdot \_\_} = \_\_\_\_\_\_\_\_$$

☑ Completion Example 10

$\dfrac{\frac{4}{9}}{\frac{4}{9}}$

$$\dfrac{\frac{4}{9}}{\frac{4}{9}} = \frac{4}{9} \cdot \_\_\_\_\_\_ = \frac{4 \cdot \_\_}{9 \cdot \_\_} = \_\_\_\_\_\_\_\_$$

Now Work Exercises 4–7 in the Margin.

Divide as indicated and reduce to lowest terms.

4. $\frac{3}{4} \div \frac{5}{8}$

5. $\frac{1}{2} \div 2$

6. $\frac{1}{2} \div \frac{1}{2}$

7. $\dfrac{\frac{9}{10}}{\frac{3}{4}}$

EXAMPLE 11

The result of multiplying two numbers is $\frac{7}{16}$. If one of the numbers is $\frac{3}{4}$, what is the other number? (**Hint:** Think in terms of whole numbers to convince yourself that this is a division problem. If the product of two numbers is 24 and one of the numbers is 6, what is the other number? You would divide 24 by 6.)

$$\begin{array}{r} 4 \\ 6\overline{)24} \\ \underline{24} \\ 0 \end{array}$$

The other number is 4.

Solution

$\frac{7}{16}$ is the result of multiplying two numbers. Divide $\frac{7}{16}$ by $\frac{3}{4}$ to find the second number.

$$\frac{7}{16} \div \frac{3}{4} = \frac{7}{16} \cdot \frac{4}{3} = \frac{7 \cdot \cancel{4}}{\cancel{4} \cdot 4 \cdot 3} = \frac{7}{12}$$

The other number is $\frac{7}{12}$.

Check by multiplying: $\frac{3}{4} \cdot \frac{7}{12} = \frac{\cancel{3} \cdot 7}{4 \cdot 4 \cdot \cancel{3}} = \frac{7}{16}$

EXAMPLE 12

If the product of $\frac{3}{2}$ with another number is $\frac{5}{18}$, what is the other number?

Solution

As in Example 11, we know the product of two numbers. So, we divide the product by the given number to find the other number.

$$\frac{5}{18} \div \frac{3}{2} = \frac{5}{18} \cdot \frac{2}{3} = \frac{5 \cdot \cancel{2}}{\cancel{2} \cdot 9 \cdot 3} = \frac{5}{27}$$

$\frac{5}{27}$ is the other number.

Check by multiplying: $\frac{3}{2} \cdot \frac{5}{27} = \frac{\cancel{3} \cdot 5}{2 \cdot \cancel{3} \cdot 9} = \frac{5}{18}$

Completion Example Answers

9. $\frac{13}{4} \div \frac{39}{5} = \frac{13}{4} \cdot \mathbf{\frac{5}{39}} = \frac{\cancel{13} \cdot \mathbf{5}}{4 \cdot \mathbf{3} \cdot \cancel{\mathbf{13}}} = \mathbf{\frac{5}{12}}$

10. $\dfrac{\frac{4}{9}}{\frac{4}{9}} = \frac{4}{9} \cdot \mathbf{\frac{9}{4}} = \frac{4 \cdot \mathbf{9}}{9 \cdot \mathbf{4}} = \mathbf{1}$

NAME ______________ SECTION ______________ DATE ______________

ANSWERS

1. ______________
2. ______________
3. ______________
4. ______________
5. ______________
6. ______________
7. ______________
8. ______________
9. ______________
10. ______________
11. ______________
12. ______________
13. ______________
14. ______________
15. ______________
16. ______________
17. ______________
18. ______________
19. ______________

Exercises 3.3

1. What is the reciprocal of $\frac{12}{13}$?

2. To divide by any nonzero number, multiply by its ______________.

3. The quotient of $0 \div \frac{5}{6}$ is ______________.

4. The quotient of $\frac{5}{6} \div 0$ is ______________.

Find the following quotients. Reduce to lowest terms whenever possible.

5. $\frac{2}{3} \div \frac{3}{4}$

6. $\frac{1}{5} \div \frac{3}{4}$

7. $\frac{3}{7} \div \frac{3}{5}$

8. $\frac{2}{11} \div \frac{2}{3}$

9. $\frac{3}{5} \div \frac{3}{7}$

10. $\frac{2}{3} \div \frac{2}{11}$

11. $\frac{5}{16} \div \frac{15}{16}$

12. $\frac{7}{18} \div \frac{3}{9}$

13. $\frac{3}{14} \div \frac{2}{7}$

14. $\frac{13}{40} \div \frac{26}{35}$

15. $\frac{5}{12} \div \frac{15}{16}$

16. $\frac{12}{27} \div \frac{10}{18}$

17. $\frac{17}{48} \div \frac{51}{90}$

18. $\frac{3}{5} \div \frac{7}{8}$

19. $\frac{13}{16} \div \frac{2}{3}$

20. ______
21. ______
22. ______
23. ______
24. ______
25. ______
26. ______
27. ______
28. ______
29. ______
30. ______
31. ______
32. ______
33. ______
34. ______
35. ______
36. ______
37. ______
38. ______
39. ______
40. ______
41. ______
42. ______
43. ______

20. $\frac{5}{6} \div \frac{3}{4}$

21. $\frac{3}{4} \div \frac{5}{6}$

22. $\frac{14}{15} \div \frac{21}{25}$

23. $\frac{3}{7} \div \frac{3}{7}$

24. $\frac{6}{13} \div \frac{6}{13}$

25. $\frac{16}{27} \div \frac{7}{18}$

26. $\frac{20}{21} \div \frac{15}{42}$

27. $\frac{25}{36} \div \frac{5}{24}$

28. $\frac{17}{20} \div \frac{3}{14}$

29. $\frac{26}{35} \div \frac{39}{40}$

30. $\frac{5}{6} \div \frac{13}{4}$

31. $\frac{7}{8} \div \frac{15}{2}$

32. $\frac{29}{50} \div \frac{31}{10}$

33. $\frac{21}{5} \div \frac{10}{3}$

34. $\frac{35}{17} \div \frac{5}{4}$

35. $\frac{21}{5} \div 3$

36. $\frac{41}{6} \div 2$

37. $3 \div \frac{21}{5}$

38. $2 \div \frac{41}{6}$

39. $5 \div \frac{15}{8}$

40. $14 \div \frac{1}{7}$

41. $\frac{\frac{15}{8}}{5}$

42. $\frac{\frac{1}{7}}{14}$

43. $\frac{\frac{3}{4}}{\frac{1}{2}}$

NAME ______________ SECTION ______________ DATE ______________

ANSWERS

44. $\dfrac{\frac{3}{4}}{2}$

45. $\dfrac{\frac{3}{4}}{3}$

46. $\dfrac{\frac{3}{4}}{\frac{1}{3}}$

47. $\dfrac{\frac{92}{7}}{\frac{46}{11}}$

48. $\dfrac{\frac{33}{32}}{\frac{11}{4}}$

49. $\dfrac{56}{\frac{1}{8}}$

50. $\dfrac{16}{\frac{4}{3}}$

51. The product of $\frac{9}{10}$ with another number is $\frac{5}{3}$. What is the other number?

52. The result of multiplying two numbers is 150. If one of the numbers is $\frac{5}{7}$, what is the other number?

44. ______________

45. ______________

46. ______________

47. ______________

48. ______________

49. ______________

50. ______________

51. ______________

52. ______________

53. ________

54. ________

55. ________

56. ________

57. [Respond below exercise.]

53. A small private college has determined that about $\frac{44}{100}$ of the students that it accepts will actually enroll. If the college wants 550 freshmen to enroll, how many should it accept?

54. The floor of the Atlantic Ocean is spreading apart at an average rate of $\frac{3}{50}$ of a meter per year. About how long will it take for the sea floor to spread 12 meters?

Check Your Number Sense

55. A bus has 45 passengers. This is $\frac{3}{4}$ of its capacity. Is the capacity of the bus more or less than 45? What is the capacity of the bus?

56. A computer printer can print eight pages in one minute. Will the printer print more or fewer than eight pages in 30 seconds? How many pages will the printer print in 15 seconds?

57. If $\frac{2}{3}$ of your salary is \$1200 per month, is your monthly salary more or less than \$1200? We know that $\frac{1}{10}$ of \$1200 is \$120. If you get a raise of $\frac{1}{10}$ of your monthly salary, will your new monthly salary be more than, less than, or equal to \$1320? Why?

NAME ______________________ SECTION ________________ DATE __________

ANSWERS

58. [Respond below exercise.]

59. [Respond below exercise.]

60. [Respond below exercise.]

61. [Respond below exercise.]

Writing and Thinking about Mathematics

58. Show that the phrases "6 divided by two" and "6 divided by one-half" have different meanings.

59. Show that the phrases "15 divided by three" and "15 divided by one-third" have different meanings.

60. Show that the phrases "12 divided by three" and "12 times one-third" have the same meaning.

61. Show that the phrases "20 divided by four" and "20 times one-fourth" have the same meaning.

62. [Respond below exercise.]

62. Is division a commutative operation? Explain briefly and give three examples using fractions to help justify your answer.

RECYCLE BIN

1. ______
2. ______
3. ______
4. ______
5. ______
6. ______
7a. ______
b. ______
8a. ______
b. ______
9a. ______
b. ______
10a. ______
b. ______
11. ______

The Recycle Bin (from Section 2.5)

Find the least common multiple (LCM) of the following sets of numbers.

1. 30, 65

2. 28, 36

3. 10, 20, 50

4. 39, 51

5. 15, 25, 100

6. 44, 88, 121

For each of the following sets of numbers, (a) find the LCM, and (b) state the number of times each number divides into the LCM.

7. 50, 125

8. 20, 24, 30

9. 12, 35, 70

10. 45, 63, 99

11. Two people meet in the optometrist's office and have a pleasant conversation. They agree to have lunch together the next time they are in the optometrist's office on the same day. If their appointments are once every 30 days for one person and once every 45 days for the other person, in how many days will they have lunch together?

3.4 Addition and Subtraction

OBJECTIVES

1. *Be able to add and subtract fractions with the same denominator.*
2. *Recall how to find the LCM.*
3. *Know how to add and subtract fractions with different denominators.*

Addition with Fractions

To add two (or more) fractions with the same denominator, think of the common denominator as the "name" of each fraction. The sum has this common name. Just as 3 pears plus 2 pears gives a total of 5 pears, 3 sevenths plus 2 sevenths gives a total of 5 sevenths. Figure 3.2 illustrates how the **sum** of $\frac{3}{7}$ and $\frac{2}{7}$ might be diagrammed.

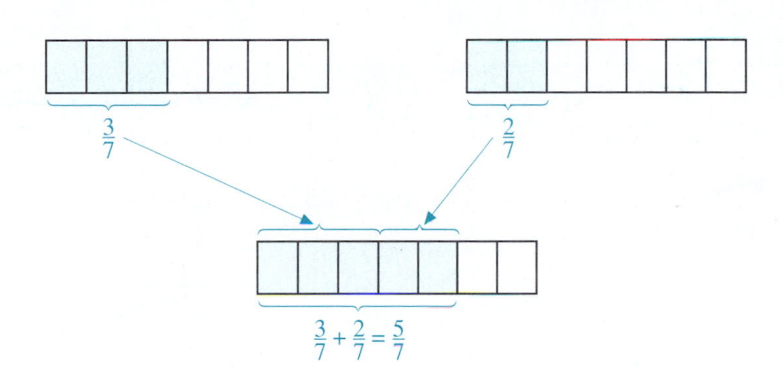

Figure 3.2

To Add Two (or More) Fractions with the Same Denominator:

1. Add the numerators.
2. Keep the common denominator.

$$\frac{a}{b} + \frac{c}{b} = \frac{a+c}{b}$$

Example 1

Perform each indicated addition.

(a) $\frac{1}{3} + \frac{1}{3} = \frac{1+1}{3} = \frac{2}{3}$

(b) $\frac{4}{5} + \frac{3}{5} = \frac{4+3}{5} = \frac{7}{5}$

(c) You can add any number of fractions with the same denominator.

$$\frac{2}{7} + \frac{3}{7} + \frac{1}{7} + \frac{6}{7} = \frac{2+3+1+6}{7} = \frac{12}{7}$$

(d) You may be able to reduce after adding.

$$\frac{4}{15} + \frac{6}{15} = \frac{4+6}{15} = \frac{10}{15} = \frac{2 \cdot \cancel{5}}{3 \cdot \cancel{5}} = \frac{2}{3}$$

Now Work Exercises 1–4 in the Margin.

Add and reduce if possible.

1. $\frac{1}{7} + \frac{4}{7} =$

2. $\frac{3}{8} + \frac{1}{8} =$

3. $\frac{3}{16} + \frac{3}{16} + \frac{5}{16} =$

4. $\frac{7}{12} + \frac{8}{12} =$

Of course, fractions to be added will not always have the same denominator. In these cases, the smallest common denominator must be found. **The least common denominator (LCD) is the least common multiple (LCM) of the denominators.**

To Add Fractions with Different Denominators:

1. Find the least common denominator (LCD).
2. Change each fraction to an equal fraction with that denominator.
3. Add the new fractions.
4. Reduce if possible.

EXAMPLE 2

$$\frac{3}{8} + \frac{11}{22}$$

(a) Find the LCD.

$$\left.\begin{aligned} 8 &= 2 \cdot 2 \cdot 2 \\ 12 &= 2 \cdot 2 \cdot 3 \end{aligned}\right\} \quad \text{LCD} = 2 \cdot 2 \cdot 2 \cdot 3 = 2^3 \cdot 3 = 24$$

(b) Find equal fractions with denominator 24.

$$\frac{3}{8} = \frac{3}{8} \cdot 1 = \frac{3}{8} \cdot \frac{3}{3} = \frac{9}{24}$$

Multiply by $\frac{3}{3}$ since $8 \cdot 3 \cdot = 24$.

$$\frac{11}{12} = \frac{11}{12} \cdot 1 = \frac{11}{12} \cdot \frac{2}{2} = \frac{22}{24}$$

Multiply by $\frac{2}{2}$ since $12 \cdot 2 = 24$.

(c) Add.

$$\frac{3}{8} + \frac{11}{12} = \frac{9}{24} + \frac{22}{24} = \frac{9 + 22}{24} = \frac{31}{24}$$

EXAMPLE 3

$$\frac{5}{21} + \frac{5}{28}$$

(a) Find the LCD.

$$\left.\begin{aligned} 21 &= 3 \cdot 7 \\ 28 &= 2 \cdot 2 \cdot 7 \end{aligned}\right\} \quad \text{LCD} = 2 \cdot 2 \cdot 3 \cdot 7 = 84$$

(b) Find equal fractions with denominator 84.

$$\frac{5}{21} = \frac{5}{21} \cdot \frac{4}{4} = \frac{20}{84}$$

$$\frac{5}{28} = \frac{5}{28} \cdot \frac{3}{3} = \frac{15}{84}$$

(c) $\frac{5}{21} + \frac{5}{28} = \frac{20}{84} + \frac{15}{84} = \frac{20 + 15}{84} = \frac{35}{84}$

(d) Now reduce. (Remember: The prime factorization of 84 is known from our work in finding the LCD.)

$$\frac{35}{84} = \frac{\cancel{7} \cdot 5}{2 \cdot 2 \cdot 3 \cdot \cancel{7}} = \frac{5}{12}$$

EXAMPLE 4

$$\frac{2}{3} + \frac{1}{6} + \frac{5}{12}$$

(a) LCD = 12 You can simply observe this or use prime factorizations.

(b) Steps (b), (c), and (d) can be written together in one process.

$$\frac{2}{3} + \frac{1}{6} + \frac{5}{12} = \frac{2}{3} \cdot \frac{4}{4} + \frac{1}{6} \cdot \frac{2}{2} + \frac{5}{12}$$

$$= \frac{8}{12} + \frac{2}{12} + \frac{5}{12}$$

$$= \frac{15}{12} = \frac{\cancel{3} \cdot 5}{2 \cdot 2 \cdot \cancel{3}} = \frac{5}{4}$$

Or, the numbers can be written vertically. The process is the same.

$$\frac{2}{3} = \frac{2}{3} \cdot \frac{4}{4} = \frac{8}{12}$$

$$\frac{1}{6} = \frac{1}{6} \cdot \frac{2}{2} = \frac{2}{12}$$

$$\frac{5}{12} = \frac{5}{12} = \frac{5}{12}$$

$$\frac{15}{12} = \frac{\cancel{3} \cdot 5}{2 \cdot 2 \cdot \cancel{3}} = \frac{5}{4}$$

✓COMPLETION EXAMPLE 5

$$\frac{2}{3} + \frac{5}{8} + \frac{1}{6}$$

(a) $3 = 3$
$8 = 2 \cdot 2 \cdot 2$
$6 = 2 \cdot 3$
LCD $= 2 \cdot 2 \cdot 2 \cdot 3 = 24$

(b) $\frac{2}{3} + \frac{5}{8} + \frac{1}{6} = \frac{2}{3} \cdot \_\_\_\_ + \frac{5}{8} \cdot \_\_\_\_ + \frac{1}{6} \cdot \_\_\_\_$

$= \_\_\_\_ + \_\_\_\_ + \_\_\_\_$

$= \_\_\_\_$

Now Work Exercises 5 and 6 in the Margin.

Add and reduce if possible.

5. $\frac{1}{4} + \frac{3}{8} + \frac{7}{10} =$

6. $\frac{1}{8} + \frac{1}{9} + \frac{1}{12} =$

Common Error

The following common error must be avoided.

Find the sum $\frac{3}{2} + \frac{1}{6}$.

WRONG SOLUTION

$$\frac{\overset{1}{\cancel{3}}}{2} + \frac{1}{\underset{2}{\cancel{6}}} = \frac{1}{2} + \frac{1}{2} = 1 \quad \textbf{WRONG}$$

You **cannot** cancel across the + sign.

CORRECT SOLUTION

Use LCD = 6.

$$\frac{3}{2} + \frac{1}{6} = \frac{3}{2} \cdot \frac{3}{3} + \frac{1}{6} = \frac{9}{6} + \frac{1}{6} = \frac{10}{6}$$

NOW reduce.

$$\frac{10}{6} = \frac{5 \cdot \cancel{2}}{3 \cdot \cancel{2}} = \frac{5}{3}$$

2 is a factor in both the numerator and the denominator.

Both the commutative and the associative properties of addition apply to fractions.

Commutative Property of Addition

If $\frac{a}{b}$ and $\frac{c}{d}$ are fractions, then

$$\frac{a}{b} + \frac{c}{d} = \frac{c}{d} + \frac{a}{b}$$

Associative Property of Addition

If $\frac{a}{b}$, $\frac{c}{d}$, and $\frac{e}{f}$ are fractions, then

$$\frac{a}{b} + \frac{c}{d} + \frac{e}{f} = \frac{a}{b} + \left(\frac{c}{d} + \frac{e}{f}\right) = \left(\frac{a}{b} + \frac{c}{d}\right) + \frac{e}{f}$$

Subtraction with Fractions

Figure 3.3 shows how the **difference** of the two fractions $\frac{5}{7}$ and $\frac{1}{7}$ might be diagrammed. Just as with addition, the common denominator "names" each fraction. The difference is found by subtracting the numerators and using the common denominator.

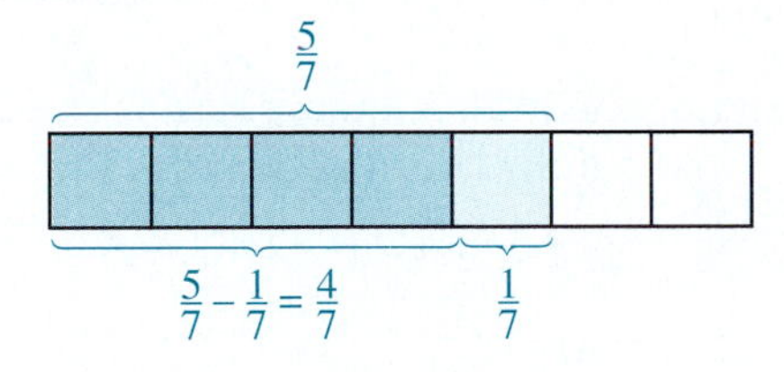

Figure 3.3

To Subtract Fractions That Have the Same Denominator:

1. Subtract the numerators.
2. Keep the common denominator.

$$\frac{a}{b} - \frac{c}{b} = \frac{a - c}{b}$$

Example 6

Perform each indicated subtraction.

(a) $\frac{5}{9} - \frac{4}{9} = \frac{5 - 4}{9} = \frac{1}{9}$

(b) You may be able to reduce after subtracting.

$$\frac{7}{8} - \frac{1}{8} = \frac{7 - 1}{8} = \frac{6}{8} = \frac{\cancel{2} \cdot 3}{\cancel{2} \cdot 4} = \frac{3}{4}$$

(c) $\frac{19}{10} - \frac{11}{10} = \frac{19 - 11}{10} = \frac{8}{10} = \frac{\cancel{2} \cdot 4}{\cancel{2} \cdot 5} = \frac{4}{5}$

Now Work Exercises 7–10 in the Margin.

To Subtract Fractions That Have Different Denominators:

1. Find the least common denominator (LCD).
2. Change each fraction to an equal fraction with that denominator.
3. Subtract the new fractions.
4. Reduce if possible.

Subtract and reduce if possible.

7. $\frac{6}{10} - \frac{3}{10} =$

8. $\frac{27}{30} - \frac{5}{30} =$

9. $\frac{7}{8} - \frac{3}{8} =$

10. $\frac{5}{6} - \frac{1}{6} =$

EXAMPLE 7

$\frac{12}{55} - \frac{2}{33}$

(a) Find the LCD.

$$\left.\begin{aligned} 55 &= 5 \cdot 11 \\ 33 &= 3 \cdot 11 \end{aligned}\right\} \quad \text{LCD} = 3 \cdot 5 \cdot 11 = 165$$

(b) Find equal fractions with denominator 165.

$$\frac{12}{55} = \frac{12}{55} \cdot \frac{3}{3} = \frac{36}{165}$$

$$\frac{2}{33} = \frac{2}{33} \cdot \frac{5}{5} = \frac{10}{165}$$

(c) Subtract.

$$\frac{12}{55} - \frac{2}{33} = \frac{36}{165} - \frac{10}{165} = \frac{36 - 10}{165} = \frac{26}{165}$$

(d) Reduce if possible.

$$\frac{26}{165} = \frac{2 \cdot 13}{3 \cdot 5 \cdot 11} = \frac{26}{165} \quad \text{cannot be reduced}$$

As with addition, steps (b), (c), and (d) can be written as one process (Completion Examples 8 and 9).

☑ COMPLETION EXAMPLE 8

$\frac{5}{18} - \frac{1}{6}$

(a) LCD = 18

(b) $\frac{5}{18} - \frac{1}{6} = \frac{5}{18} - \frac{1}{6} \cdot \frac{3}{3}$ Use $\frac{3}{3}$ since $6 \cdot 3 = 18$.

$= \frac{\_\_\_}{\_\_\_} - \frac{\_\_\_}{\_\_\_}$

$= \frac{\_\_\_}{\_\_\_} = \frac{\_\_\_}{\_\_\_}$

☑ COMPLETION EXAMPLE 9

$\frac{7}{12} - \frac{3}{20}$

(a) $\left.\begin{aligned} 12 &= 2 \cdot 2 \cdot 3 \\ 20 &= 2 \cdot 2 \cdot 5 \end{aligned}\right\} \quad \text{LCD} = \_\_\_\_\_\_$

(b) $\frac{7}{12} - \frac{3}{20} = \frac{7}{12} \cdot \_\_\_ - \frac{3}{20} \cdot \_\_\_$

$= \frac{\_\_\_}{\_\_\_} - \frac{\_\_\_}{\_\_\_}$

$= \frac{\_\_\_}{\_\_\_} = \frac{\_\_\_}{\_\_\_}$

Now Work Exercises 11 and 12 in the Margin.

Example 10

The Narragansett Grays baseball team lost 90 games in one season. If $\frac{1}{5}$ of their losses were by 1 or 2 runs and $\frac{4}{9}$ of their losses were by 3 or fewer runs, what fraction of their losses was by exactly 3 runs?

Solution

Before answering the question, we need to analyze the information and decide which arithmetic operation is needed. The losses that were by 3 or fewer runs include those that were by 1 or 2 runs. So, if we take away the losses that were by 1 or 2 runs from the losses that were by 3 or fewer runs, we will be left with the losses that were by exactly 3 runs. That is, we want to subtract. Since we are dealing with fractions and not the actual numbers of losses, we subtract $\frac{1}{5}$ from $\frac{4}{9}$. (The LCD is 45.)

$$\frac{4}{9} - \frac{1}{5} = \frac{4}{9} \cdot \frac{5}{5} - \frac{1}{5} \cdot \frac{9}{9} = \frac{20}{45} - \frac{9}{45} = \frac{11}{45}$$

Thus, the Grays lost $\frac{11}{45}$ of their games by exactly 3 runs.

Subtract and reduce if possible.

11. $\frac{7}{10} - \frac{2}{15} =$

12. $\frac{15}{16} - \frac{5}{12} =$

Completion Example Answers

5. $$\frac{2}{3} + \frac{5}{8} + \frac{1}{6} = \frac{2}{3} \cdot \mathbf{\frac{8}{8}} + \frac{5}{8} \cdot \mathbf{\frac{3}{3}} + \frac{1}{6} \cdot \mathbf{\frac{4}{4}}$$
$$= \mathbf{\frac{16}{24}} + \mathbf{\frac{15}{24}} + \mathbf{\frac{4}{24}}$$
$$= \mathbf{\frac{35}{24}}$$

8. (b) $$\frac{5}{18} - \frac{1}{6} = \frac{5}{18} - \frac{1}{6} \cdot \frac{3}{3}$$
$$= \mathbf{\frac{5}{18}} - \mathbf{\frac{3}{18}}$$
$$= \mathbf{\frac{2}{18}} = \mathbf{\frac{1}{9}}$$

9. (a) $$\left.\begin{aligned} 12 &= 2 \cdot 2 \cdot 3 \\ 20 &= 2 \cdot 2 \cdot 5 \end{aligned}\right\} \quad \text{LCD} = \mathbf{2 \cdot 2 \cdot 3 \cdot 5 = 60}$$

(b) $$\frac{7}{12} - \frac{3}{20} = \frac{7}{12} \cdot \mathbf{\frac{5}{5}} - \frac{3}{20} \cdot \mathbf{\frac{3}{3}}$$
$$= \mathbf{\frac{35}{60}} - \mathbf{\frac{9}{60}}$$
$$= \mathbf{\frac{26}{60}} = \mathbf{\frac{13}{30}}$$

NAME ______________ SECTION ______________ DATE ______________

ANSWERS

1. ______________
2. ______________
3. ______________
4. ______________
5. ______________
6. ______________
7. ______________
8. ______________
9. ______________
10. ______________
11. ______________
12. ______________

Exercises 3.4

Add and reduce if possible.

1. $\frac{6}{10} + \frac{4}{10}$

2. $\frac{3}{14} + \frac{2}{14}$

3. $\frac{1}{20} + \frac{3}{20}$

4. $\frac{3}{4} + \frac{3}{4}$

5. $\frac{5}{6} + \frac{4}{6}$

6. $\frac{7}{5} + \frac{3}{5}$

7. $\frac{11}{15} + \frac{7}{15}$

8. $\frac{7}{9} + \frac{8}{9}$

9. $\frac{3}{25} + \frac{12}{25}$

10. $\frac{7}{90} + \frac{37}{90} + \frac{21}{90}$

11. $\frac{11}{75} + \frac{12}{75} + \frac{62}{75}$

12. $\frac{14}{32} + \frac{7}{32} + \frac{1}{32}$

13. ______
14. ______
15. ______
16. ______
17. ______
18. ______
19. ______
20. ______
21. ______
22. ______
23. ______
24. ______
25. ______
26. ______
27. ______
28. ______
29. ______
30. ______
31. ______
32. ______
33. ______

13. $\frac{4}{100} + \frac{35}{100} + \frac{76}{100}$

14. $\frac{21}{95} + \frac{33}{95} + \frac{3}{95}$

15. $\frac{1}{200} + \frac{17}{200} + \frac{25}{200}$

16. $\frac{1}{12} + \frac{2}{3} + \frac{1}{4}$

17. $\frac{3}{8} + \frac{5}{16}$

18. $\frac{2}{5} + \frac{3}{10} + \frac{3}{20}$

19. $\frac{3}{4} + \frac{1}{16} + \frac{6}{32}$

20. $\frac{2}{7} + \frac{4}{21} + \frac{1}{3}$

21. $\frac{1}{6} + \frac{1}{4} + \frac{1}{3}$

22. $\frac{2}{39} + \frac{1}{3} + \frac{4}{13}$

23. $\frac{1}{2} + \frac{3}{10} + \frac{4}{5}$

24. $\frac{1}{27} + \frac{4}{18} + \frac{1}{6}$

25. $\frac{2}{7} + \frac{3}{20} + \frac{9}{14}$

26. $\frac{1}{8} + \frac{1}{12} + \frac{1}{9}$

27. $\frac{2}{5} + \frac{4}{7} + \frac{3}{8}$

28. $\frac{2}{3} + \frac{3}{4} + \frac{5}{6}$

29. $\frac{1}{5} + \frac{7}{30} + \frac{1}{6}$

30. $\frac{1}{5} + \frac{2}{15} + \frac{1}{6}$

31.
$$\begin{array}{r} \frac{3}{4} \\ \frac{1}{2} \\ +\frac{5}{12} \\ \hline \end{array}$$

32.
$$\begin{array}{r} \frac{1}{5} \\ \frac{2}{15} \\ +\frac{5}{6} \\ \hline \end{array}$$

33.
$$\begin{array}{r} \frac{7}{8} \\ \frac{2}{3} \\ +\frac{1}{9} \\ \hline \end{array}$$

NAME ______________ SECTION ______________ DATE ______________

34. $\frac{1}{27} + \frac{1}{18} + \frac{4}{9}$

35. $\frac{3}{20} + \frac{1}{100} + \frac{3}{100}$

36. $\frac{13}{100} + \frac{4}{10} + \frac{1}{1000}$

37. $\frac{7}{12} + \frac{1}{9} + \frac{2}{3}$

38. $\frac{1}{3} + \frac{8}{15} + \frac{7}{10}$

39. $\frac{9}{16} + \frac{5}{48} + \frac{3}{32}$

40. $\frac{3}{10} + \frac{1}{20} + \frac{1}{25}$

Subtract and reduce if possible.

41. $\frac{4}{7} - \frac{1}{7}$

42. $\frac{5}{7} - \frac{3}{7}$

43. $\frac{9}{10} - \frac{3}{10}$

44. $\frac{11}{10} - \frac{7}{10}$

45. $\frac{5}{8} - \frac{1}{8}$

46. $\frac{7}{8} - \frac{5}{8}$

47. $\frac{11}{12} - \frac{7}{12}$

48. $\frac{7}{12} - \frac{3}{12}$

49. $\frac{13}{15} - \frac{4}{15}$

ANSWERS

34. ______________

35. ______________

36. ______________

37. ______________

38. ______________

39. ______________

40. ______________

41. ______________

42. ______________

43. ______________

44. ______________

45. ______________

46. ______________

47. ______________

48. ______________

49. ______________

50. ______
51. ______
52. ______
53. ______
54. ______
55. ______
56. ______
57. ______
58. ______
59. ______
60. ______
61. ______
62. ______
63. ______
64. ______
65. ______
66. ______
67. ______
68. ______
69. ______
70. ______
71. ______
72. ______
73. ______
74. ______
75. ______
76. ______

50. $\frac{21}{15} - \frac{11}{15}$

51. $\frac{5}{6} - \frac{1}{3}$

52. $\frac{5}{6} - \frac{1}{2}$

53. $\frac{11}{15} - \frac{3}{10}$

54. $\frac{8}{10} - \frac{3}{15}$

55. $\frac{3}{4} - \frac{2}{3}$

56. $\frac{2}{3} - \frac{1}{4}$

57. $\frac{15}{16} - \frac{21}{32}$

58. $\frac{3}{8} - \frac{1}{16}$

59. $\frac{5}{4} - \frac{3}{5}$

60. $\frac{5}{12} - \frac{1}{6}$

61. $\frac{14}{27} - \frac{7}{18}$

62. $\frac{25}{18} - \frac{21}{27}$

63. $\frac{8}{45} - \frac{11}{72}$

64. $\frac{46}{55} - \frac{10}{33}$

65. $\frac{5}{36} - \frac{1}{45}$

66. $\frac{5}{1} - \frac{3}{4}$

67. $\frac{4}{1} - \frac{5}{8}$

68. $2 - \frac{9}{16}$

69. $\frac{31}{40} - \frac{5}{8}$

70. $\frac{14}{35} - \frac{12}{30}$

71. $\frac{20}{35} - \frac{24}{42}$

72. $\frac{3}{10} - \frac{298}{1000}$

73. $1 - \frac{9}{10}$

74. $1 - \frac{7}{8}$

75. $1 - \frac{2}{3}$

76. $1 - \frac{1}{16}$

NAME ______________ SECTION ______________ DATE ______________

ANSWERS

77. $1 - \frac{3}{20}$

78. $1 - \frac{4}{9}$

79. $\frac{7}{8} - \frac{2}{3}$

80. $\frac{14}{15} - \frac{3}{10}$

81. $\frac{1}{10} - \frac{8}{100}$

82. $\frac{3}{100} - \frac{1}{1000}$

83. Three letters weigh $\frac{1}{2}$ oz, $\frac{1}{5}$ oz, and $\frac{3}{10}$ oz. What is the total weight of the letters?

84. Using a microscope, a scientist measures the diameters of three hairs and finds them to be $\frac{1}{1000}$ in., $\frac{3}{1000}$ in., and $\frac{1}{100}$ in. What is the total of these three diameters?

85. A machinist drills four holes in a straight line. Each hole has a diameter of $\frac{1}{10}$ in. and there is $\frac{1}{4}$ in. between the holes. What is the distance between the outer edges of the first and last holes?

77. ______________

78. ______________

79. ______________

80. ______________

81. ______________

82. ______________

83. ______________

84. ______________

85. ______________

86. ______________

87. ______________

88. ______________

89. ______________

86. A notebook contains 30 sheets of paper $\left(\text{each } \frac{1}{100} \text{ in. thick}\right)$, 2 pieces of cardboard $\left(\text{each } \frac{1}{16} \text{ in. thick}\right)$, and a front and back cover $\left(\text{each } \frac{1}{4} \text{ in. thick}\right)$. What is the total thickness of the notebook?

87. Find the sum of $\frac{1}{4}$ and $\frac{3}{16}$. Subtract $\frac{1}{8}$ from the sum. What is the difference?

88. Find the difference between $\frac{2}{3}$ and $\frac{5}{9}$. Add $\frac{1}{12}$ to the difference. What is the sum?

89. Find the sum of $\frac{3}{4}$ and $\frac{5}{8}$. Multiply the sum by $\frac{2}{3}$. What is the product?

NAME ________ SECTION ________ DATE ________

ANSWERS

90. ________

91. ________

92. [Respond below exercise.] ________

90. Find the product of $\frac{9}{10}$ and $\frac{2}{3}$. Divide the product by $\frac{5}{3}$. What is the quotient?

91. About $\frac{1}{2}$ of all incoming solar radiation is absorbed by the earth, $\frac{1}{5}$ is absorbed by the atmosphere, and $\frac{1}{20}$ is scattered by the atmosphere. The rest is reflected by the earth and clouds. What fraction of solar radiation is reflected?

Collaborative Learning Exercise

92. With the class separated into teams of two to four, each team is to write one or two paragraphs explaining which topic in this chapter they found to be the most difficult and which techniques they used to learn the related material. The team leader is to read the paragraph(s) with classroom discussion to follow.

RECYCLE BIN

1. ______________

2. ______________

3. ______________

4. ______________

5. ______________

6. ______________

7. ______________

8. ______________

9. ______________

10. ______________

The Recycle Bin (from Section 2.1)

Find the value of each of the following squares. Write as many as you can from memory.

1. 9^2 **2.** 13^2 **3.** 15^2 **4.** 18^2

Find the value of each of the following expressions by using the rules for order of operations.

5. $8 \div 2 + 6 - 5 \cdot 2$

6. $8 \div (2 + 6) + 14 \cdot 2$

7. $10^2 - 9 \cdot 6 \div 2 + 1^3$

8. $3(4 + 7) - 4 \cdot 3 - 3 \cdot 7$

9. $2 \cdot 5^2 + 3(16 - 2 \cdot 8) + 5 \cdot 2^2$

10. $(5^2 + 7) \div 4 - (10 \div 5 \cdot 2)$

3.5 Comparisons and Order of Operations

OBJECTIVES

1. *Compare fractions by finding a common denominator and comparing the numerators.*
2. *Follow the rules for order of operations with fractions.*

Comparing Two or More Fractions

Many times we want to compare two (or more) fractions to see which is smaller or larger. Then we can subtract the smaller from the larger, or possibly make a decision based on the relative sizes of the fractions. Related word problems will be discussed in detail in later chapters.

To Compare Two Fractions (to Find Which is Larger or Smaller):

1. Find the least common denominator (LCD).
2. Change each fraction to an equal fraction with that denominator.
3. Compare the numerators.

Example 1

Which is larger: $\frac{5}{6}$ or $\frac{7}{8}$? How much larger?

(a) Find the LCD for 6 and 8.

$$\left.\begin{array}{l} 6 = 2 \cdot 3 \\ 8 = 2 \cdot 2 \cdot 2 \end{array}\right\} \quad \text{LCD} = 2 \cdot 2 \cdot 2 \cdot 3 = 24$$

(b) Find equal fractions with denominator 24.

$$\frac{5}{6} = \frac{5}{6} \cdot \frac{4}{4} = \frac{20}{24} \quad \text{and} \quad \frac{7}{8} = \frac{7}{8} \cdot \frac{3}{3} = \frac{21}{24}$$

(c) $\frac{7}{8}$ is larger than $\frac{5}{6}$ since 21 is larger than 20.

(d) $\frac{7}{8} - \frac{5}{6} = \frac{21}{24} - \frac{20}{24} = \frac{1}{24}$

$\frac{7}{8}$ is larger by $\frac{1}{24}$.

1. Which is larger: $\frac{9}{22}$ or $\frac{10}{33}$? How much larger?

2. Arrange $\frac{7}{12}$, $\frac{5}{9}$, and $\frac{2}{3}$ in order, from smallest to largest.

EXAMPLE 2

Which is larger: $\frac{8}{9}$ or $\frac{11}{12}$? How much larger?

(a) LCD $= 2 \cdot 2 \cdot 3 \cdot 3 = 36$

(b) $\frac{8}{9} = \frac{8}{9} \cdot \frac{4}{4} = \frac{32}{36}$ and $\frac{11}{12} = \frac{11}{12} \cdot \frac{3}{3} = \frac{33}{36}$

(c) $\frac{11}{12}$ is larger than $\frac{8}{9}$ since 33 is larger than 32.

(d) $\frac{11}{12} - \frac{8}{9} = \frac{33}{36} - \frac{32}{36} = \frac{1}{36}$

$\frac{11}{12}$ is larger by $\frac{1}{36}$.

EXAMPLE 3

Arrange $\frac{2}{3}$, $\frac{7}{10}$, and $\frac{9}{15}$ in order, from smallest to largest.

(a) LCD $= 30$

(b) $\frac{2}{3} = \frac{2}{3} \cdot \frac{10}{10} = \frac{20}{30}$; $\frac{7}{10} = \frac{7}{10} \cdot \frac{3}{3} = \frac{21}{30}$; $\frac{9}{15} = \frac{9}{15} \cdot \frac{2}{2} = \frac{18}{30}$

(c) Smallest to largest: $\frac{9}{15}$, $\frac{2}{3}$, $\frac{7}{10}$

Now Work Exercises 1 and 2 in the Margin.

Using the Rules for Order of Operations with Fractions

An expression with fractions may involve more than one arithmetic operation. To simplify such expressions, use the rules for order of operations just as they were discussed in Chapter 2 for whole numbers. Of course, all the rules for fractions must be followed, too. That is, to add or subtract, you need a common denominator; to divide, you multiply by the reciprocal of the divisor.

Rules for Order of Operations

1. First, simplify within grouping symbols, such as parentheses ( ), brackets [ ], and braces { }. Start with the innermost grouping.
2. Second, find any powers indicated by exponents.
3. Third, moving from **left to right,** perform any multiplications or divisions in the order in which they appear.
4. Fourth, moving from **left to right,** perform any additions or subtractions in the order in which they appear.

EXAMPLE 4

Evaluate the expression $\frac{1}{2} \div \frac{3}{4} + \frac{5}{6} \cdot \frac{1}{5}$.

Solution

$$\frac{1}{2} \div \frac{3}{4} + \frac{5}{6} \cdot \frac{1}{5}$$

$$= \frac{1}{\cancel{2}} \cdot \frac{\overset{2}{\cancel{4}}}{3} + \frac{5}{6} \cdot \frac{1}{5} \quad \text{Divide first.}$$

$$= \frac{2}{3} + \frac{\cancel{5}}{6} \cdot \frac{1}{\cancel{5}} \quad \text{Now multiply.}$$

$$= \frac{2}{3} + \frac{1}{6} \quad \text{Now add. (LCD = 6)}$$

$$= \frac{2}{3} \cdot \frac{2}{2} + \frac{1}{6}$$

$$= \frac{4}{6} + \frac{1}{6}$$

$$= \frac{5}{6}$$

EXAMPLE 5

Evaluate the expression $\left(\frac{3}{4} - \frac{5}{8}\right) \div \left(\frac{15}{16} - \frac{1}{2}\right)$.

$$\left(\frac{3}{4} - \frac{5}{8}\right) \div \left(\frac{15}{16} - \frac{1}{2}\right) \quad \text{Work inside the parentheses.}$$

$$= \left(\frac{6}{8} - \frac{5}{8}\right) \div \left(\frac{15}{16} - \frac{8}{16}\right)$$

$$= \left(\frac{1}{8}\right) \div \left(\frac{7}{16}\right) \quad \text{Now divide.}$$

$$= \frac{1}{\cancel{8}} \cdot \frac{\overset{2}{\cancel{16}}}{7}$$

$$= \frac{2}{7}$$

Evaluate each of the following expressions.

3. $\frac{5}{4} \div \left(1 - \frac{1}{3}\right) =$

4. $\frac{1}{4} + \left(\frac{1}{3}\right)^2 \div \frac{7}{3} =$

Example 6

Evaluate the expression $\frac{9}{10} - \left(\frac{1}{4}\right)^2 + \frac{1}{2}$.

$$\frac{9}{10} - \left(\frac{1}{4}\right)^2 + \frac{1}{2} = \frac{9}{10} - \frac{1}{16} + \frac{1}{2} \quad \text{Use the exponent first.}$$

$$= \frac{9}{10} \cdot \frac{8}{8} - \frac{1}{16} \cdot \frac{5}{5} + \frac{1}{2} \cdot \frac{40}{40} \quad \text{Now add and subtract. (LCD = 80)}$$

$$= \frac{72}{80} - \frac{5}{80} + \frac{40}{80}$$

$$= \frac{107}{80}$$

Completion Example 7

Evaluate the expression $\frac{1}{2} \cdot \frac{5}{6} + \frac{7}{15} \div 2$.

$$\frac{1}{2} \cdot \frac{5}{6} + \frac{7}{15} \div 2 = \frac{5}{12} + \frac{7}{15} \div 2$$

$$= \frac{5}{12} + \frac{7}{15} \cdot \_\_\_\_$$

$$= \frac{5}{12} + \_\_\_\_$$

$$= \frac{5}{12} \cdot \_\_\_\_ + \_\_\_\_ \cdot \_\_\_\_ \quad \text{(LCD = \_\_\_\_)}$$

$$= \_\_\_\_ + \_\_\_\_$$

$$= \_\_\_\_ = \_\_\_\_ = \_\_\_\_$$

Completion Example 8

Evaluate the expression $\left(\frac{7}{8} - \frac{7}{10}\right) \div \frac{7}{2}$.

$$\left(\frac{7}{8} - \frac{7}{10}\right) \div \frac{7}{2} = \left(\frac{35}{40} - \_\_\_\_\right) \div \frac{7}{2}$$

$$= \left(\_\_\_\_\right) \div \frac{7}{2}$$

$$= \left(\_\_\_\_\right) \cdot \_\_\_\_$$

$$= \_\_\_\_ = \_\_\_\_$$

Now Work Exercises 3 and 4 in the Margin.

Complex Fractions

A **complex fraction** is a form of a fraction in which the numerator and/or denominator are themselves fractions or the sums or differences of fractions. To simplify a complex fraction, follow the rules for order of operations by treating the numerator and denominator as if they were surrounded by parentheses.

To Simplify a Complex Fraction:

1. Simplify the numerator (add or subtract as indicated).
2. Simplify the denominator (add or subtract as indicated).
3. Divide the numerator by the denominator and reduce if possible.

EXAMPLE 9

Simplify the complex fraction $\dfrac{\frac{3}{4}+\frac{1}{2}}{1-\frac{1}{3}}$.

Solution

Simplifying the numerator gives

$$\frac{3}{4}+\frac{1}{2}=\frac{3}{4}+\frac{2}{4}=\frac{5}{4}.$$

Simplifying the denominator gives

$$1-\frac{1}{3}=\frac{3}{3}-\frac{1}{3}=\frac{2}{3}.$$

Therefore, $\dfrac{\frac{3}{4}+\frac{1}{2}}{1-\frac{1}{3}}=\dfrac{\frac{5}{4}}{\frac{2}{3}}=\dfrac{5}{4}\cdot\dfrac{3}{2}=\dfrac{15}{8}$.

Note: We could also write $\dfrac{\frac{3}{4}+\frac{1}{2}}{1-\frac{1}{3}}=\left(\frac{3}{4}+\frac{1}{2}\right)\div\left(1-\frac{1}{3}\right)$, and simplify by following the rules for order of operations.

Completion Example Answers

7. $\frac{1}{2} \cdot \frac{5}{6} + \frac{7}{15} \div 2 = \frac{5}{12} + \frac{7}{15} \div 2$

$$= \frac{5}{12} + \frac{7}{15} \cdot \mathbf{\frac{1}{2}}$$

$$= \frac{5}{12} + \mathbf{\frac{7}{30}}$$

$$= \frac{5}{12} \cdot \mathbf{\frac{5}{5}} + \mathbf{\frac{7}{30}} \cdot \mathbf{\frac{2}{2}} \quad (\text{LCD} = \mathbf{60})$$

$$= \mathbf{\frac{25}{60}} + \mathbf{\frac{14}{60}}$$

$$= \mathbf{\frac{39}{60}} = \mathbf{\frac{\not{3} \cdot 13}{\not{3} \cdot 20}} = \mathbf{\frac{23}{20}}$$

8. $\left(\frac{7}{8} - \frac{7}{10}\right) \div \frac{7}{2} = \left(\frac{35}{40} - \mathbf{\frac{28}{40}}\right) \div \frac{7}{2}$

$$= \left(\mathbf{\frac{7}{40}}\right) \div \frac{7}{2}$$

$$= \left(\mathbf{\frac{7}{40}}\right) \cdot \mathbf{\frac{2}{7}}$$

$$= \mathbf{\frac{\not{7} \cdot \not{2}}{\not{2} \cdot 20 \cdot \not{7}}} = \mathbf{\frac{1}{20}}$$

NAME ____________ SECTION ____________ DATE ____________

ANSWERS

Exercises 3.5

Find the larger number of each pair and state how much larger it is.

1. $\frac{2}{3}, \frac{3}{4}$

2. $\frac{5}{6}, \frac{7}{8}$

3. $\frac{4}{5}, \frac{17}{20}$

4. $\frac{4}{10}, \frac{3}{8}$

5. $\frac{13}{20}, \frac{5}{8}$

6. $\frac{13}{16}, \frac{21}{25}$

7. $\frac{14}{35}, \frac{12}{30}$

8. $\frac{10}{36}, \frac{7}{24}$

9. $\frac{17}{80}, \frac{11}{48}$

10. $\frac{37}{100}, \frac{24}{75}$

Arrange the numbers in order, from smallest to largest. Then find the difference between the largest and smallest numbers.

11. $\frac{2}{3}, \frac{3}{5}, \frac{7}{10}$

12. $\frac{8}{9}, \frac{9}{10}, \frac{11}{12}$

13. $\frac{7}{6}, \frac{11}{12}, \frac{19}{20}$

14. $\frac{1}{3}, \frac{5}{42}, \frac{3}{7}$

15. $\frac{1}{2}, \frac{1}{3}, \frac{1}{4}$

16. $\frac{2}{3}, \frac{3}{4}, \frac{5}{8}$

17. $\frac{7}{9}, \frac{31}{36}, \frac{13}{18}$

18. $\frac{17}{12}, \frac{40}{36}, \frac{31}{24}$

1. ____________
2. ____________
3. ____________
4. ____________
5. ____________
6. ____________
7. ____________
8. ____________
9. ____________
10. ____________
11. [Respond below exercise.]
12. [Respond below exercise.]
13. [Respond below exercise.]
14. [Respond below exercise.]
15. [Respond below exercise.]
16. [Respond below exercise.]
17. [Respond below exercise.]
18. [Respond below exercise.]

19. [Respond below exercise.] ______

20. [Respond below exercise.] ______

21. ______

22. ______

23. ______

24. ______

25. ______

26. ______

27. ______

28. ______

29. ______

30. ______

31. ______

32. ______

33. ______

34. ______

19. $\frac{1}{100}, \frac{3}{1000}, \frac{20}{10{,}000}$

20. $\frac{32}{100}, \frac{298}{1000}, \frac{3}{10}$

Evaluate each expression using the rules for order of operations.

21. $\frac{1}{2} \div \frac{7}{8} + \frac{1}{7} \cdot \frac{2}{3}$

22. $\frac{3}{5} \cdot \frac{1}{6} + \frac{1}{5} \div 2$

23. $\frac{1}{2} \div \frac{1}{2} + \frac{2}{3} \cdot \frac{2}{3}$

24. $5 - \frac{3}{4} \div 3$

25. $6 - \frac{5}{8} \div 4$

26. $\frac{2}{15} \cdot \frac{1}{4} \div \frac{3}{5} + \frac{1}{25}$

27. $\frac{5}{8} \cdot \frac{1}{10} \div \frac{3}{4} + \frac{1}{6}$

28. $\left(\frac{7}{15} + \frac{8}{21}\right) \div \frac{3}{35}$

29. $\left(\frac{1}{2} - \frac{1}{3}\right) \div \left(\frac{5}{8} + \frac{3}{16}\right)$

30. $\left(\frac{1}{3} + \frac{1}{5}\right) \cdot \left(\frac{3}{4} - \frac{1}{6}\right)$

31. $\left(\frac{1}{2}\right)^2 - \left(\frac{1}{4}\right)^3$

32. $\frac{2}{3} + \frac{3}{4} + \left(\frac{1}{2}\right)^2$

33. $\left(\frac{1}{3}\right)^2 + \left(\frac{1}{6}\right)^2 + \frac{2}{3}$

34. $\frac{1}{2} \div \frac{2}{3} + \left(\frac{1}{3}\right)^2$

NAME ____________ SECTION ____________ DATE ____________

ANSWERS

Simplify the following complex fractions.

35. $\dfrac{\frac{3}{4} - \frac{1}{2}}{1 + \frac{1}{3}}$

36. $\dfrac{\frac{1}{8} + \frac{1}{2}}{1 - \frac{2}{5}}$

37. $\dfrac{\frac{1}{5} + \frac{1}{6}}{2 + \frac{1}{3}}$

38. $\dfrac{\frac{5}{6} - \frac{1}{3}}{\frac{1}{2} + \frac{1}{5}}$

39. $\dfrac{\frac{2}{3} - \frac{1}{4}}{\frac{3}{5} - \frac{1}{4}}$

40. $\dfrac{\frac{5}{6} - \frac{2}{3}}{\frac{5}{8} - \frac{1}{16}}$

41. $\dfrac{\frac{7}{8} - \frac{3}{16}}{\frac{1}{3} - \frac{1}{4}}$

42. $\dfrac{\frac{3}{5} + \frac{4}{7}}{\frac{3}{8} + \frac{1}{10}}$

35. ____________

36. ____________

37. ____________

38. ____________

39. ____________

40. ____________

41. ____________

42. ____________

43. [Respond below exercise.] ____________

Check Your Number Sense

43. (a) If two fractions are less than 1, can their sum be more than 1? Explain.
(b) If two fractions are less than 1, can their product be more than 1? Explain.

44. [Respond below exercise.] ______________

45. [Respond below exercise.] ______________

46. [Respond below exercise.] ______________

44. Consider the fraction $\frac{1}{2}$.

(a) If this fraction is divided by 2, will the quotient be more or less than $\frac{1}{2}$?

(b) If this fraction is divided by 3, will the quotient be more or less than the quotient in part (a)?

45. Will the quotient always get smaller and smaller when a nonzero number is divided by larger and larger numbers? Can you think of a case in which this is not true? What happens when 0 is divided by larger and larger numbers?

46. Consider any fraction between 0 and 1, not including 0 or 1. If you square this number, will the result be larger or smaller than the original number? Is this always the case? Explain your answer.

3 Index of Key Ideas and Terms

NAME ______________ SECTION ______________ DATE ________

Chapter 3 Test

1. The reciprocal of $\frac{5}{8}$ is ________.

2. The fraction equivalent to $\frac{7}{16}$ with denominator 80 is ________.

3. The equation $\frac{3}{4} \cdot \frac{7}{2} = \frac{7}{2} \cdot \frac{3}{4}$ illustrates the ____________ property of ____________.

Reduce to lowest terms.

4. $\frac{90}{108}$

5. $\frac{77}{55}$

6. $\frac{117}{156}$

7. Find the product of $\frac{2}{7}$ and $\frac{14}{15}$.

8. From the sum of $\frac{3}{8}$ and $\frac{5}{12}$ subtract the sum of $\frac{1}{6}$ and $\frac{5}{9}$.

9. Find the quotient when 4 is divided by $\frac{4}{9}$.

In Exercises 10–24, perform the indicated operations and reduce all answers to lowest terms. Follow the rules for order of operations when they apply.

10. $\frac{3}{7} \cdot \frac{14}{27}$

11. $10 \div \frac{2}{5}$

12. $\frac{3}{5} \cdot \frac{1}{2} \cdot \frac{3}{8}$

13. $\frac{5}{11} \div \frac{4}{5}$

14. $2 - \frac{14}{11}$

15. $\frac{4}{35} + \frac{2}{7} + \frac{1}{10}$

ANSWERS

1. ________
2. ________
3. ________
4. ________
5. ________
6. ________
7. ________
8. ________
9. ________
10. ________
11. ________
12. ________
13. ________
14. ________
15. ________

16. ____________

17. ____________

18. ____________

19. ____________

20. ____________

21. ____________

22. ____________

23. ____________

24. ____________

25. [Respond below exercise.] ____________

26. ____________

16. $\frac{5}{16} + \left(\frac{1}{4}\right)^2$

17. $\dfrac{\frac{51}{16} - 3}{\frac{3}{8}}$

18. $\dfrac{\frac{54}{17} - \frac{3}{17}}{\frac{1}{2} + \frac{3}{4}}$

19. $\frac{17}{19} + \frac{6}{19} + \frac{15}{19}$

20. $\frac{2}{7} \cdot \frac{3}{4} \cdot \frac{7}{9}$

21. $\left(\frac{1}{2}\right)^3 - \left(\frac{1}{4}\right)^2$

22. $\frac{4}{5} - \left(\frac{1}{2}\right)^2 + \frac{1}{10}$

23. $\left(\frac{3}{4} - \frac{1}{5}\right) \div \frac{3}{2}$

24. $\frac{1}{3} \cdot \frac{5}{6} + \left(\frac{1}{3}\right)^2 \div 2$

25. Arrange the numbers $\frac{7}{8}$, $\frac{3}{4}$, and $\frac{2}{3}$ in order, from smallest to largest. Show how you arrived at this ordering.

26. The result of multiplying two numbers is $\frac{3}{5}$. If one of the numbers is $\frac{9}{10}$, what is the other number?

NAME ______________ SECTION ______________ DATE ______________

ANSWERS

Cumulative Review

1. Round off 2,549,700 to the nearest hundred thousand.

2. Give two illustrations of each of the named properties.
(a) Associative property of addition (b) Identity property for multiplication

In Exercises 3–6, (a) first estimate each answer and (b) then find the answer by performing the indicated operation.

3.
$$\begin{array}{r} 961 \\ 1745 \\ 87 \\ +\ 620 \\ \hline \end{array}$$

4.
$$\begin{array}{r} 224 \\ \times\ 108 \\ \hline \end{array}$$

5.
$$\begin{array}{r} 9048 \\ -\ 8052 \\ \hline \end{array}$$

6. $15\overline{)4750}$

7. Multiply mentally: 50×7000.

8. Without actually dividing, determine which of the numbers 2, 3, 4, 5, 9, and 10 will divide into 8190. Give a brief reason for each decision.

1. ______________

2. [Respond below exercise.]

3. ______________

4. ______________

5. ______________

6. ______________

7. ______________

8. [Respond below exercise.]

9. [Respond below exercise.] ______

10. [Respond below exercise.] ______

11. [Respond below exercise.] ______

12. [Respond below exercise.] ______

13. ______

14. ______

15. ______

9. Determine whether or not the number 431 is prime. Show all the steps you use.

10. Find the prime factorization of 780.

In Exercises 11 and 12, find the LCM of each set of numbers and tell how many times each number divides into the LCM.

11. 18, 42, 90

12. 36, 60, 84, 96

13. Use the rules for order of operations to evaluate the expression

$12 \cdot 9 \div 3^2 + 2(5^2 - 3 \cdot 5)$

Perform the indicated operations and reduce if possible.

14. $\frac{5}{9}\left(\frac{3}{28}\right)\left(\frac{14}{20}\right)$

15. $\frac{3}{20} + \frac{7}{8} + \frac{2}{15}$

NAME ______________________ SECTION ______________ DATE ________

ANSWERS

16. $\frac{27}{34} - \frac{1}{51}$

17. $\left(\frac{3}{4}\right)^2 + \frac{1}{2} \div \frac{8}{3} \cdot 3$

18. Find $\frac{3}{4}$ of $\frac{12}{7}$.

19. Find the average of the numbers 97, 85, 76, and 122.

20. Samantha opened her checking account with a deposit of $5380. She wrote checks for $95, $265, $107, and $1573 and made another deposit of $340. What was the new balance in her account?

21. If the quotient of 119 and 17 is subtracted from the product of 23 and 34, what is the difference?

16. ____________

17. ____________

18. ____________

19. ____________

20. ____________

21. ____________

22. ______________

23. ______________

24. ______________

22. The base of a laser jet printer is in the shape of a rectangle 25 centimeters wide and 32 centimeters long. What area of a desk top does the printer cover?

23. The distance between New York and Berlin is approximately $\frac{2}{3}$ the distance between Mexico City and Berlin. If the distance between Mexico City and Berlin is approximately 6000 miles, about how far apart are New York and Berlin?

24. The area of Switzerland is approximately 16,000 square miles; the area of France is approximately 210,000 square miles; and the area of Germany is approximately 140,000 square miles. (a) About what fraction (reduced) of the area of France is the area of Switzerland? (b) About what fraction (reduced) of the area of Germany is the area of Switzerland? (c) About what fraction (reduced) of the area of France is the area of Germany?

4

Mixed Numbers and Denominate Numbers

What to Expect in Chapter 4

In Chapter 4 we will discuss the meaning of mixed numbers and develop techniques for operating with mixed numbers. All of the topics developed in Chapter 1 (Whole Numbers), Chapter 2 (Prime Numbers), and Chapter 3 (Fractions) are an integral part of Chapter 4 and should be reviewed on a regular basis.

As we will see in Section 4.1, a mixed number indicates the sum of a whole number and a fraction. For example, the mixed number $5\frac{1}{3}$ is shorthand for $5 + \frac{1}{3}$. This idea leads to an understanding of how to change mixed numbers to improper fractions and improper fractions to mixed numbers. Section 4.1 also makes the distinction between reducing an improper fraction and changing an improper fraction to a mixed number.

The importance of the relationship between improper fractions and mixed numbers is emphasized in Section 4.2, where multiplication and division with complex numbers are accomplished by using the numbers in improper fraction form. Sections 4.3 and 4.4 deal with addition and subtraction with mixed numbers. In subtraction, the fraction part of a mixed number may sometimes be larger than 1. Section 4.5 explains how the rules for order of operations can be applied with mixed numbers.

The chapter ends with a section on addition and subtraction with denominate numbers, that is, numbers that are labeled with units, such as 3 hours 15 minutes and 6 feet 4 inches.

Mathematics at Work!

Everyone knows how to program a VCR, right? Not necessarily. Did you know that VCRs can tape at different speeds and that, depending upon what speed you use to tape a program, you can get up to 6 hours of programming on one tape? Generally, VCRs have three speeds for taping television programs: standard speed gives 2 hours of programming on one tape; long-play speed gives 4 hours of programming on one tape; and super-long-play speed gives 6 hours of programming on one tape. The advantage of the faster speed (shorter playing time) is that it provides a much better resolution of the picture.

Suppose you would like to make the most efficient use of a tape but, at the same time, you want your favorite program to be taped at the standard speed for better picture resolution. If you have used a tape for 1 hour at standard speed and 1 hour at long-play speed, how many hours of taping do you have left at the super-long-play speed? (Refer to Exercise 42 in Section 4.3.)

4.1 Introduction to Mixed Numbers

OBJECTIVES

1. *Understand that a mixed number is the sum of a whole number and a fraction.*
2. *Be able to change a mixed number to an improper fraction.*
3. *Be able to change an improper fraction to a mixed number.*
4. *Know the difference between reducing an improper fraction and changing it to a mixed number.*

Changing Mixed Numbers to Fraction Form

A **mixed number** is the sum of a whole number and a fraction with the fraction part less than 1. By convention, we usually write the whole number and the fraction side by side without the plus sign. For example,

$$7 + \frac{3}{4} = 7\frac{3}{4}$$ Read "seven and three-fourths."

$$10 + \frac{1}{2} = 10\frac{1}{2}$$ Read "ten and one-half."

Most people are familiar with mixed numbers and use them daily, as in: "I rode my bicycle $10\frac{1}{2}$ miles today" or, "This recipe calls for $2\frac{1}{4}$ cups of flour." However, as we will see in Section 4.2, this form of mixed numbers is not convenient for multiplication and division with mixed numbers. These operations are more easily accomplished by changing the mixed numbers to improper fractions first.

Note: To change a mixed number to an improper fraction, add the whole number and the fraction. Remember that a whole number can be written in fraction form with denominator 1.

Example 1

Change each mixed number to an improper fraction.

(a) $3\frac{4}{5} = 3 + \frac{4}{5} = \frac{3}{1} \cdot \frac{5}{5} + \frac{4}{5} = \frac{15}{5} + \frac{4}{5} = \frac{19}{5}$

(b) $2\frac{7}{8} = 2 + \frac{7}{8} = \frac{2}{1} \cdot \frac{8}{8} + \frac{7}{8} = \frac{16}{8} + \frac{7}{8} = \frac{23}{8}$

There is a pattern to changing mixed numbers to improper fractions that leads to a familiar shortcut. Since the denominator of the whole number is always 1, the LCD is always the denominator of the fraction part. Therefore, in Example 1(a), the LCD was 5 and we multiplied the whole number 3 by the common denominator 5. Similarly, in Example 1(b), we multiplied the whole number 2 by the common denominator 8. After each multiplication, we added that product to the numerator of the fraction part and used the common denominator. This process is summarized as follows.

Change each mixed number to improper fraction form.

1. $7 + \frac{1}{8}$

2. $11 + \frac{3}{4}$

3. $8\frac{1}{9}$

4. $5\frac{1}{2}$

Shortcut for Changing Mixed Numbers to Fraction Form

1. Multiply the whole number by the denominator of the fraction part.
2. Add the numerator of the fraction part to this product.
3. Write this sum over the denominator of the fraction.

With Example 1(b) as a guide, this shortcut can be diagrammed as follows:

$$2\frac{7}{8} = \frac{2 \cdot 8 + 7}{8} = \frac{16 + 7}{8} = \frac{23}{8}$$

(Diagram labels: $16 + 7$ above; $2 \cdot 8$ below.)

EXAMPLE 2

Change $5\frac{2}{3}$ to an improper fraction.

Solution

Multiply $5 \cdot 3 = 15$ and add 2: $15 + 2 = 17$

Write 17 over the denominator 3: $5\frac{2}{3} = \frac{17}{3}$

EXAMPLE 3

Change $6\frac{9}{10}$ to an improper fraction.

Solution

Multiply $10 \cdot 6 = 60$ and add 9: $60 + 9 = 69$

Write 69 over the denominator 10: $6\frac{9}{10} = \frac{69}{10}$

☑ COMPLETION EXAMPLE 4

Change $11\frac{5}{6}$ to an improper fraction.

Solution

Multiply $11 \cdot 6 =$ ______ and add 5: ______ + ______ = ______.

Write this sum, ______, over the denominator ______. Therefore,

$$11\frac{5}{6} = \frac{71}{6}$$

Now Work Exercises 1–4 in the Margin.

Changing Improper Fractions to Mixed Numbers

To reverse the process (that is, to change an improper fraction to a mixed number), we use the fact that division is one of the meanings of a fraction.

To Change an Improper Fraction to a Mixed Number:

1. Divide the numerator by the denominator to find the whole number part of the mixed number.
2. Write the remainder over the denominator as the fraction part of the mixed number. (Note that this fraction part will always be less than 1.)

EXAMPLE 5

Change $\frac{29}{4}$ to a mixed number.

Solution

Divide 29 by 4:

$$\begin{array}{r} 7 \\ 4\,\overline{)\,29} \\ \underline{28} \\ 1 \end{array} \qquad \frac{29}{4} = 7 + \frac{1}{4} = 7\frac{1}{4}$$

EXAMPLE 6

Change $\frac{59}{3}$ to a mixed number.

Solution

Divide 59 by 3:

$$\begin{array}{r} 19 \\ 3\,\overline{)\,59} \\ \underline{3} \\ 29 \\ \underline{27} \\ 2 \end{array} \qquad \frac{59}{3} = 19 + \frac{2}{3} = 19\frac{2}{3}$$

Now Work Exercises 5–8 in the Margin.

Change each improper fraction to mixed number form with the fraction part reduced.

5. $\frac{18}{7}$

6. $\frac{20}{16}$

7. $\frac{25}{2}$

8. $\frac{35}{15}$

Note: Changing an improper fraction to a mixed number is not the same as reducing an improper fraction. **Reducing** involves finding common factors in the numerator and denominator. **Changing to a mixed number** involves division of the numerator by the denominator. Common factors are not involved. In any case, the fraction part of a mixed number should be in reduced form. To ensure this, we can follow either of the following two procedures:

1. Reduce the improper fraction first, and then change this fraction to a mixed number.
2. Change the improper fraction to a mixed number first, and then reduce the fraction part.

Example 7

A customer at a supermarket deli ordered the following amounts of sliced meats: $\frac{1}{3}$ pound of roast beef, $\frac{3}{4}$ pound of turkey, $\frac{3}{8}$ pound of salami, and $\frac{1}{2}$ pound of boiled ham. What was the total amount of meat purchased? (Express the answer as a mixed number.)

Solution

To find the total amount of meat purchased, we **add** the individual amounts. From Section 3.4, we know that to add fractions, we need a common denominator. The LCD is the least common multiple of the numbers 3, 4, 8, and 2: LCD $= 2 \cdot 2 \cdot 2 \cdot 3 = 24$.

$$\frac{1}{3} + \frac{3}{4} + \frac{3}{8} + \frac{1}{2} = \frac{1}{3} \cdot \frac{8}{8} + \frac{3}{4} \cdot \frac{6}{6} + \frac{3}{8} \cdot \frac{3}{3} + \frac{1}{2} \cdot \frac{12}{12}$$

$$= \frac{8}{24} + \frac{18}{24} + \frac{9}{24} + \frac{12}{24}$$

$$= \frac{47}{24} = 1\frac{23}{24}$$

The total purchase was $1\frac{23}{24}$ pounds of meat.

Completion Example Answer

4. Multiply $11 \cdot 6 = \mathbf{66}$ and add 5: $\mathbf{66} + \mathbf{5} = \mathbf{71}$.

   Write this sum, **71**, over the denominator, **6**.

   Therefore,

   $$11\frac{5}{6} = \frac{\mathbf{71}}{\mathbf{6}}$$

NAME ____________ SECTION ____________ DATE ____________

Exercises 4.1

Write in improper fraction form reduced to lowest terms.

1. $\frac{24}{18}$ **2.** $\frac{25}{10}$ **3.** $\frac{16}{12}$ **4.** $\frac{10}{8}$

5. $\frac{39}{26}$ **6.** $\frac{48}{32}$ **7.** $\frac{35}{25}$ **8.** $\frac{18}{16}$

9. $\frac{80}{64}$ **10.** $\frac{75}{60}$

Change each of the following to a mixed number with the fraction part reduced.

11. $\frac{100}{24}$ **12.** $\frac{25}{10}$ **13.** $\frac{16}{12}$ **14.** $\frac{10}{8}$

15. $\frac{39}{26}$ **16.** $\frac{42}{8}$ **17.** $\frac{43}{7}$ **18.** $\frac{34}{16}$

19. $\frac{45}{6}$ **20.** $\frac{75}{12}$ **21.** $\frac{56}{18}$ **22.** $\frac{31}{15}$

23. $\frac{36}{12}$ **24.** $\frac{48}{16}$ **25.** $\frac{72}{16}$ **26.** $\frac{70}{34}$

ANSWERS

1. ____________
2. ____________
3. ____________
4. ____________
5. ____________
6. ____________
7. ____________
8. ____________
9. ____________
10. ____________
11. ____________
12. ____________
13. ____________
14. ____________
15. ____________
16. ____________
17. ____________
18. ____________
19. ____________
20. ____________
21. ____________
22. ____________
23. ____________
24. ____________
25. ____________
26. ____________

27. ______________
28. ______________
29. ______________
30. ______________
31. ______________
32. ______________
33. ______________
34. ______________
35. ______________
36. ______________
37. ______________
38. ______________
39. ______________
40. ______________
41. ______________
42. ______________
43. ______________
44. ______________
45. ______________
46. ______________
47. ______________
48. ______________
49. ______________
50. ______________
51. ______________

27. $\frac{45}{15}$ **28.** $\frac{60}{36}$ **29.** $\frac{35}{20}$ **30.** $\frac{185}{100}$

Change to improper fraction form and reduce.

31. $4\frac{5}{8}$ **32.** $3\frac{3}{4}$ **33.** $5\frac{1}{15}$ **34.** $1\frac{3}{5}$

35. $4\frac{2}{11}$ **36.** $2\frac{11}{44}$ **37.** $2\frac{9}{27}$ **38.** $4\frac{6}{7}$

39. $10\frac{8}{12}$ **40.** $11\frac{3}{8}$ **41.** $6\frac{8}{10}$ **42.** $14\frac{1}{5}$

43. $16\frac{2}{3}$ **44.** $12\frac{4}{8}$ **45.** $20\frac{3}{15}$ **46.** $9\frac{4}{10}$

47. $13\frac{1}{7}$ **48.** $49\frac{0}{12}$ **49.** $17\frac{0}{3}$ **50.** $3\frac{1}{50}$

In each of the following exercises, write the answer in mixed number form.

51. A tree in Yosemite National Forest grew $\frac{2}{3}$ foot, $\frac{3}{4}$ foot, $\frac{7}{8}$ foot, and $\frac{1}{2}$ foot in four consecutive years. How many feet did the tree grow during these four years?

NAME ______________ SECTION ______________ DATE ______________

ANSWERS

52. ______________

53. ______________

54. ______________

55. [Respond below exercise.]

52. A baby grew $\frac{1}{4}$ inch, $\frac{3}{8}$ inch, $\frac{3}{4}$ inch, and $\frac{9}{16}$ inch over a four-month period. How many inches did the baby grow over this time period?

53. During five days in one week the price of Xerox stock rose $\frac{1}{4}$, rose $\frac{7}{8}$, rose $\frac{3}{4}$, fell $\frac{1}{2}$, and rose $\frac{3}{8}$ of a dollar. What was the net gain (or loss) in the price of Xerox stock over this five-day period?

54. In typing the manuscript for this textbook, Dr. Wright measured the height (or thickness) of the stack of pages for each of the first four chapters as $\frac{7}{8}$ inch, $\frac{1}{2}$ inch, $\frac{3}{4}$ inch, and $\frac{5}{8}$ inch. What was the total height of the first four chapters in typed form?

Writing and Thinking about Mathematics

55. You were probably familiar with the shortcut for changing a mixed number to an improper fraction (page 226) before you read this section. In your own words, explain how this method works and why it gives the correct improper fraction every time.

[Respond below exercise.]

56. ____________

56. Explain, in your own words, why the use of the word **improper** is somewhat misleading when referring to improper fractions. Can you think of another word used in a special sense (other than its normal English usage) that could be misleading to a person not familiar with the new meaning? (**Note:** This could lead to a lively class discussion about how and why some words are given special meanings, other than their normal meanings, in technical fields.)

RECYCLE BIN

1. ____________

2. ____________

3. ____________

4. ____________

5. ____________

6. ____________

7. ____________

8. ____________

The Recycle Bin (from Sections 3.2 and 3.3)

Reduce each fraction to lowest terms.

1. $\frac{35}{45}$ **2.** $\frac{128}{320}$ **3.** $\frac{300}{100}$ **4.** $\frac{102}{221}$

Find each of the following products reduced to lowest terms.

5. $\frac{2}{3} \cdot \frac{15}{16} \cdot \frac{8}{25}$ **6.** $\frac{13}{24} \cdot \frac{15}{39} \cdot \frac{18}{20}$

Find each of the following quotients reduced to lowest terms.

7. $\frac{25}{36} \div \frac{35}{28}$ **8.** $\frac{65}{22} \div \frac{91}{33}$

4.2 Multiplication and Division

OBJECTIVES

1. *Learn how to multiply with mixed numbers.*
2. *Learn how to divide with mixed numbers.*

Multiplication with Mixed Numbers

In Sections 4.3 and 4.4, we will see that addition and subtraction with mixed numbers both rely on the fact that (as we discussed in Section 4.1) a mixed number is the sum of a whole number and a fraction. However, in this section, we will see that multiplication and division are easier to perform if we change the mixed numbers to improper fraction form rather than treat them as sums. Thus, multiplication and division with mixed numbers are the same as multiplication and division with fractions. Just as with fractions, we use prime factorizations and reduce.

To Multiply Mixed Numbers:

1. Change each number to fraction form.
2. Multiply by factoring numerators and denominators; then reduce.
3. Change the answer to a mixed number or leave it in fraction form. (The choice sometimes depends on what use is to be made of the answer.)

EXAMPLE 1

Find the product: $\frac{5}{6} \cdot 3\frac{3}{10}$.

Solution

$$\frac{5}{6} \cdot 3\frac{3}{10} = \frac{5}{6} \cdot \frac{33}{10} = \frac{\cancel{5} \cdot \cancel{3} \cdot 11}{2 \cdot \cancel{3} \cdot 2 \cdot \cancel{5}} = \frac{11}{4} \quad \text{or} \quad 2\frac{3}{4}$$

EXAMPLE 2

Multiply and reduce to lowest terms: $4\frac{1}{2} \cdot 1\frac{1}{6} \cdot 3\frac{1}{3}$.

Solution

$$4\frac{1}{2} \cdot 1\frac{1}{6} \cdot 3\frac{1}{3} = \frac{9}{2} \cdot \frac{7}{6} \cdot \frac{10}{3} = \frac{\cancel{3} \cdot \cancel{3} \cdot 7 \cdot \cancel{2} \cdot 5}{\cancel{2} \cdot 2 \cdot \cancel{3} \cdot \cancel{3}} = \frac{35}{2} \quad \text{or} \quad 17\frac{1}{2}$$

Large mixed numbers can be multiplied in the same way that we multiplied in Examples 1 and 2. The numerators and their products will be relatively large, but the technique of changing to improper fractions and then multiplying and reducing is the same.

Example 3

Linda is framing a rectangular antique circus poster that measures $24\frac{3}{8}$ inches wide by $45\frac{1}{4}$ inches long. What is the area of the glass needed to cover the poster?

Solution

Find the area (measured in square inches) of the rectangle by multiplying the width times the length.

(a) To multiply $24\frac{3}{8} \cdot 45\frac{1}{4}$, first change both numbers to improper fractions.

$$\begin{array}{r} 24 \\ \times\ 8 \\ \hline 192 \end{array} \qquad \begin{array}{r} 192 \\ +\ 3 \\ \hline 195 \end{array} \qquad 24\frac{3}{8} = \frac{195}{8}$$

$$\begin{array}{r} 45 \\ \times\ 4 \\ \hline 180 \end{array} \qquad \begin{array}{r} 180 \\ +\ 1 \\ \hline 181 \end{array} \qquad 45\frac{1}{4} = \frac{181}{4}$$

(b) Now multiply the improper fractions; then change the product back to a mixed number.

$$24\frac{3}{8} \cdot 45\frac{1}{4} = \frac{195}{8} \cdot \frac{181}{4} = \frac{35{,}295}{32} = 1102\frac{31}{32}$$

$$\begin{array}{r} 195 \\ \times\ 181 \\ \hline 195 \\ 15\ 60 \\ 19\ 5 \\ \hline 35{,}295 \end{array} \qquad \begin{array}{r} 1102\frac{31}{32} \\ 32\overline{)35{,}295} \\ \underline{32} \\ 3\ 2 \\ \underline{3\ 2} \\ 09 \\ \underline{0} \\ 95 \\ \underline{64} \\ 31 \end{array}$$

Thus, the area of the glass is $1102\frac{31}{32}$ square inches.

☑ ***Completion Example 4***

Find the product and write it as a mixed number with the fraction part in lowest terms: $\left(4\frac{3}{8}\right)\left(3\frac{1}{5}\right)\left(\frac{1}{34}\right)$.

Solution

$$\left(4\frac{3}{8}\right)\left(3\frac{1}{5}\right)\left(\frac{1}{34}\right) = \frac{35}{8} \cdot \underline{\quad\quad} \cdot \underline{\quad\quad}$$

$$= \frac{5 \cdot 7 \cdot \underline{\quad}}{2 \cdot 2 \cdot 2 \cdot \underline{\quad}} = \underline{\quad\quad}$$

☑ COMPLETION EXAMPLE 5

Multiply: $20\frac{3}{5} \cdot 18\frac{1}{4}$.

Solution

$$20\frac{3}{5} \cdot 18\frac{1}{4} = \frac{103}{5} \cdot \underline{\quad\quad}$$

$$= \underline{\quad\quad\quad} = \underline{\quad\quad\quad}$$

Now Work Exercises 1–3 in the Margin.

Find each product and write the product in mixed number form.

1. $\left(2\frac{1}{2}\right)\left(5\frac{3}{3}\right) =$

2. $4\frac{1}{5} \cdot 2\frac{3}{4} \cdot \frac{10}{33} =$

3. $10\frac{1}{12} \cdot 18\frac{3}{4} =$

In Section 3.1 we discussed the fact that finding a fractional part of another number requires multiplication. The key word is **of,** and we will find in later chapters that this is also true for decimals and percents. We emphasize this concept again here with mixed numbers.

NOTE: To find a fraction **of** a number means to **multiply** the number by the fraction.

EXAMPLE 6

Find $\frac{3}{4}$ of 40.

Solution

$$\frac{3}{4} \cdot 40 = \frac{3}{\cancel{4}_{1}} \cdot \frac{\cancel{40}^{10}}{1} = 30$$

EXAMPLE 7

Find $\frac{2}{3}$ of $5\frac{1}{4}$.

Solution

$$\frac{2}{3} \cdot 5\frac{1}{4} = \frac{2}{3} \cdot \frac{21}{4} = \frac{\cancel{2} \cdot \cancel{3} \cdot 7}{\cancel{3} \cdot \cancel{2} \cdot 2} = \frac{7}{2} \text{ or } 3\frac{1}{2}$$

4. Find $\frac{3}{10}$ of 60.

5. Find $\frac{1}{2}$ of $3\frac{1}{5}$.

Now Work Exercises 4 and 5 in the Margin.

Division with Mixed Numbers

Division with mixed numbers is the same as division with fractions, as discussed in Section 3.3. Simply change each mixed number to an improper fraction before dividing. Recall that, **to divide by any nonzero number, we multiply by its reciprocal.** That is, for $c \neq 0$ and $d \neq 0$, the **reciprocal** of $\frac{c}{d}$ is $\frac{d}{c}$ and

$$\frac{a}{b} \div \frac{c}{d} = \frac{a}{b} \cdot \frac{d}{c}$$

To Divide with Mixed Numbers:

1. Change each number to fraction form.
2. Write the reciprocal of the divisor.
3. Multiply by factoring numerators and denominators; then reduce.

Example 8

Divide: $3\frac{1}{4} \div 7\frac{4}{5}$.

Solution

$$3\frac{1}{4} \div 7\frac{4}{5} = \frac{13}{4} \div \frac{39}{5} = \frac{13}{4} \cdot \frac{5}{39}$$

Note that the divisor is $\frac{39}{5}$, and we multiply by its reciprocal, $\frac{5}{39}$.

$$= \frac{\cancel{13} \cdot 5}{4 \cdot \cancel{13} \cdot 3} = \frac{5}{12}$$

Example 9

Find the quotient: $7\frac{7}{8} \div 6$.

Solution

$$7\frac{7}{8} \div 6 = \frac{63}{8} \div \frac{6}{1} = \frac{63}{8} \cdot \frac{1}{6}$$

$$= \frac{\cancel{3} \cdot 3 \cdot 7}{8 \cdot 2 \cdot \cancel{3}} = \frac{21}{16} \quad \text{or} \quad 1\frac{5}{16}$$

COMPLETION EXAMPLE 10

Divide: $10\frac{1}{2} \div 3\frac{3}{4}$.

Solution

$$10\frac{1}{2} \div 3\frac{3}{4} = \frac{21}{2} \div \_\_\_\_ = \frac{21}{2} \cdot \_\_\_\_$$

$$= \frac{3 \cdot 7 \cdot \_\_}{2 \cdot \_\_} = \_\_\_\_ \quad \text{or} \quad \_\_\_\_$$

Now Work Exercises 6–8 in the Margin.

Find each quotient and write the quotient in the form of a mixed number.

6. $3\frac{3}{4} \div 2\frac{2}{5} =$

7. $4\frac{2}{3} \div 4\frac{2}{3} =$

8. $16 \div 2\frac{2}{3} =$

If we know two numbers, say 2 and 7, we can find their product, 14, by multiplying the two numbers. But, if we know that the product of two numbers is 24 and one of the numbers is 3, how do we find the other number? Recall from our work with whole numbers in Chapter 1 that the answer is found by dividing the product by the known number. Thus, the missing number is $24 \div 3 = 8$. Example 11 illustrates this same idea (finding a number by dividing the product of two numbers by the known number) with mixed numbers.

EXAMPLE 11

The product of $2\frac{1}{3}$ with another number is $5\frac{1}{6}$. What is the other number?

Solution

$$2\frac{1}{3} \cdot ? = 5\frac{1}{6}$$

To find the missing number, *divide* $5\frac{1}{6}$ by $2\frac{1}{3}$.

$$5\frac{1}{6} \div 2\frac{1}{3} = \frac{31}{6} \div \frac{7}{3} = \frac{31}{\underset{2}{\cancel{6}}} \cdot \frac{\overset{1}{\cancel{3}}}{7} = \frac{31}{14} \quad \text{or} \quad 2\frac{3}{14}$$

The other number is $2\frac{3}{14}$.

CHECK: $2\frac{1}{3} \cdot 2\frac{3}{14} = \frac{\cancel{7}}{3} \cdot \frac{31}{\underset{2}{\cancel{14}}} = \frac{31}{6} = 5\frac{1}{6}$

Completion Example Answers

4. $\left(4\frac{3}{8}\right)\left(3\frac{1}{5}\right)\left(\frac{1}{34}\right) = \frac{35}{8} \cdot \mathbf{\frac{16}{5}} \cdot \mathbf{\frac{1}{34}}$

$$= \frac{5 \cdot 7 \cdot \mathbf{2} \cdot \mathbf{2} \cdot \mathbf{2} \cdot \mathbf{2} \cdot \mathbf{1}}{2 \cdot 2 \cdot 2 \cdot \mathbf{5} \cdot \mathbf{2} \cdot \mathbf{17}} = \mathbf{\frac{7}{17}}$$

5. $20\frac{3}{5} \cdot 18\frac{1}{4} = \frac{103}{5} \cdot \mathbf{\frac{73}{4}}$

$$= \mathbf{\frac{7519}{20}} = \mathbf{375\frac{19}{20}}$$

10. $10\frac{1}{2} \div 3\frac{3}{4} = \frac{21}{2} \div \mathbf{\frac{15}{4}} = \frac{21}{2} \cdot \mathbf{\frac{4}{15}}$

$$= \frac{3 \cdot 7 \cdot \mathbf{2} \cdot \mathbf{2}}{2 \cdot \mathbf{3} \cdot \mathbf{5}} = \mathbf{\frac{14}{5}} \quad \text{or} \quad \mathbf{2\frac{4}{5}}$$

NAME ____________ SECTION ____________ DATE ____________

Exercises 4.2

Find the indicated products.

1. $\left(2\frac{1}{3}\right)\left(3\frac{1}{4}\right)$

2. $\left(1\frac{1}{5}\right)\left(1\frac{1}{7}\right)$

3. $4\frac{1}{2}\left(2\frac{1}{3}\right)$

4. $3\frac{1}{3}\left(2\frac{1}{5}\right)$

5. $6\frac{1}{4}\left(3\frac{3}{5}\right)$

6. $5\frac{1}{3}\left(2\frac{1}{4}\right)$

7. $\left(8\frac{1}{2}\right)\left(3\frac{2}{3}\right)$

8. $\left(9\frac{1}{3}\right)2\frac{1}{7}$

9. $\left(6\frac{2}{7}\right)1\frac{3}{11}$

10. $\left(11\frac{1}{4}\right)1\frac{1}{15}$

11. $6\frac{2}{3} \cdot 4\frac{1}{2}$

12. $4\frac{3}{8} \cdot 2\frac{2}{7}$

13. $9\frac{3}{4} \cdot 2\frac{6}{26}$

14. $7\frac{1}{2} \cdot \frac{2}{15}$

15. $\frac{3}{4} \cdot 1\frac{1}{3}$

16. $3\frac{4}{5} \cdot 2\frac{1}{7}$

17. $12\frac{1}{2} \cdot 2\frac{1}{5}$

18. $9\frac{3}{5} \cdot 1\frac{1}{16}$

ANSWERS

1. ____________
2. ____________
3. ____________
4. ____________
5. ____________
6. ____________
7. ____________
8. ____________
9. ____________
10. ____________
11. ____________
12. ____________
13. ____________
14. ____________
15. ____________
16. ____________
17. ____________
18. ____________

19. ______________
20. ______________
21. ______________
22. ______________
23. ______________
24. ______________
25. ______________
26. ______________
27. ______________
28. ______________
29. ______________
30. ______________
31. ______________
32. ______________
33. ______________
34. ______________
35. ______________
36. ______________

19. $6\frac{1}{8} \cdot 3\frac{1}{7}$

20. $5\frac{1}{4} \cdot 11\frac{1}{3}$

21. $\frac{1}{4} \cdot \frac{2}{3} \cdot \frac{6}{7}$

22. $\frac{7}{8} \cdot \frac{24}{25} \cdot \frac{5}{21}$

23. $\frac{3}{16} \cdot \frac{8}{9} \cdot \frac{3}{5}$

24. $\frac{2}{5} \cdot \frac{1}{5} \cdot \frac{4}{7}$

25. $2\frac{1}{4} \cdot 6\frac{3}{8} \cdot 1\frac{5}{27}$

26. $1\frac{3}{32} \cdot 1\frac{1}{7} \cdot 1\frac{1}{25}$

27. $1\frac{5}{16} \cdot 1\frac{1}{3} \cdot 1\frac{1}{5}$

28. $24\frac{1}{5} \cdot 35\frac{1}{6}$

29. $72\frac{3}{5} \cdot 25\frac{1}{6}$

30. $42\frac{5}{6} \cdot 30\frac{1}{7}$

31. $75\frac{1}{3} \cdot 40\frac{1}{25}$

32. $36\frac{3}{4} \cdot 17\frac{5}{12}$

33. Find $\frac{2}{3}$ of 60

34. Find $\frac{1}{4}$ of 80.

35. Find $\frac{1}{5}$ of 100

36. Find $\frac{3}{5}$ of 100.

NAME ________________ SECTION ________________ DATE ________

ANSWERS

37. Find $\frac{1}{2}$ of $2\frac{5}{8}$

38. Find $\frac{1}{6}$ of $1\frac{3}{4}$.

39. Find $\frac{9}{10}$ of $3\frac{5}{7}$

40. Find $\frac{7}{8}$ of $6\frac{4}{5}$.

Find the indicated quotients.

41. $\frac{2}{21} \div \frac{2}{7}$

42. $\frac{9}{32} \div \frac{5}{8}$

43. $\frac{5}{12} \div \frac{3}{4}$

44. $\frac{6}{17} \div \frac{6}{17}$

45. $\frac{5}{6} \div 3\frac{1}{4}$

46. $\frac{7}{8} \div 7\frac{1}{2}$

47. $\frac{29}{50} \div 3\frac{1}{10}$

48. $4\frac{1}{5} \div 3\frac{1}{3}$

49. $2\frac{1}{17} \div 1\frac{1}{4}$

50. $5\frac{1}{6} \div 3\frac{1}{4}$

51. $2\frac{2}{49} \div 3\frac{1}{14}$

52. $6\frac{5}{6} \div 2$

37. ________

38. ________

39. ________

40. ________

41. ________

42. ________

43. ________

44. ________

45. ________

46. ________

47. ________

48. ________

49. ________

50. ________

51. ________

52. ________

53. ______________

54. ______________

55. ______________

56. ______________

57. ______________

58. ______________

59. ______________

60. ______________

61. ______________

62. ______________

63. ______________

53. $4\frac{1}{5} \div 3$

54. $4\frac{5}{8} \div 4$

55. $6\frac{5}{6} \div \frac{1}{2}$

56. $4\frac{5}{8} \div \frac{1}{4}$

57. $4\frac{1}{5} \div \frac{1}{3}$

58. $1\frac{1}{32} \div 3\frac{2}{3}$

59. $7\frac{5}{11} \div 4\frac{1}{10}$

60. $13\frac{1}{7} \div 4\frac{2}{11}$

61. A man drives $17\frac{7}{10}$ miles one way to work, five days a week. How many miles does he drive each week going to and from work?

62. A length of pipe is $27\frac{3}{4}$ feet. What would be the total length if $36\frac{1}{2}$ of these pipe sections were laid end-to-end?

63. A woman reads $\frac{1}{6}$ of a book in 3 hours. If the book contains 540 pages, how many pages does she read in 3 hours? How long will she take to read the entire book?

NAME ______________________ SECTION ______________ DATE __________

64. Three towns (A, B, and C) are located on the same highway (assume a straight section of highway). Towns A and B are 53 kilometers apart. Town C is $45\frac{9}{10}$ kilometers from town B. How far apart are towns A and C? (The sketch shows that there are two possible situations to consider.)

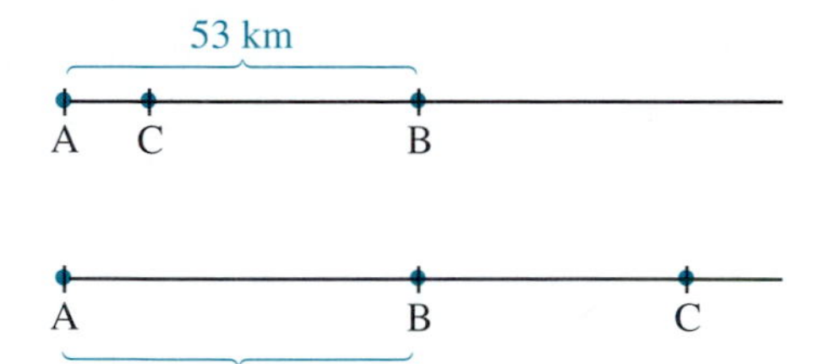

65. A telephone pole is 32 feet long. If $\frac{5}{16}$ of the pole must be underground and $\frac{11}{16}$ of the pole aboveground, how much of the pole is underground? How much is aboveground?

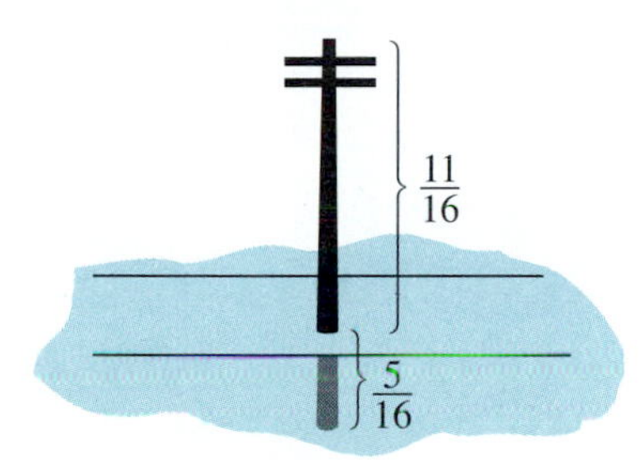

66. The total distance around a square (its perimeter) is found by multiplying the length of one side by 4. Find the perimeter of a square if the length of one side is $5\frac{1}{16}$ inches.

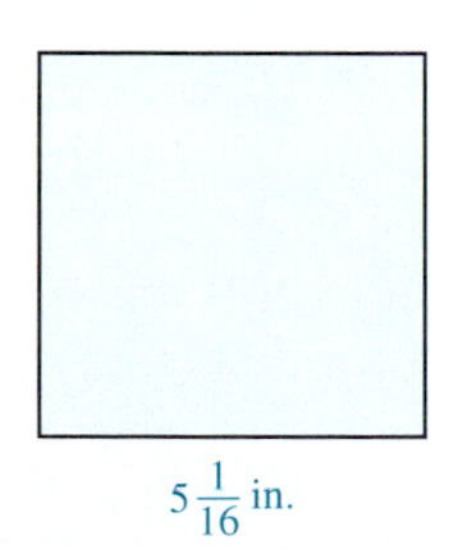

67. The product of $\frac{9}{10}$ with another number is $1\frac{2}{3}$. What is the other number?

ANSWERS

64. [Respond below exercise.]

65. ______________

66. ______________

67. ______________

68. ______________

69. ______________

70. ______________

71. ______________

72. ______________

68. The result of multiplying two numbers is $10\frac{1}{3}$. If one of the numbers if $7\frac{1}{6}$, what is the other number?

69. An airplane is carrying 150 passengers. This is $\frac{6}{7}$ of its capacity. What is the capacity of the airplane?

70. The sale price of a coat is $135. This is $\frac{3}{4}$ of the original price. What was the original price (price before the sale)?

71. The sale price of a new computer system (including a printer) is $2400. This is $\frac{3}{4}$ of the original price. What was the original price (price before the sale)?

72. A used car is advertised at a special price of $3500. If this sale price is $\frac{4}{5}$ of the original price, what was the original price (price before the sale)?

NAME ______________ SECTION ______________ DATE ______________

ANSWERS

73a. ______________

b. ______________

74. ______________

75. ______________

76. ______________

73. Your car averages $26\frac{3}{10}$ miles per gallon of gas. (a) How many miles can your car travel on $17\frac{1}{2}$ gallons of gas? (b) If each gallon costs $132\frac{9}{10}$ cents, what would you pay (in cents) for $17\frac{1}{2}$ gallons?

74. An equilateral triangle is one in which all three sides are the same length. What is the perimeter of an equilateral triangle with sides of length $15\frac{7}{10}$ centimeters?

75. On a road map, 1 inch represents 50 miles. How many miles are represented by $3\frac{1}{4}$ inches?

76. One tablespoon of butter contains 36 milligrams of cholesterol. How many milligrams of cholesterol are there in $2\frac{3}{4}$ tablespoons of butter?

77. [Respond below exercise.] ____________

Check Your Number Sense

77. Suppose the product of $5\frac{7}{10}$ and some other number is $10\frac{1}{2}$. Answer the following questions without doing any calculations:

(a) Do you think that this other number is more than 1 or less than 1? Why?

(b) Do you think that this other number is more than 2 or less than 2? Why? Find the other number.

RECYCLE BIN

1. ____________
2. ____________
3. ____________
4. ____________
5. ____________
6. ____________
7. ____________
8. ____________

The Recycle Bin (from Section 3.4)

Find each of the following sums and reduce to lowest terms.

1. $\frac{2}{3} + \frac{1}{7}$

2. $\frac{5}{8} + \frac{9}{10}$

3. $\frac{3}{5} + \frac{7}{20} + \frac{1}{12}$

4. $\frac{11}{18} + \frac{2}{5} + \frac{3}{10}$

Find each of the following differences and reduce to lowest terms.

5. $\frac{7}{8} - \frac{9}{20}$

6. $\frac{19}{21} - \frac{3}{28}$

7. $\frac{17}{32} - \frac{17}{64}$

8. $\frac{20}{39} - \frac{9}{26}$

4.3 Addition

OBJECTIVES

1. *Know how to add mixed numbers when the fraction parts have the same denominator.*
2. *Know how to add mixed numbers when the fraction parts have different denominators.*
3. *Be able to write the sum of mixed numbers in the form of a mixed number with the fraction part less than 1.*

Addition with Mixed Numbers

From Section 4.1, we know that a mixed number is the sum of a whole number and a fraction. This means that the sum of mixed numbers is the sum of whole numbers and fractions. Keeping this relationship and the commutative and associative properties of addition in mind, we add mixed numbers by treating the whole numbers and the fraction parts separately.

To Add Mixed Numbers:

1. Add the fraction parts.
2. Add the whole numbers.
3. Write the sum as a mixed number so that the fraction part is less than 1. (If the sum of the fraction parts is more than 1, rewrite it as a mixed number and add it to the sum of the whole numbers.)

Example 1

Find the sum: $4\frac{2}{7} + 6\frac{3}{7}$.

Solution

We can write each number as a sum and then use the commutative and associative properties of addition to treat the whole numbers and fraction parts separately.

$$\begin{aligned} 4\frac{2}{7} + 6\frac{3}{7} &= 4 + \frac{2}{7} + 6 + \frac{3}{7} \\ &= (4 + 6) + \left(\frac{2}{7} + \frac{3}{7}\right) \\ &= 10 + \frac{5}{7} \\ &= 10\frac{5}{7} \end{aligned}$$

Find each sum and write the sum in the form of a mixed number.

1. $2\frac{1}{2} + 3\frac{1}{3} =$

2. $14\frac{7}{10} + 22\frac{1}{5} =$

3. $4\frac{5}{8} + 10\frac{7}{8} =$

4. $\begin{array}{r} 8 \\ +\ 4\frac{3}{10} \\ \hline \end{array}$

EXAMPLE 2

Add: $25\frac{1}{6} + 3\frac{7}{18}$.

Solution

$$\begin{aligned} 25\frac{1}{6} + 3\frac{7}{18} &= 25 + \frac{1}{6} + 3 + \frac{7}{18} \\ &= (25 + 3) + \left(\frac{1}{6} + \frac{7}{18}\right) \\ &= 28 + \left(\frac{1}{6} \cdot \frac{3}{3} + \frac{7}{18}\right) \\ &= 28 + \left(\frac{3}{18} + \frac{7}{18}\right) \\ &= 28\frac{10}{18} = 28\frac{5}{9} \end{aligned}$$

Or, vertically,

$$\begin{array}{rcrcr} 25\frac{1}{6} & = & 25\frac{1}{6} \cdot \frac{3}{3} & = & 25\frac{3}{18} \\ +\ 3\frac{7}{18} & = & 3\frac{7}{18} & = & 3\frac{7}{18} \\ \hline & & & & 28\frac{10}{18} = 28\frac{5}{9} \end{array}$$

LCD = 18

EXAMPLE 3

Add the mixed numbers: $7\frac{2}{3} + 9\frac{4}{5}$.

Solution

$$\begin{array}{rcrcr} 7\frac{2}{3} & = & 7\frac{2}{3} \cdot \frac{5}{5} & = & 7\frac{10}{15} \\ +\ 9\frac{4}{5} & = & 9\frac{4}{5} \cdot \frac{3}{3} & = & 9\frac{12}{15} \\ \hline & & & & 16\frac{22}{15} = 16 + 1\frac{7}{15} = 17\frac{7}{15} \end{array}$$

LCD = 15

↑ Fraction part is greater than 1. ↑ Change it to a mixed number.

Now Work Exercises 1–4 in the Margin.

EXAMPLE 4

A triangle has sides measuring $25\frac{1}{2}$ meters, $32\frac{3}{10}$ meters, and $41\frac{7}{10}$ meters. Find the perimeter of (total distance around) the triangle.

Solution

We find the perimeter by adding the lengths of the three sides.

$$25\frac{1}{2} = 25\frac{5}{10}$$

$$32\frac{3}{10} = 32\frac{3}{10}$$

$$41\frac{7}{10} = 41\frac{7}{10}$$

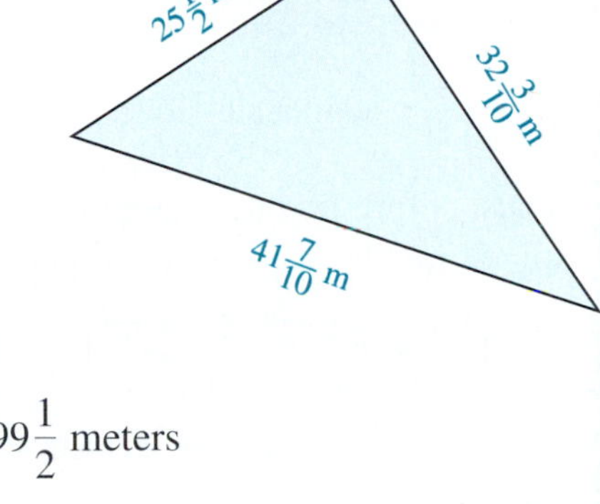

$$98\frac{15}{10} = 98 + 1\frac{5}{10} = 99\frac{1}{2} \text{ meters}$$

The perimeter is $99\frac{1}{2}$ meters.

Estimating with Mixed Numbers

Sometimes we make major errors when operating with mixed numbers because we are so intent on remembering some rule or technique for operating with fractions. The fact is that the fraction parts of mixed numbers have a much lesser effect on the answer than the whole number parts have. Thus, to avoid some errors when operating with mixed numbers, we can estimate the answer simply by using the whole number parts and ignoring the fraction parts. In such cases, we say that the mixed numbers have been **truncated.**

To Estimate an Answer with Mixed Numbers:

1. Perform the related operations with truncated mixed numbers.
2. If one of the numbers is a fraction only, replace this fraction with the whole number 1. (This replacement can be used for estimating as long as no denominator or divisor is 0.)

(**Note:** This technique is illustrated with addition in Example 5. However, it applies to *all* operations: addition, subtraction, multiplication, and division.)

EXAMPLE 5

Estimate the value of the following expression by using truncated mixed numbers: $14\frac{3}{5} + 6\frac{1}{3} + 2\frac{1}{4}$.

Solution

$14\frac{3}{5}$ truncates to 14.

$6\frac{1}{3}$ truncates to 6.

$2\frac{1}{4}$ truncates to 2.

Thus, before finding a common denominator for the fraction parts in the mixed numbers, we can estimate the sum of

$$14\frac{3}{5} + 6\frac{1}{3} + 2\frac{1}{4} \quad \text{as} \quad 14 + 6 + 2 = 22$$

Therefore, we would expect the sum of the mixed numbers to be approximately 22.

NAME ____________ SECTION ____________ DATE ____________

ANSWERS

1a. ______ b. ______
2a. ______ b. ______
3a. ______ b. ______
4a. ______ b. ______
5a. ______ b. ______
6a. ______ b. ______
7a. ______ b. ______
8a. ______ b. ______
9a. ______ b. ______
10a. ______ b. ______
11a. ______ b. ______
12a. ______ b. ______

Exercises 4.3

First (a) estimate each sum by using truncated mixed numbers; then (b) find each sum.

1. $10 + 4\frac{5}{7}$

2. $6\frac{1}{2} + 3\frac{1}{2}$

3. $5\frac{1}{3} + 2\frac{2}{3}$

4. $8 + 7\frac{3}{11}$

5. $12 + \frac{3}{4}$

6. $15 + \frac{9}{10}$

7. $3\frac{1}{4} + 5\frac{5}{12}$

8. $5\frac{11}{12} + 3\frac{5}{6}$

9. $21\frac{1}{3} + 13\frac{1}{18}$

10. $9\frac{2}{3} + \frac{5}{6}$

11. $4\frac{1}{2} + 3\frac{1}{6}$

12. $3\frac{1}{4} + 7\frac{1}{8}$

13a. ____________
b. ____________
14a. ____________
b. ____________
15a. ____________
b. ____________
16a. ____________
b. ____________
17a. ____________
b. ____________
18a. ____________
b. ____________
19a. ____________
b. ____________
20a. ____________
b. ____________
21a. ____________
b. ____________
22a. ____________
b. ____________
23a. ____________
b. ____________
24a. ____________
b. ____________
25a. ____________
b. ____________
26a. ____________
b. ____________
27a. ____________
b. ____________
28a. ____________
b. ____________
29a. ____________
b. ____________
30a. ____________
b. ____________

13. $25\frac{1}{10} + 17\frac{1}{4}$

14. $5\frac{1}{7} + 3\frac{1}{3}$

15. $6\frac{5}{12} + 4\frac{1}{3}$

16. $5\frac{3}{10} + 2\frac{1}{14}$

17. $8\frac{2}{9} + 4\frac{1}{27}$

18. $11\frac{3}{4} + 2\frac{5}{16}$

19. $6\frac{4}{9} + 12\frac{1}{15}$

20. $4\frac{1}{6} + 13\frac{9}{10}$

21. $21\frac{3}{4} + 6\frac{3}{4}$

22. $3\frac{5}{8} + 3\frac{5}{8}$

23. $7\frac{3}{5} + 2\frac{1}{8}$

24. $9\frac{1}{8} + 3\frac{7}{12}$

25. $3\frac{1}{3} + 4\frac{1}{4} + 5\frac{1}{5}$

26. $\frac{3}{7} + 2\frac{1}{14} + 2\frac{1}{6}$

27. $20\frac{5}{8} + 42\frac{5}{6}$

28. $25\frac{2}{3} + 1\frac{1}{16}$

29. $32\frac{1}{64} + 4\frac{1}{24} + 17\frac{3}{8}$

30. $3\frac{1}{20} + 7\frac{1}{15} + 2\frac{3}{10}$

NAME ____________ SECTION ____________ DATE ____________

31. $\begin{array}{r} 4\frac{1}{3} \\ 8\frac{3}{8} \\ +\ 6\frac{1}{6} \\ \hline \end{array}$

32. $\begin{array}{r} 3\frac{2}{3} \\ 14\frac{1}{10} \\ +\ 5\frac{1}{5} \\ \hline \end{array}$

33. $\begin{array}{r} 13\frac{5}{8} \\ 13\frac{1}{12} \\ +\ 10\frac{1}{4} \\ \hline \end{array}$

34. $\begin{array}{r} 5\frac{1}{8} \\ 1\frac{1}{5} \\ +\ 3\frac{1}{40} \\ \hline \end{array}$

35. $\begin{array}{r} 27\frac{2}{3} \\ 30\frac{5}{8} \\ +\ 31\frac{5}{6} \\ \hline \end{array}$

36. A bus trip is made in three parts. The first part takes $2\frac{1}{3}$ hours, the second takes $2\frac{1}{2}$ hours, and the part takes $3\frac{3}{4}$ hours. How long does the entire trip take?

37. A construction company was contracted to build three sections of highway. One section was $20\frac{7}{10}$ kilometers, the second section was $3\frac{4}{10}$ kilometers, and the third section was $11\frac{6}{10}$ kilometers. What was the total length of highway built?

38. A triangle (three-sided figure) has sides that measure $42\frac{3}{4}$ feet, $23\frac{1}{2}$ feet, and $22\frac{7}{8}$ feet. What is the perimeter of (total distance around) the triangle?

ANSWERS

31a. ____________

b. ____________

32a. ____________

b. ____________

33a. ____________

b. ____________

34a. ____________

b. ____________

35a. ____________

b. ____________

36. ____________

37. ____________

38. ____________

39. ______________

40. ______________

41. ______________

42. ______________

43a. ______________

b. ______________

c. ______________

44. [Respond below exercise.] ______________

39. A quadrilateral (four-sided figure) has sides that measure $3\frac{1}{2}$ inches, $2\frac{1}{4}$ inches, $3\frac{5}{8}$ inches, and $2\frac{3}{4}$ inches. What is the total distance around the quadrilateral?

40. The United States Postal Service sells bulk-rate postage stamps with denominations of $7\frac{3}{5}$¢, $13\frac{1}{5}$¢, $16\frac{7}{10}$¢, $17\frac{1}{2}$¢, and $24\frac{1}{10}$¢. If a company bought 1000 of each of these stamps, what would be the total amount paid?

41. Among the top 50 movie rentals in 1990, "The Hunt for Red October" earned $\$58\frac{1}{2}$ million, "Presumed Innocent" earned $\$43\frac{4}{5}$ million, "Robocop 2" earned $\$22\frac{3}{10}$ million, and "Goodfellas" earned $\$18\frac{1}{5}$ million. What was the total of the rental earnings of these four films in 1990?

42. A video cassette tape has a running time of 2 hours at standard speed, 4 hours at long-play speed, and 6 hours at super-long-play speed. If $\frac{1}{2}$ of the tape is recorded at standard speed, $\frac{1}{4}$ at long-play speed, and $\frac{1}{4}$ at super-long-play speed, what is the total playing time of the entire recording?

Check Your Number Sense

43. Using the idea of truncated mixed numbers, estimate each of the following products mentally.

(a) $2\frac{1}{7} \cdot 3\frac{5}{8}$ (b) $20\frac{1}{7} \cdot 30\frac{5}{8}$ (c) $200\frac{1}{7} \cdot 300\frac{5}{8}$

44. Of your three estimated answers in Exercise 43, which one do you think is closest to the actual product? Why?

4.4 Subtraction

Subtraction with Mixed Numbers

Subtraction with mixed numbers involves working with the whole numbers and the fraction parts separately in a manner similar to addition. With subtraction, though, we have another concern related to the sizes of the fraction parts. We will see that when the fraction part being subtracted is larger than the other fraction part, the smaller fraction part must be changed to an improper fraction by **borrowing** 1 from its whole number.

To Subtract Mixed Numbers:

1. Subtract the fraction parts.
2. Subtract the whole numbers.

EXAMPLE 1

Find the difference: $4\frac{3}{5} - 1\frac{2}{5}$.

Solution $4\frac{3}{5} - 1\frac{2}{5} = (4 - 1) + \left(\frac{3}{5} - \frac{2}{5}\right)$

$$= 3 + \frac{1}{5}$$

$$= 3\frac{1}{5}$$

Or, vertically,

$$\begin{array}{r} 4\frac{3}{5} \\ -1\frac{2}{5} \\ \hline 3\frac{1}{5} \end{array}$$

EXAMPLE 2

Subtract: $10\frac{3}{5} - 6\frac{3}{20}$.

Solution

$$\begin{array}{rcr} 10\frac{3}{5} & = & 10\frac{12}{20} \\ -\ 6\frac{3}{20} & = & 6\frac{3}{20} \\ \hline & & 4\frac{9}{20} \end{array} \quad \text{LCD} = 20$$

Now Work Exercises 1–3 in the Margin.

OBJECTIVES

1. *Know how to subtract mixed numbers when the fraction part of the number being subtracted is smaller than or equal to the fraction part of the other number.*
2. *Know how to subtract mixed numbers when the fraction part of the number being subtracted is larger than the fraction part of the other number.*

Find each difference.

1. $9\frac{7}{10} - 3\frac{1}{5} =$

2. $\begin{array}{r} 15\frac{2}{3} \\ -\ 7\frac{1}{10} \\ \hline \end{array}$

3. $6\frac{3}{4} - 2\frac{3}{14} =$

When the fraction part of the number being subtracted is larger than the fraction part of the first number, we **borrow** 1 from the whole-number part and rewrite the first number as a whole number plus an improper fraction. Then the subtraction of the fraction parts can proceed as before.

If the Fraction Part Being Subtracted Is Larger than the First Fraction:

1. **Borrow** the whole number 1 from the first whole number.
2. Add this 1 to the first fraction. (This will always result in an improper fraction that is larger than the fraction being subtracted.)
3. Then subtract.

EXAMPLE 3

Find the difference: $7\frac{2}{5} - 4\frac{3}{5}$.

Solution

Since $\frac{3}{5}$ is larger than $\frac{2}{5}$, **borrow** the whole number 1 from 7. We write 7 as 6 + 1; then we write $1 + \frac{2}{5}$ as $\frac{7}{5}$.

$$\begin{aligned} 7\frac{2}{5} &= 6 + 1 + \frac{2}{5} = 6 + 1\frac{2}{5} = 6\frac{7}{5} \\ -4\frac{3}{5} &= 4 + \phantom{1 +{}} \frac{3}{5} = 4 + \frac{3}{5} = 4\frac{3}{5} \\ & \qquad\qquad\qquad\qquad\qquad\quad\; 2\frac{4}{5} \end{aligned}$$

EXAMPLE 4

Subtract $19\frac{2}{3} - 5\frac{3}{4}$.

Solution

First change the fraction parts so that they have the same denominator, 12. Then **borrow** the whole number 1 from 19.

$$\begin{aligned} 19\frac{2}{3} &= 19\frac{8}{12} = 18\frac{20}{12} \qquad 1 + \frac{8}{12} = \frac{12}{12} + \frac{8}{12} = \frac{20}{12} \\ -\;5\frac{3}{4} &= 5\frac{9}{12} = 5\frac{9}{12} \\ & \qquad\qquad\; 13\frac{11}{12} \end{aligned}$$

EXAMPLE 5

Find the difference: $7 - 4\frac{2}{3}$.

Solution

In this case, the whole number 7 has no fraction part. We still **borrow** the whole number 1. We write 1 as $\frac{3}{3}$ so that its denominator will be the same as that of the other fraction part, $\frac{2}{3}$.

$$\begin{array}{rcl} 7 & = & 6\frac{3}{3} \\ -4\frac{2}{3} & = & 4\frac{2}{3} \\ \hline & & 2\frac{1}{3} \end{array}$$

Here, $7 = 6 + 1 = 6 + \frac{3}{3} = 6\frac{3}{3}$.

COMPLETION EXAMPLE 6

Subtract: $10 - 6\frac{5}{8}$.

Solution

The whole number 10 has no fraction part. We still need to **borrow** the whole number 1 from 10.

$$\begin{array}{rcl} 10 & = & 9\,\underline{\quad} \\ -\;6\frac{5}{8} & = & 6\,\frac{5}{8} \\ \hline & & \underline{\quad\quad} \end{array}$$

COMPLETION EXAMPLE 7

Find the difference as indicated.

$$\begin{array}{rcl} 14\frac{3}{4} & = 14\,\underline{\quad} & = 13\,\underline{\quad} \\ -\;5\frac{9}{10} & = \;\;5\,\underline{\quad} & = \;\;5\,\underline{\quad} \\ \hline & & \underline{\quad\quad} \end{array}$$

Now Work Exercises 4 and 5 in the Margin.

Subtract.

4. $17 - 9\frac{6}{11} =$

5. $\begin{array}{r} 13\frac{1}{2} \\ -\;8\frac{5}{6} \\ \hline \end{array}$

Example 8

When he was first using the new word processor on his computer, Karl found that he was taking about $25\frac{1}{2}$ minutes to type 4 pages of a term paper. Now he takes about $18\frac{3}{5}$ minutes to type 4 pages. By how many minutes has he improved in typing 4 pages? in typing 1 page?

Solution

(a) To find by how many minutes Karl has improved in typing 4 pages, we simply find the difference between the two times.

$$\begin{array}{rcrcr} 25\frac{1}{2} & = & 25\frac{5}{10} & = & 24\frac{15}{10} \\ -18\frac{3}{5} & = & 18\frac{6}{10} & = & 18\frac{6}{10} \\ \hline & & & & 6\frac{9}{10} \end{array}$$

So, in typing 4 pages, Karl has improved by about $6\frac{9}{10}$ minutes.

(**Note:** With truncated numbers, we could have estimated his improvement time as $25 - 18 = 7$ minutes.)

(b) To find his improvement time in typing one page, we divide the answer in part (a) by 4.

$$6\frac{9}{10} \div 4 = \frac{69}{10} \div \frac{4}{1} = \frac{69}{10} \cdot \frac{1}{4} = \frac{69}{40} = 1\frac{29}{40}$$

He has improved by about $1\frac{29}{40}$ minutes per page.

Completion Example Answers

6. $$\begin{array}{rcr} 10 & = & 9\mathbf{\frac{8}{8}} \\ -\ 6\frac{5}{8} & = & 6\frac{5}{8} \\ \hline & & \mathbf{3\frac{3}{8}} \end{array}$$

7. $$\begin{array}{rcrcr} 14\frac{3}{4} & = & 14\mathbf{\frac{15}{20}} & = & 13\mathbf{\frac{35}{20}} \\ -\ 5\frac{9}{10} & = & 5\mathbf{\frac{18}{20}} & = & 5\mathbf{\frac{18}{20}} \\ \hline & & & & \mathbf{8\frac{17}{20}} \end{array}$$

NAME ______________ SECTION ______________ DATE ______________

ANSWERS

Exercises 4.4

Find each of the following differences.

1. $5\frac{1}{2} - 1$

2. $7\frac{3}{4} - 2$

3. $4\frac{5}{12} - 3$

4. $3\frac{5}{8} - 2$

5. $6\frac{1}{2} - 2\frac{1}{2}$

6. $9\frac{1}{4} - 5\frac{1}{4}$

7. $5\frac{3}{4} - 2\frac{1}{4}$

8. $7\frac{9}{10} - 3\frac{3}{10}$

9. $14\frac{5}{8} - 11\frac{3}{8}$

10. $20\frac{7}{16} - 15\frac{5}{16}$

11. $4\frac{7}{8} - 1\frac{1}{4}$

12. $9\frac{5}{16} - 2\frac{1}{4}$

13. $5\frac{11}{12} - 1\frac{1}{4}$

14. $10\frac{5}{6} - 4\frac{2}{3}$

15. $8\frac{5}{6} - 2\frac{1}{4}$

16. $15\frac{5}{8} - 11\frac{3}{4}$

17. $14\frac{6}{10} - 3\frac{4}{5}$

18. $8\frac{3}{32} - 4\frac{3}{16}$

1. ______________
2. ______________
3. ______________
4. ______________
5. ______________
6. ______________
7. ______________
8. ______________
9. ______________
10. ______________
11. ______________
12. ______________
13. ______________
14. ______________
15. ______________
16. ______________
17. ______________
18. ______________

19. ______________
20. ______________
21. ______________
22. ______________
23. ______________
24. ______________
25. ______________
26. ______________
27. ______________
28. ______________
29. ______________
30. ______________
31a. ______________
b. ______________
32. ______________
33. ______________

19. $5\frac{9}{10} - 2$ **20.** $7 - 6\frac{2}{3}$ **21.** $12 - 4\frac{1}{5}$

22. $75 - 17\frac{5}{6}$ **23.** $4\frac{9}{16} - 2\frac{7}{8}$ **24.** $3\frac{7}{10} - 2\frac{5}{6}$

25. $20\frac{3}{6} - 3\frac{4}{8}$ **26.** $17\frac{3}{12} - 12\frac{2}{8}$ **27.** $18\frac{2}{7} - 4\frac{1}{3}$

28. $13\frac{5}{8} - 6\frac{11}{20}$ **29.** $18\frac{7}{8} - 2\frac{2}{3}$ **30.** $10\frac{3}{10} - 2\frac{1}{2}$

31. Sara can paint a room in $3\frac{3}{5}$ hours, and Emily can paint a room of the same size in $4\frac{1}{5}$ hours. (a) How many hours are saved by having Sara paint a room of this size? (b) How many minutes are saved?

32. A teacher graded two sets of test papers. The first set took $3\frac{3}{4}$ hours to grade, and the second set took $2\frac{3}{5}$ hours. How much faster did she grade the second set?

33. Mike takes $1\frac{1}{2}$ hours to clean a pool, and Tom takes $2\frac{1}{3}$ hours to clean the same pool. How much longer does Tom take?

NAME ______________ SECTION ______________ DATE ______________

ANSWERS

34. ______________

35. ______________

36. ______________

37. ______________

38. ______________

39. ______________

34. A long-distance runner ran 10 miles in $70\frac{3}{10}$ minutes. Three months later, she ran the same 10 miles in $63\frac{7}{10}$ minutes. By how much did her time improve?

35. A certain stock was selling for $43\frac{7}{8}$ dollars per share. One month later, it was selling for $48\frac{1}{2}$ dollars per share. By how much did the stock increase in price?

36. You need to lose 10 pounds. If you weigh 180 pounds now and you lose $3\frac{1}{4}$ pounds during the first week and $3\frac{1}{2}$ pounds during the second week, how much more weight do you need to lose?

37. Mr. Johnson originally weighed 240 pounds. During each week of six weeks of dieting, he lost $5\frac{1}{2}$ pounds, $2\frac{3}{4}$ pounds, $4\frac{5}{16}$ pounds, $1\frac{3}{4}$ pounds, $2\frac{5}{8}$ pounds, and $3\frac{1}{4}$ pounds. If he was 35 years old, what did he weigh at the end of the six weeks?

38. A salesman drove $5\frac{3}{4}$ hours one day and $6\frac{1}{2}$ hours the next day. How much more time did he spend driving on the second day?

39. On average, the air that we inhale includes $1\frac{1}{4}$ parts water, and the air we exhale includes $5\frac{9}{10}$ parts water. How many more parts water are in exhaled air?

40. ______________

[Respond below exercise.]
41. ______________

42a. ______________

b. ______________

c. ______________

[Respond below exercise.]
43. ______________

40. A person who is running will burn about $14\frac{7}{10}$ calories each minute, and a person who is walking will burn about $5\frac{1}{2}$ calories each minute. How many more calories does a runner burn in a minute than a walker?

Writing and Thinking about Mathematics

41. In your own words, describe the technique for estimating answers when mixed numbers are involved in calculations.

42. Using the rule you just described in Exercise 41, mentally estimate each of the following differences.

(a) $15\frac{6}{7} - 11\frac{3}{4}$ (b) $136\frac{17}{40} - 125\frac{23}{30}$ (c) $945\frac{1}{10} - 845\frac{3}{100}$

43. Consider the following problem:

The product of two numbers is $16\frac{1}{2}$. If one of the numbers is $2\frac{3}{4}$, what is the other number?

(a) Rewrite the problem using truncated mixed numbers and solve this new problem. (b) Does this technique make it easier for you to understand the original problem? Why or why not? (c) What is the solution to the original problem?

RECYCLE BIN

1. ______________

2. ______________

3. ______________

4. ______________

The Recycle Bin (from Sections 2.1 and 3.5)

Use the rules for order of operations to simplify each of the following expressions.

1. $6 + 2(13 - 5 \cdot 2) + 12 \div 2$

2. $5^3 + 6^2 - 3 \cdot 5 \cdot 2^2$

3. $\left(\frac{1}{3}\right)^2 + \frac{5}{8} \div \frac{1}{4}$

4. $\frac{3}{4} \div \frac{1}{2} \cdot 6 - \frac{1}{5}(45)$

4.5 Complex Fractions and Order of Operations

OBJECTIVES

1. *Recognize and know how to simplify complex fractions.*
2. *Be able to follow the rules for order of operations with mixed numbers.*

Simplifying Complex Fractions

In Section 3.5 we defined a **complex fraction** as the form of a fraction in which the numerator and/or denominator are themselves fractions or the sums or differences of fractions. Since a mixed number is the sum of a whole number and a fraction, and since whole numbers can be written in fraction form, we can now adjust the definition of a complex fraction to include mixed numbers. Thus, we are led to the following more complete definition of a complex fraction.

Definition

A **complex fraction** is a fraction in which the numerator and/or denominator are themselves fractions or mixed numbers or the sums or differences of fractions or mixed numbers.

To Simplify a Complex Fraction:

1. Simplify the numerator so that it is a single fraction, possibly an improper fraction.
2. Simplify the denominator so that it also is a single fraction, possibly an improper fraction.
3. Divide the numerator by the denominator and reduce if possible.

Example 1

Simplify the complex fraction: $\dfrac{3\frac{2}{5}}{\frac{1}{4}+\frac{3}{5}}$.

Solution

$3\frac{2}{5} = \frac{17}{5}$ Change the mixed number to an improper fraction.

$\frac{1}{4} + \frac{3}{5} = \frac{5}{20} + \frac{12}{20} = \frac{17}{20}$

So,

$$\frac{3\frac{2}{5}}{\frac{1}{4}+\frac{3}{5}} = \frac{\frac{17}{5}}{\frac{17}{20}} = \frac{17}{5} \div \frac{17}{20} = \frac{17}{5} \cdot \frac{20}{17} = \frac{17 \cdot 4 \cdot 5}{5 \cdot 17} = \frac{5}{1} = 5$$

Simplify each of the following complex fractions.

1. $\dfrac{3\frac{1}{4}}{\frac{3}{4}+\frac{7}{8}}$

2. $\dfrac{5\frac{1}{2}+\frac{1}{2}}{\frac{7}{10}-\frac{1}{2}}$

Example 2

Simplify the complex fraction: $\dfrac{5\frac{2}{3}-2\frac{1}{3}}{1\frac{1}{2}+\frac{1}{2}}$.

Solution

$$5\frac{2}{3}-2\frac{1}{3}=3\frac{1}{3}=\frac{10}{3} \quad \text{and} \quad 1\frac{1}{2}+\frac{1}{2}=2$$

So,

$$\frac{5\frac{2}{3}-2\frac{1}{3}}{1\frac{1}{2}+\frac{1}{2}}=\frac{\frac{10}{3}}{2}=\frac{\frac{10}{3}}{\frac{2}{1}}=\frac{10}{3}\div\frac{2}{1}=\frac{10}{3}\cdot\frac{1}{2}=\frac{5\cdot\cancel{2}\cdot 1}{3\cdot\cancel{2}}=\frac{5}{3}=1\frac{2}{3}$$

Now Work Exercises 1 and 2 in the Margin.

Order of Operations

With complex fractions, we know to simplify the numerator and denominator separately; that is, we treat the fraction bar as a grouping symbol. More generally, expressions that involve more than one operation are evaluated (or simplified) by using the rules for order of operations. These rules were discussed in Sections 2.1 and 3.5 and are listed again here for easy reference.

Rules for Order of Operations

1. First, simplify within grouping symbols, such as parentheses ( ), brackets [ ], and braces { }. Start with the innermost grouping.
2. Second, find any powers indicated by exponents.
3. Third, moving from **left to right,** perform any multiplications or divisions in the order in which they appear.
4. Fourth, moving from **left to right,** perform any additions or subtractions in the order in which they appear.

Example 3

Use the rules for order of operations to simplify the following expression:

$$2\frac{1}{2}\cdot 1\frac{1}{6}+7\div\frac{3}{4}.$$

Solution

$$2\frac{1}{2}\cdot 1\frac{1}{6}+7\div\frac{3}{4}=\frac{5}{2}\cdot\frac{7}{6}+\frac{7}{1}\cdot\frac{4}{3}$$ Multiply and divide from left to right.

$$=\frac{35}{12}+\frac{28}{3}$$ Now add.

$$=\frac{35}{12}+\frac{28}{3}\cdot\frac{4}{4}$$

$$=\frac{35}{12}+\frac{112}{12}$$

$$=\frac{147}{12}=12\frac{3}{12}=12\frac{1}{4}$$

Or, working with separate parts, we can write

$$2\frac{1}{2}\cdot 1\frac{1}{6}=\frac{5}{2}\cdot\frac{7}{6}=\frac{35}{12}=2\frac{11}{12}$$ multiplying

$$7\div\frac{3}{4}=\frac{7}{1}\cdot\frac{4}{3}=\frac{28}{3}=9\frac{1}{3}$$ dividing

$$2\frac{11}{12}=2\frac{11}{12}$$ adding the results

$$+\,9\frac{1}{3}=9\frac{4}{12}$$

$$11\frac{15}{12}=12\frac{3}{12}=12\frac{1}{4}$$

Example 4

Evaluate the following expression by using the rules for order of operations:

$$\left(2\frac{1}{2}\right)^2+\frac{1}{5}\cdot\frac{1}{6}\div\frac{1}{15}.$$

Solution

$$\left(2\frac{1}{2}\right)^2+\frac{1}{5}\cdot\frac{1}{6}\div\frac{1}{15}=\left(\frac{5}{2}\right)^2+\frac{1}{5}\cdot\frac{1}{6}\div\frac{1}{15}$$

$$=\frac{25}{4}+\frac{1}{30}\cdot\frac{15}{1}$$

$$=\frac{25}{4}+\frac{1}{2}$$

$$=\frac{25}{4}+\frac{2}{4}$$

$$=\frac{27}{4}$$

$$=6\frac{3}{4}$$

Use the rules for order of operations to evaluate each of the following expressions.

3. $\frac{3}{10} \cdot \frac{1}{6} + \frac{2}{5} \div 2$

4. $3\frac{5}{7} \div (2^2 + 3^2)^2$

Now Work Exercises 3 and 4 in the Margin.

EXAMPLE 5

Find the average of the mixed numbers $1\frac{1}{2}$, $2\frac{3}{4}$, and $3\frac{5}{8}$.

Solution

We find the sum first and then divide by 3.

$$\begin{aligned} 1\frac{1}{2} &= 1\frac{4}{8} \\ 2\frac{3}{4} &= 2\frac{6}{8} \\ +3\frac{5}{8} &= 3\frac{5}{8} \\ \hline & 6\frac{15}{8} = 7\frac{7}{8} \end{aligned}$$

$$7\frac{7}{8} \div 3 = \frac{63}{8} \cdot \frac{1}{3} = \frac{\cancel{3} \cdot 3 \cdot 7}{8 \cdot \cancel{3}} = \frac{21}{8} = 2\frac{5}{8}$$

Therefore, the average is $2\frac{5}{8}$.

NAME ______________ SECTION ______________ DATE ______________

ANSWERS

1. ______
2. ______
3. ______
4. ______
5. ______
6. ______
7. ______
8. ______
9. ______
10. ______
11. ______
12. ______

Exercises 4.5

Simplify each of the following complex fractions.

1. $\dfrac{2 + \frac{1}{5}}{1 + \frac{1}{4}}$

2. $\dfrac{\frac{2}{3} + \frac{1}{5}}{4\frac{1}{2}}$

3. $\dfrac{4 + \frac{1}{3}}{6 + \frac{1}{4}}$

4. $\dfrac{2 - \frac{1}{3}}{1 - \frac{1}{3}}$

5. $\dfrac{\frac{2}{3} - \frac{1}{4}}{\frac{3}{5} - \frac{1}{4}}$

6. $\dfrac{\frac{5}{6} - \frac{2}{3}}{\frac{5}{8} - \frac{1}{16}}$

7. $\dfrac{1}{\frac{1}{12} + \frac{1}{6}}$

8. $\dfrac{1}{\frac{1}{5} + \frac{2}{15}}$

9. $\dfrac{\frac{3}{10} + \frac{1}{6}}{3}$

10. $\dfrac{\frac{7}{12} + \frac{1}{15}}{5}$

11. $\dfrac{7\frac{1}{3} + 2\frac{1}{5}}{6\frac{1}{9} + 2}$

12. $\dfrac{5\frac{2}{3} - 1\frac{1}{6}}{3\frac{1}{2} + 3\frac{1}{6}}$

13. ______________

14. ______________

15. ______________

16. ______________

17. ______________

18. ______________

19. ______________

20. ______________

21. ______________

22. ______________

13. $\dfrac{6\frac{1}{100} + 5\frac{3}{100}}{2\frac{1}{2} + 3\frac{1}{10}}$

14. $\dfrac{4\frac{7}{10} - 2\frac{9}{10}}{5\frac{1}{100}}$

Evaluate the following expressions using the rules for order of operations.

15. $\frac{3}{5} \cdot \frac{1}{6} + \frac{1}{5} \div 2$

16. $\frac{1}{2} \div \frac{1}{2} + 1 - \frac{2}{3} \cdot 3$

17. $3\frac{1}{2} \cdot 5\frac{1}{3} + \frac{5}{12} \div \frac{15}{16}$

18. $2\frac{1}{4} + 1\frac{1}{5} + 2 \div \frac{20}{21}$

19. $\frac{5}{8} - \frac{1}{3} \cdot \frac{2}{5} + 6\frac{1}{10}$

20. $1\frac{1}{6} \cdot 1\frac{2}{19} \div \frac{7}{8} + \frac{1}{38}$

21. $\frac{3}{10} + \frac{5}{6} \div \frac{1}{4} \cdot \frac{1}{8} - \frac{7}{60}$

22. $5\frac{1}{7} \div (2 + 1)^2$

NAME ______________________ SECTION ______________ DATE __________

ANSWERS

23. $\left(2 - \frac{1}{3}\right) \div \left(1 - \frac{1}{3}\right)^2$

24. $\left(2\frac{4}{9} + 1\frac{1}{18}\right) \div \left(1\frac{2}{9} - \frac{1}{6}\right)$

25. If the product of $5\frac{1}{2}$ and $2\frac{1}{4}$ is added to the quotient of $\frac{9}{10}$ and $\frac{3}{4}$, what is the sum?

26. If the quotient of $\frac{5}{8}$ and $\frac{1}{2}$ is subtracted from the product of $2\frac{1}{4}$ and $3\frac{1}{5}$, what is the difference?

27. If $\frac{9}{10}$ of 70 is divided by $\frac{3}{4}$ of 10, what is the quotient?

28. If $\frac{2}{3}$ of $4\frac{1}{4}$ is added to $\frac{5}{8}$ of $6\frac{1}{3}$, what is the sum?

23. ____________

24. ____________

25. ____________

26. ____________

27. ____________

28. ____________

29. ______________

30. ______________

31. ______________

32. ______________

33. [Respond below exercise.] ______________

29. Find the average of the numbers $\frac{7}{8}$, $\frac{9}{10}$, and $1\frac{3}{4}$.

30. Find the average of the numbers $\frac{5}{6}$, $\frac{1}{15}$, and $\frac{17}{30}$.

31. Find the average of the numbers $5\frac{1}{8}$, $7\frac{1}{2}$, $4\frac{3}{4}$, and $10\frac{1}{2}$.

32. Find the average of the numbers $4\frac{7}{10}$, $3\frac{9}{10}$, $5\frac{1}{100}$, and $11\frac{3}{20}$.

Collaborative Learning Exercise

33. With the class separated into teams of two to four students, each team is to write one or two paragraphs on the following two topics. Then the team leader is to read the paragraphs with classroom discussion to follow.

(a) What topic have you found to be the most interesting in the text so far? Why?

(b) What topic have you found to be the most useful in the text so far? Why?

4.6 Denominate Numbers

Simplifying Mixed Denominate Numbers

Abstract numbers are numbers with no units of measure attached—for example, 25, 1, $\frac{1}{2}$, and $5\frac{3}{4}$. **Denominate numbers** are numbers with units of measure attached—for example, 25 centimeters, 1 ounce, $\frac{1}{2}$ inch, and $5\frac{3}{4}$ feet.

Denominate numbers with two or more related units are called **mixed denominate numbers.** Examples of commonly used mixed denominate numbers are

5 ft 8 in., 3 lb 4 oz, and 1 hr 45 min

With mixed numbers we make sure that the fraction part is less than 1. Thus, we write $6\frac{1}{2}$, not $5\frac{3}{2}$. Similarly, in **simplified mixed denominate numbers,** the number of smaller units is less than 1 of the larger unit. Table 4.1 shows equivalent measures of length, weight, liquid volume, and time used in the U.S. customary system of measure. (Metric tables are included inside the back cover of the text for interested students. Metric measures are discussed in detail in the Measurement chapter, which may or may not appear in your edition of this textbook.)

Table 4.1 U.S. Customary Units of Measure

Length		*Liquid Volume*	
1 foot (ft)	= 12 inches (in.)	1 pint (pt)	= 16 fluid ounces (fl oz)
1 yard (yd)	= 3 ft	1 quart (qt)	= 2 pt = 32 fl oz
1 mile (mi)	= 5280 ft	1 gallon (gal)	= 4 qt
Weight		*Time*	
1 pound (lb)	= 16 ounces (oz)	1 minute (min)	= 60 seconds (sec)
1 ton (t)	= 2000 lb	1 hour (hr)	= 60 min
		1 day	= 24 hr

To simplify mixed denominate numbers, either you must have a table of equivalent values (such as Table 4.1) with you or you must memorize the basic equivalent values. Most people know at least some of these values from experience. The following examples illustrate the technique for simplifying mixed denominate numbers. Remember that the number of smaller units must be less than 1 of the next larger unit.

OBJECTIVES

1. *Know the difference between abstract numbers and denominate numbers.*
2. *Be able to simplify mixed denominate numbers.*
3. *Know how to add like denominate numbers.*
4. *Know how to subtract like denominate numbers.*

Simplify each of the following mixed denominate numbers.

1. 3 hr 75 min

2. 10 ft 13 in.

EXAMPLE 1

Simplify the mixed denominate number 3 ft 14 in.

Solution

Since 14 in. is more than 1 ft (there are 12 in. in 1 ft), we write

$$\begin{aligned} 3\text{ ft }14\text{ in.} &= 3\text{ ft} + 12\text{ in.} + 2\text{ in.} \\ &= \underbrace{3\text{ ft} + 1}\text{ ft} + 2\text{ in.} \\ &= 4\text{ ft} + 2\text{ in.} \\ &= 4\text{ ft }2\text{ in.} \end{aligned}$$

EXAMPLE 2

Simplify the mixed denominate number 5 lb 30 oz.

Solution

Since 30 oz is more than 1 lb, we write

$$\begin{aligned} 5\text{ lb }30\text{ oz} &= 5\text{ lb} + 16\text{ oz} + 14\text{ oz} \\ &= \underbrace{5\text{ lb} + 1}\text{ lb} + 14\text{ oz} \\ &= 6\text{ lb} + 14\text{ oz} \\ &= 6\text{ lb }14\text{ oz} \end{aligned}$$

EXAMPLE 3

Simplify the mixed denominate number 2 hr 70 min.

Solution

Since 70 min is more than 1 hr, we write

$$\begin{aligned} 2\text{ hr }70\text{ min} &= 2\text{ hr} + 60\text{ min} + 10\text{ min} \\ &= \underbrace{2\text{ hr} + 1\text{ hr}} + 10\text{ min} \\ &= 3\text{ hr} + 10\text{ min} \\ &= 3\text{ hr }10\text{ min} \end{aligned}$$

Now Work Exercises 1 and 2 in the Margin.

Adding and Subtracting Like Denominate Numbers

Understanding how to simplify mixed denominate numbers helps in both adding and subtracting such numbers. **Like denominate numbers** are denominate numbers that have the same units or that can be written with the same units. For example,

5 ft 10 in. and 2 ft 3 in. are like denominate numbers,

and

3 hr 5 min and 4 hr 15 min are like denominate numbers.

To Add Like Denominate Numbers:

1. Write the numbers in column form so that like units are aligned.
2. Add the numbers in each column.
3. Simplify the resulting mixed denominate number, if necessary.

Examples 4–6 illustrate addition with like denominate numbers and simplifying the sum.

EXAMPLE 4

$$\begin{array}{r} 3 \text{ ft} \quad 2 \text{ in.} \\ 2 \text{ ft} \quad 8 \text{ in.} \\ + \quad 5 \text{ ft} \quad 5 \text{ in.} \\ \hline 10 \text{ ft } 15 \text{ in.} \end{array} = 10 \text{ ft} + 12 \text{ in.} + 3 \text{ in.}$$
$$= 11 \text{ ft } 3 \text{ in.}$$

EXAMPLE 5

$$\begin{array}{r} 2 \text{ hr } 15 \text{ min} \\ + \ 4 \text{ hr } 50 \text{ min} \\ \hline 6 \text{ hr } 65 \text{ min} \end{array} = 6 \text{ hr} + 60 \text{ min} + 5 \text{ min}$$
$$= 7 \text{ hr } 5 \text{ min}$$

EXAMPLE 6

In three different trips to the supermarket, Dan bought the following amounts of milk: 2 gal 2 qt, 1 gal 3 qt, and 1 gal 2 qt. What was the total amount of milk he bought on these three trips?

Solution

To find the total amount, we add the denominate numbers and simplify.

$$\begin{array}{r} 2 \text{ gal } 2 \text{ qt} \\ 1 \text{ gal } 3 \text{ qt} \\ + \ 1 \text{ gal } 2 \text{ qt} \\ \hline 4 \text{ gal } 7 \text{ qt} \end{array} = 4 \text{ gal} + 4 \text{ qt} + 3 \text{ qt}$$
$$= 5 \text{ gal } 3 \text{ qt}$$

The total amount of milk that Dan bought was 5 gal 3 qt.

To Subtract Like Denominate Numbers:

1. Write the numbers in column form so that like units are aligned.
2. If necessary for subtraction, borrow 1 of the larger units and rewrite the top number.
3. Subtract the like units.

3. Add and simplify the sum.

$$\begin{array}{r} 20 \text{ min } 45 \text{ sec} \\ +\ 13 \text{ min } 25 \text{ sec} \\ \hline \end{array}$$

4. Subtract.

$$\begin{array}{r} 5 \text{ gal } 1 \text{ qt } \ 2 \text{ fl oz} \\ -\ 2 \text{ gal } 3 \text{ qt } 15 \text{ fl oz} \\ \hline \end{array}$$

Examples 7–9 illustrate subtraction with like denominate numbers.

EXAMPLE 7

$$\begin{array}{r} 8 \text{ lb } 14 \text{ oz} \\ -\ 3 \text{ lb } 10 \text{ oz} \\ \hline 5 \text{ lb } \ \ 4 \text{ oz} \end{array}$$

EXAMPLE 8

$$\begin{array}{r} 6 \text{ ft } 5 \text{ in.} \\ -\ 2 \text{ ft } 8 \text{ in.} \\ \hline \end{array}$$

Here, 5 in. is smaller than 8 in., and we cannot subtract directly. So, we **borrow** 1 ft = 12 in. from 6 ft.

$$\begin{array}{rcr} 6 \text{ ft } 5 \text{ in.} & = & 5 \text{ ft } 17 \text{ in.} \\ -\ 2 \text{ ft } 8 \text{ in.} & = & 2 \text{ ft } \ \ 8 \text{ in.} \\ \hline & & 3 \text{ ft } \ \ 9 \text{ in.} \end{array}$$

(12 in. + 5 in. = 17 in.)

EXAMPLE 9

On the first day of a long business trip (including both airplane and car travel) a salesman traveled for 13 hr 20 min. On the second day he traveled for 10 hr 50 min. What was the difference in his travel time for the two days?

Solution

To find the difference, we subtract the two times. In this case, we need to know that there are 60 minutes in 1 hour.

$$\begin{array}{r} 13 \text{ hr } 20 \text{ min} \\ -\ 10 \text{ hr } 50 \text{ min} \\ \hline \end{array}$$

Since 20 min is smaller than 50 min, we cannot subtract directly. So, we **borrow** 1 hr = 60 min from 13 hr.

$$\begin{array}{rcr} 13 \text{ hr } 20 \text{ min} & = & 12 \text{ hr } 80 \text{ min} \\ -\ 10 \text{ hr } 50 \text{ min} & = & 10 \text{ hr } 50 \text{ min} \\ \hline & & 2 \text{ hr } 30 \text{ min} \end{array}$$

(60 min + 20 min = 80 min)

So, the difference in travel time for the two days was 2 hr 30 min.

Now Work Exercises 3 and 4 in the Margin.

NAME ____________ SECTION ____________ DATE ____________

ANSWERS

1. ______
2. ______
3. ______
4. ______
5. ______
6. ______
7. ______
8. ______
9. ______
10. ______
11. ______
12. ______
13. ______
14. ______
15. ______
16. ______
17. ______
18. ______

Exercises 4.6

Simplify the following mixed denominate numbers.

1. 3 ft 20 in.

2. 4 ft 18 in.

3. 6 lb 20 oz

4. 3 lb 24 oz

5. 5 min 80 sec

6. 14 min 90 sec

7. 2 days 30 hr

8. 5 days 36 hr

9. 8 gal 5 qt

10. 2 gal 6 qt

11. 4 pt 20 fl oz

12. 3 pt 24 fl oz

Add and simplify if necessary.

13.
```
   2 ft 8 in.
   5 ft 4 in.
 + 1 ft 7 in.
```

14.
```
   3 ft 5 in.
   6 ft 5 in.
 + 2 ft 3 in.
```

15.
```
   10 lb 10 oz
 +  7 lb  8 oz
```

16.
```
   4 lb 5 oz
 + 4 lb 7 oz
```

17.
```
   8 min 35 sec
 + 9 min 35 sec
```

18.
```
    5 min 10 sec
 + 14 min 35 sec
```

19. ________
20. ________
21. ________
22. ________
23. ________
24. ________
25. ________
26. ________
27. ________
28. ________
29. ________
30. ________
31. ________
32. ________
33. ________
34. ________
35. ________
36. ________

19. 2 hr 15 min 45 sec
+ 1 hr 55 min 30 sec

20. 5 hr 20 min 30 sec
+ 2 hr 35 min 40 sec

21. 2 days 20 hr 50 min
+ 3 days 5 hr 45 min

22. 1 day 15 hr 40 min
+ 2 days 10 hr 20 min

23. 5 gal 2 qt
3 gal 2 qt
+ 4 gal 3 qt

24. 4 gal 3 qt
1 gal 2 qt
+ 3 gal 1 qt

25. 4 gal 3 qt 10 fl oz
+ 2 gal 3 qt 10 fl oz

26. 5 gal 3 qt 15 fl oz
+ 1 gal 2 qt 8 fl oz

27. 5 yd 2 ft 8 in.
+ 6 yd 2 ft 10 in.

28. 3 yd 1 ft 7 in.
+ 2 yd 2 ft 3 in.

Subtract.

29. 5 yd 1 ft 7 in.
− 2 yd 2 ft 5 in.

30. 8 yd 2 ft 3 in.
− 7 yd 1 ft 8 in.

31. 9 gal 2 qt 4 fl oz
− 5 gal 3 qt 6 fl oz

32. 20 gal 1 qt 13 fl oz
− 14 gal 2 qt 10 fl oz

33. 15 hr 30 min
− 12 hr 45 min

34. 6 hr 20 min
− 4 hr 40 min

35. 15 min 20 sec
− 10 min 30 sec

36. 30 min 15 sec
− 20 min 25 sec

NAME ____________ SECTION ____________ DATE ____________

ANSWERS

37. 8 lb 4 oz
− 3 lb 12 oz

38. 20 lb 10 oz
− 10 lb 14 oz

39. 6 ft 5 in.
− 2 ft 9 in.

40. 3 ft 8 in.
− 1 ft 10 in.

41. What is 2 ft 5 in. more than 4 ft 8 in.?

42. Find the sum of 6 hr 30 min, 9 hr 15 min, and 4 hr 45 min.

43. What is the difference between 14 hr and 10 hr 20 min?

44. What is 3 lb 11 oz less than 6 lb 8 oz?

45. What is 7 qt more than 5 gal 2 qt?

37. ____________
38. ____________
39. ____________
40. ____________
41. ____________
42. ____________
43. ____________
44. ____________
45. ____________

46. ______________

47. ______________

48. ______________

49. [Respond below exercise.]

50. [Respond below exercise.]

46. What is the sum of 4 gal 3 qt, 2 gal 3 qt, and 1 gal 2 qt?

47. The plans for a new bookcase call for 3 shelves that are each 3 ft 8 in. long and 4 shelves that are each 1 ft 5 in. long. What total length of shelf board is required for these shelves?

48. Mr. and Mrs. Gonzalez spent the following amounts of time race-walking for their health during one week: 1 hr 10 min, 1 hr 30 min, 45 min, 1 hr 5 min, 50 min, 1 hr 40 min, and 2 hr. How much time did they spend race-walking that week?

Writing and Thinking about Mathematics

49. In your own words, explain the difference between abstract numbers and denominate numbers. Write a sentence in which each type of number is used at least once.

50. Explain how to simplify the denominate number 23 hr 59 min 60 sec.

4 Index of Key Ideas and Terms

NAME ______________ SECTION ______________ DATE ______________

ANSWERS

1. ______________

2a. ______________

b. ______________

3a. ______________

b. ______________

4. ______________

5. ______________

6. ______________

7. ______________

8. ______________

9. ______________

10. ______________

11. ______________

12. ______________

Chapter 4 Test

1. The expression $\frac{13}{0}$ is undefined, but $\frac{0}{13} = $ ______.

2. Change each improper fraction to a mixed number.

(a) $\frac{17}{5}$ (b) $\frac{100}{33}$

3. Change each mixed number to an improper fraction.

(a) $6\frac{5}{8}$ (b) $4\frac{3}{10}$

In the remaining exercises, express your answers as either whole numbers or mixed numbers with the fraction part reduced.

4. Find the sum of $2\frac{3}{4}$ and $3\frac{5}{6}$.

5. Find the difference between $7\frac{1}{10}$ and $4\frac{4}{15}$.

6. Find the sum of the product of $3\frac{3}{5}$ and $6\frac{2}{3}$ with the quotient of $2\frac{1}{3}$ and 14.

In Exercises 7–18, perform the indicated operations.

7. $\begin{array}{r} 9\frac{3}{8} \\ + \frac{5}{6} \\ \hline \end{array}$

8. $\begin{array}{r} 7 \\ - 5\frac{3}{5} \\ \hline \end{array}$

9. $4\frac{3}{10} + 2\frac{3}{4}$

10. $6\frac{1}{6} - 4\frac{2}{7}$

11. $\frac{3}{5} + 1\frac{7}{10} + 2\frac{1}{8}$

12. $6\frac{2}{5} \cdot 3\frac{1}{8}$

13. ______________

14. ______________

15. ______________

16. ______________

17. ______________

18. ______________

19. ______________

20. ______________

21. ______________

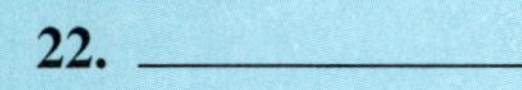

13. $2\frac{2}{3} \cdot \frac{5}{8} \cdot 2\frac{1}{4}$

14. $\frac{5}{6} \div 2\frac{1}{2}$

15. $4\frac{7}{8} \div 3\frac{1}{4}$

16. $\left(3\frac{1}{4}\right)\left(1\frac{2}{13}\right)\left(3\frac{1}{9}\right)$

17. $\frac{4}{5} \div 1\frac{1}{3} + \frac{2}{15} \cdot \frac{10}{7}$

18. $\dfrac{6\frac{1}{5} + 3\frac{7}{10}}{5\frac{3}{10} - 4\frac{1}{2}}$

19. A businesswoman flew to Japan in 14 hours and 10 minutes. Her flight to Europe lasted 8 hours and 30 minutes. What was the difference in flight times?

20. A carpenter measured the lengths of three boards. The lengths were 2 ft 3 in., 6 ft 8 in., and 10 ft 4 in. What was the total length of the three boards?

21. The lengths of the three sides of a triangle are $5\frac{9}{10}$ centimeters, $3\frac{7}{10}$ centimeters, and $7\frac{3}{10}$ centimeters. What is the perimeter of the triangle?

22. A box has a square bottom (and top) that measures $8\frac{1}{2}$ inches along each edge. The height of the box measures $12\frac{1}{4}$ inches. What is the total area of the six faces of the box? (**Note:** Each face of the box is a rectangle. A square is a rectangle with all four sides equal.)

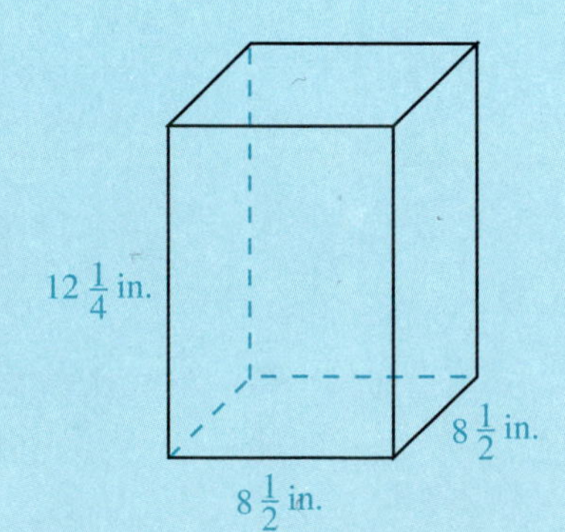

NAME ______________________ SECTION ________________ DATE __________

ANSWERS

1. [Respond below exercise.] __________

2. __________

3. __________

4. [Respond in exercise.] __________

5. __________

6. __________

7. __________

8a. __________

b. __________

Cumulative Review

1. Write 53,460 in expanded notation and in its English word equivalent.

2. Round off 265,400 to the nearest ten thousand.

3. Name the property of addition illustrated: $16 + 35 = 35 + 16$.

4. Match each expression with its best estimate.

______	(a) $165 + 92 + 86 + 131$	A. 200
______	(b) $8467 \div 43$	B. 500
______	(c) $33 \cdot 46$	C. 1000
______	(d) $6476 - 5392$	D. 1500

5. Multiply mentally: 800(7000).

6. Use the rules for order of operations to evaluate the following expression:
$7^2 + 2(12 \cdot 4 \div 2 \cdot 3) - 7 \cdot 5$

7. List all the prime numbers less than 30.

8. Find the prime factorization of each number:
(a) 170 (b) 305

9. ____________

10. ____________

11. ____________

12. ____________

13. ____________

14. ____________

15. ____________

16. ____________

17. ____________

18. ____________

19. ____________

20. ____________

21. ____________

22. ____________

23. ____________

9. The value of $0 \div 36$ is _____, while $36 \div 0$ is _____.

10. Find the average of 44, 35, 53, and 40.

Perform the indicated operations. Reduce all fractions to lowest terms.

11. $\begin{array}{r} 8597 \\ +\ 4653 \\ \hline \end{array}$

12. $\begin{array}{r} 3782 \\ -\ 1255 \\ \hline \end{array}$

13. $\begin{array}{r} 732 \\ \times\ 36 \\ \hline \end{array}$

14. $13\overline{)2639}$

15. $\frac{7}{8} \div 6$

16. $\frac{7}{18}\left(\frac{3}{10}\right)\left(\frac{12}{28}\right)$

17. $\frac{7}{8} - \frac{7}{12}$

18. $\frac{9}{10} + \frac{4}{5} + \frac{1}{6}$

19. $\begin{array}{r} 13\frac{4}{5} \\ -\ 8\frac{4}{7} \\ \hline \end{array}$

20. $4\frac{3}{8} \div 2\frac{1}{2}$

21. $5\left(5\frac{2}{5}\right)\left(5\frac{5}{6}\right)$

22. $\dfrac{1\frac{3}{10} + 2\frac{3}{5}}{2\frac{4}{5} - 1\frac{1}{2}}$

23. $\left(1\frac{1}{3}\right)^2 + 6\frac{4}{5} \div 2\frac{1}{8} - 2\frac{3}{10}$

NAME ______________________ SECTION ________________ DATE __________

ANSWERS

24. [Respond below exercise.]

25a. ____________

b. ____________

26. ____________

27. ____________

24. Arrange the following fractions in order from smallest to largest.

$\frac{7}{10}, \frac{3}{4}, \frac{5}{6}$

25. Find the following sums and simplify if possible.

(a)
$$\begin{array}{r} 3 \text{ hr } 15 \text{ min } 35 \text{ sec} \\ +\ 1 \text{ hr } 55 \text{ min } 40 \text{ sec} \\ \hline \end{array}$$

(b)
$$\begin{array}{r} 20 \text{ lb } 10 \text{ oz} \\ +\ 7 \text{ lb } 12 \text{ oz} \\ \hline \end{array}$$

26. On Tuesday morning, Kevin had $453 in his checking account. That day, he deposited $1500 in his account and wrote checks for the following amounts: $87, $250, and $675. What was his new balance on Wednesday morning?

27. If a car has a 22-gallon gas tank, how many gallons of gas will it take to fill the tank when it is $\frac{1}{4}$ full?

28. ______________

29. ______________

30. ______________

28. Paula bought two pairs of shoes for \$37 per pair, three blouses for \$49 each, a pair of slacks for \$29, and a sweater for \$58. What was the average price per item?

29. The discount price of a new VCR is \$180. If this price is $\frac{4}{5}$ of the original price, what was the original price?

30. In November 1990, men in the age group 25–54 watched TV for an average of 26 hours and 44 minutes per week. During the same month, women in the same age group watched TV for an average of 30 hours and 34 minutes per week. What was the difference between the women's and the men's TV viewing times for one week in November 1990? What was the difference for the entire month?

5

Decimal Numbers

What to Expect in Chapter 5

In Chapter 5, the concept of decimal numbers is developed and students will learn to work with the basic operations with decimal numbers. Two related topics—scientific notation and square roots—are included in this chapter because of the current emphasis on working with calculators. In Section 5.1, the techniques involved in reading, writing, and rounding off decimal numbers are discussed, along with the use of the word **and** to indicate placement of the decimal point. Operations with decimal numbers are discussed in Sections 5.2 through 5.4, as well as the use of rounded-off numbers to estimate answers.

Scientific notation, which involves positive and negative exponents, is introduced in Section 5.5 so that students can read results on their calculators when very large or very small decimal numbers are involved. In Section 5.6, we see that fractions and decimals are closely related and can be used together by changing all the numbers in one problem to one form or the other. In Section 5.7, square roots and the Pythagorean Theorem are discussed, providing valuable understanding and interesting applications of decimal numbers.

Mathematics at Work!

View of Wrigley Field, home of the Chicago Cubs.

As illustrated in the figure below, the shape of a baseball infield is a square 90 feet on each side. (A square is a four-sided plane figure in which all four sides are the same length and adjoining sides meet at 90° angles.) Do you think that the distance from home plate to second base is more than 180 feet or less than 180 feet? more than 90 feet or less than 90 feet? The distance from the pitcher's mound to home plate is $60\frac{1}{2}$ feet. Is the pitcher's mound exactly halfway between home plate and second base? Do the two diagonals of the square (shown as dashed lines in the figure) intersect at the pitcher's mound? What is the distance from home plate to second base (to the nearest tenth of a foot)? (See Exercise 61 in Section 5.7.)

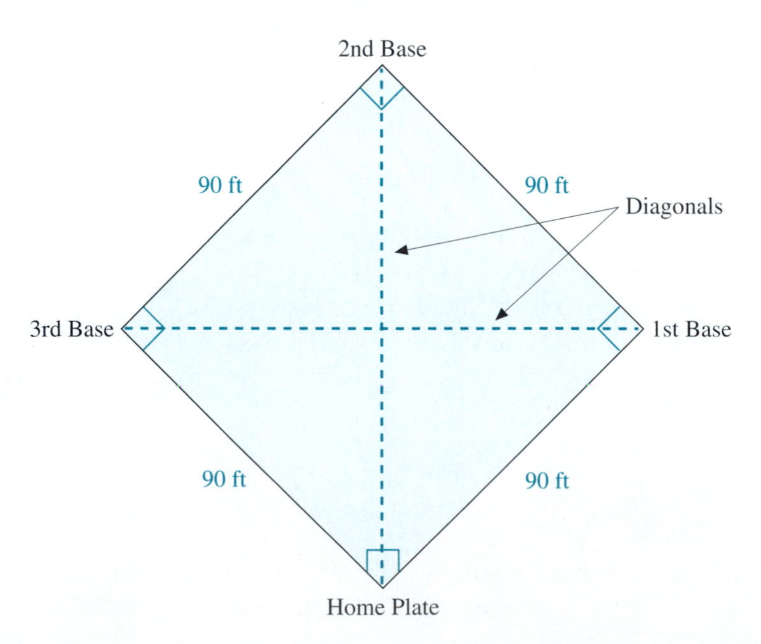

5.1 Reading, Writing, and Rounding Off Decimal Numbers

OBJECTIVES

1. *Learn to read and write decimal numbers.*
2. *Realize the importance of the word* **and** *in reading and writing decimal numbers.*
3. *Understand that* **th** *at the end of a word indicates a fraction part.*
4. *Know how to round off decimal numbers to indicated places of accuracy.*

Reading and Writing Decimal Numbers

In Section 1.1, we discussed whole numbers and their representation using a **place value system** and the ten digits 0, 1, 2, 3, 4, 5, 6, 7, 8, and 9. The value of each place (for whole numbers) in the system is a power of ten:

1, 10, 100, 1000, 10,000, 100,000, and so on.

In this section, we show how this **decimal system** can be extended to include fractions as well as whole numbers. We will need the following two concepts:

1. A **finite** amount is an amount that can be counted.
2. An **infinite** amount is an amount that cannot be counted.

Definition

A **finite decimal number** (or **terminating decimal number**) is any rational number (fraction or mixed number) with a power of ten in the denominator of the fraction.

While we generally operate only with finite decimal numbers in arithmetic, you should be aware that there are three classifications of decimal numbers:

1. finite (or terminating) decimals,
2. infinite repeating decimals, and
3. infinite nonrepeating decimals.

Infinite repeating decimals and infinite nonrepeating decimals will be discussed in Sections 5.6 and 5.7. **In Sections 5.1–5.5, we use the term *decimal number* to refer only to finite (or terminating) decimals.**

Note that the whole numbers can also be classified as decimal numbers since any whole number can be written in fraction form with denominator 1. Examples of decimal numbers are

$$\frac{3}{10}, \quad \frac{75}{100}, \quad \frac{349}{1000}, \quad \frac{7}{1}, \quad 8\frac{1}{10}, \quad 21\frac{16}{100}$$

The common **decimal notation** uses a decimal point, with whole numbers written to the left of the decimal point and fractions written to the right of the decimal point. The values of several places in this decimal system are shown in Figure 5.1.

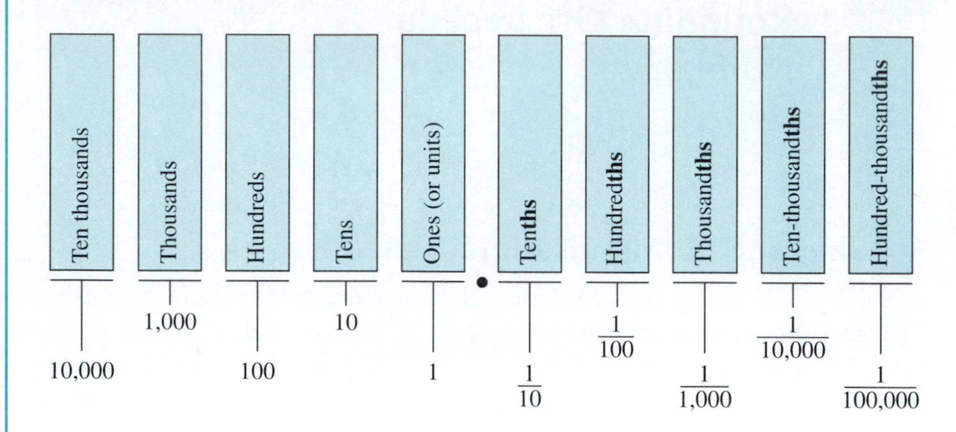

Figure 5.1

In reading a fraction such as $\frac{346}{1000}$, we read the numerator as a whole number ("three hundred forty-six"), then attach the name of the denominator ("thousandths"). The same procedure is followed with numbers written in decimal notation.

$$\frac{346}{1000} = 0.346$$ Read "three hundred forty-six thousandths."

If there is no whole number part, then 0 is commonly written to the left of the decimal point. In general, decimal numbers are read (and written) according to the following convention.

To Read or Write a Decimal Number:

1. Read (or write) the whole number.
2. Read (or write) the word **and** in place of the decimal point.
3. Read (or write) the fraction part as a whole number with the name of the place of the last digit on the right.

EXAMPLE 1

Write the decimal number $48\frac{6}{10}$ in decimal notation and in words.

Solution

4 8 . 6 in decimal notation

forty-eight and six tenths in words

And indicates the decimal point; the digit 6 is in the tenths position.

EXAMPLE 2

Write the decimal number $5\frac{398}{1000}$ in decimal notation and in words.

Solution

5 . 3 9 8 in decimal notation

five and three hundred ninety-eight thousandths in words

And indicates the decimal point; the digit 8 is in the thousandths position.

EXAMPLE 3

Write the decimal number $12\frac{75}{10,000}$ in decimal notation and in words.

Solution

Two 0's must be inserted as placeholders.

1 2 . 0 0 7 5 in decimal notation

twelve and seventy-five ten-thousandths in words

And indicates the decimal point; the digit 5 is in the ten-thousandths position.

Special Notes:

1. The letters **th** at the end of a word indicate a fraction part (a part to the right of the decimal point).

   six hundred = 600
   six hundred**ths** = 0.06

2. The hyphen (-) indicates one word.

   four hundred thousand = 400,000
   four hundred-thousand**ths** = 0.00004

Now Work Exercises 1–4 in the Margin.

EXAMPLE 4

Write **seventeen thousandths** in decimal notation.

Solution

0 is inserted as a placeholder.

0 . 0 1 7

The digit 7 is in the thousandths position.

Compare Examples 5 and 6 very carefully. Note how deletion of the word **and** in Example 6 completely changes the number from that in Example 5.

Write each decimal number in words.

1. 20.7

2. 9.04

3. 18.651

4. 2.0008

Write each number in decimal notation.

5. ten and eleven hundredths

6. four and five thousandths

7. eight hundred and three tenths

8. one thousand six hundred and two hundred sixty-four thousandths

EXAMPLE 5 Write **six hundred and five thousandths** in decimal notation.

Solution

600.005

Two 0's are inserted as placeholders.

The digit 5 is in the thousandths position.

EXAMPLE 6 Write **six hundred five thousandths** in decimal notation.

Solution

0.605

Now Work Exercises 5–8 in the Margin.

Rounding Off Decimal Numbers

Measuring devices such as rulers, metersticks, speedometers, micrometers, and surveying transits give only approximate measurements. (See Figure 5.2.) This is because, whether the units are large (such as miles and kilometers) or small (such as inches and centimeters), there are always smaller, more accurate units (such as eighths of an inch and millimeters) that could be used to indicate a particular measurement. We are constantly dealing with approximate (or rounded-off) numbers in our daily lives. If a recipe calls for 1.5 cups of flour and the cook puts in 1.47 cups (or 1.53 cups), the result will be about the same. In fact, the cook will never put exactly the same amount of flour in the mixture, no matter how many times the recipe is used.

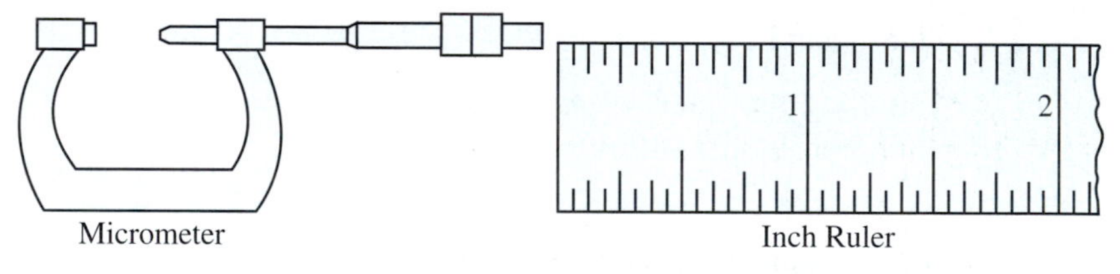

(a) The micrometer is marked to give approximate measures of small objects.

(b) The ruler is marked to give approximate measures of lengths in fourths, eighths, and sixteenths of an inch.

Figure 5.2

Rounding off a given number means finding another number close to the given number. The desired place of accuracy must be stated. For example, looking at the number line below, we can see that 5.76 is closer to 5.80 than it is to 5.70. So (to the nearest tenth), 5.76 rounds off to 5.80, or 5.8.

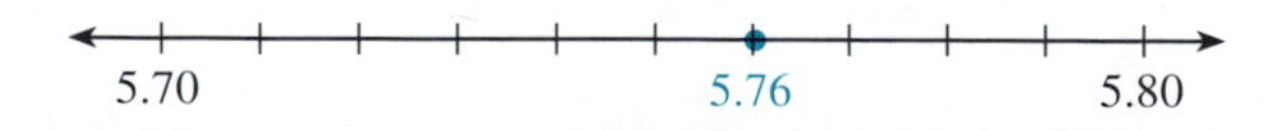

There are several generally accepted rules for rounding off decimal numbers. For example, the IRS allows rounding off to the nearest dollar on income tax forms. Sometimes, in cases where many numbers are involved, a decimal number is rounded to the nearest even digit at some particular place value. The rule chosen depends on the use of the numbers and whether or not there might be a penalty for an error.

In this text, we will use the following rules for rounding off, with one exception: in the case of dollars and cents, we will follow the common business practice of always rounding up to the next higher cent.

Rules for Rounding Off Decimal Numbers

1. Look at the single digit just to the right of the place of desired accuracy.
2. If this digit is 5 or greater, make the digit in the desired place of accuracy one larger and replace all digits to the right with 0's. If a 9 is made one larger, then the next digit to the left is increased by 1. Otherwise, all digits to the left remain unchanged.
3. If this digit is less than 5, leave the digit in the desired place of accuracy as it is and replace all digits to the right with 0's. All digits to the left remain unchanged.

Note: Trailing 0's to the right of the decimal point must be dropped so that the place of accuracy is clearly understood. If a rounded-off number has a 0 in the desired place of accuracy, then that 0 remains. In whole numbers, all 0's must remain.

Example 7 Round off 6.749 to the nearest tenth.

Solution

(a) 6 . 7 4 9

7 is in the tenths position. The next digit is 4.

(b) Since 4 is less than 5, leave the 7 and replace 4 and 9 with 0's.
(c) 6.749 rounds off to **6.7** (to the nearest tenth).

Note that even though 6.7 = 6.700, the trailing 0's in 6.700 are dropped to indicate the position of accuracy.

Example 8 Round off 13.73962 to the nearest thousandth.

Solution

(a) 1 3 . 7 3 9 6 2

9 is in the thousandths position. The next digit is 6.

(b) Since 6 is greater than 5, make 9 one larger and replace 6 and 2 with 0's. (Making 9 one larger gives 10, which affects the digit 3 as well as the 9.)
(c) 13.73962 rounds off to **13.740** (to the nearest thousandth). (Note that in 13.74000 only two training 0's are dropped. The remaining 0 is written to indicate that the accuracy is to the thousandths place.)

Round off each number as indicated.

9. 3.349 (nearest tenth)

10. 0.07921 (nearest thousandth)

11. 7558 (nearest thousand)

12. 0.0006873 (nearest hundred-thousandth)

✓ COMPLETION EXAMPLE 9

Round off 8.00241 to the nearest ten-thousandth.

Solution

(a) The digit in the ten-thousandth position is _____.

(b) The next digit to the right is _____.

(c) Since _____ is less than 5, leave _____ as it is and replace _____ with a 0.

(d) 8.00241 rounds off to _______________ (to the nearest _______________).

✓ COMPLETION EXAMPLE 10

Round off 7361 to the nearest hundred.

Solution

(a) The decimal point is understood to be to the right of _____.

(b) The digit in the hundreds position is _____.

(c) The next digit to the right is _____.

(d) Since _____ is greater than 5, change the _____ to _____ and replace _____ and _____ with 0's.

(e) So, 7361 rounds off to _____________ (to the nearest hundred).

Now Work Exercises 9–12 in the Margin.

Completion Example Answers

9. (a) The digit in the ten-thousandth position is **4.**
   (b) The next digit to the right is **1.**
   (c) Since **1** is less than 5, leave **4** as it is and replace **1** with a 0.
   (d) 8.00241 rounds off to **8.0024** (to the nearest **ten-thousandth**).

10. (a) The decimal point is understood to be to the right of **1.**
    (b) The digit in the hundreds position is **3.**
    (c) The next digit to the right is **6.**
    (d) Since **6** is greater than 5, change the **3** to **4** and replace **6** and **1** with 0's.
    (e) So, 7361 rounds off to **7400** (to the nearest hundred).

NAME ______________ SECTION ______________ DATE ______________

ANSWERS

Exercises 5.1

Write the following mixed numbers in decimal notation.

1. $37\frac{498}{1000}$ **2.** $18\frac{76}{100}$ **3.** $4\frac{11}{100}$

4. $87\frac{3}{1000}$ **5.** $95\frac{2}{10}$ **6.** $56\frac{3}{100}$

7. $62\frac{7}{10}$ **8.** $100\frac{25}{100}$

Write the following decimal numbers in mixed number form.

9. 82.56 **10.** 93.07 **11.** 10.576

12. 100.6 **13.** 65.003 **14.** 172.35

Write the following numbers in decimal notation.

15. three tenths **16.** fourteen thousandths

17. seventeen hundredths **18.** six and twenty-eight hundredths

19. sixty and twenty-eight thousandths

20. seventy-two and three hundred ninety-two thousandths

21. eight hundred fifty and thirty-six ten-thousandths

1. ______________
2. ______________
3. ______________
4. ______________
5. ______________
6. ______________
7. ______________
8. ______________
9. ______________
10. ______________
11. ______________
12. ______________
13. ______________
14. ______________
15. ______________
16. ______________
17. ______________
18. ______________
19. ______________
20. ______________
21. ______________

22. ______________

[Respond below exercise.]
23. ______________

[Respond below exercise.]
24. ______________

[Respond below exercise.]
25. ______________

[Respond below exercise.]
26. ______________

[Respond below exercise.]
27. ______________

[Respond below exercise.]
28. ______________

[Respond below exercise.]
29. ______________

[Respond below exercise.]
30. ______________

[Respond in exercise.]
31a. ______________

[Respond in exercise.]
b. ______________

[Respond in exercise.]
c. ______________

[Respond in exercise.]
d. ______________

[Respond in exercise.]
e. ______________

[Respond in exercise.]
32a. ______________

[Respond in exercise.]
b. ______________

[Respond in exercise.]
c. ______________

[Respond in exercise.]
d. ______________

33. ______________

34. ______________

35. ______________

36. ______________

37. ______________

38. ______________

22. seven hundred and seventy-seven thousandths

Write the following decimal numbers in words.

23. 0.5 **24.** 0.93 **25.** 5.06

26. 35.078 **27.** 7.003 **28.** 607.607

29. 10.4638 **30.** 600.615

Fill in the blanks to correctly complete each statement.

31. Round off 8472 to the nearest hundred.

(a) The decimal point is understood to be to the right of ______.

(b) The digit in the hundredths position is ______.

(c) The next digit to the right is ______.

(d) Since ______ is greater than 5, change the ______ to ______ and replace ______ and ______ with 0's.

(e) So, 8472 rounds off to ______________ (to the nearest hundred).

32. Round off 1.00643 to the nearest ten-thousandth.

(a) The digit in the ten-thousandths position is ______.

(b) The next digit to the right is ______.

(c) Since ______ is less than 5, leave ______ as it is and replace ______ with a 0.

(d) So, 1.00643 rounds off to ______________ (to the nearest ______________).

Round off each of the following numbers as indicated.

To the nearest tenth:

33. 4.763 **34.** 5.031 **35.** 76.349

36. 76.352 **37.** 89.015 **38.** 7.555

NAME ______________________ SECTION ______________________ DATE ______________

39. 18.009 **40.** 14.33382

To the nearest hundredth:

41. 0.385 **42.** 0.296 **43.** 5.7226

44. 8.9874 **45.** 6.99613 **46.** 13.13465

47. 0.0782 **48.** 6.0035

To the nearest thousandth:

49. 0.0672 **50.** 0.05550 **51.** 0.6338

52. 7.6666 **53.** 32.4785 **54.** 9.4302

55. 0.00191 **56.** 20.76962

To the nearest whole number (or nearest unit, or nearest one):

57. 479.23 **58.** 6.8 **59.** 19.999

60. 382.48 **61.** 649.66 **62.** 439.78

63. 6333.11 **64.** 8122.825

ANSWERS

39. ______
40. ______
41. ______
42. ______
43. ______
44. ______
45. ______
46. ______
47. ______
48. ______
49. ______
50. ______
51. ______
52. ______
53. ______
54. ______
55. ______
56. ______
57. ______
58. ______
59. ______
60. ______
61. ______
62. ______
63. ______
64. ______

65. ______________
66. ______________
67. ______________
68. ______________
69. ______________
70. ______________
71. ______________
72. ______________
73. ______________
74. ______________
75. ______________
76. ______________
77. ______________
78. ______________
79. ______________
80. ______________
81. ______________
82. ______________
83. ______________
84. ______________
85. ______________
86. ______________
87. ______________
88. ______________

To the nearest ten:

65. 5163 **66.** 6475 **67.** 495

68. 572.5 **69.** 998.5 **70.** 378.92

71. 92,540.9 **72.** 7007.7

To the nearest thousand:

73. 7398 **74.** 62,275 **75.** 47,823.4

76. 103,499 **77.** 217,480.2 **78.** 9872.5

79. 4,500,762 **80.** 573,333.3

81. 0.0005783 (nearest hundred-thousandth) **82.** 0.5449 (nearest hundredth)

83. 473.8 (nearest ten) **84.** 5.00632 (nearest thousandth)

85. 473.8 (nearest hundred) **86.** 5750 (nearest thousand)

87. 3.2296 (nearest thousandth) **88.** 15.548 (nearest tenth)

NAME ____________ SECTION ____________ DATE ____________

ANSWERS

89. 78,419 (nearest ten thousand)

90. 78,419 (nearest ten)

In each of the following exercises, write the decimal numbers that are not whole numbers in words.

91. One inch is equal to 2.54 centimeters.

92. One gallon of water weighs 8.33 pounds.

93. The winning car in the 1991 Indianapolis 500 race averaged 176.457 mph.

94. In 1990, the U.S. dollar was worth 1.3892 Swiss francs.

95. The number π is approximately equal to 3.14159.

96. The number **e** (used in higher level mathematics) is approximately equal to 2.71828.

97. One acre is approximately equal to 0.405 hectares.

98. One liter is approximately equal to 0.264 gallons.

89. ____________

90. ____________

91. [Respond below exercise.] ____________

92. [Respond below exercise.] ____________

93. [Respond below exercise.] ____________

94. [Respond below exercise.] ____________

95. [Respond below exercise.] ____________

96. [Respond below exercise.] ____________

97. [Respond below exercise.] ____________

98. [Respond below exercise.] ____________

99. [Respond below exercise.] ____________

100. [Respond below exercise.] ____________

Writing and Thinking about Mathematics

99. In your own words, state why the word **and** is so commonly misused when numbers are spoken and/or written. Bring an example of this from a newspaper, magazine, or television show, to class for discussion.

100. Why are hyphens used when some words are written? Give four examples of the use of hyphens when writing decimal numbers in word form.

RECYCLE BIN

1a. ____________

b. ____________

2a. ____________

b. ____________

3a. ____________

b. ____________

4a. ____________

b. ____________

5a. ____________

b. ____________

6a. ____________

b. ____________

7. ____________

The Recycle Bin (from Sections 1.2 and 1.3)

First (a) estimate each sum, then (b) find each sum.

1. $\begin{array}{r} 2093 \\ +7014 \\ \hline \end{array}$

2. $\begin{array}{r} 875 \\ 573 \\ +107 \\ \hline \end{array}$

3. $\begin{array}{r} 127{,}484 \\ 382{,}567 \\ +\ 52{,}661 \\ \hline \end{array}$

First (a) estimate each difference, then (b) find each difference.

4. $\begin{array}{r} 357 \\ -169 \\ \hline \end{array}$

5. $\begin{array}{r} 7844 \\ -3843 \\ \hline \end{array}$

6. $\begin{array}{r} 90{,}006 \\ -\ 5{,}115 \\ \hline \end{array}$

7. What number should be added to 850 in order to get a sum of 1200?

5.2 Addition and Subtraction (and Estimating)

OBJECTIVES

1. *Be able to add with decimal numbers.*
2. *Be able to subtract with decimal numbers.*
3. *Know how to estimate sums and differences with rounded-off decimal numbers.*

Addition with Decimal Numbers

We can add decimal numbers by writing each number in expanded notation and adding the whole numbers and fractions separately. Thus, we can emphasize that fractions with common denominators are added. For example,

$$\begin{aligned} 6.15 + 3.42 &= 6 + \frac{1}{10} + \frac{5}{100} + 3 + \frac{4}{10} + \frac{2}{100} \\ &= (6 + 3) + \left(\frac{1}{10} + \frac{4}{10}\right) + \left(\frac{5}{100} + \frac{2}{100}\right) \\ &= 9 + \frac{5}{10} + \frac{7}{100} \\ &= 9 + \frac{50}{100} + \frac{7}{100} \\ &= 9 + \frac{57}{100} \\ &= 9.57 \end{aligned}$$

Of course, addition of decimal numbers can be accomplished in a much easier way by writing the decimal numbers one under the other and keeping the decimal points aligned vertically. In this way, whole numbers are added to whole numbers, tenths to tenths, hundredths to hundredths, and so on (as we did with the fractions in the example just given). The decimal point in the sum is in line with the decimal points in the addends. Thus,

Decimal points are aligned vertically.

$$\begin{array}{r} 6.15 \\ +3.42 \\ \hline 9.57 \end{array}$$

Any number of 0's may be written to the right of the last digit in the fraction part of a number to help keep the digits in the correct alignment. This will not change the value of any number or the value of the sum.

To Add Decimal Numbers:

1. Write the addends in a vertical column.
2. Keep the decimal points aligned vertically.
3. Keep digits with the same position value aligned. (Zeros may be written in to help keep the digits aligned properly.)
4. Add the numbers, keeping the decimal point in the sum aligned with the other decimal points.

Find each sum.

1. 45.2 + 2.08 + 3.5 =

2. 4 + 5.7 + 0.63 =

3.
```
   17
    8.61
    5.004
+  29.19
```

EXAMPLE 1

Find the sum 6.3 + 5.42 + 14.07.

Solution

```
   6.30   ← 0 may be written in to help keep the digits aligned.
   5.42
+ 14.07
-------
  25.79
```

EXAMPLE 2

Find the sum 9 + 4.86 + 37.479 + 0.6.

Solution

```
   9.000   ← The decimal point is understood to be to the right of 9,
               as in 9.0, 9.00, or 9.000.
   4.860
  37.479
+  0.600   ← 0's are written in to help keep the digits aligned properly.
--------
  51.939
```

EXAMPLE 3

Add:
```
  56.2
  85.75
+29.001
```

Solution

You can write
```
   56.200   ← 0's written in to keep the digits in line.
   85.750
+  29.001
---------
  170.951
```

Now Work Exercises 1–3 in the Margin.

Subtraction with Decimal Numbers

To Subtract Decimal Numbers:

1. Write the addends in a vertical column.
2. Keep the decimal points aligned vertically.
3. Keep digits with the same position value aligned. (Zeros may be written in as aids.)
4. Subtract, keeping the decimal point in the difference aligned with the other decimal points.

EXAMPLE 4

Find the difference 16.715 − 4.823.

Solution

$$\begin{array}{r} 16.715 \\ -\ \ 4.823 \\ \hline 11.892 \end{array}$$

EXAMPLE 5

Find the difference 21.2 − 13.716.

Solution

$$\begin{array}{r} 21.200 \\ -13.716 \\ \hline 7.484 \end{array}$$ ← Write in 0's.

Now Work Exercises 4 and 5 in the Margin.

Estimating Sums and Differences

We can estimate a sum (or difference) by rounding off each number to the place of the **leftmost nonzero digit** and then adding (or subtracting) these rounded-off numbers. This technique of estimating answers is especially helpful when working with decimal numbers, where the placement of the decimal point is so important.

Note that, depending on the position of the leftmost nonzero digit, the rounded-off numbers may be whole numbers or decimal fractions. Also, different numbers within one problem might be rounded off to different places of accuracy.

Find each difference.

4. 50.036 − 47.58 =

5. 14 − 4.176 =

EXAMPLE 6

First (a) estimate the sum 74 + 3.529 + 52.61; then (b) find the actual sum.

Solution

(a) Estimate by adding rounded-off numbers.

74	rounds to →	70
3.529	rounds to →	4
52.61	rounds to →	+ 50
		124 ← estimate

(b) Find the actual sum.

$$\begin{array}{r} 74.000 \\ 3.529 \\ +\ 52.610 \\ \hline 130.139 \end{array} \leftarrow \text{actual sum}$$

NAME ______________ SECTION ______________ DATE ______________

ANSWERS

Exercises 5.2

Find each of the indicated sums. Estimate your answers, either mentally or on paper, before doing the actual calculations. Check to see that your sums are close to the estimated values.

1. 0.6 + 0.4 + 1.3

2. 5 + 6.1 + 0.4

3. 0.59 + 6.91 + 0.05

4. 3.488 + 16.593 + 25.002

5. 37.02 + 25 + 6.4 + 3.89

6. 4.0086 + 0.034 + 0.6 + 0.05

7. 43.766 + 9.33 + 17 + 206

8. 52.3 + 6 + 21.01 + 4.005

9. 2.051 + 0.2006 + 5.4 + 37

10. 5 + 2.37 + 463 + 10.88

1. ______________
2. ______________
3. ______________
4. ______________
5. ______________
6. ______________
7. ______________
8. ______________
9. ______________
10. ______________

11. ____________
12. ____________
13. ____________
14. ____________
15. ____________
16. ____________
17. ____________
18. ____________
19. ____________
20. ____________
21. ____________
22. ____________
23. ____________
24. ____________
25. ____________
26. ____________
27. ____________
28. ____________
29. ____________
30. ____________

11. $\begin{array}{r} 47.3 \\ 42.03 \\ +\ 29.003 \\ \hline \end{array}$

12. $\begin{array}{r} 1.007 \\ 20.063 \\ +\ 0.49 \\ \hline \end{array}$

13. $\begin{array}{r} 4.128 \\ 0.02 \\ +\ 3. \\ \hline \end{array}$

14. $\begin{array}{r} 5.0015 \\ 2.443 \\ +0.0469 \\ \hline \end{array}$

15. $\begin{array}{r} 75.2 \\ 3.682 \\ +14.995 \\ \hline \end{array}$

16. $\begin{array}{r} 107.39 \\ 5.061 \\ 23.54 \\ +\ 64.9801 \\ \hline \end{array}$

17. $\begin{array}{r} 34.967 \\ 50.6 \\ 8.562 \\ +\ 9.3 \\ \hline \end{array}$

18. $\begin{array}{r} 4.156 \\ 3.7 \\ 25.682 \\ +\ 13.405 \\ \hline \end{array}$

19. $\begin{array}{r} 74. \\ 3.529 \\ 52.62 \\ +\ 7.001 \\ \hline \end{array}$

20. $\begin{array}{r} 983.4 \\ 47.518 \\ 805.411 \\ +300.766 \\ \hline \end{array}$

Find each of the indicated differences. First estimate the differences mentally.

21. $5.2 - 3.76$

22. $17.83 - 8.9$

23. $29.5 - 13.61$

24. $1.0057 - 0.03$

25. $78.015 - 13.068$

26. $\begin{array}{r} 22.418 \\ -\ 17.523 \\ \hline \end{array}$

27. $\begin{array}{r} 4.8 \\ -\ 0.0026 \\ \hline \end{array}$

28. $\begin{array}{r} 31.009 \\ -\ 0.534 \\ \hline \end{array}$

29. $\begin{array}{r} 4. \\ -\ 1.0566 \\ \hline \end{array}$

30. $\begin{array}{r} 40.718 \\ -\ 6.532 \\ \hline \end{array}$

NAME ______________ SECTION ______________ DATE ______________

ANSWERS

31. ______________

32. ______________

33a. ______________

b. ______________

34. ______________

31. Theresa got a haircut for $30.00 and a manicure for $10.50. If she tipped the stylist $5, how much change did she receive from a $50 bill?

32. The inside radius of a pipe is 2.38 inches and the outside radius is 2.63 inches, as shown in the figure. What is the thickness of the pipe? (**Note:** The radius of a circle is the distance from the center of the circle to a point on the circle.)

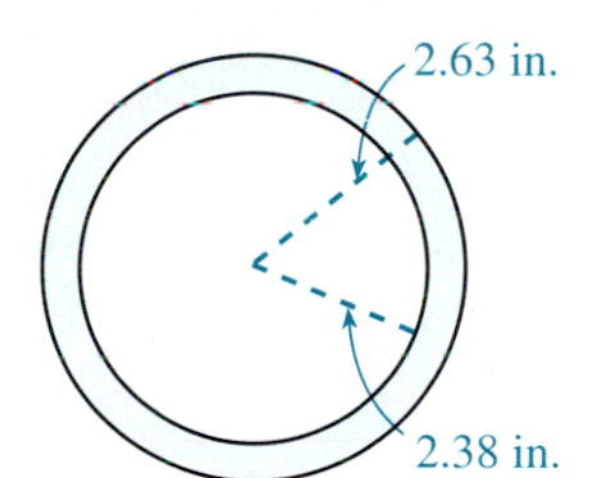

33. Mr. Johnson bought the following items at a department store: slacks, $32.50; shoes, $43.75; shirt, $18.60. (a) How much did he spend? (b) What was his change if he gave the clerk a $100 bill? (Tax was included in the prices.)

34. An architect's scale drawing shows a rectangular lot 2.38 inches on one side and 3.76 inches on the other side. What is the perimeter of (distance around) the rectangle in the drawing?

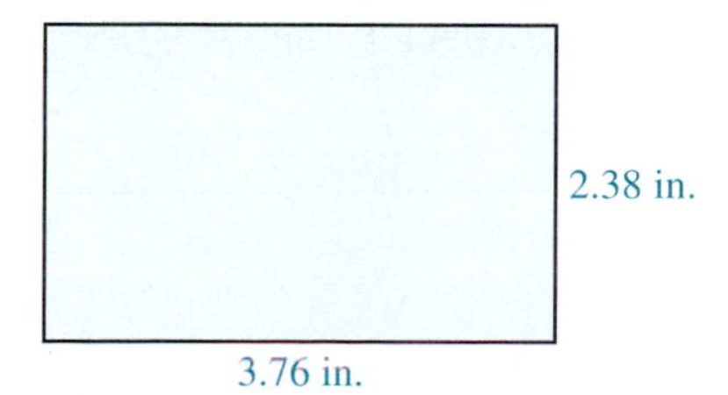

35a. ______________

b. ______________

36a. ______________

b. ______________

37. ______________

38. ______________

39. ______________

35. Mrs. Johnson bought the following items at a department store: dress, $47.25; shoes, $35.75; purse, $12.50. (a) How much did she spend? (b) What was her change if she gave the clerk a $100 bill? (Tax was included in the prices.)

36. Suppose your checking account shows a balance of $280.35 at the beginning of the month. During the month, you make deposits of $310.50, $200, and $185.50, and you write checks for $85.50, $210.20, and $600. (a) Estimate the balance at the end of the month. (b) Find the balance at the end of the month.

37. Your income for one month was $1800. You paid $570 for rent, $40 for electricity, $20 for water, and $35 for gas. **Estimate** what remained of your month's income after you paid those bills.

38. Martin wants to buy a car for $15,000. His credit union will loan him $10,500, but he must pay $280 for a license fee and $900 for taxes. **Estimate** the amount of cash he needs in order to buy the car.

39. In 1991 U.S. farmers produced 242.526 million bushels of oats, 464.495 million bushels of barley, and 1980.704 million bushels of wheat. What was the total combined amount of these grains produced in 1991?

NAME ______________ SECTION ______________ DATE ______________

ANSWERS

40. ______________

41. ______________

42. ______________

43. [Respond below exercise.] ______________

40. In 1990 Procter & Gamble spent $2284.5 million on advertising. In the same year, Johnson & Johnson spent $653.7 million and Bristol-Myers spent $428.7 million on advertising. How much more did Procter & Gamble spend on advertising than the other two companies combined?

41. The eccentricity of a planet's orbit is the measure of how much the orbit varies from a perfectly circular pattern. Mercury's orbit has an eccentricity of 0.205630, while Earth's orbit has an eccentricity of 0.016711. How much greater is Mercury's eccentricity of orbit than Earth's?

42. In a recent year, Boston, MA received 42.25 inches of rain and 23.7 inches of snow, while Burlington, VT received 32.52 inches of rain and 42.1 inches of snow. How much more rain was there in Boston than in Burlington? How much more snow was there in Burlington than in Boston?

Check Your Number Sense

43. Suppose that you were to add two decimal numbers with 0 as their whole number parts. Can their sum possibly be (a) more than 2? (b) more than 1? (c) less than 1? Explain briefly.

[Respond below exercise.]

44. ______

44. Suppose that you were to subtract two decimal numbers with 0 as their whole number parts. Can their difference possibly be (a) more than 1? (b) less than 1? (c) equal to 1? Explain briefly.

RECYCLE BIN

1. ______

2. ______

3. ______

4. ______

5. ______

The Recycle Bin (from Section 1.4)

Find each of the following products.

1. $50 \cdot 7000$

2. $\begin{array}{r} 35 \\ \times\ 15 \\ \hline \end{array}$

3. $\begin{array}{r} 195 \\ \times\ \ 36 \\ \hline \end{array}$

4. First estimate the following product, and then find the product.

$\begin{array}{r} 4571 \\ \times\ \ 375 \\ \hline \end{array}$

5. Find the sum of seventy-five and ninety-three and the difference between one hundred sixty-four and eighty-five. Then, find the product of the sum and the difference.

5.3 Multiplication (and Estimating)

OBJECTIVES

1. *Be able to multiply decimal numbers and place the decimal point correctly in the product.*
2. *Be able to multiply decimal numbers mentally by powers of 10.*
3. *Know how to estimate products with rounded-off decimal numbers.*

Multiplying Decimal Numbers

The method for multiplying decimal numbers is similar to that for multiplying whole numbers. The only difference is the need to determine how to place the decimal point in the product when multiplying decimal numbers. To illustrate how to place the decimal point in a product, several products are shown here in both fraction form and decimal form. The denominators in the fractions are all powers of ten.

Products in Fraction Form	**Products in Decimal Form**
$\frac{1}{10} \cdot \frac{1}{100} = \frac{1}{1000}$	.1 ← 1 place; × .01 ← 2 places; total of 3 places (thousandths); = .001
$\frac{3}{10} \cdot \frac{5}{100} = \frac{15}{1000}$	.3 ← 1 place; × .05 ← 2 places; total of 3 places (thousandths); = .015
$\frac{6}{100} \cdot \frac{4}{1000} = \frac{24}{100{,}000}$	.004 ← 3 places; × .06 ← 2 places; total of 5 places (hundred thousandths); = .00024

As these illustrations indicate, there is **no need to keep the decimal points lined up for multiplication.** The following rule states how to multiply two decimal numbers and place the decimal point in the product.

To Multiply Decimal Numbers:

1. Multiply the two numbers as if they were whole numbers.
2. Count the total number of places to the right of the decimal points in both numbers being multiplied.
3. Place the decimal point in the product so that the number of places to the right of it is the same as that found in Step 2.

EXAMPLE 1

```
     2.432   ← 3 places }
  ×    5.1   ← 1 place  } total of 4 places
  --------
      2432
    12 160
  --------
   12.4032   ← 4 places in the product
```

Find each product.

1. (0.8)(0.2) =

2. (5.6)(0.04) =

3. 3.781 × 3.01

Example 2

Multiply 4.35 × 12.6.

$$\begin{array}{r} 4.35 \\ \times\ 12.6 \\ \hline 2\,610 \\ 8\,70 \\ 43\,5 \\ \hline 54.810 \end{array}$$

4.35 ← 2 places; 12.6 ← 1 place; total of 3 places

54.810 ← 3 places in the product

Example 3

Multiply (0.046)(0.007).

$$\begin{array}{r} 0.046 \\ \times\ 0.007 \\ \hline 0.000322 \end{array}$$

0.046 ← 3 places; 0.007 ← 3 places; total of 6 places

0.000322 ← 6 places in the product

This means that three 0's had to be inserted between the 3 and the decimal point.

Completion Example 4

$$\begin{array}{r} 3.4 \\ \times\ 5.8 \\ \hline 2\,72 \\ 17\,0 \\ \hline \_\_\_\_ \end{array}$$

3.4 ← ____ place; 5.8 ← ____ place; total of ____ places

________ ____ places in the product

Completion Example 5

Multiply 0.003 × 0.03.

$$\begin{array}{r} 0.003 \\ \times\ 0.03 \\ \hline \_\_\_\_ \end{array}$$

0.003 ← ____ places; 0.03 ← ____ places; total of ____ places

________ ____ places in the product

____ 0's had to be inserted to place the decimal point.

Now Work Exercises 1–3 in the Margin.

Multiplying Decimal Numbers by Powers of 10

In Section 1.4, we discussed multiplication of whole numbers by powers of ten by placing 0's to the right of the number. Now that decimal numbers have been introduced, we can see that inserting 0's in a whole number has the effect of moving the decimal point to the right. (**Note:** Even though we do not always write the decimal point in a whole number, it is understood to be just to the right of the rightmost digit.) The following more general guidelines can be used to multiply all decimal numbers by powers of 10.

To Multiply a Decimal Number by a Power of 10:

1. Move the decimal point to the right.
2. Move it the same number of places as the number of 0's in the power of 10.

Multiplication by **10** moves the decimal point **one** place **to the right.**

Multiplication by **100** moves the decimal point **two** places **to the right.**

Multiplication by **1000** moves the decimal point **three** places **to the right.**

And so on.

***Example* 6**

The following products illustrate multiplication powers of 10.

(a) $10(2.68) = 26.8$ — Move the decimal point 1 place to the right.

(b) $100(2.68) = 268. = 268$ — Move the decimal point 2 places to the right.

(c) $1000(0.9653) = 965.3$ — Move the decimal point 3 places to the right.

(d) $1000(7.2) = 7200. = 7200$ — Move the decimal point 3 places to the right.

(e) $10^2(3.5149) = 351.49$ — Move the decimal point 2 places to the right. The exponent tells how many places to move the decimal point.

Now Work Exercises 4–6 in the Margin.

Find each product mentally by using your knowledge of multiplication by powers of 10.

4. $8.75 \times 10 =$

5. $100(6.3) =$

6. $1000 \times 0.1894 =$

In the metric system of measurement, units of length are set up so that there are 10 of one unit in the next larger unit. For example, there are 10 millimeters (mm) in 1 centimeter (cm), and there are 10 cm in 1 decimeter (dm), and there are 10 dm in 1 meter (m). Therefore, to change from any number of a particular unit of length in the metric system to a smaller unit of measure, we multiply by some power of 10 (or simply move the decimal point to the right). Table 5.1 illustrates some basic relationships between units of length in the metric system. (**Note:** There are other units of measure in the metric system for weight, area, and volume. The metric system is discussed in detail later in the text.

Table 5.1 Basic Units of Measure of Length in the Metric System

1 m = 10 dm	1 dm = 10 cm	1 cm = 10 mm
1 m = 100 cm	1 m = 1000 mm	
1 m = 10 dm = 100 cm = 1000 mm		
1 km = 1000 m		

Find the equivalent measures in the metric system.

7. 5.6 m = ______ cm

8. 35.25 cm = ______ mm

9. 16.43 km = ______ m

Changing Metric Measures of Length

To change to a measure that is

one unit smaller, multiply by 10.	3 cm = 30 mm
two units smaller, multiply by 100.	5 m = 500 cm
three units smaller, multiply by 1000.	14 m = 14 000 mm

And so on.

Example 7

The following examples illustrate how to change from larger to smaller units of length in the metric system. (Refer to Table 5.1 if you need help.)

(a) 4.32 m = 100(4.32) cm = 432 cm

(b) 4.32 m = 1000(4.32) mm = 4320 mm

(c) 14.6 cm = 10(14.6) mm = 146 mm

(d) 3.51 km = 1000(3.51) m = 3510 m

Now Work Exercises 7–9 in the Margin.

As an aid in understanding relative lengths in the metric system, we show some comparisons to the U.S. customary system here.

1 meter is about 39.36 inches.

1 meter is about 3.28 feet.

1 centimeter is about 0.394 inches.

1 kilometer is about 0.62 miles.

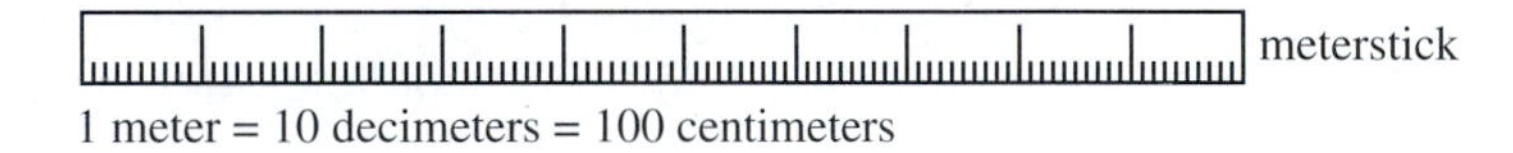

1 meter = 10 decimeters = 100 centimeters

yardstick
1 yard = 3 feet = 36 inches

ruler
1 foot = 12 inches

Estimating Products of Decimal Numbers

Estimating products can be done by rounding off each number to the place of the last nonzero digit on the left and multiplying these rounded-off numbers. This technique is particularly helpful in correctly placing the decimal point in the actual product.

EXAMPLE 8

First (a) estimate the product (0.356)(6.1); then (b) find the product and use the estimation to help place the decimal point.

Solution

(a) Estimate by multiplying rounded-off numbers.

$$\begin{array}{r l} 0.4 & \text{(0.356 rounded off)} \\ \times\ 6 & \text{(6.1 rounded off)} \\ \hline 2.4 & \leftarrow \text{estimate} \end{array}$$

(b) Find the actual product.

$$\begin{array}{r l} 0.356 & \\ \times\ \ 6.1 & \\ \hline 356 & \\ 2\,136 & \\ \hline 2.1716 & \leftarrow \text{actual product} \end{array}$$

The estimated product 2.4 helps place the decimal point correctly in the product 2.1716.

EXAMPLE 9

First (a) find the product (19.9)(2.3); then (b) estimate the product and use this estimation as a check.

Solution

(a) Find the actual product.

$$\begin{array}{r l} 19.9 & \\ \times\ \ 2.3 & \\ \hline 5\,97 & \\ 39\,8 & \\ \hline 45.77 & \leftarrow \text{actual product} \end{array}$$

(b) Estimate the product and use this estimation to check that the actual product is reasonable.

$$\begin{array}{r l} 20.0 & \text{(19.9 rounded off)} \\ \times\ \ \ 2 & \text{(2.3 rounded off)} \\ \hline 40.0 & \leftarrow \text{estimated product} \end{array}$$

The estimated product of 40.0 indicates that the actual product is reasonable and the placement of the decimal point is correct.

Note that even though estimated products (or estimated sums or estimated differences) are close to the actual value, they do not guarantee the absolute accuracy of this actual value. The estimates help only in placing decimal points and checking the reasonableness of answers. Experience and understanding are needed to judge whether or not a particular answer is reasonably close to an estimate or vice versa.

Word Problems

Some word problems may involve several operations with decimal numbers. The words do not usually say directly to add, subtract, or multiply. Experience and reasoning abilities are needed to decide which operation (if any) to perform with the given numbers. In Example 10, we illustrate a problem that involves several steps, and we show how estimating can provide a check for a reasonable answer.

EXAMPLE 10

You can buy a car for \$7500 cash or you can make a down payment of \$1875 and then pay \$546.67 each month for 12 months. How much can you save by paying cash?

Solution

(a) Find the amount paid in monthly payments by **multiplying** the amount of each payment by 12. In this case, judgment dictates that we use 12 and do not round off to 10, since we do not want to lose two full monthly payments in our estimate. (See Exercise 78 to understand what happens if 10 is used in the estimated calculations.)

Estimate

$$\begin{array}{r} \$500 \\ \times \quad 12 \\ \hline 1000 \\ 500 \\ \hline \$6000 \end{array}$$ estimated monthly payments

Actual Amount

$$\begin{array}{r} \$546.67 \\ \times \quad 12 \\ \hline 1093\ 34 \\ 5466\ 7 \\ \hline \$6560.04 \end{array}$$ paid in monthly payments

(b) Find the total amount paid by **adding** the down payment to the answer in part (a).

Estimate

$$\begin{array}{rl} \$2000 & \text{down payment} \\ +\ 6000 & \text{monthly payments} \\ \hline \$8000 & \text{estimated total} \end{array}$$

Actual Amount

$$\begin{array}{rl} \$1875.00 & \text{down payment} \\ +\ 6560.04 & \text{monthly payments} \\ \hline \$8435.04 & \text{total paid} \end{array}$$

(c) Find the savings by **subtracting** \$7500 (the cash price) from the answer in part (b).

Estimate

$$\begin{array}{rl} \$8000 & \text{estimated total} \\ -\ 7500 & \text{cash price} \\ \hline \$\ 500 & \text{estimated savings} \end{array}$$

Actual Amount

$$\begin{array}{rl} \$8435.04 & \text{total paid} \\ -\ 7500.00 & \text{cash price} \\ \hline \$\ 935.04 & \text{savings by paying cash} \end{array}$$

The estimated \$500 saved by paying cash is reasonably close to the actual savings of \$935.04.

Completion Example Answers

4.
$$\begin{array}{rl} 3.4 & \leftarrow \textbf{1} \text{ place} \\ \times\ 5.8 & \leftarrow \textbf{1} \text{ place} \end{array} \Big\} \text{ total of } \textbf{2} \text{ places}$$
$$\begin{array}{r} 2\ 72 \\ 17\ 0 \\ \hline \mathbf{19.72} \end{array}$$ **2** places in the product

5.
$$\begin{array}{rl} 0.003 & \leftarrow \textbf{3} \text{ places} \\ \times\ 0.03 & \leftarrow \textbf{2} \text{ places} \\ \hline \mathbf{0.00009} & \end{array} \Big\} \text{ total of } \textbf{5} \text{ places}$$
5 places in the product
4 0's had to be inserted to place the decimal point.

NAME ______________ SECTION ______________ DATE ______________

Exercises 5.3

1. Match each indicated product with the best estimate of that product.

______	(a) (0.7)(0.8)	A. 0.06
______	(b) (34.5)(0.11)	B. 1.0
______	(c) (0.63)(9.81)	C. 3.0
______	(d) (0.34)(0.18)	D. 5.0
______	(e) (4.6)(1.2)	E. 6.0

2. Match each indicated product with the best estimate of that product.

______	(a) 1.75(0.04)	A. 0.008
______	(b) 1.75(0.004)	B. 0.08
______	(c) 17.5(0.04)	C. 0.8
______	(d) 1.75(4)	D. 8.0

Find each of the indicated products.

3. (0.6)(0.7)
4. (0.3)(0.8)
5. (0.2)(0.2)
6. (0.3)(0.3)
7. 8(2.7)
8. 4(9.6)
9. 1.4(0.3)
10. 1.5(0.6)
11. (0.2)(0.02)
12. (0.3)(0.03)
13. 5.4(0.02)
14. 7.3(0.01)
15. 0.23×0.12
16. 0.15×0.15
17. 8.1×0.006
18. 7.1×0.008
19. 0.06×0.01
20. 0.25×0.01

ANSWERS

1a. [Respond in exercise.]
b. [Respond in exercise.]
c. [Respond in exercise.]
d. [Respond in exercise.]
e. [Respond in exercise.]
2a. [Respond in exercise.]
b. [Respond in exercise.]
c. [Respond in exercise.]
d. [Respond in exercise.]
3. ______
4. ______
5. ______
6. ______
7. ______
8. ______
9. ______
10. ______
11. ______
12. ______
13. ______
14. ______
15. ______
16. ______
17. ______
18. ______
19. ______
20. ______

21. ______
22. ______
23. ______
24. ______
25. ______
26. ______
27. ______
28. ______
29. ______
30. ______
31. ______
32. ______
33. ______
34. ______
35. ______
36. ______
37. ______
38. ______
39. ______
40. ______
41. ______
42. ______
43. ______
44. ______
45. ______

21. 3(0.125)

22. 4(0.375)

23. 1.6(0.875)

24. 5.3(0.75)

25. 6.9(0.25)

26. 4.8(0.25)

27. 0.83(6.1)

28. 0.27(0.24)

29. 0.16(0.5)

30. 0.28(0.5)

Find each of the following products mentally by using your knowledge of multiplication by powers of ten.

31. 100(3.46)

32. 100(20.57)

33. 100(7.82)

34. 100(6.93)

35. 100(16.1)

36. 100(38.2)

37. 10(0.435)

38. 10(0.719)

39. 10(1.86)

40. 1000(4.1782)

41. 1000(0.38)

42. 1000(0.47)

43. 10,000 × 0.005

44. 10,000 × 0.00615

45. 10,000 × 7.4

NAME ____________ SECTION ____________ DATE ____________

Find the equivalent measures in the metric system. (Refer to Table 5.1 for help.)

46. 5 cm = ______ mm **47.** 13 cm = ______ mm **48.** 3 m = ______ cm

49. 15 m = ______ cm **50.** 3.2 m = ______ mm **51.** 6.17 m = ______ mm

52. 6.5 km = ______ m **53.** 16 km = ______ m **54.** 0.5 km = ______ m

55. 0.6 km = ________ m = ________ cm = ________ mm

56. 2.53 km = ________ m = ________ cm = ________ mm

57. 0.02 km = ________ m = ________ cm = ________ mm

58. 10.7 km = ________ m = ________ cm = ________ mm

In Exercises 59–70, first estimate the product, and then find the actual product.

59. $\begin{array}{r} 0.106 \\ \times\ 0.09 \\ \hline \end{array}$ **60.** $\begin{array}{r} 1.07 \\ \times\ 0.5 \\ \hline \end{array}$ **61.** $\begin{array}{r} 5.08 \\ \times\ 0.4 \\ \hline \end{array}$

62. $\begin{array}{r} 0.0106 \\ \times\ 0.087 \\ \hline \end{array}$ **63.** $\begin{array}{r} 0.0213 \\ \times\ 0.065 \\ \hline \end{array}$ **64.** $\begin{array}{r} 83.105 \\ \times\ 0.111 \\ \hline \end{array}$

ANSWERS

46. [Respond in exercise.] ____________

47. [Respond in exercise.] ____________

48. [Respond in exercise.] ____________

49. [Respond in exercise.] ____________

50. [Respond in exercise.] ____________

51. [Respond in exercise.] ____________

52. [Respond in exercise.] ____________

53. [Respond in exercise.] ____________

54. [Respond in exercise.] ____________

55. [Respond in exercise.] ____________

56. [Respond in exercise.] ____________

57. [Respond in exercise.] ____________

58. [Respond in exercise.] ____________

59. ____________

60. ____________

61. ____________

62. ____________

63. ____________

64. ____________

65. ____________

66. ____________

67. ____________

68. ____________

69. ____________

70. ____________

71. ____________

72a. ____________

b. ____________

73a. ____________

b. ____________

74. ____________

65. $\begin{array}{r} 17.002 \\ \times\ 0.101 \\ \hline \end{array}$

66. $\begin{array}{r} 86.1 \\ \times\ 0.057 \\ \hline \end{array}$

67. $\begin{array}{r} 7.83 \\ \times\ 0.18 \\ \hline \end{array}$

68. $\begin{array}{r} 95.62 \\ \times\ 0.57 \\ \hline \end{array}$

69. $\begin{array}{r} 6.02 \\ \times\ 0.57 \\ \hline \end{array}$

70. $\begin{array}{r} 8.034 \\ \times\ 0.29 \\ \hline \end{array}$

71. To buy a car, you may pay $2036.50 in cash, or you may put down $400 and make 18 monthly payments of $104.30. How much would you save by paying cash?

72. Suppose a tax assessor figures the tax at 0.07 of the assessed value of a home. (a) If the assessed value is determined at a rate of 0.32 times the market value, what taxes are paid on a home with a market value of $136,500? (b) Estimate these taxes and check to see that the actual taxes are close to the estimate.

73. (a) If the sale price of a new refrigerator is $583 and sales tax is figured at 0.06 times the price, approximately what amount is paid for the refrigerator? (b) What is the exact amount paid for the refrigerator?

74. If you were paid a salary of $350 per week and $13.75 for each hour you worked over 40 hours in a week, how much would you make if you worked 45 hours in one week?

NAME ______________ SECTION ______________ DATE ______________

ANSWERS

75. ______________

76. ______________

77. ______________

[Respond below exercise.]
78. ______________

75. Find the perimeter and area of a square with sides 3 ft long.

76. In 1990 New York state led the nation in per capita funding for its public libraries with $30.42 spent for each person. How much funding would have been received that year by a library that served a town of 23,500 people?

77. In 1991 the average price per pound received by U.S. farmers for cattle was $0.726. At this price, what would be the value of 42,500 pounds of cattle?

Writing and Thinking about Mathematics

78. In Example 10, we stated the following problem:

> You can buy a car for $7500 cash or you can make a down payment of $1875 and pay $546.67 each month for 12 months. How much can you save by paying cash?

Estimate the savings by using rounded-off numbers. (This includes rounding off 12 months to 10 months.) Explain why this estimated savings does not seem reasonable, and explain why we must be careful about using rounded-off numbers in practical applications.

79. [Respond below exercise.]

80. [Respond below exercise.]

79. Suppose you are interested only in a rounded-off answer for a product. Would there be any difference in the products produced by the following two procedures?

(a) First multiply the two numbers as they are and then round off the product to the desired place of accuracy.
(b) First round off each number to the desired place of accuracy and then multiply the rounded-off numbers.

Explain, in your own words, why you think these two procedures would produce the same result or different results.

80. We stated in the text that 1 meter is about 39.36 inches. Use each of the techniques discussed in Exercise 79 to find (to the nearest tenth of an inch) how many inches are in 17.523 meters. Discuss why the results are different (or the same).

RECYCLE BIN

1. ______
2. ______
3. ______
4. ______
5. [Respond below exercise.]

The Recycle Bin (from Section 1.5)

Find the whole number quotients and remainders for each of the following.

1. $15\overline{)120}$ **2.** $20\overline{)315}$ **3.** $50\overline{)4057}$ **4.** $230\overline{)46790}$

5. The state of Alaska has a population of about 550,000 people and a land area of about 570,000 square miles (or about 1,480,000 square kilometers). (a) Approximately how many square miles are there in Alaska for each person? (b) About how many square kilometers are there for each person?

5.4 Division (and Estimating)

OBJECTIVES

1. *Be able to divide decimal numbers and place the decimal point correctly in the quotient.*
2. *Be able to divide decimal numbers mentally by powers of 10.*
3. *Know how to estimate quotients with rounded-off decimal numbers.*

Dividing Decimal Numbers

The process of division (called the division algorithm) with decimal numbers is, in effect, the same as that for division with whole numbers. This is reasonable because whole numbers are decimal numbers; that is, we can always write the decimal point to the right of a whole number if we so choose. As the following example illustrates, division with whole numbers gives a quotient and, possibly, a remainder.

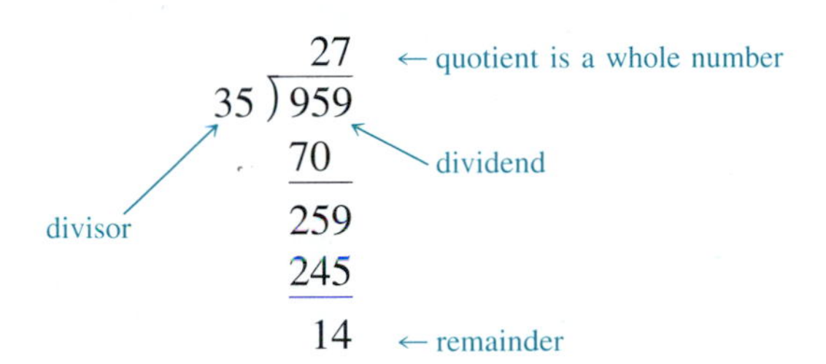

Now that we have decimal numbers, we can continue to divide and get a quotient that is a decimal number other than a whole number. Zeros are added onto the dividend if they are needed.

$$\begin{array}{r} 27.4 \\ 35\,\overline{)\,959.0} \\ \underline{70} \\ 259 \\ \underline{245} \\ 14\ 0 \\ \underline{14\ 0} \\ 0 \end{array} \quad \begin{array}{l} \leftarrow \text{quotient is a decimal number} \\ \leftarrow 0 \text{ added on} \end{array}$$

If the divisor is a decimal number other than a whole number, multiply both the divisor and the dividend by a power of 10 so that the divisor is a whole number. For example, we can write

$$4.9\,\overline{)\,51.45} \quad \text{as} \quad \frac{51.45}{4.9} \cdot \frac{10}{10} = \frac{514.5}{49} \quad \text{or} \quad 49\,\overline{)\,514.5}$$

By multiplying both the divisor (4.9) and the dividend (51.45) by 10, we have a new divisor (49) that is a whole number. In a similar manner, we can write

$$1.36\,\overline{)\,5.1} \quad \text{as} \quad \frac{5.1}{1.36} \cdot \frac{100}{100} = \frac{510}{136} \quad \text{or} \quad 136\,\overline{)\,510.}$$

In this case, both the divisor (1.36) and the dividend (5.1) are multiplied by 100 so that we have a new whole number divisor (136).

This discussion leads to the following procedure for dividing decimal numbers.

To Divide Decimal Numbers:

1. Move the decimal point in the divisor to the right so that the divisor is a whole number.
2. Move the decimal point in the dividend the same number of places to the right.
3. Place the decimal point in the quotient directly above the new decimal point in the dividend. **(Note: Be sure to do this before dividing.)**
4. Divide just as you would with whole numbers. (**Note:** 0's may be added in the dividend as needed to be able to continue the division process.)

EXAMPLE 1

Divide: 51.45 ÷ 4.9.

Solution

(a) Write the problem as follows:

$4.9\overline{)51.45}$

(b) To move the decimal point so that the divisor is a whole number, move each decimal point one place. This makes the whole number 49 the divisor. Place the decimal point in the quotient before dividing.

. ← decimal point in quotient

$4.9.\overline{)51.4.5}$

(c) Divide as with whole numbers.

$$
\begin{array}{r}
10.5 \\
49.\overline{)514.5} \\
\underline{49} \\
24 \\
\underline{0} \\
24\ 5 \\
\underline{24\ 5} \\
0
\end{array}
$$

EXAMPLE 2

Divide: 5.1 ÷ 1.36.

Solution

(a) Write the problem as follows:

$1.36\overline{)5.1}$

(b) Move the decimal points so the divisor is a whole number. Add 0's in the dividend if needed. Place the decimal point in the quotient before dividing.

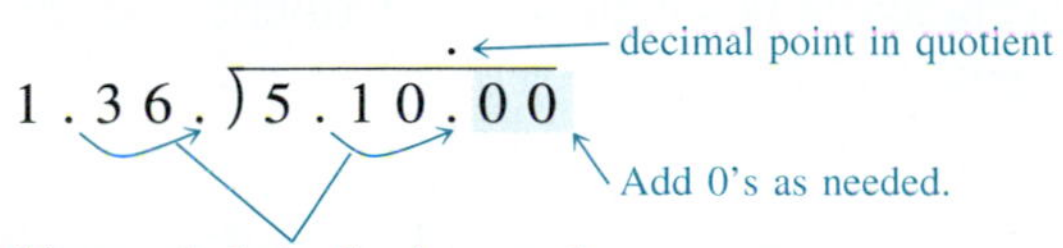

(c) Divide.

```
       3.75
136. )510.00
      408
      102 0
       95 2
        6 80
        6 80
           0
```

EXAMPLE 3

Divide: $6.3252 \div 6.3$.

Solution

```
      1.004
6.3. )6.3.252
      6 3
        0 2
          0
         25
          0
        252
        252
          0
```

Note: There **must** be a digit to the right of the decimal point in the quotient above every digit to the right of the decimal point in the dividend.

COMPLETION EXAMPLE 4

Divide: $24.225 \div 4.25$.

```
          5.
4.25. ) 24.22.5
        ______

          ____
```

Now Work Exercises 1–3 in the Margin.

Find each quotient.

1. $9.87 \div 2.1$

2. $9.393 \div 9.3$

3. $10.01 \div 2.86$

In each of Examples 1–4, the remainder was 0. Certainly, this will not always be the case. In general, some place of accuracy for the quotient is agreed upon before the division is performed. If the remainder is not 0 by the time this place of accuracy is reached, then we divide one more place and round off the quotient.

When the Remainder is not 0:

1. Decide first how many decimal places are to be in the quotient.
2. Divide until the quotient is **one digit past the place of desired accuracy.**
3. Using this last digit, round off the quotient to the desired place of accuracy.

EXAMPLE 5 Find $8.24 \div 2.9$ to the nearest tenth.

Solution

Divide until the quotient is in hundredths (one more place than tenths); then round off to tenths.

```
             hundredths
               read approximately
        2.84 ≈ 2.8
2.9. ) 8.2.40     rounded off to tenths
       5 8
       2 4 4
       2 3 2
         1 20
         1 16
            4
```

$8.24 \div 2.9 \approx 2.8$ accurate to the nearest tenth

EXAMPLE 6 Find $1.83 \div 4.1$ to the nearest hundredth.

Solution

Divide until the quotient is in thousandths (one more place than hundredths); then round off to hundredths.

```
                          read approximately
            0 . 4 4 6 ≈ 0 . 4 5
4 . 1 . ) 1 . 8 . 3 0 0
          1   6   4      Add as many 0's as needed.
              1   9 0
              1   6 4
                  2 6 0
                  2 4 6
                    1 4
```

$1.83 \div 4.1 \approx 0.45$ accurate to the nearest hundredth

Example 7

Find 17 ÷ 3.3 to the nearest thousandth.

Solution

Divide until the quotient is in ten-thousandths; then round off to thousandths.

```
              5.1515 ≈ 5.152
3.3.)17.0.0000
      16 5           Add as many 0's as needed.
         5 0
         3 3
         1 70
         1 65
            50
            33
            170
            165
              5
```

$17 \div 3.3 \approx 5.152$ accurate to thousandths

Now Work Exercises 4 and 5 in the Margin.

4. Find 1.83 ÷ 7 (to the nearest hundredth).

5. Find 43.721 ÷ 0.06 (to the nearest thousandth).

Dividing Decimal Numbers by Powers of 10

In Section 5.3, we found that multiplication by powers of 10 can be accomplished by moving the decimal point to the right (making a number larger than the original number). Division by powers of 10 can be accomplished by moving the decimal point to the left (making a number smaller than the original number).

To Divide a Decimal Number by a Power of 10:

1. Move the decimal point to the left.
2. Move it the same number of places as the number of 0's in the power of 10.

Division by **10** moves the decimal point **one** place **to the left.**

Division by **100** moves the decimal point **two** places **to the left.**

Division by **1000** moves the decimal point **three** places **to the left.**

And so on.

Find each quotient by using your knowledge of division by powers of 10.

6. $42.31 \div 100$

7. $328 \div 10$

8. $\dfrac{57.6}{10{,}000}$

Two general guidelines will help you understand work with powers of 10:

1. Multiplication by a power of 10 will make a number larger, so move the decimal point to the right.
2. Division by a power of 10 will make a number smaller, so move the decimal point to the left.

Example 8

The following quotients illustrate division by powers of 10.

(a) $4.16 \div 100 = \dfrac{4.16}{100} = 0.0416$ Move the decimal point two places to the left.

(b) $782 \div 10 = \dfrac{782.}{10} = 78.2$ Move the decimal point one place to the left.

(c) $5.933 \div 1000 = \dfrac{5.933}{1000} = 0.005933$ Move the decimal point three places to the left.

Now Work Exercises 6–8 in the Margin.

As we stated in Section 5.3, the units of length in the metric system are set up so that there are 10 of one unit in the next larger unit. (This relationship is also true for measures of weight and liquid volume.) To change from larger units of length to smaller units of length, we multiply by an appropriate power of 10 (as we did in Section 5.3). Now, to reverse this process (that is, to change from smaller units of length to larger units of length), we divide by an appropriate power of 10. (Refer to Table 5.1 to review the basic relationships of length in the metric system.)

Changing Metric Measures of Length

To change to a measure that is	**Example**
one unit larger, divide by 10.	4 mm = 0.4 cm
two units larger, divide by 100.	6 mm = 0.06 dm
three units larger, divide by 1000.	8 mm = 0.008 m
And so on.	

Example 9

The following examples illustrate changing from smaller to larger units of length in the metric system. (Refer to Table 5.1 if you need help.)

(a) $564 \text{ cm} = \dfrac{564}{100} \text{ m} = 5.64 \text{ m}$

(b) $564 \text{ mm} = \dfrac{564}{1000} \text{ m} = 0.564 \text{ m}$

(c) $1030 \text{ m} = \dfrac{1030}{1000} \text{ km} = 1.03 \text{ km}$

Now Work Exercises 9–11 in the Margin.

Find the equivalent measures in the metric system.

9. 350 cm = ______ m

10. 68 mm = ______ cm

11. 3952 mm = ______ m

Estimating Quotients of Decimal Numbers

As with addition, subtraction, and multiplication, we can use estimating with division to help in placing the decimal point in the quotient and to verify the reasonableness of the quotient. The technique is used to round off both the divisor and the dividend to the place of the last nonzero digit on the left and then divide with these rounded-off values.

EXAMPLE 10

First (a) estimate the quotient $6.2 \div 0.302$, and then (b) find the quotient to the nearest tenth.

Solution

(a) Estimate the quotient using $6.2 \approx 6$ and $0.302 \approx 0.3$.

```
       2 0.
0.3 ) 6.0
      6
      0 0
        0
        0
```

(b) Find the quotient to the nearest tenth.

```
           20.52 ≈ 20.5
0.302 ) 6.200 00
        6 04
          160
            0
          160 0
          151 0
            9 00
            6 04
```

The estimated value 20 is very close to the rounded-off quotient 20.5.

We know from Section 1.6 that the **average** (or **arithmetic average**) of a set of numbers can be found by adding the numbers, then dividing the sum by the number of addends. The term **average** is also used in phrases such as "average speed of 43 miles per hour" or "the average price of a pair of shoes." These averages can also be found by division. If the total amount of a quantity (distance, dollars, gallons of gas, etc.) and a number of units (time, items bought, miles, etc.) are known, then we can find the **average amount per unit** by dividing the amount by the number of units.

Example 11

The gas tank of a car holds 17 gallons of gasoline. Approximately how many miles per gallon does the car average if it will go 470 miles on one tank of gas?

Solution

The question calls only for an approximate answer. Thus, we can use rounded-off values:

$$17 \approx 20 \text{ gal} \quad \text{and} \quad 470 \approx 500$$

Now divide to approximate the **average** number of miles per gallon:

$$\begin{array}{r} 25 \\ 20\overline{)500} \\ \underline{40} \\ 100 \\ \underline{100} \\ 0 \end{array} \quad \text{miles per gallon}$$

The car averages about 25 miles per gallon.

If an average amount per unit is known, then a corresponding total amount can be found by multiplying. For example, if you ride your bicycle at an average speed of 15.2 miles per hour, then the distance you travel can be found by multiplying your average speed by the time you spend riding.

Example 12

If you ride your bicycle at an average speed of 15.2 miles per hour, how far will you ride in 3.5 hours?

Solution

Multiply the average speed by the number of hours.

$$\begin{array}{rl} 15.2 & \text{miles per hour} \\ \times \quad 3.5 & \text{hours} \\ \hline 7\,60 & \\ 45\,6 & \\ \hline 53.20 & \text{miles} \end{array}$$

You will ride 53.2 miles in 3.5 hours.

Completion Example Answer

4.

$$\begin{array}{r} 5.7 \\ 4.25.\overline{)24.22.5} \\ \underline{\mathbf{21\ 25}} \\ \mathbf{2\ 975} \\ \underline{\mathbf{2\ 975}} \\ \mathbf{0} \end{array}$$

NAME ______________ SECTION ______________ DATE ______________

Exercises 5.4

1. Match the indicated quotient with the best estimate of that quotient.

_______ (a) $3.1\overline{)6.386}$ A. 0.02

_______ (b) $0.1\overline{)216.5}$ B. 2

_______ (c) $3.7\overline{)281.6}$ C. 5

_______ (d) $18.5\overline{)127.9}$ D. 75

_______ (e) $4.1\overline{)0.0884}$ E. 2000

2. Match the indicated quotient with the best estimate of that quotient.

_______ (a) $27.58 \div 0.003$ A. 10

_______ (b) $27.58 \div 0.03$ B. 100

_______ (c) $27.58 \div 0.3$ C. 1000

_______ (d) $27.58 \div 3$ D. 10,000

Divide.

3. $4.68 \div 2$

4. $1.71 \div 3$

5. $4.95 \div 5$

6. $1.62 \div 9$

7. $0.064 \div 0.8$

8. $0.63 \div 0.7$

9. $82.24 \div 0.04$

10. $16.02 \div 0.03$

11. $48 \div 2.4$

12. $28 \div 5.6$

Find each quotient to the nearest tenth.

13. $8\overline{)455}$

14. $4\overline{)263}$

15. $9.4\overline{)6.538}$

ANSWERS

1a. [Respond in exercise.]

b. [Respond in exercise.]

c. [Respond in exercise.]

d. [Respond in exercise.]

e. [Respond in exercise.]

2a. [Respond in exercise.]

b. [Respond in exercise.]

c. [Respond in exercise.]

d. [Respond in exercise.]

3. ______________

4. ______________

5. ______________

6. ______________

7. ______________

8. ______________

9. ______________

10. ______________

11. ______________

12. ______________

13. ______________

14. ______________

15. ______________

16. ______________
17. ______________
18. ______________
19. ______________
20. ______________
21. ______________
22. ______________
23. ______________
24. ______________
25. ______________
26. ______________
27. ______________
28. ______________
29. ______________
30. ______________
31. ______________
32. ______________
33. ______________
34. ______________

16. $4.6\overline{)5}$ **17.** $7.05\overline{)0.4977}$ **18.** $0.37\overline{)4.683}$

19. $1.62\overline{)34}$ **20.** $1.33\overline{)75}$

Find each quotient to the nearest hundredth.

21. $24\overline{)0.1463}$ **22.** $1.23\overline{)14.91129}$ **23.** $0.075\overline{)0.42753}$

24. $2.7\overline{)2.583}$ **25.** $23\overline{)62.949}$ **26.** $9\overline{)2}$

27. $13\overline{)65.476}$ **28.** $3.181\overline{)6}$

Find each quotient mentally by using your knowledge of division by powers of 10.

29. $78.4 \div 100$ **30.** $16.4963 \div 100$ **31.** $50.36 \div 100$

32. $45.621 \div 1000$ **33.** $73.85 \div 1000$ **34.** $18.6 \div 1000$

35. $\frac{167}{10}$ **36.** $\frac{138.1}{10}$ **37.** $\frac{7.85}{10}$

NAME ____________ SECTION ____________ DATE ____________

38. $\frac{1.54}{10,000}$ **39.** $\frac{169.9}{10,000}$ **40.** $\frac{10.413}{10,000}$

Find the equivalent measures in the metric system.

41. 5 mm = ______ cm

42. 11 mm = ______ cm

43. 83 cm = ______ m

44. 95 cm = ______ m

45. 344 mm = ______ m

46. 255 mm = ______ m

47. 1500 m = ______ km

48. 2400 m = ______ km

49. 97.2 mm = ______ cm

50. 18.5 mm = ______ cm

51. 32 mm = ______ cm = ______ dm = ______ m

52. 560 mm = ______ cm = ______ dm = ______ m

In Exercises 53–58, first estimate each quotient; then find the quotient to the nearest thousandth.

53. $23\overline{)71}$ **54.** $69\overline{)293}$ **55.** $85.3\overline{)24.31}$

56. $2.57\overline{)0.4961}$ **57.** $16.2\overline{)0.11623}$ **58.** $25.7\overline{)6.27}$

ANSWERS

35. ____________

36. ____________

37. ____________

38. ____________

39. ____________

40. ____________

41. [Respond in exercise.]

42. [Respond in exercise.]

43. [Respond in exercise.]

44. [Respond in exercise.]

45. [Respond in exercise.]

46. [Respond in exercise.]

47. [Respond in exercise.]

48. [Respond in exercise.]

49. [Respond in exercise.]

50. [Respond in exercise.]

51. [Respond in exercise.]

52. [Respond in exercise.]

53. ____________

54. ____________

55. ____________

56. ____________

57. ____________

58. ____________

59a. ____________

b. ____________

60a. ____________

b. ____________

61a. ____________

b. ____________

62a. ____________

b. ____________

63. ____________

64. ____________

Read each problem carefully before you decide which operation(s) are required.

59. (a) If a car averages 24.6 miles per gallon, about how far will it go on 18 gallons of gas? (b) Exactly how many miles will it go on 18 gallons of gas?

60. (a) If a bicyclist rode 250.6 miles in 13.2 hours, about how fast did she ride per hour? (b) What was her average speed in miles per hour (to the nearest tenth)?

61. A quarter section of beef can be bought cheaper than the same amount of meat purchased a few pounds at a time. (a) Estimate the cost per pound if 150 pounds costs $187.50. (b) What is the cost per pound?

62. (a) If you drive 9.5 hours at an average speed of 52.2 miles per hour, about how far will you drive? (b) Exactly how far will you drive?

63. If new tires cost $56.50 per tire and tax is figured at 0.06 times the cost of each tire, what will you pay for four new tires?

64. If you bought 10 books for a total price of $225 plus tax, and tax is figured at 0.06 times the price, what average amount did you pay per book, including tax?

NAME ______________ SECTION ______________ DATE ______________

ANSWERS

65. ______________

66. ______________

67. ______________

68. ______________

69. ______________

65. If the total price of a stereo was $312.70 including tax at 0.06 times the list price, you can find the list price by dividing the total price by 1.06. What was the list price? (**Note:** 1.06 represents the list price plus 0.06 times the list price.)

66. Suppose that the total interest paid on a 30-year mortgage for a home loan of $60,000 is going to be $189,570. What will be the payment each month if the payments are to pay off both the loan and the interest?

67. In 1992 the Cleveland Indians had a team batting average of 0.266 and had 1495 base hits. Find the number of team at bats, to the nearest whole number, by dividing base hits by batting average.

68. In a recent year Mike Mussina of the Baltimore Orioles led all American League pitchers with 18 wins and a 0.783 winning percentage. Find Mussina's total number of pitching decisions (games won or lost), to the nearest whole number, by dividing wins by winning percentage.

69. In a recent year Mark Price of the Cleveland Cavaliers basketball team led the NBA with a 0.947 free-throw percentage. He successfully made 270 free throws. Find his number of attempted free throws, to the nearest whole number, by dividing free throws made by free-throw percentage.

70. ______________

71. ______________

72. [Respond below exercise.]

73a. ______________

b. ______________

74. [Respond below exercise.]

70. At Olympic Stadium in Montreal, home of the Expos baseball team, the outfield wall indicates distances from home plate in both feet and meters. The distance down the foul lines is 325 feet. Convert this distance to meters, to the nearest tenth of a meter. (Use 1 m = 3.28 ft.)

71. A marathon footrace is 26.219 miles in length. Convert this distance to kilometers, to the nearest tenth of a kilometer. (Use 1 km = 0.621 mi.)

Check Your Number Sense

72. When a textbook is made, it is often printed and bound in sets of 16 pages, called signatures. (Each signature is actually one large sheet of paper with the individual pages laid out in a rectangular arrangement. It is then folded several times to create the "booklet," or signature, of 16 pages.) Which of the following are possible textbook lengths if only whole 16-page signatures are used? (a) 256 pages (b) 500 pages (c) 368 pages (d) 1264 pages (e) 648 pages. Explain, briefly, why you chose the values you did.

73. The public address announcer at Candlestick Park in San Francisco earns $75 for announcing each Giants baseball game. (a) If she announced 72 games in one particular season, about how much did she earn: $3500, $5000, $6500, or $8000? (b) If there are 81 home games in one year, approximately what is the maximum amount that she can earn announcing games: $3500, $5000, $6500, or $8000?

Writing and Thinking about Mathematics

74. In your own words, explain why, when converting units in the metric system, we **multiply** by a power of 10 to change to a **smaller** unit of measure, and we **divide** by a power of 10 to change to a **larger** unit of measure.

5.5 Scientific Notation

OBJECTIVES

1. *Know how to read and write very large numbers in scientific notation.*
2. *Know how to read and write very small numbers in scientific notation.*
3. *Be able to read the results on a calculator set in scientific notation mode.*

Scientific Notation and Calculators

In scientific applications in fields of study such as physics, chemistry, biology, and astronomy the numbers used are sometimes either very large or very small. Also, any operations with these numbers are likely to yield numbers even larger or smaller. For example, light travels about 6,000,000,000,000 (6 trillion) miles in one year. (This distance is known as a light-year.) The distance from the earth to the sun is about 93,000,000 (93 million) miles. Due to the large number of 0's, errors can be made when writing down and operating with such numbers. Therefore, mathematicians and scientists have developed a type of shorthand notation for writing such numbers called **scientific notation.**

Also, since hand-held calculators are so widely used now, scientific notation is even more useful because of the limited number of digits on a calculator's display screen. Most calculators display, at most, seven to nine digits. Scientific calculators can be set to display decimal numbers in scientific notation.

Scientific Notation

In **scientific notation,** decimal numbers are written as the product of a decimal number between 1 and 10 and a power of 10.
(**Note:** In this section, we will see that the exponent on 10 can be either a positive or a negative number. Positive and negative numbers will be discussed in detail in Chapter 9.)

To write 93,000,000 in scientific notation, we write the number as follows:

$$93{,}000{,}000 = \underbrace{9.3}_{\text{between 1 and 10}} \times 10^{7} \quad \text{(distance from the earth to the sun in miles)}$$

The exponent 7 tells how many places the decimal point in 9.3 is to be moved to the right so that the product will be equal to 93,000,000.

To write 6,000,000,000,000 in scientific notation, we write the number as follows:

$$6{,}000{,}000{,}000{,}000 = \underbrace{6.0}_{\text{between 1 and 10}} \times 10^{12} \quad \text{(a light-year in miles)}$$

The exponent 12 tells how many places the decimal point in 6.0 is to be moved to the right so that the product will be equal to 6,000,000,000,000.

Now, if a calculator is used to find the product $93{,}000{,}000 \times 40{,}000$, the display will be similar to one of the following:

3.72 12 or 3.72 $\underline{12}$ or 3.72 E12

In each case, 12 is understood to be the exponent on 10 in scientific notation:

$$93{,}000{,}000 \times 40{,}000 = 3{,}720{,}000{,}000{,}000 = 3.72 \times 10^{12}$$

Thus, multiplication by 10^{12} moves the decimal point 12 places to the right:

$$3.72 \times 10^{12} = 3.720{,}000{,}000{,}000.$$

The decimal point is moved 12 places to the right.

EXAMPLE 1

Each of the following numbers is written in both decimal notation and scientific notation.

Decimal Notation		Scientific Notation
(a) 75,000	=	7.5×10^4
(b) 978,000	=	9.78×10^5
(c) 12,340,000	=	1.234×10^7

Scientific Notation with Negative Exponents

Because **negative numbers** can be used as exponents in scientific notation, we present a brief introduction to the concept of negative numbers here. A more detailed and complete discussion on negative numbers is given in Chapter 9.

Temperature readings below 0 are common in many parts of the world. Mathematically, we can indicate the idea of a reading of **10° below 0** with the number **−10°** (read **negative 10 degrees**). Negative numbers are numbers that are less than 0, and these numbers are indicated with a negative sign (−). Number lines such as the one in Figure 5.3 are frequently used to illustrate the idea of negative numbers.

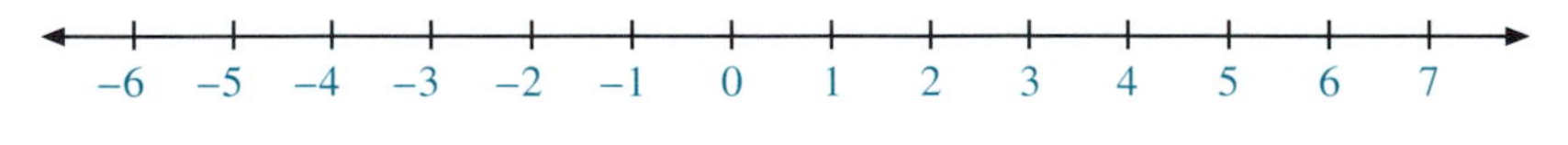

Figure 5.3

At this point, we need only know that **a negative exponent in scientific notation is used to indicate that the decimal point is to be moved to the left.**

Scientific notation makes use of negative exponents in the representations of very small numbers. For example,

$0.0008 = 8.0 \times 10^{-4}$ (the diameter of a red blood cell in centimeters)

Or, in another sense,

$8.0 \times 10^{-4} = 0.0008.0 = 0.0008$

The decimal point is moved 4 places to the left.

Similarly,

$9.234 \times 10^{-6} = 0.000009.234 = 0.000009234$

The decimal point is moved 6 places to the left.

Example 2

Each of the following numbers is written in both decimal notation and scientific notation.

Decimal Notation		Scientific Notation
(a) 0.00075	=	7.5×10^{-4}
(b) 0.00632	=	6.32×10^{-3}
(c) 0.0000000004	=	4.0×10^{-10}

Now Work Exercises 1–4 in the Margin.

Write each number in scientific notation.

1. 783,200,000

2. 0.00039

Write each of the following numbers in decimal notation without using exponents.

3. 2.37×10^{5}

4. 6.5×10^{-4}

NAME ______________ SECTION ______________ DATE ______________

ANSWERS

1. ______
2. ______
3. ______
4. ______
5. ______
6. ______
7. ______
8. ______
9. ______
10. ______
11. ______
12. ______
13. ______
14. ______
15. ______
16. ______
17. ______
18. ______
19. ______
20. ______
21. ______
22. ______
23. ______
24. ______
25. ______
26. ______

Exercises 5.5

Write each of the following numbers in scientific notation.

1. 5,000,000 **2.** 4,300,000 **3.** 750,000

4. 890,000 **5.** 67,000,000 **6.** 45,100,000

7. 175,000,000 **8.** 732,000,000,000 **9.** 213,700,000,000

10. 824,500,000,000 **11.** 0.00062 **12.** 0.00057

13. 0.000025 **14.** 0.000034 **15.** 0.0000008

16. 0.0000002 **17.** 0.00000000671 **18.** 0.00000000255

19. 0.00000000000321 **20.** 0.00000000000786

Write each of the following numbers in decimal notation without using exponents.

21. 5.7×10^3 **22.** 6.3×10^4 **23.** 7.54×10^4

24. 1.26×10^7 **25.** 4.72×10^6 **26.** 8.99×10^8

27. ____________
28. ____________
29. ____________
30. ____________
31. ____________
32. ____________
33. ____________
34. ____________
35. ____________
36. ____________
37. ____________
38. ____________
39. ____________
40. ____________
41a. ____________
b. ____________
42a. ____________
b. ____________
43a. ____________
b. ____________
44a. ____________
b. ____________
45a. ____________
b. ____________
46a. ____________
b. ____________
47a. ____________
b. ____________
48a. ____________
b. ____________
49a. ____________
b. ____________
50a. ____________
b. ____________

27. 5.7×10^{-3} **28.** 6.3×10^{-4} **29.** 1.84×10^{-5}

30. 3.17×10^{-13} **31.** 5.24×10^{-8} **32.** 2.155×10^{-10}

State whether or not each number given is written in scientific notation. If not, rewrite the number in scientific notation.

33. 56.71×10^3 **34.** 19.823×10^4 **35.** 1.44×10^6

36. 1.85×10^5 **37.** 474.3×10^{-5} **38.** 32.1×10^{-3}

39. 17.86×10^{-6} **40.** 25.9×10^{-7}

Set your calculator in scientific mode and use it to perform the following operations. Write your answers (a) in the scientific notation you read on the display and (b) in decimal notation.

41. $120 \div 0.003$ **42.** $155 \div 0.0005$ **43.** $57{,}000 \times 94{,}000$

44. $125{,}000 \times 32{,}000$ **45.** $88{,}000 \times 3500$ **46.** $45{,}000 \times 2500$

47. $0.00036 \div 1800$ **48.** $0.00024 \div 600$

49. $5000 \times 65{,}000 + 7000 \times 2000$

50. $4000 \times 27{,}000 + 800 \times 30{,}000$

NAME ____________ SECTION ____________ DATE ____________

ANSWERS

51. [Respond below exercise.]

52. [Respond below exercise.]

53. [Respond below exercise.]

54. ____________

51. Light travels approximately 3×10^8 meters per second. (a) How fast does light travel in centimeters per second? (b) Write both of these numbers in decimal notation.

52. An atom of gold weighs approximately 3.25×10^{-22} grams. Write this number in decimal notation.

53. The atomic weight of carbon-12 is 1.9926×10^{-26} kilograms. Write this number in decimal notation.

54. Nematode sea worms are the most numerous of all sea and land animals, with an estimated population of 40,000,000,000,000,000,000,000,000. Write this number in scientific notation.

55. ______________

56. ______________

57. [Respond below exercise.]

58. [Respond below exercise.]

55. The Environmental Protection Agency estimates that, by the year 2010, the amount of municipal solid waste that we produce each year in the United States will reach 2.5×10^8 tons. Write this number in standard decimal notation.

56. A typical slime mold spore measures about 0.000015 meter in diameter. Write this number in scientific notation.

57. In a recent year, the United States consumed 8.117×10^{16} Btu (British thermal units) of energy and produced only 6.747×10^{16} Btu of energy. Write each of these numbers in standard decimal notation.

58. As of 1991, the natural gas reserves of Canada, the United States, and Mexico were estimated at 9.76×10^{13} cubic feet, 1.693×10^{14} cubic feet, and 7.27×10^{13} cubic feet, respectively. Write each of these numbers in standard decimal notation.

5.6 Decimals and Fractions

OBJECTIVES

1. *Know how to change decimal numbers to fraction form and/or mixed number form.*
2. *Know how to change fractions and mixed numbers to decimal form.*
3. *Recognize both terminating and nonterminating decimals.*
4. *Understand that working with decimals and fractions in the same problem may involve rounded-off numbers and approximate answers.*

Changing from Decimals to Fractions

In Section 5.1, we discussed the fact that whole numbers, finite decimal numbers, fractions, and mixed numbers are different forms of the same type of number—namely, **rational numbers.** In this section, we will introduce the concept of **infinite** (or **nonterminating**) decimals and show how to add, subtract, multiply, and divide with various combinations of fractions, mixed numbers, and decimal numbers.

Finite (or terminating) decimal numbers can be written in fraction form with denominators that are powers of ten. For example,

$$0.25 = \frac{25}{100} \quad \text{and} \quad 0.025 = \frac{25}{1000}$$

The rightmost digit, 5, is in the hundredths position, so the fraction has 100 in the denominator.

The rightmost digit, 5, is in the thousandths position, so the fraction has 1000 in the denominator.

Changing from Decimals to Fractions

A finite (or terminating) decimal number can be written in fraction form by writing a fraction with the following:

1. A **numerator** that consists of the whole number formed by all the digits of the decimal number and
2. A **denominator** that is the power of ten that names the position of the rightmost digit.

In the following examples, each decimal number is changed to fraction form and then reduced, if possible, by using the factoring techniques discussed in Chapter 3 for reducing fractions.

Example 1

$$0.25 = \frac{25}{100} = \frac{\cancel{5} \cdot \cancel{5} \cdot 1}{2 \cdot \cancel{5} \cdot 2 \cdot \cancel{5}} = \frac{1}{4}$$

↑ hundredths

Example 2

$$0.32 = \frac{32}{100} = \frac{\cancel{4} \cdot 8}{\cancel{4} \cdot 25} = \frac{8}{25}$$

↑ hundredths

EXAMPLE 3

$$0.131 = \frac{131}{1000}$$

↑ thousandths

EXAMPLE 4

$$0.075 = \frac{75}{1000} = \frac{\cancel{25} \cdot 3}{\cancel{25} \cdot 40} = \frac{3}{40}$$

↑ thousandths

EXAMPLE 5

$$2.6 = \frac{26}{10} = \frac{\cancel{2} \cdot 13}{\cancel{2} \cdot 5} = \frac{13}{5}$$

↑ tenths

or, as a mixed number,

$$2.6 = 2\frac{6}{10} = 2\frac{3}{5}$$

EXAMPLE 6

$$1.42 = \frac{142}{100} = \frac{\cancel{2} \cdot 71}{\cancel{2} \cdot 50} = \frac{71}{50}$$

↑ hundredths

or, as a mixed number,

$$1.42 = 1\frac{42}{100} = 1\frac{21}{50}$$

Changing from Fractions to Decimals

Changing from Fractions to Decimals

A fraction can be written in decimal form by dividing the numerator by the denominator.

1. If the remainder is 0, the decimal is said to be **terminating.**
2. If the remainder is not 0, the decimal is said to be **nonterminating.**

The following examples illustrate fractions that convert to terminating decimals.

EXAMPLE 7

$\frac{3}{8}$

$$\begin{array}{r} .375 \\ 8\overline{)3.000} \\ \underline{2\,4} \\ 60 \\ \underline{56} \\ 40 \\ \underline{40} \\ 0 \end{array}$$

$\frac{3}{8} = 0.375$

EXAMPLE 8

$\frac{3}{4}$

$$\begin{array}{r} .75 \\ 4\overline{)3.00} \\ \underline{2\,8} \\ 20 \\ \underline{20} \\ 0 \end{array}$$

$\frac{3}{4} = 0.75$

EXAMPLE 9

$\frac{4}{5}$

$$\begin{array}{r} .8 \\ 5\overline{)4.0} \\ \underline{4\,0} \\ 0 \end{array}$$

$\frac{4}{5} = 0.8$

Nonterminating decimals can be **repeating** or **nonrepeating.** A nonterminating repeating decimal has a repeating pattern to its digits. Every fraction with a whole number numerator and nonzero denominator is either a terminating decimal or a repeating decimal. (Such numbers are called **rational numbers.**) Nonterminating, nonrepeating decimals are called **irrational numbers** and are discussed in Section 5.7 with square roots.

The following examples illustrate fractions that convert to nonterminating repeating decimals.

EXAMPLE 10

$\frac{1}{3}$

$$\begin{array}{r} .333 \\ 3\overline{)1.000} \\ \underline{9} \\ 10 \\ \underline{9} \\ 10 \\ \underline{9} \\ 1 \end{array}$$

← The 3 will repeat without end.

← Continuing to divide will give a remainder of 1 each time.

We write $\frac{1}{3} = 0.333\ldots$ The three dots mean "and so on" or to continue without stopping.

EXAMPLE 11

$\frac{7}{12}$

$$\begin{array}{r} .5833 \\ 12\overline{)7.0000} \\ \underline{6\,0} \\ 1\,00 \\ \underline{96} \\ 40 \\ \underline{36} \\ 40 \\ \underline{36} \\ 4 \end{array}$$

← The 3 will repeat without end.

← Continuing to divide will give a remainder of 4 each time.

We write $\frac{7}{12} = 0.58333\ldots$

EXAMPLE 12

$\frac{1}{7}$

$$\begin{array}{r} .142857 \\ 7\overline{)1.000000} \\ \underline{7} \\ 30 \\ \underline{28} \\ 20 \\ \underline{14} \\ 60 \\ \underline{56} \\ 40 \\ \underline{35} \\ 50 \\ \underline{49} \\ 1 \end{array}$$

← The six digits will repeat in the same pattern without end.

The remainder will repeat in sequence 1, 3, 2, 6, 4, 5, 1, and so on. Therefore, the digits in the quotient will also repeat.

We write $\frac{1}{7} = 0.142857142857142857\ldots$

Another way of writing repeating decimals is to write a **bar** over the repeating digits. Thus, in Examples 10, 11, and 12, we can write

$$\frac{1}{3} = 0.\overline{3} \quad \text{and} \quad \frac{7}{12} = 0.58\overline{3} \quad \text{and} \quad \frac{1}{7} = 0.\overline{142857}$$

We may choose to round off the quotient to some decimal place just as we did with division in Section 5.5. Perform the division one place past the desired round-off position.

EXAMPLE 13

$$\frac{5}{11} \qquad \begin{array}{r} .454 \\ 11\overline{)5.000} \\ \underline{4\,4} \\ 60 \\ \underline{55} \\ 50 \\ \underline{44} \end{array} \approx 0.45 \quad \text{(nearest hundredth)}$$

EXAMPLE 14

$$\frac{5}{6} \qquad \begin{array}{r} .833 \\ 6\overline{)5.000} \\ \underline{4\,8} \\ 20 \\ \underline{18} \\ 20 \\ \underline{18} \end{array} \approx 0.83 \quad \text{(nearest hundredth)}$$

Operating with Both Fractions and Decimals

As the following examples illustrate, we can perform operations and comparisons with both fractions and decimal numbers by changing the fractions to decimal form. (**Note:** In some cases, this may involve rounding off the decimal form of a number and settling for an approximate answer. To obtain a more accurate answer, we may need to change the decimals to fraction form and then perform the operations.)

EXAMPLE 15

Find the sum $10\frac{1}{2} + 7.32 + 5\frac{3}{5}$ in decimal form.

Solution

$$\begin{array}{rcr l} 10\frac{1}{2} & = & 10.50 & \left(\frac{1}{2} = 0.50\right) \\ 7.32 & = & 7.32 & \\ +\ 5\frac{3}{5} & = & 5.60 & \left(\frac{3}{5} = 0.60\right) \\ \hline & & 23.42 & \end{array}$$

EXAMPLE **16**

Determine whether $\frac{3}{16}$ is larger than 0.18 by changing $\frac{3}{16}$ to decimal form and then comparing the two numbers. Find the difference.

Solution

Divide first.

$$\begin{array}{r} .1875 \\ 16\overline{)3.0000} \\ \underline{1\,6} \\ 1\,40 \\ \underline{1\,28} \\ 120 \\ \underline{112} \\ 80 \\ \underline{80} \\ 0 \end{array}$$

So, $\frac{3}{16} = 0.1875$.

Now subtract.

$$\begin{array}{r} 0.1875 \\ -0.1800 \\ \hline 0.0075 \end{array} \quad \text{difference}$$

Thus, $\frac{3}{16}$ is larger than 0.18 and their difference is 0.0075.

NAME ____________ SECTION ____________ DATE ____________

ANSWERS

1. ____________
2. ____________
3. ____________
4. ____________
5. ____________
6. ____________
7. ____________
8. ____________
9. ____________
10. ____________
11. ____________
12. ____________
13. ____________
14. ____________
15. ____________
16. ____________
17. ____________
18. ____________
19. ____________
20. ____________
21. ____________
22. ____________
23. ____________
24. ____________

Exercises 5.6

Change each decimal to fraction form. Do not reduce.

1. 0.9 **2.** 0.3 **3.** 0.5 **4.** 0.8

5. 0.62 **6.** 0.38 **7.** 0.57 **8.** 0.41

9. 0.526 **10.** 0.625 **11.** 0.016 **12.** 0.012

13. 5.1 **14.** 7.2 **15.** 8.15 **16.** 6.35

Change each decimal to fraction form (or mixed number form) and reduce if possible.

17. 0.125 **18.** 0.36 **19.** 0.18 **20.** 0.375

21. 0.225 **22.** 0.455 **23.** 0.17 **24.** 0.029

25. 3.2 **26.** 1.25 **27.** 6.25 **28.** 2.75

Change each fraction to decimal form. If the decimal is nonterminating, write it with the bar notation over the repeating pattern of digits.

29. $\frac{2}{3}$ **30.** $\frac{5}{16}$ **31.** $\frac{5}{11}$ **32.** $\frac{3}{11}$

33. $\frac{11}{16}$ **34.** $\frac{9}{16}$ **35.** $\frac{3}{7}$ **36.** $\frac{5}{7}$

37. $\frac{1}{6}$ **38.** $\frac{5}{18}$ **39.** $\frac{5}{9}$ **40.** $\frac{2}{9}$

Change each fraction to decimal form rounded off to the nearest thousandth.

41. $\frac{7}{24}$ **42.** $\frac{16}{33}$ **43.** $\frac{5}{12}$ **44.** $\frac{13}{16}$

25. ____________
26. ____________
27. ____________
28. ____________
29. ____________
30. ____________
31. ____________
32. ____________
33. ____________
34. ____________
35. ____________
36. ____________
37. ____________
38. ____________
39. ____________
40. ____________
41. ____________
42. ____________
43. ____________
44. ____________

NAME ______________________ SECTION ______________ DATE __________

ANSWERS

45. $\frac{1}{32}$ **46.** $\frac{1}{14}$ **47.** $\frac{16}{13}$ **48.** $\frac{20}{9}$

49. $\frac{30}{32}$ **50.** $\frac{40}{3}$

Perform the indicated operations by writing all the numbers in decimal form. Round off to the nearest thousandth if necessary.

51. $\frac{1}{4} + 0.25 + \frac{1}{5}$ **52.** $\frac{3}{4} + \frac{1}{10} + 3.55$

53. $\frac{5}{8} + \frac{3}{5} + 0.41$ **54.** $6 + 2\frac{37}{100} + 3\frac{11}{50}$

55. $2\frac{53}{100} + 5\frac{1}{10} + 7.35$ **56.** $37.02 + 25 + 6\frac{2}{5} + 3\frac{89}{100}$

45. ________
46. ________
47. ________
48. ________
49. ________
50. ________
51. ________
52. ________
53. ________
54. ________
55. ________
56. ________

57. ______

58. ______

59. ______

60. ______

61. ______

62. ______

63. ______

64. ______

65. ______

66. ______

67. ______

68. ______

69. ______

70. ______

71. ______

72. ______

57. $1\frac{1}{4} - 0.125$

58. $2\frac{1}{2} - 1.75$

59. $36.71 - 23\frac{1}{5}$

60. $3.1 - 2\frac{1}{100}$

61. $\left(\frac{35}{100}\right)(0.73)$

62. $\left(5\frac{1}{10}\right)(2.25)$

63. $\left(1\frac{3}{8}\right)(3.1)(2.6)$

64. $\left(1\frac{3}{4}\right)\left(2\frac{1}{2}\right)(5.35)$

65. $5\frac{54}{100} \div 2.1$

66. $72.16 \div \frac{2}{5}$

67. $13.65 \div \frac{1}{2}$

68. $91.7 \div \frac{1}{4}$

In each of the following exercises, change any fraction to decimal form; then determine which number is larger. Find the difference.

69. $2\frac{1}{4}$, 2.3

70. $\frac{7}{8}$, 0.878

71. 0.28, $\frac{3}{11}$

72. $\frac{1}{3}$, 0.3

NAME ______________ SECTION ______________ DATE ______________

ANSWERS

73. ______________

74. ______________

75. ______________

76. ______________

77. ______________

In Exercises 73–78, change each decimal number (except the year 1990) into fraction or mixed number form.

73. According to the 1990 census, there are 70.3 people per square mile in the United States.

74. The average weight for a one-year-old girl is 9.1 kg.

75. The median age for men at first marriage is 26.3 years. The median age for women at first marriage is 24.1 years.

76. The maximum speed of a giant tortoise on land is about 0.17 mph.

77. There are about 22.8 students per teacher in California public schools.

78. ______________

79. ______________

80. ______________

81. ______________

82. ______________

78. The surface gravity on Mars is about 0.38 times the gravity on Earth. The atmospheric pressure on Mars is about 0.01 times the atmospheric pressure on Earth.

In Exercises 79–82, change each fraction or mixed number (but not years) into decimal form. Round to the nearest thousandth if necessary.

79. By 1990, one estimation was that $\frac{3}{50}$ of U.S. households received 20 or more TV stations.

80. In a recent year, about $\frac{8}{57}$ of the advertising budgets in the automotive industry was spent on newspaper ads.

81. Since 1970, the average annual tuition cost for college has increased $5\frac{18}{25}$ times.

82. In the 1992 presidential election, Bill Clinton received $\frac{1081}{2500}$ of the popular vote.

5.7 Square Roots and the Pythagorean Theorem

OBJECTIVES

1. *Memorize the squares of the whole numbers from 1 to 20.*
2. *Memorize the perfect square whole numbers from 1 to 400.*
3. *Understand the terms* **square root, radical sign, radicand,** *and* **radical.**
4. *Know how to use a calculator to find square roots.*
5. *Be able to simplify radical expressions.*
6. *Understand the terms* **right triangle, hypotenuse,** *and* **leg.**
7. *Know and be able to use the Pythagorean Theorem.*

Square Roots and Irrational Numbers

A number is **squared** when it is multiplied by itself. If a whole number is squared, the result is called a **perfect square.** For example, squaring 7 gives $7^2 = 49$, and 49 is a perfect square. Table 5.2 shows the perfect square numbers found by squaring the whole numbers from 1 to 20. A more complete table (of powers, roots, and prime factorizations) is located in the back of the book.

Table 5.2 Squares of Whole Numbers from 1 to 20

$1^2 = 1$	$5^2 = 25$	$9^2 = 81$	$13^2 = 169$	$17^2 = 289$
$2^2 = 4$	$6^2 = 36$	$10^2 = 100$	$14^2 = 196$	$18^2 = 324$
$3^2 = 9$	$7^2 = 49$	$11^2 = 121$	$15^2 = 225$	$19^2 = 361$
$4^2 = 16$	$8^2 = 64$	$12^2 = 144$	$16^2 = 256$	$20^2 = 400$

Since $5^2 = 25$, the number 5 is called the **square root** of 25. We write $\sqrt{25} = 5$. (Read "the square root of 25 is 5.") Similarly,

$$\sqrt{49} = 7 \quad \text{since} \quad 7^2 = 49$$
$$\sqrt{1} = 1 \quad \text{since} \quad 1^2 = 1$$
$$\sqrt{0} = 0 \quad \text{since} \quad 0^2 = 0$$

The symbol $\sqrt{}$ is called a **radical sign.** The number under the radical sign is called the **radicand.** The complete expression, such as $\sqrt{25}$, is called a **radical.**

Table 5.3 contains the square roots of the perfect square numbers from 1 to 400. (Notice that Table 5.3 is just another way of looking at Table 5.2.) Both tables should be memorized.

Table 5.3 Square Roots of Perfect Squares from 1 to 400

$\sqrt{1} = 1$	$\sqrt{25} = 5$	$\sqrt{81} = 9$	$\sqrt{169} = 13$	$\sqrt{289} = 17$
$\sqrt{4} = 2$	$\sqrt{36} = 6$	$\sqrt{100} = 10$	$\sqrt{196} = 14$	$\sqrt{324} = 18$
$\sqrt{9} = 3$	$\sqrt{49} = 7$	$\sqrt{121} = 11$	$\sqrt{225} = 15$	$\sqrt{361} = 19$
$\sqrt{16} = 4$	$\sqrt{64} = 8$	$\sqrt{144} = 12$	$\sqrt{256} = 16$	$\sqrt{400} = 20$

The square roots of some numbers are not found as easily as those in the preceding tables. In fact, most square roots are irrational numbers **(nonrepeating infinite decimals);** that is, most square roots can only be approximated with decimals.

Use a calculator to find the value of each of the following radicals accurate to four decimal places.

1. $\sqrt{7}$

2. $\sqrt{70}$

3. $\sqrt{600}$

4. $\sqrt{9000}$

Decimal approximations of $\sqrt{2}$ are given here to emphasize this idea and to help you understand that there is no finite decimal number whose square is 2.

$$\begin{array}{r} 1.4 \\ \times\ 1.4 \\ \hline 56 \\ 14 \\ \hline 1.96 \end{array} \qquad \begin{array}{r} 1.414 \\ \times\ 1.414 \\ \hline 5656 \\ 1414 \\ 5656 \\ 1414 \\ \hline 1.999396 \end{array} \qquad \begin{array}{r} 1.41421 \\ \times\ 1.41421 \\ \hline 141421 \\ 282842 \\ 565684 \\ 141421 \\ 565684 \\ 141421 \\ \hline 1.9999899241 \end{array} \qquad \begin{array}{r} 1.41422 \\ \times\ 1.41422 \\ \hline 282844 \\ 282844 \\ 565688 \\ 141422 \\ 565688 \\ 141422 \\ \hline 2.0000182084 \end{array}$$

So, $\sqrt{2}$ is between 1.41421 and 1.41422.

To find (or approximate) a square root with a calculator, use the key labeled $\boxed{\sqrt{x}}$.

EXAMPLE 1

Use a calculator to find $\sqrt{2}$ accurate to nine decimal places.

Solution

STEP 1: Enter 2.

STEP 2: Press the key $\boxed{\sqrt{x}}$.

STEP 3: The display will show 1.414213562.

Similarly, with a calculator,

$\sqrt{3} = 1.732050808$ accurate to nine decimal places

and

$\sqrt{5} = 2.236067977$ accurate to nine decimal places

Now Work Exercises 1–4 in the Margin.

Simplifying Radical Expressions

The square roots just discussed and many others are irrational numbers, and their decimal forms are infinite nonrepeating decimals. However, there are situations, particularly in algebra, in which we are not interested in the decimal representation of a square root, but in the simplified form of a radical expression. To simplify radicals in general, we need the following property of square roots.

Property of Square Roots

For decimal numbers a and b,

$$\sqrt{ab} = \sqrt{a} \cdot \sqrt{b}$$

With this property, we can simplify a radical expression by finding a perfect square factor. For example,

$$\sqrt{18} = \sqrt{9 \cdot 2} = \sqrt{9} \cdot \sqrt{2} = 3\sqrt{2} \quad \text{(9 is a square factor.)}$$

The expression $3\sqrt{2}$ is simplified because the radicand, 2, has no perfect square factor. That is, **a radical is considered to be in simplest form when the radicand has no square number factor.**

Simplify the following radicals.

EXAMPLE 2

$\sqrt{50}$

Solution

$$\sqrt{50} = \sqrt{25 \cdot 2} = \sqrt{25} \cdot \sqrt{2} = 5\sqrt{2} \quad \text{(25 is a square factor.)}$$

EXAMPLE 3

$\sqrt{75}$

Solution

$$\sqrt{75} = \sqrt{25 \cdot 3} = \sqrt{25} \cdot \sqrt{3} = 5\sqrt{3}$$

EXAMPLE 4

$\sqrt{450}$

Solution

$$\sqrt{450} = \sqrt{9 \cdot 50} = \sqrt{9 \cdot 25 \cdot 2}$$
$$= \sqrt{9} \cdot \sqrt{25} \cdot \sqrt{2} = 3 \cdot 5\sqrt{2} = 15\sqrt{2}$$

or

$$\sqrt{450} = \sqrt{225 \cdot 2} = \sqrt{225} \cdot \sqrt{2} = 15\sqrt{2}$$

Now Work Exercises 5–7 in the Margin.

Simplify each of the following radicals.

5. $\sqrt{80}$

6. $\sqrt{90}$

7. $\sqrt{3000}$

The Pythagorean Theorem

The following discussion involving right triangles serves as an application of squares and square roots and uses the terms defined below.

Right triangle: A triangle containing a right angle (90°).

Hypotenuse: The longest side of a right triangle; the side opposite the right angle.

Leg: Each of the other two sides of a right triangle.

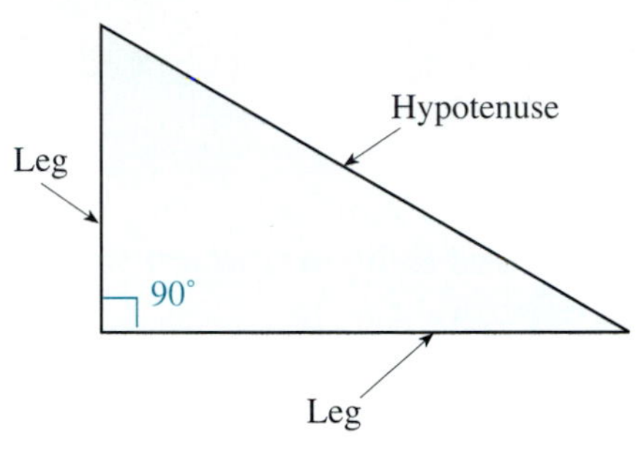

Pythagoras (c. 585–501 B.C.), a famous Greek mathematician, is given credit for discovering the following theorem (even though historians have found that the facts of the theorem were known before the time of Pythagoras).

The Pythagorean Theorem

In a right triangle, the square of the hypotenuse is equal to the sum of the squares of the two legs:

$$c^2 = a^2 + b^2$$

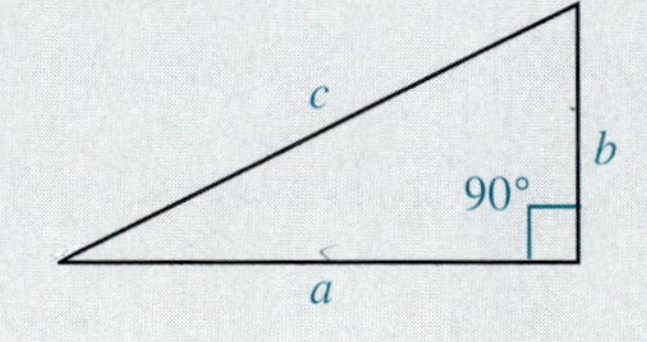

Example 5

Show that a triangle with sides of 3 inches, 4 inches, and 5 inches must be a right triangle.

Solution

If the triangle is a right triangle, then its three sides must satisfy the property stated in the Pythagorean Theorem: $c^2 = a^2 + b^2$. Or, in this case, $5^2 = 3^2 + 4^2$. Since $25 = 9 + 16$ is a true statement, the triangle is a right triangle.

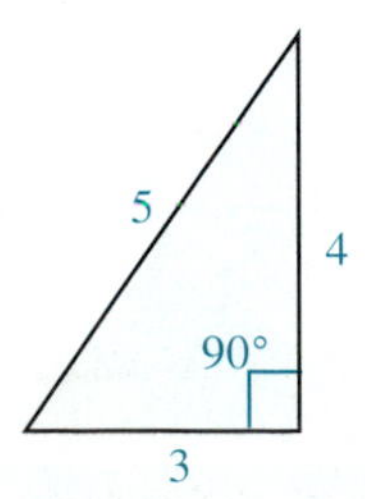

EXAMPLE 6

Find the length of the hypotenuse of a right triangle with legs of length 12 cm and 5 cm.

Solution

Let c = the length of the hypotenuse.
Now, by the Pythagorean Theorem,

$$c^2 = 12^2 + 5^2$$

$$c^2 = 144 + 25$$

$$c^2 = 169$$

$$c = \sqrt{169} = 13$$

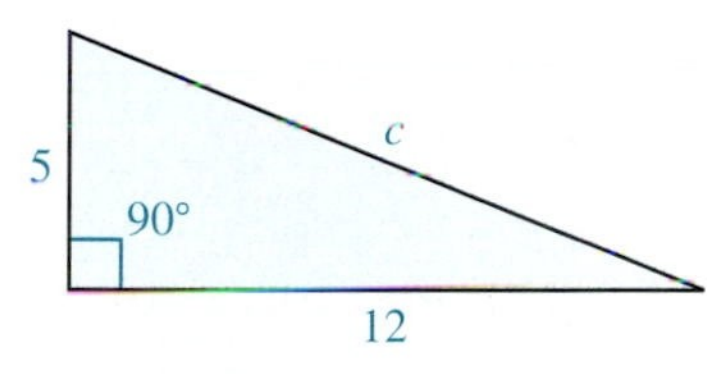

The length of the hypotenuse is 13 centimeters.

Rarely do all three sides of a right triangle have whole number values. Examples 7 and 8 illustrate right triangles with irrational numbers as the lengths of the hypotenuses.

EXAMPLE 7

Find the length of the hypotenuse of a right triangle in which both legs have a length of 1 meter.

Solution

$$c^2 = 1^2 + 1^2$$

$$c^2 = 1 + 1 = 2$$

$$c = \sqrt{2}$$

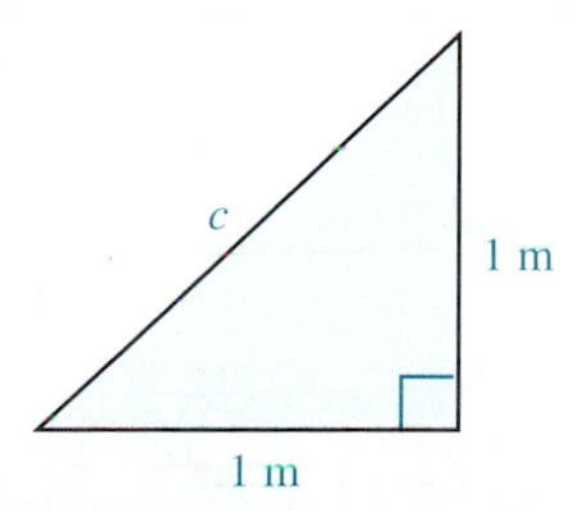

The length of the hypotenuse is $\sqrt{2}$ meters (or about 1.41 meters).

Use the Pythagorean Theorem to determine whether or not each of the following sets of three numbers could be the lengths of the sides of a right triangle.

8. 16 cm, 30 cm, 34 cm

9. 4 ft, 5 ft, 6 ft

EXAMPLE 8

A guy wire is attached to the top of a telephone pole and anchored to the ground 10 feet from the base of the pole. If the pole is 20 feet high, what is the length of the guy wire?

Solution

Let $x =$ the length of the guy wire.
Then, by the Pythagorean Theorem,

$$x^2 = 10^2 + 20^2$$

$$x^2 = 100 + 400$$

$$x^2 = 500$$

$$x = \sqrt{500} = \sqrt{100 \cdot 5} = \sqrt{100} \cdot \sqrt{5} = 10\sqrt{5}$$

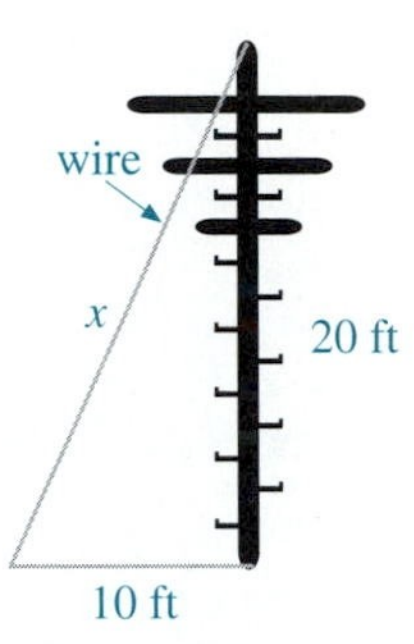

The guy wire is $10\sqrt{5}$ feet long (or about 22.4 feet long).

Now Work Exercises 8 and 9 in the Margin.

NAME ______________ SECTION ______________ DATE ______________

ANSWERS

Exercises 5.7

State whether or not each number is a perfect square in Exercises 1–10.

1. 144 **2.** 169 **3.** 81 **4.** 16 **5.** 400

6. 225 **7.** 242 **8.** 48 **9.** 45 **10.** 40

11. Show by squaring that $\sqrt{3}$ is between 1.732 and 1.733.

12. Show by squaring that $\sqrt{5}$ is between 2.236 and 2.237.

Simplify the radical expressions in Exercises 13–36.

13. $\sqrt{12}$ **14.** $\sqrt{28}$ **15.** $\sqrt{24}$

16. $\sqrt{32}$ **17.** $\sqrt{48}$ **18.** $\sqrt{288}$

1. ______________
2. ______________
3. ______________
4. ______________
5. ______________
6. ______________
7. ______________
8. ______________
9. ______________
10. ______________
11. [Respond below exercise.]
12. [Respond below exercise.]
13. ______________
14. ______________
15. ______________
16. ______________
17. ______________
18. ______________

19. ______________
20. ______________
21. ______________
22. ______________
23. ______________
24. ______________
25. ______________
26. ______________
27. ______________
28. ______________
29. ______________
30. ______________
31. ______________
32. ______________
33. ______________
34. ______________
35. ______________
36. ______________
37. ______________
38. ______________
39. ______________
40. ______________

19. $\sqrt{363}$ **20.** $\sqrt{242}$ **21.** $\sqrt{500}$

22. $\sqrt{300}$ **23.** $\sqrt{128}$ **24.** $\sqrt{125}$

25. $\sqrt{72}$ **26.** $\sqrt{98}$ **27.** $\sqrt{605}$

28. $\sqrt{150}$ **29.** $\sqrt{169}$ **30.** $\sqrt{196}$

31. $\sqrt{800}$ **32.** $\sqrt{80}$ **33.** $\sqrt{90}$

34. $\sqrt{40}$ **35.** $\sqrt{256}$ **36.** $\sqrt{361}$

Use the Pythagorean Theorem to determine whether or not each of the triangles in Exercises 37 and 38 is a right triangle.

37.

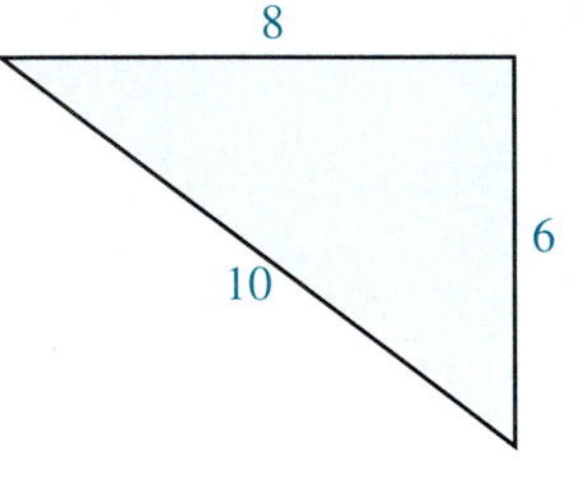

38.

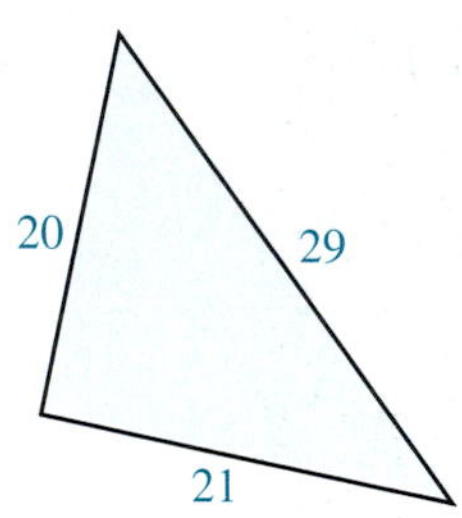

Find the length of the hypotenuse of each of the following right triangles.

39.

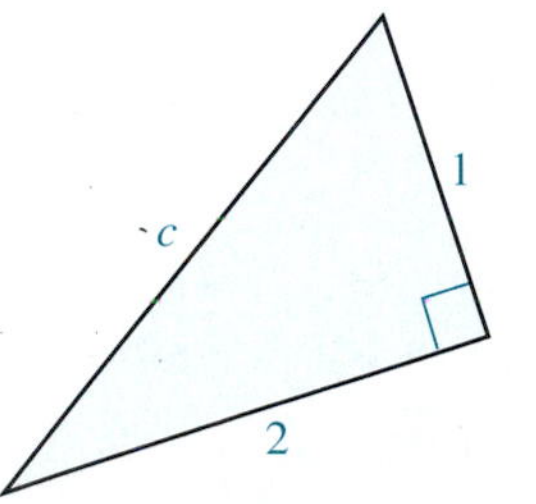

40.

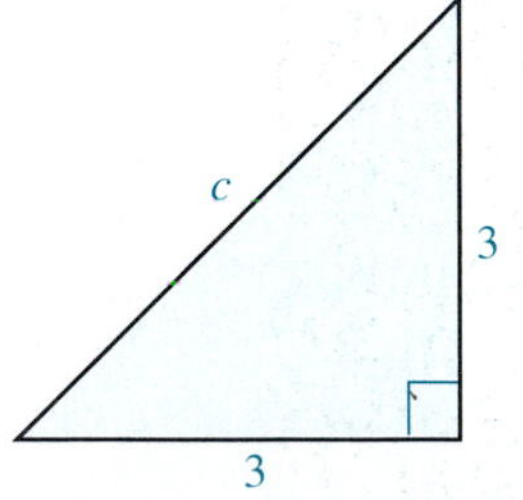

NAME ________________ SECTION ________________ DATE ________

41.

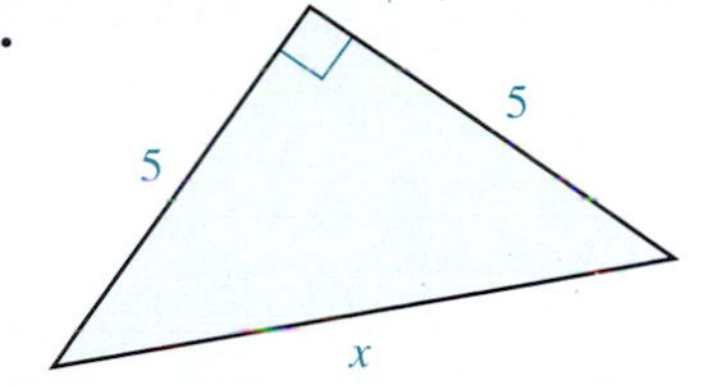

42.

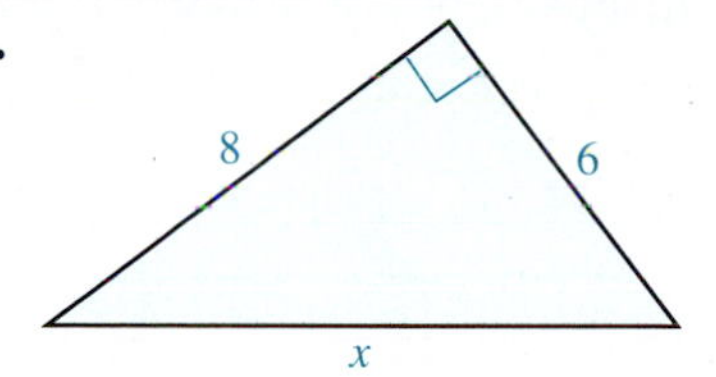

43.

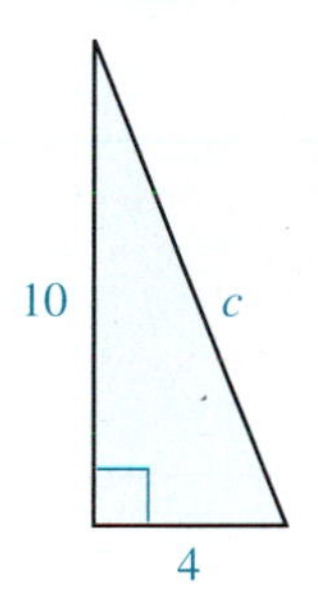

44.

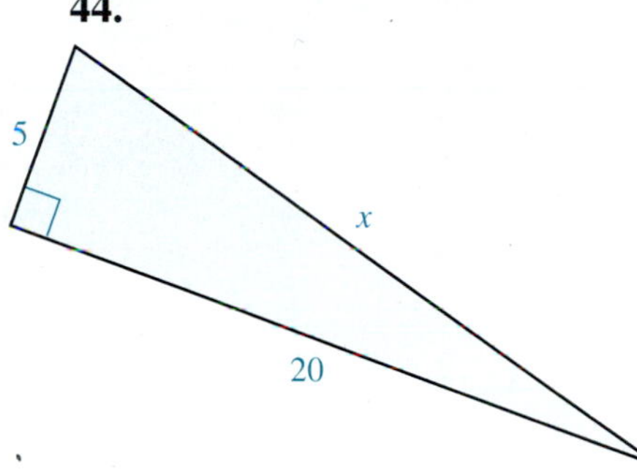

First, mentally estimate the value of each of the following radicals and write down your estimate. Then use a calculator to find the value accurate to four decimal places.

45. $\sqrt{8}$ **46.** $\sqrt{34}$ **47.** $\sqrt{45}$

48. $\sqrt{10}$ **49.** $\sqrt{40}$ **50.** $\sqrt{20}$

51. $\sqrt{75}$ **52.** $\sqrt{200}$ **53.** $\sqrt{300}$

54. $\sqrt{80}$ **55.** $\sqrt{500}$ **56.** $\sqrt{95}$

ANSWERS

41. ________
42. ________
43. ________
44. ________
45. ________
46. ________
47. ________
48. ________
49. ________
50. ________
51. ________
52. ________
53. ________
54. ________
55. ________
56. ________

57. ______________

58. ______________

59. ______________

60. ______________

57. The base of a fire-engine ladder is 20 feet from a building and reaches to a fourth floor window 60 feet above ground level. How far is the ladder extended (to the nearest tenth of a foot)?

58. A forestay that helps support a ship's mast reaches from the top of the mast, which is 20 meters high, to a point on the deck 10 meters from the base of the mast. What is the length of the stay (to the nearest tenth of a meter)?

59. The Xerox Center building in Chicago is 500 feet tall. At a certain time of day, it casts a shadow that is 150 feet long. At that time of day, what is the distance (to the nearest tenth of a foot) from the tip of the shadow to the top of the Xerox building?

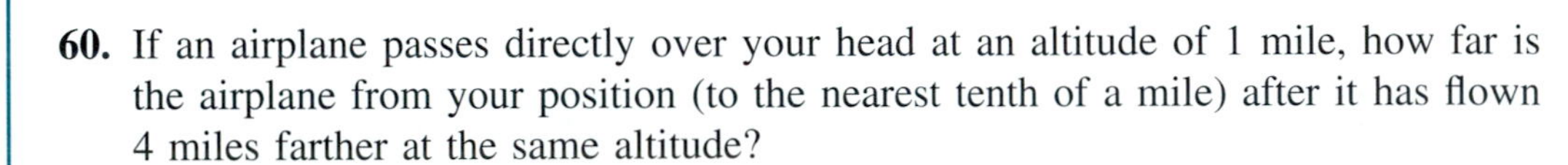

60. If an airplane passes directly over your head at an altitude of 1 mile, how far is the airplane from your position (to the nearest tenth of a mile) after it has flown 4 miles farther at the same altitude?

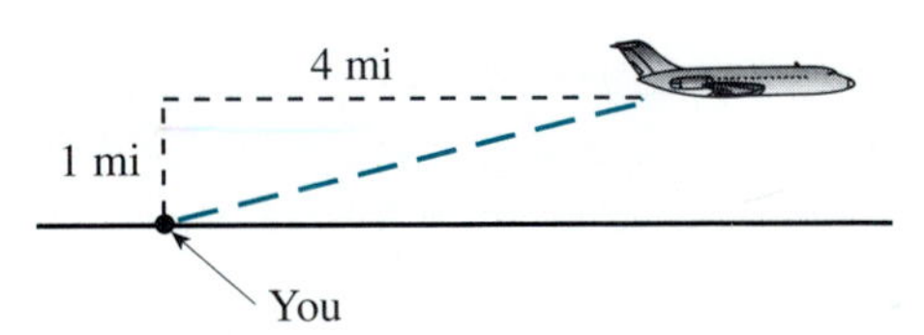

NAME ______________________ SECTION ______________ DATE __________

ANSWERS

61a. ______________

b. ______________

62a. ______________

b. ______________

63. ______________

64. ______________

61. The shape of a baseball infield is a square with sides 90 feet long. (a) Find the distance (to the nearest tenth of a foot) from home plate to second base. (b) The diagonals of the square intersect halfway between home plate and second base. If the pitcher's mound is $60\frac{1}{2}$ feet from home plate, is the pitcher's mound closer to home plate or to second base?

62. A square is said to be inscribed in a circle if each corner of the square lies on the circle. (a) Find the circumference and area of a circle with diameter 20 mm. (b) Find the perimeter and area of a square inscribed in the circle.

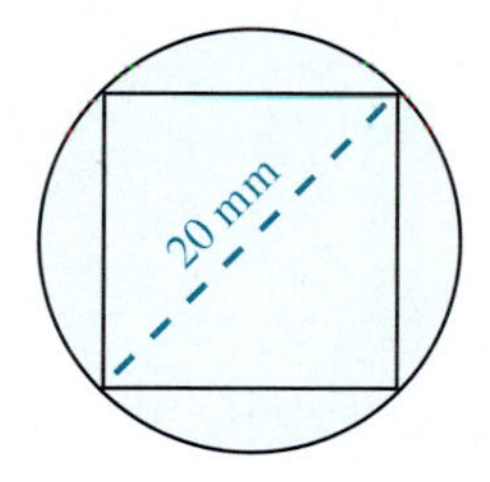

63. Before painting a picture on canvas, an artist must stretch the canvas on a rectangular wooden frame. To be sure that the corners of the canvas are true right angles, the artist can measure the diagonals of the stretched canvas. What should be the diagonal measure, to the nearest tenth of an inch, of a canvas whose sides are 24 inches and 38 inches in length?

64. While installing windows in a new home, a builder measures the diagonals of rectangular window casements to verify that their corners are true right angles. What should be the diagonal measure, to the nearest tenth of an inch, of a window casement with dimensions 36 inches by 54 inches?

65. ______________

66. ______________

67. [Respond below exercise.] ______________

65. To create a square inside a square, a quilting pattern requires four triangular pieces like the one shaded in the figure shown here. If the square in the center measures 10 cm on a side, and the two legs of each triangle are of equal length, how long are the legs of each triangle, to the nearest tenth of a centimeter?

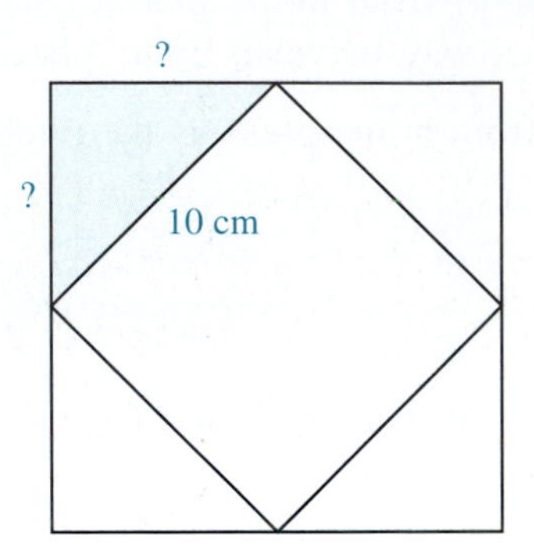

66. The shape of home plate in the game of baseball can be created by cutting off two triangular pieces at the corners of a square, as shown in the figure. If each of the triangles has a hypotenuse of 12 inches and legs of equal length, what is the length of one side of the original square, to the nearest tenth of an inch?

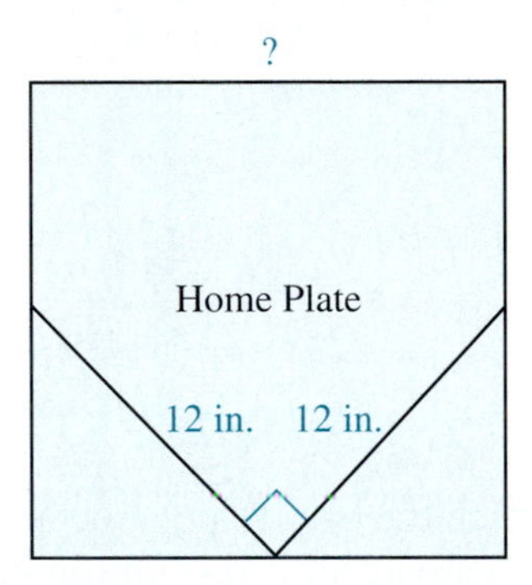

Collaborative Learning Exercise

67. [**Note:** In algebra, the fraction $\frac{1}{2}$ is used as an exponent to indicate square root. That is, we can write $\sqrt{a} = a^{\frac{1}{2}}$. Similarly, a cube root is indicated as $\sqrt[3]{a} = a^{\frac{1}{3}}$ and a fourth root is indicated as $\sqrt[4]{a} = a^{\frac{1}{4}}$.]

With the class separated into teams of two to four, each team is to write what they think is a good definition of the terms **cube root** and **fourth root.** Using a calculator, they are to find both the cube root and the fourth root (accurate to four decimal places) of each of the following numbers:

(a) 8 (b) 16 (c) 27 (d) 125

The team leader is to read the results with classroom discussion to follow.

5 Index of Key Ideas and Terms

NAME ______________________ SECTION ________________ DATE ________

ANSWERS

1. [Respond below exercise.] ____________
2. ____________
3. ____________
4. ____________
5. ____________
6. ____________
7. ____________
8. ____________
9. ____________
10. ____________
11. ____________
12. ____________
13. ____________
14. ____________
15. ____________
16. ____________

Chapter 5 Test

1. Write the decimal number 30.657 in words.

2. Write the decimal number 0.375 in fraction form reduced to lowest terms.

3. Change $\frac{5}{16}$ to decimal form.

In Exercises 4–6, round off as indicated.

4. 203.018 (to the nearest hundredth)

5. 1972.8046 (to the nearest thousand)

6. 0.0693 (to the nearest tenth)

In Exercises 7–16, perform the indicated operations.

7. $85.815 + 17.943$

8. $95.6 - 93.712$

9. $82 + 14.91 + 25.2$

10. $100.64 - 82.495$

11. $13\frac{2}{5} + 6 + 17.913$

12. $1\frac{1}{1000} - 0.09705$

13. $(0.35)(0.84)$

14. $\begin{array}{r} 16.31 \\ \times\ 0.785 \\ \hline \end{array}$

15. $(1.92)(1000)$

16. $3.614 \div 100$

17. ____________

18. ____________

19. [Respond in exercise.]

20. [Respond in exercise.]

21. ____________

22. ____________

23. ____________

24. ____________

25a. ____________

b. ____________

26. ____________

Find each quotient to the nearest hundredth.

17. $82 \div 4.6$

18. $0.13\overline{)8.617}$

Find the equivalent measures in the metric system.

19. 0.681 km = __________ m = __________ cm = __________ mm

20. 355 mm = __________ cm = __________ m

Write each of the following numbers in scientific notation.

21. 86,500,000

22. 0.000372

Simplify the following radical expressions. Then use a calculator to find the value of each expression accurate to four decimal places.

23. $\sqrt{800}$

24. $\sqrt{125}$

25. (a) If an automobile gets 18.3 miles per gallon and the tank holds 21.4 gallons, approximately how far can the car travel on a full tank? (b) To the nearest mile, how far can the car actually travel on a full tank?

26. A football field is rectangular in shape—100 yards long and 40 yards wide. To the nearest tenth of a yard, what is the distance from one corner of the field to another corner? (That is, what is the length of a diagonal of the rectangle?)

NAME ______________________ SECTION ________________ DATE __________

ANSWERS

Cumulative Review

1. Name the property illustrated.

(a) $15 + 6 = 6 + 15$ (b) $2(6 \cdot 7) = (2 \cdot 6)7$ (c) $51 + 0 = 51$

Evaluate the expressions in Exercises 2–5.

2. $2^2 + 3^2$

3. $18 + 5(2 + 3^2) \div 11 \cdot 2$

4. $24 + [6 + (5^2 \cdot 1 - 3)]$

5. $36 \div 4 + 9 \cdot 2^2$

6. Find the prime factorization of 420.

7. Find the LCM for each set of numbers.

(a) 28, 70 (b) 8, 16, 32, 64

In Exercises 8–14, perform the indicated operations.

8. $\frac{14}{15} - \frac{9}{10}$

9. $4\frac{5}{8} + 3\frac{9}{10}$

10. $306\frac{3}{8} - 250\frac{13}{20}$

11. $10\frac{1}{2} + 5\frac{3}{4} + 16\frac{1}{10}$

12. $\left(\frac{3}{4} - \frac{1}{8}\right) \div \left(\frac{13}{16} - \frac{1}{2}\right)$

13. $\left(\frac{1}{2}\right)^2\left(\frac{14}{15}\right) + \frac{4}{15} \div 2$

14. $22\frac{1}{2} \div 3\frac{3}{4}$

1a. ________
b. ________
c. ________
2. ________
3. ________
4. ________
5. ________
6. ________
7a. ________
b. ________
8. ________
9. ________
10. ________
11. ________
12. ________
13. ________
14. ________

15. ____________

16. ____________

17. ____________

18. ____________

19a. ____________

b. ____________

20. [Respond below exercise.]

21a. ____________

b. ____________

22. ____________

15. Simplify the complex fraction: $\dfrac{\frac{1}{2} + \frac{5}{6}}{1 - \frac{5}{8}}$

16. Find $\frac{3}{4}$ of 96.

17. Find the quotient: $32.1 \div 1.7$ (to the nearest tenth).

18. Find the product of 21.6 and 0.35 and subtract the sum of 1.375 and 4.

19. Write the following numbers in scientific notation.
(a) 18,700,000 (b) 0.000000632

20. Use the Pythagorean Theorem to show that a triangle with sides of 12 cm, 16 cm, and 20 cm is a right triangle.

21. The business office at a university purchased five vans at a price of $21,540 each, including tax. (a) Approximately how much was paid for the vans? (b) Exactly how much was paid for the vans?

22. The sale price of a washing machine was $240. This was $\frac{4}{5}$ of the original price. What was the original price?

6

Ratios and Proportions

What to Expect in Chapter 6

Chapter 6 introduces another meaning for fractions (ratios) and the very important topic of solving equations. Section 6.1 gives the definition of a ratio and shows three different ways in which to write a ratio. In Section 6.2, ratios are used in a way familiar to shoppers—price per unit. Proportions (equations stating that two ratios are equal) are introduced in Section 6.3. In Section 6.4, we show how to find the value of one unknown term in a proportion. The unknown term is called a **variable,** and the methods for finding the value of this term provide an introduction to equation solving. (**Note:** Solving different forms of equations will be discussed in Chapter 9 and in other courses in mathematics.) The chapter closes with a section on problem solving with the use of proportions.

Mathematics at Work!

Home owners are familiar with trips to the store to buy such things as paint, fertilizer, and grass seed for repairs and maintenance. In many cases, the purchase must be for more than is needed at one time for one particular job because the units do not work out exactly and the store will not sell part of a can of paint or part of a bag of fertilizer. A little application of mathematics can salve many financial wounds by minimizing any excess amounts that must be bought. Consider the following situation.

You want to treat your lawn with fungicide/insecticide/fertilizer mix. The local nursery sells bags of this mix for $14 per bag, and each bag contains 16 pounds of mix. The recommended coverage for one bag is 2000 square feet. If your lawn consists of two rectangular shapes, one 30 feet by 100 feet and the other 50 feet by 80 feet, how many pounds of fertilizer mix would it take to cover your lawn? How many bags do you need to buy? How much would you pay (not including tax)? (See Exercise 41, Section 6.5.)

Lawn and garden care are perhaps the most routine maintenance tasks faced by every home owner.

6.1 Ratios

OBJECTIVES

1. *Understand the meaning of a ratio.*
2. *Know several notations for ratios.*

Understanding Ratios

We know two meanings for fractions.

1. To indicate a part of a whole:

$\frac{7}{8}$ means $\frac{\text{7 pieces of cherry pie}}{\text{8 pieces in the whole pie}}$

2. To indicate division:

$\frac{3}{8}$ means $3 \div 8$ or

$$\begin{array}{r} .375 \\ 8\,\overline{)\,3.000} \\ \underline{2\,4} \\ 60 \\ \underline{56} \\ 40 \\ \underline{40} \\ 0 \end{array}$$

A third use of fractions is to compare two quantities. Such a comparison is called a **ratio.** For example,

$\frac{3}{4}$ might mean $\frac{\text{3 dollars}}{\text{4 dollars}}$ or $\frac{\text{3 hours}}{\text{4 hours}}$

Definition

A **ratio** is a comparison of two quantities by division. The ratio of a to b can be written as

$$\frac{a}{b} \quad \text{or} \quad a : b \quad \text{or} \quad a \text{ to } b$$

Ratios have the following characteristics:

1. Ratios can be reduced, just as fractions can be reduced.

2. Whenever the units of the numbers in a ratio are the same, then the ratio has no units. We say that the ratio is an **abstract number.**

3. When the numbers in a ratio have different units, then the numbers must be labeled to clarify what is being compared. Such a ratio is called a **rate.**

For example, the ratio of 55 miles : 1 hour $\left(\text{or } \frac{\text{55 miles}}{\text{1 hour}}\right)$ is a rate of 55 miles per hour (or 55 mph).

Write each comparison as a ratio reduced to lowest terms.

1. 3 quarters to 1 dollar

2. 36 inches to 5 feet

3. Inventory shows 5000 washers and 4000 bolts. What is the ratio of washers to bolts?

Examples of Ratios

EXAMPLE 1 Compare the quantities 30 students and 40 chairs as a ratio.

Solution

Since the units (students and chairs) are not the same, the units must be written in the ratio. We can write

(a) $\dfrac{30 \text{ students}}{40 \text{ chairs}}$ (b) 30 students : 40 chairs

(c) 30 students **to** 40 chairs

Furthermore, the same ratio can be simplified by reducing. Since

$\dfrac{30}{40} = \dfrac{3}{4}$, we can write the ratio as $\dfrac{3 \text{ students}}{4 \text{ chairs}}$.

The reduced ratio can also be written as

3 students **:** 4 chairs or 3 students **to** 4 chairs.

EXAMPLE 2 Write the comparison of 2 feet to 3 yards as a ratio.

Solution

(a) We can write the ratio as $\dfrac{2 \text{ feet}}{3 \text{ yards}}$.

(b) However, a better procedure is to change to common units. Since 1 yard contains 3 feet, 3 yards = 9 feet. Now we can write the ratio as an abstract number:

$$\frac{2 \text{ feet}}{3 \text{ yards}} = \frac{2 \text{ feet}}{9 \text{ feet}} = \frac{2}{9} \quad \text{or} \quad 2 : 9 \quad \text{or} \quad 2 \textbf{ to } 9$$

EXAMPLE 3

Baseball players' batting averages are published in the newspapers. Suppose a player has a batting average of .250. What does this mean?

Solution

A batting average is a ratio of hits to times at bat. Thus, a batting average of .250 means

$$.250 = \frac{250 \text{ hits}}{1000 \text{ times at bat}}$$

Reducing gives

$$.250 = \frac{250}{1000} = \frac{250 \cdot 1}{250 \cdot 4} = \frac{1}{4} = \frac{1 \text{ hit}}{4 \text{ times at bat}}$$

This means that we can expect this player to hit safely at a rate of 1 hit for every 4 times he comes to bat.

Now Work Exercises 1–3 in the Margin.

NAME ______________ SECTION ______________ DATE ______________

ANSWERS

Exercises 6.1

Write the following comparisons as ratios reduced to lowest terms. Use common units in the numerator and denominator whenever possible.

1. 1 dime to 4 nickels
2. 5 nickels to 3 quarters
3. 5 dollars to 5 quarters
4. 6 dollars to 50 dimes
5. 250 miles to 5 hours
6. 270 miles to 4.5 hours
7. 50 miles to 2 gallons of gas
8. 60 miles to 5 gallons of gas
9. 30 chairs to 25 people
10. 25 people to 30 chairs
11. 18 inches to 2 feet
12. 36 inches to 2 feet
13. 8 days to 1 week
14. 21 days to 4 weeks

1. ______
2. ______
3. ______
4. ______
5. ______
6. ______
7. ______
8. ______
9. ______
10. ______
11. ______
12. ______
13. ______
14. ______

15. ______________

16. ______________

17. ______________

18. ______________

19. ______________

20. ______________

21. ______________

22. ______________

15. $200 in profit to $500 invested

16. $200 in profit to $1000 invested

17. 100 centimeters to 1 meter

18. 10 centimeters to 1 millimeter

19. 125 hits to 500 times at bat

20. 100 hits to 500 times at bat

21. About 28 out of every 100 African-Americans have type-A blood. Express this fact as a ratio in lowest terms.

22. A serving of four home-baked chocolate chip cookies weighs 40 grams and contains 12 grams of fat. What is the ratio of fat grams to total grams, in lowest terms?

NAME ______________________ SECTION ______________ DATE ________

ANSWERS

23. ______________

24. ______________

25. [Respond below exercise.]

23. In recent years, 18 out of every 100 students taking the SAT (Scholastic Aptitude Test) have scored 600 or above on the mathematics portion of the test. Write the ratio, in lowest terms, of the number of scores 600 or above to the number of scores below 600.

24. In a recent year, Albany, NY reported a total of 60 clear days, the rest being cloudy or partly cloudy. For a 365-day year, write the ratio of clear days to cloudy or partly cloudy days in lowest terms.

Writing and Thinking about Mathematics

25. About 8 out of every 100 men exhibit dichromatism, or partial color blindness. What is the ratio of partially color-blind men to men who are unaffected? The ratio of all people affected by dichromatism to those who are not is about 1 to 22. Is this ratio the same as the ratio you determined for men? What does this tell you about dichromatism in the female population?

RECYCLE BIN

1. ______________

2. ______________

3. ______________

4. ______________

5. ______________

6. ______________

7. ______________

8. ______________

The Recycle Bin (from Sections 3.2, 4.2, 5.3, and 5.4)

Perform the indicated operations and simplify if possible.

1. $3\frac{1}{2} \cdot 3\frac{1}{7}$

2. $4\frac{5}{8} \div 7\frac{3}{5}$

3. $\frac{14}{32} \cdot \frac{12}{28} \cdot \frac{4}{15}$

4. $20 \div \frac{1}{4}$

5. (19.2)(7.8)

6. 14.12(.0015)

7. $13.4\overline{)73.7}$

8. $0.25\overline{)100}$

6.2 Price per Unit

OBJECTIVES

1. *Understand the concept of price per unit.*
2. *Know how to set up a price per unit as a ratio.*
3. *Be able to calculate a price per unit by using division.*

Understanding Price per Unit

When you buy a new battery for your car, you usually buy just one battery; or, if you buy flashlight batteries, you probably buy them in sets of two or four. In cases such as these, the price for one unit (one battery) is clearly marked or understood. However, when you buy groceries, the same item may be packaged in more than one size. Since you want to get the most for your money, you want the better (or best) buy. To calculate the **price per unit** (or **unit price**), you can set up the ratio of price to units and divide.

To Find the Price per Unit:

1. Set up a ratio (usually in fraction form) of **price to units.**
2. Divide the price by the number of units.

Note: In the past, many consumers did not understand the concept of price per unit or know how to determine such a number. Because of this, consumers were not always aware of what they were buying, and most states now have a law that grocery stores must display the price per unit for certain goods they sell.

Examples of Unit Pricing

In the following examples, we show the division process as you would write it on paper. However, you may choose to use a calculator to perform these operations. In either case, your first step should be to write each ratio so you can clearly see what you are dividing and that all ratios are comparing the same types of units.

Also note that the comparisons being made in the examples and exercises are with different amounts of the same brand and quality of goods. Any comparison of a relatively expensive brand of high quality with a cheaper brand of lower quality would be meaningless because factors other than price would be involved.

EXAMPLE 1

A 12-ounce can of beans is priced at 80¢ while an 18-ounce can of the same beans is $1.10. Which is the better buy?

Solution

In order to decide which buy is better, we write two ratios of price to units, divide, and compare the results. In this problem, we must convert

\$1.10 to 110¢ so that both ratios will be comparing cents to ounces.

(a) $\frac{80¢}{12\text{ oz}}$

$$
\begin{array}{r}
6.66 \\
12\overline{)80.00} \\
\underline{72} \\
8\,0 \\
\underline{7\,2} \\
80 \\
\underline{72}
\end{array}
$$

or 6.7¢ per ounce

(b) $\frac{110¢}{18\text{ oz}}$

$$
\begin{array}{r}
6.11 \\
18\overline{)110.00} \\
\underline{108} \\
2\,0 \\
\underline{1\,8} \\
20 \\
\underline{18}
\end{array}
$$

or 6.2¢ per ounce

(**Note:** 6.11 is rounded up to 6.2 because we are dealing with money.)

Thus, the larger can (18 ounces for \$1.10) is the better buy, since the price per ounce is smaller.

Example 2

Pancake syrup comes in three different-sized bottles:

36 fluid ounces for \$3.29,

24 fluid ounces for \$2.49, and

12 fluid ounces for \$1.59.

Find the price per fluid ounce for each size of bottle and tell which is the best buy.

Solution

After each ratio is set up, the numerator is changed from dollars and cents to just cents so that the division will yield cents per ounce.

(a) $\frac{\$3.29}{36\text{ oz}} = \frac{329¢}{36\text{ oz}}$

$$
\begin{array}{r}
9.11 \\
36\overline{)329.00} \\
\underline{324} \\
5\,0 \\
\underline{3\,6} \\
40 \\
\underline{36}
\end{array}
$$

or 9.2¢ per ounce or 9.2¢/oz

(**Note:** "9.2¢/oz" is read "9.2¢ per ounce.")

(b) $\frac{\$2.49}{24 \text{ oz}} = \frac{249¢}{24 \text{ oz}}$

$$\begin{array}{r} 10.37 \\ 24 \overline{)\,249.00} \\ \underline{24} \\ 09 \\ \underline{0} \\ 9\,0 \\ \underline{7\,2} \\ 1\,80 \\ \underline{1\,68} \end{array}$$

or 10.4¢/oz

(c) $\frac{\$1.59}{12 \text{ oz}} = \frac{159¢}{12 \text{ oz}}$

$$\begin{array}{r} 13.25 \\ 12 \overline{)\,159.00} \\ \underline{12} \\ 39 \\ \underline{36} \\ 3\,0 \\ \underline{2\,4} \\ 60 \\ \underline{60} \end{array}$$

or 13.3¢/oz

The largest container (36 fluid ounces) is the best buy at 9.2¢/oz.

In each of the examples, the largest number of units was the best buy. In general, this is true because the manufacturer wants you to buy more of the product. However, the largest amount may not always be the best buy for every individual. For example, people who do not use much pancake syrup may want to buy a smaller bottle. Even though they pay more per unit, they do not end up throwing any into the trash, which would be more expensive in the long run.

Special comment on the term *per*: The student should be aware that the term **per** can be interpreted to mean **divided by.** For example,

cents **per** ounce means cents **divided by** ounces;

dollars **per** pound means dollars **divided by** pounds;

miles **per** hour means miles **divided by** hours; and

miles **per** gallon means miles **divided by** gallons.

NAME ______________ SECTION ______________ DATE ______________

ANSWERS

1. ______________
2. ______________
3. ______________
4. ______________
5. ______________
6. ______________
7. ______________
8. ______________
9. ______________
10. ______________
11. ______________
12. ______________
13. ______________
14. ______________

Exercises 6.2

Find the unit price of each of the following items and tell which is the better (or best) buy.

1. boxed rice
16 oz (1 lb) at 89¢
32 oz (2 lb) at $1.69

2. noodles
42 oz at $2.99
14 oz at $1.39

3. sour cream
8 oz $\left(\frac{1}{2}\text{ pt}\right)$ at 69¢
11 oz at $1.29

4. coffee
48 oz (3 lb) at $7.29
13 oz at $2.45

5. instant coffee
8 oz at $3.49
2 oz at $1.29

6. instant coffee
12 oz at $4.49
8 oz at $3.69

7. frozen orange juice
16 fl oz at $1.69
12 fl oz at 99¢
6 fl oz at 69¢

8. tortilla chips
11 oz at $2.32
16 oz at $2.74
7.5 oz at $1.72

9. boxed dry milk
9.6 oz at $1.49
25.6 oz at $3.59

10. cookies
16 oz at $1.99
20 oz at $2.59

11. saltine crackers
8 oz at 99¢
16 oz at $1.19

12. peanut butter
18 oz at $1.85
28 oz at $2.69

13. aluminum foil
200 sq ft at $4.19
75 sq ft at $1.59
25 sq ft at 69¢

14. liquid dish soap
32 oz at $2.29
22 oz at $1.69
12 oz at $1.09

15. ______________

16. ______________

17. ______________

18. ______________

19. ______________

20. ______________

21. ______________

22. ______________

23. ______________

24. ______________

25. ______________

26. ______________

27. ______________

28. ______________

29. ______________

30. ______________

15. sliced bologna
8 oz at $1.09
12 oz at $1.59

16. sliced ham
16 oz at $4.69
8 oz at $2.59

17. laundry detergent
9 lb 3 oz (147 oz) at $5.99
72 oz at $3.99
42 oz at $2.39
17 oz at $1.23

18. peanut butter
12 oz at $1.29
18 oz at $1.89
28 oz at $2.79
40 oz at $3.89

19. bottled bleach
1 gal (128 fl oz) at $1.14
$\frac{1}{2}$ gal (64 fl oz) at 85¢
1 qt (32 fl oz) at 59¢

20. jelly
18 oz at $1.29
32 oz at $1.69
48 oz at $2.49

21. boxed doughnuts
14 oz at $1.49
9 oz at $1.09
5 oz at 69¢

22. mustard
8 oz at 65¢
12 oz at $1.09
24 oz at $1.19

23. mayonnaise
16 oz at $1.23
32 oz at $1.69

24. salad dressing
8 oz at $1.09
16 oz at $1.69

25. mustard
8 oz at 69¢
16 oz at $1.09

26. catsup
14 oz at 83¢
32 oz at 99¢

27. chili with beans
40 oz at $2.29
15 oz at 89¢

28. dill pickles
32 oz at $1.59
16 oz at $1.19

29. chili beans
30 oz at 89¢
15.5 oz at 59¢

30. baked beans
31 oz at 79¢
16 oz at 53¢

6.3 Proportions

Understanding Proportions

Consider the equation $\frac{3}{6} = \frac{4}{8}$. This statement (or equation) says that two ratios are equal. Such an equation is called a **proportion.** As we will see, proportions may be true or false.

Definition

A **proportion** is a statement that two ratios are equal.
In symbols,

$$\frac{a}{b} = \frac{c}{d} \quad \text{is a proportion.}$$

A proportion has four **terms:**

first term → a, second term → b, third term → c, fourth term → d

$$\frac{a}{b} = \frac{c}{d}$$

The first and fourth terms (a and d) are called the **extremes.**

The second and third terms (b and c) are called the **means.**

To help in remembering which terms are the extremes and which terms are the means, think of a general proportion written with colons, as shown here.

means (b, c)

$$a : b = c : d$$

extremes (a, d)

With this form, you can see that a and d are the two end terms; thus, the name **extremes** might seem more reasonable.

OBJECTIVES

1. *Understand that a proportion is an equation.*
2. *Be familiar with the terms* **means** *and* **extremes.**
3. *Know that in a true proportion, the product of the means is equal to the product of the extremes.*

EXAMPLE 1

In the proportion $\frac{8.4}{4.2} = \frac{10.2}{5.1}$, tell which numbers are the extremes and which are the means.

Solution

8.4 and 5.1 are the extremes.

4.2 and 10.2 are the means.

✔ COMPLETION EXAMPLE 2

In the proportion $\frac{2\frac{1}{2}}{10} = \frac{3\frac{1}{4}}{13}$, tell which numbers are the extremes and which are the means.

Solution

__________ and __________ are the extremes.

__________ and __________ are the means.

Identifying True Proportions

In a true proportion, the product of the extremes is equal to the product of the means. In symbols,

$$\frac{a}{b} = \frac{c}{d} \quad \text{if and only if} \quad a \cdot d = b \cdot c$$

where $b \neq 0$ and $d \neq 0$.

Note that in a proportion the terms can be any of the types of numbers that we have studied: whole numbers, fractions, mixed numbers, or decimals. Thus, all the related techniques for multiplying these types of numbers should be reviewed at this time.

EXAMPLE 3

Determine whether the proportion $\frac{9}{13} = \frac{4.5}{6.5}$ is true or false.

Solution

```
  6.5                              4.5
×   9                            × 13
-----                            ----
 58.5  ← product of extremes      13 5
                                  45
                                 ----
                                  58.5  ← product of means
```

Since $9(6.5) = 13(4.5)$, the proportion is true.

EXAMPLE 4

Is the proportion $\frac{5}{8} = \frac{7}{10}$ true or false?

Solution

$$5 \cdot 10 = 50 \quad \text{and} \quad 8 \cdot 7 = 56$$

Since $50 \neq 56$, the proportion is false.

EXAMPLE 5

Is the proportion $\frac{\frac{3}{5}}{\frac{3}{4}} = \frac{12}{15}$ true or false?

Solution

The extremes are $\frac{3}{5}$ and 15, and their product is

$$\frac{3}{5} \cdot 15 = \frac{3}{5} \cdot \frac{15}{1} = 9.$$

The means are $\frac{3}{4}$ and 12, and their product is

$$\frac{3}{4} \cdot 12 = \frac{3}{4} \cdot \frac{12}{1} = 9.$$

Since the product of the extremes is equal to the product of the means, the proportion is true.

COMPLETION EXAMPLE 6

Is the proportion $\frac{1}{4} : \frac{2}{3} = 9 : 24$ true or false?

Solution

(a) The extremes are __________ and __________.

The means are __________ and __________.

(b) The product of the extremes is __________.

The product of the means is __________.

(c) The proportion is __________ because the products are __________.

Now Work Exercises 1–3 in the Margin.

Determine whether each proportion is true or false.

1. $\frac{15}{20} = \frac{21}{28}$

2. $\frac{5\frac{1}{2}}{2} = \frac{6\frac{1}{2}}{3}$

3. $\frac{1.3}{1.5} = \frac{1.82}{2.1}$

Completion Example Answers

2. $\mathbf{2\frac{1}{2}}$ and **13** are the extremes.

   10 and $\mathbf{3\frac{1}{4}}$ are the means.

6. (a) The extremes are $\mathbf{\frac{1}{4}}$ and **24.**

   The means are $\mathbf{\frac{2}{3}}$ and **9.**

   (b) The product of the extremes is

   $\mathbf{\frac{1}{4} \cdot 24 = 6}$

   The product of the means is

   $\mathbf{\frac{2}{3} \cdot 9 = 6}$

   (c) The proportion is **true** because the products are **equal.**

NAME ______________ SECTION ______________ DATE __________

ANSWERS

1. ______________
2. ______________
3. ______________
4. ______________
5. ______________
6. ______________
7. ______________
8. ______________
9. ______________
10. ______________
11. ______________
12. ______________
13. ______________
14. ______________

Exercises 6.3

1. In the proportion $\frac{7}{8} = \frac{476}{544}$

(a) the extremes are ______ and ______;

(b) the means are ______ and ______.

2. In the proportion $\frac{x}{y} = \frac{w}{z}$

(a) the extremes are ______ and ______;

(b) the means are ______ and ______;

(c) the two terms ______ and ______ cannot be 0.

Determine whether each proportion is true or false by comparing the products of the means and extremes.

3. $\frac{5}{6} = \frac{10}{12}$

4. $\frac{2}{7} = \frac{5}{17}$

5. $\frac{7}{21} = \frac{4}{12}$

6. $\frac{3}{5} = \frac{60}{100}$

7. $\frac{125}{1000} = \frac{1}{8}$

8. $\frac{3}{8} = \frac{375}{1000}$

9. $\frac{1}{4} = \frac{25}{100}$

10. $\frac{7}{8} = \frac{875}{1000}$

11. $\frac{3}{16} = \frac{9}{48}$

12. $\frac{2}{3} = \frac{66}{100}$

13. $\frac{1}{3} = \frac{33}{100}$

14. $\frac{14}{6} = \frac{21}{8}$

15. ______________

16. ______________

17. ______________

18. ______________

19. ______________

20. ______________

21. ______________

22. ______________

23. ______________

24. ______________

25. [Respond below exercise.] ______________

15. $\frac{12}{18} = \frac{14}{21}$

16. $\frac{5}{6} = \frac{7}{8}$

17. $\frac{7.5}{10} = \frac{3}{4}$

18. $\frac{6.2}{3.1} = \frac{10.2}{5.1}$

19. $\frac{8\frac{1}{2}}{2\frac{1}{3}} = \frac{4\frac{1}{4}}{1\frac{1}{6}}$

20. $\frac{6\frac{1}{5}}{1\frac{1}{7}} = \frac{3\frac{1}{10}}{\frac{8}{14}}$

21. $\frac{6}{24} = \frac{10}{48}$

22. $\frac{7}{16} = \frac{3\frac{1}{2}}{8}$

23. $\frac{10}{17} = \frac{5}{8\frac{1}{2}}$

24. $\frac{9}{20} = \frac{2\frac{1}{4}}{5}$

Writing and Thinking about Mathematics

25. Consider the proportion $\frac{16}{5} = \frac{22.4}{7}$. Show why multiplying both sides of the equation by 35 has the same effect as setting the product of the extremes equal to the product of the means. Do you think that this same technique (multiplying both sides of the equation by the least common multiple of the denominators) will work with all true proportions? Why or why not?

6.4 Finding the Unknown Term in a Proportion

OBJECTIVES

1. *Learn how to find the unknown term in a proportion.*
2. *Recall the skills used in multiplying and dividing with whole numbers, fractions, mixed numbers, and decimals.*

Understanding the Meaning of an Unknown Term

Proportions can be used to solve certain types of word problems. In these problems, a proportion is set up in which one of the terms in the proportion is not known and the solution to the problem is the value of this unknown term. In this section, we will use the fact that in a true proportion the product of the extremes is equal to the product of the means as a method for finding the unknown term in a proportion.

Definition

A **variable** is a symbol (generally a letter of the alphabet) that is used to represent an unknown number or any one of several numbers. The set of possible values for a variable is called its **replacement set.**

Finding the Unknown Term in a Proportion

To Find the Unknown Term in a Proportion:

1. In the proportion, the unknown term is represented with a variable (some letter such as x, y, w, A, B, etc.).
2. Write an equation that sets the product of the extremes equal to the product of the means.
3. Divide both sides of the equation by the number multiplying the variable. (This number is called the **coefficient** of the variable.)

The resulting equation will have a coefficient of 1 for the variable and will give the missing value for the unknown term in the proportion.

Important Notes: As you work through each problem, be sure to write each new equation below the previous equation in the same format shown in the examples. (Arithmetic that cannot be done mentally should be performed to one side, and the results written in the next equation.) This format carries over into solving all types of equations at all levels of mathematics. Also, when a number is written next to a variable, such as in $3x$ or $4y$, the meaning is to multiply the number times the value of the variable. That is, $3x = 3 \cdot x$ and $4y = 4 \cdot y$. The number 3 is the coefficient of x and 4 is the coefficient of y.

EXAMPLE 1 Find the value of x if $\frac{3}{6} = \frac{5}{x}$.

Solution

In this case, you may be able to see that the correct value of x is 10 since

$\frac{3}{6}$ reduces to $\frac{1}{2}$ and $\frac{1}{2} = \frac{5}{10}$.

However, not all proportions involve such simple ratios, and the following general method of solving for the unknown is important.

(a) $\frac{3}{6} = \frac{5}{x}$ Write the proportion.

(b) $3 \cdot x = 6 \cdot 5$ Write the product of the extremes equal to the product of the means.

(c) $\frac{\cancel{3} \cdot x}{\cancel{3}} = \frac{30}{3}$ Divide both sides by the coefficient of the variable, 3.

(d) $x = 10$ Reduce both sides to find the solution.

EXAMPLE 2 Find the value of y if $\frac{6}{16} = \frac{y}{24}$.

Solution

Note that the variable may appear on the right side of the equation [as shown here in step (b)] as well as on the left side of the equation (as shown in Example 1). In either case, we **divide both sides of the equation by the coefficient of the variable.**

(a) $\frac{6}{16} = \frac{y}{24}$ Write the proportion.

(b) $6 \cdot 24 = 16 \cdot y$ Write the product of the extremes equal to the product of the means.

(c) $\frac{6 \cdot 24}{16} = \frac{\cancel{16} \cdot y}{\cancel{16}}$ Divide both sides by 16, the coefficient of y.

(d) $9 = y$ Reduce both sides to find the value of y.

Second Solution

Reduce the fraction $\frac{6}{16}$ before solving the proportion.

(a) $\frac{6}{16} = \frac{y}{24}$ Write the proportion.

(b) $\frac{3}{8} = \frac{y}{24}$ Reduce the fraction: $\frac{6}{16} = \frac{3}{8}$.

(c) $3 \cdot 24 = 8 \cdot y$ Proceed to solve as before.

(d) $\frac{3 \cdot 24}{8} = \frac{\cancel{8} \cdot y}{\cancel{8}}$

(e) $9 = y$

EXAMPLE 3 Find w if $\dfrac{w}{7} = \dfrac{20}{\frac{2}{3}}$.

Solution

(a) $\dfrac{w}{7} = \dfrac{20}{\frac{2}{3}}$ Write the proportion.

(b) $\dfrac{2}{3} \cdot w = 7 \cdot 20$ Set the product of the extremes equal to the product of the means.

(c) $\dfrac{\frac{2}{3} \cdot w}{\frac{2}{3}} = \dfrac{7 \cdot 20}{\frac{2}{3}}$ Divide each side by the coefficient $\frac{2}{3}$.

(d) $w = \dfrac{7}{1} \cdot \dfrac{20}{1} \cdot \dfrac{3}{2}$ Simplify. Remember: to divide by a fraction, multiply by its reciprocal.

(e) $w = 210$

COMPLETION EXAMPLE 4 Find A if $\dfrac{A}{3} = \dfrac{7.5}{6}$.

Solution

(a) $\dfrac{A}{3} = \dfrac{7.5}{6}$

(b) $6 \cdot A =$ ________

(c) $\dfrac{6 \cdot A}{6} =$ ________

(d) $A =$ ________

As illustrated in Examples 5 and 6, when the coefficient of the variable is a fraction, we can multiply both sides of the equation by the reciprocal of this coefficient. This is the same as dividing by the coefficient, but fewer steps are involved.

EXAMPLE 5 Find x if $\dfrac{x}{1\frac{1}{2}} = \dfrac{1\frac{2}{3}}{3\frac{1}{3}}$.

Solution

(a) $\dfrac{x}{1\frac{1}{2}} = \dfrac{1\frac{2}{3}}{3\frac{1}{3}}$

Solve each proportion for the unknown term.

1. $\frac{3}{10} = \frac{R}{100}$

2. $\frac{\frac{1}{2}}{6} = \frac{5}{x}$

3. $\frac{x}{1.5} = \frac{11}{2.75}$

(b) $\frac{10}{3} \cdot x = \frac{\cancel{3}}{2} \cdot \frac{5}{\cancel{3}}$ Write each mixed number as an improper fraction.

(c) $\frac{10}{3} \cdot x = \frac{5}{2}$

(d) $\frac{\cancel{3}}{\cancel{10}} \cdot \frac{\cancel{10}}{\cancel{3}} \cdot x = \frac{3}{10} \cdot \frac{5}{2}$ Multiply each side by $\frac{3}{10}$, the reciprocal of $\frac{10}{3}$.

(e) $x = \frac{3}{\underset{2}{\cancel{10}}} \cdot \frac{\cancel{5}}{2} = \frac{3}{4}$

COMPLETION EXAMPLE 6

Find y if $\frac{2\frac{1}{2}}{6} = \frac{3}{y}$.

Solution

(a) $\frac{2\frac{1}{2}}{6} = \frac{3}{y}$

(b) $\frac{5}{2} \cdot y =$ __________ $\left(2\frac{1}{2} = \frac{5}{2}\right)$

(c) $\frac{\cancel{2}}{\cancel{5}} \cdot \frac{\cancel{5}}{\cancel{2}} \cdot y = \frac{2}{5} \cdot$ __________

(d) $y =$ __________

Now Work Exercises 1–3 in the Margin.

Completion Example Answers

4. (a) $\frac{A}{3} = \frac{7.5}{6}$

(b) $6 \cdot A = \mathbf{3 \cdot 7.5}$

(c) $\frac{\cancel{6} \cdot A}{\cancel{6}} = \frac{\mathbf{22.5}}{\mathbf{6}}$

(d) $A = \mathbf{3.75}$

6. (a) $\frac{2\frac{1}{2}}{6} = \frac{3}{y}$

(b) $\frac{5}{2} \cdot y = \mathbf{3 \cdot 6}$

(c) $\frac{\cancel{2}}{\cancel{5}} \cdot \frac{\cancel{5}}{\cancel{2}} \cdot y = \frac{2}{5} \cdot \mathbf{18}$

(d) $y = \frac{\mathbf{36}}{\mathbf{5}} \left(\textbf{or } \mathbf{7\frac{1}{5}}\right)$

NAME ______________________ SECTION ______________ DATE ________

Exercises 6.4

Solve for the variable in each of the following proportions.

1. $\frac{3}{6} = \frac{6}{x}$

2. $\frac{7}{21} = \frac{y}{6}$

3. $\frac{5}{7} = \frac{x}{28}$

4. $\frac{4}{10} = \frac{5}{x}$

5. $\frac{8}{B} = \frac{6}{30}$

6. $\frac{7}{B} = \frac{5}{15}$

7. $\frac{1}{2} = \frac{x}{100}$

8. $\frac{3}{4} = \frac{x}{100}$

9. $\frac{A}{3} = \frac{7}{2}$

10. $\frac{x}{100} = \frac{1}{20}$

11. $\frac{3}{5} = \frac{60}{D}$

12. $\frac{3}{16} = \frac{9}{x}$

13. $\frac{\frac{1}{2}}{x} = \frac{5}{10}$

14. $\frac{\frac{2}{3}}{3} = \frac{y}{127}$

15. $\frac{\frac{1}{3}}{x} = \frac{5}{9}$

16. $\frac{\frac{3}{4}}{7} = \frac{3}{z}$

17. $\frac{\frac{1}{8}}{6} = \frac{\frac{1}{2}}{w}$

18. $\frac{\frac{1}{6}}{5} = \frac{5}{w}$

ANSWERS

1. ______
2. ______
3. ______
4. ______
5. ______
6. ______
7. ______
8. ______
9. ______
10. ______
11. ______
12. ______
13. ______
14. ______
15. ______
16. ______
17. ______
18. ______

19. ______________

20. ______________

21. ______________

22. ______________

23. ______________

24. ______________

25. ______________

26. ______________

27. ______________

28. ______________

29. ______________

30. ______________

31. ______________

32. ______________

33. ______________

34. ______________

35. ______________

36. ______________

19. $\frac{1}{4} = \frac{1\frac{1}{2}}{y}$

20. $\frac{1}{5} = \frac{x}{2\frac{1}{2}}$

21. $\frac{1}{5} = \frac{x}{7\frac{1}{2}}$

22. $\frac{2}{5} = \frac{R}{100}$

23. $\frac{3}{5} = \frac{R}{100}$

24. $\frac{A}{4} = \frac{75}{100}$

25. $\frac{A}{4} = \frac{50}{100}$

26. $\frac{20}{B} = \frac{1}{4}$

27. $\frac{30}{B} = \frac{25}{100}$

28. $\frac{A}{20} = \frac{15}{100}$

29. $\frac{1}{3} = \frac{R}{100}$

30. $\frac{2}{3} = \frac{R}{100}$

31. $\frac{9}{x} = \frac{4\frac{1}{2}}{11}$

32. $\frac{y}{6} = \frac{2\frac{1}{2}}{12}$

33. $\frac{x}{4} = \frac{1\frac{1}{4}}{5}$

34. $\frac{5}{x} = \frac{2\frac{1}{4}}{27}$

35. $\frac{x}{3} = \frac{16}{3\frac{1}{5}}$

36. $\frac{6.2}{5} = \frac{x}{15}$

NAME ______________ SECTION ______________ DATE ______________

ANSWERS

37. $\frac{3.5}{2.6} = \frac{10.5}{B}$ **38.** $\frac{4.1}{3.2} = \frac{x}{6.4}$ **39.** $\frac{7.8}{1.3} = \frac{x}{0.26}$

40. $\frac{7.2}{y} = \frac{4.8}{14.4}$ **41.** $\frac{150}{300} = \frac{R}{100}$ **42.** $\frac{19.2}{96} = \frac{R}{100}$

43. $\frac{12}{B} = \frac{25}{100}$ **44.** $\frac{13.5}{B} = \frac{15}{100}$ **45.** $\frac{A}{42} = \frac{65}{100}$

46. $\frac{A}{244} = \frac{18}{100}$ **47.** $\frac{A}{850} = \frac{30}{100}$ **48.** $\frac{A}{595} = \frac{6}{100}$

49. $\frac{5684}{B} = \frac{98}{100}$ **50.** $\frac{24}{27} = \frac{R}{100}$

Check Your Number Sense

Now that you are familiar with proportions and the techniques for finding the unknown term in a proportion, check your general understanding by choosing the answer (using mental calculations only) that seems the most reasonable to you in each of the following exercises. After you have checked the answers in the back of the text, work out each problem that you missed to help you develop a better understanding of proportions.

51. Given the proportion $\frac{x}{100} = \frac{1}{4}$, which of the following values seems the most reasonable for x?

(a) 10 (b) 25 (c) 50 (d) 75

37. ______________

38. ______________

39. ______________

40. ______________

41. ______________

42. ______________

43. ______________

44. ______________

45. ______________

46. ______________

47. ______________

48. ______________

49. ______________

50. ______________

51. ______________

52. ____________

53. ____________

54. ____________

55. ____________

56. ____________

52. Given the proportion $\frac{x}{200} = \frac{1}{10}$, which of the following values seems the most reasonable for x?

(a) 10 (b) 20 (c) 30 (d) 40

53. Given the proportion $\frac{3}{5} = \frac{60}{x}$, which of the following values seems the most reasonable for x?

(a) 50 (b) 80 (c) 100 (d) 150

54. Given the proportion $\frac{4}{10} = \frac{20}{x}$, which of the following values seems the most reasonable for x?

(a) 10 (b) 30 (c) 40 (d) 50

55. Given the proportion $\frac{1.5}{3} = \frac{x}{6}$, which of the following values seems the most reasonable for x?

(a) 1.5 (b) 2.5 (c) 3.0 (d) 4.5

56. Given the proportion $\frac{2\frac{1}{3}}{x} = \frac{4\frac{2}{3}}{10}$, which of the following values seems the most reasonable for x?

(a) $4\frac{2}{3}$ (b) 5 (c) 20 (d) 40

RECYCLE BIN

1. ____________

2. ____________

3. ____________

4. ____________

5. ____________

6. ____________

The Recycle Bin (from Section 5.6)

Perform the indicated operations by using decimal forms of the fractions.

1. $\frac{3}{4} + \frac{3}{5} + 7.16$

2. $5\frac{1}{2} + 20.3 + 16.8$

3. $27\frac{5}{8} - 13.925$

4. $8.13\left(2\frac{1}{4}\right)$

5. $6\frac{7}{10} + 2\frac{1}{2}(6.1)$

6. $5.8\left(3\frac{1}{10}\right) - 3 \div \frac{1}{5}$

6.5 Problem Solving with Proportions

OBJECTIVES

1. *Learn how to identify the unknown quantity in a problem.*
2. *Recognize the type of problem that can be solved by using a proportion.*
3. *Be able to set up a proportion with an unknown term in one of the appropriate patterns that will lead to a solution of the problem.*

When to Use Proportions

A proportion is a statement that two ratios are equal. Thus, problems that involve two ratios are the type that can be solved by using proportions.

To Solve a Word Problem Using Proportions:

1. Identify the unknown quantity and use a variable to represent this quantity.
2. Set up a proportion in which the units are compared as in Pattern A or Pattern B shown here.

   Suppose that a motorcycle will travel 352 miles on 11 gallons of gas. How far would you expect it to travel on 15 gallons of gas? (Let x = the unknown number of miles.)

Pattern A Each ratio has different units, but they are in the same order. For example,

$$\frac{352 \textbf{ miles}}{11 \text{ gallons}} = \frac{x \textbf{ miles}}{15 \text{ gallons}}$$

Pattern B Each ratio has the same units, the numerators correspond, and the denominators correspond. For example,

$$\frac{352 \textbf{ miles}}{x \textbf{ miles}} = \frac{11 \text{ gallons}}{15 \text{ gallons}}$$

(352 miles corresponds to 11 gallons, and x miles corresponds to 15 gallons.)

3. Solve the proportion.

Problem Solving with Proportions

Example 1

You drove your car 500 miles and used 20 gallons of gasoline. How many miles would you expect to drive on 30 gallons of gasoline?

Solution

(a) Let x represent the unknown number of miles.

(b) Set up a proportion using either Pattern A or Pattern B. **Label the numerators and denominators to be sure the units are in the same order.**

$$\frac{500 \text{ miles}}{20 \text{ gallons}} = \frac{x \text{ miles}}{30 \text{ gallons}}$$

Pattern A Each ratio has different units but the numerators are the same units and the denominators are the same units.

(c) Solve the proportion.

$$\frac{500}{20} = \frac{x}{30}$$

$$500 \cdot 30 = 20 \cdot x$$

$$\frac{15{,}000}{20} = \frac{\cancel{20} \cdot x}{\cancel{20}}$$

$$750 = x$$

You would expect to drive 750 miles on 30 gallons of gas.

Note: *Any* of the following proportions yield the same answer.

$$\frac{20 \text{ gallons}}{500 \text{ miles}} = \frac{30 \text{ gallons}}{x \text{ miles}}$$

Pattern A Each ratio has different units but the numerators are the same units and the denominators are the same units.

$$\frac{500 \text{ miles}}{x \text{ miles}} = \frac{20 \text{ gallons}}{30 \text{ gallons}}$$

Pattern B Each ratio has the same units, numerators correspond, and denominators correspond.

$$\frac{x \text{ miles}}{500 \text{ miles}} = \frac{30 \text{ gallons}}{20 \text{ gallons}}$$

Pattern B Each ratio has the same units, numerators correspond, and denominators correspond.

EXAMPLE 2

An architect draws the plans for a building using a scale of $\frac{1}{2}$ inch to represent 10 feet. How many feet would 6 inches represent?

Solution

(a) Let y represent the unknown number of feet.

(b) Set up a proportion labeling the numerators and denominators so that $\frac{1}{2}$ inch corresponds to 10 feet.

$$\frac{\frac{1}{2} \text{ inch}}{6 \text{ inches}} = \frac{10 \text{ feet}}{y \text{ feet}}$$

(c) Solve the proportion. $\frac{\frac{1}{2}}{6} = \frac{10}{y}$

$$\frac{1}{2} \cdot y = 6 \cdot 10$$

$$\frac{\cancel{\frac{1}{2}} \cdot y}{\cancel{\frac{1}{2}}} = \frac{60}{\frac{1}{2}}$$

$$y = \frac{60}{1} \cdot \frac{2}{1}$$

$$y = 120$$

6 inches would represent 120 feet.

✔ COMPLETION EXAMPLE 3

A recommended mixture of weed killer is 3 capfuls for 2 gallons of water. How many capfuls should be mixed with 5 gallons of water?

Solution

(a) Let x = unknown number of capfuls of weed killer.

(b) $\frac{x \text{ capfuls}}{5 \text{ gallons}} = \frac{3 \text{ capfuls}}{2 \text{ gallons}}$

$$\_\_\_\_ \cdot x = 5 \cdot \_\_\_\_$$

$$\frac{\_\_\_\_ \cdot x}{?} = \frac{15}{?}$$

$$x = \_\_\_\_$$

____ capfuls of weed killer should be mixed with 5 gallons of water.

✔ COMPLETION EXAMPLE 4

A jelly manufacturer puts 2.5 ounces of sugar into every 6-ounce jar of jelly. How many ounces of jelly can be made with 300 ounces of sugar?

Solution

(a) Let A = unknown amount of jelly.

(b) $\frac{2.5 \text{ oz. sugar}}{6 \text{ oz. jelly}} = \frac{300 \text{ oz. sugar}}{A \text{ oz. jelly}}$

$$\_\_\_\_ \cdot \_\_\_\_ = 1800$$

$$\frac{\_\_\_\_ \cdot \_\_\_\_}{?} = \frac{1800}{?}$$

$$A = \_\_\_\_$$

300 ounces of sugar will make ____ ounces of jelly.

Now Work Exercises 1 and 2 in the Margin.

Nurses and doctors work with proportions when prescribing medicine and giving injections. Medical texts write proportions in the form

$$2 : 40 :: x : 100 \quad \text{instead of} \quad \frac{2}{40} = \frac{x}{100}$$

Using either notation, the solution is found by setting the product of the extremes equal to the product of the means and solving the equation for the unknown quantity.

1. One pound of candy costs $4.75. How many pounds of candy can be bought for $28.50?

2. Precinct 1 has 520 registered voters and Precinct 2 has 630 registered voters. If the ratio of actual voters to registered voters was the same in both precincts, how many voted in Precinct 2 in the last election if 104 voted in Precinct 1?

EXAMPLE 5 Solve the proportion.

2 ounces : 40 grams :: x ounces : 100 grams

Solution

$$\overbrace{2 : \underbrace{40 :: x}_{\text{means}} : 100}^{\text{extremes}}$$

$$2 \cdot 100 = 40 \cdot x$$

$$\frac{200}{40} = \frac{\cancel{40} \cdot x}{\cancel{40}}$$

$$5 = x$$

The solution is $x = 5$ ounces.

Completion Example Answers

3. (a) Let x = unknown number of capfuls of weed killer.

(b) $\dfrac{x \text{ capfuls}}{5 \text{ gallons}} = \dfrac{3 \text{ capfuls}}{2 \text{ gallons}}$

$$\mathbf{2} \cdot x = 5 \cdot \mathbf{3}$$

$$\frac{\cancel{\mathbf{2}} \cdot x}{\cancel{\mathbf{2}}} = \frac{15}{\mathbf{2}}$$

$$x = \mathbf{7.5}$$

7.5 capfuls of weed killer should be mixed with 5 gallons of water.

4. (a) Let A = unknown amount of jelly.

(b) $\dfrac{2.5 \text{ oz sugar}}{6 \text{ oz jelly}} = \dfrac{300 \text{ oz sugar}}{A \text{ oz jelly}}$

$$\mathbf{2.5 \cdot A} = 1800$$

$$\frac{\cancel{\mathbf{2.5}} \cdot \mathbf{A}}{\cancel{\mathbf{2.5}}} = \frac{1800}{\mathbf{2.5}}$$

$$A = \mathbf{720}$$

300 ounces of sugar will make **720** ounces of jelly.

NAME ______________ SECTION ______________ DATE ______________

ANSWERS

1. ______________

2. ______________

3. ______________

4. ______________

5. ______________

6. ______________

7. ______________

Exercises 6.5

Solve the following word problems using proportions.

1. A mapmaker uses a scale of 2 inches to represent 30 miles. How many miles do 3 inches represent?

2. An investor thinks she should make $12 for every $100 she invests. How much would she expect to make on a $1500 investment?

3. If one dozen (12) eggs costs $1.09, what do three dozen eggs cost?

4. The price of a certain fabric is $1.75 per yard. How many yards can be bought with $35 (not including tax)?

5. A baseball team bought 8 bats for $96. What would the team pay for 10 bats?

6. A store owner expects to make a profit of $2 on an item that sells for $10. How much profit can he expect to make on a similar item that sells for $60?

7. A saleswoman makes $8 for every $100 worth of the product that she sells. What will she make if she sells $5000 worth of the product?

8. ______________

9. ______________

10. ______________

11. 

12. ______________

13. ______________

8. If property taxes are figured at \$1.50 for every \$100 in evaluation, what taxes will be paid on a home valued at \$85,000?

9. A condominium owner pays property taxes of \$2000 per year. If taxes are figured at a rate of \$1.25 for every \$100 in value, what is the value of his condominium?

10. Sales tax is figured at 6¢ for every \$1.00 of merchandise purchased. What was the purchase price on an item that had a sales tax of \$2.04?

11. An architect drew plans for a city park using a scale of $\frac{1}{4}$ inch to represent 25 feet. How many feet would 2 inches represent?

12. A building 14 stories high casts a shadow 30 feet long at a certain time of day. What is the length of the shadow of a 20-story building at the same time of day in the same city?

13. Two numbers are in the ratio of 4 to 3. The number 10 is in that same ratio to a fourth number. What is the fourth number?

NAME ______________________ SECTION ______________ DATE ________

ANSWERS

14. ____________

15. ____________

16. ____________

17. ____________

18. ____________

19. ____________

14. A salesman figured he drove 560 miles every two weeks. How far would he drive in three months (12 weeks)?

15. Driving steadily, a woman made a 200-mile trip in $4\frac{1}{2}$ hours. How long would it take her to drive 500 miles at the same rate of speed?

16. If you can drive 286 miles in $5\frac{1}{2}$ hours, how long will it take you to drive 468 miles at the same rate of speed?

17. What will 21 gallons of gasoline cost if gasoline costs $1.25 per gallon?

18. If diesel fuel costs $1.10 per gallon, how much diesel fuel will $24.53 buy?

19. An electric fan makes 180 revolutions per minute. How many revolutions will the fan make if it runs for 24 hours?

20. ______________

21. ______________

22. ______________

23. ______________

24. ______________

25. ______________

26. ______________

20. An investor made $144 in one year on a $1000 investment. What would she have earned if her investment had been $4500?

21. A typist can type 8 pages of manuscript in 56 minutes. How long will this typist take to type 300 pages?

22. On a map, $1\frac{1}{2}$ inches represent 40 miles. How many inches represent 50 miles?

23. In the metric system, there are 2.54 centimeters in one inch. How many centimeters are there in one foot?

24. If 40 pounds of fertilizer are needed to cover 2400 square feet of lawn, how many pounds of fertilizer are needed for a lawn of 5400 square feet?

25. An English teacher must read and grade 27 essays. If the teacher takes 20 minutes to read and grade 3 essays, how much time will he need to grade all 27 essays?

26. If 2 cups of flour are needed to make 12 biscuits, how much flour is needed to make 9 of the same kind of biscuits?

NAME ______________________ SECTION ______________________ DATE ______________

ANSWERS

27. If 2 cups of flour are needed to make 12 biscuits, how many of the same kind of biscuits can be made with 3 cups of flour?

The following exercises are examples of proportions in medicine. Solve for x and be sure to label your answers. (The abbreviations are from the metric system.)

28. 1 liter : 1000 mL :: x liter : 5000 mL

29. 1 kg : 1000 g :: x kg : 2700 g

30. 1 mg : 1000 mcg :: x mg : 4 mcg

31. 1 g : 1000 mg :: 0.5 g : x mg

32. 1 dram : 60 grains :: x dram : 90 grains

33. $\frac{\text{1 ounce}}{\text{8 drams}} = \frac{\text{0.5 ounce}}{x}$

34. $\frac{\text{1 dram}}{\text{60 minims}} = \frac{x}{\text{180 minims}}$

35. $\frac{\text{1 ounce}}{\text{480 minims}} = \frac{\text{4 ounces}}{x}$

36. $\frac{\text{1 g}}{\text{15 grains}} = \frac{x}{\text{60 grains}}$

37. $\frac{\text{1 ounce}}{\text{30 g}} = \frac{\text{2 ounces}}{x}$

38. $\frac{\text{1 tsp}}{\text{4 mL}} = \frac{x}{\text{12 mL}}$

39. $\frac{\text{1 ounce}}{\text{30 mL}} = \frac{x}{\text{15 mL}}$

40. $\frac{\text{1 pint}}{\text{500 mL}} = \frac{\text{1.5 pints}}{x}$

27. ______________

28. ______________

29. ______________

30. ______________

31. ______________

32. ______________

33. ______________

34. ______________

35. ______________

36. ______________

37. ______________

38. ______________

39. ______________

40. ______________

41. ______________

42. ______________

43. ______________

44. ______________

41. You want to treat your lawn with a fungicide/insecticide/fertilizer mix. The local nursery sells bags of this mix for $14 per bag, and each bag contains 16 pounds of mix. The recommended coverage for one bag is 2000 square feet. (a) If your lawn consists of two rectangular shapes, one 30 feet by 100 feet and the other 50 feet by 80 feet, how many pounds of fertilizer mix would it take to cover your lawn? (b) How many bags do you need to buy? (c) How much would you pay (not including tax)?

42. One bag of Weed & Fertilizer contains 18 pounds of fertilizer and weed treatment with a recommended coverage of 5000 square feet. (a) If your lawn is in the shape of a rectangle 150 feet by 220 feet, how many pounds of Weed & Fertilizer do you need to cover your lawn? (b) If the cost of one bag is $12, what will you pay for the amount of fertilizer you need (not including tax)?

43. The local paint store recommends that you use one gallon of paint to cover 400 square feet of properly prepared wall space in your home. You want to paint the walls of three rooms—one room measuring 10 feet by 12 feet, another 12 feet by 13 feet, and the third 16 feet by 15 feet. (a) If each room has 8-foot ceilings, how many gallons of paint do you need to buy? (b) Will you have any paint left over? (c) How much? (**Note:** For this exercise, ignore window space and door area.)

44. One bag of dichondra lawn food contains 20 pounds of fertilizer, and its recommended coverage is 4000 square feet. (a) If you want to cover a lawn that is in the shape of a rectangle 120 feet by 160 feet, how many pounds of lawn food do you need? (b) How many bags of lawn food would you need to buy?

NAME ________________ SECTION ________________ DATE ________

ANSWERS

45. ____________

46. ____________

47. ____________

48. ____________

49. [Respond below exercise.]

Check Your Number Sense

In each of the following exercises, use mental calculations and your judgment to choose the best answer. After you have checked your answers in the back of the text, work out any problems that you missed to help reinforce your understanding.

45. A computer microchip manufacturer expects that 3 out of every 100 microchips it produces will be defective. If 5000 microchips are produced in one production run, about how many would be expected to be defective in that run?

(a) 15 (b) 50 (c) 150 (d) 300

46. A professional baseball player has hit 12 home runs in the first 55 games of the season. At this rate, about how many home runs would you expect him to hit over a complete season of 162 games?

(a) 15 (b) 25 (c) 35 (d) 45

47. A food distributor is mailing a survey to shoppers to study their grocery purchasing habits. From experience, the company expects about 6 out of every 100 addressees to return the survey. If they would like at least 1000 responses, what is the minimum number of surveys that they should send out?

(a) 6000 (b) 10,000 (c) 20,000 (d) 60,000

48. The world's tropical rain forests are being destroyed at a rate of 96,000 acres per day. About how many acres of rain forest are being destroyed every hour?

(a) 4000 (b) 8000 (c) 500,000 (d) 1,000,000

Collaborative Learning Exercises

With the class separated into teams of two to four, each team is to analyze the following two exercises and decide on the best of the given choices for the answer. The team leader is to discuss the team's choice and the reason for this choice with the class.

49. Marine biologists at a killer whale feeding ground photograph and identify 35 whales during a research trip. The next year they return to the same location and identify 40 whales, 28 of which they had identified the year before. About how many whales would you estimate are in the group that these biologists are studying?

(a) 40 (b) 50 (c) 75 (d) 100

[Respond below exercise.]
50. ____________

50. Forest rangers are concerned about the deer population in a certain region. To get a good estimate of the deer population, they find, tranquilize, and tag 50 deer. One month later, they again locate 50 deer, and of these 50, 5 are deer that were previously tagged. On the basis of these procedures and results, what do the rangers estimate the deer population to be for that region?

(a) 100 (b) 150 (c) 500 (d) 1000

RECYCLE BIN

1. ____________

2. ____________

3. ____________

4. ____________

The Recycle Bin (from Sections 4.2–4.4)

1. Consider the product $4\frac{1}{2} \cdot 1\frac{1}{6} \cdot 1\frac{1}{5}$. Do you think that this product is (a) close to 6, (b) close to 20, (c) close to 100? Find the product.

2. Consider the sum $6\frac{1}{3} + 5\frac{3}{10} + 8\frac{1}{15}$. Do you think that this sum is (a) close to 10, (b) close to 20, (c) close to 100? Find the sum.

3. Consider the quotient $6\frac{2}{3} \div 1\frac{1}{3}$. Do you think that this quotient is (a) more than 6 or (b) less than 6? Find the quotient.

4. Use the rules for order of operations to find the value of the following expression: $\frac{1}{2} \cdot 3\frac{1}{2} + 5\frac{1}{10} \cdot \frac{10}{17} - 3\frac{5}{12}$.

6 Index of Key Ideas and Terms

NAME ________________ SECTION ________________ DATE ________

Chapter 6 Test

1. The ratio 7 to 6 can be expressed as the fraction ______.

2. In the proportion $\frac{3}{4} = \frac{75}{100}$, 3 and 100 are called the ______________.

3. Is the proportion $\frac{4}{6} = \frac{9}{14}$ true or false? Give a reason for your answer.

Write the following comparisons as ratios reduced to lowest terms. Use common units whenever possible.

4. 3 weeks to 35 days

5. 6 nickels to 3 quarters

6. 220 miles to 4 hours

7. Find the unit prices and tell which is the better buy for a carton of cottage cheese:
8 oz at 54¢ or 32 oz at $2.05

Solve the following proportions.

8. $\frac{9}{17} = \frac{x}{51}$

9. $\frac{3}{5} = \frac{10}{y}$

10. $\frac{50}{x} = \frac{0.5}{0.75}$

11. $\frac{2.25}{y} = \frac{1.5}{13}$

12. $\frac{\frac{3}{8}}{x} = \frac{9}{20}$

13. $\frac{\frac{1}{3}}{x} = \frac{5}{\frac{1}{6}}$

14. $9 : 4 :: y : 2$

15. 9 is to 4 as x is to 16

ANSWERS

1. ______________

2. ______________

3. ______________

4. ______________

5. ______________

6. ______________

7. ______________

8. ______________

9. ______________

10. ______________

11. ______________

12. ______________

13. ______________

14. ______________

15. ______________

16. ______________

17. ______________

18. ______________

19. ______________

20. ______________

21. ______________

22. ______________

16. How far (to the nearest tenth of a mile) can you drive on 5 gallons of gas if you can drive 120 miles on 3.5 gallons of gas?

17. On a certain map, 2 inches represents 15 miles. How far apart are two towns that are $3\frac{1}{5}$ inches apart on the map?

18. If you can buy 4 tires for $236, what will be the cost of 5 of the same type of tires?

19. If you drive 165 miles in $3\frac{2}{3}$ hours, what is your average speed in miles per hour?

20. A house painter knows that he can paint 3 houses every five weeks. At this rate, how long will it take him to paint 15 houses?

21. If 2 units of water weigh 124.8 pounds, what is the weight of 5 of these units of water?

22. A manufacturing company expects to make a profit of $3 on a product that it sells for $8. How much profit does the company expect to make on a similar product that it sells for $20?

NAME ____________ SECTION ____________ DATE ____________

Cumulative Review

1. Write five thousand seven hundred forty in the form of a decimal number.

2. Write the decimal number 3.075 in words.

3. Find the prime factorization of 4510.

Evaluate each of the following expressions.

4. $3 \cdot 5^2 + 20 \div 5$

5. $(80 + 10) \div 5 \cdot 2 - 14 + 2^2$

6. Find the LCM for 30, 35, and 42.

7. Find the average of 50, 54, 60, and 76.

Perform the indicated operations and reduce all answers.

8. $1 - \frac{15}{16}$

9. $4\frac{5}{8} \div 2$

10. $\left(1\frac{2}{3}\right)\left(2\frac{4}{5}\right)\left(3\frac{1}{7}\right)$

11. $\dfrac{\frac{1}{5} + \frac{1}{10}}{\frac{1}{3} - \frac{1}{4}}$

12. $\begin{array}{r} 6 \\ -2\frac{3}{4} \\ \hline \end{array}$

13. $\begin{array}{r} 15\frac{1}{2} \\ +22\frac{3}{4} \\ \hline \end{array}$

14. Find $\frac{3}{4}$ of 76.

ANSWERS

1. ____________

2. [Respond below exercise.] ____________

3. ____________

4. ____________

5. ____________

6. ____________

7. ____________

8. ____________

9. ____________

10. ____________

11. ____________

12. ____________

13. ____________

14. ____________

15. ______________

16. [Respond below exercise.]

17. ______________

18. ______________

19. ______________

20. ______________

21. ______________

22. ______________

23. ______________

24. ______________

25. ______________

15. Round off 0.03572 to the nearest thousandth.

16. Write each of the numbers in words.

(a) 600.006 (b) 0.606

Perform the indicated operations.

17. $\begin{array}{r} 847.8 \\ 436.92 \\ +354.718 \\ \hline \end{array}$

18. $\begin{array}{r} 32.007 \\ -15.835 \\ \hline \end{array}$

19. $\begin{array}{r} 566.3 \\ \times \ 7.2 \\ \hline \end{array}$

20. Divide $1.028 \div 1.3$ and find the quotient to the nearest hundredth.

21. Write each of the following numbers in scientific notation.

(a) 93,500,000 (b) 0.000000482

Solve each of the following proportions.

22. $\frac{5}{7} = \frac{x}{3\frac{1}{2}}$

23. $\frac{0.5}{1.6} = \frac{0.1}{A}$

24. (a) If you drove your car 250 miles in 5.5 hours, what was your average speed per hour (to the nearest tenth of a mile)? (b) How far could you drive at this average speed in 7 hours?

25. In December, Mario was told he would receive a bonus of \$4000. However, $\frac{1}{10}$ was withheld for state taxes, $\frac{1}{5}$ was withheld for federal taxes, and \$200 was withheld for Social Security. How much cash did Mario receive after these deductions?

7

Percent (Calculators Recommended)

What to Expect in Chapter 7

Chapter 7 deals with ideas and applications related to percent, one of the most useful and important mathematical concepts in our daily lives. Section 7.1 introduces percent by explaining that percent means hundredths (or per hundred) and that, therefore, percents can be represented in fraction form with denominator 100. Percent of profit and the relationship between percents and decimals are also included in Section 7.1. Section 7.2 discusses the relationships among percents, fractions, and decimals. Sections 7.3 and 7.4 lay the groundwork for applications with percents in Sections 7.5 and 7.6 by helping students realize that all percent problems are basically one of three types of problems. Both the method using the proportion $\frac{R}{100} = \frac{A}{B}$ (Section 7.3) and the method using the formula $R \times B = A$ (Section 7.4) are used in the development.

In Sections 7.5 and 7.6, we emphasize Pólya's four-step process for solving problems:

1. Understand the problem.
2. Devise a plan.
3. Carry out the plan.
4. Look back over the results.

Percents are often used to measure and report the attitudes and characteristics of the general population.

Mathematics at Work!

The concept of percent is an important part of our daily lives. Income tax is based on a percent of income; property tax is based on a percent of the value of a home; a professional athlete's skills are measured as percents (batting percentage, free-throw percentage, pass completion percentage); profit is calculated as a percent of investment; commissions are calculated as percents of sales; and tips for service are figured as a percent of the bill.

Even reading a newspaper or magazine with understanding requires some knowledge of percents. Below are excerpts from front pages of the *Los Angeles Times* during one week in March 1994.

In reference to a meeting of top business leaders:

. . . 35% admitted eating at a fast-food restaurant during the previous week.

German unemployment rose to 10.5%, . . .

In reference to the reformed KGB:

. . . a 30% cut in staff, . . .

In reference to the economy in India:

. . . about 40% of revenue comes from import fees.

Discussing results of new state tests in California:

Only 7% of fourth graders showed a substantial grasp of mathematical concepts; . . .

. . . and 30% of the high school sophomores demonstrated only a "superficial understanding" of what they read.

7.1 Decimals and Percents

OBJECTIVES

1. *Understand that percent means hundredths.*
2. *Relate percent to fractions with denominator 100.*
3. *Compare profit to investment as a percent.*
4. *Be able to change percents to decimals and decimals to percents.*

Understanding Percent

The word **percent** comes from the Latin *per centum*, meaning **per hundred.** So, **percent means hundredths,** or the **ratio of a number to 100.** The symbol % is called the **percent sign.** As we shall see, this sign has the same meaning as the fraction $\frac{1}{100}$. For example,

$$\frac{35}{100} = 35\left(\frac{1}{100}\right) = 35\% \quad \text{and} \quad \frac{60}{100} = 60\left(\frac{1}{100}\right) = 60\%$$

In Figure 7.1, the large square is partitioned into 100 small squares, and each small square represents $\frac{1}{100}$, or 1%, of the large square. Therefore, in Figure 7.1, the portion of the large square that is shaded is $\frac{27}{100} = 27\left(\frac{1}{100}\right) = 27\%$.

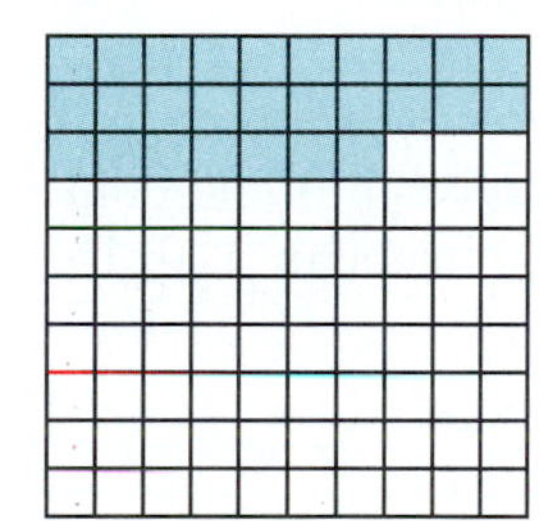

Figure 7.1

If a fraction has a denominator of 100, then (with no change in the numerator) the numerator can be read as a percent by dropping the denominator and adding the % sign.

EXAMPLE 1

Each fraction is changed to a percent.

(a) $\frac{20}{100} = 20 \cdot \frac{1}{100} = 20\%$ Remember that **percent** means **hundredths,** and the % sign indicates hundredths, or $\frac{1}{100}$.

(b) $\frac{85}{100} = 85\%$

(c) $\frac{6.4}{100} = 6.4\%$ Note that the decimal point is not moved; that is, the numerator is unchanged.

(d) $\frac{3\frac{1}{2}}{100} = 3\frac{1}{2}\%$ Note that since the denominator is 100, the numerator is unchanged even though it contains a fraction.

(e) $\frac{240}{100} = 240\%$ If the numerator is larger than 100, then the number is larger than 1 and the percent is more than 100%.

Change each fraction to a percent.

1. $\frac{9}{100}$

2. $\frac{1.25}{100}$

3. $\frac{125}{100}$

4. $\frac{6\frac{1}{4}}{100}$

Now Work Exercises 1–4 in the Margin.

Percent of Profit

Percent of profit is the ratio of money made to money invested. Generally, two investments do not involve the same amount of money and, therefore, the comparative success of each investment cannot be based on the **amount** of profit. In comparing investments, the investment with the greater percent of profit is considered the better investment.

The use of percent gives an effective method of comparison because each ratio of money made (profit) to money invested has the same denominator (100).

Example 2

Calculate the percent of profit for both (a) and (b) and tell which is the better investment.

(a) \$150 profit made by investing \$300

(b) \$200 profit made by investing \$500

Solution

In each case, find the ratio of dollars profit to dollars invested and reduce the ratio so that it has a denominator of 100. Do not reduce to lowest terms.

(a) $\frac{\$150 \text{ profit}}{\$300 \text{ invested}} = \frac{3 \cdot 50}{3 \cdot 100} = \frac{50}{100} = 50\%$

(b) $\frac{\$200 \text{ profit}}{\$500 \text{ invested}} = \frac{5 \cdot 40}{5 \cdot 100} = \frac{40}{100} = 40\%$

Investment (a) is better than investment (b) because 50% is larger than 40%. Obviously, \$200 profit is more than \$150 profit, but the \$500 risked as an investment in (b) is considerably more than the \$300 risked in (a).

Completion Example 3

Which is the better investment?

(a) Making \$40 profit by investing \$200.

(b) Making \$75 profit by investing \$300.

Solution

Write each ratio as hundredths and compare the percents.

(a) $\frac{\$40 \text{ profit}}{\$200 \text{ invested}} = \frac{2 \cdot \underline{\quad}}{2 \cdot 100} = \frac{\underline{\quad}}{100} = \underline{\qquad}\%$

(b) $\frac{\$75 \text{ profit}}{\$300 \text{ invested}} = \frac{3 \cdot \underline{\quad}}{3 \cdot 100} = \frac{\underline{\quad}}{100} = \underline{\qquad}\%$

Investment ______ is better since ______% is more than ______%.

Decimals and Percents

One way to change a decimal to a percent is to change the decimal to fraction form with denominator 100 and then change the fraction to percent form. For example,

Decimal Form		Fraction Form		Percent Form
0.47	=	$\frac{47}{100}$	=	47%
0.93	=	$\frac{93}{100}$	=	93%

In both of these illustrations the decimals are in hundredths and changing to fraction form is relatively easy. If the decimal is not in hundredths, we can multiply by 1 in the form of $\frac{100}{100}$, as follows:

$$0.325 = 0.325 \cdot \frac{100}{100} = 0.325(100) \cdot \frac{1}{100} = 32.5 \cdot \frac{1}{100} = 32.5\%$$

Now, looking at the fact that 0.325 = 32.5%, we can see that the decimal point was moved two places to the right and the % sign was written. This leads to the following general rule.

To Change a Decimal to a Percent:

STEP 1: Move the decimal point two places to the right.

STEP 2: Write the % sign.

(These two steps have the effect of multiplying by 100 and then dividing by 100.)

The following relationships between decimals and percents will serve as helpful guidelines.

A decimal number that is

1. less than 0.01 is less than 1%.
2. between 0.01 and 0.10 is between 1% and 10%.
3. between 0.10 and 1.00 is between 10% and 100%.
4. more than 1 is more than 100%.

Example 4

Change each decimal to an equivalent percent.

(a) 0.253 (b) 0.905 (c) 2.65

(d) 0.7 (e) 0.002

Change each decimal to a percent.

5. 0.04

6. 1.35

7. 0.936

Change each percent to a decimal.

8. 12%

9. 18.2%

10. 0.7%

Solution

(a) 0.253 = 25.3% — Decimal point moved two places to the right; % sign added

(b) 0.905 = 90.5%

(c) 2.65 = 265% Note that this is more than 100%.

(d) 0.7 = 70% The decimal point is not written here. We could write 70% or 70.0%.

(e) 0.002 = 0.2% Note that this is less than 1%. — Decimal point moved two places to the right; % sign added

To change percents to decimals, we reverse the procedure for changing decimals to percents. For example,

$$39\% = 39\left(\frac{1}{100}\right) = \frac{39}{100} = 0.39$$

Remember, the decimal point is moved two places to the left when dividing by 100.

To Change a Percent to a Decimal:

STEP 1: Move the decimal point two places to the left.

STEP 2: Delete the % sign.

Example 5 Change each percent to an equivalent decimal.

(a) 76% (b) 18.5% (c) 100% (d) 1.3% (e) 0.25%

Solution

(a) 76% = 0.76 ← % sign deleted — Understood decimal point; Decimal point moved two places left

(b) 18.5% = 0.185

(c) 100% = 1.00 = 1

(d) 1.3% = 0.013

(e) 0.25% = 0.0025 Note that 0.25% is less than 1% and its decimal equivalent is less than 0.01.

Now Work Exercises 5–10 in the Margin.

Completion Example Answers

3. (a) $\frac{\$40 \text{ profit}}{\$200 \text{ invested}} = \frac{2 \cdot \mathbf{20}}{2 \cdot 100} = \frac{\mathbf{20}}{100} = \mathbf{20\%}$

(b) $\frac{\$75 \text{ profit}}{\$300 \text{ invested}} = \frac{3 \cdot \mathbf{25}}{3 \cdot 100} = \frac{\mathbf{25}}{100} = \mathbf{25\%}$

Investment **(b)** is better since **25%** is more than **20%**.

NAME ______________ SECTION ______________ DATE ______________

Exercises 7.1

What percent of each square is shaded?

1. 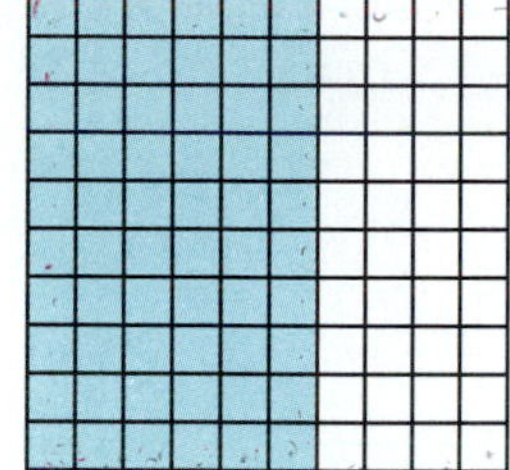

2. 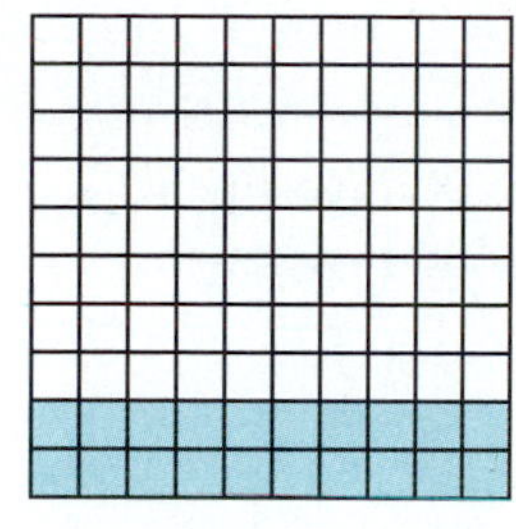

3. 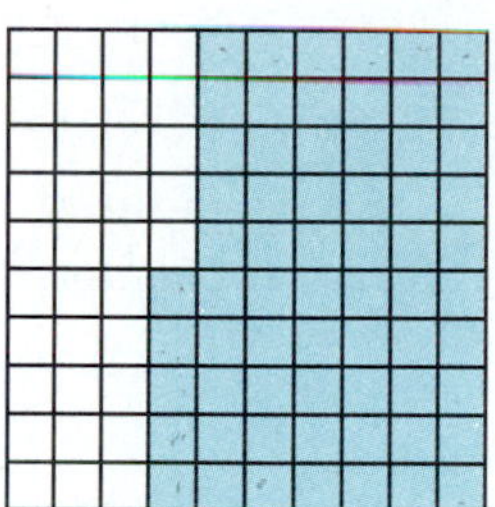

4. 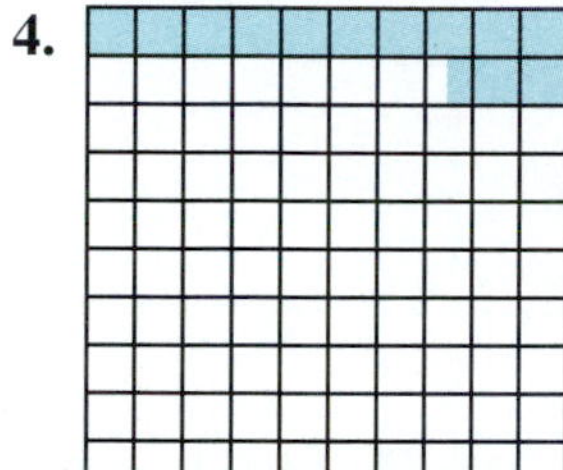

5. 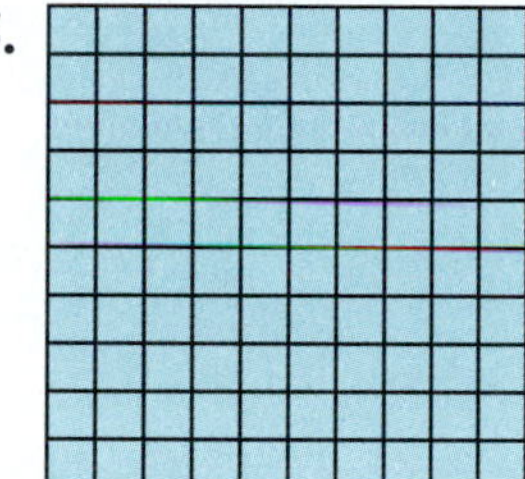

6. 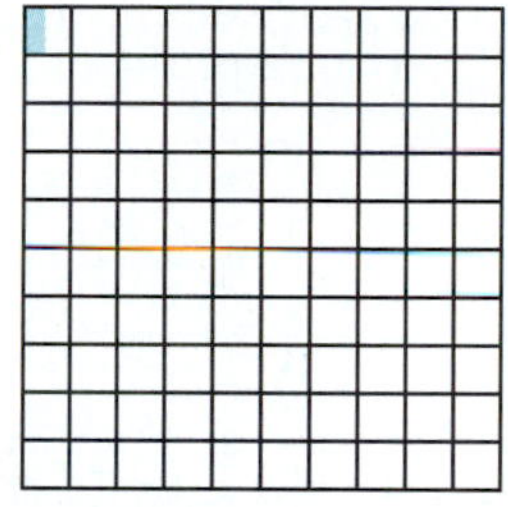

Change the following fractions to percents.

7. $\frac{20}{100}$

8. $\frac{9}{100}$

9. $\frac{15}{100}$

10. $\frac{62}{100}$

11. $\frac{53}{100}$

12. $\frac{68}{100}$

13. $\frac{125}{100}$

14. $\frac{200}{100}$

15. $\frac{336}{100}$

16. $\frac{13.4}{100}$

17. $\frac{0.48}{100}$

18. $\frac{0.5}{100}$

19. $\frac{2.14}{100}$

20. $\frac{1.62}{100}$

ANSWERS

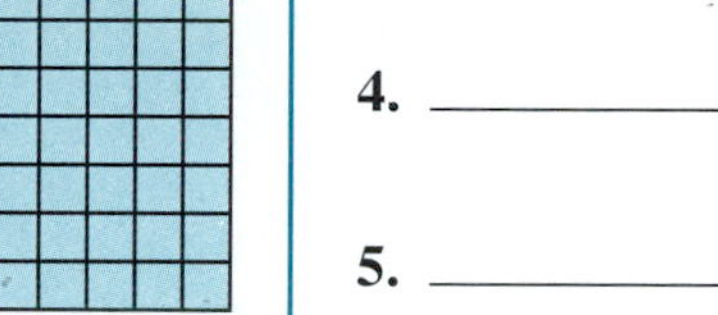

1. ______
2. ______
3. ______
4. ______
5. ______
6. ______
7. ______
8. ______
9. ______
10. ______
11. ______
12. ______
13. ______
14. ______
15. ______
16. ______
17. ______
18. ______
19. ______
20. ______

21. [Respond below exercise.] ____________

22. [Respond below exercise.] ____________

23. [Respond below exercise.] ____________

24. [Respond below exercise.] ____________

25. ____________

26. ____________

27. ____________

28. ____________

29. ____________

30. ____________

31. ____________

32. ____________

33. ____________

34. ____________

35. ____________

36. ____________

37. ____________

38. ____________

39. ____________

40. ____________

In Exercises 21–24, write the ratio of profit to investment as hundredths and then as percents. Compare the percents and tell which investment is better, (a) or (b).

21. (a) A profit of $36 on a $200 investment
(b) A profit of $51 on a $300 investment

22. (a) A profit of $40 on a $400 investment
(b) A profit of $60 on a $600 investment

23. (a) A profit of $150 on a $1500 investment
(b) A profit of $200 on a $2000 investment

24. (a) A profit of $300 on a $2000 investment
(b) A profit of $360 on a $3000 investment

Change the following decimals to percents.

25. 0.02 **26.** 0.09 **27.** 0.1 **28.** 0.7

29. 0.36 **30.** 0.52 **31.** 0.40 **32.** 0.65

33. 0.025 **34.** 0.035 **35.** 0.055 **36.** 0.004

37. 1.10 **38.** 1.75 **39.** 2 **40.** 2.3

NAME ______________ SECTION ______________ DATE ______________

ANSWERS

Change the following percents to decimals.

41. 2% **42.** 7% **43.** 18% **44.** 20%

45. 30% **46.** 80% **47.** 0.26% **48.** 0.52%

49. 125% **50.** 120% **51.** 232% **52.** 215%

53. 17.3% **54.** 10.1% **55.** 13.2% **56.** 6.5%

57. Suppose that sales tax is figured at 7.25%. Change 7.25% to a decimal.

58. The interest rate on a loan is 6.4%. Change 6.4% to a decimal.

59. The sales commission for the clerk in a retail store is figured at 8.5%. Change 8.5% to a decimal.

60. In calculating his sales commission, Mr. Howard multiplies by the decimal 0.12. Change 0.12 to a percent.

61. To calculate what your maximum house payment should be, a banker multiplied your income by 0.28. Change 0.28 to a percent.

62. The discount you earn by paying cash is found by multiplying the amount of your purchase by 0.02. Change 0.02 to a percent.

41. ______________
42. ______________
43. ______________
44. ______________
45. ______________
46. ______________
47. ______________
48. ______________
49. ______________
50. ______________
51. ______________
52. ______________
53. ______________
54. ______________
55. ______________
56. ______________
57. ______________
58. ______________
59. ______________
60. ______________
61. ______________
62. ______________

63. ______________

64. ______________

65. ______________

66. ______________

67. ______________

68. ______________

63. The rate of profit based on sales price is 45%. Change 45% to a decimal.

64. The discount during a special sale on dresses is 30%. Change 30% to a decimal.

65. Suppose the state license fee is figured by multiplying the cost of your car by 0.065. Change 0.065 to a percent.

66. With 22.2% of its residents having no health coverage, New Mexico was recently ranked as the state with the highest percentage of health care uninsured. Write 22.2% as a decimal.

67. Per capita property tax collections in California increased by 133% between 1980 and 1991. Write 133% as a decimal.

68. According to the Aluminum Association, the percentage of aluminum cans recycled in 1990 was 63.6%, and this percentage dropped to 62.4% in 1991. Write 63.6% and 62.4% as decimals.

RECYCLE BIN

1. ______________

2. ______________

3. ______________

4. ______________

5. ______________

6. ______________

The Recycle Bin (from Section 6.4)

Solve for the variable in each of the following proportions.

1. $\frac{R}{100} = \frac{4}{5}$ **2.** $\frac{R}{100} = \frac{2}{3}$ **3.** $\frac{23}{100} = \frac{A}{500}$

4. $\frac{75}{100} = \frac{A}{60}$ **5.** $\frac{90}{100} = \frac{36}{B}$ **6.** $\frac{10}{100} = \frac{7.4}{B}$

7.2 Fractions and Percents

OBJECTIVES

1. *Be able to change fractions and mixed numbers to percents.*
2. *Be able to change percents to mixed numbers and fractions.*

Changing Fractions and Mixed Numbers to Percents

As we discussed in Section 7.1, if a fraction has denominator 100, we can change it to a percent by writing the numerator and adding the % sign. If the denominator is a factor of 100 (2, 4, 5, 10, 20, 25, or 50), we can write the fraction in an equivalent form with denominator 100 and then change it to a percent. For example,

$$\frac{1}{4} = \frac{1}{4} \cdot \frac{25}{25} = \frac{25}{100} = 25\%$$

$$\frac{3}{5} = \frac{3}{5} \cdot \frac{20}{20} = \frac{60}{100} = 60\%$$

$$\frac{17}{50} = \frac{17}{50} \cdot \frac{2}{2} = \frac{34}{100} = 34\%$$

However, most fractions do not have denominators that are factors of 100. A more general approach (easily applied with calculators) is to change the fraction to decimal form by dividing the numerator by the denominator, and then change the decimal to a percent.

To Change a Fraction to a Percent:

STEP 1: Change the fraction to a decimal.
(Divide the numerator by the denominator.)

STEP 2: Change the decimal to a percent.

Example 1

Change $\frac{5}{8}$ to a percent.

Solution

(a) Divide:

$$\begin{array}{r} 0.625 \\ 8\overline{)5.000} \\ \underline{4\,8} \\ 20 \\ \underline{16} \\ 40 \\ \underline{40} \\ 0 \end{array}$$

(b) Change 0.625 to a percent: $\frac{5}{8} = 0.625 = 62.5\%$

Change each fraction or mixed number to a percent.

1. $\frac{13}{20}$

2. $1\frac{1}{2}$

3. $\frac{1}{4}$

EXAMPLE 2

Change $\frac{18}{20}$ to a percent.

Solution

(a) Divide:

$$\begin{array}{r} 0.9 \\ 20\,\overline{)\,18.0} \\ \underline{18\,0} \\ 0 \end{array}$$

(b) Change 0.9 to a percent: $\frac{18}{20} = 0.9 = 90\%$

EXAMPLE 3

Change $2\frac{1}{4}$ to a percent.

Solution

(a) $2\frac{1}{4} = \frac{9}{4}$

$$\begin{array}{r} 2.25 \\ 4\,\overline{)\,9.00} \\ \underline{8} \\ 1\,0 \\ \underline{8} \\ 20 \\ \underline{20} \\ 0 \end{array}$$

(b) $2\frac{1}{4} = \frac{9}{4} = 2.25 = 225\%$

Now Work Exercises 1–3 in the Margin.

Agreement for Rounding Off Decimal Quotients

In this text,

1. Decimal quotients that are exact with four decimal places (or less) will be written with four decimal places (or less).
2. Decimal quotients that are not exact will be divided to the fourth place, and the quotient will be rounded off to the third place (thousandths).

Using a Calculator

Most calculators will give answers accurate to 8 or 9 decimal places. So, if you use a calculator to perform the long division when changing a fraction to a decimal, be sure to follow the agreement in Statement 2 in the preceding box.

EXAMPLE 4

Using a calculator, $\frac{1}{3} = 0.3333333$.

(a) Rounding off the decimal quotient to the third decimal place:

$\frac{1}{3} = 0.333 = 33.3\%$ The answer is rounded off and not exact.

(b) Without a calculator, we can divide and use fractions:

$$\begin{array}{r} 0.33\frac{1}{3} \\ 3\overline{)1.00} \\ \underline{9} \\ 10 \\ \underline{9} \\ 1 \end{array}$$

$\frac{1}{3} = 0.33\frac{1}{3} = 33\frac{1}{3}\%$ or $33.\overline{3}\%$

$33\frac{1}{3}\%$ and $33.\overline{3}\%$ are exact, and 33.3% is rounded off.

Note: Any of these answers is acceptable, but be aware that 33.3% is a rounded-off answer.

EXAMPLE 5

During the years 1921 to 1981, the New York Yankees baseball team played in 33 World Series Championships and won 22 of them. What percent of these championships did the Yankees win?

Solution

(a) The percent won can be found by using a calculator and changing the fraction $\frac{22}{33}$ to decimal form and then changing the decimal to a percent. Using a calculator,

$\frac{22}{33} = 0.6666666 \approx 0.667 = 66.7\%$

(b) Without a calculator, we can divide and use fractions:

$$\begin{array}{r} 0.66\frac{2}{3} \\ 3\overline{)2.00} \\ \underline{1\,8} \\ 20 \\ \underline{18} \\ 2 \end{array}$$

$\frac{22}{33} = \frac{\cancel{11} \cdot 2}{\cancel{11} \cdot 3} = \frac{2}{3} = 0.66\frac{2}{3} = 66\frac{2}{3}\%$ or $66.\overline{6}\%$

The Yankees won $66\frac{2}{3}\%$ of these championships.

Change each fraction to a percent. A calculator may be used.

4. $\frac{3}{8}$

5. $\frac{3}{40}$

6. $\frac{7}{9}$

Note: Any one of the answers—$66\frac{2}{3}\%$, $66.\bar{6}\%$, or 66.7%—is acceptable.

EXAMPLE 6

Using a calculator, $\frac{1}{7} = 0.1428571$.

(a) Rounding off the decimal quotient to the third decimal place:

$$\frac{1}{7} = 0.143 = 14.3\%$$

(b) Using long division, you can write the answer with a fraction or continue to divide and round off the decimal answer. The choice is yours.

$$\begin{array}{r} 0.14\frac{2}{7} = 14\frac{2}{7}\% \\ 7\overline{)1.00} \\ \underline{7} \\ 30 \\ \underline{28} \\ 2 \end{array} \quad \text{or} \quad \begin{array}{r} 0.1428 = 14.3\% \\ 7\overline{)1.0000} \\ \underline{7} \\ 30 \\ \underline{28} \\ 20 \\ \underline{14} \\ 60 \\ \underline{56} \\ 4 \end{array}$$

by rounding off the decimal quotient to the third decimal place

Now Work Exercises 4–6 in the Margin.

Changing Percents to Fractions and Mixed Numbers

To Change a Percent to a Fraction or a Mixed Number

STEP 1: Write the percent as a fraction with denominator 100 and drop the % sign.

STEP 2: Reduce the fraction.

EXAMPLE 7

Change each percent to an equivalent fraction or mixed number in reduced form.

(a) 60% (b) $7\frac{1}{4}\%$ (c) 130%

Solution

(a) $60\% = \frac{60}{100} = \frac{3 \cdot \cancel{20}}{5 \cdot \cancel{20}} = \frac{3}{5}$

(b) $7\frac{1}{4}\% = \frac{7\frac{1}{4}}{100} = \frac{\frac{29}{4}}{100} = \frac{29}{4} \cdot \frac{1}{100} = \frac{29}{400}$

(c) $130\% = \frac{130}{100} = \frac{13 \cdot \cancel{10}}{10 \cdot \cancel{10}} = \frac{13}{10} = 1\frac{3}{10}$

Now Work Exercises 7–10 in the Margin.

A Common Misunderstanding

The fractions $\frac{1}{4}$ and $\frac{1}{2}$ are often confused with the percents $\frac{1}{4}\%$ and $\frac{1}{2}\%$. The differences can be clarified by using decimals.

PERCENT	DECIMAL	FRACTION
$\frac{1}{4}\%$ (or 0.25%)	0.0025	$\frac{1}{400}$
$\frac{1}{2}\%$ (or 0.5%)	0.005	$\frac{1}{200}$
25%	0.25	$\frac{1}{4}$
50%	0.50	$\frac{1}{2}$

Thus,

$\frac{1}{4} = 0.25$ and $\frac{1}{4}\% = 0.0025$

$0.25 \neq 0.0025$

Similarly,

$\frac{1}{2} = 0.50$ and $\frac{1}{2}\% = 0.005$

$0.50 \neq 0.005$

You can think of $\frac{1}{4}$ as being one-fourth of a dollar (a quarter) and $\frac{1}{4}\%$ as being one-fourth of a penny. $\frac{1}{2}$ can be thought of as one-half of a dollar and $\frac{1}{2}\%$ as one-half of a penny.

Change each percent to a fraction or mixed number.

7. 80%

8. 16%

9. $5\frac{1}{2}\%$

10. 235%

Some percents are so common that their decimal and fraction equivalents should be memorized. Their fractional values are particularly easy to work with, and many times calculations involving these fractions can be done mentally.

Common Percent—Decimal—Fraction Equivalents

$1\% = 0.01 = \frac{1}{100}$ $\quad 33\frac{1}{3}\% = 0.33\frac{1}{3} = \frac{1}{3}$ $\quad 12\frac{1}{2}\% = 0.125 = \frac{1}{8}$

$25\% = 0.25 = \frac{1}{4}$ $\quad 66\frac{2}{3}\% = 0.66\frac{2}{3} = \frac{2}{3}$ $\quad 37\frac{1}{2}\% = 0.375 = \frac{3}{8}$

$50\% = 0.50 = \frac{1}{2}$ $\quad 62\frac{1}{2}\% = 0.625 = \frac{5}{8}$

$75\% = 0.75 = \frac{3}{4}$ $\quad 87\frac{1}{2}\% = 0.875 = \frac{7}{8}$

$100\% = 1.00 = 1$

NAME ______ SECTION ______ DATE ______

ANSWERS

1. ______
2. ______
3. ______
4. ______
5. ______
6. ______
7. ______
8. ______
9. ______
10. ______
11. ______
12. ______
13. ______
14. ______
15. ______
16. ______
17. ______
18. ______
19. ______
20. ______
21. ______
22. ______
23. ______
24. ______
25. ______
26. ______
27. ______
28. ______
29. ______
30. ______
31. ______
32. ______

Exercises 7.2

Change the following fractions and mixed numbers to percents.

1. $\frac{3}{100}$ **2.** $\frac{16}{100}$ **3.** $\frac{7}{100}$ **4.** $\frac{29}{100}$

5. $\frac{1}{2}$ **6.** $\frac{3}{4}$ **7.** $\frac{1}{4}$ **8.** $\frac{1}{20}$

9. $\frac{11}{20}$ **10.** $\frac{7}{10}$ **11.** $\frac{3}{10}$ **12.** $\frac{3}{5}$

13. $\frac{1}{5}$ **14.** $\frac{2}{5}$ **15.** $\frac{4}{5}$ **16.** $\frac{1}{50}$

17. $\frac{13}{50}$ **18.** $\frac{1}{25}$ **19.** $\frac{12}{25}$ **20.** $\frac{24}{25}$

21. $\frac{1}{8}$ **22.** $\frac{5}{8}$ **23.** $\frac{7}{8}$ **24.** $\frac{1}{9}$

25. $\frac{5}{9}$ **26.** $\frac{2}{7}$ **27.** $\frac{3}{7}$ **28.** $\frac{5}{6}$

29. $\frac{7}{11}$ **30.** $\frac{5}{11}$ **31.** $1\frac{1}{14}$ **32.** $1\frac{1}{6}$

33. ______
34. ______
35. ______
36. ______
37. ______
38. ______
39. ______
40. ______
41. ______
42. ______
43. ______
44. ______
45. ______
46. ______
47. ______
48. ______
49. ______
50. ______
51. ______
52. ______
53. ______
54. ______
55. ______
56. ______
57. ______
58. ______
59. ______
60. ______
61. ______
62. ______
63. ______
64. ______
65. ______

33. $1\frac{1}{20}$ **34.** $1\frac{1}{4}$ **35.** $1\frac{3}{4}$ **36.** $1\frac{1}{5}$

37. $1\frac{3}{8}$ **38.** $2\frac{1}{2}$ **39.** $2\frac{1}{10}$ **40.** $2\frac{1}{15}$

Change the following percents to fractions or mixed numbers.

41. 10% **42.** 5% **43.** 15% **44.** 17%

45. 25% **46.** 30% **47.** 50% **48.** $12\frac{1}{2}\%$

49. $37\frac{1}{2}\%$ **50.** $16\frac{2}{3}\%$ **51.** $33\frac{1}{3}\%$

52. $66\frac{2}{3}\%$ **53.** 33% **54.** $\frac{1}{2}\%$

55. $\frac{1}{4}\%$ **56.** 1% **57.** 100%

58. 125% **59.** 120% **60.** 150%

61. 0.3% **62.** 2.5% **63.** 62.5%

64. 0.2% **65.** 0.75%

NAME ____________ SECTION ____________ DATE ____________

Find the missing forms of each number.

	Fraction Form	Decimal Form	Percent Form
66.	$\frac{5}{8}$	(a) ______	(b) ______
67.	$\frac{11}{20}$	(a) ______	(b) ______
68.	(a) ______	0.09	(b) ______
69.	(a) ______	1.75	(b) ______
70.	(a) ______	(b) ______	36%
71.	(a) ______	(b) ______	10.5%

72. A department store offers a 30% discount during a special sale on men's suits. Change 30% to a fraction reduced to lowest terms.

73. In a sophomore class of 250 students, 10 represent the sophomore class on the student council. What percent of the class is on the student council?

74. Malcolm planned to drive 300 miles on a trip. After 3 miles, what percent of the trip had he driven?

75. (a) There are 12 inches in a foot. What percent of a foot is an inch?

(b) There are 4 quarts in a gallon. What percent of a gallon is a quart?

(c) There are 16 ounces in a pound. What percent of a pound is an ounce?

76. A desk once belonging to George Washington was recently sold at auction. It was expected to bring $80,000 but actually sold for $165,000. Write the actual selling price as a percent of the expected price.

ANSWERS

66a. ______
b. ______
67a. ______
b. ______
68a. ______
b. ______
69a. ______
b. ______
70a. ______
b. ______
71a. ______
b. ______
72. ______
73. ______
74. ______
75a. ______
b. ______
c. ______
76. ______

77. ______________

78a. [Respond below exercise.] ______________
b. [Respond below exercise.] ______________
c. [Respond below exercise.] ______________

77. According to an article in *The Boston Globe* (January 23, 1994), only about 250 of the nation's 175,000 dairy farmers still deliver milk door-to-door. What percent of dairy farmers still deliver milk?

Check Your Number Sense

78. Three pairs of fractions are given. In each case, tell which fraction would be easier to change to a percent mentally and explain your reasoning.

(a) $\frac{7}{20}$ or $\frac{2}{9}$ (b) $\frac{3}{10}$ or $\frac{5}{16}$ (c) $\frac{1}{12}$ or $\frac{1}{25}$

RECYCLE BIN

1. ______________
2. ______________
3. ______________
4a. ______________
b. ______________

The Recycle Bin (from Section 6.5)

Solve the following problems.

1. A salesman makes \$9 for every \$100 worth of merchandise he sells. What will he make if he sells \$80,000 worth of merchandise?

2. An architect drew up plans for a new building using a scale of $\frac{1}{2}$ inch to represent 36 feet. How many feet would 3 inches represent?

3. If gasoline costs \$1.12 per gallon, how many gallons can be bought for \$16.80?

4. A certain type of fertilizer is sold in bags containing 10 pounds. The recommended coverage is 10 pounds per 3000 square feet. (a) If your lawn is in the shape of a rectangle 90 feet by 80 feet, how many pounds of this fertilizer do you need to cover your lawn? (b) How many bags do you need to buy?

7.3 Solving Percent Problems Using the Proportion $\frac{R}{100} = \frac{A}{B}$ (Optional)

OBJECTIVES

1. *Recognize three types of percent problems.*
2. *Solve percent problems using the proportion $\frac{R}{100} = \frac{A}{B}$.*

The Basic Proportion $\frac{R}{100} = \frac{A}{B}$

Many people have difficulty working with percent simply because they do not know whether to add, subtract, multiply, or divide. **There are only three basic types of problems using percent.** In this section, we will develop a technique for solving all three types of problems by using just one basic proportion of the form $\frac{R}{100} = \frac{A}{B}$. The terms represented in this basic proportion are explained in the following box.

$R\%$ = Rate or percent

B = Base (number we are finding the percent of)

A = Amount or percentage (a part of the base)

The relationship among R, B, and A is given in the basic proportion

$$\frac{R}{100} = \frac{A}{B}$$

As an example of how to set up a basic proportion, consider the statement

25% of 60 is 15.

We can set up this statement in the form of the basic proportion as follows:

25%	of	60	is	15.
↑ Rate		↑ Base		↑ Amount

The basic proportion

$$\frac{R}{100} = \frac{A}{B} \quad \text{becomes} \quad \frac{25}{100} = \frac{15}{60}.$$

Using the Basic Proportion

If any two of the values R, A, and B are known, then the third can be found by substituting the known values into the basic proportion and solving for the one unknown term.

The Three Basic Types of Percent Problems and the Proportion $\frac{R}{100} = \frac{A}{B}$

TYPE 1: Find the Amount, given the Base and the Percent.

What is 65% of 500? $\frac{65}{100} = \frac{A}{500}$

R and B are known. The object is to find A.

TYPE 2: Find the Base, given the Percent and the Amount.

57% of what number is 51.3? $\frac{57}{100} = \frac{51.3}{B}$

R and A are known. The object is to find B.

TYPE 3: Find the Percent, given the Base and the Amount.

What percent of 170 is 204? $\frac{R}{100} = \frac{204}{170}$

A and B are known. The object is to find R.

The following examples illustrate how to substitute into the basic proportion and how to solve the resulting proportion for the unknown term by using the methods discussed in Chapter 6. **Remember that *B* is the number you are taking the percent of.** *B* is the number that follows the word **of.**

Example 1

What is 65% of 500?

Solution

In this problem, $R\% = 65\%$ and $B = 500$. We want to find the value of A. Substitution in the basic proportion gives

$$\frac{65}{100} = \frac{A}{500}$$

$$65 \cdot 500 = 100 \cdot A$$

The product of the extremes must equal the product of the means.

$$\frac{32{,}500}{100} = \frac{\cancel{100} \cdot A}{\cancel{100}}$$

$$325 = A$$

So, 65% of 500 is $\underline{\ 325\ }$.

EXAMPLE 2

57% of what number is 51.3?

Solution

In this problem, $R\% = 57\%$ and $A = 51.3$. We want to find the value of B. Substitution in the basic proportion gives

$$\frac{57}{100} = \frac{51.3}{B}$$

$$57 \cdot B = 100 \cdot 51.3$$ The product of the extremes must equal the product of the means.

$$\frac{\cancel{57} \cdot B}{\cancel{57}} = \frac{5130}{57}$$

$$B = 90$$

So, 57% of $\underline{\ 90\ }$ is 51.3.

EXAMPLE 3

What percent of 170 is 204?

Solution

In this problem, $B = 170$ and $A = 204$. We want to find the value of R. (Do you expect $R\%$ to be more than 100% or less than 100%?) Substitution in the basic proportion gives

$$\frac{R}{100} = \frac{204}{170}$$

$$170 \cdot R = 100 \cdot 204$$

$$\frac{\cancel{170} \cdot R}{\cancel{170}} = \frac{20{,}400}{170}$$

$$R = 120 \quad (\text{or } R\% = 120\%)$$

So, $\underline{\ 120\%\ }$ of 170 is 204.

Now Work Exercises 1–3 in the Margin.

Find the unknown quantity.

1. 15% of 80 is ______.

2. 86% of ______ is 430.

3. ______% of 90 is 19.8.

EXAMPLE 4

Many food product labels now list total calories and number of calories from fat. Dietary experts believe that a healthy diet has at most 30% of its calories derived from fat. Following this guideline, if an adult consumes 2500 calories per day, how many calories can be from fat?

Solution

In this problem, $R\% = 30\%$ and $B = 2500$. We want to find the value of A. Substitution in the basic proportion gives

$$\frac{30}{100} = \frac{A}{2500}$$

$$30 \cdot 2500 = 100 \cdot A$$

$$\frac{75{,}000}{100} = \frac{100 \cdot A}{100}$$

$$750 = A$$

So, in a healthy diet of 2500 calories, no more than 750 calories should be derived from fat.

NAME ______ SECTION ______ DATE ______

Exercises 7.3

Use the basic proportion $\frac{R}{100} = \frac{A}{B}$ to solve each of the following problems for the unknown quantity.

1. 10% of 90 is ______.
2. 5% of 72 is ______.
3. 15% of 50 is ______.
4. 25% of 60 is ______.
5. 75% of 32 is ______.
6. 60% of 40 is ______.
7. 100% of 47 is ______.
8. 80% of 80 is ______.
9. 2% of ______ is 5.
10. 20% of ______ is 14.
11. 3% of ______ is 27.
12. 30% of ______ is 45.
13. 100% of ______ is 62.
14. 50% of ______ is 35.
15. 150% of ______ is 69.
16. 110% of ______ is 440.
17. ______% of 50 is 75.
18. ______% of 120 is 48.

ANSWERS

1. ______
2. ______
3. ______
4. ______
5. ______
6. ______
7. ______
8. ______
9. ______
10. ______
11. ______
12. ______
13. ______
14. ______
15. ______
16. ______
17. ______
18. ______

19. ____________
20. ____________
21. ____________
22. ____________
23. ____________
24. ____________
25. ____________
26. ____________
27. ____________
28. ____________
29. ____________
30. ____________
31. ____________
32. ____________
33. ____________
34. ____________
35. ____________
36. ____________
37. ____________
38. ____________
39. ____________
40. ____________
41. ____________
42. ____________

19. ________% of 70 is 21.

20. ________% of 15 is 5.

21. ________% of 44 is 66.

22. ________% of 50 is 10.

23. ________% of 54 is 18.

24. ________% of 100 is 38.

Each of the following problems is one of the three types discussed, with slightly changed wording. Remember that B (the base) is the number you are finding the percent **of.**

25. ________ is 50% of 35.

26. ________ is 31% of 85.

27. 32 is 20% of ________.

28. 79 is 100% of ________.

29. 23 is ________% of 10.

30. 18 is ________% of 10.

31. 40 is $33\frac{1}{3}$% of ________.

32. 76.8 is 15% of ________.

33. 142.6 is 23% of ________.

34. 20.25 is 25% of ________.

35. 48 is ________% of 24.

36. 8 is ________% of 12.

37. ________ is 96% of 35.

38. ________ is 84% of 52.

39. ________ is 18% of 425.

40. ________ is 28% of 640.

41. Find 18% of 345.

42. Find 13.5% of 95.

NAME ______ SECTION ______ DATE ______

ANSWERS

43. Find 150% of 70.

44. What percent of 32 is 24?

45. 200 is 125% of what number?

46. 450 is 150% of what number?

47. What percent of 100 is 66.5?

48. Find 86.5% of 100.

49. What percent of 96 is 16?

50. 160 is 80% of what number?

51. According to U.S. Census information from 1980 and 1990, Mesa, Arizona had the largest population growth among the nation's 100 most populated cities. Between 1980 and 1990, the population of Mesa increased by 89%, to 288,091. What was Mesa's population in 1980 (to the nearest whole number)?

52. Due to the increasing cost of breakfast cereals, more and more people are buying private-label brands rather than national brands. In a recent year, the sale of private-label cereals rose from 170 million boxes to 180 million boxes. What was the percent increase in sales (to the nearest tenth of a percent)?

43. ______

44. ______

45. ______

46. ______

47. ______

48. ______

49. ______

50. ______

51. ______

52. ______

53. ______________

54. ______________

53. A Massachusetts Highway Department study done in the summer of 1993 noted that 68% of the fans attending Red Sox games at Fenway Park travel there by private automobile. The park seats 34,142 people at a sold-out game. About how many people (to the nearest ten) would you expect to arrive by private automobile at a sold-out game?

54. The same study noted in Exercise 53 found that 40% of the fans at Yankee Stadium travel there by car. Yankee Stadium seats 57,545 at sold-out games. About how many people (to the nearest ten) would you expect to arrive at a sold-out Yankee game by some means other than automobile?

7.4 Solving Percent Problems Using the Equation $R \times B = A$

OBJECTIVES

1. *Recognize three types of percent problems.*
2. *Solve percent problems using the equation $R \times B = A$.*
3. *Estimate percents and amounts related to percents.*

The Basic Equation $R \times B = A$

In Section 7.3, we introduced the three types of percent problems and the use of the basic proportion $\frac{R}{100} = \frac{A}{B}$ to solve these problems. If both sides of this proportion are multiplied by B, and R is changed to decimal form (or fraction form), we have the following result.

$$\frac{R}{100} = \frac{A}{B}$$

$$\frac{R}{100} \times B = \frac{A}{B} \times B$$

or

$$R \times B = A,$$

with R the value of $R\%$ **in decimal form.**

We will call this last equation, $\boldsymbol{R \times B = A}$, the basic equation for solving percent problems.

Now, consider the statement

25% of 60 is 15.

We can set this statement in the form of the basic equation as follows:

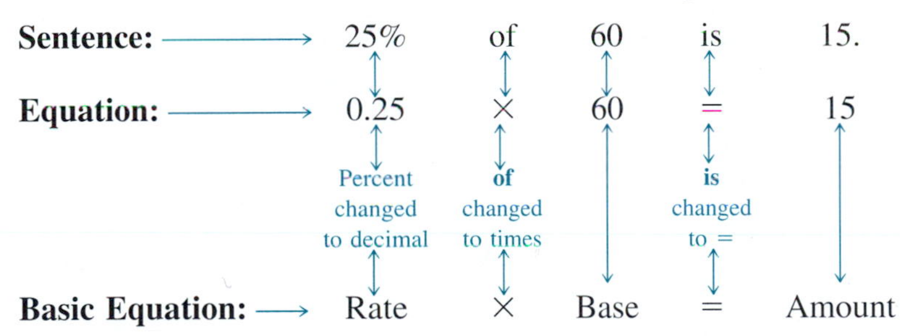

The terms represented in the basic equation are explained in the following box.

> R = Rate or percent (as a decimal or fraction)
>
> B = Base (number we are finding the percent of)
>
> A = Amount or percentage (a part of the base)
>
> **of** means to multiply (the cross sign $\times$ is used in the equation)
>
> **is** means equals (=)
>
> The relationship among R, B, and A is given in the basic equation
>
> $$R \times B = A$$

Using the Basic Equation

Even though there are just **three basic types of problems that involve percent,** many people have a difficult time differentiating among them and determining whether to multiply or divide in a particular problem. Most of these difficulties can be avoided by using the basic equation $R \times B = A$ and the equation-solving skills learned in Chapter 6. If the values of two of the quantities in the equation $R \times B = A$ are known, then these values can be substituted into the equation, and the third value can be determined by solving the resulting equation.

The Three Basic Types of Percent Problems and the Equation $R \times B = A$

TYPE 1: Find the Amount, given the Base and the Percent.

What is 45% of 70?

$$\begin{array}{c} R \times B = A \\ \swarrow \quad \downarrow \quad \downarrow \\ 0.445 \times 70 = A \end{array}$$

R and B are known. The object is to find A.

TYPE 2: Find the Base, given the Percent and the Amount.

30% of what number is 18.36?

$$\begin{array}{c} R \times B = A \\ \swarrow \quad \downarrow \quad \searrow \\ 0.330 \times B = 18.36 \end{array}$$

R and A are known. The object is to find B.

TYPE 3: Find the Percent, given the Base and the Amount.

What percent of 84 is 16.8?

$$\begin{array}{c} R \times B = A \\ \downarrow \quad \downarrow \quad \searrow \\ R \times 84 = 16.8 \end{array}$$

A and B are known. The object is to find R.

Remember:

(a) **Of** means to multiply when used with decimals or fractions.

(b) **Is** means =.

(c) The percent is changed to decimal or fraction form.

Problem Type 1: Finding the Value of A

Example 1

What is 45% of 70?

Solution

Here, $R = 45\% = 0.45$ and $B = 70$.

Substituting into the equation $R \times B = A$ gives

$$0.45 \times 70 = A$$
$$31.5 = A$$

$$\begin{array}{r} 70 \\ \times\ 0.45 \\ \hline 3\ 50 \\ 28\ 0\ \ \\ \hline 31.50 \end{array}$$

So, 45% of 70 is 31.5.

Note: The multiplication can be done by hand, as shown here, or with a calculator. In either case, the equations should be written so the = signs are aligned one above the other, as shown in the solution.

EXAMPLE 2

What is 18% of 200?

Solution

In this problem, $R = 18\% = 0.18$ and $B = 200$.

Substituting into the equation $R \times B = A$ gives

$$0.18 \times 200 = A$$
$$36 = A$$

So, 18% of 200 is 36.

Type 1 problems are the most common and, generally, the easiest to solve. The rate (R) and base (B) are known. The unknown amount (A) is found by multiplying the rate times the base.

Problem Type 2: Finding the Value of *B*

EXAMPLE 3

30% of what number is 18.36?

Solution

In this case, $R = 30\% = 0.30$ and $A = 18.36$.

Substituting into the equation $R \times B = A$ gives

$$0.30 \times B = 18.36$$

$$\frac{\cancel{0.30} \times B}{\cancel{0.30}} = \frac{18.36}{0.30}$$ Divide both sides by 0.30, the coefficient of B.

$$B = 61.2$$

$$\begin{array}{r} 61.2 \\ .30.\overline{)\,18.36.0} \\ 18\ 0\ \ \ \ \\ \hline 36\ \ \\ 30\ \ \\ \hline 60 \\ 60 \\ \hline 0 \end{array}$$

So, 30% of 61.2 is 18.36.

Find the unknown quantity.

1. 10% of 137 is ______.

2. 50% of ______ is 73.

3. ______% of 56 is 67.2.

EXAMPLE 4 82% of ________ is 246.

Solution

Here, $R = 82\% = 0.82$ and $A = 246$.

Substituting into the equation $R \times B = A$ gives

$$0.82 \times B = 246$$

$$\frac{\cancel{0.82} \times B}{\cancel{0.82}} = \frac{246}{0.82}$$ Divide both sides by 0.82, the coefficient of B. The division can be performed with a calculator.

$$B = 300$$

So, 82% of ___300___ is 246.

Problem Type 3: Finding the Value of R

EXAMPLE 5 What percent of 84 is 16.8?

Solution

In this problem, $B = 84$ and $A = 16.8$.

Substituting into the equation $R \times B = A$ gives

$$R \times 84 = 16.8$$

$$\frac{R \times \cancel{84}}{\cancel{84}} = \frac{16.8}{84}$$ Divide both sides by 84, the coefficient of R.

$$\begin{array}{r} 0.2 \\ 84 \overline{)16.8} \\ \underline{16.8} \\ 0 \end{array}$$

$$R = .2 = 20\%$$

So, ___20%___ of 84 is 16.8.

EXAMPLE 6 ________% of 195 is 234.

Solution

In this case, $B = 195$ and $A = 234$.

Substituting into the equation $R \times B = A$ gives

$$R \times 195 = 234$$

$$\frac{R \times \cancel{195}}{\cancel{195}} = \frac{234}{195}$$ Divide both sides by 195, the coefficient of R.

$$R = 1.2 = 120\%$$ R is more than 100% because A is larger than B.

So, ___120%___ of 195 is 234.

Now Work Exercises 1–3 in the Margin.

When working with applications, the wording of a problem is unlikely to be exactly as shown in Examples 1–6. The following examples show some alternative wordings, but each problem is still one of the three basic types.

EXAMPLE 7

Find 65% of 42.

Solution

Here the Rate = 65% and the Base = 42. This is a Type 1 Problem:

65% of 42 is ______.

$.65 \times 42 = A$

$27.3 = A$

$$\begin{array}{r} .65 \\ \times \quad 42 \\ \hline 1.30 \\ 26.0 \\ \hline 27.30 \end{array}$$

EXAMPLE 8

What percent of 125 gives a result of 31.25?

Solution

Here the Rate is unknown. The Base = 125 and the Amount = 31.25. This is a Type 3 Problem:

______% of 125 is 31.25.

$R \times 125 = 31.25$

$$\frac{R \times \cancel{125}}{\cancel{125}} = \frac{31.25}{125}$$

$R = .25 = 25\%$

$$\begin{array}{r} 0.25 \\ 125 \overline{)31.25} \\ 25\,0 \\ \hline 6\,25 \\ 6\,25 \\ \hline 0 \end{array}$$

Now Work Exercises 4 and 5 in the Margin.

The following examples show how to use fraction equivalents for percents to simplify your work.

EXAMPLE 9

What is 25% of 24?

$$A = \frac{1}{\cancel{4}} \times \overset{6}{\cancel{24}} = 6$$

Type 1 Problem

$25\% = \frac{1}{4}$

EXAMPLE 10

Find 75% of 36.

$$A = \frac{3}{\cancel{4}} \times \overset{9}{\cancel{36}} = 27$$

Type 1 Problem

$75\% = \frac{3}{4}$

4. Find 33% of 180.

5. Find 150% of 60.

EXAMPLE 11

What is the value of $33\frac{1}{3}\%$ of 72?

$$A = \frac{1}{\cancel{3}} \times \overset{24}{\cancel{72}} = 24$$

Type 1 Problem

$33\frac{1}{3}\% = \frac{1}{3}$

EXAMPLE 12

$37\frac{1}{2}\%$ of what number is 300?

$$\frac{3}{8} \times B = 300$$

Type 2 Problem

$37\frac{1}{2}\% = \frac{3}{8}$

$$\frac{\cancel{8}}{\cancel{3}} \times \frac{\cancel{3}}{\cancel{8}} \times B = \frac{8}{\cancel{3}} \times \overset{100}{\cancel{300}}$$

$\frac{8}{3}$ is the reciprocal of $\frac{3}{8}$.

$$B = 800$$

Remember that a percent is just another form of a fraction and that, when you find a percent of a given number, the amount will be:

(a) smaller than the given number if the percent is less than 100%;

(b) greater than the given number if the percent is more than 100%.

NAME ______________ SECTION ______________ DATE ______________

ANSWERS

Exercises 7.4

Solve each problem for the unknown quantity. You may use a calculator to perform calculations; however, you should first write the basic equation $R \times B = A$ and other resulting equations one below the other with = signs aligned.

1. 10% of 70 is ______.

2. 5% of 62 is ______.

3. 15% of 60 is ______.

4. 25% of 72 is ______.

5. 75% of 12 is ______.

6. 60% of 30 is ______.

7. 100% of 36 is ______.

8. 80% of 50 is ______.

9. 2% of ______ is 3.

10. 20% of ______ is 17.

11. 3% of ______ is 21.

12. 30% of ______ is 21.

13. 100% of ______ is 75.

14. 50% of ______ is 42.

15. 150% of ______ is 63.

16. 110% of ______ is 330.

17. ______% of 60 is 90.

18. ______% of 150 is 60.

19. ______% of 75 is 15.

20. ______% of 12 is 4.

21. ______% of 34 is 17.

22. ______% of 30 is 6.

23. ______% of 48 is 16.

24. ______% of 100 is 35.

Each of the following problems is one of the three types discussed written in a different way. Remember that B is the number you are finding the percent of.

25. ______ is 50% of 25.

26. ______ is 31% of 76.

27. 22 is 20% of ______.

28. 86 is 100% of ______.

1. ______
2. ______
3. ______
4. ______
5. ______
6. ______
7. ______
8. ______
9. ______
10. ______
11. ______
12. ______
13. ______
14. ______
15. ______
16. ______
17. ______
18. ______
19. ______
20. ______
21. ______
22. ______
23. ______
24. ______
25. ______
26. ______
27. ______
28. ______

29. ______________
30. ______________
31. ______________
32. ______________
33. ______________
34. ______________
35. ______________
36. ______________
37. ______________
38. ______________
39. ______________
40. ______________
41. ______________
42. ______________
43. ______________
44. ______________
45. ______________
46. ______________
47. ______________
48. ______________
49. ______________
50. ______________
51. ______________
52. ______________
53. ______________
54. ______________
55. ______________

29. 18 is ______% of 10.

30. 15 is ______% of 10.

31. 24 is $33\frac{1}{3}\%$ of ______.

32. 92.1 is 15% of ______.

33. 119.6 is 23% of ______.

34. 9.5 is 25% of ______.

35. 36 is ______% of 18.

36. 60 is ______% of 40.

37. ______ is 96% of 17.

38. ______ is 84% of 32.

39. ______ is 18% of 325.

40. ______ is 28% of 460.

41. Find 18% of 244.

42. Find 15.2% of 75.

43. Find 120% of 60.

44. What percent of 32 is 8?

45. 100 is 125% of what number?

Use fractions to solve the following problems. Do the work mentally, if you can, after you have set up the related equation.

46. Find 50% of 32.

47. Find $66\frac{2}{3}\%$ of 60.

48. What is $12\frac{1}{2}\%$ of 80?

49. What is $62\frac{1}{2}\%$ of 16?

50. $33\frac{1}{3}\%$ of 75 is what number?

51. 25% of 150 is what number?

52. 75% of what number is 21?

53. 50% of what number is 35?

54. $37\frac{1}{2}\%$ of what number is 61.2?

55. 100% of what number is 76.3?

NAME ________________ SECTION ________________ DATE ________

ANSWERS

56. ________________

57. ________________

58. ________________

59. ________________

60. ________________

56. Magazine publishers will often give significant discounts to subscribers. The cover price of *Newsweek* is $79.65 for 6 months, but a 6-month subscription costs only $18.57! What is the percent savings (to the nearest tenth) for someone who subscribes to *Newsweek*?

57. Only 27% of American deaths in the Revolutionary War occurred in battle. All other mortalities were the result of things such as exposure, disease, and starvation. Of the estimated 25,300 deaths, how many men died in battle (to the nearest hundred)?

58. During his presidency from 1945 to 1953, Harry Truman vetoed 250 congressional bills, and 12 of those vetoes were overridden. What percent of Truman's vetoes were overridden?

59. William O. Douglas served as an associate justice on the Supreme Court for 36 years, from 1939 to 1975, the longest tenure of any Supreme Court justice in history. He lived to be 82. What percent of his life (to the nearest one percent) did he spend on the Supreme Court?

60. The decade from 1960 to 1970 saw the largest 10-year increase in the number of male elementary and high school teachers in our nation's history—from 402,000 to 691,000. What was the percent increase from 1960 to 1970 (to the nearest one percent)?

61. ____________

62. ____________

63. ____________

64. ____________

65. ____________

66. ____________

67. ____________

68. ____________

69. ____________

70. ____________

Check Your Number Sense

In each of the following exercises, one of the choices is the correct answer to the problem. Work through all the exercises by reasoning and calculating mentally. After you have checked your answers, you should work out the answers to any problems you missed and try to understand why your reasoning was incorrect.

61. Write 7% as a fraction.

(a) $\frac{7}{1}$ (b) $\frac{7}{10}$ (c) $\frac{7}{100}$ (d) $\frac{7}{1000}$

62. Find 10% of 730.

(a) 7.3 (b) 73 (c) 14.6 (d) 146

63. Find 20% of 730.

(a) 7.3 (b) 73 (c) 14.6 (d) 146

64. What is 100% of 9.03?

(a) 0.0903 (b) 0.903 (c) 9.03 (d) 90.3

65. 10% of what number is 29?

(a) 2.9 (b) 29 (c) 290 (d) 2900

66. 1% of what number is 29?

(a) 0.29 (b) 2.9 (c) 290 (d) 2900

67. What percent of 100 is 25?

(a) 0.25% (b) 2.5% (c) 25% (d) 250%

68. What percent of 1000 is 25?

(a) 0.25% (b) 2.5% (c) 25% (d) 250%

69. 245 is approximately 50% of what number?

(a) 24 (b) 120 (c) 500 (d) 1200

70. 72 is approximately 30% of what number?

(a) 25 (b) 75 (c) 150 (d) 250

7.5 Applications with Percent: Discount, Sales Tax, and Tipping

OBJECTIVES

1. *Know Pólya's Four-Step Process for Solving Problems.*
2. *Be familiar with the terms discount and sales tax.*
3. *Apply skills related to types of percent problems to solving applications.*

Pólya's Four-Step Process for Solving Problems

George Pólya (1887–1985), a famous professor at Stanford University, studied the process of discovery learning. Among his many accomplishments, he developed a four-step process as an approach to problem solving:

1. Understand the problem. **2.** Devise a plan.

3. Carry out the plan. **4.** Look back over the results.

There are a variety of types of applications discussed throughout this text and in subsequent courses in mathematics, and you will find these four steps helpful as guidelines for understanding and solving all of them. Applying the necessary skills to solve exercises, such as adding fractions or solving equations, is not the same as accumulating the knowledge with which to solve problems. Problem solving can involve careful reading, reflection, and some original or independent thought.

Basic Steps for Problem Solving

1. Understand the problem. For example,
   (a) Read the problem carefully and identify the key words.
   (b) Understand what information is given and what is to be found.

2. Devise a plan using, for example, one or all of the following:
   (a) Guess or estimate or make a list of possibilities.
   (b) Draw a picture or diagram.
   (c) Use a variable and form an equation.

3. Carry out the plan. For example,
   (a) Try all the possibilities you have listed.
   (b) Solve any equation that you may have set up.

4. Look back over the results. For example,
   (a) Can you see an easier way to solve the problem?
   (b) Does your solution actually work? Does it make sense?
   (c) If there is an equation, check your answer in the equation.

In this section, we will be concerned with applications that involve percent and the types of percent problems that have been discussed in the first four sections of Chapter 7. The use of a calculator is recommended, but keep in mind that it is a tool to enhance, not replace, the necessary skills and abilities related to problem solving. Study the following examples carefully, as they illustrate some basic plans for problem solving with percent.

Discount

A **discount** is a reduction in the **original price** (or **marked price**) of an item. The new reduced price is called the **sale price,** and the discount is the difference between the original price and the sale price. The **rate of discount** (or **percent of discount**) is a percent that represents what part the discount is of the original price.

> **Note:** In this text, when solving money problems, rounded-off answers will be rounded to the next higher cent even if the next digit is less than 5.

EXAMPLE 1

A refrigerator that regularly sells for \$975 is on sale at a 30% discount. (a) What is the amount of the discount? (b) What is the sale price?

Solution

First find the amount of the discount, and then find the sale price by subtracting the discount from the original price.

(a) Since the rate of discount is 30%, we find 30% of \$975.

30% of \$975 is ________.

$0.30 \times 975 = A$

$292.50 = A$

	\$97 5	original price
×	.30	rate of discount
	\$292.50	discount

The discount is \$292.50.

(b) Find the sale price by subtracting the discount from the original price.

\$975.00	original price
− 292.50	discount
\$682.50	sales price

The sale price is \$682.50.

Sales Tax

Sales tax is a tax charged on goods sold by retailers, and it is assessed by states for income to operate state services. The rate of the sales tax varies from state to state (or even city to city in some cases). In fact, some states do not have a sales tax.

EXAMPLE 2

If the sales tax rate is 6%, what would be the final cost of the refrigerator in Example 1?

Solution

First find the amount of the sales tax, and then add this tax to the sale price found in Example 1 to find the final cost.

(a) Find the amount of the sales tax.

6% of \$682.50 is ________.

$0.06 \times 692.50 = A$

$40.95 = A$

	\$692.50	sale price
×	0.06	tax rate
	\$40.9500	sales tax

The sales tax is \$40.95.

(b) Add the sales tax to the sale price to find the final cost.

	\$682.50	sale price
+	40.95	sales tax
	\$723.45	final cost

The final cost of the refrigerator is \$723.45.

Example 3

An auto dealer paid \$7566 for a large order of a special part. This was **not** the original price. He received a 3% discount off the original price because he paid cash. What was the original price?

Solution

Since \$7566 is not the original price, do **not** take 3% of \$7566. In fact, we are to find the original price. Therefore, we must reason that if a price is discounted by 3%, then 97% of the original price is what remains (100% − 3% = 97%). Thus, \$7566 represents 97% of the original price.

97% of ______ is \$7566

$.97 \times B = 7566$

$$\frac{.97 \times B}{.97} = \frac{7566}{.97}$$

$B = 7800$

$$
\begin{array}{r}
7800. \\
.97.\overline{)\,7566.00.} \\
\underline{679} \\
776 \\
\underline{776} \\
0\,0 \\
\underline{0} \\
00 \\
\underline{0} \\
0
\end{array}
$$

The original price was \$7800.

Note: To check this result, take 3% of \$7800 to find the discount. Subtract this discount from \$7800 to get the discounted price of \$7566.

Tipping

Tipping (or **leaving a tip**) is the custom of leaving a percent of a bill (usually at a restaurant) as a payment to the waiter or waitress for providing good service. The amount of the tip is usually 10–20% of the total amount charged (including tax), depending on the quality of the service. The "usual" tip is 15%, but in the case of particularly bad service, no tip is required.

Since most people do not carry a calculator with them when dining out, the calculation of a 15% tip can be an interesting exercise in percents. The following rule of thumb uses the basic facts that 5% is half of 10% and that 10% of a decimal number can be found by moving the decimal point 1 place to the left.

Rule of Thumb for Calculating a 15% Tip:

1. For ease of calculation, round off the amount of the bill to the nearest whole dollar.
2. Find 10% of the rounded-off amount by moving the decimal point 1 place to the left.
3. Divide the answer in Step 2 by 2. (This represents 5% of the rounded-off amount, or one-half of 10%.)
4. Add the two amounts found in Steps 2 and 3. This sum is the amount of the tip.

Note: There are other rules of thumb that people use to calculate the amount of a tip. For example, you might always round up instead of rounding off to the nearest whole dollar, or you might round off to the nearest ten dollars, or you might leave a tip that makes the total of your expenses a whole-dollar amount. Consider two bills of $7.95 and $7.20. Some might round off each bill to $10.00 and leave a tip of 15% as $1.00 + $0.50 = $1.50 in each case. However, according to the rule in the text, we will calculate the tips as follows:

Round off $7.95 to $8.00 and calculate 15% as $0.80 + $0.40 = $1.20.

Round off $7.20 to $7.00 and calculate 15% as $0.70 + $0.35 = $1.05.

Example 4

You and a friend dine out and the bill comes to $28.40, including tax. If you plan to leave a 15% tip, what should be the amount of the tip?

Solution

For ease of computation, round off $28.40 to $28.00.
Find 10% of $28.00 by moving the decimal point one place to the left:

$$10\% \text{ of } 28.00 = 2.80$$

Now find 5% of $28.00 by dividing $2.80 by 2:

$$2.80 \div 2 = 1.40$$

Adding gives:

$$\begin{array}{r} \$2.80 \\ +\ 1.40 \\ \hline \$4.20 \end{array} \quad \text{amount of tip}$$

We will say that the tip should be $4.20 as a textbook answer. However, practically speaking (depending on things such as how much change you have in your pocket and how good the service was), you might leave either $4.25 or $4.50.

NAME ______________ SECTION ______________ DATE ______________

ANSWERS

1a. ______________

b. ______________

2a. ______________

b. ______________

3a. ______________

b. ______________

4a. ______________

b. ______________

Exercises 7.5

The following problems may involve several calculations, and calculators are recommended. Follow Pólya's four-step process for problem solving. After you have read the problem so that you understand it, write down the known information and your plan for solving the problem. Then follow through with your plan.

1. A store owner received a 3% discount from the manufacturer when she bought $15,500 worth of dresses. (a) What was the amount of the discount? (b) What did she pay for the dresses?

2. If the sales tax in a certain state is figured at 6%: (a) How much tax is there on a purchase of $30.20? (b) What is the total amount paid for the purchase?

3. If sales tax was figured at 6%: (a) How much tax was paid on the purchase of three textbooks priced at $55.00, $25.50, and $34.95? (b) What would be the total cost of all three books?

4. A new briefcase was priced at $275. If it was to be marked down 30%: (a) What would be the discount? (b) What would be the new price?

5a. ______________

b. ______________

c. ______________

6a. ______________

b. ______________

7a. ______________

b. ______________

8. ______________

9. ______________

5. The discount on a fur coat was \$150. This was a 20% discount. (a) What was the original selling price of the coat? (b) What was the sale price? (c) What was the total amount paid for the coat if a 6% sales tax was added to the sale price?

6. The discount on men's suits was \$50, and they were on sale for \$200. (a) What was the original selling price? (b) What was the rate of discount?

7. In order to get more subscribers, a book club offered three books for a total price of \$7.02. The total selling price was originally \$17.55 for all three books. (a) What was the amount of the discount? (b) Based on the original selling price, what was the rate of the discount on these three books?

8. An auto supply store received a shipment of auto parts and a bill for \$845.30. Some of the parts were not as ordered, and they were returned immediately. The value of the parts returned was \$175.50. The terms of the billing provided the store with a 2% discount if it paid cash (for the parts it kept) within two weeks. What did the store pay for the parts it kept if it paid cash within two weeks?

9. Towels were on sale at a discount of 30%. If the sale price was \$3.01, what was the original price?

NAME ______________________ SECTION ______________ DATE __________

ANSWERS

10. Computer disks were on sale for \$5.24 per box. What was the original price per box if the sale price represents a discount of 20%?

10. ______________

11. ______________

12. ______________

13. ______________

In Exercises 11–16, use the rule of thumb stated in the text to calculate a 15% tip.

11. You invited your friend to lunch at the local coffeehouse, and the bill totaled \$12.60. Your friend offered to leave the tip. What amount did he leave as a 15% tip?

12. A lawyer took four clients to dinner. The bill for the meals and drinks was \$150.00 plus a 6% sales tax. What amount did she leave as a tip (rounding up to the nearest dollar) if she left 15%?

13. A bill for a family dining at a restaurant was \$38.40. What would a 15% tip have been? What total amount should they have left on the table if they wanted to go before the waiter came to pick up the money?

14. ______________

15. ______________

16. ______________

17a. ______________

b. ______________

14. On a recent business trip, Ken and Joe had breakfast at the restaurant next to the motel where they were staying. Ken's breakfast bill was $6.95 and Joe's was $8.75. How much should each man have left as a tip if each tipped 15%?

15. Mrs. Chung had two large pizzas delivered to her home for $26.00. If she tipped the delivery person 15%, what total amount did she pay the driver?

16. Juan took his date out for dinner before the senior prom. The total bill, including tax, was $42.00. How much did he tip the waitress if he left 15%? What were his total expenses for the meal?

17. Linda is enrolled in freshman calculus. She has the choice of buying the text in hardback form for $60.00 or in paperback form for $46.50. Tax is figured at 5% of the selling price. The bookstore buys back hardback books for 40% of the selling price and paperback books for 30% of the selling price. (a) Which book is the more economical buy for Linda if she sells her book back to the bookstore at the end of the semester? (b) How much does she save?

NAME ______________ SECTION ______________ DATE ______________

ANSWERS

18. ______________

19. ______________

20. ______________

21a. ______________

b. ______________

18. A student missed 3 problems on a mathematics test and received a grade of 85%. If all the problems were of equal value, how many problems were on the test?

19. Suppose you sell your home for $100,000 and you owe the Savings and Loan $60,000 on the first trust deed. You pay a real estate agent a commission of 6% of the selling price and other fees and taxes totaling $1200. How much cash do you have from the sale?

20. Sheets are marked $22.50 and pillowcases $7.50. What is the sale price of each item if each item is discounted 25% off the marked price?

21. In the roofing business, shingles are sold by the "square," which is enough material to cover a 10 ft by 10 ft square (or 100 square feet). A roofing supplier has a closeout on shingles at a 30% discount. (a) If the original price was $230 per square, what is the sale price per square? (b) How much would a roofer pay for 34 squares?

22. [Respond below exercise.]

23a. [Respond below exercise.]

b. [Respond below exercise.]

24. [Respond below exercise.]

Writing and Thinking about Mathematics

22. Make up your own rule of thumb for leaving a 20% tip.

23. A man weighed 220 pounds. He lost 22 pounds in three months. Then he gained back 22 pounds over the next four months. (a) What percent of his weight did he lose in the first three months? (b) What percent of his weight did he gain back? The loss and gain are the same amount, but the two percents are different. Explain why.

Collaborative Learning Exercise

24. With the class separated into teams of two to four students, each team is to analyze the following discussion on discount and decide how to answer the related questions. Then each team leader is to present the team's answers and related ideas to the class for general discussion.

In Example 1 on page 460, we calculated a 30% discount to find the sale price of a refrigerator. Then, in Example 2, we calculated the sales tax at 6% of the sale price. The final cost of the refrigerator was then found by adding the sales tax to the sale price. Could most of these calculations have been avoided by simply discounting the original price by 24% to determine the final cost (30% − 6% = 24%)? Explain your reasoning.

7.6 Applications with Percent: Commission, Profit, and Others

OBJECTIVES

1. *Be familiar with and understand the term* **commission.**
2. *Understand that percent of profit is a ratio and can be based on cost or on selling price.*
3. *Know how to calculate percent of profit.*

Commission

A **commission** is a fee paid to an agent or salesperson for a service. Commissions are usually a percent of a negotiated contract or a percent of sales.

Example 1

A saleswoman earns a fixed salary of $900 a month plus a commission of 8% on whatever amount she sells over $6500 in merchandise. What did she earn the month she sold $11,800 in merchandise?

Solution

First find the base of her commission by subtracting $6500 from the total amount she sold. Then find 8% of this base and add her monthly salary.

(a) Subtract $6500 from $11,800.

$11,800	total amount she sold
− 6,500	amount on which she does not earn a commission
$ 5,300	base of commission

(b) Find 8% of this base.

8% of $5300 is ________

$0.08 \times 5300 = A$

$424 = A$

$5300	base of commission
× .08	rate of commission
$ 424	commission

(c) Her monthly pay is the sum of her salary and her commission.

$ 900.00	fixed salary
+ 424.00	commission
$1324.00	total pay for the month

She earned $1324 for the month.

Percent of Profit

In business, a company's **profit** is the difference between income (or revenue) and costs, such as wages, materials, and rent. However, manufacturers and retailers are also concerned with the profit on each item produced or sold, which is simply the difference between its selling price to the customer and the cost to the company. This is the type of profit that will be discussed in this section.

Terms Related to Profit

Profit: The difference between selling price and cost (profit = selling price − cost)

Percent of profit: There are two types; both are ratios with profit in the numerator.

1. Percent of profit **based on cost** is the ratio of profit to cost:

$$\frac{\text{profit}}{\text{cost}} = \%\text{ of profit based on cost}$$

2. Percent of profit **based on selling price** is the ratio of profit to selling price:

$$\frac{\text{profit}}{\text{selling price}} = \%\text{ of profit based on selling price}$$

EXAMPLE 2

A company manufactures and sells plastic boxes that cost $21 each to produce, and that sell for $28 each. (a) What is the profit on each box? (b) What is the percent of profit based on cost? (c) What is the percent of profit based on selling price?

Solution

(a) To find the profit, subtract the cost from the selling price.

$$\begin{array}{rl} \$28 & \text{selling price} \\ -\ 21 & \text{cost} \\ \hline \$\ 7 & \text{profit} \end{array}$$

(b) To find the percent of profit based on cost, divide the profit by the cost.

$$\frac{\$7\text{ profit}}{\$21\text{ cost}} = \frac{1}{3} = 0.33\frac{1}{3} = 33\frac{1}{3}\%\text{ profit based on cost}$$

(c) To find the percent of profit based on selling price, divide the profit by the selling price.

$$\frac{\$7\text{ profit}}{\$28\text{ selling price}} = \frac{1}{4} = 0.25 = 25\%\text{ profit based on selling price}$$

Note that in parts (b) and (c) the profit is $7 and this does not change. What changes, and gives different percents, is the denominator.

Percent of profit **based on cost** is higher than percent of profit **based on selling price.** The business community reports whichever percent serves its purposes better. Your responsibility as an investor or consumer is to know which percent is reported and what it means to you.

✔ COMPLETION EXAMPLE 3

Women's coats were on sale for $250. (a) If the coats cost the store owner $200, what was his percent of profit based on cost? (b) What was his percent of profit based on selling price?

Solution

(a) First find the profit.

$$\begin{array}{rl} \$250 & \text{selling price} \\ -\underline{\quad\quad} & \text{cost} \\ \$\underline{\quad\quad} & \text{profit} \end{array}$$

(b) Find the percent of profit based on cost.

$$\frac{\underline{\quad\quad}\ \text{profit}}{\underline{\quad\quad}\ \text{cost}} = \underline{\quad\quad}\%\ \text{profit based on cost}$$

(c) Find the percent of profit based on selling price.

$$\frac{\underline{\quad\quad}\ \text{profit}}{\underline{\quad\quad}\ \text{selling price}} = \underline{\quad\quad}\%\ \text{profit based on selling price}$$

Completion Example Answers

3. (a) First find the profit.

$$\begin{array}{rl} \$250 & \text{selling price} \\ -\ \mathbf{200} & \text{cost} \\ \hline \$\ \mathbf{50} & \text{profit} \end{array}$$

(b) Find the percent of profit based on cost.

$$\frac{\mathbf{\$50}\text{ profit}}{\mathbf{\$200}\text{ cost}} = \mathbf{25\%}\text{ profit based on cost}$$

(c) Find the percent of profit based on selling price.

$$\frac{\mathbf{\$50}\text{ profit}}{\mathbf{\$250}\text{ selling price}} = \mathbf{20\%}\text{ profit based on selling price}$$

NAME ______________ SECTION ______________ DATE ______________

ANSWERS

1. ______________

2. ______________

3. ______________

4. ______________

Exercises 7.6

1. A realtor works on a 6% commission. What is his commission on a house he sold for $95,000?

2. A realtor selling commercial property works on a 4% commission. What is her commission on a building she sold for $875,000?

3. A sales clerk receives a monthly salary of $500 plus a commission of 6% on all sales over $2500. What did the clerk earn the month she sold $16,000 in merchandise?

4. A shoe saleswoman works on a fixed salary of $300 per month plus a 5% commission. How much did she make during the month in which she sold $7500 worth of shoes?

5. ______________

6a. ______________

b. ______________

7. ______________

8a. ______________

b. ______________

c. ______________

9a. ______________

b. ______________

c. ______________

5. If a salesman works on a 10% commission only (no monthly salary), how much merchandise will he have to sell to earn $2800 in one month?

6. A basketball player made 120 of 300 shots she attempted. (a) What percent of her shots did she make? (b) What percent did she miss?

7. In one season, a basketball player missed 15% of his free throws. How many free throws did he make if he attempted 180?

8. A set of golf clubs cost the golf pro $400, and he sold them in the pro shop for $550. (a) What was his profit? (b) What was his percent of profit based on cost? (c) What was his percent of profit based on selling price?

9. Men's suits were on sale for $300. Each one cost the store owner $250. (a) What was the profit for the store? (b) What was the store's percent of profit based on cost? (c) What was the store's percent of profit based on selling price?

NAME ______________ SECTION ______________ DATE ______________

ANSWERS

10. The cost of a television set to a store owner was $350, and he sold the set for $490. (a) What was his profit? (b) What was his percent of profit based on cost? (c) What was his percent of profit based on selling price?

10a. ______________

b. ______________

c. ______________

11a. ______________

b. ______________

11. A car dealer bought a used car for $1500. He marked up the price so he would make a profit of 25% based on his cost. (a) What was the selling price? (b) If the customer paid 8% of the selling price in taxes and fees, what did the customer pay for his car?

12. ______________

13. ______________

12. The property taxes on a house were $750. What was the tax rate if the house was valued at $25,000 for tax purposes?

13. A computer programmer was told he would be given a bonus of 5% of any money his programs could save the company. How much would he have to save the company to earn a bonus of $6000?

14. ______________

15a. ______________

b. ______________

c. ______________

16. ______________

17. ______________

18. ______________

14. You want to purchase a new home for $98,000. The bank will loan you 80% of the purchase price, but you must pay a loan fee of 2% of the amount of the loan, plus other fees totaling $850. How much cash do you need in order to purchase the home?

15. The author of a book was told she would have to cut the number of pages by 12% in order for the book to sell at a competitive price and still show a profit for the publisher. (a) What percent of the pages were in the final form of the book? (b) If the book contained 220 pages in its final form, how many pages did the original form contain? (c) How many pages were cut?

16. A pigeon egg incubates in 18 days, while a duck egg requires 30 days to hatch. Expressed as a percent, how much longer does it take for the duck egg to hatch?

17. In 1991, Ford Motor Company produced 81,594 Mustangs, which was about 10.6% of its total production. About how many cars did Ford produce in 1991 (to the nearest thousand)?

18. Along with cattle, sheep, and poultry, Alaskan farmers also raise reindeer as livestock. In fact, in a recent year, 37,000 reindeer accounted for 71% of Alaska's livestock population. To the nearest thousand, what was the total livestock population in Alaska that year?

NAME ______________ SECTION ______________ DATE ______________

ANSWERS

19. ______________

20. ______________

21a. [Respond below exercise.]

b. [Respond below exercise.]

22. [Respond below exercise.]

19. In 1980, there were 11.3 billion shares of stock traded on the New York Stock Exchange. That volume increased steadily through 1987, when 47.8 billion shares were traded. Write the number of shares traded in 1987 as a percent of the number traded in 1980 (to the nearest one percent).

20. In 1980, the U.S. government's public debt per capita was $3,985, and by 1990 that debt had increased to $13,000 per capita. Write the per capita debt in 1990 as a percent of the per capita dept in 1980, to the nearest one percent.

Writing and Thinking about Mathematics

21. Joel worked in a men's clothing store on a straight 8% commission. His friend, who worked at the same store, earned a monthly salary of $500 plus a 4% commission. (a) How much did each make during the month in which each sold $18,500 worth of clothing? (b) What percent more did Joel make than his friend? Explain why there is more than one answer to part (b).

22. (a) What topic (or topics) discussed in Chapter 7 have you found to be the most interesting? Briefly, explain why. (b) What topic (or topics) have you found to be the most difficult to learn? Briefly, explain why.

23. [Respond below exercise.]

24. [Respond below exercise.]

23. (a) What topic (or topics) discussed in the text to this point have you found to be most interesting? Briefly, explain why. (b) What topic (or topics) have you found to be the most difficult to learn? Briefly, explain why.

Collaborative Learning Exercise

24. With the class separated into teams of two to four students, each team is to analyze the following discussion of atomic half-life and answer the related questions as best they can. A general classroom discussion should follow with the class coming to an understanding of the concepts involved.

The half-life of a radioactive chemical element is the time that it takes for 50% of the atoms in the element sample to decay. Scientists can use methods such as carbon-14 dating to determine the ages of archaeological artifacts by comparing the amount of carbon-14 remaining in an artifact with the level of carbon-14 that would have been present originally. The half-life of carbon-14 is 5730 years. What percent of 10 grams would be left after 5730 years? How many grams? What percent would be left after 11,460 years? How many grams? What percent would be left after 17,190 years? How many grams? Describe a pattern in your calculations over these periods of time. What percent would be left after six such periods of time?

7 Index of Key Ideas and Terms

NAME ______________________ SECTION ________________ DATE ________

Chapter 7 Test

In Exercises 1–5, change each number to a percent.

1. $\frac{101}{100}$ **2.** 0.003 **3.** 0.173

4. $\frac{8}{25}$ **5.** $2\frac{3}{8}$

In Exercises 6–8, change each percent to a decimal.

6. 70% **7.** 180% **8.** 9.3%

In Exercises 9–11, change each percent to a fraction or mixed number with the fraction part reduced.

9. 130% **10.** $35\frac{1}{2}\%$ **11.** 8.6%

In Exercises 12–17, find the unknown quantity.

12. 10% of 56 is _____. **13.** 12 is _____% of 36.

14. 33 is 110% of _____. **15.** _____ is 62% of 475.

16. 6.765 is 33% of _____. **17.** 16.55 is _____% of 50.

Choose the correct answer by reasoning and calculating mentally.

18. Write 11% as a fraction.

(a) $\frac{11}{1}$ (b) $\frac{11}{10}$ (c) $\frac{11}{100}$ (d) $\frac{11}{1000}$

ANSWERS

1. ________
2. ________
3. ________
4. ________
5. ________
6. ________
7. ________
8. ________
9. ________
10. ________
11. ________
12. ________
13. ________
14. ________
15. ________
16. ________
17. ________
18. ________

19. ____________

20. ____________

21. ____________

22a. ____________

b. ____________

c. ____________

23a. ____________

b. ____________

c. ____________

24a. ____________

b. ____________

25. ____________

19. Write $\frac{2}{3}$ as a percent.

(a) 66% (b) 67% (c) $33\frac{1}{3}\%$ (d) $66\frac{2}{3}\%$

20. Find 25% of 100.

(a) 2.5 (b) 25 (c) 250 (d) 2500

21. 10% of what number is 62?

(a) 6.2 (b) 62 (c) 620 (d) 6200

22. Using the rule of thumb stated in the text, calculate the 15% tip for each of the following amounts.

(a) $6.72 (b) $25.35 (c) $17.95

23. A hardware store has light fixtures on sale for $50.00. This represents a 20% discount. (a) What was the original selling price? (b) If the fixtures cost the store $40, what was the store's percent of profit based on cost? (c) What was the store's percent of profit based on selling price?

24. A customer received a 2% discount on the purchase of a new dining room set because she paid in cash. (a) If she paid $1176, what was the original selling price? (b) What was the amount of the discount?

25. A salesman earns $900 each month plus a commission of 8% of his sales over $20,000. How much did he earn for the month in which his sales were $50,000?

NAME ______________________ SECTION ________________ DATE __________

ANSWERS

Cumulative Review

1. The number 0 is called the additive ____________.

2. Name the property of addition that is illustrated.

   $17 + 15 = 15 + 17$

3. Round off as indicated:

   (a) 12,943 (nearest thousand) (b) 3.09672 (nearest thousandth)

First estimate the result, then perform the indicated operations to find the actual result.

4. $\begin{array}{r} 8695 \\ 457 \\ +1206 \\ \hline \end{array}$

5. $\begin{array}{r} 8500 \\ -4675 \\ \hline \end{array}$

6. $72\overline{)6696}$

7. $\begin{array}{r} 182 \\ \times\ 36 \\ \hline \end{array}$

8. Evaluate the expression $3(5 + 2^2) - 6 - 2 \cdot 3^2$.

9. List the squares of the prime numbers less than 30.

10. (a) Find the LCM, and (b) state the number of times each number divides into the LCM for the following set of numbers: 56, 60, 75.

Evaluate each of the following expressions. Reduce all fractions.

11. $\frac{15}{28} \cdot \frac{7}{25}$

12. $\frac{3}{4} + \frac{7}{8} \div \frac{1}{2}$

13. $5.6 + 7\frac{1}{2} - 3\frac{5}{8}$

14. $\dfrac{\frac{1}{5} + \frac{7}{20}}{2\frac{1}{2} - 1\frac{3}{5}}$

15. Solve the proportion: $\dfrac{7}{x} = \dfrac{3\frac{1}{2}}{11}$

1. ____________

2. ____________

3a. ____________

b. ____________

4. ____________

5. ____________

6. ____________

7. ____________

8. ____________

9. [Respond below exercise.]

10a. ____________

b. ____________

11. ____________

12. ____________

13. ____________

14. ____________

15. ____________

16. ______________

17. ______________

18a. ______________

b. ______________

19a. ______________

b. ______________

c. ______________

d. ______________

20a. ______________

b. ______________

21. ______________

22a. ______________

b. ______________

c. ______________

23a. ______________

b. ______________

24a. ______________

b. ______________

c. ______________

d. ______________

16. Find 75% of 104.

17. 82% of ______ is 328.

18. Write each of the following numbers in scientific notation.

(a) 563,000,000 (b) 0.0000942

19. Find the equivalent measures in the metric system.

(a) 345 cm = __________ mm (b) 1.6 m = __________ cm

(c) 830 mm = __________ cm (d) 5200 m = __________ km

20. Simplify the following square roots.

(a) $\sqrt{175}$ (b) $\sqrt{108}$

21. Find the length of the hypotenuse of a right triangle that has sides of length 15 feet and 20 feet.

22. A clothing store has dresses on sale for $80.00. This represents a 20% discount. (a) What was the original selling price? (b) If the dresses cost the store $60, what was the store's percent of profit based on cost? (c) What was the store's percent of profit based on selling price?

23. You ordered two pizzas to be delivered to your home. One was for $11.75 and the other was for $15.80. (a) If you wanted to give the driver a 15% tip, what would be the amount of the tip? (b) How much would you pay the driver?

24. You paid $4500 as a down payment on a new car and made 48 monthly payments of $350 each. (a) Approximately how much did you pay for the car? (b) Exactly how much did you pay for the car? (c) How much more did you pay for the car than you would have if you had bought the car originally for $17,040 in cash? (d) What percent more did you pay by making monthly payments?

8

Consumer Applications

What to Expect in Chapter 8

Interest is money paid for the use of money, and the related concepts of simple interest (Section 8.1) and compound interest (Section 8.2) provide important applications of percent and interest that every adult in our society should understand.

Some of the many expenses in becoming a homeowner and other expenses in maintaining a home once you own it are discussed in Section 8.3. This discussion and related problems are designed to be informative regardless of whether you own a home, rent an apartment, or anticipate owning a home. Section 8.4 presents a similar discussion of the expenses involved in buying a car and owning a car.

Note that all the consumer applications in this chapter involve percent in some way.

Mathematics at Work!

The interest rate that you pay on a loan will depend on the bank from which you borrow and the purpose of the loan.

At some time in your life, you will probably borrow money (this is the same as using a credit card), lend money, or have a savings account. In any case, you will need to understand the concept of **interest,** money paid for the use of money. If you have a loan (or a savings account) with a bank, you should know what type of interest you are paying (or getting)—**simple interest** or **compound interest**—and what the rates are. You should know that the terms for loans and investments are negotiable, so it may pay you to shop around with different banks and savings and loans for the best terms.

The formula for calculating **simple interest** (involving only one payment at the end of the term of the loan) is

$$I = P \times r \times t$$

where

I = Interest (earned or paid)

P = Principal (the amount invested or borrowed)

r = Rate of interest

t = Time (in years)

Interest paid on interest earned is called **compound interest.** When interest is compounded, the total amount A accumulated (including principal and interest) is given by the formula

$$A = P\left(1 + \frac{r}{n}\right)^{nt}$$

where n is the number of compounding periods in one year. For example, if interest is compounded quarterly, then $n = 4$. If interest is compounded daily, then $n = 360$. (**Note:** When calculating interest, we will use 30 days per month and 360 days per year, a common practice in business.)

Suppose that \$5000 is invested at 8% for 4 years and interest is compounded. Do you think there will be a difference in the accumulated amount if (a) interest is compounded quarterly and (b) interest is compounded daily? Will this difference will be (a) about \$5, (b) about \$20, (c) about \$50, or (d) over \$100? (See Exercise 13 in Section 8.2.)

8.1 Simple Interest

OBJECTIVES

1. *Understand the concept of* **simple interest.**
2. *Find simple interest using the formula* $I = P \times r \times t$.

Understanding Simple Interest

Interest is money paid for the use of money. The money that is invested or borrowed is called the **principal.** The **rate** is the **percent of interest** and is almost always stated as an **annual (yearly) rate.**

Interest is either paid or earned, depending on whether you are the borrower or the lender. In either case, the calculations involved are the same. Although interest rates can vary from year to year (or even daily, as in the case of home loans) and from one part of the world to another, the concept of interest is the same everywhere.

Some loans (called **notes**) are based on **simple interest** and involve only one payment (including interest and principal) at the end of the term of the loan. Such loans are usually made for a period of one year or less. **Compound interest,** which involves interest paid on interest, will be discussed in Section 8.2.

The following formula is used to calculate simple interest.

Formula for Calculating Simple Interest

$$I = P \times r \times t$$

where

I = Interest (earned or paid)

P = Principal (the amount invested or borrowed)

r = Rate of interest (stated as an annual, or yearly, rate) and used in decimal or fraction form

t = Time (in years or fraction of a year)

Note: For calculation purposes, we will use 360 days in one year (30 days in a month). This is a common practice in business and banking; however, with the advent of computers, many lending institutions now base their calculations on 365 days per year and pay or charge interest on a daily basis.

Example 1

If you were to borrow \$2000 at 12% for one year, how much interest would you pay?

Solution

Use the formula for simple interest: $I = P \times r \times t$ with

$P = \$2000, \quad r = 12\% = 0.12, \quad \text{and} \quad t = 1 \text{ year}$

$I = \$2000 \times 0.12 \times 1 = \$240$

You would pay \$240 interest.

1. Ralph borrowed $3000 at 7% for one year. How much interest did he pay?

2. Mrs. Rice loaned her daughter $1500 at 10% interest for 9 months. How much interest did she earn?

Example 2

If you decided you would need the $2000 for only 90 days, how much interest would you pay? (The interest rate would still be stated at 12%, the annual rate.)

Solution

Use the formula for simple interest: $I = P \times r \times t$ with

$$P = \$2000, \quad r = 12\% = 0.12, \quad \text{and}$$

$$t = 90 \text{ days}$$

$$= \frac{90}{360} \text{ year} = \frac{1}{4} \text{ year}$$

$$I = \$2000 \times 0.12 \times \frac{1}{4} = \frac{240}{4} = \$60$$

or $$I = \$2000 \times 0.12 \times 0.25 = \$60$$

You would pay $60 interest if you borrowed the money for 90 days.

✓ Completion Example 3

Sylvia borrowed $2400 at 10% interest for 30 days. How much interest did she have to pay?

Solution

$P =$ __________, $r =$ __________% $=$ __________,

$t = 30$ days $=$ __________ year

$I =$ __________ $\times$ __________ $\times$ __________ $=$ __________

Now Work Exercises 1 and 2 in the Margin.

If you know the values of any three of the variables in the formula $I = P \times r \times t$, you can find the value of the fourth variable by substituting into the formula and solving the resulting equation for the unknown variable. This procedure is illustrated in Examples 4 and 5.

Example 4

How much (what principal) would you need to invest if your investment returned 9% interest and you wanted to make $100 in interest in 30 days?

Solution

Here the principal is unknown, while the interest ($I = \$100$), the rate of interest ($r = 9\% = 0.09$), and the time ($t = 30$ days) are all known. However, before substituting into the formula, we must change t to a fraction of a year: $t = \frac{30}{360} = \frac{1}{12}$ year.

$$I = P \times r \times t$$

$$100 = P \times 0.09 \times \frac{1}{12}$$

$$100 = P \times \frac{9}{100} \times \frac{1}{12}$$

$$100 = P \times \frac{3}{400}$$

$$\frac{400}{3} \times \frac{100}{1} = P \times \frac{\cancel{3}}{\cancel{400}} \times \frac{\cancel{400}}{\cancel{3}} \quad \text{Multiply both sides by } \frac{400}{3}.$$

$$\frac{40{,}000}{3} = P$$

$$P = \$13{,}333.34$$

or using $\frac{1}{12} = 0.083$ as a rounded-off decimal,

$$100 = P \times 0.09 \times 0.083$$

$$100 = P \times 0.00747$$

$$\frac{100}{0.00747} = \frac{P \times \cancel{0.00747}}{\cancel{0.00747}}$$

$$P = \$13{,}386.89$$

Using the rounded-off decimal 0.083 leads to a "round off error." You should understand that once a rounded-off value is used in any calculation there will be some error. The correct result can be found by using 0.083333333, which is the display on your calculator for 1 ÷ 12.

EXAMPLE 5

Stuart wants to borrow $1500 at 10% and is willing to pay $250 in simple interest. How long can he keep the money?

Solution

Here time is unknown and the principal is $1500, the interest is $250, and the rate is 10% = 0.10. Substituting into the formula $I = P \times r \times t$ gives

$$250 = 1500 \times 0.10 \times t$$

$$250 = 150 \times t$$

$$\frac{250}{150} = \frac{\cancel{150} \times t}{\cancel{150}}$$

$$\frac{5}{3} = t$$

or, $t = 1\frac{2}{3}$ years or 1 year 8 months

Stuart can borrow the money for $1\frac{2}{3}$ years.

Now Work Exercise 3 in the Margin.

3. What interest rate would you be paying if you borrowed $1000 for 6 months and paid $60 in interest?

Completion Example Answers

3. $P =$ **\$2400**, $r =$ **10%** $=$ **0.10**, $t =$ 30 days $= \mathbf{\frac{1}{12}}$ year

   $I = \mathbf{2400 \times 0.10 \times \frac{1}{12}} = \mathbf{\$20}$

NAME ____________ SECTION ____________ DATE ________

ANSWERS

1. ____________

2. ____________

3. ____________

4. ____________

5. ____________

6. ____________

Exercises 8.1

1. What is the simple interest paid on $500 at 6% for one year?

2. What is the simple interest paid on $2000 at 8% for one year?

3. How much interest would be paid on a loan of $1000 at 18% for 6 months? (**Note:** This interest rate may seem high, but the interest rates on some credit cards are even higher.)

4. How much interest would be paid on a loan of $3000 at 12% for 9 months?

5. You invested $2000 at 8% for 60 days. How much interest did your money earn?

6. Stacey loaned her brother $1500 for 8 months at 10% interest. How much interest did she earn?

7. ______________

8. ______________

9. ______________

10. ______________

11. ______________

12. ______________

13. ______________

7. What principal will earn $50 in interest if it is invested at 8% for 90 days?

8. What principal will earn $75 in interest if it is invested for 60 days at 9%?

9. How long will it take for $1000 invested at 5% to earn $50 in simple interest?

10. What length of time will it take to earn $70 in simple interest if $2000 is invested at 7%?

11. What will be the interest earned in one year on a savings account of $800 if the bank pays 4% interest?

12. If interest is paid at 6% for one year, what will a principal of $1800 earn?

13. If a principal of $900 is invested at a rate of 9% for 90 days, what will be the interest earned?

NAME ______________ SECTION ______________ DATE ______________

ANSWERS

14. ______________

15. ______________

16. ______________

17. ______________

18a. ______________

b. ______________

19. ______________

14. A loan of \$5000 is made at 8% for a period of 6 months. How much interest is paid?

15. If you borrow \$750 for 30 days at 18%, how much interest will you pay?

16. How much interest is paid on a 60-day loan of \$500 at 12%?

17. Find the simple interest paid on a savings account of \$2800 for 120 days at 3.5%.

18. A savings account of \$5300 is left for 90 days drawing interest at a rate of 5%. (a) How much interest is earned? (b) What is the amount in the account at the end of 90 days?

19. Every 6 months a stock pays 10% in dividends (interest on investment). What will be the earnings of \$14,600 invested for 6 months? (Remember that rates of interest are given as annual rates.)

20. ______________

21a. ______________

b. ______________

22a. ______________

b. ______________

23. ______________

24. ______________

25. ______________

20. If you charge \$1000 worth of merchandise at a local department store at 18% interest, how much will you owe at the end of 60 days?

21. You buy an oven on sale from \$500 to \$450, but you don't make a payment for 60 days and are charged interest at a rate of 18%. (a) How much do you pay for the oven by waiting 60 days to pay? (b) How much do you save by buying the oven on sale? (Sales tax is not included here.)

22. A friend borrows \$500 from you for a period of 8 months and pays you interest at 6%. (a) How much interest are you paid? (b) If you had asked 8%, how much more interest would you have earned?

23. What principal would have to be invested at 8% for 60 days to earn interest of \$500?

24. What rate of interest is charged if a loan of \$2500 for 90 days is paid off with \$2562.50?

25. How many days must you leave \$1000 in a savings account at 5.5% to earn \$11.00?

NAME ______________ SECTION ______________ DATE ______________

ANSWERS

26a. ______________
b. ______________
c. ______________
d. ______________

27a. ______________
b. ______________
c. ______________
d. ______________

28a. ______________
b. ______________

29. ______________

30. ______________

26. Determine the missing item in each row.

Principal	Rate	Time	Interest
\$ 400	16%	90 days	\$ (a)
\$ (b)	15%	120 days	\$ 5.00
\$ 560	12%	(c)	\$ 5.60
\$2700	(d)	40 days	\$25.50

27. Determine the missing item in each row.

Principal	Rate	Time	Interest
\$ 500	18%	30 days	\$ (a)
\$ 500	18%	(b)	\$15.00
\$ 500	(c)	90 days	\$22.50
\$ (d)	18%	30 days	\$ 1.50

28. (a) If Carlos has a savings account of \$25,000 drawing interest at 8%, how much interest will he earn in 6 months? (b) How long must he leave the money in the account to earn \$1500?

29. Ms. Lee has accumulated \$240,000 and she wants to live on the interest each year. If she needs \$2000 a month to live on, what interest rate must she earn on her money?

30. Mr. Smith has a savings account of \$2500 that draws 4.5% interest. How many days will it take for him to earn \$56.25?

31. ______________

32. ______________

33. ______________

34. ______________

35a. ______________

b. ______________

c. ______________

d. ______________

31. A bank decides to loan $5 million to a contractor to build new homes. How much interest will the bank earn in one year if the interest rate is 9.2%?

32. A credit card company has $120 million loaned to its customers at 18.9%. How much interest will it earn in one month?

33. A small airline company borrowed $7.5 million to buy some new airplanes. The loan rate was 7.5%, and the airline paid $562,500 in interest. What was the length of time of the loan?

34. A department store keeps $15 million in merchandise in stock. If the store pays interest at 9% on a bank loan for this stock, how much interest will the store pay in 3 months' time?

35. Determine the missing item in each row.

Principal	Rate	Time	Interest
$1000	$10\frac{1}{2}\%$	60 days	$\underline{\text{(a)}}$
$ 800	$13\frac{1}{2}\%$	__(b)__	$ 18.00
$2000	__(c)__	9 months	$172.50
$\underline{\text{(d)}}$	$7\frac{1}{2}\%$	1 year	$ 85.00

8.2 Compound Interest

OBJECTIVES

1. *Understand the concept of* **compound interest.**
2. *Find interest compounded periodically with repeated calculations.*
3. *Find compound interest using the formula* $A = P\left(1 + \frac{r}{n}\right)^{nt}$.

Understanding Compound Interest

Interest paid on interest is called **compound interest.** To calculate compound interest, we can calculate the simple interest for each period of time that interest is now compounded, using a **new principal for each calculation.** This new principal is the **previous principal plus the earned interest.** The calculations can be performed in a step-by-step manner, as indicated in the following outline.

To Calculate Compound Interest:

STEP 1: Using the formula for simple interest, $I = P \times r \times t$, calculate the simple interest where $t = \frac{1}{n}$ and n is the number of periods per year for compounding. For example,

for compounding annually, $n = 1$ and $t = \frac{1}{1} = 1$

for compounding semiannually, $n = 2$ and $t = \frac{1}{2}$

for compounding quarterly, $n = 4$ and $t = \frac{1}{4}$

for compounding monthly, $n = 12$ and $t = \frac{1}{12}$

for compounding daily, $n = 360$ and $t = \frac{1}{360}$

STEP 2: Add this interest to the principal to create a new value for the principal.

STEP 3: Repeat Steps 1 and 2 however many times the interest is to be compounded.

In Examples 1 and 3, we show how compound interest can be calculated in the step-by-step manner just outlined. This process serves to develop a basic understanding of the concept of compound interest. (Remember that if calculations with dollars and cents involve three or more decimal places, round answers up to the next higher cent regardless of the digit in the thousandths place.)

After Example 3 we will discuss another formula and show how to calculate compound interest with a calculator and the key marked $\boxed{y^x}$ (or $\boxed{x^y}$).

EXAMPLE 1

You deposit \$1000 in an account that pays 8% interest compounded quarterly (every 3 months). How much interest will you earn in one year?

Solution

Use $t = \frac{1}{n} = \frac{1}{4}$ and the formula $I = P \times r \times t$ and calculate the interest four times.

(a) First period: $P = \$1000$.

$$I = \mathbf{\$1000} \times 0.08 \times \frac{1}{4} = \$20 \text{ interest}$$

(b) Second period: $P = \$1000 + \$20 = \$1020$.

$$I = \mathbf{\$1020} \times 0.08 \times \frac{1}{4} = \$20.40 \text{ interest}$$

(c) Third period: $P = \$1020 + \$20.40 = \$1040.40$.

$$I = \mathbf{\$1040.40} \times 0.08 \times \frac{1}{4} = \$20.81 \text{ interest}$$

(d) Fourth period: $P = \$1040.40 + \$20.81 = \$1061.21$.

$$I = \mathbf{\$1061.21} \times 0.08 \times \frac{1}{4} = \$21.23 \text{ interest}$$

$$\begin{array}{r} \$20.00 \\ 20.40 \\ 20.81 \\ +\ 21.23 \\ \hline \$82.44 \end{array} \quad \text{total interest earned in 1 year}$$

The balance in the account will be \$1000 + \$82.44 = \$1082.44.

EXAMPLE 2

In Example 1, how much more interest will you earn by having the interest compounded quarterly for one year rather than calculated just as simple interest for the year?

Solution

Simple interest for one year would be $I = \$1000 \times 0.08 \times 1 = \$80.00$. The difference is

$$\begin{array}{rl} \$82.44 & \text{compound interest} \\ -\ 80.00 & \text{simple interest} \\ \hline \$\ 2.44 & \text{more by compounding quarterly} \end{array}$$

EXAMPLE 3

If an account is compounded monthly at 10%, how much interest will \$6000 earn in three months?

Solution

Use $t = \frac{1}{12}$ and calculate the interest three times.

(a) First period: $P = \$6000$.

$$I = \mathbf{\$6000} \times 0.10 \times \frac{1}{12} = \$50 \text{ interest}$$

(b) Second period: $P = \$6000 + \$50 = \$6050$.

$$I = \mathbf{\$6050} \times 0.10 \times \frac{1}{12} = \$50.42 \text{ interest}$$

(c) Third period: $P = \$6050 + \$50.42 = \$6100.42$.

$$I = \mathbf{\$6100.42} \times 0.10 \times \frac{1}{12} = \$50.84 \text{ interest}$$

The total interest earned will be

$$\begin{array}{r} \$\ 50.00 \\ 50.42 \\ +\ \ 50.84 \\ \hline \$151.26 \end{array}$$

Now Work Exercise 1 in the Margin.

1. George deposits \$500 in a savings account that pays 6% interest compounded quarterly. How much interest will he earn in 9 months?

Finding Compound Interest Using the Formula $A = P\left(1 + \frac{r}{n}\right)^{nt}$

The steps outlined in Examples 1 and 3 illustrate how the principal is adjusted for each time period of the compounding process and show the interest earned over each time period. The following compound interest formula can be used to find the total **amount** accumulated (also called the **future value** of the principal). To work with this formula, use a calculator and follow the process outlined in Example 4. **Note that the steps follow the rules for order of operations.**

Compound Interest Formula

When interest is compounded, the total **amount** A accumulated (including principal and interest) is given by the formula

$$A = P\left(1 + \frac{r}{n}\right)^{nt}$$

where

P = the principal

r = the annual interest rate (in decimal or fraction form)

t = the length of time in years

n = the number of compounding periods in one year

EXAMPLE 4

Mike invests $4000 at 6% interest to be compounded monthly. What will be the amount in his account in 5 years?

Solution

$P = \$4000, \quad r = 6\% = 0.06,$

$n = 12$ times per year, $\quad t = 5$ years

Substituting into the formula gives

$$\begin{aligned} A &= 4000\left(1 + \frac{0.06}{12}\right)^{12\cdot 5} \\ &= 4000(1 + 0.005)^{60} \\ &= 4000(1.005)^{60} \\ &= 4000(1.348850153) \\ &= \$5395.41 \quad \text{total amount accumulated over 5 years} \end{aligned}$$

The expression in the formula can be evaluated with a calculator by following the rules for order of operations.

STEP 1: Enter 0.06. The display should show 0.06.

STEP 2: Press the key [÷].

STEP 3: Enter 12. The display should show 12.

STEP 4: Press the key [=]. The display should show 0.005.

STEP 5: Press the key [+].

STEP 6: Enter 1.

STEP 7: Press the key [=]. The display should show 1.005. This is the base for the exponent.

STEP 8: Press the key [x^y]. (This key is marked [y^x] on some calculators.)

STEP 9: Enter 60. (This is the exponent.)

STEP 10: Press the key [=].

You should now have 1.348850153 on your calculator display. This is the future value of $1. Since $4000 was invested, we need to multiply by 4000 to find the total amount, *A*.

STEP 11: Press the key [×].

STEP 12: Enter 4000. This is the principal.

STEP 13: Press the key [=].

The display should read 5395.40061.

The amount in Mike's account after 5 years will be $5395.41.

The sequence of steps used in Example 4 can be diagrammed as follows:

(Enter) 0.06 (rate of interest) (press) [÷] (enter) 12 (number of compounding periods) (press) [=] (press) [+]

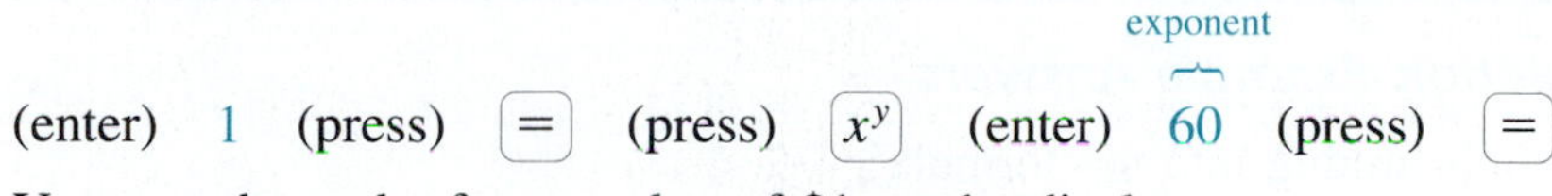

You now have the future value of $1 on the display.

(Press) × (enter) 4000 (press) =

(principal: 4000)

You now have the future value of $4000 on the display.

If we want to find the interest earned on an investment after using the compound interest formula, we subtract the original principal from the final amount:

$$I = A - P$$

EXAMPLE 5

How much interest did Mike earn in the investment described in Example 4?

Solution

$$\begin{aligned} I &= A - P \\ &= 5395.41 - 4000.00 \\ &= 1395.41 \end{aligned}$$

Mike earned $1395.41 in interest.

COMPLETION EXAMPLE 6

(a) Use the compound interest formula to find the value of $10,000 invested for 3 years if it is compounded daily at 12%. (b) Find the amount of interest earned.

Solution

(a) Follow the steps outlined in Example 4 with

$P = \$10{,}000,\quad r = 12\% = 0.12,\quad n = 360,\quad t = 3$

Substituting into the formula gives

$$\begin{aligned} A &= 10{,}000(1 + \_\_\_\_\_\_\_\_)^{360(\_\_\_\_)} \\ &= 10{,}000(1 + \_\_\_\_\_\_\_\_)^{\_\_\_\_} \\ &= 10{,}000(1.\_\_\_\_\_\_\_\_)^{\_\_\_\_} \\ &= 10{,}000(\_\_\_\_\_\_\_\_) \\ &= \_\_\_\_\_\_\_\_ \end{aligned}$$

The value (or Amount) will be $________.

(b) The interest earned will be

$I = \$\_\_\_\_\_\_\_\_ - \$10{,}000.00 = \$\_\_\_\_\_\_\_\_.$

Now Work Exercise 2 in the Margin.

2. Use the compound interest formula (and a calculator) to find the value of $4000 invested for 10 years compounded daily at 5%.

Completion Example Answers

6. (a) Substituting into the formula gives

$$A = 10{,}000\left(1 + \frac{\mathbf{0.12}}{\mathbf{360}}\right)^{360(3)}$$

$$= 10{,}000(1 + \mathbf{0.000333333})^{\mathbf{1080}}$$

$$= 10{,}000(1.\mathbf{000333333})^{\mathbf{1080}}$$

$$= 10{,}000(\mathbf{1.433243431})$$

$$= \mathbf{14{,}332.44}$$

The value (or Amount) will be **$14,332.44.**

(b) I = **$14,332.44** − $10,000.00 = **$4332.44**

NAME ______ SECTION ______ DATE ______

ANSWERS

Exercises 8.2

In Exercises 1–8, use the formula for simple interest, $I = P \times r \times t$, repeatedly, as shown in Examples 1 and 3 in the text. Show the calculations for each period of compounding.

1. (a) If a bank compounds interest quarterly at 4% on a certificate of deposit, what will an investment of $13,000 be worth in 6 months? (b) in one year?

2. If a $9000 deposit in a savings account earns 5% compounded monthly, what will be the balance in the account in 6 months?

3. (a) If an account is compounded quarterly at a rate of 6%, how much interest will be earned on $5000 in one year? (b) What will be the total amount in the account? (c) How much more interest will be earned in the first year because the compounding is done quarterly rather than annually?

4. (a) How much interest will be earned in 2 years on a loan of $4000 compounded semiannually at 6%? (b) How much will be owed at the end of the 2-year period?

5. (a) How much interest will be earned on a savings account of $3000 in two years if interest is compounded annually at 5.5%? (b) if interest is compounded semiannually?

1a. ______

b. ______

2. ______

3a. ______

b. ______

c. ______

4a. ______

b. ______

5a. ______

b. ______

6. ______________

7. ______________

8. ______________

9. ______________

10a. ______________

b. ______________

11. [Respond below exercise.]

6. If interest is calculated at 10% compounded quarterly, what will be the value of $15,000 in 9 months?

7. Calculate the interest earned in six months on $10,000 compounded monthly at 8%.

8. Calculate the interest you will pay on a loan of $6500 for one year if the interest is compounded every 3 months at 18% and you make no monthly payments.

Note that in Exercises 9 and 10, payments are made so that the principal actually decreases.

9. Suppose that you borrow $4000 and agree to make four equal payments of $1000 each plus interest over the next four years. Interest is to be calculated at a rate of 6% based only on what you owe. How much interest will you pay?

10. Linda agreed to loan her son $400 under the following terms: he is to make payments of $100 plus interest every 60 days, and the interest rate is to be 6.6% annually. (a) How long will it take him to repay the loan? (b) How much interest will he pay?

In Exercises 11–20, use the formula for compound interest, $A = P\left(1 + \frac{r}{n}\right)^{nt}$, and the formula for finding interest, $I = A - P$, whenever each applies.

11. (a) Calculate the interest in one year on $5000 compounded monthly at 12%. (b) Suppose the interest is compounded semiannually. Is the accumulated value the same? (c) If not, explain why not in your own words.

NAME ______________ SECTION ______________ DATE ______________

ANSWERS

12. (a) What will be the interest on $10,000 compounded daily at 10% for one year? (b) What is the difference between this and simple interest at 10% for one year?

13. (a) Find the value of $5000 compounded quarterly at 8% for 4 years. (b) What do you think the difference in interest would be if the money were compounded daily: about $5, $20, $50, or over $100? (c) Find the exact difference in interest.

14. (a) What would be the value of a $20,000 savings account at the end of 5 years if interest were calculated at 7% compounded annually? (b) How much more would be earned if the interest were compounded daily?

15. (a) Suppose that $3000 is invested at 5% and compounded monthly for one year. Find the accumulated value. (b) Is the accumulated amount the same if the original principal of $3000 is compounded annually for 12 years? If not, what is the difference?

16. (a) Find the value of $25,000 compounded daily at 5% for 20 years. (b) Do you think that the amount will be doubled or more than doubled if the rate is doubled to 10%? (c) Find the amount if the rate is 10%.

In Exercises 17–20, find the amount A and the interest earned I for the given information.

	Compounding Period	Principal	Annual Rate	Time	A	$I = A - P$
17.	Quarterly	$1000	10%	5 yr	(a) ____	(b) ____
18.	Monthly	$1000	10%	5 yr	(a) ____	(b) ____
19.	Daily	$2000	5%	10 yr	(a) ____	(b) ____
20.	Daily	$7500	8%	20 yr	(a) ____	(b) ____

12a. ____
b. ____

13a. ____
b. ____
c. ____

14a. ____
b. ____

15. [Respond below exercise.]

16. [Respond below exercise.]

17a. ____
b. ____

18a. ____
b. ____

19a. ____
b. ____

20a. ____
b. ____

21. [Respond below exercise.] ______

22. [Respond below exercise.] ______

Writing and Thinking about Mathematics

21. Use your calculator and choose the values of t (in years) to use in the formula for compound interest until you find how many years of daily compounding at 6% are needed for an investment of $6000 to approximately double in value. Write down the values you chose for t, why you chose those particular values, and the corresponding accumulated values of money. Explain why you agree or disagree with the idea that $10,000 would double in a shorter time period.

t	$\left(1 + \frac{.06}{360}\right)^{360t}$	A

22. Use your calculator and choose the values of t (in years) to use in the formula for compound interest until you find how many years of daily compounding at 6% are needed for an investment of $10,000 to approximately triple in value. Write down the values you chose for t, why you chose those particular values, and the corresponding accumulated values of money. Explain why you agree or disagree with the idea that $30,000 would triple in a shorter time period.

t	$\left(1 + \frac{.06}{360}\right)^{360t}$	A

8.3 Buying and Owning a Home

OBJECTIVES

1. *Become aware of and learn how to calculate the expenses involved in buying a home.*
2. *Know how to calculate the percent of your income that is spent on your home.*

Buying a Home

Not everyone can or should own a home. In many cases, renting an apartment or home is wise and perfectly satisfactory. However, for those people who are considering buying a home, this financial investment will probably be the largest one of their lives and they should be informed buyers. This section provides a base for understanding some of the terminology used by realtors and bankers in selling and financing homes and for developing a "feeling" for the amount of cash needed to actually buy a home. Some terms related to home purchasing are explained briefly in the following box.

Expenses in Buying a Home

Purchase price: The selling price (what you have agreed to pay)

Down payment: Cash you pay to the seller (usually 20% to 30% of the purchase price)

Mortgage loan (1st trust deed): Loan to you by bank or savings and loan (difference between purchase price and down payment)

Mortgage fee (or points): Loan fee charged by the lender (usually 1% to 3% of the mortgage loan)

Fire insurance: Insurance against the loss of your home by fire (required by almost all lenders)

Recording fees: Fees for recording you as the legal owner

Property taxes: Taxes that must be prepaid before the lender will give you the loan (usually six months in advance)

Legal fees: Fees charged by a lawyer or escrow company for completing all forms in a legal manner

Example 1

You buy a home for $150,000. Your down payment is 20% of the selling price, and the mortgage fee is 2 points (2% of the new mortgage). You also have to pay $500 for fire insurance, $350 for taxes, $50 for recording fees, and $310 for legal fees. (a) What is the amount of your mortgage? (b) How much cash must you provide to complete the purchase?

Solution

(a) To find the amount of the mortgage, find 80% of the selling price. (Since the down payment is 20%, the mortgage will be $100\% - 20\% = 80\%$ of the selling price.)

$$\begin{array}{rl} \$150{,}000 & \text{selling price} \\ \times \quad\quad 0.80 & \\ \hline \$120{,}000.00 & \text{mortgage (or 1st trust deed)} \end{array}$$

(b) Add the mortgage fee, the down payment, and all the other fees.

$120,000	mortgage	$30,000	down payment
× 0.02		2,400	mortgage fees
$2400.00	mortgage fee	500	fire insurance
		350	property taxes
$150,000	selling price	50	recording fees
× 0.20		+ 310	legal fees
$30,000.00	down payment	$33,610	cash needed to complete purchase

The mortgage will be $120,000 and you will need $33,610 in cash.

Owning a Home

After you have bought your home you must make the payments on the mortgage, pay property taxes, pay for utilities (water, electricity, and gas), and pay for repairs. (Don't be too discouraged. One advantage of this is that many of these expenses are deductible from your income taxes.)

Expenses in Owning a Home

Monthly mortgage payment: Payment to mortgage holder includes both principal and interest

Property taxes: May be paid monthly, semiannually, or annually

Homeowner's insurance: Can be included with your fire insurance; includes insurance against theft and liability

Utilities: Monthly payments for water, electricity, and gas

Maintenance: Repairs, yardwork, painting, and so on

EXAMPLE 2

(a) What were the total expenses for your home the month you paid $1250 on your mortgage loan, $100 in taxes, $50 for fire insurance, $200 for all utilities, and $75 for maintenance? (b) If your salary was $3650, what percentage of your salary was spent on your home?

Solution

(a) Find your total expenses.

$1250	loan payment
100	taxes
50	fire insurance
200	utilities
+ 75	maintenance
$1675	total spent on home

(b) Calculate the percent of your salary spent on your home.

$$\frac{1675}{3650} = 0.4589 = 45.89\%$$

NAME ______________ SECTION ______________ DATE ______________

ANSWERS

Exercises 8.3

1. A home is sold for $162,500. The buyer has to make a down payment of 25% of the selling price, pay a loan fee of 2% of the mortgage, $200 for fire insurance, $50 for recording fees, $580 for taxes, and $570 for legal fees. (a) What is the amount of the down payment? (b) What is the amount of the mortgage? (c) How much cash does the buyer need to complete the purchase?

1a. ______ b. ______ c. ______

2. The purchase price on a home is $98,000 and the buyer makes a down payment of $9,800 (10% of the selling price) and pays a loan fee of $2\frac{1}{2}\%$ of the new mortgage. He also pays $250 for legal fees, $320 for taxes, and $425 for fire insurance. (The seller agrees to pay all recording fees.) (a) What is the amount of the loan fee? (b) How much does the buyer owe in order to complete the purchase?

2a. ______ b. ______

3. You bought a home for $85,000 and made a down payment of $17,000. If you paid a mortgage fee (loan fee) of $1360, (a) what percent of the first trust deed was this fee? (b) You also paid $250 for fire insurance, $35 for recording fees, $170 for taxes, and $195 for legal fees. How much cash did you need to complete the purchase?

3a. ______ b. ______

4. A house is sold for $125,000 and the buyer makes a 30% down payment and is charged a 1% loan fee on the first trust deed. The buyer is also charged $220 for recording fees, $345 for legal fees, $450 for taxes, and $520 for fire insurance. (a) How much is the down payment? (b) How much is the first trust deed? (c) How much is the loan fee? (d) How much cash does the buyer need?

4a. ______ b. ______ c. ______ d. ______

5. A condominium was sold for $96,000. The buyer made a down payment of 25% of the selling price and paid a loan fee of $1\frac{1}{2}\%$ of the amount of the mortgage. She also paid $420 for fire insurance, $35 for recording fees, and $380 for taxes. (The seller paid all legal fees.) (a) What was the amount of the down payment? (b) What was the amount of the mortgage? (c) How much was the loan fee? (d) How much cash did she need to complete the purchase?

5a. ______ b. ______ c. ______ d. ______

6a. ______________

b. ______________

7a. ______________

b. ______________

8a. ______________

b. ______________

9a. ______________

b. ______________

10. ______________

6. In March, Ms. Smith made a mortgage payment of $625 and paid $25 for taxes, $10 for water, $35 for electricity, $40 for gas, and $55 for a plumber's bill. (a) What were her home expenses in March? (b) If her income was $1975, what percent of her income did she spend on her home?

7. During July, the Johnsons made a loan payment of $875 and paid $90 for taxes, $85 for utilities, $70 for fire insurance, and $285 for a painter and other repairs. (a) How much did the Johnsons spend on their home in July? (b) If the combined income of Mr. and Mrs. Johnson was $4620, what percent of their income did they spend on their home?

8. In one month Sam paid $150 for home repairs, $390 for his home loan, $60 for utilities, $20 for fire insurance, and $40 for taxes. (a) What were his home expenses for that month? (b) What percent of his $2000 income did he spend on his home?

9. Your income for one month was $1800. (a) If you made a mortgage payment of $578 and paid $25 in taxes, a water bill of $15, an electric bill of $35, a gas bill of $45, and a fire insurance premium of $40, how much did you spend on your home? (b) What percent of your income was this?

10. Your income was $2800 per month and you figured you could afford to spend 30% of this each month on a home. What mortgage payment could you make if you estimated taxes at $60 per month, utilities at $80 per month, fire insurance at $65 per month, and repairs at 7% of your income per month?

8.4 Buying and Owning a Car

OBJECTIVES

1. *Become aware of and learn how to calculate the expenses involved in buying a car.*
2. *Know how to calculate the percent of your income that is spent on your car.*

Buying a Car

Buying a car is not as expensive or complicated as buying a home. However, more people buy cars than homes and, in many cases, buying a car is the most expensive purchase of a person's life. As with finances in general, paying cash for a car is cheaper than financing the car with a bank or savings and loan. If you are going to finance the purchase of a car, at least be aware of the expenses involved and study all the papers so that you know the total amount that you are paying for the car.

Expenses in Buying a Car

Purchase price: The selling price agreed on by the seller and the buyer

Sales tax: A fixed percent that varies from state to state

License fee: Fixed by the state, often based on the type of car and its value

EXAMPLE 1

You are going to buy a new car for \$18,500. The bank will loan you 70% of all the related expenses, including taxes and fees. If sales tax is figured at 8% and there is a license fee of \$250, how much cash do you need in order to buy the car?

Solution

(a) First find the total of all the related expenses.

\$18,500	selling price	\$18,500	selling price
× 0.08	tax rate	1,480	sales tax
\$ 1,480	sales tax	+ 250	license fee
		\$20,230	total expenses

(b) Find 30% of all the expenses. (Since the bank will loan you 70%, you must provide 100% − 70% = 30% of the total expenses in cash.)

\$20,230	total expenses
× 0.30	
\$6,069.00	cash

Owning a Car

Actually, the bank owns your car until all payments are made on the loan. The car is their security for the loan. However, in addition to the monthly payments, you must pay for insurance, any necessary repairs, and general maintenance costs.

Expenses in Owning a Car

Monthly payments: Payments made if you borrowed money to buy the car

Auto insurance: Covers a variety of situations (liability, collision, towing, theft, and so on)

Operating costs: Basic items such as gasoline, oil, tires, tune-ups

Repairs: Replacing worn or damaged parts

EXAMPLE 2

In one month, Bonnie's car expenses were as follows: loan payment, $350; insurance, $80; gasoline, $100; oil and filter, $22; new headlight, $32. (a) What were her total car expenses for that month? (b) What percent of her car expenses was for the loan payment?

Solution

(a) Find her total expenses.

$350	loan payment
80	insurance
100	gasoline
22	oil and filter
+ 32	headlight
$584	total expenses

(b) Find the percent of the total spent on the loan payment.

$$\frac{350}{584} \approx 0.5993 \approx 60\%$$

Bonnie spent $584 on her car, and the loan payment was about 60% of her expenses.

NAME ______________________ SECTION ________________ DATE __________

ANSWERS

1. ____________

2. ____________

3. ____________

4. ____________

5. ____________

Exercises 8.4

1. To buy a used car for $6800, you must pay a 6% sales tax and a license fee of $120. If the bank will loan you 85% of your expenses, how much cash do you need to buy the car?

2. John wants to buy a new convertible for $25,000. His credit union will loan him 80% of his expenses. What amount of cash does he need to buy the car if the sales taxes are 8.5% and the license fee is $250?

3. How much cash do you need to buy a car for $18,000 if the sales tax is calculated at 6%, the license fee is $200, and the loan company will let you borrow 75% of your expenses?

4. A used car is priced at $4500. Your old car is worth $800 on a trade-in. The sales tax is figured at 7.5% of the selling price, and the license fee is $80. If the savings and loan will lend you $3000, how much cash do you need to buy the car?

5. Your old car is worth $1500 if you trade it in for a new car priced at $11,200. Sales tax is 6% and the license fee is $225. If the bank will loan you 80% of your expenses, how much cash do you need to buy the new car?

6a. ______________

b. ______________

7. ______________

8a. ______________

b. ______________

[Respond below
9. exercise.]

10a. ______________

b. ______________

c. ______________

6. (a) If your car expenses for one month were $325 for the loan payment, $35 for insurance, $120 for gasoline, and $145 for two new tires, what were your total car expenses for the month? (b) If your income was $2000, what percent of your income was used for car expenses?

7. Nancy decided her old car needed painting. If the paint job was priced at $1650, including repairing some dents, and she figured that driving her car cost an average of 24¢ per mile, including gas, oil, and insurance, what were her car expenses that month if she drove 1200 miles?

8. Art owns his car, but it needs a new transmission for $1200 (installed). (a) What were his car expenses the month that he had the new transmission installed if he also spent $65 for insurance, $75 for gas, $15 for oil and filter, and $300 for a tune-up? (b) If he took $1000 from his savings account to help pay for the transmission, what percent of his income of $2600 was used for the remainder of his car expenses?

9. Danielle decided she would like to have a new car, but she could not afford one if the car expenses averaged more than 20% of her monthly income. If she figured the expenses would average $300 for a loan payment, $70 for insurance, $65 for gas, $8 for oil, $10 for tire wear, and $40 for general repairs, could she afford the car with a monthly income of $2100?

10. Suppose that you owned a car, your monthly income was $1800, and you figured you could spend 15% of this to operate a car. (a) What would you be able to spend on gas if insurance cost $85, oil and filter cost $20, and you estimated $60 per month for other expenses? (b) How many gallons of gas could you buy if gas cost $1.25 per gallon? (c) How many miles could you drive if your car averaged 19 miles per gallon?

8 Index of Key Ideas and Terms

NAME ______________ SECTION ______________ DATE ______________

ANSWERS

1. ______________

2. ______________

3. ______________

4. ______________

5a. ______________

b. ______________

6. ______________

Chapter 8 Test

1. How much simple interest will be earned in one year on a savings account of $1500 if the bank pays 4.5% interest?

2. You made $42 in interest on an investment at 7% for 9 months. What principal did you invest?

3. If Ms. King had a savings account of $3000 earning 5.5% simple interest, how long would it take her to earn $82.50 in interest?

4. If an investment of $7500 earns simple interest of $131.25 in 45 days, what is the rate of interest?

5. A certificate of deposit is earning 6% compounded quarterly. (a) If $1000 is deposited in the account, what will be the balance in the account at the end of one year? (b) at the end of five years?

6. How much more would be in the balance of the account in five years in Exercise 5 if the money were compounded daily?

7. ______________

8. ______________

9a. ______________

b. ______________

10. ______________

11a. ______________

b. ______________

12. ______________

7. Fred is buying a car for $17,500. Sales tax is 6% and his license and transfer fees total $513. If he receives a $4000 trade-in allowance and a $1200 factory rebate, how much cash will he need to buy the car?

8. Referring to Exercise 7, if Fred finances 70% of his expenses before the trade-in and the rebate, how much cash will he need?

9. The purchase price of a home is $120,000. The buyer makes a down payment of 20% of the purchase price and pays a loan fee of 1 point. (a) What is the amount of the down payment? (b) What is the amount of the loan fee?

10. Referring to Exercise 9, if the buyer pays $50 for recording fees, $450 for property taxes, and $350 for escrow fees, how much cash does the buyer need to complete the transaction?

11. In one month, Janice paid $200 for repairs, $665 for her home loan, $80 for utilities, $60 for fire insurance, and $85 for taxes. (a) What were her home expenses for that month? (b) What percent of her $3000 monthly income did she spend on her home?

12. A home was sold for $350,000 in Palm Springs. The buyer made a down payment of 25% of the selling price and paid a loan fee of 1.5% of the amount of the mortgage. He also paid $820 for fire insurance, $1200 for property taxes, and $550 for other fees. How much cash did he need to complete the purchase?

NAME ______ SECTION ______ DATE ______

ANSWERS

1. ______

2. ______

3a. ______

b. ______

4. ______

5. ______

6. ______

7. ______

8. ______

9. ______

10. ______

Cumulative Review

1. Write the following whole number in standard notation:
two hundred sixty-one thousand, seventy-five

2. Write the following decimal number in standard decimal notation:
three hundred and four thousandths

3. Round off as indicated:
(a) 16.938 (to the nearest hundredth) (b) 539.673 (to the nearest hundred)

4. Find the decimal equivalent to $\frac{14}{35}$.

5. Find the decimal equivalent to $\frac{21}{40}$.

6. Write $\frac{9}{5}$ as a percent.

7. Write $1\frac{1}{2}\%$ in decimal form.

Perform the indicated operations for Exercises 8–19. Reduce all fractions to lowest terms.

8. $\frac{2}{15} + \frac{11}{15} + \frac{7}{15}$

9. $4 - \frac{3}{11}$

10. $\frac{2}{5} \cdot \frac{1}{3} \cdot \frac{4}{7}$

11. ______________

12. ______________

13. ______________

14. ______________

15. ______________

16. ______________

17. ______________

18. ______________

19. ______________

20. ______________

21. ______________

22. ______________

23. ______________

11. $2\frac{4}{15} + 3\frac{1}{6} + 4\frac{7}{10}$

12. $70\frac{1}{4} - 23\frac{5}{6}$

13. $4\frac{5}{7} \cdot 2\frac{6}{11}$

14. $6 \div 3\frac{1}{3}$

15. $(700)(8000)$

16. $403 - 4.012$

17. $71 + 0.354 + 4.39$

18. $(0.27)(0.043)$

19. $27.404 \div 0.34$

20. Evaluate: $50 + 2(36 \div 3^2 \cdot 2) + 12 \div 4 - 2^2$

21. Use the tests for divisibility to determine if 732 can be divided exactly by 2, 3, 4, 5, 9, or 10.

22. Find the prime factorization of 396.

23. Find the LCM of the numbers 14, 21, and 30.

NAME ______________________ SECTION ______________ DATE __________

ANSWERS

24. Division by _____ is undefined.

25. In the proportion $\frac{7}{8} = \frac{21}{24}$, the extremes are _____ and _____, and the means are _____ and _____.

26. 15% of ________ is 7.5.

27. 9.25% of 200 is ________.

28. 65 is _____% of 26.

29. Solve the following proportion for x: $\frac{1\frac{2}{3}}{x} = \frac{10}{2\frac{1}{4}}$

30. The sum of two numbers is 521. If one of the numbers is 196, what is the other number?

24. ____________

25. ____________

26. ____________

27. ____________

28. ____________

29. ____________

30. ____________

31. ____________

32. ____________

33a. ____________

b. ____________

34. ____________

35a. ____________

b. ____________

31. An investment pays 6.25% simple interest. What is the interest on $4800 invested for 6 months?

32. What principal is invested at 5.5% if, after 100 days, the interest earned is $69.30?

33. Five thousand dollars is invested in a savings account. (a) What is in the balance of the account after 9 years if interest is earned at 8% compounded quarterly? (b) if interest is earned at 8% compounded daily?

34. Find the length (to the nearest tenth of a centimeter) of the hypotenuse of a right triangle with legs of length 10 centimeters and 20 centimeters.

35. Write each of the following numbers in scientific notation.

(a) 73,800,000

(b) 0.00045

9

Introduction to Algebra

What to Expect in Chapter 9

Chapter 9 provides an introduction to several topics from algebra. The concept of negative numbers is introduced, along with the rules for operating with a new type of number (called an **integer**). In Section 9.1, number lines are used to provide a "picture" of integers and their relationships. Integers are graphed (plotted) on number lines, and number lines are used to give an intuitive understanding of the very important idea of the magnitude (**absolute value**) of an integer. This idea is the foundation for the rules of addition and subtraction with integers developed in Sections 9.2 and 9.3.

Multiplication and division with integers are discussed in Section 9.4, and the fact that division by 0 is undefined is reinforced—this time in terms of integers. Section 9.5 discusses methods for simplifying and evaluating algebraic expressions by using the distributive property in combining like terms. Section 9.6 provides a discussion of translating English phrases into algebraic expressions and presents basic techniques for solving equations with integers as solutions.

The chapter closes with a section on problem solving by using the translating skills developed in Section 9.6 and then setting up and solving the related equations.

Mathematics at Work!

A car rental counter is a common sight at most airport terminals.

Suppose that you are a traveling salesperson and your company has allowed you \$50 per day for car rental. You rent a car for \$30 per day plus 5¢ per mile. How many miles can you drive for the \$50 allowed?

Plan: Set up an equation relating the amount of money spent and the money allowed in your budget and solve the equation.

Solution: Let x = number of miles driven.

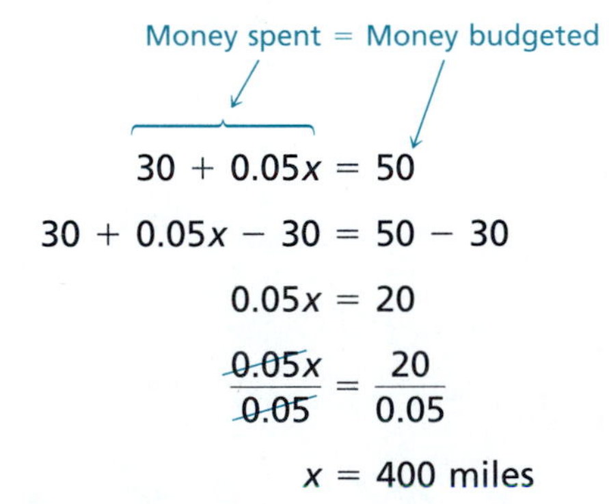

$$30 + 0.05x = 50$$
$$30 + 0.05x - 30 = 50 - 30$$
$$0.05x = 20$$
$$\frac{0.05x}{0.05} = \frac{20}{0.05}$$
$$x = 400 \text{ miles}$$

You would like to get the most miles for your money. The car rental agency down the street charges only \$25 per day and 8¢ per mile. How many miles would you get for your \$50 with this price structure? Is this a better deal for you? (See Exercise 43, Section 9.7.)

Integers (Number Lines and Absolute Value)

OBJECTIVES

1. *Understand what the integers are.*
2. *Know how to find the opposite of an integer.*
3. *Be able to graph sets of integers on number lines.*
4. *Know how to use the inequality symbols* <, >, ≤, *and* ≥.
5. *Learn the definition of absolute value.*

Number Lines

The concepts of positive and negative numbers occur frequently in our daily lives:

	Negative	**Zero**	**Positive**
Temperatures are recorded as:	below zero	zero	above zero
Businesses report:	losses	no gain	profits
Golf scores are recorded as:	below par	par	above par

In this chapter, we will develop techniques for understanding and operating with positive and negative numbers. We begin with the graphs of numbers on horizontal lines called **number lines.** For example, choose some point on a horizontal line and label it with the number 0 (Figure 9.1).

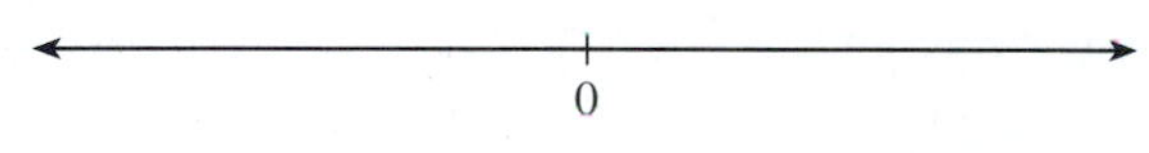

Figure 9.1

Now choose another point on the line to the right of 0 and label it with the number 1 (Figure 9.2).

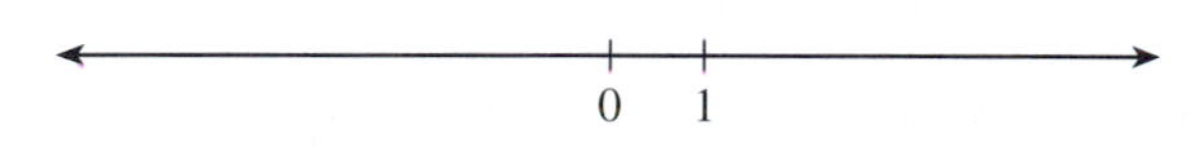

Figure 9.2

Points corresponding to all remaining whole numbers are now determined and are all to the right of 0. That is, the point for 2 is the same distance from 1 as 1 is from 0, and so on (Figure 9.3).

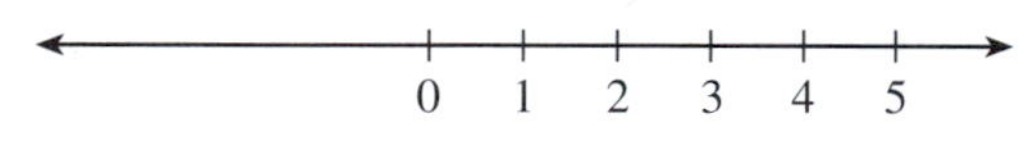

Figure 9.3

The **graph** of a number is the point on a number line that corresponds to that number, and the number is called the **coordinate** of the point. The terms **number** and **point** are used interchangeably. Thus, we might refer to the **point** 6. The graphs of numbers are indicated by marking the corresponding points with large dots (Figure 9.4).

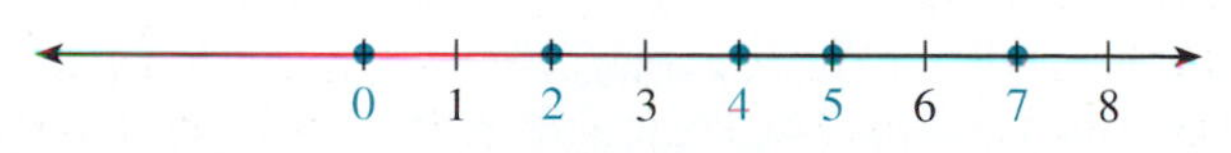

The graph of the set of numbers $A = \{0, 2, 4, 5, 7\}$

Figure 9.4

The point one unit to the left of 0 is the **opposite of 1.** It is also called **negative 1** and is symbolized as **-1.** Similarly, the **opposite of 2** is called **negative 2** and is symbolized as **-2,** the opposite of 3 is -3, and so on (Figure 9.5).

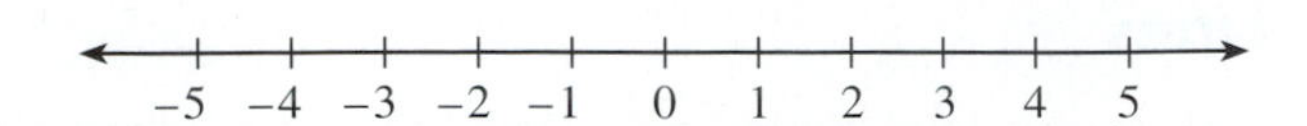

Figure 9.5

Definition

The set of **integers** is the set of whole numbers and their opposites:

Integers: $\ldots, -4, -3, -2, -1, 0, 1, 2, 3, 4, \ldots$

The counting numbers (all whole numbers except 0) are called **positive integers** and may be written as

$+1, +2, +3, +4,$ and so on.

The **opposites** of the counting numbers are called **negative integers** and are written as

$-1, -2, -3, -4,$ and so on.

The number 0 is neither positive nor negative (Figure 9.6).

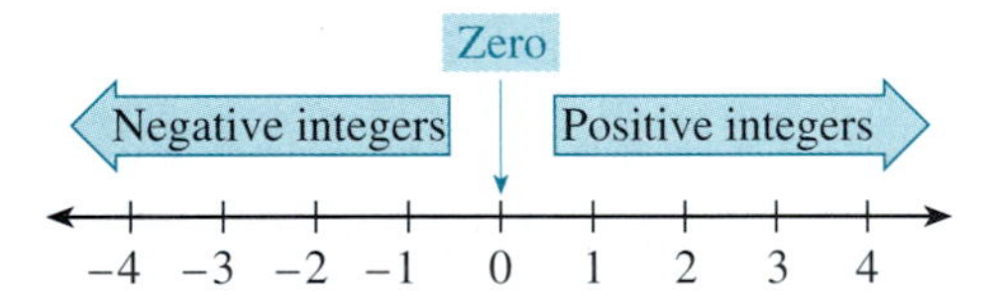

Figure 9.6

Note the following facts about integers:

1. The opposite of a positive integer is a negative integer. For example,

   $-(+2) = -2$ and $-(+7) = -7$. (opposite)

2. The opposite of a negative integer is a positive integer. For example,

   $-(-3) = +3$ and $-(-4) = +4$. (opposite)

3. The opposite of 0 is 0. (That is, $-0 = 0$.)

Note: Integers are not the only type of number that can be graphed on number lines. We will see later in this chapter that fractions, mixed numbers, decimals, square roots, and the opposites of all these types of numbers can also be graphed on number lines. So, you should be aware that the set of integers does not include all positive numbers or all negative numbers.

Example 1

Find the opposite of each of the following numbers.

(a) -6 (b) -20 (c) $+8$

Solution

(a) $-(-6) = 6$ (b) $-(-20) = 20$ (c) $-(+8) = -8$

Example 2

Graph the set of integers $B = \{-4, -2, 0, 1, 2\}$.

Solution

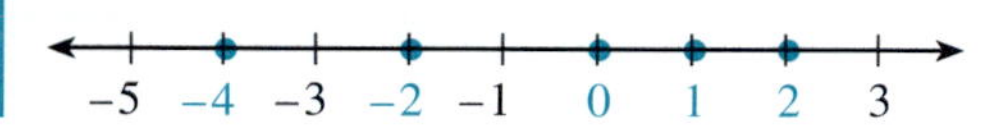

Example 3

Graph the set of positive odd integers $C = \{1, 3, 5, 7, 9, \ldots\}$.

Solution

The three dots above the number line indicate that the pattern in the graph continues without end. The set of positive odd integers is an **infinite** set.

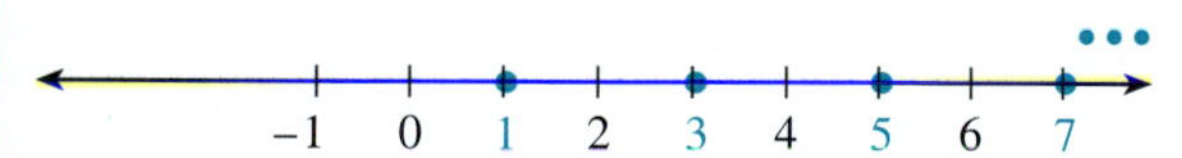

Now Work Exercises 1–3 in the Margin.

1. Find the opposite of each number.

   (a) -5 (b) $+12$

2. Graph the set of integers:

   $\{-3, 0, 1, 2\}$

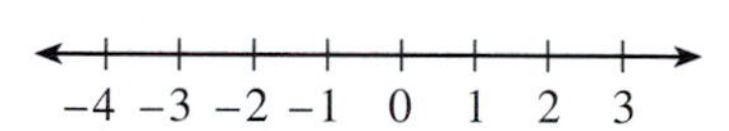

3. Graph the set of odd integers:

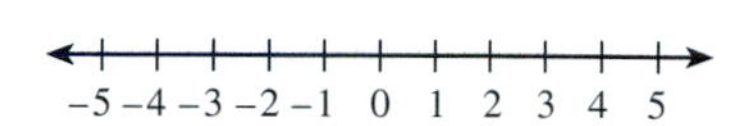

Inequality Symbols

On a horizontal number line, **smaller numbers are always to the left of larger numbers.** Each number is smaller than any number to its right and larger than any number to its left. We use the following **inequality symbols** to indicate the order of numbers on the number line and their relative sizes.

Symbols for Order

$<$ is read "is less than"
$>$ is read "is greater than"
$\leq$ is read "is less than or equal to"
$\geq$ is read "is greater than or equal to"

The following relationships can be observed on the number line in Figure 9.7.

Using <		or	Using >	
$2 < 5$	2 is less than 5		$5 > 2$	5 is greater than 2
$-3 < 0$	-3 is less than 0		$0 > -3$	0 is greater than -3
$-4 < -2$	-4 is less than -2		$-2 > -4$	-2 is greater than -4

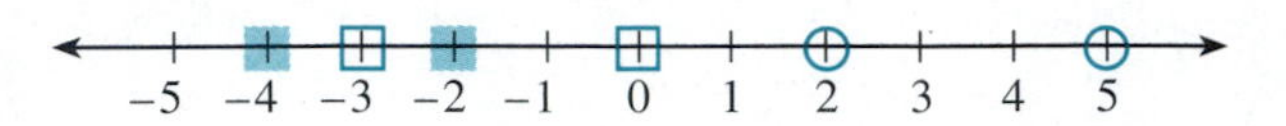

Figure 9.7

One useful idea implied by the previous discussion is that expressions containing the symbols < and > can be read either from right to left or from left to right. For example, we might read

$1 < 8$ as "1 is less than 8" (reading from left to right)

or

$1 < 8$ as "8 is greater than 1." (reading from right to left)

Also note that for the symbol $\geq$ (or the symbol $\leq$) we can write an expression such as

$5 \geq -10$ since 5 is greater than -10

and

$5 \geq 5$ since 5 is equal to 5.

The symbol $\geq$ is read **greater than or equal to,** and a true relationship is represented if either

the first number is **greater** than the second

or

the first number is **equal** to the second.

EXAMPLE 4

Determine whether the following statements are true or false. Rewrite any false statement so that it is true.

(a) $3 \leq 11$ (b) $6 > -1$ (c) $-7 > 0$ (d) $0 \geq 0$

(e) $-6 \leq -6$

Solution

(a) $3 \leq 11$ is true since 3 is less than 11.

(b) $6 > -1$ is true since 6 is greater than -1.

(c) $-7 > 0$ is false.
We can change the inequality to read $-7 < 0$ or $0 > -7$.

(d) $0 \geq 0$ is true since $0 = 0$.

(e) $-6 \leq -6$ is true since $-6 = -6$.

Now Work Exercises 4–6 in the Margin.

Absolute Value

Another concept related to **signed numbers** (positive numbers, negative numbers, and zero) is that of **absolute value,** symbolized by two vertical bars, $| \;|$. (**Note:** The definition given here for the absolute value of an integer is valid for any type of number on a number line.)

Absolute value is used to indicate the distance a number is from 0. For example, we know that -4 and $+4$ are both 4 units from 0. Thus, we can write $|-4| = |+4| = 4$. (See Figure 9.8.)

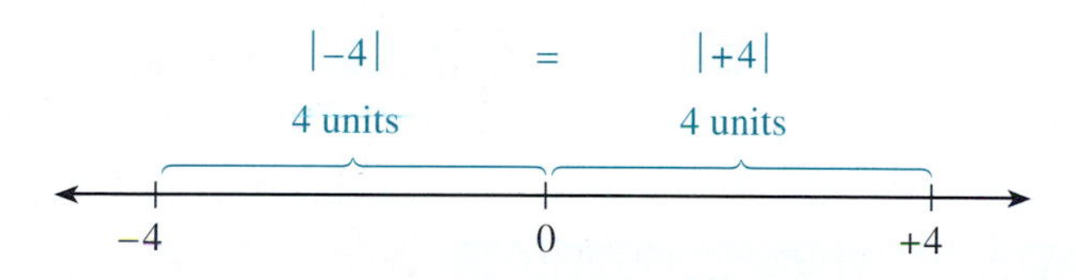

Figure 9.8

Definition

The **absolute value** of an integer is its distance from 0.
The absolute value of an integer is never negative.
We can express the definition symbolically, for any integer a, as follows:

If a is a positive integer or 0, $|a| = a$.

If a is a negative integer, $|a| = -a$.

Note: When $\boldsymbol{a}$ represents a negative number, the symbol $\boldsymbol{-a}$ represents a positive number. **That is, the opposite of a negative number is a positive number.** For example,

$$\text{If } a = -8, \text{ then } -a = -(-8) = +8.$$

And, for $a = -8$, we can write

$$|a| = -a$$

$$|-8| = -(-8) = +8$$

The absolute value of -8 is the opposite of -8.

EXAMPLE 5

$|-3| = |+3| = 3$

3 units 3 units

-3 0 $+3$

Determine whether each of the following statements is true or false. Rewrite any false statement so that it is true.

4. $-5 < 0$

5. $-4 \geq 4$

6. $-10 \leq -10$

7. Find the value of each absolute value.
(a) $|-15|$ (b) $|9|$ (c) $|-4|$

Determine whether each of the following statements is true or false. Rewrite any false statement so that it is true.

8. $|-5| \geq |+5|$

9. $|-38| > |-39|$

EXAMPLE 6

$|0| = 0$

No units

0

EXAMPLE 7

True or false: $|-12| \geq 12$.

Solution

True, since $|-12| = 12$ and $12 \geq 12$. (Remember that the symbol $\geq$ is read **greater than or equal to** so that "equal to" is valid with this symbol.)

Now Work Exercises 7–9 in the Margin.

NAME ______________ SECTION ______________ DATE ______________

Exercises 9.1

Graph each of the following sets of numbers on a number line.

1. $\{0, 1, 2\}$

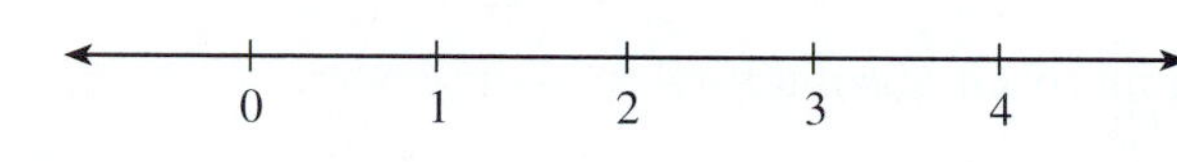

2. $\{0, 2, 4\}$

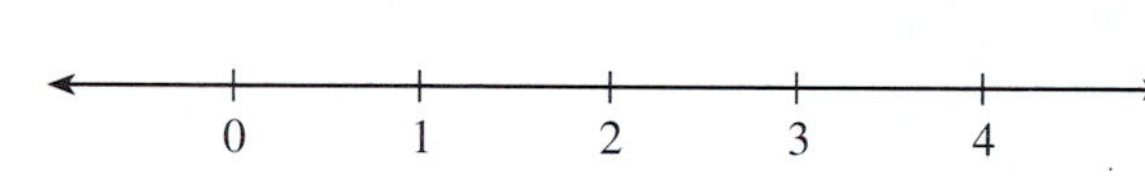

3. $\{-3, -1, 1\}$

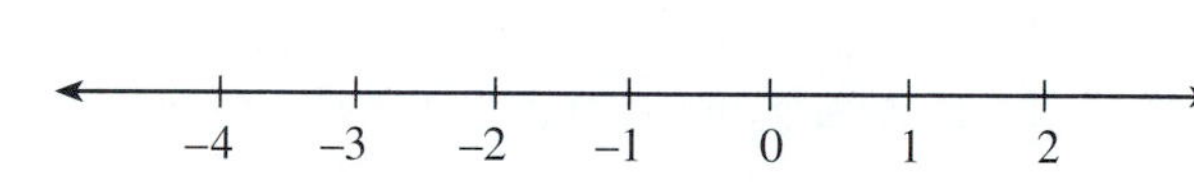

4. $\{-3, -2, 0\}$

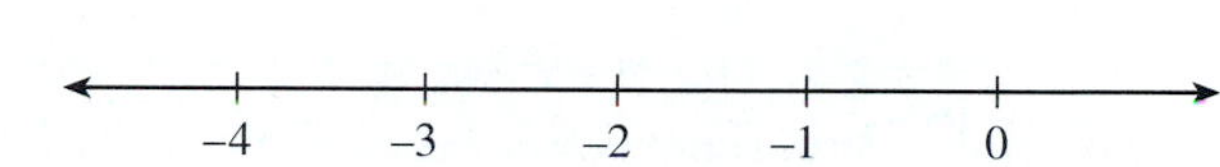

5. $\{-10, -9, -8, -7\}$

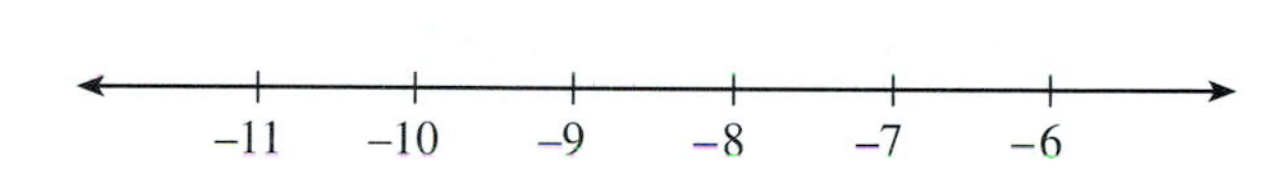

6. $\{-5, -4, -2, -1\}$

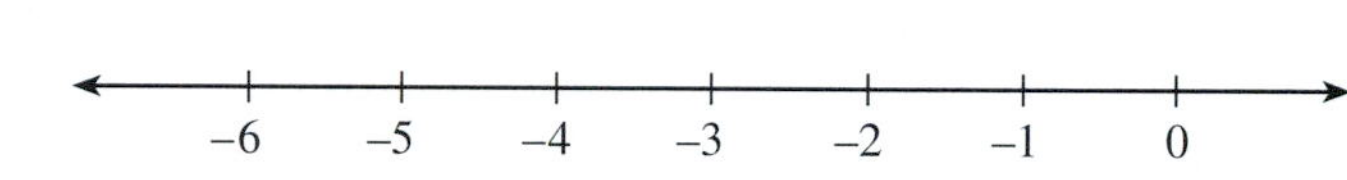

7. $\{-5, 0, 5\}$

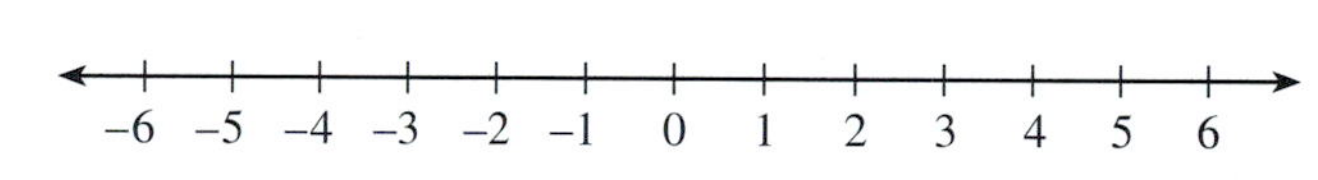

8. $\{-3, -1, 0, 1, 3\}$

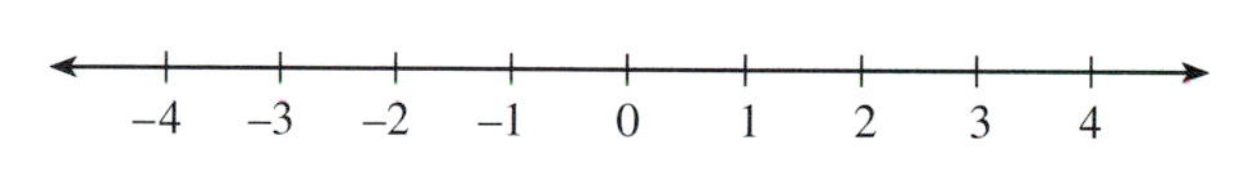

9. $\{1, 2, 3, \ldots, 10\}$

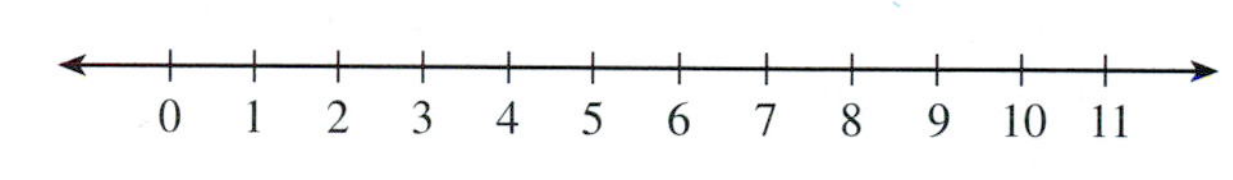

10. $\{-2, -1, 0, \ldots, 5\}$

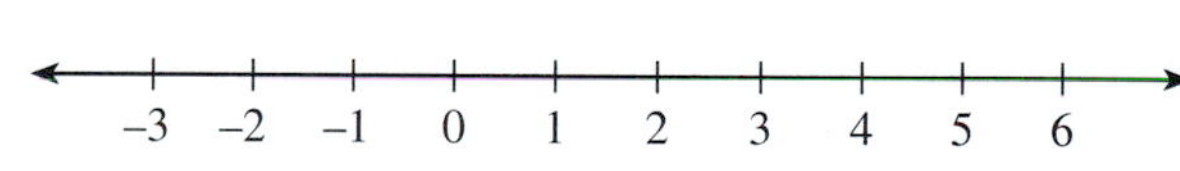

11. $\{0, 2, 4, 6, 8, \ldots\}$

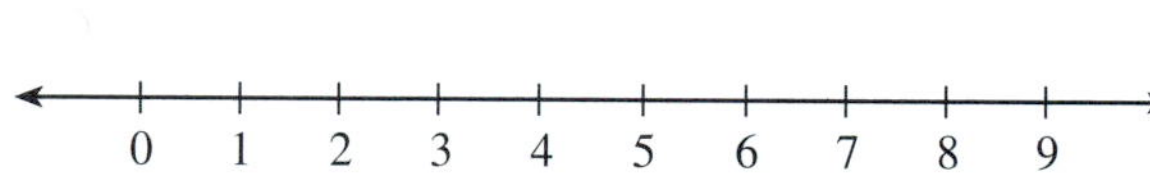

12. $\{\ldots, -7, -5, -3, -1\}$

–8 –7 –6 –5 –4 –3 –2 –1 0

13. All whole numbers less than 4

14. All integers less than 4

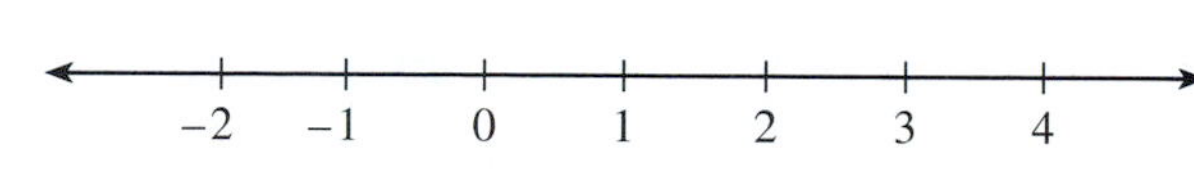

15. All integers less than 0

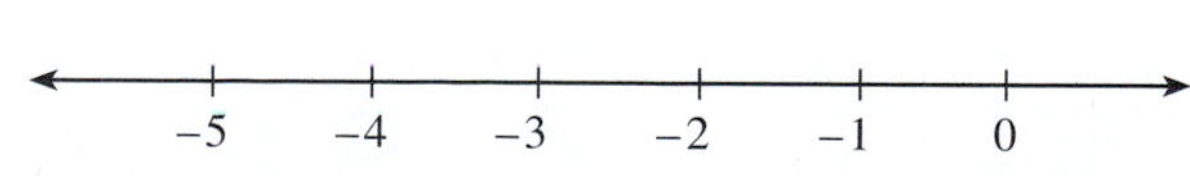

16. All negative integers greater than -5

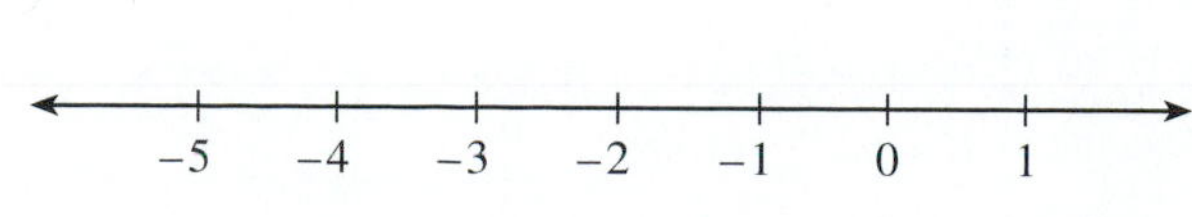

ANSWERS

1. [Respond below exercise.]

2. [Respond below exercise.]

3. [Respond below exercise.]

4. [Respond below exercise.]

5. [Respond below exercise.]

6. [Respond below exercise.]

7. [Respond below exercise.]

8. [Respond below exercise.]

9. [Respond below exercise.]

10. [Respond below exercise.]

11. [Respond below exercise.]

12. [Respond below exercise.]

13. [Respond below exercise.]

14. [Respond below exercise.]

15. [Respond below exercise.]

16. [Respond below exercise.]

17. ______
18. ______
19. ______
20. ______
21. ______
22. ______
23. ______
24. ______
25. ______
26. ______
27. ______
28. ______
29. ______
30. ______
31. ______
32. ______
33. ______
34. ______
35. ______
36. ______
37. ______
38. ______
39. ______
40. ______
41. ______
42. ______
43. ______
44. ______
45. ______
46. ______
47. ______
48. ______
49. ______
50. ______

Find the opposite of each number.

17. -10 **18.** -9 **19.** 14

20. -6 **21.** 0 **22.** 40

Find each absolute value as indicated.

23. $|-6|$ **24.** $|-10|$ **25.** $|+24|$ **26.** $|+16|$

27. $|-20|$ **28.** $|-50|$ **29.** $|0|$ **30.** $|27|$

Identify the appropriate symbol that will make each statement true: $<$, $>$, or $=$.

31. 5 ______ 0 **32.** 7 ______ -2 **33.** -57 ______ -50

34. -30 ______ -29 **35.** $|19|$ ______ 19 **36.** $|-3|$ ______ 3

37. $|-21|$ ______ -21 **38.** -63 ______ $|-63|$

Determine whether each statement is true or false. If the statement is false, change the inequality or equality symbol so that the statement is true. (There may be more than one change that will make a false statement correct.)

39. $0 = -0$ **40.** $-10 < -11$ **41.** $21 > |-21|$

42. $|-5| < 5$ **43.** $-6 = |-6|$ **44.** $19 = |-19|$

45. $-4 \geq 4$ **46.** $-12 \geq 12$ **47.** $|-3| \leq 3$

48. $-45 < |-45|$ **49.** $0 < |0|$ **50.** $|-4| > 0$

9.2 Addition with Integers

OBJECTIVES

1. *Develop an intuitive understanding of adding integers along a number line.*
2. *Learn how to add with integers.*

An Intuitive Approach

We have learned how to add, subtract, multiply, and divide with whole numbers, fractions, mixed numbers, and decimal numbers. In the next three sections, we will discuss the basic rules and techniques for these same four operations with integers.

As an intuitive approach to addition with integers, consider an open field with a straight line marked with integers (much like a football field is marked every 10 yards). Imagine that a ball player stands at the point marked 0 and throws the ball to the point marked +5 and then stands at +5 and throws the ball 3 more units in the positive direction (to the right). Where will the ball land? (See Figure 9.9.)

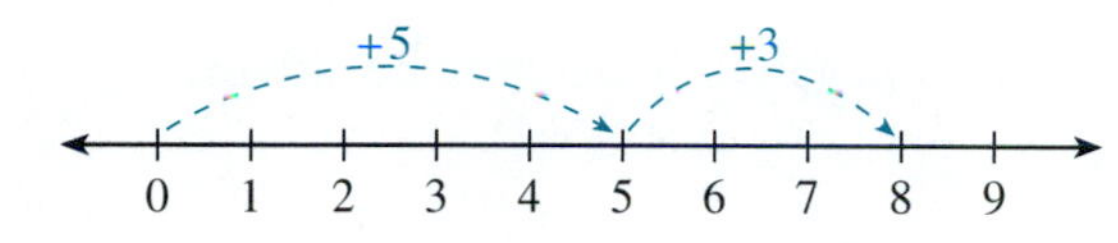

Figure 9.9

The ball will land at the point marked +8. We have essentially added the two positive integers +5 and +3.

$$(+5) + (+3) = +8 \qquad \text{or} \qquad (5) + (3) = 8$$

Now, if the same player stands at 0 and throws the ball the same two distances in the opposite direction (to the left), where will the ball land on the second throw? The ball will land at −8. We have just added two negative integers, −5 and −3. (See Figure 9.10.)

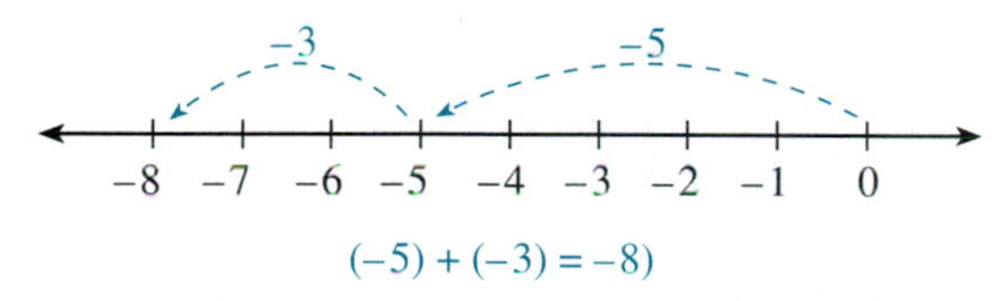

Figure 9.10

Sums involving both positive and negative integers are illustrated in Figures 9.11 and 9.12.

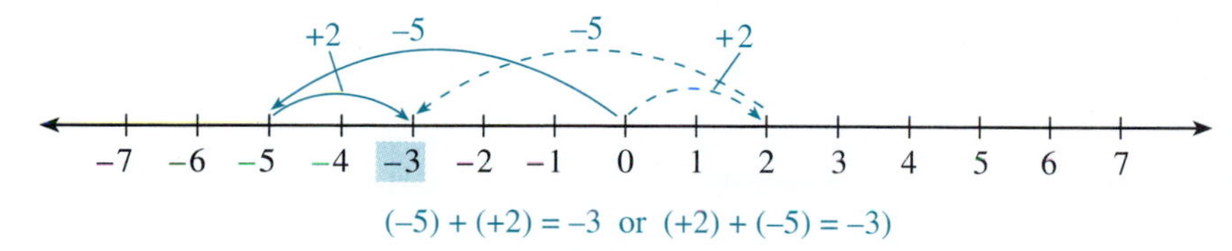

Figure 9.11

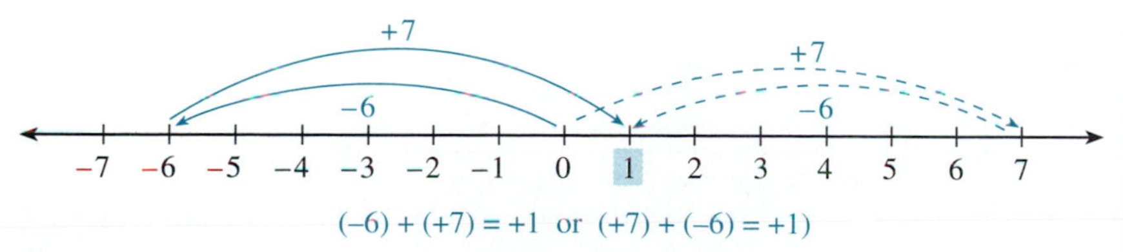

Figure 9.12

Although we are concerned here mainly with developing the techniques for adding integers, we note that addition with integers is **commutative,** as illustrated in Figures 9.11 and 9.12, and **associative.**

To help you understand the rules for addition, we make the following suggestions for reading expressions with + and − signs:

+ used as the sign of a number is read **positive.**

+ used as an operation is read **plus.**

− used as the sign of a number is read **negative.**

− used as an operation is read **minus.** (See Section 9.3.)

In summary:

1. The sum of two positive integers is positive:

$(+6) + (+5) = +11$

positive plus positive positive

2. The sum of two negative integers is negative:

$(-4) + (-3) = -7$

negative plus negative negative

3. The sum of a positive and a negative integer may be negative or positive (or zero), depending on which number is farther from 0.

$(+5) + (-7) = -2$

positive plus negative negative

$(+9) + (-3) = +6$

positive plus negative positive

EXAMPLE 1

Find each of the following sums.

(a) $(-5) + (+4)$ (b) $(-3) + (-8)$

(c) $(+6) + (-10)$ (d) $(-7) + (+7)$

Solution

(a) 4 units right

−5 −4 −3 −2 −1 0

$(-5) + (+4) = -1$

(b)

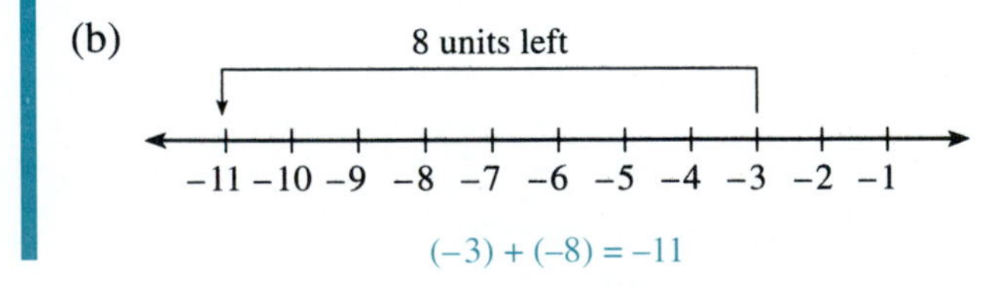

$(-3) + (-8) = -11$

(c)

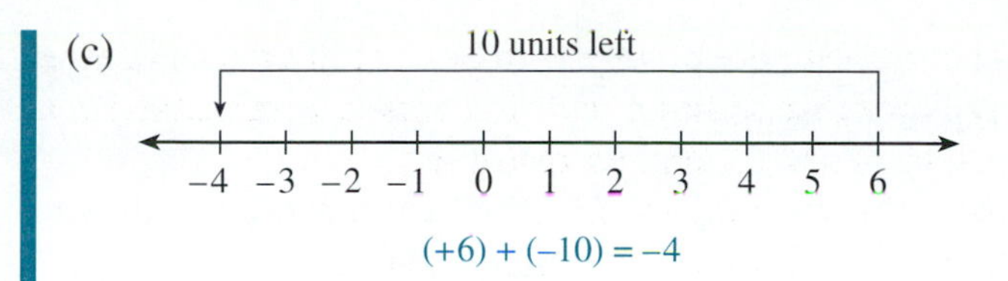

$(+6) + (-10) = -4$

(d)

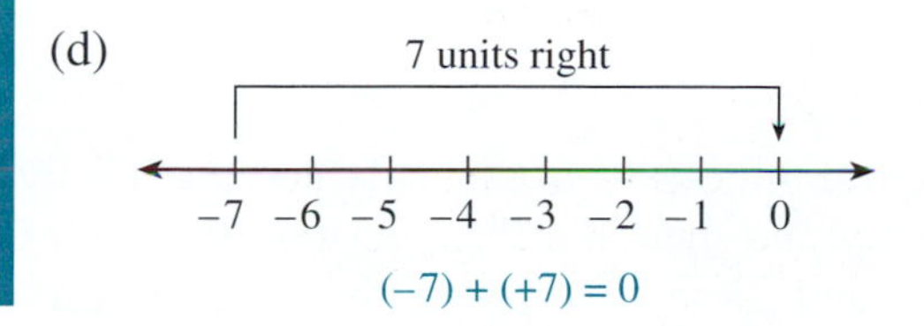

$(-7) + (+7) = 0$

Now Work Exercises 1–4 in the Margin.

Rules for Addition with Integers

Saying that one number is farther from 0 than another number is the same as saying that the first number has a larger absolute value. With this basic understanding, the rules for adding integers can be written out formally in terms of absolute value as follows.

Rules for Addition with Integers

1. To add two integers with like signs, add their absolute values and use the common sign.

common sign

$$(+7) + (+2) = +(|+7| + |+2|) = +(7 + 2) = +9$$

common sign

$$(-7) + (-2) = -(|-7| + |-2|) = -(7 + 2) = -9$$

2. To add two integers with unlike signs, subtract their absolute values (the smaller from the larger) and use the sign of the integer with the larger absolute value.

$$(-16) + (+10) = -(|-16)| - |+10|) = -(16 - 10) = -6$$

$$(+16) + (-10) = +(|+16| - |-10|) = +(16 - 10) = +6$$

$$(-30) + (+30) = (|-30| - |+30|) = (30 - 30) = 0$$

Equations in algebra are almost always written horizontally, so addition and subtraction with integers are done in the horizontal format much of the time. However, there are situations (such as in long division) where sums and differences are written in a vertical format with one number directly under another, as illustrated in Example 2.

Find each of the following sums.

1. $(-6) + (-10) =$ ______

2. $13 + (-3) =$ ______

3. $(-21) + (+8) =$ ______

4. $(-30) + (-5) =$ ______

Find each of the following sums.

5. $\begin{array}{r} -10 \\ +3 \\ \hline \end{array}$

6. $\begin{array}{r} -5 \\ +9 \\ \hline \end{array}$

7. $\begin{array}{r} -2 \\ -3 \\ -15 \\ \hline \end{array}$

8. $\begin{array}{r} -20 \\ +6 \\ -1 \\ \hline \end{array}$

EXAMPLE 2

Find each of the following sums.

(a) $\begin{array}{r} -30 \\ 8 \\ -12 \\ \hline \end{array}$ (b) $\begin{array}{r} 52 \\ -20 \\ -15 \\ \hline \end{array}$ (c) $\begin{array}{r} -10 \\ -11 \\ -9 \\ \hline \end{array}$

Solution

One technique for adding several integers is to mentally add the positive and negative integers separately and then add these results. (We are, in effect, using the commutative and associative properties of addition.)

(a) $\begin{array}{r} -30 \\ 8 \\ -12 \\ \hline -34 \end{array}$ or $\begin{array}{r} -42 \\ 8 \\ \hline -34 \end{array}$ (b) $\begin{array}{r} 52 \\ -20 \\ -15 \\ \hline 17 \end{array}$ or $\begin{array}{r} 52 \\ -35 \\ \hline 17 \end{array}$ (c) $\begin{array}{r} -10 \\ -11 \\ -9 \\ \hline -30 \end{array}$

Now Work Exercises 5–8 in the Margin.

Note: The positive sign (+) may be omitted when writing positive numbers, but the negative sign (−) must always be written for negative numbers. Thus, **if there is no sign in front of an integer, the integer is understood to be positive.**

NAME ______________ SECTION ______________ DATE ______________

ANSWERS

Exercises 9.2

Find each of the indicated sums.

1. $(+6) + (-4)$

2. $(+8) + (-7)$

3. $(+4) + (+6)$

4. $(+5) + (-8)$

5. $(+16) + (+3)$

6. $(-8) + (-2)$

7. $(-3) + (-6)$

8. $(-2) + (+2)$

9. $(+4) + (-4)$

10. $(+13) + (+12)$

11. $(+6) + (-10)$

12. $(+14) + (-17)$

13. $(+5) + (-3)$

14. $(+15) + (-18)$

15. $(-4) + (-12)$

16. $(-8) + (+8)$

17. $(+2) + (-6)$

18. $(-9) + (+5)$

19. $(-16) + (+3) + (+13)$

20. $(-5) + (+5) + (14)$

1. ______________

2. ______________

3. ______________

4. ______________

5. ______________

6. ______________

7. ______________

8. ______________

9. ______________

10. ______________

11. ______________

12. ______________

13. ______________

14. ______________

15. ______________

16. ______________

17. ______________

18. ______________

19. ______________

20. ______________

21. ______
22. ______
23. ______
24. ______
25. ______
26. ______
27. ______
28. ______
29. ______
30. ______
31. ______
32. ______
33. ______
34. ______
35. ______
36. ______
37. ______
38. ______
39. ______
40. ______

21. $(-1) + (-2) + (+7)$

22. $(+3) + (-4) + (-5)$

23. $(+6) + (+3) + (+5)$

24. $(-18) + (-5) + (-7)$

25. $(-1) + (+2) + (-4) + (+2)$

26. $\begin{array}{r} -4 \\ \underline{+8} \end{array}$

27. $\begin{array}{r} -5 \\ \underline{-10} \end{array}$

28. $\begin{array}{r} -13 \\ \underline{-6} \end{array}$

29. $\begin{array}{r} +16 \\ \underline{+25} \end{array}$

30. $\begin{array}{r} +14 \\ \underline{-8} \end{array}$

31. $\begin{array}{r} +20 \\ \underline{-7} \end{array}$

32. $\begin{array}{r} +2 \\ -5 \\ \underline{-3} \end{array}$

33. $\begin{array}{r} +8 \\ +3 \\ \underline{-1} \end{array}$

34. $\begin{array}{r} +10 \\ -4 \\ \underline{+2} \end{array}$

35. $\begin{array}{r} -16 \\ -8 \\ \underline{+12} \end{array}$

36. $\begin{array}{r} -15 \\ -20 \\ \underline{-6} \end{array}$

37. $\begin{array}{r} -4 \\ -17 \\ \underline{+11} \end{array}$

38. $\begin{array}{r} +13 \\ -5 \\ +17 \\ \underline{-25} \end{array}$

39. $\begin{array}{r} +14 \\ -14 \\ +37 \\ \underline{-37} \end{array}$

40. $\begin{array}{r} -8 \\ -5 \\ -13 \\ \underline{-22} \end{array}$

NAME ________________ SECTION ________________ DATE ________

ANSWERS

Choose the response that correctly completes each statement.

41. If x and y are integers, then $x + y$ is (never, sometimes, always) equal to 0.

42. If x and y are integers, then $x + y$ is (never, sometimes, always) positive.

43. If x and y are integers, then $x + y$ is (never, sometimes, always) negative.

44. If x is a positive integer and y is a negative integer, then $x + y$ is (never, sometimes, always) equal to 0.

45. If x and y are both negative integers, then $x + y$ is (never, sometimes, always) equal to 0.

Use a calculator to find the value of each expression. (**Note:** On a calculator, the key marked [+/-] is used to change the sign of a number from positive to negative or from negative to positive. The two keys marked [+] and [−] are used for the operations of addition and subtraction, respectively.)

46. $8305 + (-4783) + (-5400)$

47. $-4785 + (-7300) + (-2540)$

48. $-11{,}970 + (-4375) + (25{,}000) + (15{,}000)$

49. $39{,}500 + (-34{,}100) + (-53{,}700) + (60{,}000)$

50. $-35{,}632 + (-10{,}658) + (17{,}344) + (-3479)$

41. ________

42. ________

43. ________

44. ________

45. ________

46. ________

47. ________

48. ________

49. ________

50. ________

51. [Respond below exercise.]

52. [Respond below exercise.]

53. [Respond below exercise.]

Writing and Thinking about Mathematics

51. Give three illustrations of the associative property of addition with integers.

52. Describe in your own words the conditions under which the sum of two integers will be 0.

53. Explain in your own words how the expression $-x$ can possibly represent a positive integer.

9.3 Subtraction with Integers

OBJECTIVES

1. *Understand the concept of an additive inverse.*
2. *Learn how to subtract with integers.*

Additive Inverse

Subtraction with whole numbers is defined in terms of addition. For example, the difference $10 - 7$ is equal to 3 because $7 + 3 = 10$. If we are restricted to whole numbers, then the difference $10 - 15$ cannot be found because there is no whole number that we can add to 15 to get 10. However, now that we are familiar with integers, we can reason that

$$15 + (-5) = 10 \quad \text{and therefore} \quad 10 - 15 = -5$$

In this manner, we can now define subtraction in terms of addition with integers so that a larger number may be subtracted from a smaller number.

To define subtraction with integers, we need to emphasize and clarify the relationship between any integer and its opposite.

Definition

The **opposite** of an integer is called its **additive inverse.** The sum of an integer and its additive inverse is 0. Symbolically, for any integer a,

$$a + (-a) = 0$$

Note: The symbol $-a$ should be read as **the opposite of *a*.** Since a is a variable, $-a$ might be positive, negative, or 0. For example,

if $a = 13$, then $-a = -13$ and $-a$ is **negative.**

However,

if $a = -4$, then $-a = -(-4) = +4$ and $-a$ is **positive.**

↖ ↗ opposite of ↑ negative 4

Example 1

Find the additive inverse (opposite) of each integer.

(a) 7 (b) -3 (c) -22

Solution

(a) The additive inverse of 7 is -7, and

$7 + (-7) = 0$

(b) The additive inverse of -3 is $+3$, and

$(-3) + (+3) = 0$

(c) The additive inverse of -22 is $+22$, and

$(-22) + (+22) = 0$

Find each of the following differences.

1. $+2 - (-3) =$ ______

2. $-5 - (+2) =$ ______

3. $(-8) - (-4) =$ ______

4. $(-4) - (-5) =$ ______

Subtraction with Integers

Just as subtraction with whole numbers is defined in terms of addition, subtraction with integers is also defined in terms of addition. The distinction is that now we can subtract larger numbers from smaller numbers, and we can subtract negative numbers and positive numbers from each other. In effect, we can now study differences that result in negative numbers. For example,

$$4 - 9 = -5 \quad \text{because} \quad 9 + (-5) = 4$$
$$10 - 13 = -3 \quad \text{because} \quad 13 + (-3) = 10$$
$$8 - (-6) = 14 \quad \text{because} \quad (-6) + 14 = 8$$
$$-7 - (-5) = -2 \quad \text{because} \quad (-5) + (-2) = -7$$

Definition

For any integers a and b, we define the difference between a and b as follows:

$$a - b = a + (-b)$$

This equation is read

"a minus b is equal to a plus the opposite of b."

Thus, as we have discussed and as the following examples illustrate, subtraction is accomplished by adding the opposite of the subtrahend.

EXAMPLE 2

Find the following differences.

(a) $(+2) - (-6)$ (b) $(-3) - (-7)$

(c) $(-5) - (-2)$ (d) $(-9) - (-9)$

Solution

(a) $(+2) - (-6) = (+2) + (+6) = +8$

↖ minus changed to plus ↗ ↑ opposite of -6

(b) $(-3) - (-7) = (-3) + (+7) = +4$

↖ minus changed to plus ↗ ↑ opposite of -7

(c) $(-5) - (-2) = (-5) + (+2) = -3$

↖ minus changed to plus ↗ ↑ opposite of -2

(d) $(-9) - (-9) = (-9) + (+9) = 0$

Now Work Exercises 1–4 in the Margin.

The interrelationship between addition and subtraction with integers allows us to eliminate the use of so many parentheses. This simpler notation is illustrated as follows:

$$9 - 12 = 9 - (+12) = 9 + (-12) = -3$$

9 minus positive 12 — 9 plus negative 12

Thus, the expression $9 - 12$ can be thought of as subtraction or addition. The result is the same in either case. Understanding this notation takes some practice, but it is quite important since it is commonly used in all mathematics textbooks. Study the following examples carefully.

EXAMPLE 3

(a) $7 - 13 = -6$

(b) $10 - 8 = 2$

(c) $-4 - 8 = -12$

(d) $6 - 9 = -3$

(e) $-20 + 15 - 10 = -5 - 10 = -15$

The following examples show that subtraction is **not commutative** and **not associative.**

$12 - 5 = 7$ and $5 - 12 = -7$

So, $12 - 5 \neq 5 - 12$ and, in general,

$$\boldsymbol{a - b \neq b - a}$$

Similarly,

$9 - (4 - 1) = 9 - (3) = 6$ and $(9 - 4) - 1 = 5 - 1 = 4$

So, $9 - (4 - 1) \neq (9 - 4) - 1$ and, in general,

$$\boldsymbol{a - (b - c) \neq (a - b) - c}$$

As with addition, integers can be written vertically in subtraction. One number is written underneath the other, and the sign of the integer being subtracted (the minuend) is changed and the resulting numbers are added. That is, we add the opposite of the integer that is subtracted.

EXAMPLE 4

To Subtract:	Add the opposite of the bottom number.	
−35	−35	
−22	+22	sign is changed
	−13	difference

EXAMPLE 5

To Subtract:	Add the opposite of the bottom number.	
−16	−16	
+14	−14	sign is changed
	−30	difference

Example 6

To Subtract: Add the opposite of the bottom number.

$$\begin{array}{r} -20 \\ \underline{-87} \end{array} \qquad \begin{array}{rl} -20 & \\ \underline{+87} & \text{sign is changed} \\ +67 & \text{difference} \end{array}$$

Now Work Exercises 5–8 in the Margin.

Find each of the following differences.

5. Subtract: $\begin{array}{r} -19 \\ \underline{-13} \end{array}$

6. Subtract: $\begin{array}{r} -2 \\ \underline{-7} \end{array}$

Evaluate each of the following expressions.

7. $-12 - 8 =$ ______

8. $-2 + 1 - 6 - 5 =$ ______

NAME ______________________ SECTION ______________________ DATE ____________

ANSWERS

1. ____________
2. ____________
3. ____________
4. ____________
5. ____________
6. ____________
7. ____________
8. ____________
9. ____________
10. ____________
11. ____________
12. ____________
13. ____________
14. ____________
15. ____________
16. ____________
17. ____________
18. ____________
19. ____________
20. ____________
21. ____________
22. ____________
23. ____________
24. ____________
25. ____________
26. ____________
27. ____________
28. ____________
29. ____________
30. ____________

Exercises 9.3

Find the additive inverse of each number.

1. 16 **2.** 26 **3.** -45

4. -75 **5.** -8 **6.** 37

Perform the indicated operations.

7. $(+5) - (+2)$ **8.** $(+16) - (+3)$ **9.** $(+8) - (-3)$

10. $(+12) - (-4)$ **11.** $(-5) - (+2)$ **12.** $(-10) - (+3)$

13. $(-10) - (-1)$ **14.** $(-15) - (-1)$ **15.** $(-3) - (-7)$

16. $(-2) - (-12)$ **17.** $(-4) - (+6)$ **18.** $(-9) - (+13)$

19. $(-13) - (-14)$ **20.** $(-12) - (-15)$ **21.** $(+9) - (-9)$

22. $(+11) - (-11)$ **23.** $(+15) - (-2)$ **24.** $(+20) - (-3)$

25. $(-7) - (-7)$ **26.** $(-5) - (-5)$ **27.** $(+8) - (+8)$

28. $(+5) - (+5)$ **29.** $(-16) - (+10)$ **30.** $(-17) - (+14)$

31. ____________
32. ____________
33. ____________
34. ____________
35. ____________
36. ____________
37. ____________
38. ____________
39. ____________
40. ____________
41. ____________
42. ____________
43. ____________
44. ____________
45. ____________
46. ____________
47. ____________
48. ____________
49. ____________
50. ____________
51. ____________
52. ____________
53. ____________
54. ____________
55. ____________
56. ____________
57. ____________
58. ____________
59. ____________
60. ____________
61. ____________

Subtract the bottom number from the top number.

31. $\begin{array}{r} 18 \\ \underline{-12} \end{array}$ **32.** $\begin{array}{r} 24 \\ \underline{16} \end{array}$ **33.** $\begin{array}{r} -8 \\ \underline{-12} \end{array}$ **34.** $\begin{array}{r} -13 \\ \underline{-18} \end{array}$

35. $\begin{array}{r} -4 \\ \underline{+5} \end{array}$ **36.** $\begin{array}{r} 32 \\ \underline{-48} \end{array}$ **37.** $\begin{array}{r} -6 \\ \underline{-30} \end{array}$ **38.** $\begin{array}{r} -25 \\ \underline{-13} \end{array}$

39. $\begin{array}{r} -45 \\ \underline{-16} \end{array}$ **40.** $\begin{array}{r} 28 \\ \underline{-15} \end{array}$

Evaluate each of the following expressions.

41. $6 + 2$ **42.** $4 + 8$ **43.** $7 - 1$

44. $9 - 4$ **45.** $4 + 6$ **46.** $8 + 9$

47. $-3 - 1$ **48.** $-2 - 6$ **49.** $12 - 6$

50. $-13 + 4$ **51.** $-20 + 14$ **52.** $-10 + 9$

53. $24 - 32$ **54.** $14 - 17$ **55.** $-12 - 6$

56. $-2 - 8$ **57.** $-6 + 6$ **58.** $-7 + 7$

59. $-5 + 12 - 3$ **60.** $-20 - 2 + 6$ **61.** $-4 - 10 - 7$

NAME ______________ SECTION ______________ DATE ______________

ANSWERS

62. $13 - 4 + 6 - 5$

63. $-4 + 10 - 12 + 1$

64. $19 - 5 - 8 - 6$

65. $-8 + 14 - 10 + 4$

Identify the correct symbol that will make each statement true: $<$, $>$, or $=$.

66. $-8 + (-2)$ _____ $-5 - 6$

67. $-7 + (-4)$ _____ $-3 - 3$

68. $-10 - 10$ _____ $0 - 20$

69. $-9 - 9$ _____ $0 - 18$

70. $5 - (-1)$ _____ $5 - 1$

71. Beginning with a temperature of 10° above zero, the temperature was measured hourly for five hours. It rose 4°, dropped 3°, dropped 2°, rose 1°, and dropped 5°. What was the final temperature recorded?

72. In a five-day week, the stock market showed a gain of 28 points, a gain of 12 points, a loss of 19 points, a loss of 3 points, and a loss of 16 points. What was the net change in the stock market for the week?

73. In seven running plays in a football game, the quarterback threw passes that gained 12 yards, lost 2 yards, lost 5 yards, gained 3 yards, gained 35 yards, lost 4 yards, and gained 15 yards. What was his net passing yardage for these plays?

62. ______________

63. ______________

64. ______________

65. ______________

66. ______________

67. ______________

68. ______________

69. ______________

70. ______________

71. ______________

72. ______________

73. ______________

74. ______________

75. ______________

76. ______________

77. ______________

78. ______________

79. [Respond below exercise.]

80. [Respond below exercise.]

81. [Respond below exercise.]

74. Suppose that par for a certain golf course is 72. In a four-day tournament, Nancy scored 3 under par, even par, 2 under par, and 1 over par. What was her total number of strokes for the tournament?

Use a calculator to find the value of each of the following expressions.

75. $15{,}855 - 18{,}273$

76. $-302{,}500 - 257{,}600 + 207{,}300$

77. $400{,}000 - 1{,}780{,}350 + 542{,}000$

78. $354{,}750 - 425{,}792 - 356{,}425 + 321{,}273$

Writing and Thinking about Mathematics

79. Give two examples that illustrate why subtraction is not commutative.

80. Give two examples that illustrate why subtraction is not associative.

81. Explain, in your own words, how the difference of two negative integers might be a positive integer.

9.4 Multiplication and Division with Integers

OBJECTIVES

1. *Learn how to multiply with integers.*
2. *Learn how to divide with integers.*

Multiplication with Integers

Multiplication with whole numbers is a shorthand form of repeated addition. Multiplication with integers can be thought of in the same manner. For example,

$$4 + 4 + 4 + 4 + 4 + 4 = 6 \cdot 4 = 24$$

and

$$(-4) + (-4) + (-4) + (-4) + (-4) + (-4) = 6(-4) = -24$$

and

$$(-2) + (-2) + (-2) = 3(-2) = -6$$

Repeated addition with a negative integer results in a product of a positive integer and a negative integer. Therefore, because the sum of negative integers is negative, we can reason that **the product of a positive integer and a negative integer is negative.**

EXAMPLE 1

(a) $5(-5) = -25$

(b) $3(-20) = -60$

(c) $8(-11) = -88$

(d) $-1(7) = -7$

The product of two negative integers can be explained in terms of opposites and the fact that, for any integer a,

$$\mathbf{-a = 1(a)}$$

That is, the opposite of a can be treated as -1 times a.

Now consider the product $(-3)(-5)$ and the following procedure:

$$\underset{\uparrow}{(-3)}(-5) = \underbrace{-1(3)}_{-1 \cdot 3}(-5) = -1 \cdot [(3)(-5)]$$

opposite → of 3

$$= \underbrace{-1 \cdot [-15]}_{-1 \cdot [-15]} = \underset{\uparrow}{-}\,[-15] = +15$$

$-1 \cdot [-15]$ → opposite of -15

Although one example does not prove a rule, this process is such that we can use it as a general procedure and come to the following correct conclusion: **The product of two negative integers is positive.**

EXAMPLE 2

(a) $(-7)(-2) = +14$

(b) $-8(-5) = +40$

(c) $-9(-11) = +99$

(d) $0(-12) = 0$ and
$0(+12) = 0$ and
$0(0) = 0$

As Example 2(d) illustrates, the product of 0 and any integer is 0. The rules for multiplication with integers can be stated as follows.

Find the following products.

1. $6(-3) =$ ______

2. $(-8)(-2) =$ ______

3. $(-14)(0) =$ ______

4. $(-2)(-2)(-2) =$ ______

5. $(-3)(-3)(0) =$ ______

6. $(-6)(-1)(4)(-3) =$ ______

Rules for Multiplication with Integers

If a and b are positive integers, then

1. The product of two positive integers is positive:

   $a \cdot b = ab$

2. The product of two negative integers is positive:

   $(-a)(-b) = ab$

3. The product of a positive integer and a negative integer is negative:

   $a(-b) = -ab$

4. The product of 0 and any integer is 0:

   $a \cdot 0 = 0$ and $(-a) \cdot 0 = 0$

The **commutative** and **associative** properties of multiplication are valid for multiplication with integers, just as they are with whole numbers. Thus, to find the product of more than two integers, we can multiply any two, then continue to multiply until all the integers have been multiplied.

EXAMPLE 3

(a) $(-4)(-2)(+5) = [(-4)(-2)](+5) = [+8](+5) = +40$

(b) $(-2)(-2)(-3)(-6) = +4(-3)(-6) = -12(-6) = +72$

(c) $(-1)(5)(-2)(-1)(7)(3) = -5(-2)(-1)(7)(3) = +10(-1)(7)(3)$
$= -10(7)(3) = -70(3) = -210$

Now Work Exercises 1–6 in the Margin.

Division with Integers

Remember that division can be indicated in the form of a fraction with the numerator to be divided by the denominator. That is,

$$a \div b = \frac{a}{b}$$

Thus, we can write $63 \div 7$ as $\frac{63}{7}$. Now, since the operation of division is defined in terms of multiplication, we know that

$$\frac{63}{7} = 9 \quad \text{because} \quad 63 = 7 \cdot 9$$

This relationship between division and multiplication is true for integers as well as whole numbers.

Definition

For integers a, b, and x (where $b \neq 0$),

$$\frac{a}{b} = x \quad \text{means that} \quad a = b \cdot x$$

Special Note: In this text, we are emphasizing the rules for signs when operating (adding, subtracting, multiplying, and dividing) with integers. With this emphasis in mind, the problems are set up in such a way that the results are integers. However, you should be aware of the fact that the rules for operating with integers are valid for operating with any type of signed number, including positive and negative fractions, mixed numbers, and decimal numbers. Operations with these types of signed numbers are discussed in later courses.

EXAMPLE 4

(a) $\frac{+28}{+4} = +7$ because $+28 = (+4)(+7)$.

(b) $\frac{-28}{-4} = +7$ because $-28 = (-4)(+7)$.

(c) $\frac{+28}{-4} = -7$ because $+28 = (-4)(-7)$.

(d) $\frac{-28}{+4} = -7$ because $-28 = (+4)(-7)$.

The rules for division with integers can be stated as follows.

Rules for Division with Integers

If a and b are positive integers, then

1. The quotient of two positive integers is positive:

$$\frac{a}{b} = +\frac{a}{b}$$

2. The quotient of two negative integers is positive:

$$\frac{-a}{-b} = +\frac{a}{b}$$

3. The quotient of a positive integer and a negative integer is negative:

$$\frac{-a}{b} = -\frac{a}{b} \quad \text{and} \quad \frac{a}{-b} = -\frac{a}{b}$$

Find the following quotients.

7. $\frac{-20}{10} =$ ______

8. $\frac{-40}{-2} =$ ______

9. $\frac{0}{-8} =$ ______

10. $\frac{36}{-2} =$ ______

11. $\frac{-42}{0} =$ ______

12. $\frac{-50}{-25} =$ ______

The rules for multiplication and division with nonzero integers can be summarized in the following two statements:

1. If two nonzero integers have the same sign, then both their product and their quotient will be positive.
2. If two nonzero integers have unlike signs, then both their product and their quotient will be negative.

Now Work Exercises 7–12 in the Margin.

In Section 1.5, we stated that division by 0 is undefined. For completeness and understanding, we explain this fact again here.

Division by 0 is not Defined

Case 1: Suppose that $a \neq 0$ and $\frac{a}{0} = x$. Then, by the meaning of division, $a = 0 \cdot x$. But this is not possible since $0 \cdot x = 0$ for any value of x and we stated that $a \neq 0$.

Case 2: Suppose that $\frac{0}{0} = x$. Then $0 = 0 \cdot x$, which is true for all values of x. But this is not allowed since we must have a unique answer for division.

Therefore, we conclude that, in every case, **division by 0 is not defined.**

EXAMPLE 5

(a) $\frac{8}{0}$ is undefined. If $\frac{8}{0} = x$, then $8 = 0 \cdot x$, which is not possible.

(b) $\frac{-17}{0}$ is undefined. If $\frac{-17}{0} = x$, then $-17 = 0 \cdot x$, which is not possible.

NAME ______________ SECTION ______________ DATE ______________

ANSWERS

1. ______
2. ______
3. ______
4. ______
5. ______
6. ______
7. ______
8. ______
9. ______
10. ______
11. ______
12. ______
13. ______
14. ______
15. ______
16. ______
17. ______
18. ______
19. ______
20. ______
21. ______
22. ______
23. ______
24. ______
25. ______
26. ______
27. ______
28. ______
29. ______
30. ______

Exercises 9.4

Find the following products.

1. $5(-3)$ **2.** $4(-6)$ **3.** $-6(-4)$

4. $-2(-7)$ **5.** $-5(4)$ **6.** $-8(3)$

7. $14(2)$ **8.** $13(3)$ **9.** $-10(5)$

10. $-11(3)$ **11.** $(-7)3$ **12.** $(-2)9$

13. $6(-8)$ **14.** $9(-4)$ **15.** $-7(-9)$

16. $-8(-9)$ **17.** $0(-6)$ **18.** $0(-4)$

19. $(-6)(-5)(3)$ **20.** $(-2)(-1)(7)$ **21.** $4(-2)(-3)$

22. $5(-6)(-1)$ **23.** $(-5)(3)(-4)$ **24.** $(-3)(7)(-5)$

25. $(-7)(-2)(-3)$ **26.** $(-4)(-4)(-4)$ **27.** $(-3)(-3)(-5)$

28. $(-2)(-2)(-8)$ **29.** $(-3)(-4)(-5)$ **30.** $(-2)(-5)(-7)$

31. ____________
32. ____________
33. ____________
34. ____________
35. ____________
36. ____________
37. ____________
38. ____________
39. ____________
40. ____________
41. ____________
42. ____________
43. ____________
44. ____________
45. ____________
46. ____________
47. ____________
48. ____________
49. ____________
50. ____________
51. ____________
52. ____________
53. ____________
54. ____________
55. ____________
56. ____________

31. $(-5)(0)(-6)$

32. $(-6)(0)(-2)$

33. $(-1)(-1)(-1)$

34. $(-3)(-3)(-3)$

35. $(-2)(-2)(-2)(-2)$

36. $(-2)(-4)(-4)$

37. $(-1)(-4)(-7)(+3)$

38. $(-5)(-2)(-1)(+5)$

39. $(-2)(-3)(-10)(-5)$

40. $(-11)(-2)(-4)(-1)$

Find the following quotients.

41. $\frac{-12}{4}$

42. $\frac{-18}{2}$

43. $\frac{-14}{7}$

44. $\frac{-28}{7}$

45. $\frac{-20}{-5}$

46. $\frac{-30}{-3}$

47. $\frac{-50}{-10}$

48. $\frac{-30}{-5}$

49. $\frac{30}{-6}$

50. $\frac{40}{-8}$

51. $\frac{75}{-25}$

52. $\frac{80}{-4}$

53. $\frac{12}{6}$

54. $\frac{24}{8}$

55. $\frac{36}{9}$

56. $\frac{22}{11}$

NAME ____________ SECTION ____________ DATE ____________

ANSWERS

57. $\frac{-39}{13}$ **58.** $\frac{27}{-9}$ **59.** $\frac{32}{-4}$ **60.** $\frac{23}{-23}$

61. $\frac{-34}{-17}$ **62.** $\frac{-60}{-15}$ **63.** $\frac{-8}{-8}$ **64.** $\frac{26}{-13}$

65. $\frac{-31}{0}$ **66.** $\frac{17}{0}$ **67.** $\frac{0}{-20}$ **68.** $\frac{0}{-16}$

69. $\frac{35}{0}$ **70.** $\frac{0}{25}$

71. If the product of -12 and -3 is added to the product of -25 and 2, what is the sum?

72. If the quotient of -35 and 7 is subtracted from the product of -11 and -5, what is the difference?

73. What number should be added to -15 to get a sum of -29?

74. What number should be added to -40 to get a sum of -10?

75. What is the difference between the product of -16 and 2 and the quotient of -45 and 5?

57. ____________

58. ____________

59. ____________

60. ____________

61. ____________

62. ____________

63. ____________

64. ____________

65. ____________

66. ____________

67. ____________

68. ____________

69. ____________

70. ____________

71. ____________

72. ____________

73. ____________

74. ____________

75. ____________

76. [Respond below exercise.]

77. [Respond below exercise.]

78. [Respond below exercise.]

Writing and Thinking about Mathematics

76. If you multiply an odd number of negative numbers together, do you think that the product will be positive or negative? Explain your reasoning.

77. If you multiply an even number of negative numbers together, do you think that the product will be positive or negative? Explain your reasoning.

78. In your own words, explain why division by 0 is not possible in arithmetic.

9.5 Combining Like Terms and Evaluating Algebraic Expressions

Combining Like Terms

A single number is called a **constant,** and a symbol (or letter) used to indicate more than one number is called a **variable.** Any constant or variable or the indicated product of constants and powers of variables is called a **term.** Examples of terms are

$$18, \quad \frac{3}{4}, \quad 5xy, \quad -4x^2, \quad -10x, \quad \text{and} \quad \frac{5}{9}x^3y$$

As we discussed in Section 6.2, a number written next to a variable (as in $-10x$) or a variable written next to another variable (as in xy) indicates multiplication. In the term $-4x^2$, the constant -4 is called the **numerical coefficient** of x^2 (or simply the **coefficient** of x^2). If a term has only one variable, then the exponent of the variable is called the **degree** of the term. Constants are said to be of zero degree.

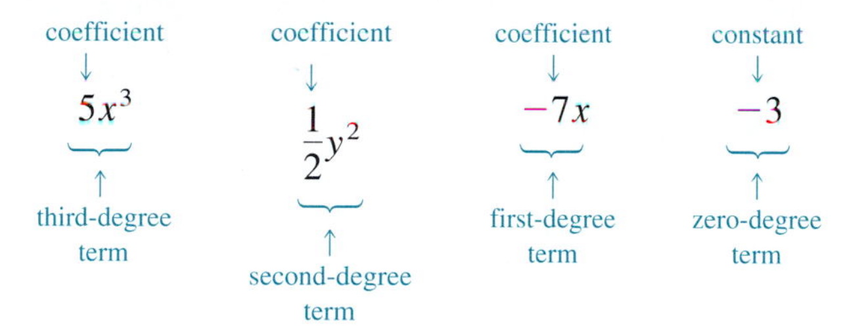

Like terms (or **similar terms**) are terms that contain the same variables (if any) raised to the same powers. Whatever power a variable is raised to in one term, it is raised to the same power in other like terms. Constants are like terms.

Like Terms

-7, 19, and $\frac{7}{8}$	are like terms because each term is a constant.
$-3a$, $14a$, and $5a$	are like terms because each term contains the same variable, a, raised to the same power, 1.
$4xy^2$ and $8xy^2$	are like terms because each term contains the same two variables, x and y, with x first-degree in both terms and y second-degree in both terms.

Unlike Terms

$9x$ and $3x^2$	are unlike terms (**not** like terms) because x is not of the same power in both terms.
$15ab^2$ and $-4a^2b$	are unlike terms since the variables are **not** of the same power in both terms.

OBJECTIVES

1. Know the meanings of the terms **constant, variable, term,** and **coefficient.**
2. Recognize like terms and be able to combine like terms.
3. Understand the distributive property of multiplication over addition.
4. Be able to evaluate algebraic expressions for given values of the variables by using the rules for order of operations.

If no number is written next to a variable, then the coefficient is understood to be 1. If a $(-)$ sign is written next to a variable, then the coefficient is understood to be -1. For example,

$$x = 1 \cdot x, \quad a^3 = 1 \cdot a^3, \quad \text{and} \quad y = 1 \cdot y$$
$$-x = -1 \cdot x, \quad -a^3 = -1 \cdot a^3, \quad \text{and} \quad -y = -1 \cdot y$$

EXAMPLE 1

From the following list of terms, pick out the like terms.

$$5, \quad 4y, \quad -12, \quad -y, \quad 3x^2z, \quad -\frac{1}{2}y, \quad -2x^2z, \quad 0$$

Solution

(a) 5, -12, and 0 are like terms. All are constants.

(b) $4y$, $-y$, and $-\frac{1}{2}y$ are like terms. All have the same variable factor, y.

(c) $3x^2z$ and $-2x^2z$ are like terms. All have the same variable factor, x^2z.

Algebraic expressions involving the sums or differences of terms, such as

$$8 + x, \quad 5y^2 - 11y, \quad \text{and} \quad 2x^2 - 5x - 2$$

are called **polynomials.** To simplify this type of algebraic expression, we **combine like terms** whenever possible. For example,

$$7x + 2x = 9x \quad \text{and} \quad 8n + 3n + 2n = 13n$$

The technique that we use for combining like terms is based on the following **distributive property of multiplication over addition** (or simply the **distributive property**).

Distributive Property of Multiplication over Addition

For any numbers a, b, and c,

$$a(b + c) = ab + ac$$

Another form of the distributive property is

$$ba + ca = (b + c)a$$

This form is particularly useful when b and c are numerical coefficients because it leads directly to the explanation of combining like terms. Thus,

$$ba + ca = (b + c)a$$
$$7x + 2x = (7 + 2)x = 9x \quad \text{by the distributive property}$$

(Intuitively, we can think of x as a label or name. Thus, just as 7 oranges + 2 oranges is 9 oranges, $7x + 2x$ is $9x$.) Similarly,

$$6n - n + 3n = (6 - 1 + 3)n = 8n \quad \text{Note that } -1 \text{ is the coefficient in } -n.$$

Example 2

Simplify each expression by combining like terms whenever possible.

(a) $6x + 10x + 3$ (b) $4x^2 + x^2 - 8x^2$

(c) $2(x + 3) + 4(x - 7)$ (d) $3a - 52b + 9$

Solution

(a) $6x + 10x + 3 = (6 + 10)x + 3$
$= 16x + 3$

Use the distributive property with $6x + 10x$. Note that the constant 3 is not combined with $16x$ because they are not like terms.

(b) $4x^2 + x^2 - 8x^2 = (4 + 1 - 8)x^2$
$= -3x^2$

Note that $+1$ is the coefficient of x^2.

(c) $2(x + 3) + 4(x - 7) = 2x + 6 + 4x - 28$ First use the distributive property directly.
$= 2x + 4x + 6 - 28$
$= (2 + 4)x - 22$ Combine like terms.
$= 6x - 22$

(d) $3a - 52b + 9$ This expression is already simplified since it has no like terms.

Now Work Exercises 1–4 in the Margin.

Simplify each expression by combining like terms.

1. $5x + 6x =$ ________

2. $-3y - 4y =$ ________

3. $6x + 6 + x - 8 =$ ________

4. $2(n + 4) + 3(n - 3) =$ ________

Evaluating Algebraic Expressions

Algebraic expressions can be evaluated for given values of the variables by substituting one value for each variable into the expression and then following the rules for order of operations. When like terms are present, the process of evaluation can be made easier by first combining like terms.

To Evaluate an Algebraic Expression:

1. Combine like terms.
2. Substitute the values given for any variables. (**Note:** To indicate multiplication, enclose the numbers substituted in parentheses.)
3. Follow the rules for order of operations.

The rules for order of operations are restated here for easy reference.

Rules for Order of Operations:

1. First, simplify within grouping symbols, such as parentheses ( ), brackets [ ], or braces { }. Start with the innermost grouping.
2. Second, find any powers indicated by exponents.
3. Third, moving from **left to right,** perform any multiplications or divisions in the order in which they appear.
4. Fourth, moving from **left to right,** perform any additions or subtractions in the order in which they appear.

Evaluate each expression for $x = 4$ and $a = -5$.

5. $2x - 8 =$ ______

6. $3a + 14 =$ ______

7. $6x^2 - 5x^2 + 3x - 2x - 7 + 8 =$ ______

8. $7a + 3(x - 1) =$ ______

EXAMPLE 3

Evaluate the expression $6ab + ab + 4a - a$ for $a = -2$ and $b = +3$.

Solution

Combining like terms gives

$$\begin{aligned} 6ab + ab + 4a - a &= (6 + 1)ab + (4 - 1)a \\ &= 7ab + 3a \end{aligned}$$

Substituting $a = -2$ and $b = +3$ into the simplified expression and following the rules for order of operations gives the value of the expression.

$$\begin{aligned} &7ab + 3a \\ &= 7(-2)(3) + 3(-2) \\ &= -42 - 6 \\ &= -48 \end{aligned}$$

Note that the substituted numbers must be in parentheses to indicate multiplication.

☑ *COMPLETION EXAMPLE 4*

Evaluate the expression $7x^2 - 2x^2 + 2x - x^2 - 14x$ for $x = 3$.

Solution

Combining like terms

$$\begin{aligned} &7x^2 - 2x^2 + 2x - x^2 - 14x \\ &= (\_\_\_\_\_\_\_\_\_\_\_\_)x^2 + (\_\_\_\_\_\_\_\_)x \\ &= \_\_\_\_\_x^2 - \_\_\_\_\_x \end{aligned}$$

Substituting $x = 3$ and evaluating:

$$\begin{aligned} 4x^2 - 12x &= 4(\_\_\_\_\_)^2 - 12(\_\_\_\_\_) \\ &= \_\_\_\_\_ - \_\_\_\_\_ = \_\_\_\_\_ \end{aligned}$$

Now Work Exercises 5–8 in the Margin.

Completion Example Answers

4. $7x^2 - 2x^2 + 2x - x^2 - 14x = (\mathbf{7 - 2 - 1})x^2 + (\mathbf{2 - 14})x$
$= \mathbf{4}x^2 - \mathbf{12}x$

Substituting $x = 3$ and evaluating:

$$\begin{aligned} 4x^2 - 12x &= 4(\mathbf{3})^2 - 12(\mathbf{3}) \\ &= \mathbf{36} - \mathbf{36} = \mathbf{0} \end{aligned}$$

NAME ______________ SECTION ______________ DATE ______________

ANSWERS

Exercises 9.5

Simplify each of the following algebraic expressions by combining like terms whenever possible.

1. $6x + 2x$
2. $4x - 3x$
3. $5x + x$
4. $7x - 3x$
5. $-10a + 3a$
6. $-11y + 4y$
7. $-18y + 6y$
8. $-2x - 5x$
9. $-5x - 4x$
10. $-x - 2x$
11. $-7x - x$
12. $2x - 2x$
13. $3x - 5x + 12x$
14. $2a + 14a - 25a$
15. $6c - 13c + 5c$
16. $40x - 30x - 10x$
17. $16n - 15n - 3n$
18. $12y - 12y + 4y$
19. $5x^2 - 3x^2 + 2x$
20. $-2x^2 - x^2 - x$
21. $7x^2 - 4x^2 + 20$
22. $4x + 7 - 8 + 3x$
23. $-5x - 1 + 8 + 9x$
24. $10y + 3 - 4 - 6y$
25. $2(x + 1) + 3(x - 1)$
26. $4(x - 1) + 5(x - 2)$
27. $3ab + 6a + 2b$
28. $9xy - 2x + 5y$
29. $x^2 - 5x + 6$
30. $x^2 - 7x + 12$

1. ______
2. ______
3. ______
4. ______
5. ______
6. ______
7. ______
8. ______
9. ______
10. ______
11. ______
12. ______
13. ______
14. ______
15. ______
16. ______
17. ______
18. ______
19. ______
20. ______
21. ______
22. ______
23. ______
24. ______
25. ______
26. ______
27. ______
28. ______
29. ______
30. ______

31. ______
32. ______
33. ______
34. ______
35. ______
36. ______
37. ______
38. ______
39. ______
40. ______
41. ______
42. ______
43. ______
44. ______
45. ______
46. ______
47. ______
48. ______
49. ______
50. ______

31. $-4n + n + 1 - 3$

32. $-2c + 5c + 6 - 5$

33. $3x^2 - 4x^2 + 2x - 1$

34. $-5x^2 + 4x^2 - 17x + 20x + 42 + 3$

35. $12x^2 - 2x^2 + 15x + 13x - 35 - 41$

Evaluate each of the following expressions for $x = -3$, $y = 2$, $z = 3$, $a = -1$, and $c = -2$.

36. $x - 2$

37. $y - 2$

38. $z - 3$

39. $2x + 3x - 7$

40. $7a - a + 3$

41. $-3y - 4y + 6 - 2$

42. $-2c - 3c + 1 - 4$

43. $5y - 2y - 3y + 4$

44. $2x - 3x + x - 8$

45. $3a + 2x + 7x - a + x$

46. $6 + 2(a + 1) + 3z$

47. $5y + 14 + 3(c - 1) - 4$

48. $5x^2 - 8x + 2x + 7 - 9$

49. $4z^2 + 2z - 5z + 2 + 10$

50. $2c^2 + 8 + 2c + 3c + c^2$

9.6 Translating English Phrases and Solving Equations

OBJECTIVES

1. *Be able to translate English phrases into algebraic expressions.*
2. *Be able to translate algebraic expressions into English phrases.*
3. *Know how to solve linear equations that are of the form $ax + b = c$ when the constants a, b, and c are integers.*

Translating English Phrases

Many word problems can be solved by using the following three basic skills:

1. Know how to translate English phrases into algebraic expressions.
2. Be able to set up an equation using the translated expressions.
3. Solve the resulting equation.

In this section, we will learn how to translate English phrases by finding key words and how to develop techniques for solving equations. In Section 9.7, we will see how these skills can be used to set up equations using translated expressions to problem solve.

The following list contains some of the key words that appear in word problems. These words indicate what operations are to be performed either with known numbers (called **constants**) or with variables and constants.

Key Words (that indicate operations)

Addition	Subtraction	Multiplication	Division
add	subtract (from)	multiply	divide
sum	difference	product	quotient
plus	minus	times	
more than	less than	twice, double	
increased by	decreased by	of (with fractions)	

The following examples illustrate how English phrases can be translated into algebraic expressions. Different letters are used as variables to represent the unknown numbers, and the key words are in boldface print.

English Phrase	Algebraic Expression
1. A number **plus** 7 The **sum** of a number and 7 7 **more than** a number	$x + 7$
2. 8 **times** a number The **product** of a number and 8 A number **multiplied by** 8	$8n$
3. The **quotient** of a number and 6 A number **divided by** 6	$\frac{n}{6}$
4. **Twice** a number **decreased by** 51 51 **less than twice** a number 51 **subtracted from two times** a number	$2y - 51$

Write an English phrase that means the same as each expression.

1. $13x$

2. $n - 45$

3. $2(y + 7)$

4. $\frac{50}{x}$

The words **difference** and **quotient** deserve special mention because their use implies that the numbers are to be operated on in the order given. For example,

"the **difference** between x and 10" means $x - 10$, not $10 - x$;

"the **quotient** of y and 20" means $\frac{y}{20}$, not $\frac{20}{y}$.

A skill closely related to translating English phrases into algebraic expressions is the reverse process—that is, looking at an algebraic expression and creating an English phrase that has the same meaning. In such cases, there is often more than one correct phrase.

EXAMPLE 1

For each expression, write an English phrase that has the same meaning.

(a) $14x$ (b) $n + 35$

(c) $2x + 5x$ (d) $\frac{y}{3}$

Solution

	Algebraic Expression	Possible English Phrase
(a)	$14x$	The product of 14 and a number
(b)	$n + 35$	35 more than a number
(c)	$2x + 5x$	Twice a number plus 5 times the number
(d)	$\frac{y}{3}$	A number divided by 3

Now Work Exercises 1–4 in the Margin.

Solving Equations ($ax + b = c$)

If an equation contains a variable, such as the equation $5x + 1 = 16$, then any number may be substituted for x. For example,

Substituting $x = 8$ gives $5 \cdot 8 + 1 = 16$ a false statement

Substituting $x = 3$ gives $5 \cdot 3 + 1 = 16$ a true statement

We can substitute as many values for x as we choose. However, the object is not just to substitute values for the variable x, but to find the value for x that results in a true statement when this value is substituted for x in the equation. This procedure is called **solving the equation.** The value found is called the **solution** of the equation.

We just found that substituting $x = 3$ in the equation $5x + 1 = 16$ gives a true statement, and thus, $x = 3$ is a solution of the equation. In fact, $x = 3$ is the *only* solution.

Definition

A **first-degree equation in x** (or **linear equation in x**) is any equation that can be written in the form

$$ax + b = c \quad \text{where } a, b, \text{ and } c \text{ are constants and } a \neq 0.$$

(**Note:** A variable other than x may be used.)

EXAMPLE 2

Given the equation $3n + 15 = 27$, (a) show that 4 is a solution, and (b) show that 2 is not a solution.

Solution

(a) Since $3 \cdot 4 + 15 = 27$ is a true statement, 4 is a solution.

(b) Since $3 \cdot 2 + 15 = 27$ is a false statement, 2 is not a solution.

A fundamental fact of algebra, stated here without proof, is that **every first-degree equation has exactly one solution.** Sometimes, with simple equations, this solution can be found by observation or by trial and error. However, by using the following principles, we can guarantee finding the solution in an efficient, organized manner that works, regardless of how complicated the equation is. The basic idea is that **the same operation is to be performed on both sides of the equal sign (that is, both sides of the equation).**

For example, if you add 17 to one side of an equation, you must add 17 to the other side as well. You cannot add 17 to one side and subtract 17 from the other side. As illustrated in Figure 9.13, you can think of the equal sign as the fulcrum (or balance point) on a balance scale.

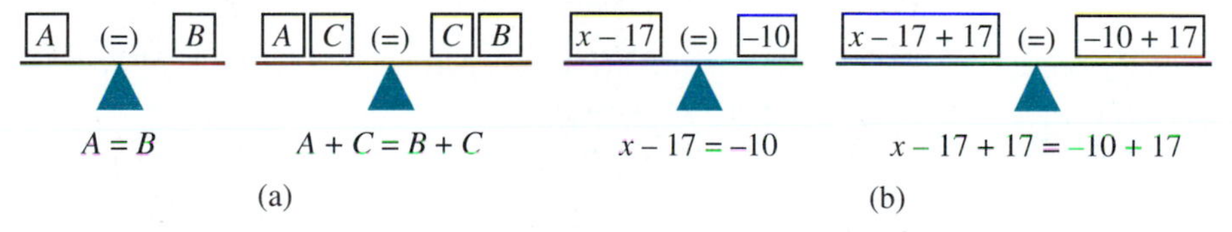

Figure 9.13

Principles Used in Solving a First-Degree Equation

In the four basic principles stated here, A and B represent algebraic expressions or constants.

1. **The Addition Principle:**
   If $A = B$ is true, then $A + C = B + C$ is also true for any number C. (The same number may be added to both sides of an equation.)

2. **The Subtraction Principle:**
   If $A = B$ is true, then $A - C = B - C$ is also true for any number C. (The same number may be subtracted from both sides of an equation.)

continued

3. The Multiplication Principle:
If $A = B$ is true, then $C \cdot A = C \cdot B$ is also true for any number C. (Both sides of an equation may be multiplied by the same number.)

4. The Division Principle:
If $A = B$ is true, then $\frac{A}{C} = \frac{B}{C}$ is also true for any nonzero number C. (Both sides of an equation may be divided by the same nonzero number.)

Note that in the Division Principle, division by C is the same as multiplication by the reciprocal of C, namely $\frac{1}{C}$. Thus, we could list the Division and Multiplication Principles as one principle. The use of any of these principles with an equation gives a new equation with the same solution. Such equations are said to be **equivalent.** The process of solving equations involves applying the principles just listed to find equivalent equations until the solution is obvious, as in $x = 3$, $y = 15$, or $700 = A$.

EXAMPLE 3 Solve the equation $x + 13 = 25$.

Solution

$x + 13 = 25$ Write the equation.

$x + 13 - 13 = 25 - 13$ Use the Subtraction Principle and subtract 13 from both sides.

$x + 0 = 12$ Simplify both sides.

$x = 12$

EXAMPLE 4 Solve the equation $5y = 180$.

Solution

$5y = 180$ Write the equation.

$\frac{\cancel{5}y}{\cancel{5}} = \frac{180}{5}$ Use the Division Principle and divide both sides by 5. (Or use the Multiplication Principle and multiply both sides by $\frac{1}{5}$.)

$y = 36$ Simplify both sides.

EXAMPLE 5

Solve the equation $\frac{n}{6} = 31$.

Solution

$\frac{n}{6} = 31$ Write the equation.

$\frac{\cancel{6}}{1} \cdot \frac{n}{\cancel{6}} = \frac{31}{1} \cdot \frac{6}{1}$ Use the Multiplication Principle and multiply both sides by $\frac{6}{1}$.

$n = 186$ Simplify both sides.

Now Work Exercises 5 and 6 in the Margin.

More difficult problems may involve several steps. Keep in mind the following general guidelines.

General Guidelines for Solving Equations

1. The goal is to isolate the variable on one side of the equation (either the right side or the left side).
2. First use the Addition and Subtraction Principles whenever numbers are subtracted or added to the variable.
3. Use the Multiplication and Division Principles to make 1 (or +1) the coefficient of the variable.

EXAMPLE 6

Solve the equation $4x - 16 = 64$.

Solution

$4x - 16 = 64$	Write the equation.
$4x - 16 + 16 = 64 + 16$	Use the Addition Principle and add 16 to both sides.
$4x + 0 = 80$	Simplify both sides.
$4x = 80$	
$\frac{1}{\cancel{4}} \cdot \cancel{4}x = \frac{1}{4} \cdot 80$	Use the Multiplication Principle and multiply both sides by $\frac{1}{4}$. (Or use the Division Principle and divide both sides by 4.)
$1 \cdot x = 20$	Simplify both sides.
$x = 20$	

Checking Solutions to Equations:

Checking can be done by substituting the solution found back into the original equation to see if the resulting statement is true.

CHECKING EXAMPLE 6

$$4x - 16 = 64$$

$$4 \cdot 20 - 16 \stackrel{?}{=} 64$$

$$80 - 16 \stackrel{?}{=} 64$$

$$64 = 64 \quad \text{a true statement}$$

Solve each of the following equations.

5. $3x = -12$

6. $y - 2 = -5$

7. Solve the following equation.

$3x - 35 = 1$

✓ Completion Example 7

Explain each step in the solution process shown here.

Equation	Explanation
$13 = 5n - 7$	Write the equation.
$13 + 7 = 5n - 7 + 7$	______________
$20 = 5n$	______________
$\frac{20}{5} = \frac{\cancel{5}n}{\cancel{5}}$	______________
$4 = n$	______________

Now Work Exercise 7 in the Margin.

Solving Equations Involving Integers

The same principles used for solving first-degree equations with whole-number constants and coefficients can be used to solve equations with integer (or decimal or fraction) constants and coefficients. The following examples illustrate solving equations with integer constants and coefficients and integer solutions.

Example 8

Solve the equation $x + 17 = 14$.

Solution

$x + 17 = 14$ Write the equation.

$x + 17 - 17 = 14 - 17$ Use the Subtraction Principle and subtract 17 from both sides. (This is the same as adding -17 to both sides.)

$x + 0 = -3$ Simplify both sides.

$x = -3$

Example 9

Solve the equation $-5y = 80$.

Solution

$-5y = 80$ Write the equation.

$\frac{\cancel{-5}y}{\cancel{-5}} = \frac{80}{-5}$ Use the Division Principle and divide both sides by -5. (We want the coefficient of y to be $+1$.)

$y = -16$ Simplify both sides.

EXAMPLE 10

Solve the equation $21 = -4x + 16$.

Solution

$12 = -4x + 16$	Write the equation.
$12 - 16 = -4x + 16 - 16$	Use the Subtraction Principle and subtract 16 from both sides. (This is the same as adding -16 to both sides.)
$-4 = -4x$	Simplify both sides.
$\frac{-4}{-4} = \frac{\cancel{-4}x}{\cancel{-4}}$	Use the Division Principle and divide both sides by -4. (We want the coefficient of x to be $+1$.)
$1 = x$	Simplify both sides.

EXAMPLE 11

Solve the equation $5n + 2 = 3n - 8$.

Solution

$5n + 2 = 3n - 8$	Write the equation.
$5n + 2 - 2 = 3n - 8 - 2$	Add -2 to both sides.
$5n = 3n - 10$	Simplify both sides.
$5n - 3n = 3n - 10 - 3n$	Add $-3n$ to both sides.
$2n = -10$	Simplify both sides.
$\frac{1}{\cancel{2}} \cdot \cancel{2}n = \frac{1}{2} \cdot (-10)$	Multiply both sides by $\frac{1}{2}$. (This is the same as dividing both sides by 2.)
$n = -5$	Simplify both sides.

Now Work Exercises 8–10 in the Margin.

If an expression on one side or the other of an equation contains a number times a quantity in parentheses (or other grouping symbol), we use the distributive property to **clear** the parentheses, and then we proceed as before.

EXAMPLE 12

Solve the equation $3(x + 2) = 18 - x$.

Solution

$3(x + 2) = 18 - x$	Write the equation.
$3x + 6 = 18 - x$	Use the distributive property.
$3x + 6 - 6 = 18 - x - 6$	Subtract 6 from both sides.
$3x = 12 - x$	Simplify both sides.
$3x + x = 12 - x + x$	Add x to both sides.
$4x = 12$	Simplify both sides.
$\frac{1}{\cancel{4}} \cdot \cancel{4}x = \frac{1}{4} \cdot 12$	Multiply both sides by $\frac{1}{4}$.
$x = 3$	Simplify both sides.

Solve each of the following equations.

8. $4x - 10 = -38$

9. $14 = -2n + 16$

10. $3x - 4 = 5x + 14$

Solve each of the following equations.

11. $2(x + 1) = 17 - x$

12. $y + 9 = 3(y + 7)$

✓ Completion Example 13

Complete the solution process for the equation $y - 1 = 2(y - 5)$.

Solution

Equation	**Explanation**
$y - 1 = 2(y - 5)$	Write the equation.
______	Use the distributive property.
______	______
______	______
______	______

Now Work Exercises 11 and 12 in the Margin.

Completion Example Answers

7.

Equation	**Explanation**
$13 = 5n - 7$	Write the equation.
$13 + 7 = 5n - 7 + 7$	**Add 7 to both sides.**
$20 = 5n$	**Simplify.**
$\frac{20}{5} = \frac{\cancel{5}n}{\cancel{5}}$	**Divide both sides by 5.**
$4 = n$	**Simplify.**

13.

Equation	**Explanation**
$y - 1 = 2(y - 5)$	Write the equation.
$\mathbf{y - 1 = 2y - 10}$	Use the distributive property.
$\mathbf{y - 1 + 10 = 2y - 10 + 10}$	**Add 10 to both sides.**
$\mathbf{y + 9 = 2y}$	**Simplify.**
$\mathbf{y + 9 - y = 2y - y}$	**Subtract *y* from both sides. (Or add −*y* to both sides.)**
$\mathbf{9 = y}$	**Simplify.**

NAME ______________ SECTION ______________ DATE ______________

ANSWERS

1. ______
2. ______
3. ______
4. ______
5. ______
6. ______
7. ______
8. ______
9. ______
10. ______
11. ______
12. ______
13. ______
14. ______
15. ______
16. ______
17. ______
18. ______
19. ______
20. ______
21. [Respond below exercise.]
22. [Respond below exercise.]
23. [Respond below exercise.]
24. [Respond below exercise.]
25. [Respond below exercise.]
26. [Respond below exercise.]

Exercises 9.6

Write an algebraic expression described by each of the following English phrases. (Any variable may be used to represent the unknown number.)

1. a number plus 5

2. 8 more than a number

3. the sum of a number and 9

4. the product of a number and 9

5. the quotient of a number and 9

6. the difference of a number and 9

7. 13 less than a number

8. 55 decreased by a number

9. 13 decreased by a number

10. 16 plus a number

11. 4 more than twice a number

12. 15 increased by twice a number

13. 8 times a number plus 6

14. 6 more than 5 times a number

15. 18 less than the quotient of a number and 2

16. 25 more than the product of a number and 8

17. the difference of 3 and twice a number

18. the sum of 1 and three times a number

19. the product of 6 and a number increased by 4 times the number

20. 7 times a number decreased by the product of the number and 5

Write an English phrase that means the same as each expression. (There may be more than one correct phrase.)

21. $5x$

22. $n + 15$

23. $x - 6$

24. $y - 32$

25. $y + 491$

26. $7y$

27. [Respond below exercise.] ______

28. [Respond below exercise.] ______

29. [Respond below exercise.] ______

30. [Respond below exercise.] ______

31. [Respond below exercise.] ______

32. [Respond below exercise.] ______

33. [Respond below exercise.] ______

34. [Respond below exercise.] ______

35. [Respond below exercise.] ______

36. ______

37. ______

38. ______

39. ______

40. ______

41. [Respond in exercise.] ______

42. [Respond in exercise.] ______

27. $\frac{n}{17}$ **28.** $\frac{x}{10}$ **29.** $2x + 1$

30. $3x - 1$ **31.** $\frac{y}{3} + 20$ **32.** $\frac{a}{5} + 12$

33. $15x - 15$ **34.** $\frac{6}{x}$ **35.** $13 - x$

A set of numbers follows the equation in each problem below. Determine which number in the set is the solution of the equation by substituting the numbers into the equation.

36. $x - 17 = 10$; $\{27, 30, 33\}$ **37.** $2n + 3 = 13$; $\{0, 2, 5, 7\}$

38. $5n + 8 = 23$; $\{0, 1, 2, 3\}$ **39.** $\frac{n}{3} + 6 = 8$; $\{3, 6, 9, 12\}$

40. $\frac{x}{2} - 4 = 0$; $\{2, 4, 6, 8, 10\}$

Give an explanation (or a reason) for each step in the solution process.

41.

$3x + 10 = 22$ ______

$3x + 10 - 10 = 22 - 10$ ______

$3x = 12$ ______

$\frac{3x}{3} = \frac{12}{3}$ ______

$x = 4$ ______

42.

$2x - 15 = -35$ ______

$2x - 15 + 15 = -35 + 15$ ______

$2x = -20$ ______

$\frac{2x}{2} = \frac{-20}{2}$ ______

$x = -10$ ______

NAME ____________ SECTION ____________ DATE ____________

Solve each of the following equations.

43. $x - 12 = 6$

44. $x - 10 = 9$

45. $n + 3 = -13$

46. $n + 7 = -15$

47. $7y = 42$

48. $9y = 72$

49. $\frac{n}{7} = 32$

50. $\frac{x}{3} = 10$

51. $2x + 1 = -11$

52. $3x - 1 = -16$

53. $18 = -4y - 6$

54. $28 = -5n - 7$

55. $\frac{n}{2} + 3 = 15$

56. $\frac{x}{5} - 9 = 6$

57. $\frac{x}{3} + 12 = 20$

58. $\frac{x}{4} - 35 = 15$

59. $70 = 2x - 6$

60. $82 = 3x + 4$

61. $3x + 5 = 2x - 5$

62. $7x - 10 = 5x + 20$

ANSWERS

43. ____________

44. ____________

45. ____________

46. ____________

47. ____________

48. ____________

49. ____________

50. ____________

51. ____________

52. ____________

53. ____________

54. ____________

55. ____________

56. ____________

57. ____________

58. ____________

59. ____________

60. ____________

61. ____________

62. ____________

63. ______________

64. ______________

65. ______________

66. ______________

67. ______________

68. ______________

69. ______________

70. ______________

71. [Respond below exercise.]

63. $4n + 7 = 6n + 11$

64. $8n - 14 = 11n - 5$

65. $2(x + 3) = x - 1$

66. $3(x - 1) = x + 13$

67. $3(n - 5) = 2(n + 2)$

68. $2(n + 1) = 4(n - 1)$

69. $5x - 10 = 2(x - 20)$

70. $9x + 30 = 7(x - 2)$

Writing and Thinking about Mathematics

71. The solution of an equation is the same whether the variable appears on the right side of the equation or on the left side of the equation. For example, the solution of the equation $3x + 1 = 25$ is the same as the solution of the equation $25 = 3x + 1$. Does this seem reasonable to you? Explain briefly.

9.7 Problem Solving

OBJECTIVES

1. Be able to solve number problems by using equations.
2. Be able to solve problems involving geometric concepts by using equations.

Solving Number Problems

The word problems discussed here can be solved by translating English phrases into algebraic expressions—a skill that we developed in Section 9.6. The four-step problem-solving approach is outlined below:

Basic Steps for Solving Number Problems:

1. Read the problem carefully at least twice. Look for key words and phrases that can be translated into algebraic expressions.
2. Assign a variable as the unknown quantity and form an equation using the expressions you translated.
3. Solve the equation.
4. Look back over the problem and check that the answer is reasonable.

EXAMPLE 1

Four less than twice a number is equal to 20. Find the number.

Solution

STEP 1: Let n = the unknown number.

STEP 2: Translate "four less than twice a number" to $\underline{2n - 4}$.

STEP 3: Form the equation: $\underline{2n - 4 = 20}$.

STEP 4: Solve the equation:

$$2n - 4 = 20$$
$$2n - 4 + 4 = 20 + 4$$
$$2n = 24$$
$$\frac{1}{2} \cdot 2n = \frac{1}{2} \cdot 24$$
$$1 \cdot n = 12$$
$$n = 12$$

CHECK:

$$2(12) - 4 \stackrel{?}{=} 20$$
$$24 - 4 \stackrel{?}{=} 20$$
$$20 = 20$$

The number is 12.

COMPLETION EXAMPLE 2

A number is increased by 45 and the result is 30. What is the number?

Solution

STEP 1: Let x = the unknown number.

STEP 2. Translate "a number is increased by 45" to ________________.

STEP 3: Form the equation: ______________.

STEP 4: Solve the equation:

$$\_\_\_\_\_\_\_\_ = \_\_\_\_\_\_\_\_$$

$$x + 45 - \_\_\_\_ = 30 - \_\_\_\_$$

$$x + \_\_\_\_ = \_\_\_\_\_\_$$

$$x = \_\_\_\_\_\_$$

CHECK:

$$\_\_\_\_ + 45 \stackrel{?}{=} 30$$

$$\_\_\_\_ = 30$$

So ______ is the correct number.

EXAMPLE 3

A span of a suspension bridge is the distance between its supports. The longest span on the Tacoma Narrows bridge in Washington State is 2800 feet. This is 40 feet more than twice the longest span of the Triboro bridge in New York City. What is the longest span of the Triboro bridge?

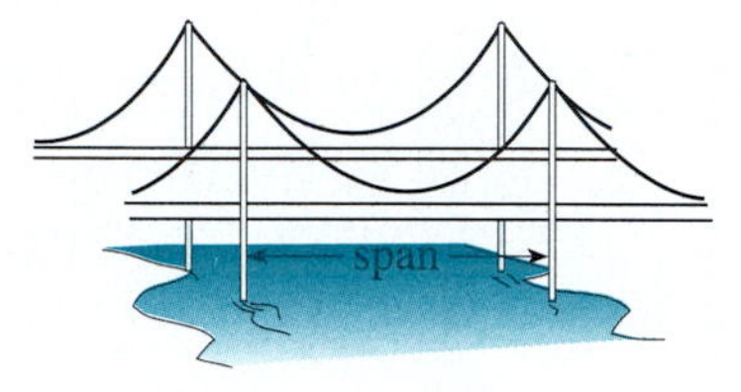

Solution

STEP 1: Let $x =$ the length of the longest span of the Triboro bridge.

STEP 2: Translate "40 feet more than twice the longest span of the Triboro bridge" to $\underline{2x + 40}$.

STEP 3: Form the equation: $\underline{2x + 40 = 2800}$.

STEP 4: Solve the equation:

$$2x + 40 = 2800$$

$$2x + 40 - 40 = 2800 - 40$$

$$2x + 0 = 2760$$

$$2x = 2760$$

$$\frac{\cancel{2}x}{\cancel{2}} = \frac{2760}{2}$$

$$x = 1380$$

The longest span of the Triboro birdge is 1380 feet.

(The student should check this result in the original equation.)

Solving Geometry Problems

The concepts of perimeter, area, and volume of geometric figures are related to various formulas in the form of equations. In this section, the corresponding formula will be provided in each problem and all the values except one will be known. By substituting the known values, the unknown value can be found by using the techniques we have learned for solving equations.

For easy reference, Table 9.1 contains the formulas for perimeter (P) and area (A) of six geometric figures.

Table 9.1 Formulas for Finding Perimeter and Area of Six Geometric Figures

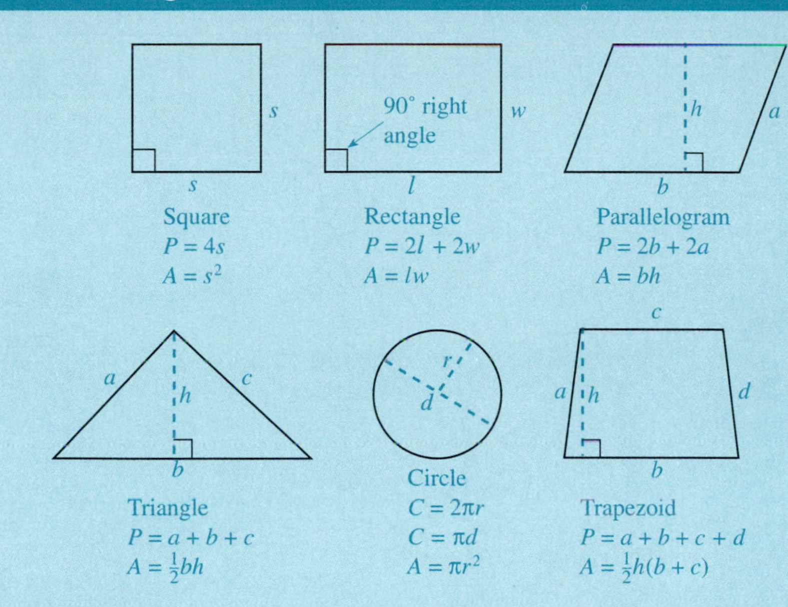

Note: The Greek letter π is the symbol used for the constant number 3.1415926535 . . . . This constant is an infinite decimal with no pattern to its digits. For our purposes, we will use $\pi = 3.14$, but you should be aware that 3.14 is only an approximation to π.

Example 4

A rectangular swimming pool has a perimeter of 160 meters and a length that is 20 meters more than its width. Find its width and length.

Solution

Use the formula for the perimeter of a rectangle: $P = 2l + 2w$. Sketch a picture and label the sides where

$$w = \text{the width}$$

$$w + 20 = \text{the length}$$

Substituting into the formula gives the equation to be solved:

$$160 = 2(w + 20) + 2w$$

$$160 = 2w + 40 + 2w$$

$$160 = 4w + 40$$

$$160 - 40 = 4w + 40 - 40$$

$$120 = 4w$$

$$\frac{120}{4} = \frac{4w}{4}$$

$$30 = w$$

The width of the pool is 30 meters, and the length is $w + 20 = 50$ meters.

EXAMPLE 5

Suppose that the area of a triangle is 45 square feet. If the base is 10 feet long, what is the height of the triangle?

Solution

The formula for the area of a triangle is $A = \frac{1}{2} \cdot b \cdot h$. In this problem, we know that $A = 45$ and $b = 10$. Substitution gives the equation

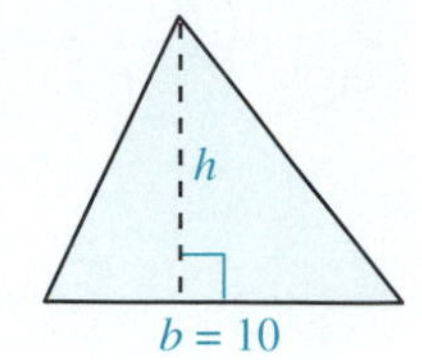

$$45 = \frac{1}{2} \cdot 10 \cdot h$$

Now, we want to solve this equation for h.

$$45 = \frac{1}{2} \cdot 10 \cdot h$$

$$45 = 5 \cdot h$$

$$\frac{45}{5} = \frac{\cancel{5} \cdot h}{\cancel{5}}$$

$$9 = h$$

The height of the triangle is 9 feet.

Completion Example Answers

2. STEP 1: Let $x =$ the unknown number.

   STEP 2: Translate "a number is increased by 45" to $\mathbf{x + 45}$.

   STEP 3: Form the equation: $\mathbf{x + 45 = 30}$.

   STEP 4: Solve the equation:

$$\mathbf{x + 45 = 30}$$

$$x + 45 - \mathbf{45} = 30 - \mathbf{45}$$

$$x + \mathbf{0} = \mathbf{-15}$$

$$x = \mathbf{-15}$$

   CHECK:

$$\mathbf{-15} + 45 \stackrel{?}{=} 30$$

$$\mathbf{30} = 30$$

   So $\mathbf{-15}$ is the correct number.

NAME ______________________ SECTION ______________ DATE ________

ANSWERS

1. [Respond below exercise.]

2. [Respond below exercise.]

3. [Respond below exercise.]

4. [Respond below exercise.]

5. [Respond below exercise.]

6. [Respond below exercise.]

7. [Respond in exercise.]

8. [Respond in exercise.]

Exercises 9.7

(**Note:** Some of the exercises in this section involve the use of decimal numbers. You should understand that the rules and methods for solving equations with decimal numbers are the same as those for solving equations involving only integers.)

Translate each of the following equations into a sentence in English. Do not solve the equation.

1. $n + 5 = 16$

2. $x - 7 = 8$

3. $4x - 8 = 20$

4. $\frac{y}{2} = 61$

5. $\frac{n}{3} + 1 = 13$

6. $15 = 25 - 2x$

Follow the steps outlined in this section to solve Exercises 7–10.

7. $\underbrace{\text{The difference between a number and twelve}}_{\text{(a)}}$ is equal to 16. What is the number?

Let x = the unknown number.

Translation of (a): ______________________

Equation: ______________________

Solve the equation.

8. $\underbrace{\text{The sum of a number and fifteen}}_{\text{(a)}}$ is equal to 35. What is the number?

Let n = the unknown number.

Translation of (a): ______________________

Equation: ______________________

Solve the equation.

9. [Respond in exercise.] ________

10. [Respond in exercise.] ________

11. ________

12. ________

13. ________

14. ________

9. $\underbrace{\text{Four less than three times a number}}_{\text{(a)}}$ is equal to 26. What is the number?

Let y = the unknown number.

Translation of (a): ________________

Equation: ________________

Solve the equation.

10. Twenty is equal to $\underbrace{\text{one more than twice a number.}}_{\text{(a)}}$ What is the number?

Let x = the unknown number.

Translation of (a): ________________

Equation: ________________

Solve the equation.

In working out Exercises 11–23, make a guess as to what you think the answer is. [For example, do you think the number is positive (greater than 0) or negative (less than 0)? Do you think the number is greater than 50 or less than 50?] Do not be afraid to make a "wrong" guess. Then, translate the phrases, form an equation, and solve the equation.

11. The sum of a number and 45 is equal to 56. What is the number?

12. The difference between a number and 14 is 27. What is the number?

13. What is the number whose product with 3 is equal to 51?

14. If the quotient of a number and 6 is 24, what is the number?

NAME ______________ SECTION ______________ DATE ______________

ANSWERS

15. ______________

16. ______________

17. ______________

18. ______________

19. ______________

20. ______________

21. ______________

15. If the product of a number and four is increased by 3, the result is 23. Find the number.

16. If the product of a number and 7 is decreased by 5, the result is 44. Find the number.

17. Eleven is the result if 9 is subtracted from the quotient of a number and 5. What is the number?

18. Ten and eight tenths is equal to four less than twice a number. What is the number?

19. The product of a number and 8 increased by 24 is equal to twice the number. Find the number.

20. Find a number such that three times the sum of the number and 4 is equal to -60.

21. Five more than twice a number is equal to 20 more than the number. What is the number?

22. ____________

23. ____________

24. ____________

25. ____________

26. ____________

27. ____________

22. Twenty plus a number is equal to the sum of twice the number and three times the same number. Find the number.

23. Twice a number decreased by 6 is equal to 5 less than the number. What is the number?

Use your knowledge of geometric figures to set up an equation to solve each of the following problems.

24. The perimeter of a rectangle is 40 meters. If the length is 15 meters, what is the width of the rectangle? ($P = 2l + 2w$)

25. If the base of a triangle is 18 inches long and its area is 54 square inches, what is the height of the triangle? $\left(A = \frac{1}{2}bh\right)$

26. A parallelogram is a four-sided figure in which opposite sides are parallel. We can also show (in geometry) that these opposite sides have the same length. Suppose that the perimeter of a parallelogram is 180 yards. If one of two equal sides has a length of 50 yards, what is the length of each of the other two sides? ($P = 2a + 2b$)

27. A trapezoid is a four-sided figure in which just two sides are parallel. If a trapezoid has an area of 195 square millimeters and parallel sides of length 20 millimeters and 10 millimeters, what is the height of the trapezoid? $\left[A = \frac{1}{2}h(b + c)\right]$

NAME ________________ SECTION ________________ DATE ________________

ANSWERS

28. ________

29. ________

30. ________

31. ________

32. ________

33. ________

28. The perimeter of a circle is called its circumference. If the circumference of a circle is 18.84 inches, what is the radius of the circle? ($C = 2\pi r$)(Use $\pi = 3.14$.)

29. In any triangle, the sum of the measures of the angles equals 180°. If two of the angles have measures of 30° and 50°, what is the measure of the third angle? ($\alpha + \beta + \gamma = 180°$)

30. A rectangular solid is in the shape of a box. Consider a rectangular solid with a volume of 60 cubic inches. If the box is 3 inches wide and 4 inches long, what is its height? ($V = lwh$)

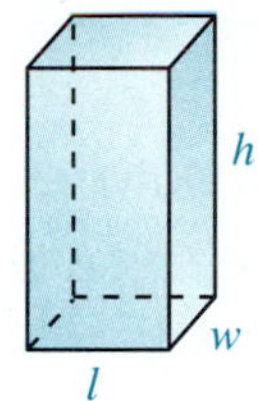

31. A rectangular solid is in the shape of a box. If the box has a volume of 360 cubic feet and is 4 feet high and 15 feet long, what is its width? ($V = lwh$)

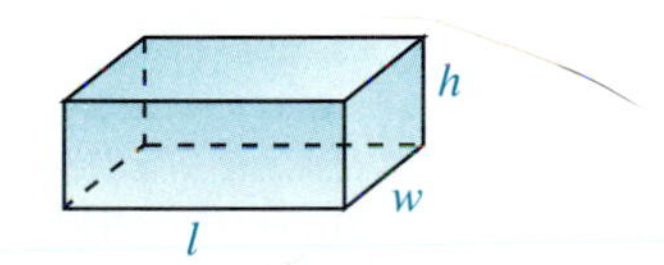

(**Note:** For Exercises 32–35, refer to Section 5.7 for the discussion of square roots. Use a calculator to find the answers to the nearest tenth of a unit.)

32. If the area of a square is 28 square meters, what is the length of one side of the square? ($A = s^2$)

33. Find the length of one side of a square with an area of 450 square inches. ($A = s^2$)

34. ______________

35. ______________

36. ______________

37. ______________

38. ______________

39. ______________

34. Find the radius of a circle that encloses an area of 28.26 square feet. ($A = \pi r^2$) (Use $\pi = 3.14$.)

35. What is the radius of a circle that has an area of 37.68 square centimeters? ($A = \pi r^2$)(Use $\pi = 3.14$.)

36. The perimeter of a triangle is 36 centimeters. If two sides are equal in length and the third side is 8 centimeters long, what is the length of each of the other two sides? ($P = a + b + c$)

37. A real estate agent says that the current value of a home is $40,000 more than twice its value when it was new. If the current value is $160,000, what was the value of the home when it was new?

38. The BART Trans-Bay Tubes, underwater rapid transit tunnels below San Francisco Bay, are 19,008 feet long and are the longest underwater tunnels in North America. They are 2058 feet longer than three times the length of the Sumner Tunnel under Boston Harbor. How long is the Sumner Tunnel?

39. According to the U.S. Department of Agriculture, the projected average annual expenditure by a middle-income family to raise a child born in 1990 to age 17 is $12,357. What is the projected total expenditure for raising a child to age 17?

NAME ______________ SECTION ______________ DATE ______________

ANSWERS

40. ______________

41. ______________

42. ______________

43. [Respond below exercise.]

40. According to the National Association of Realtors, in April 1992 the median price of a single family home in Honolulu, Hawaii was \$342,000. This was \$32,000 more than four times the median price of a home in Detroit at that time. What was the median price of a home in Detroit?

41. In the National Park System, the smallest National Seashore is Fire Island in New York, which is 19,579 acres. That is 3024 acres less than one sixth of the number of acres in the Gulf Islands in Florida and Mississippi, which make up the largest National Seashore Park. How many acres are there in the Gulf Islands?

42. A saltbox-style house (see diagram) is characterized by a long sloping roof in back. If such a roof measures 60 feet along its peak, 25 feet along one edge from top to front, and has a total area of 3900 square feet, what is the length along one edge from top to back?

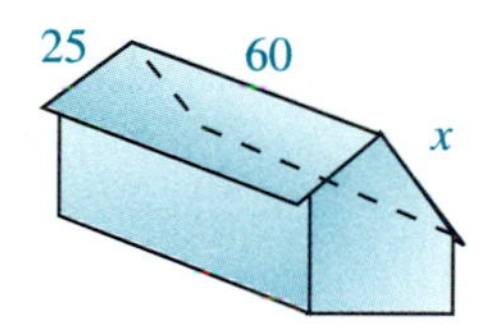

43. Suppose that you are a traveling salesperson and your company has allowed you \$50 per day for car rental. You rent a car for \$30 per day plus 5¢ per mile. How many miles can you drive for the \$50 allowed?

Plan: Set up an equation relating the amount of money spent and the money allowed in your budget and solve the equation.

Solution: Let x = number of miles driven.

Money spent = Money budgeted

$$30 + 0.05x = 50$$

$$30 + 0.05x - 30 = 50 - 30$$

$$0.05x = 20$$

$$\frac{0.05x}{0.05} = \frac{20}{0.05}$$

$$x = 400 \text{ miles}$$

You would like to get the most miles for your money. The car rental agency down the street charges only \$25 per day and 8¢ per mile. How many miles would you get for your \$50 with this price structure? Is this a better deal for you?

44. [Respond below exercise.]

45. [Respond below exercise.]

46. [Respond below exercise.]

44. Under which of the following programs for a truck rental would you get the greatest number of miles for your $200 budget if you want the truck for two days?
(a) A fee of $40 per day plus 15¢ per mile
(b) A fee of $60 per day plus 10¢ per mile

Writing and Thinking about Mathematics

45. If you planned to travel (a) less than 800 miles or (b) more than 800 miles with a truck under the rental conditions discussed in Exercise 44, which of the two programs would you use? Explain why.

Collaborative Learning Exercise

46. As a result of his studies in discovery learning, George Pólya (1887–1985), a famous professor at Stanford University, developed the following four-step process as an approach to problem solving:

(1) Understand the problem.
(2) Devise a plan.
(3) Carry out the plan.
(4) Look back over the results.

With the class separated into teams of three to four students, each team is to discuss some nonmathematical problem that a member of the team solved today (for example: what to have for breakfast, where to study, what route to take to school). Can this person's thoughts and actions be described in terms of Pólya's four steps? As the team looks back over the solution, would there have been a better or more efficient way in which to solve this person's problem? The team leader is to report the results of this discussion to the class.

9 Index of Key Ideas and Terms

NAME ______________ SECTION ______________ DATE ______________

Chapter 9 Test

1. Graph the following set of integers on a number line.

$\{-3, -1, 0, |-2|, |-4|\}$

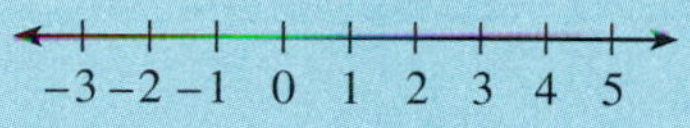

2. The opposite of -11 is ______.

3. State whether each of the following statements is true or false. If the statement is false, rewrite it in a true form.

(a) $-12 > -15$

(b) $|-15| \leq |-12|$

Perform the indicated operations.

4. $(+14) + (-6)$

5. $(-9) - (+3)$

6. $(-3) + (+7) + (-4) + (-9)$

7. $-10 - 5 - 1$

8. $17 - 4 - 12 + 15$

9. $(-4)(-7)$

10. $(-3)(-5)(2)(-3)$

11. $\frac{-56}{-7}$

12. $\frac{-3}{0}$

13. $\frac{65}{-13}$

14. Simplify each of the following expressions by combining like terms.

(a) $4x - 5 - x + 10$

(b) $5x^2 + x - 3 + x^2 - 4x + 8$

15. (a) Simplify the following expression, and (b) evaluate it for $x = 2$ and $y = -1$.

$2x + 2(y + 3) - (3y + 4) + 3(x - 3)$

ANSWERS

1. [Respond below exercise.] ______
2. ______
3a. ______
b. ______
4. ______
5. ______
6. ______
7. ______
8. ______
9. ______
10. ______
11. ______
12. ______
13. ______
14a. ______
b. ______
15a. ______
b. ______

16. ______________

17. ______________

18. ______________

19. ______________

20. ______________

21. ______________

22. ______________

23. ______________

24. ______________

Solve each of the following equations.

16. $5x + 1 = 36$

17. $3x - 4 = -19$

18. $y - 8 = 5y - 32$

19. $3(n + 3) = 2(7 - n)$

20. If the product of a number and 4 is decreased by 10, the result is 50. Find the number.

21. Four times the difference between a number and 7 is equal to 2 plus the number. Find the number.

22. What is the number whose product with 6 is equal to twice the number minus 12?

23. The perimeter of a rectangle is 60 feet. If the length is 20 feet, what is the width of the rectangle? $(P = 2l + 2w)$

24. The base of a triangle is 30 centimeters long, and its area is 210 square centimeters. What is the height of the triangle? $\left(A = \frac{1}{2}bh\right)$

NAME ______________________ SECTION ______________________ DATE ____________

Cumulative Review

Perform the indicated operations and simplify.

1. $1 - \frac{1}{16}$

2. $\frac{3}{10} - \frac{2}{15}$

3. $25\frac{1}{6} \times 13\frac{3}{8}$

4. $340\frac{1}{10} - 150\frac{3}{5}$

5. $\dfrac{4 - \frac{1}{2}}{\frac{1}{24} + \frac{1}{6}}$

6. Multiply: (45.8)(16.73)

7. Divide (to the nearest hundredth): $176 \div 32.1$

Evaluate the following expressions. Reduce all fractions.

8. $\frac{16}{0}$

9. $\frac{1}{2} + \frac{3}{4} \cdot \left(\frac{1}{2}\right)^2 - \frac{1}{16}$

10. $(-7)^2 \cdot 5 + 2.1(-6) \div \frac{1}{2}$

In Exercises 11–14, choose the correct answer by reasoning and calculating mentally.

11. Write 7% as a fraction.

(a) $\frac{7}{1}$ (b) $\frac{7}{10}$ (c) $\frac{7}{100}$ (d) $\frac{7}{1000}$

12. Find 20% of 730.

(a) 7.3 (b) 73 (c) 14.6 (d) 146

13. What percent of 100 is 25?

(a) 0.25% (b) 2.5% (c) 25% (d) 250%

14. 245 is approximately 50% of what number?

(a) 120 (b) 500 (c) 1000 (d) 2000

15. 135% of ______ is 270.

16. 25% of 200 is ______.

ANSWERS

1. ______

2. ______

73 ______

4. ______

5. ______

6. ______

7. ______

8. ______

9. ______

10. ______

11. ______

12. ______

13. ______

14. ______

15. ______

16. ______

17. ______________

18. ______________

19. ______________

20. ______________

21. ______________

22. ______________

23. ______________

24. ______________

25. ______________

26a. ______________

b. ______________

c. ______________

27. ______________

28. ______________

29. ______________

30. ______________

17. Find the average (mean) of the numbers 5000, 6250, 9475, and 8672.

Find the equivalent metric units as indicated.

18. 3000 m = __________ km

19. 8.6 cm = __________ mm

Simplify the following radical expressions.

20. $\sqrt{600}$

21. $2\sqrt{10} + \sqrt{90}$

Write each of the following numbers in scientific notation.

22. 93,500,000

23. 0.00247

24. Find the value of $10,000 invested at 6% compounded daily for 5 years.

25. If a car averages 23.6 miles per gallon, how many miles will it go on 15 gallons of gas?

26. The cost of a sofa to a furniture store was $750. The sofa was sold for $1250. (a) What was the store's profit? (b) What was the percent of profit based on cost? (c) What was the percent of profit based on selling price?

27. The scale on a map indicates that $1\frac{3}{4}$ inches represents 50 miles. How many miles apart are two cities that are 2 inches apart on the map?

Solve each of the following equations.

28. $2x + 1 = -11$

29. $5(n - 1) = 3n + 15$

30. If one is added to a number, the result is equal to eleven more than three times the number. What is the number?

Measurement

What to Expect in this Chapter

This Chapter discusses the relationships among the units of measure in two systems: the metric system and the U.S. customary system. The emphasis in Sections M.1 and M.2 is placed on understanding the metric system and on changing from one unit of measure to an equivalent unit of measure within the metric system. To help you understand and to illustrate the ease of working within the metric system, special charts (with units of measure as headings) are provided, which clearly relate changes in units to placement of the decimal point. Section M.1 deals with units of length and units of area, and Section M.2 covers units of weight and units of volume.

Section M.3 discusses measures in the U.S. customary system, gives tables of equivalent measures in the metric system and the U.S. customary system, and shows how to change from a unit in one system to an equivalent unit in the other. The chapter ends with Section M.4, which presents units of measure used in medicine and formulas for converting an adult's dosage of medicine to a child's dosage.

Prospective home buyers will often look at many different properties before making a purchase.

Mathematics at Work!

When discussing a home with a prospective buyer, a real estate agent will usually have a sheet of paper with a picture of the home, the important features of the home (such as how many bedrooms, how many bathrooms, dining room, family room, living room, size and type of garage, whether or not there is a swimming pool, etc.), and the size of each room. Another fact of interest to many home buyers is the size of the lot, or the acreage. In the case of residential property, the lot size is generally stated in square feet; but, historically, people think of purchasing land by the acre, and many home owners would prefer to know what part of an acre the lot is (or, for larger parcels, how many acres are in the lot). In Section M.3, we state the fact that 43,560 $ft^2 = 1$ acre. Using this fact, answer the following question.

Suppose that the home you are considering buying sits on a rectangular lot that is 90 feet by 121 feet. What fraction of an acre is the lot?

(See also Exercise 73 in Section M.3.)

M.1 Metric System: Length and Area

OBJECTIVES

1. *Learn the basic prefixes in the metric system.*
2. *Know that the **meter** is the basic unit of length.*
3. *Be able to change metric measures of length without a chart.*
4. *Know that area is measured in **square units.***
5. *Be able to change metric measures of area without a chart.*
6. *Know that the **are** and **hectare** are the basic units of land area.*

About 90% of the people in the world use the metric system of measurement. The United States is the only major industrialized country still committed to the U.S. customary system (formerly called the English system). Even in the United States, the metric system has been used for years in such fields as medicine, science, and the military. Auto mechanics who work on imported cars must be familiar with the metric system because the measures of the auto parts are in metric units.

Length

The **meter** is the basic unit of length in the metric system. Smaller and larger units are named by putting a prefix in front of the basic unit—for example, **centi**meter and **kilo**meter. The prefixes we will use are listed here from smallest to largest unit size:

milli- (thousandths)	deka- (tens)
centi- (hundredths)	hecto- (hundreds)
deci- (tenths)	kilo- (thousands)

Other prefixes that indicate extremely small units are *micro-, nano-, pico-, femto-,* and *atto-*. Prefixes that indicate extremely large units are *mega-, giga-,* and *tera-*.

Some examples of metric lengths are shown in Figure M.1.

one millimeter (mm): About the width of the wire in a paper clip

one centimeter (cm): About the width of a paper clip

one meter (m): Just over 39 inches, or slightly longer than a yard
one kilometer (km): About 0.62 of a mile

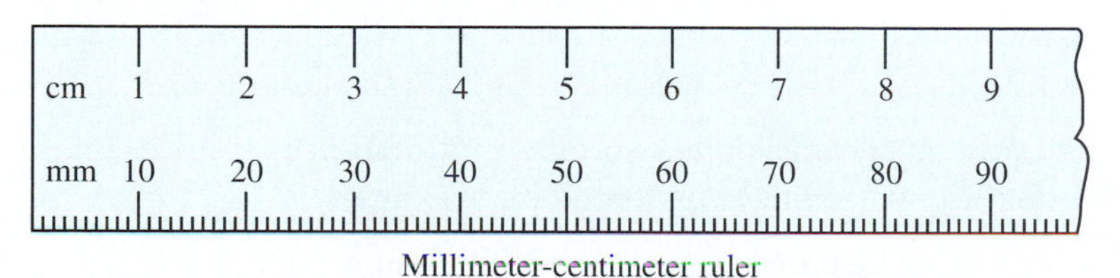

Millimeter-centimeter ruler

Figure M.1

Table M.1 lists the metric prefixes and their values in decimal form. **These prefixes must be memorized in order.** Table M.2 lists the measures of length, their relationships to the meter, and their abbreviations.

Table M.1 Metric Prefixes and Their Values

Prefix	Value	
milli-	0.001	thousandths
centi-	0.01	hundredths
deci-	0.1	tenths
basic unit	1	ones
deka-	10	tens
hecto-	100	hundreds
kilo-	1000	thousands

Table M.2 Measures of Length

1 **milli**meter (mm)	= 0.001 meter
1 **centi**meter (cm)	= 0.01 meter
1 **deci**meter (dm)	= 0.1 meter
1 meter (m)	= 1.0 meter
1 **deka**meter (dam)	= 10 meters
1 **hecto**meter (hm)	= 100 meters
1 **kilo**meter (km)	= 1000 meters

As you can tell from Tables M.1 and M.2, the metric units of length are related to each other by powers of 10. That is, you multiply by 10 to get the equivalent measure in the next lower (or smaller) unit. For example,

1 m = 10 dm and 1 cm = 10 mm

Conversely, you need only divide by 10 to get the equivalent measure in the next higher (or larger) unit. For example,

1 m = 0.1 dam and 1 dam = 0.1 hm

Thus, you will have more of a smaller unit and less of a larger unit.

The following method of using a chart makes conversion from one metric unit to another quite easy.

Changing Units of Length

A chart like the one shown here can be used to change from one unit of length to another in the metric system.

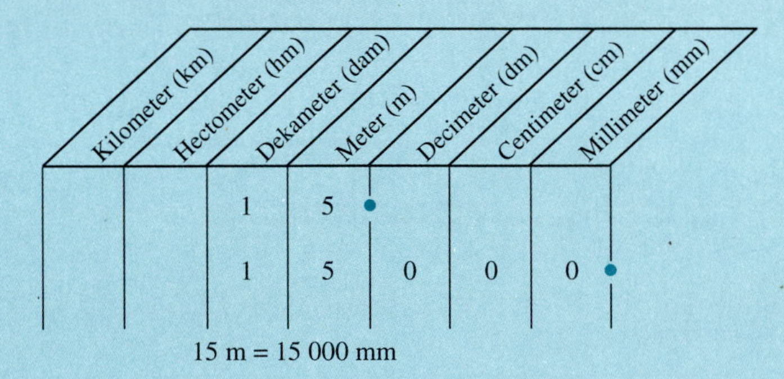

1. List each unit across the top. Memorize the unit prefixes in order.
2. Enter the given number so that each digit is in one column and the decimal point is on the given unit line.
3. Move the decimal point to the desired unit line.
4. Fill in all spaces with 0's.

In the metric system,

1. A 0 is written to the left of the decimal point if there is no whole number part (0.25 m).
2. No commas are used in writing numbers. If a number has more than four digits (left or right of the decimal point), the digits are grouped in threes from the decimal point with a space between the groups (14 000 m).

Refer to the chart below for Examples 1–5.

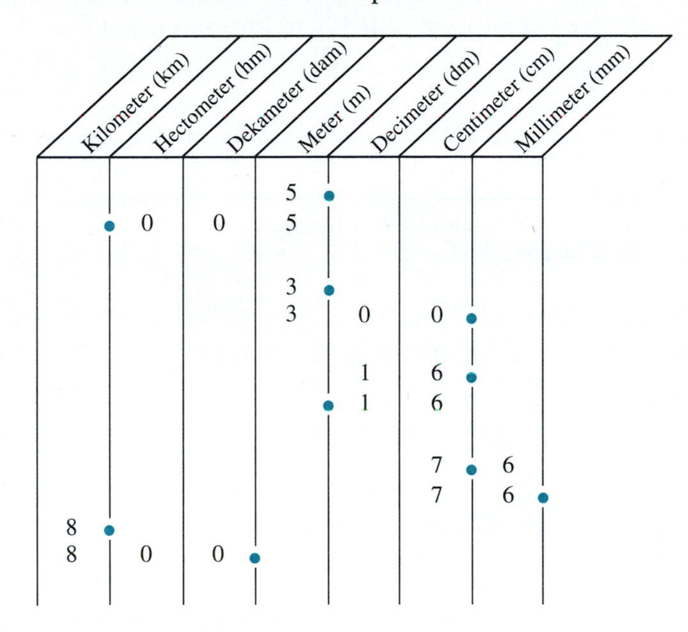

EXAMPLE 1 5m = ______ km

Solution

Move the decimal point left to the km line and fill in the columns with 0's:

5 m = 0.005 km

EXAMPLE 2 3 m = ______ cm

Solution

Move the decimal point right to the cm line and fill in the columns with 0's:

3 m = 300 cm

EXAMPLE 3 Change 16 cm to meters.

Solution

Move the decimal point left to the m line. Be sure to write 0 for the whole number part:

16 cm = 0.16 m

Make your own chart on a piece of paper and then use the chart to change the following units as indicated.

1. 5 m = __________ mm

2. 1.6 m = __________ cm

3. 83 m = __________ mm

4. 7 cm = __________ m

5. 325 mm = __________ m

EXAMPLE 4 How many millimeters are there in 7.6 cm?

Solution

Move the decimal point right to the mm line:

7.6 cm = 76 mm

EXAMPLE 5 Convert 8 km to dekameters.

Solution

Move the decimal point right to the dam line and fill in 0's:

8 km = 800 dam

Using a chart, change the units as indicated in Completion Examples 6–10.

COMPLETION EXAMPLES 6–10

6. 4 m = __________ mm

7. 3.1 cm = __________ mm

8. 50 cm = __________ m

9. 18 km = __________ m

10. 25 km = __________ m

Now Work Exercises 1–5 in the Margin.

After some practice, you may not need a chart to help in changing units. If so, you can use the following technique. However, the chart method is highly recommended.

Changing Metric Measures of Length Without a Chart

1. To change to a measure that is

One unit smaller, multiply by 10:	3 cm = 30 mm
Two units smaller, multiply by 100:	5 m = 500 cm
Three units smaller, multiply by 1000:	14 m = 14 000 mm
and so on.	

2. To change to a measure that is

One unit larger, divide by 10:	50 cm = 5 dm
Two units larger, divide by 100:	50 cm = 0.5 m
Three units larger, divide by 1000:	13 mm = 0.013 m
and so on.	

COMPLETION EXAMPLES 11–14

11. 42 m = 420 dm = 420 cm = __________ mm

12. 17 m = __________ dm = __________ cm = __________ mm

13. 6 m = 0.6 dam = 0.06 hm = __________ km

14. 112 m = __________ dam = __________ hm = __________ km

Now Work Exercises 6–10 in the Margin.

Area

Area is a measure of the interior of (or surface enclosed by) a figure in a plane. For example, the two rectangles in Figure M.2 have different areas because they have different amounts of interior space, or different amounts of space in the plane are enclosed by the sides of the figures.

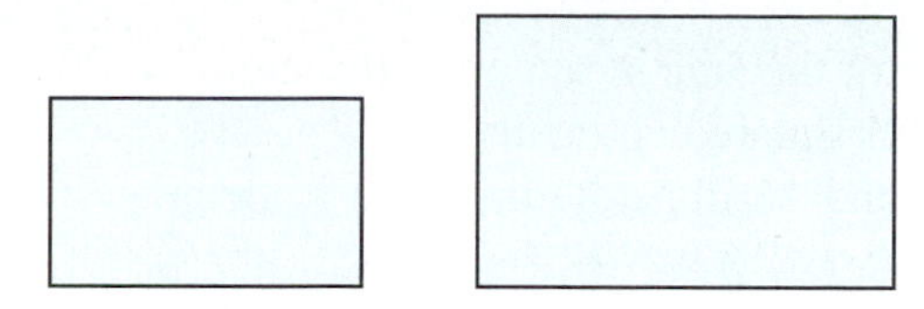

These two rectangles have different **areas**.

Figure M.2

Area is measured in **square units.** In metric units, a square that is 1 centimeter long on each side is said to have an area of 1 square centimeter, or the area is $1 cm^2$. A rectangle that has a length of 7 centimeters and a width of 4 centimeters encloses 28 squares, each of which has an area of 1 cm^2. So the rectangle is said to have an area of 28 cm^2. (See Figure M.3.)

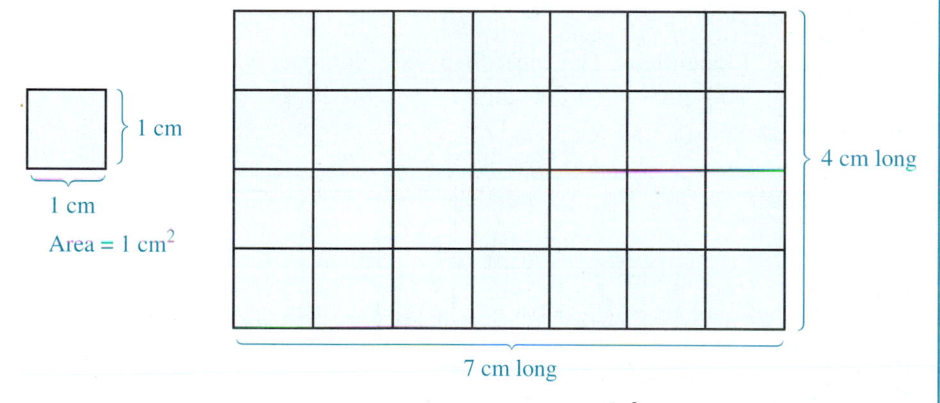

Area = 7 cm x 4 cm = 28 cm^2

There are 28 squares that are each 1 cm^2 in the large rectangle.

Figure M.3

Table M.3 shows metric area measures useful for relatively small areas. For example, the area of the floor of your classroom could be measured in square meters for carpeting, and the area of this page could be measured in square centimeters. Other measures, listed in Table M.4 on page M11, are used for measuring land area.

Table M.3 Measures of Small Area

$1 cm^2 = 100 mm^2$

$1 dm^2 = 100 cm^2 = 10\,000 mm^2$

$1 m^2 = 100 dm^2 = 10\,000 cm^2 = 1\,000\,000 mm^2$

Without a chart, change the units as indicated.

6. 35 m = __________ cm

7. 35 m = __________ mm

8. 6.4 cm = __________ mm

9. 5.9 m = __________ km

10. 8 m = __________ km

Note that each unit of area in the metric system is 100 times the next smaller unit of area—**not** just 10 times, as it is with length. For example, consider the rectangle in Figure M.3. If the sides are measured in millimeters, then the dimensions are 70 millimeters by 40 millimeters (each length is increased by a factor of 10) and the area is

$$7 \text{ cm} \times 4 \text{ cm} = 28 \text{ cm}^2$$

or

$$70 \text{ mm} \times 40 \text{ mm} = 2800 \text{ mm}^2$$

Thus, the measure of the area in square millimeters is 100 times the measure of the same area in square centimeters.

Examples 15 and 16 illustrate this idea in detail.

EXAMPLE 15

A square 1 centimeter on a side encloses 100 square millimeters.

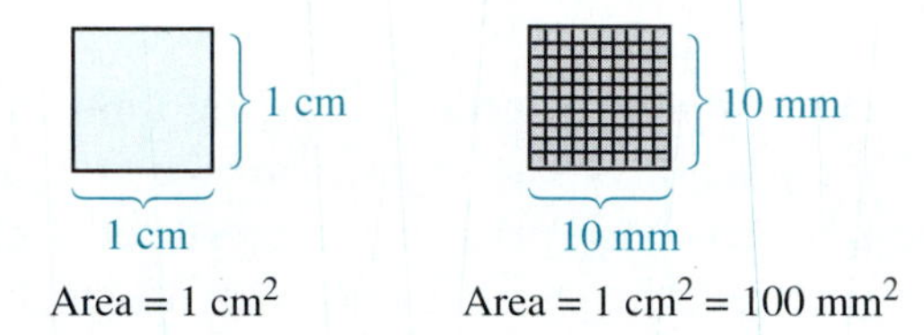

EXAMPLE 16

A square 1 decimeter (10 cm) on a side encloses 1 square decimeter. ($1 \text{ dm}^2 = 100 \text{ cm}^2 = 10\,000 \text{ mm}^2$)

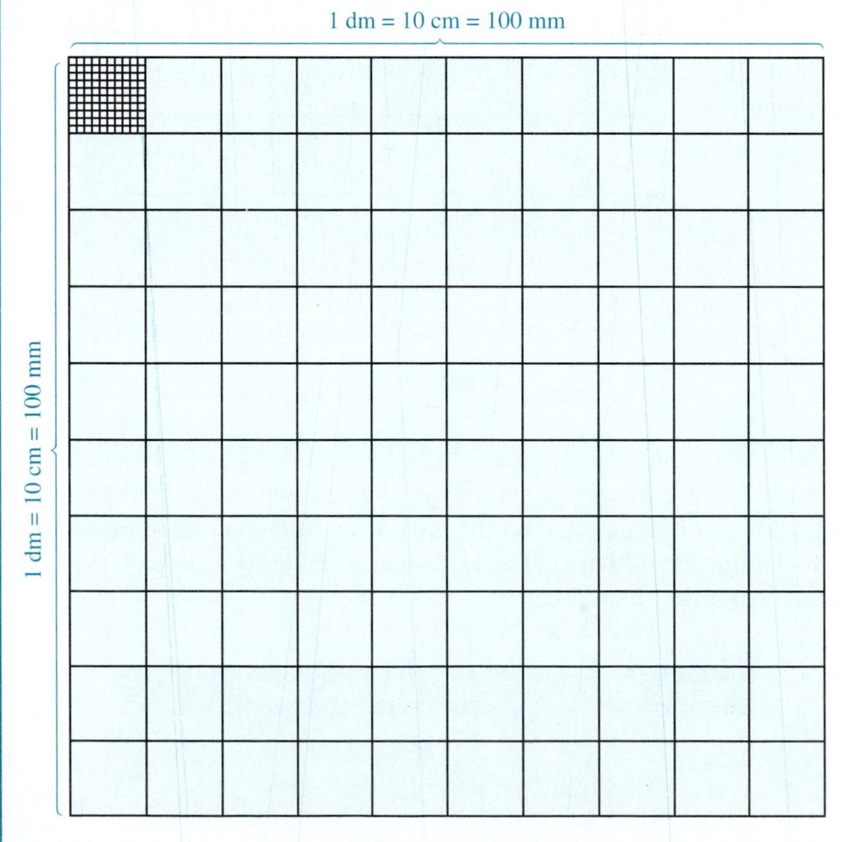

A chart similar to that used on page M4 can be used to change measures of area. **The key difference is that there must be two digits in each column.** This corresponds to multiplication of each smaller unit of area by 100.

Changing Units of Area

A chart like the one shown here can be used to change from one unit of area to another.

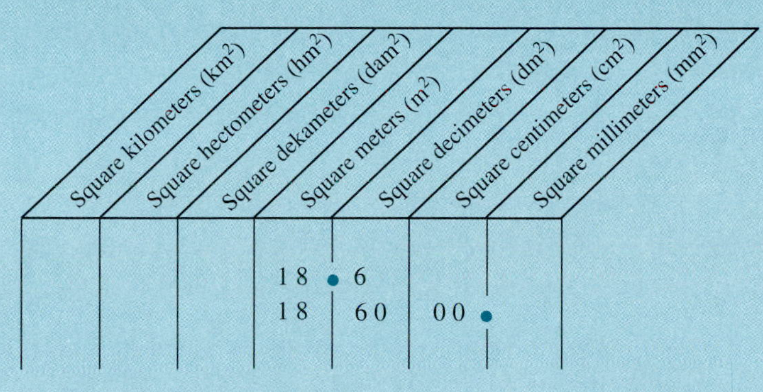

$18.6 \text{ m}^2 = 186\,000 \text{ cm}^2$

1. List each area unit across the top. (Abbreviations will do.)
2. Enter the given number so that there are two digits in each column, with the decimal point on the given unit line.
3. Move the decimal point to the desired unit line.
4. Fill in the spaces with 0's, using two digits per column.

Refer to the chart below for Examples 17–21.

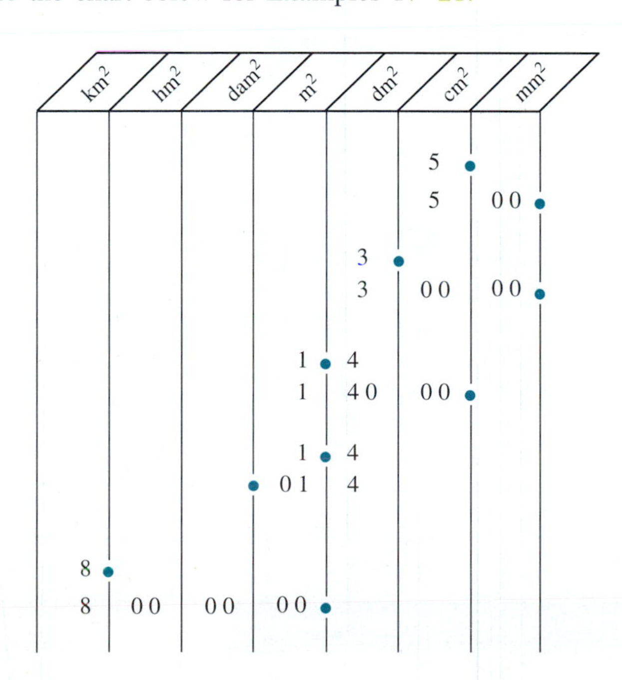

Example 17 $5\ \text{cm}^2 =$ _____ mm^2.

Solution

Move the decimal point right to the mm^2 line and fill in the columns with two 0's per column:

$$5\ \text{cm}^2 = 500\ \text{mm}^2$$

Example 18 Convert $3\ \text{dm}^2$ to mm^2.

Solution

Move the decimal point right to the mm^2 line and fill in two 0's per column:

$$3\ \text{dm}^2 = 30\ 000\ \text{mm}^2$$

Example 19 Change $1.4\ \text{m}^2$ to cm^2.

Solution

Move the decimal point right to the mm^2 line and fill in with 0's so that there are two digits per column:

$$1.4\ \text{m}^2 = 14\ 000\ \text{cm}^2$$

Example 20 How many dam^2 are there in $1.4\ \text{m}^2$?

Solution

Move the decimal point left to the dam^2 line and fill in only one 0 so that there are two digits in the column between the decimal point and the digit 1:

$$1.4\ \text{m}^2 = 0.014\ \text{dam}^2$$

Example 21 Express $8\ \text{km}^2$ as m^2.

Solution

Move the decimal point right to the m^2 line and fill in two 0's per column:

$$8\ \text{km}^2 = 8\ 000\ 000\ \text{m}^2$$

Using a chart, change the units as indicated in Completion Examples 22–24.

☑ Completion Examples 22–24

22. $8.52\ \text{m}^2 =$ ______________ $\text{dm}^2 =$ ______________ cm^2

23. $147\ \text{cm}^2 =$ ______________ mm^2

24. $3.8\ \text{cm}^2 =$ ______________ $\text{dm}^2 =$ ______________ m^2

Now Work Exercises 11–14 in the Margin.

Change the units as indicated.

11. $22\ \text{cm}^2 =$ ____________ mm^2

12. $500\ \text{mm}^2 =$ ____________ cm^2

13. $3.7\ \text{dm}^2 =$ ____________ cm^2

$=$ ____________ mm^2

14. $5200\ \text{mm}^2$

$=$ ____________ cm^2

$=$ ____________ dm^2

Land Area

A square 10 meters on a side encloses an area of 1 **are** (a). A **hectare** (ha) is 100 ares. The are and hectare are used in measuring land area. (See Table M.4.)

Table M.4 Measures of Land Area

$1 \text{ a} = 100 \text{ m}^2$
$1 \text{ ha} = 100 \text{ a} = 10\,000 \text{ m}^2$

EXAMPLE 25

A square 10 meters on a side encloses 100 m^2, or 1 are.

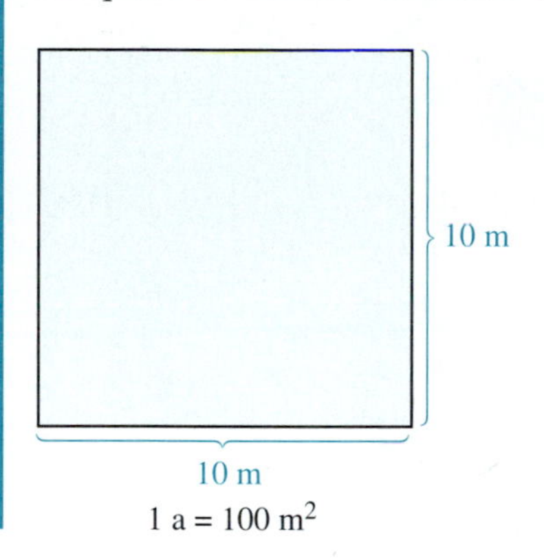

$1 \text{ a} = 100 \text{ m}^2$

EXAMPLE 26

(a) To change units from a to m^2, multiply by 100 (move the decimal point two places to the right):

$$3.2 \text{ a} = 100(3.2) \text{ m}^2 = 320 \text{ m}^2$$

(b) To change units from m^2 to a, divide by 100 (move the decimal point two places to the left):

$$65 \text{ m}^2 = \frac{65}{100} \text{a} = 0.65 \text{ a}$$

EXAMPLE 27

How many ares are in 1 km^2? (**Note:** For comparison, one km is about 0.6 mile, so 1 km^2 is about 0.6 mi $\times$ 0.6 mi = 0.36 mi^2.)

Solution

Remember that 1 km = 1000 m, so

$$\begin{aligned} 1 \text{ km}^2 &= 1000 \text{ m} \times 1000 \text{ m} \\ &= 1\,000\,000 \text{ m}^2 \\ &= 10\,000 \text{ a} \quad \text{Divide m}^2 \text{ by 100 to get ares.} \end{aligned}$$

Change the units as indicated.

15. 3.6 a = __________ m^2

16. 0.73 ha = __________ a

= __________ m^2

17. 5.4 ha = __________ a

= __________ m^2

18. 49 500 m^2 = __________ a

= __________ ha

Example 28

A farmer plants corn and beans as shown in the following figure. How many hectares are planted in corn? In beans? (From Example 27 we know that 1 km^2 = 10 000 a.)

Solution

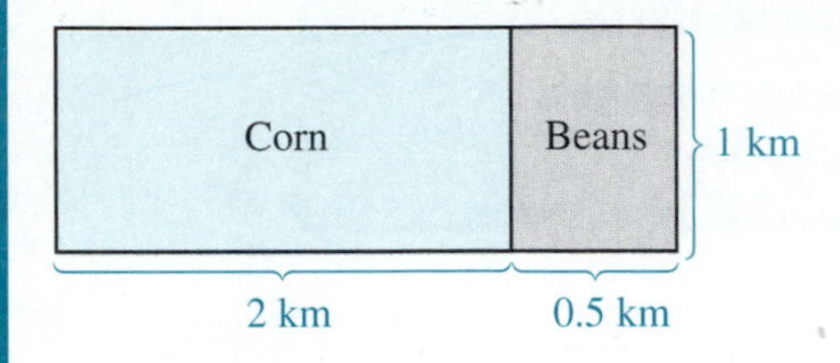

Corn: (2 km) · (1 km) = 2 km^2 = 20 000 a = 200 ha

Beans: (0.5 km) · (1 km) = 0.5 km^2 = 5000 a = 50 ha

Now Work Exercises 15–18 in the Margin.

Completion Example Answers

6. 4 m = **4000** mm 7. 3.1 cm = **31** mm 8. 50 cm = **0.5** m

9. 18 km = **18 000** m 10. 25 km = **25 000** m

11. 42 m = 420 dm = 4200 cm = **42 000** mm

12. 17 m = **170** dm = **1700** cm = **17 000** mm

13. 6 m = 0.6 dam = 0.06 hm = **0.006** km

14. 112 m = **11.2** dam = **1.12** hm = **0.112** km

22. 8.52 m^2 = **852** dm^2 = **85 200** cm^2

23. 147 cm^2 = **14 700** mm^2

24. 3.8 cm^2 = **0.038** dm^2 = **0.000 38** m^2

NAME ______________ SECTION ______________ DATE ______________

ANSWERS

1. [Respond below exercise.] ______
2. ______
3. ______
4. ______
5. ______
6. ______
7. ______
8. ______
9. ______
10. ______
11. ______
12. ______
13. ______
14. ______
15. ______
16. ______
17. ______
18. ______
19. ______
20. ______
21. ______
22. ______
23. ______
24. ______
25. ______
26. ______
27. ______
28. ______
29. ______

Exercises M.1

1. Write the six metric prefixes discussed in this section in order from largest to smallest.

Change the following units of length as indicated. Use the chart method until you become thoroughly familiar with the metric system.

2. 1 m = ______ cm

3. 5 m = ______ cm

4. 12 m = ______ cm

5. 6 m = ______ cm

6. 2 m = ______ mm

7. 0.3 m = ______ mm

8. 0.7 m = ______ mm

9. 1.4 m = ______ mm

10. 1.6 cm = ______ mm

11. 1.8 cm = ______ mm

12. 25 cm = ______ mm

13. 35 cm = ______ mm

14. 4 m = ______ dm

15. 16 m = ______ dm

16. 7 dm = ______ cm

17. 21 dm = ______ cm

18. 3 km = ______ m

19. 5 km = ______ m

20. 5.28 km = ______ m

21. 6.4 km = ______ m

22. 11 mm = ______ cm

23. 26 mm = ______ cm

24. 72 mm = ______ cm

25. 48 mm = ______ cm

26. Change 6 mm to decimeters.

27. Change 12 mm to centimeters.

28. Change 20 mm to meters.

29. Change 30 mm to meters.

30. ____________
31. ____________
32. ____________
33. ____________
34. ____________
35. ____________
36. ____________
37. ____________
38. ____________
39. ____________
40. ____________
41. ____________
42. ____________
43. ____________
44. ____________
45. ____________
46. ____________
47. ____________
48. ____________
49. ____________
50. ____________
51. ____________
52. ____________
53. ____________
54. ____________
55. ____________

30. Convert 145 mm to meters.

31. Convert 256 mm to meters.

32. Convert 25 cm to meters.

33. Convert 32 cm to meters.

34. How many meters are in 150 cm?

35. How many meters are in 170 cm?

36. How many kilometers are in 3000 m?

37. How many kilometers are in 2400 m?

38. Express 500 m in kilometers.

39. Express 400 m in kilometers.

40. Express 3.45 m in centimeters.

41. Express 4.62 m in centimeters.

42. 6.3 cm = __________ m

43. 5.2 cm = __________ m

44. 3.25 m = __________ mm

45. 6.41 m __________ mm

46. How many centimeters are in 3 mm?

47. How many centimeters are in 5 mm?

48. Change 32 mm to meters.

49. Change 57 mm to meters.

50. What number of kilometers is equivalent to 20 000 m?

51. What number of kilometers is equivalent to 35 000 m?

52. Express 1.5 km in meters.

53. Express 2.3 km in meters.

54. How many kilometers are in 0.5 m?

55. How many kilometers are in 1.5 m?

NAME ____________ SECTION ____________ DATE ____________

Change the following units of area as indicated. Use the chart method until you become thoroughly familiar with the metric system.

56. 3 cm^2 = ________ mm^2

57. 5.6 cm^2 = ________ mm^2

58. 8.7 cm^2 = ________ mm^2

59. 3.61 cm^2 = ________ mm^2

60. Express 600 mm^2 in cm^2

61. Express 28 mm^2 in cm^2.

62. How many cm^2 are in 1400 mm^2?

63. How many cm^2 are in 20 000 mm^2?

64. 4 dm^2 = ________ cm^2 = ________ mm^2

65. 7.3 dm^2 = ________ cm^2 = ________ mm^2

66. 57 dm^2 = ________ cm^2 = ________ mm^2

67. 0.6 dm^2 = ________ cm^2 = ________ mm^2

68. 17 m^2 = ________ dm^2 = ________ cm^2 = ________ mm^2

69. 2.9 m^2 = ________ dm^2 = ________ cm^2 = ________ mm^2

70. 0.03 m^2 = ________ dm^2 = ________ cm^2 = ________ mm^2

Change the following units of land area as indicated.

71. 7.8 a = ________ m^2

72. 300 a = ________ m^2

73. 0.04 a = ________ m^2

74. 0.53 a = ________ m^2

75. 8.69 ha = ________ a = ________ m^2

76. 7.81 ha = ________ a = ________ m^2

ANSWERS

56. ________

57. ________

58. ________

59. ________

60. ________

61. ________

62. ________

63. ________

64. [Respond in exercise.]

65. [Respond in exercise.]

66. [Respond in exercise.]

67. [Respond in exercise.]

68. [Respond in exercise.]

69. [Respond in exercise.]

70. [Respond in exercise.]

71. ________

72. ________

73. ________

74. ________

75. [Respond in exercise.]

76. [Respond in exercise.]

77. [Respond in exercise.]

78. [Respond in exercise.]

79. ____________

80. ____________

81. ____________

82. ____________

83. ____________

84. ____________

77. 0.16 ha = ____________ a = ____________ m^2

78. 0.02 ha = ____________ a = ____________ m^2

79. How many hectares are in 1 a?

80. How many hectares are in 15 a?

81. How many hectares are in 5 km^2?

82. Change 4.76 km^2 to hectares.

83. Change 0.3 km^2 to hectares.

84. Change 650 ha to ares.

M.2 Metric System: Weight and Volume

OBJECTIVES

1. *Become familiar with the concepts of mass and weight.*
2. *Know that the **kilogram** is the basic unit of mass.*
3. *Know that in some fields the **gram** is a more convenient unit of mass.*
4. *Be able to change metric measures of mass.*
5. *Know that volume is measured in **cubic units.***
6. *Be able to change metric measures of volume.*
7. *Know that the **liter** is the basic unit of liquid volume.*

Mass (Weight)

Mass is the amount of material in an object. Regardless of where the object is in space, its mass remains the same. (See Figure M.4.) **Weight** is the force of the earth's gravitational pull on an object. The farther an object is from earth, the less the gravitational pull of the earth. Thus, astronauts experience weightlessness in space, but their mass is unchanged.

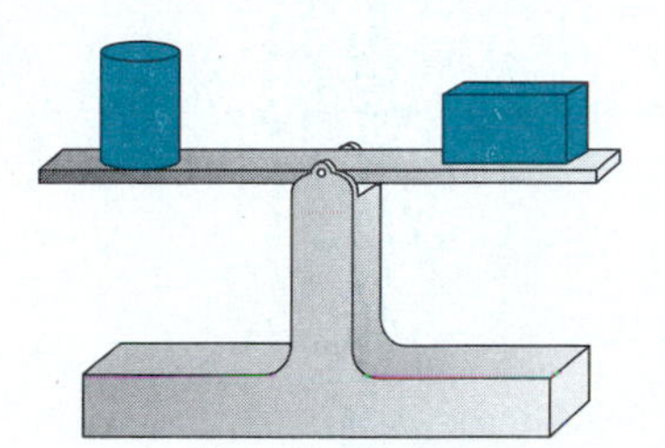

The two objects have the same **mass** and balance on an equal arm balance, regardless of their location in space.

Figure M.4

Because most of us do not stray far from the earth's surface, in this text weight and mass will be used interchangeably. Thus, a **mass** of 20 kilograms will be said to **weigh** 20 kilograms.

The basic unit of mass in the metric system is the **kilogram,*** about 2.2 pounds. In some fields, such as medicine, the **gram** (about the mass of a paper clip) is more convenient to use as a basic unit than the kilogram.

Large masses, such as loaded trucks and railroad cars, are measured by the **metric ton** (1000 kilograms, or about 2200 pounds). (See Tables M.5 and M.6.)

Table M.5 Measures of Mass

1 **milli**gram (mg)	= 0.001 gram
1 **centi**gram (cg)	= 0.01 gram
1 **deci**gram (dg)	= 0.1 gram
1 gram (g)	= 1.0 gram
1 **deka**gram (dag)	= 100 grams
1 **hecto**gram (hg)	= 1000 grams
1 **kilo**gram (kg)	= 1000 grams
1 metric ton (t)	= 1000 kilograms

Table M.6 Equivalent Measures of Mass

1000 mg = 1 g	0.001 g = 1 mg
1000 g = 1 kg	0.001 kg = 1 g
1000 kg = 1 t	0.001 t = 1 kg
1 t = 1000 kg = 1 000 000 g	
= 1 000 000 000 mg	

* Technically, a kilogram is the mass of a certain cylinder of platinum-iridium alloy kept by the International Bureau of Weights and Measures in Paris.

Originally, the basic unit was a gram, defined as the mass of 1 cm^3 of distilled water at 4° Celsius. This mass is still considered accurate for many purposes, so

1 cm^3 of water has a mass of 1 g
1 dm^3 of water has a mass of 1 kg
1 m^3 of water has a mass of 1000 kg, or 1 metric ton

The centigram, decigram, dekagram, and hectogram have little practical use and are not included in the exercises. For completeness, they are all included in the headings of the charts used to change units.

Changing Units of Mass

A chart like the one shown here can be used to change from one unit of mass to another.

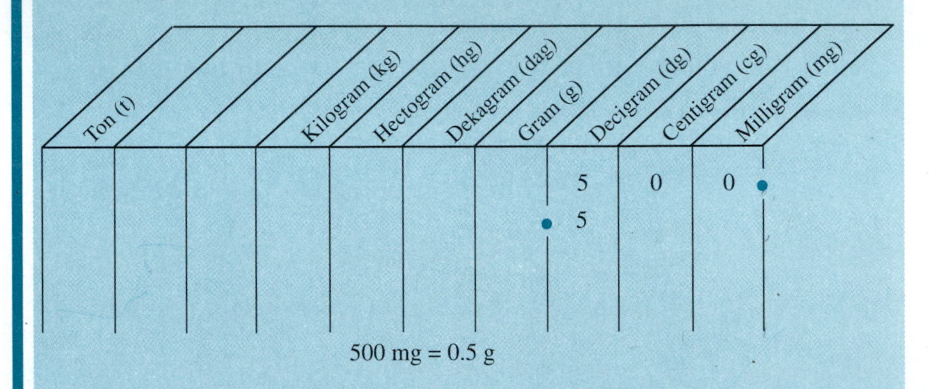

1. List each unit across the top.
2. Enter the given number so that there is one digit in each column, with the decimal point on the given unit line.
3. Move the decimal point to the desired unit line.
4. Fill in the spaces with 0's, using one digit per column.

Refer to the following chart for Examples 1–5.

Ton (t)			Kilogram (kg)	Hectogram (hg)	Dekagram (dag)	Gram (g)	Decigram (dg)	Centigram (cg)	Milligram (mg)
								2	3.
							.0	2	3
						6.			
						6	0	0	0.
		4	9.						
		4	9	0	0	0.			
5.									
5	0	0	0.						
		7	0.						
	.0	7	0						

Example 1 Express 23 mg in grams.

Solution

Move the decimal point left to the g line and fill in the dg column with 0. Write 0 for the whole number part:

23 mg = 0.023 g

Example 2 Express 6 g in milligrams.

Solution

Move the decimal point right to the mg line and fill in the columns with 0's:

6 g = 6000 mg

Example 3 Change 49 kg to grams.

Solution

Move the decimal point right to the g line and fill in the columns with 0's:

49 kg = 49 000 g

Example 4 How many kilograms are in 5 t?

Solution

Move the decimal point right to the kg line and fill in the columns with 0's:

5 t = 5000 kg

Example 5 How many tons are in 70 kg?

Solution

Move the decimal point left to the t line and fill in with one 0. The whole number part is 0:

70 kg = 0.07 t

Change the units as indicated in Completion Examples 6–10.

Completion Examples 6–10

6. 60 mg = 0.06 g = ________ kg

7. 135 mg = ________ g = ________ kg

8. 5 700 000 kg = ________ t

9. 100 g = ________ kg

10. 78 g = ________ mg

Now Work Exercises 1–5 in the Margin.

Change the units as indicated.

1. 43 g = ________ mg

2. 7.8 g = ________ mg

3. 350 mg = ________ g

4. 75 kg = ________ g

5. 3940 kg = ________ t

Volume

Volume is a measure of the space enclosed by a three-dimensional figure and is measured in **cubic units.** The volume or space contained within a cube that is 1 centimeter on each edge is **one cubic centimeter, or 1 cm^3,** as shown in Figure M.5. A cubic centimeter is about the size of a sugar cube.

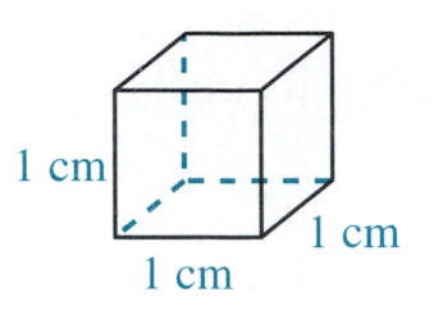

Volume = 1 cm^3

Figure M.5

A rectangular solid that has edges of 3 centimeters, 2 centimeters, and 5 centimeters has a volume of 3 cm × 2 cm × 5 cm = 30 cm^3. We can think of the rectangular solid as being three layers of 10 cubic centimeters, as shown in Figure M.6.

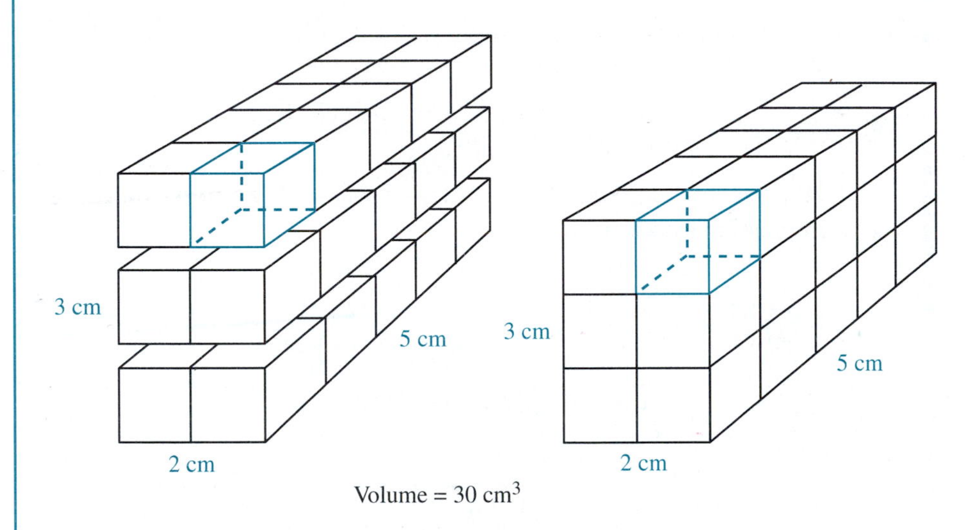

Volume = 30 cm^3

Figure M.6

If a cube is 1 decimeter along each edge, then the volume of the cube is 1 cubic decimeter (or 1dm^3). In terms of centimeters, this same cube has volume

$$10 \text{ cm} \times 10 \text{ cm} \times 10 \text{ cm} = 1000 \text{ cm}^3$$

That is, as shown in Figure M.7,

$$1 \text{ dm}^3 = 1000 \text{ cm}^3 \quad \text{or} \quad 1000 \text{ cm}^3 = 1 \text{ dm}^3$$

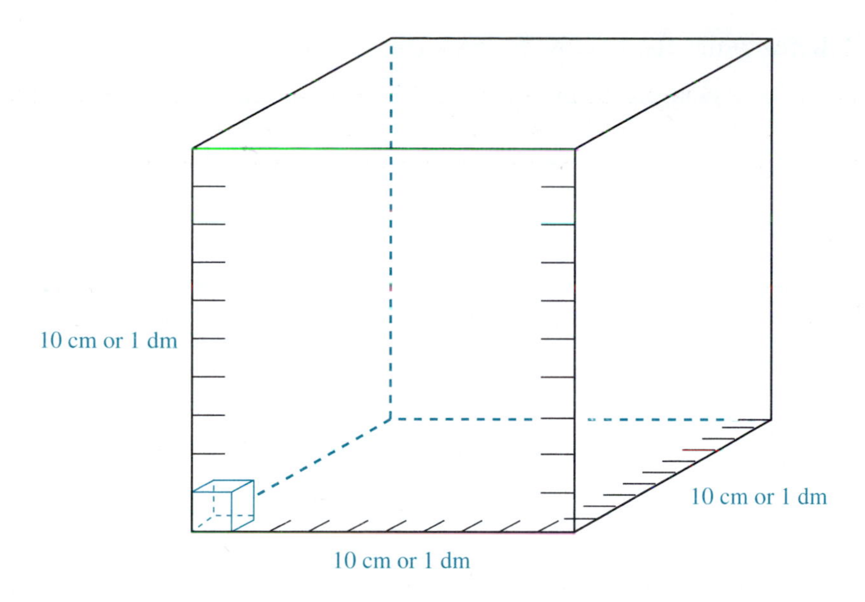

Figure M.7

The following relationships are true for cubic units in the metric system:

1. Equivalent cubic units in the next smallest unit can be found by multiplying the original number of units by 1000.
2. Equivalent cubic units in the next largest unit can be found by dividing the original number of units by 1000.

Again, we can use a chart; however, this time **there must be three digits in each column.**

Changing Units of Volume

A chart like the one shown here can be used to change from one unit of volume to another.

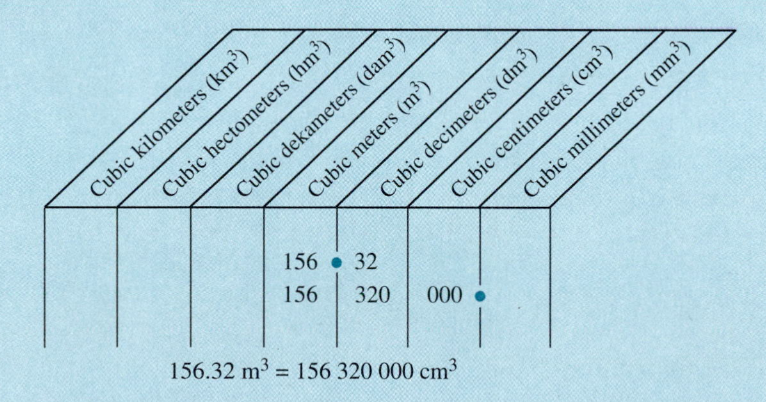

1. List each volume unit across the top. (Abbreviations will do.)
2. Enter the given number so that there are three digits in each column, with the decimal point on the given unit line.
3. Move the decimal point to the desired unit line.
4. Fill in the spaces with 0's, using three digits per column.

Refer to the chart below for Examples 11–14.

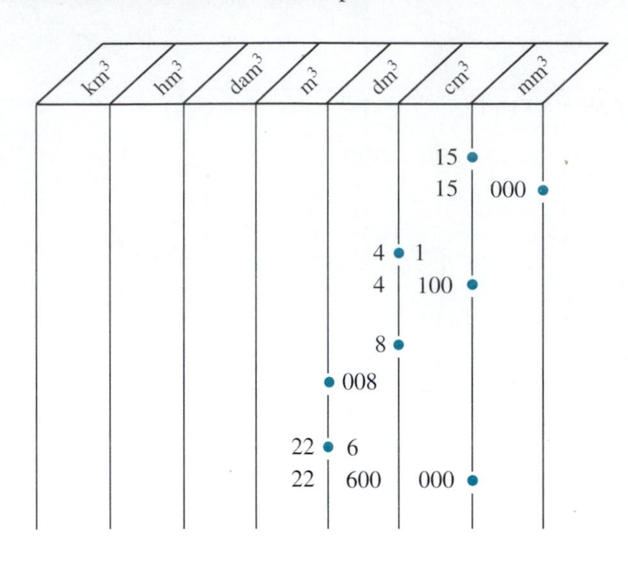

EXAMPLE 11 $15 \text{ cm}^3 =$ ________ mm^3

Solution

Move the decimal point right to the mm³ line and fill in the column with 0's so that there are three digits per column:

$15 \text{ cm}^3 = 15\ 000 \text{ mm}^3$

EXAMPLE 12 Convert 4.1 dm³ to cm³.

Solution

Move the decimal point right to the cm³ line and fill in the column with 0's so that there are three digits per column:

$4.1 \text{ dm}^3 = 4100 \text{ cm}^3$

EXAMPLE 13 How many cubic meters are there in 8 dm³?

Solution

Move the decimal point left to the m³ line and fill in with 0's. Write 0 as the whole number part:

$8 \text{ dm}^3 = 0.008 \text{ m}^3$

EXAMPLE 14 Express 22.6 m³ in cm³.

Solution

Move the decimal point right to the cm³ line and fill in with 0's:

$22.6 \text{ m}^3 = 22\ 600\ 000 \text{ cm}^3$

Volumes measured in cubic kilometers are so large that they are not used in everyday situations. Possibly some scientists work with these large volumes; however, in this text the discussion and exercises are related to the more practical units of m³, dm³, cm³, and mm³.

Using a chart, change the units as indicated in Completion Examples 15–17.

COMPLETION EXAMPLES 15–17

15. 3.7 dm^3 = ________ cm^3

16. 0.8 mm^3 = ________ cm^3

17. 4 m^3 = ________ dm^3 = ________ 3 cm^3

Liquid Volume

Liquid volume is measured in **liters** (abbreviated L). You are probably familiar with 1 L and 2 L bottles of soda on your grocer's shelf. A **liter** is the volume enclosed in a cube that is 10 centimeters on each edge. So, 1 liter is equal to

$$10 \text{ cm} \times 10 \text{ cm} \times 10 \text{ cm} = 1000 \text{ cm}^3 \quad \text{or} \quad 1 \text{ liter} = 1000 \text{ cm}^3$$

That is, the cubic box shown in Figure M.7 would hold 1 liter of liquid.

The prefixes kilo-, hecto-, deka-, deci-, centi-, and milli- all indicate the same part of a liter as they do of the meter (see Tables M.7 and M.8). The same type of chart used in Section M.1 **with one digit per column** will be helpful for changing units. The centiliter (cL), deciliter (dL), and dekaliter (daL) are not commonly used and are not included in the tables or exercises.

Table M.7 Measures of Liquid Volume

1 **milli**liter (mL)	= 0.001 liter
1 liter (L)	= 1.0 liter
1 **hecto**liter (hL)	= 100 liters
1 **kilo**liter (kL)	= 1000 liters

Table M.8 Equivalent Measures of Volume

1000 mL = 1 L	1 mL = 1 cm^3
1000 L = 1 kL	1 L = 1 dm^3
10 hL = 1 kL	1 kL = 1 m^3

The following chart shows examples of unit changes for several liquid volumes.

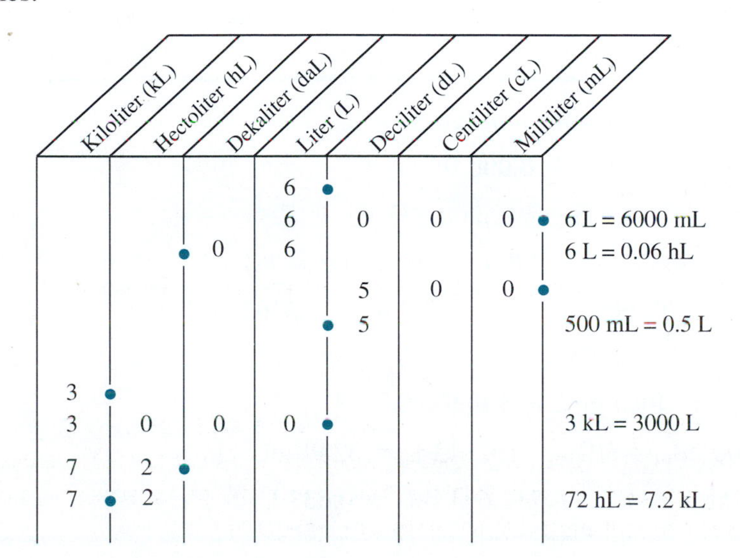

Change the units as indicated.

6. 2 mL = __________ L

7. 3.6 kL = __________ L

8. 500 mL = __________ L

9. 500 mL = __________ cm^3

10. 42 hL = __________ kL

Using the chart, change the units in Completion Examples 18–21 as indicated.

✔ Completion Examples 18–21

18. 5000 mL = __________ L

19. 3.2 L = __________ mL

20. 60 hL = __________ kL

21. 637 mL = __________ L

There is an interesting "crossover" relationship between liquid volume measures and cubic volume measures. Since

1 L = 1000 mL and 1 L = 1000 cm^3

we have

1 mL = 1 cm^3

Also,

1 kL = 1000 L = 1 000 000 cm^3 and 1 000 000 cm^3 = 1 m^3

This gives

1 kL = 1000 L = 1 m^3

Example 22

70 mL = __________ cm^3

70 mL = 70 cm^3

Example 23

3.8 kL = __________ m^3

3.8 kL = 3.8 m^3

Now Work Exercises 6–10 in the Margin.

Completion Example Answers

6. 60 mg = 0.06 g = **0.000 06** kg
7. 135 mg = **0.135** g = **0.000 135** kg
8. 5 700 000 kg = **5 700** t 9. 100 g = **0.1** kg
10. 78 g = **78 000** mg 15. 3.7 dm^3 = **3700** cm^3
16. 0.8 mm^3 = **0.0008** cm^3
17. 4 m^3 = **4000** dm^3 = **4 000 000** cm^3
18. 5000 mL = **5** L 19. 3.2 L = **3200** mL
20. 60 hL = **6** kL 21. 637 mL = **0.637** L

NAME ______________ SECTION ______________ DATE ______________

ANSWERS

1. ______
2. ______
3. ______
4. ______
5. ______
6. ______
7. ______
8. ______
9. ______
10. ______
11. ______
12. ______
13. ______
14. ______
15. ______
16. ______
17. ______
18. ______
19. ______
20. ______
21. ______
22. ______
23. ______
24. ______
25. ______
26. ______
27. ______
28. ______
29. ______
30. ______

Exercises M.2

Change the following units of mass (weight) as indicated.

1. 2 g = ______ mg

2. 7 kg = ______ g

3. 3700 kg = ______ t

4. 34.5 mg = ______ g

5. 5600 g = ______ kg

6. 4000 kg = ______ t

7. 91 kg = ______ t

8. 73 kg = ______ mg

9. 0.7 g = ______ mg

10. 0.54 g = ______ mg

11. How many kilograms are there in 5 t?

12. How many kilograms are there in 17 t?

13. Change 2 t to kilograms.

14. Change 896 mg to grams.

15. Express 986 g in milligrams.

16. Express 342 kg in grams.

17. Convert 75 000 g to kilograms.

18. Convert 3000 mg to grams.

19. Convert 7 t to grams.

20. Convert 0.4 t to grams.

21. Change 0.34 g to kilograms.

22. Change 0.78 g to milligrams.

23. How many grams are in 16 mg?

24. How many milligrams are in 2.5 g?

25. 92.3 g = ______ kg

26. 3.94 g = ______ mg

27. 7.58 t = ______ kg

28. 5.6 t = ______ kg

29. 2963 kg = ______ t

30. 3547 kg = ______ t

31. [Respond in exercise.]

32. [Respond in exercise.]

33. [Respond in exercise.]

34. [Respond in exercise.]

35. ______

36. ______

37. ______

38. ______

39. ______

40. ______

41. ______

42. ______

43. ______

44. ______

45. ______

46. ______

47. ______

48. ______

49. ______

50. ______

51. ______

52. ______

53. ______

54. ______

55. [Respond in exercise.]

56. [Respond in exercise.]

Complete the following tables.

31. 1 cm^3 = ______ mm^3
1 dm^3 = ______ cm^3
1 m^3 = ______ dm^3
1 km^3 = ______ m^3

32. 1 dm = ______ cm
1 dm = ______ mm
1 dm^2 = ______ cm^2
1 dm^2 = ______ mm^2
1 dm^3 = ______ cm^3
1 dm^3 = ______ mm^3

33. 1 m = ______ dm
1 m = ______ cm
1 m^2 = ______ dm^2
1 m^2 = ______ cm^2
1 m^3 = ______ dm^3
1 m^3 = ______ cm^3

34. 1 km = ______ m
1 km^2 = ______ m^2
1 km^3 = ______ m^3
1 km = ______ dm
1 km^2 = ______ ha
1 km^3 = ______ kL

Change the following units of volume as indicated.

35. 73 m^3 = ______ dm^3

36. 0.9 m^3 = ______ dm^3

37. 525 cm^3 = ______ m^3

38. 400 m^3 = ______ cm^3

39. 8.7 m^3 = ______ cm^3

40. 63 dm^3 = ______ m^3

41. How many cm^3 are in 45 mm^3?

42. How many mm^3 are in 3.1 cm^3?

43. Change 19 mm^3 to dm^3.

44. Change 5 cm^3 to mm^3.

45. Convert 2 dm^3 to cm^3.

46. Convert 76.4 mL to liters.

47. Change 5.3 L to milliliters.

48. Change 30 cm^3 to milliliters.

49. Change 30 cm^3 to liters.

50. Change 5.3 mL to liters.

51. 48 kL = ______ L

52. 72 000 L = ______ kL

53. 290 L = ______ kL

54. 569 mL = ______ L

55. 80 L = ______ mL = ______ cm^3

56. 7.3 L = ______ mL = ______ cm^3

M.3 U.S. Customary Measurements and Metric Equivalents

OBJECTIVES

1. *Become familiar with common units of measure in the U.S. customary system.*
2. *Be able to change measures within the U.S. customary system.*
3. *Know how to use formulas for converting between Fahrenheit and Celsius measures of temperature.*
4. *Learn how to use tables to convert units of measure between the U.S. customary system and the metric system.*

U.S. Customary Measurements

In the U.S. customary system (formerly the English system), the units are not systematically related as are the units in the metric system. Historically, some of the units were associated with parts of the body, which would vary from person to person. For example, a foot was the length of a person's foot, and a yard was the distance from the tip of one's nose to the tip of one's fingers with arm outstretched. A king might dictate his own foot to be the official "foot," but, of course, the next king might have a different-sized foot.

There is considerably more stability now because the official weights and measures are monitored by the government.

In this section, we will discuss the common units for length, area, liquid volume, weight, and time, and how to find equivalent measures. The basic relationships are listed in Table M.9. The measures of time are universal.

Table M.9 U.S. Customary Units of Measure and Equivalents

Length		Liquid Volume	
1 foot (ft)	= 12 inches (in.)	1 pint (pt)	= 16 fluid ounces (fl oz)
1 yard (yd)	= 3 ft	1 quart (qt)	= 2 pt = 32 fl oz
1 mile (mi)	= 5280 ft	1 gallon (gal)	= 4 qt
Area		**Time**	
1 ft^2	= 144 $in.^2$	1 minute (min)	= 60 seconds (sec)
1 yd^2	= 9 ft^2	1 hour (hr)	= 60 min
1 acre	= 4840 yd^2	1 day	= 24 hr
	= 43,560 ft^2		
Weight			
1 pound (lb)	= 16 ounces (oz)		
1 ton (t)	= 2000 lb		

To change units, you must either have a table of equivalent values with you or memorize the basic equivalent values. Most people know some of these values but not all.

There are several methods used to change from one unit to another. One is to use proportions and solve these proportions; another is to substitute ratios (commonly used in science courses); and another is to substitute equivalent values for just one unit and multiply. We will illustrate the third technique because it is probably the simplest.

Examples 1–9 illustrate several conversions between various commonly known U.S. customary units of measure.

EXAMPLE 1

4 ft = __________ in.

Solution

Think of 4 ft as the product of 4(1 ft) and replace 1 ft with 12 in.

$$4 \text{ ft} = 4(1 \text{ ft}) = 4(12 \text{ in.}) = 48 \text{ in.}$$

EXAMPLE 2

Convert 6 ft to inches.

Solution

$$6 \text{ ft} = 6(1 \text{ ft}) = 6(12 \text{ in.}) = 72 \text{ in.}$$

EXAMPLE 3

How many feet are there in 36 in.?

Solution

In changing from a smaller unit to a larger unit, we use a fraction. We know that 12 in. = 1 ft, so 1 in. = $\frac{1}{12}$ ft. Now,

$$36 \text{ in.} = 36(1 \text{ in.}) = 36\left(\frac{1}{12}\text{ft}\right) = 3 \text{ ft}$$

EXAMPLE 4

How many yards are there in 12 ft?

Solution

Since 3 ft = 1 yd, we have 1 ft = $\frac{1}{3}$yd, so

$$12 \text{ ft} = 12(1 \text{ ft}) = 12\left(\frac{1}{3}\text{yd}\right) = 4 \text{ yd}$$

EXAMPLE 5

How many acres are there in 87,120 ft^2?

Solution

Since 43,560 ft^2 = 1 acre, we have 1 $\text{ft}^2 = \frac{1}{43{,}560}$ acre, so

$$87{,}120 \text{ ft}^2 = 87{,}120\ (1 \text{ ft}^2) = 87{,}120\left(\frac{1}{43{,}560}\text{ acre}\right) = 2 \text{ acres}$$

Example 6

6 qt = __________ pt

Solution

6 qt = 6(1 qt) = 6(2 pt) = 12 pt

Example 7

Represent 2.5 lb in ounces.

Solution

2.5 lb = 2.5(1 lb) = 2.5(16 oz) = 40 oz

Example 8

Express 36 hr in days.

Solution

Since 24 hr = 1 day, we know that 1 hr = $\frac{1}{24}$ day.

Now we can substitute and multiply:

$$36 \text{ hr} = 36(1 \text{ hr}) = 36\left(\frac{1}{24}\text{day}\right) = 1\frac{1}{2}\text{days or } 1.5 \text{ days}$$

Example 9

How many seconds are there in 3 hr?

Solution

In this case, you must make two substitutions. First change hours to minutes, then change minutes to seconds.

3 hr = 3(1 hr) = 3(60 min) = 180 min

180 min = 180(1 min) = 180(60 sec) = 10,800 sec

So,

3 hr = 10,800 sec

Now Work Exercises 1–5 in the Margin.

U.S. Customary and Metric Equivalents

We will begin the following discussion of equivalent measures in the U.S. customary and metric systems with measures of temperature.

Convert the measures as indicated.

1. 2 ft = __________ in.

2. 8 in. = __________ ft

3. 2 gal = __________ qt

4. 3 t = __________ lb

5. 45 min = __________ hr

Temperature

U.S. customary measure is in **degrees Fahrenheit**(°F).
Metric measure is in **degrees Celsius** (°C).

The two scales are shown on thermometers in the following figure. Approximate conversions can be made by reading along a ruler or the edge of a piece of paper held horizontally across the page.

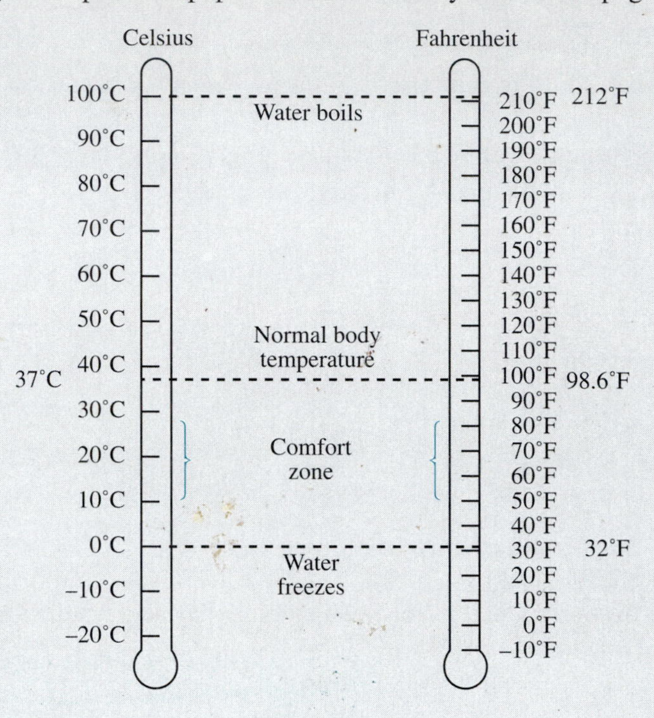

EXAMPLE 10

Hold a straightedge horizontally across the two thermometers and you will read the following equivalent degree measures:

100° C = 212° F Water boils at sea level.

40° C = 104° F Hot day in the desert

20° C = 68° F Comfortable room temperature

The two formulas that give exact conversions between degrees Celsius and degrees Fahrenheit are provided here for easy reference.

Let F = temperature in degrees Fahrenheit and C = temperature in degrees Celsius. Then,

$$C = \frac{5(F - 32)}{9} \quad \text{and} \quad F = \frac{9 \cdot C}{5} + 32$$

A calculator will give answers accurate to 8 digits (or more). Answers that are not exact may be rounded off to whatever place of accuracy you choose. As a general rule, the desired place of accuracy is stated in the problem.

EXAMPLE 11

Let $F = 86°$ and find the equivalent measure in degrees Celsius.

Solution

Using the formula for conversion from Fahrenheit to Celsius, we find

$$C = \frac{5(86 - 32)}{9} = \frac{5(54)}{9} = 30$$

So,

$$86° \text{ F} = 30° \text{ C}$$

EXAMPLE 12

Let $C = 40°$ and convert this temperature to degrees Fahrenheit.

Solution

The formula for conversion from Celsius to Fahrenheit gives

$$F = \frac{9 \cdot 40}{5} + 32 = 72 + 32 = 104$$

So,

$$40° \text{ C} = 104° \text{ F}$$

In the tables of length equivalents, area equivalents, volume equivalents, and mass equivalents (Tables M.10–M.13), the equivalent measures are rounded off. Any calculations with these measures (with or without a calculator) cannot be any more accurate than the measure in the table. Figure M.8 shows examples of some length equivalents.

In Examples 13–16, use Table M.10 to convert measurements as indicated. (Also see Figure M.8.)

Table M.10 Length Equivalents

U.S. to Metric	Metric to U.S.
1 in = 2.54 cm (exact)	1 cm = 0.394 in.
1 ft = 0.305 m	1 m = 3.28 ft
1 yd = 0.914 m	1 m = 1.09 yd
1 mi = 1.61 km	1 km = 0.62 mi

1 inch = 2.54 cm

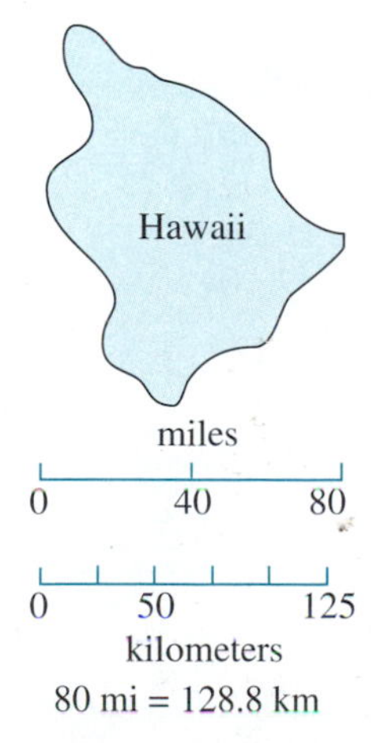

80 mi = 128.8 km

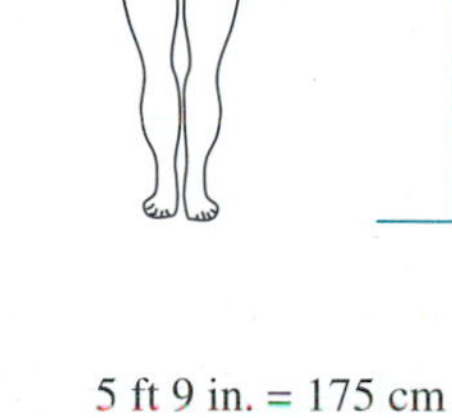

5 ft 9 in. = 175 cm

Figure M.8

Convert the measures as indicated.

6. 20 in. = __________ cm

7. 100 yd = __________ m

8. 10 m = __________ ft

9. 20 km = __________ mi

EXAMPLE 13 6 ft = __________ cm

Solution

6 ft = 72 in. = 72(2.54 cm) = 183 cm rounded off

or

6 ft = 6(0.305 m) = 1.83 m = 183 cm

EXAMPLE 14 How many kilometers is the same as 25 miles?

Solution

25 mi = 25(1.61 km) = 40.25 km

EXAMPLE 15 How many feet are there in 30 m?

Solution

30 m = 30(3.28 ft) = 98.4 ft

EXAMPLE 16 Convert 10 km to miles.

Solution

10 km = 10(0.62 mi) = 6.2 mi

Now Work Exercises 6–9 in the Margin.

Use Table M.11 in Examples 17–20 to convert the measures as indicated. (Also see Figure M.9.)

Table M.11 Area Equivalents

U.S. to Metric	Metric to U.S.
1 in.^{2} = 6.45 cm^2	1 cm^2 = 0.155 in.^{2}
1 ft^2 = 0.093 m^2	1 m^2 = 10.764 ft^2
1 yd^2 = 0.836 m^2	1 m^2 = 1.196 yd^2
1 acre = 0.405 ha	1 ha = 2.47 acres

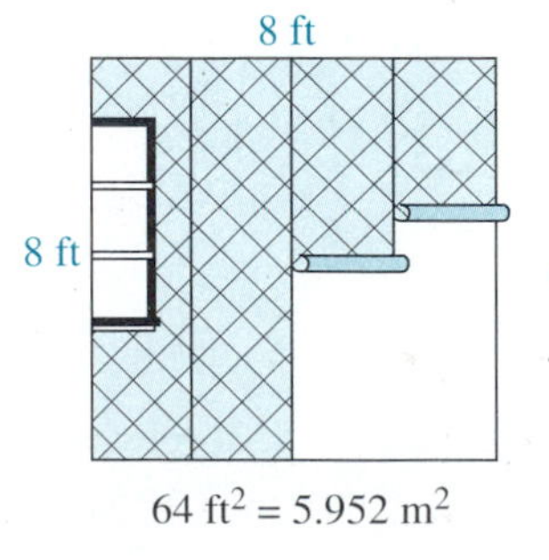

64 ft^2 = 5.952 m^2

0.875 in.^{2} = 5.64 cm^2

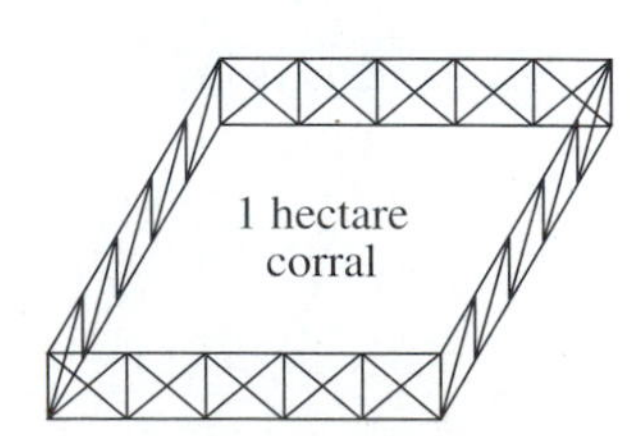

1 ha = 2.47 acres

Figure M.9

Example 17 $40 \text{ yd}^2 =$ ________ m^2

Solution

$40 \text{ yd}^2 = 40(0.836 \text{ m}^2) = 33.44 \text{ m}^2$

Example 18 How many hectares are in 5 acres?

Solution

$5 \text{ acres} = 5(0.405 \text{ ha}) = 2.025 \text{ ha}$

Example 19 Convert 5 ha to acres.

Solution

$5 \text{ ha} = 5(2.47 \text{ acres}) = 12.35 \text{ acres}$

Example 20 Change 100 cm^2 to square inches.

Solution

$100 \text{ cm}^2 = 100(0.155 \text{ in.}^2) = 15.5 \text{ in.}^2$

Now Work Exercises 10–13 in the Margin.

Convert the measures as indicated.

10. $6 \text{ in.}^2 =$ ________ cm^2

11. $625 \text{ ft}^2 =$ ________ m^2

12. $50 \text{ m}^2 =$ ________ ft^2

13. $3 \text{ ha} =$ ________ acres

Use Table M.12 to convert the measures as indicated in Examples 21–23. (Also see Figure M.10.)

Table M.12 Volume Equivalents

U.S. to Metric	Metric to U.S.
$1 \text{ in.}^3 = 16.387 \text{ cm}^3$	$1 \text{ cm}^3 = 0.06 \text{ in}^3$
$1 \text{ ft}^3 = 0.028 \text{ m}^3$	$1 \text{ m}^3 = 35.315 \text{ ft}^3$
1 qt = 0.946 L	1 L = 1.06 qt
1 gal = 3.785 L	1 L = 0.264 gal

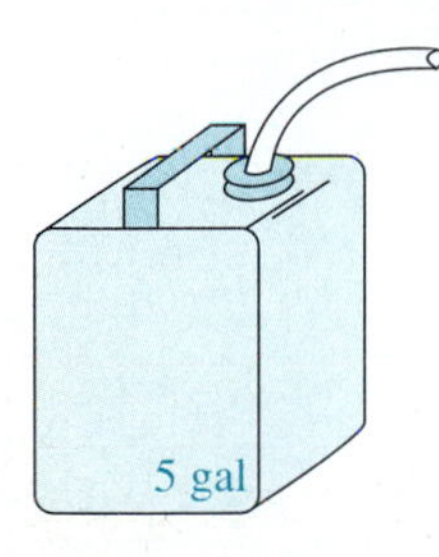

5 gal = 18.925 L

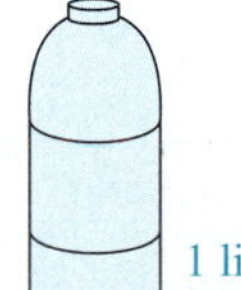

1 L = 1.06 qt

$3 \text{ in.}^3 = 49.161 \text{ cm}^3$

Figure M.10

Convert the measures as indicated.

14. 3 qt = __________ L

15. 5 gal = __________ L

16. 100 cm^3 = __________ in.^3

17. 5 m^3 = __________ ft^3

EXAMPLE 21 20 gal = __________ L

Solution

$$20 \text{ gal} = 20(3.785 \text{ L}) = 75.7 \text{ L}$$

EXAMPLE 22 Change 42 L to gallons.

Solution

$$42 \text{ L} = 42(0.264 \text{ gal}) = 11.088 \text{ gal}$$

or

$$42 \text{ L} = 11.1 \text{ gal} \quad \text{rounded off}$$

EXAMPLE 23 Express 10 cm^3 in cubic inches.

Solution

$$10 \text{ cm}^3 = 10(0.06 \text{ in.}^3) = 0.6 \text{ in.}^3$$

Now Work Exercises 14–17 in the Margin.

Use Table M.13 to convert the measures as indicated in Examples 24 and 25. (Also see Figure M.11.)

Table M.13 Mass Equivalents

U.S. to Metric	Metric to U.S.
1 oz = 28.35 g	1 g = 0.035 oz
1 lb = 0.454 kg	1 kg = 2.205 lb

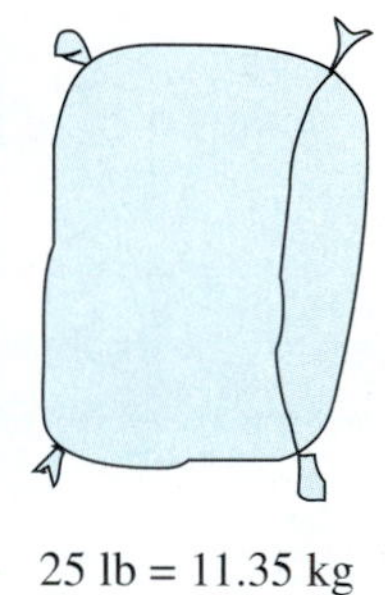

25 lb = 11.35 kg

9 kg = 19.85 lb

Figure M.11

EXAMPLE 24 5 lb = __________ kg

Solution

$$5 \text{ lb} = 5(0.454 \text{ kg}) = 2.27 \text{ kg}$$

EXAMPLE 25 Convert 15 kg to pounds.

Solution

$$15 \text{ kg} = 15(2.205 \text{ lb}) = 22.075 \text{ lb}$$

or

$$15 \text{ kg} = 33.1 \text{ lb} \quad \text{rounded off}$$

NAME ______________ SECTION ______________ DATE ________

Exercises M.3

Use Tables M.9–M.13 as references and any appropriate formulas to convert the following measures as indicated.

1. 5 ft = __________ in.

2. 3 ft = __________ in.

3. 1.5 ft = __________ in.

4. 48 in. = __________ ft

5. 30 in. = __________ ft

6. 4 yd = __________ ft

7. Change $2\frac{1}{3}$ yd to feet.

8. Change $1\frac{2}{3}$ yd to feet.

9. How many miles are in 10,560 ft?

10. Change 3 mi to feet.

11. How many yards are in 6 ft?

12. How many yards are in 9 ft?

13. Convert 2 pt to fluid ounces.

14. Convert 3 pt to fluid ounces.

15. Express 3 qt in pints.

16. Express 1.5 qt in pints.

17. Convert 5 gal to quarts.

18. Convert 3 gal to quarts.

19. 20 qt = __________ gal

20. 13 qt = __________ gal

21. 3 lb = __________ oz

22. 1.75 lb = __________ oz

23. $2\frac{1}{2}$ t = __________ lb

24. 3 t = __________ lb

25. 5 hr = __________ min

26. 4 hr = __________ min

27. How many minutes are in $1\frac{1}{2}$ hr?

28. Change 15 min to hours.

29. Express 3 days in hours.

30. Convert 0.5 hr to seconds.

31. How many seconds are in 5 min?

32. How many seconds are in 3 min?

33. Change 240 sec to minutes.

34. Change 300 sec to minutes.

35. How many days are in 48 hr?

36. How many days are in 72 hr?

37. 25° C = __________° F

38. 35° C = __________° F

39. 50° F = __________° C

40. 100° F = __________° C

41. Change 32° F to °C

42. Change 41° F to °C

ANSWERS

1. ________
2. ________
3. ________
4. ________
5. ________
6. ________
7. ________
8. ________
9. ________
10. ________
11. ________
12. ________
13. ________
14. ________
15. ________
16. ________
17. ________
18. ________
19. ________
20. ________
21. ________
22. ________
23. ________
24. ________
25. ________
26. ________
27. ________
28. ________
29. ________
30. ________
31. ________
32. ________
33. ________
34. ________
35. ________
36. ________
37. ________
38. ________
39. ________
40. ________
41. ________
42. ________

43. ____________
44. ____________
45. ____________
46. ____________
47. ____________
48. ____________
49. ____________
50. ____________
51. ____________
52. ____________
53. ____________
54. ____________
55. ____________
56. ____________
57. ____________
58. ____________
59. ____________
60. ____________
61. ____________
62. ____________
63. ____________
64. ____________
65. ____________
66. ____________
67. ____________
68. ____________
69. ____________
70. ____________
71. ____________
72. ____________
73. ____________
74. ____________
75a. ____________
b. ____________
76a. ____________
b. ____________

43. How many meters are in 3 yd?

44. How many meters are in 5 yd?

45. Change 60 mi to kilometers.

46. Change 100 mi to kilometers.

47. Convert 200 km to miles.

48. Convert 65 km to miles.

49. Express 50 cm in inches.

50. Express 100 cm in inches.

51. 3 in^2 = ____________ cm^2

52. 16 in.^2 = ____________ cm^2

53. 600 ft^2 = ____________ m^2

54. 300 ft^2 = ____________ m^2

55. 100 yd^2 = ____________ m^2

56. 250 yd^2 = ____________ m^2

57. 1000 acres = ____________ ha

58. 250 acres = ____________ ha

59. How many acres are in 300 ha?

60. How many acres are in 400 ha?

61. Change 5 m^2 to square feet.

62. Change 10 m^2 to square feet.

63. Change 30 cm^2 to square inches.

64. Change 50 cm^2 to square inches.

65. 10 qt = ____________ L

66. 20 qt = ____________ L

67. 10 L = ____________ qt

68. 50 L = ____________ gal

69. 10 lb = ____________ kg

70. 500 kg = ____________ lb

71. 16 oz = ____________ g

72. 100 g = ____________ oz

73. Suppose that the home you are considering buying sits on a rectangular lot that is 270 feet by 121 feet. What fraction of an acre is the lot?

74. A new manufacturing building covers an area of 3 acres. How many square feet of ground does the new building cover?

75. A painting of a landscape is on a rectangular canvas that measures 3 feet by 4 feet. (a) How many square inches of wall space will the painting cover when it is hanging? (b) How many square yards?

76. Mr. Zimmer decided to build a fence around a 5-acre pasture so that some of his cattle could graze there. (a) If he put 50 head of cattle in the field, how many square feet were there in the field for each head of cattle? (b) How many square yards for each head of cattle?

Applications in Medicine

Measurement and Changing Units

This section is designed to provide an introduction to measurements used in medicine, techniques (proportions) for changing units, and formulas related to dosages of medicine.

Household fluid measures are based on the uses of household cooking and eating utensils. This method of measurement is **not** accurate, but it is useful. The **apothecaries' system of fluid measure** is used by pharmacists. (See Tables M.14 and M.15 and Figure M.12.)

OBJECTIVES

1. *Become familiar with household fluid measures, metric fluid measures, and apothecaries' fluid measures.*
2. *Be able to use proportions to change units from one system to another.*
3. *Learn three formulas for calculating dosages of medicine for children.*

Table M.14 Abbreviations

Household Fluid Measure	Apothecaries' Fluid Measure
gtt = 1 drop	♏ = 1 minim (about 1 drop of water)
tsp = 1 teaspoonful	*f*ʒ = 1 fluidram (60 minims)
tbsp = 1 tablespoonful	*f*℥ = 1 fluidounce (480 minims)

Table M.15 Fluid Measure Equivalents

Household	Metric	Apothecaries'
1 gtt	= 0.06 mL	= 1 ♏
15 gtt	= 1.0 mL	= 15 ♏
75 gtt	= 5.0 mL	= 75 ♏
1 tsp	= 5.0 mL	= $1\frac{1}{4}$ *f*ʒ
3 tsp	= 15.0 mL	= 4 *f*ʒ
1 tbsp	= 15.0 mL	= 4 *f*ʒ = $\frac{1}{2}$ *f*℥
1 glassful	= 240.0 mL	= 8 *f*℥

To Use Proportions to Change from One System to Another:

1. Find the corresponding equivalent units in the table of fluid measure equivalents (Table M.15).
2. Write x for the unknown quantity.
3. Write a proportion using the ratio of equivalent measures and the ratio of x to the known measure being changed.
4. Solve the proportion.

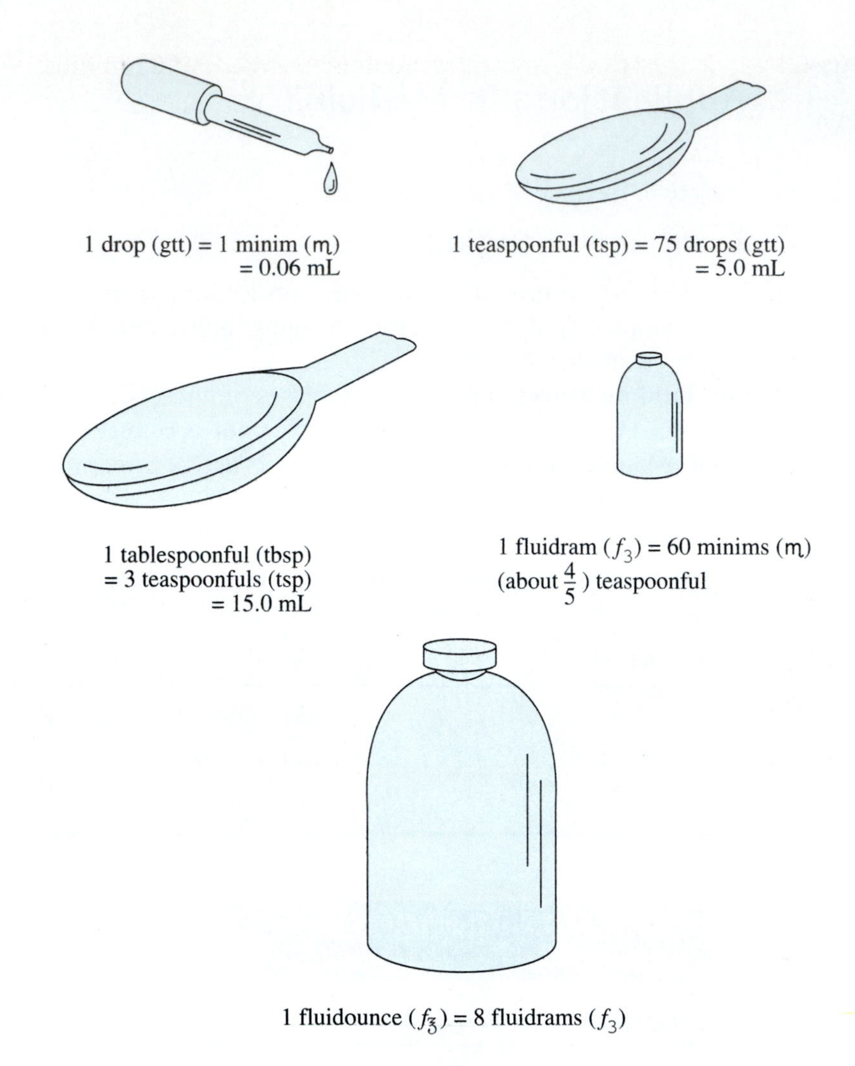

Figure M.12

EXAMPLE 1

How many milliliters (mL) are in 6 drops (gtt)?

Solution

From Table M.15, 1 gtt = 0.06 mL.
Let x = unknown number of milliliters.
Set up the proportion:

$$\frac{x \text{ mL}}{6 \text{ gtt}} = \frac{0.06 \text{ mL}}{1 \text{ gtt}}$$

$$1 \cdot x = 6(0.06)$$

$$x = 0.36 \text{ mL}$$

So, 6 drops (gtt) = 0.36 milliliters (mL), or there is 0.36 milliliter in 6 drops.

EXAMPLE 2 How many tablespoonfuls (tbsp) are in 60 milliliters?

Solution

From Table M.15, 1 tbsp = 15.0 mL.
Let x = unknown number of tablespoonfuls.
Set up the proportion:

$$\frac{x \text{ tbsp}}{60 \text{ mL}} = \frac{1 \text{ tbsp}}{15.0 \text{ mL}}$$

$$15 \cdot x = 60 \cdot 1$$

$$\frac{\cancel{15}x}{\cancel{15}} = \frac{60}{15}$$

$$x = 4 \text{ tbsp}$$

So, 60 mL = 4 tbsp, or 4 tablespoonfuls are 60 milliliters.

EXAMPLE 3 How many teaspoonfuls (tsp) are in 8 fluidrams($f ʒ$)?

Solution

From Table M.15, 1 tsp = $1\frac{1}{3} f ʒ$.

Let x = unknown number of teaspoonfuls.
Set up the proportion:

$$\frac{x \text{ tsp}}{8\, f ʒ} = \frac{1 \text{ tsp}}{1\frac{1}{3}\, f ʒ}$$

$$\frac{4}{3} \cdot x = 8 \cdot 1$$

$$\frac{\cancel{3}}{\cancel{4}} \cdot \frac{\cancel{4}}{\cancel{3}} \cdot x = \frac{3}{4} \cdot 8$$

$$x = 6 \text{ tsp}$$

There are 6 teaspoonfuls in 8 fluidrams, or $8 f ʒ$ = 6 tsp.

EXAMPLE 4 How many glassfuls are in 480 milliliters?

Solution

From Table M.15, 1 glassful = 240.0 mL.
Let x = unknown number of glassfuls.
Set up the proportion:

$$\frac{x \text{ glassfuls}}{480 \text{ mL}} = \frac{1 \text{ glassful}}{240 \text{ mL}}$$

$$240 \cdot x = 480 \cdot 1$$

$$\frac{\cancel{240} \cdot x}{\cancel{240}} = \frac{480}{240}$$

$$x = 2 \text{ glassfuls}$$

Thus, 2 glassfuls = 480 mL, or there are 2 glassfuls in 480 milliliters.

Calculating Dosages

Three Formulas Used to Calculate Dosages of Medicine for Children

1. **Fried's Rule** (birth to age 2):

$$\text{Child's dose} = \frac{\text{age in months} \times \text{adult dose}}{150}$$

2. **Young's Rule** (ages 1 to 12):

$$\text{Child's dose} = \frac{\text{age in years} \times \text{adult dose}}{\text{age in years} + 12}$$

3. **Clark's Rule** (ages 2 and over):

$$\text{Child's dose} = \frac{\text{weight in pounds} \times \text{adult dose}}{150}$$

Example 5

Use Fried's Rule to find a child's dose of paregoric, given:

Child's age = 12 months

Adult dose = 5 mL

Solution

Substituting into the formula gives

$$\text{Child's dose} = \frac{12 \text{ months} \times 5 \text{ mL}}{150}$$

$$= \frac{60}{150}$$

$$= 0.4 \text{ mL}$$

The child's dose would be 0.4 mL of paregoric.

Example 6

Use Young's Rule to find a child's dose of castor oil, given:

Child's age = 30 months (2.5 years)

Adult dose = 15 mL

Solution

Substituting into the formula gives

$$\text{Child's dose} = \frac{2.5 \text{ yr} \times 15 \text{ mL}}{2.5 + 12}$$

$$= \frac{37.5}{14.5}$$

$$= 2.6 \text{ mL} \quad \text{rounded off}$$

Using Young's Rule, the child's dose would be 2.6 mL of castor oil.

NAME ________________ SECTION ________________ DATE ________

ANSWERS

Exercises M.4

Use Table M.15 and proportions to find the following equivalent measures.

1. 5 gtt = ________ mL

2. 7 mL = ________ gtt

3. 4 tbsp = ________ mL

4. 3 tbsp = ________ *f*ʒ

5. 6*f*ʒ = ________ tsp

6. 10 mL = ________ tsp

7. 0.18 mL = ________ gtt

8. $2\frac{1}{2}$ glassfuls = ________ mL

9. 5 tbsp = ________ *f*ʒ

10. 4 tbsp = ________ *f*℥

11. 360 mL = ________ glassfuls

12. 4 gtt = ________ ♍

13. 5♍ = ________ gtt

14. 30 gtt = ________ mL

15. 30 mL = ________ tsp

16. 6*f*℥ = ________ glassfuls

17. 45 mL = ________ tbsp

18. 5 tsp = ________ *f*ʒ

19. 150♍ = ________ gtt

20. 6 glassfuls = ________ mL

Use Fried's Rule to find the child's dose from the given information.

21. Child's age: 9 months
Adult dose: 250 mL liquid ampicillin

22. Child's age: 5 months
Adult dose: 30 mL milk of magnesia

1. ________
2. ________
3. ________
4. ________
5. ________
6. ________
7. ________
8. ________
9. ________
10. ________
11. ________
12. ________
13. ________
14. ________
15. ________
16. ________
17. ________
18. ________
19. ________
20. ________
21. ________
22. ________

23. ______________

24. ______________

25. ______________

26. ______________

27. ______________

28. ______________

29. ______________

23. Child's age: 20 months
Adult dose: 15 mL castor oil

Use Young's Rule to find the child's dose from the given information.

24. Child's age: 5 years
Adult dose: 250 mL liquid ampicillin

25. Child's age: 8 years
Adult dose: 10 mL Robitussin®

26. Child's age: 6 years
Adult dose: 15 mL castor oil

Use Clark's Rule to find the child's dose from the given information.

27. Child's weight: 20 lb
Adult dose: 5 mL Polaramine® syrup

28. Child's weight: 60 lb
Adult dose: 250 mL penicillin

29. Child's weight: 75 lb
Adult dose: 15 mL castor oil

Index of Key Ideas and Terms

NAME ______________ SECTION ______________ DATE ______________

ANSWERS

Chapter Test

In Problems 1–30, change the units as indicated

1. 7 cm = _____ m

2. 15 m = _____ cm

3. 173 mm = _____ m

4. 0.3 m = _____ mm

5. 17 g = _____ kg

6. 103 kg = _____ g

7. 42 kg = _____ t

8. 6 t = _____ kg

9. 4 m^2 = _____ cm^2

10. 16 cm^2 = _____ m^2

11. 45 gtt = _____ mL

12. 12 tsp = _____ *f*ʒ

13. 2 L = _____ cm^3

14. 175 mL = _____ L

15. 2.75 ft = _____ in.

16. $6\frac{1}{2}$ yd = _____ ft

17. 74 in. = _____ ft

18. 2 ft = _____ yd

1. ______________
2. ______________
3. ______________
4. ______________
5. ______________
6. ______________
7. ______________
8. ______________
9. ______________
10. ______________
11. ______________
12. ______________
13. ______________
14. ______________
15. ______________
16. ______________
17. ______________
18. ______________

19. ______

20. ______

21. ______

22. ______

23. ______

24. ______

25. ______

26. ______

27. ______

28. ______

29. ______

30. ______

31. ______

32. ______

19. $4\frac{3}{4}$ lb = ______ oz

20. 1.3 t = ______ lb

21. 54 oz = ______ lb

22. 800 lb = ______ t

23. 3.2 hr = ______ min

24. $2\frac{5}{12}$ days = ______ hr

25. 168 sec = ______ min

26. 84 hr = ______ days

27. 1.6 pt = ______ fl oz

28. $2\frac{3}{8}$ gal = ______ qt

29. $3\frac{1}{4}$ qt = ______ gal

30. 22 fl oz ______ pt

31. Use the appropriate rule for calculating the child's dose from the following information:

Child's weight: 45 lb
Adult dose: 250 mL liquid ampicillin

32. Use the formula $C = \frac{5(F - 32)}{9}$ to convert 68° F to degrees Celsius.

NAME ______________________ SECTION ______________________ DATE ____________

ANSWERS

1. ____________
2. ____________
3. ____________
4. ____________
5. ____________
6. ____________
7. ____________
8. ____________
9. ____________
10. ____________
11. ____________
12a. ____________
b. ____________
13. ____________

Cumulative Review

1. Evaluate the expression: $16 + (12 \cdot 3 + 2^3) \div 4 - 20$.

2. Evaluate the expression: $(0.5)^2 + 2\frac{1}{5} \div \frac{11}{6} - 0.3$.

3. Evaluate the expression: $(1.5)^2 - (1.2)^2$.

4. Find the average of the following set of numbers:

33, 50, 50, 42, 41, 60, 46

5. If a savings account earned \$75 in interest at 5% for 9 months, what was the original amount in the account?

6. If an investment pays 6% interest compounded daily, what will be the value of \$8000 in 5 years?

7. If a stock pays a 10% dividend every quarter, what will be the total of the dividends paid on a \$6000 investment over 4 years? (Assume that the dividends are reinvested each quarter.)

8. 18 is __________ % of 90.

9. 70% of __________ is 91.

10. 37.5% of 103.4 is __________.

11. 86% of 14,000 is __________.

12. (a) What were your car expenses for the month in which you paid \$320 for four new tires, \$25 for oil and filter, \$80 for gas, \$80 for insurance, and \$125 for a tune-up? (b) If your income is \$2520 per month, what percent of your income did you spend on your car that month?

13. If your car averages 22.5 miles per gallon, how many miles will it go on 18 gallons of gas?

14. ____________

15a. ____________

b. ____________

16. ____________

17. ____________

18a. ____________

b. ____________

c. ____________

d. ____________

e. ____________

19. ____________

20a. ____________

b. ____________

21a. ____________

b. ____________

22a. ____________

b. ____________

23a. ____________

b. ____________

24. ____________

14. Which is the better buy of a container of frozen orange juice: (a) 16 fl oz for \$1.79 or (b) 12 fl oz for \$1.09?

15. Calculators are on sale at a discount of 28%, and the original price was \$62.50. (a) What is the approximate amount of the discount? (b) What is the sale price?

16. Add and simplify:

$$\begin{array}{r} 12 \text{ hr } 23 \text{ min } 12 \text{ sec} \\ +\ \ 6 \text{ hr } 39 \text{ min } 48 \text{ sec} \\ \hline \end{array}$$

17. Find the difference:

$$\begin{array}{r} 4 \text{ yd } 2 \text{ ft } 5 \text{ in.} \\ -\ 1 \text{ yd } 2 \text{ ft } 9 \text{ in.} \\ \hline \end{array}$$

18. Change the units as indicated.

(a) 35.6 cm = ____________ mm

(b) 18.72 cm = ____________ m

(c) 195 cm^2 = ____________ mm^2

(d) 0.54 L = ____________ mL

(e) 78 000 cm^3 = ____________ m^3

19. Simplify the expression: $\sqrt{72} + 2\sqrt{50}$.

20. (a) Find the length of the hypotenuse for the right triangle shown:

(b) Use a calculator to find this value to the nearest hundredth.

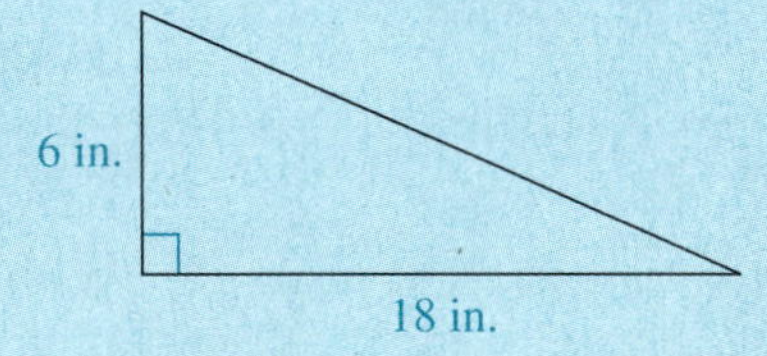

21. Write each of the following numbers in scientific notation.

(a) 890,000,000

(b) 0.0000632

22. Perform the indicated operations.

(a) $(-5) + (-4) + (-12)$

(b) $(16 - 18)(13 - 5)$

23. Solve each of the following equations.

(a) $3x + 10 = -11$

(b) $4y - 14 = 24.8$

24. Five more than six times a number is equal to 71. Find the number.

Statistics and Reading Graphs

What to Expect in this Chapter

This chapter deals with applications of mathematical concepts related to information that is likely to appear in daily newspapers, magazines, and consumer reports. The basic statistical terms **data, statistic, mean, median,** and **range** are discussed in Section S.1.

Sections S.2–S.4 discuss the methods for reading and calculating information related to the following types of graphs:

circle graphs and pictographs (Section S.2)

bar graphs and line graphs (Section S.3)

histograms and frequency polygons (Section S.4)

The knowledge and skills related to reading and understanding graphs, simple calculations involving the information in graphs, and calculating basic statistics from given lists of data are considered fundamental and necessary skills by most employers.

Mathematics at Work!

If you want to convince someone quickly and easily that you know of a trend, a percentage, or a relative size, draw a graph. As the old saying goes, "A picture is worth a thousand words." Graphs (circle graphs, pictographs, bar graphs, and line graphs) appear almost daily in newspapers, magazines, and journals and are a basic part of many textbooks. *USA Today* has a regular feature on the front page entitled "USA SNAPSHOTS"® (a look at statistics that shape the nation), which uses graphs so that information can be communicated with very few words.

Even though graphs are convenient and effective, you (as a consumer and citizen) should be aware of how some graphs can imply ideas that are not really true. In reading graphs, you should learn to look past the graphics to the actual numerical scales and other information stated in the graph. Ask yourself questions such as:

"Do the percents total 100%?"
"Is any information missing or misleading?"
"Are the figures or diagrams used in proper proportion?"

Don't become a complete pessimist; just be astute and observant.

A figure similar to the following appeared in the June, 1994 issue of *Reader's Digest* in connection with an article entitled "America's Free Market: The Global Winner." Study the three graphs carefully to see if you can understand the information being illustrated.

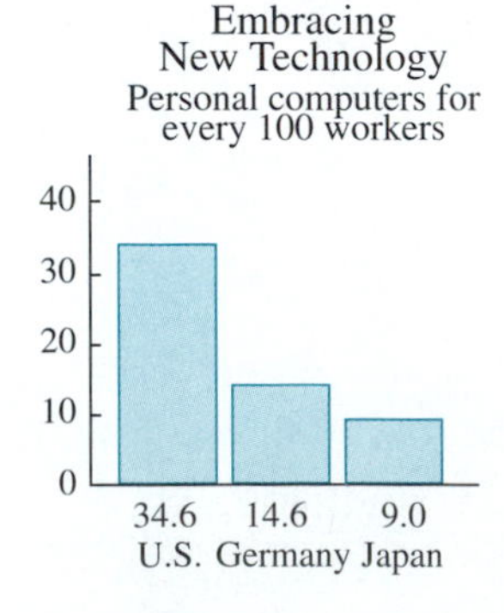

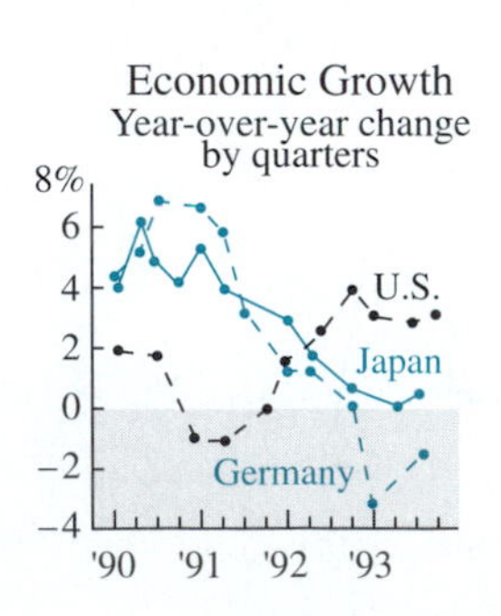

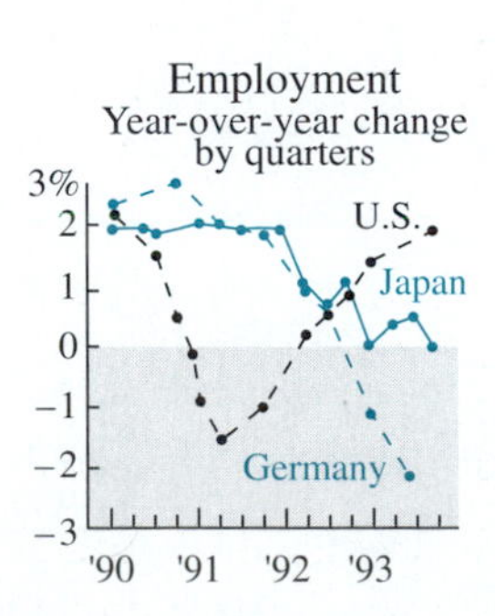

S.1 Mean, Median, and Range

OBJECTIVES

1. *Learn the meanings of the following terms:*

   data, statistic, mean, median, range

2. *Know how to find the* **mean, median,** *and* **range** *of a set of data items.*

Statistical Terms

Statistics is the study of how to gather, organize, analyze, and interpret numerical information. In this section, we will study only three measures (or three statistics) that are easily found or calculated: **mean, median,** and **range.** Other measures that you might read about or study in a statistics course are:

> **standard deviation** and **variance** (both measures of how "spread-out" data is)
>
> ***z*-score** (a measure that compares numbers that are expressed in diffferent units—for example, your score on a biology exam and your score on a mathematics exam)
>
> **correlation coefficient** (a measure of how two different types of data might be related—for example, the relationship between the amount of schooling a person has and the amount of that person's lifetime earnings)

You will need a semester or two of algebra in order to be able to study and understand these topics, so keep working hard.

The following terms and their definitions are necessary for understanding the topics and related problems in this section. They are listed here in one place for easy reference.

Terms Used in the Study of Statistics

Data: Value(s) measuring some characteristic of interest. (We will consider only numerical data.)

Statistic: A single number describing some characteristic of the data.

Mean: The arithmetic average of the data. (Find the sum of all the data and divide by the number of data items.)

Median: The middle of the data after the data has been arranged in order.

Range: The difference between the largest and the smallest data items.

Finding the Mean, Median, and Range of a Set of Data

Examples 1–3 that follow refer to the data from Group A and Group B shown at the top of the next page.

Group A: Annual Income for 8 Families

$18,000; $12,000; $15,000; $17,000

$35,000; $70,000; $15,000; $20,000

Group B: Grade Point Averages (GPA) for 11 Students

2.0; 2.0; 1.9; 3.1; 3.5; 2.9;

2.5; 3.6; 2.0; 2.4; 3.4

EXAMPLE 1

Find the mean income for the families in Group A.

Solution

The mean is the arithmetic average of the data. Therefore, we find the sum of the 8 incomes and divide by 8.

```
    $18,000            $25,250   mean income
     12,000        8 ) 202,000
     15,000            16
     17,000            --
     35,000             42
     70,000             40
     15,000             --
 +   20,000              2 0
 ----------              1 6
   $202,000              ---
                           40
                           40
                           --
                           00
                            0
                           --
                            0
```

The mean annual income is $25,250.

The **median** is another type of "average." In a set of ranked data (data arranged in order), the median is the middle value. As we will see, the determination of this value depends on whether there is an odd number of data items or an even number of data items.

To Find the Median:

1. Rank the data. (Arrange the data in order, either from smallest to largest or largest to smallest.)
2. If there is an **odd** number of items, the median is the middle item.
3. If there is an **even** number of items, the median is the value found by averaging the two middle items. (**Note:** This value may or may not be in the data.)

EXAMPLE 2

Find the median income for Group A and the median GPA for Group B.

Solution

First we rank both sets of data in order from smallest to largest.

Group A	**Group B**
1. $12,000	1. 1.9
2. $15,000	2. 2.0
3. $15,000	3. 2.0
4. $17,000	4. 2.0
5. $18,000	5. 2.4
6. $20,000	6. 2.5
7. $35,000	7. 2.9
8. $70,000	8. 3.1
	9. 3.4
	10. 3.5
	11. 3.6

Group A has 8 items (an even number), so the median is the average of the fourth and fifth items.

$$\text{Median income} = \frac{\$17{,}000 + \$18{,}000}{2} = \frac{\$35{,}000}{2} = \$17{,}500$$

Group B has 11 items (an odd number), so the median is the middle item (the sixth item).

$$\text{Median GPA} = 2.5$$

Once the data has been ranked (as was done in Example 2), the range is easily determined. Remember that the range is the difference between the largest and the smallest items in the set of data.

EXAMPLE 3

Find the range for both Group A and Group B.

Solution

From the ranked data in Example 2, we can calculate each range as follows:

$$\text{Group A range} = \$70{,}000 - \$12{,}000 = \$58{,}000$$

$$\text{Group B range} = 3.6 - 1.9 = 1.7$$

EXAMPLE 4

On an English exam, two students scored 95 points each, five students scored 86 points each, one student scored 82 points, another student scored 78 points, and six students scored 75 points each. What is (a) the class mean, (b) the class median, and (c) the range of the scores?

Solution

(a) *To find the mean:* Multiply as shown below, add the products, and then divide by 15 because the sum represents the sum of the scores of the 15 students.

$$\begin{array}{r} 95 \\ \times\ 2 \\ \hline 190 \end{array} \quad \begin{array}{r} 86 \\ \times\ 5 \\ \hline 430 \end{array} \quad \begin{array}{r} 82 \\ \times\ 1 \\ \hline 82 \end{array} \quad \begin{array}{r} 78 \\ \times\ 1 \\ \hline 78 \end{array} \quad \begin{array}{r} 75 \\ \times\ 6 \\ \hline 450 \end{array}$$

Now add these five products and divide the sum by 15.

$$\begin{array}{r} 190 \\ 430 \\ 82 \\ 78 \\ +\ 450 \\ \hline 1230 \end{array} \qquad \begin{array}{r} 82 \\ 15\overline{)1230} \\ \underline{120} \\ 30 \\ \underline{30} \\ 0 \end{array}$$ class mean

(b) *To find the median:* Arrange the scores in order and pick the middle (8th) score. (**Note:** The arrangement may be from largest to smallest or smallest to largest.)

95, 95, 86, 86, 86, 86, 86, 82, 78, 75, 75, 75, 75, 75, 75

↑ median (82)

(c) *To find the range:* Subtract the lowest score from the highest score.

Range $= 95 - 75 = 20$

Thus, the class mean is 82, the median is 82, and the range is 20. (**Note:** In this example, the mean and median happen to be the same. In general, they are not necessarily the same.)

☑ COMPLETION EXAMPLE 5

If you had scores of 82, 75, and 72 on three mathematics tests, what is the lowest score you could get on the fourth test (the final exam) and still earn a grade of B? (For a B you must average between 80 and 89.)

Solution

We can proceed in the following manner by dealing with the total number of points needed to earn a B.

(a) If 4 exams are to average 80 (or more), then the total number of points must be

$$\begin{array}{r} 80 \\ \times\ \ 4 \\ \hline \_\_\_\_ \end{array}$$ Total points (or more) needed

(b) On the first three exams you have accumulated the following numbers of points:

$$\begin{array}{r} 82 \\ 75 \\ +\ 72 \\ \hline \end{array}$$

_____ Points accumulated

(c) To average 80 (or more) on your four exams, you need

$$\begin{array}{r} 320 \\ - \quad\quad \\ \hline \end{array}$$

_____ Points (or more) needed on final exam

(d) Thus, you need a score of at least _____ on your final exam to earn a B for the course.

Note: Many people believe that the mean (or arithmetic average) is relied on too much in reporting central tendencies for data such as income, housing costs, and taxes, because a few very high or very low items can distort the picture of a central tendency. For example, the median of $17,500 for the incomes in Group A is probably more representative of the data than the mean of $25,250. This is because the one high income of $70,000 raises the mean considerably, whereas the median is not affected by this one extreme **outlier value.** When you read an article in a magazine or newspaper that reports means or medians, you should now have a better understanding of the implications of these statistical measures.

Completion Example Answers

5. (a)

$$\begin{array}{r} 80 \\ \times \quad 4 \\ \hline \mathbf{320} \end{array}$$

Total points (or more) needed

(b)

$$\begin{array}{r} 82 \\ 75 \\ +\ 72 \\ \hline \mathbf{229} \end{array}$$

Points accumulated

(c)

$$\begin{array}{r} 320 \\ -\ \mathbf{229} \\ \hline \mathbf{91} \end{array}$$

Points (or more) needed on final exam

(d) Thus, you need a score of at least **91** on your final exam to earn a B for the course.

NAME ________________ SECTION ________________ DATE ________

ANSWERS

1a. ________
b. ________
c. ________

2a. ________
b. ________
c. ________

3a. ________
b. ________
c. ________

4a. ________
b. ________
c. ________

Exercises S.1

In Exercises 1–12, find the mean, median, and range of the given data.

1. Ten geology students had the following scores on a final exam:

75, 83, 93, 65, 85, 85, 88, 90, 55, 71

(a) Mean = ______ (b) Median = ______ (c) Range = ______

2. Joe did the following numbers of sit-ups each day in one week:

25, 52, 48, 42, 38, 58, 52

(a) Mean = ______ (b) Median = ______ (c) Range = ______

3. The local college's men's basketball team scored the following points per game during the 20-game season:

85, 60, 62, 70, 75, 52, 88, 50, 80, 72
90, 85, 85, 93, 70, 75, 68, 73, 65, 82

(a) Mean = ______ (b) Median = ______ (c) Range = ______

4. In a study of sleeping habits, 15 adults reported the following hours of sleep the night before an airplane trip:

4, 6, 6, 7, 6.5, 6.5, 7.5, 8.5
5, 6, 4.5, 5.5, 9, 3, 8

(a) Mean = ______ (b) Median = ______ (c) Range = ______

5a. ______________

b. ______________

c. ______________

6a. ______________

b. ______________

c. ______________

7a. ______________

b. ______________

c. ______________

8a. ______________

b. ______________

c. ______________

5. Vera kept track of her golf scores for twelve 18-hole rounds. Her scores were:

85, 90, 82, 85, 87, 80, 78, 82, 88, 82, 86, 81

(a) Mean = ______ (b) Median = ______ (c) Range = ______

6. Joanne went to six different repair shops to get the following estimates for repair of her car:

\$425, \$525, \$325, \$300, \$500, \$325

(a) Mean = ______ (b) Median = ______ (c) Range = ______

7. The fire department reported the following mileage for tires used on their nine fire engines:

14,000; 14,000; 11,000; 15,000; 9,000;
14,000; 12,000; 10,000; 9,000

(a) Mean = ______ (b) Median = ______ (c) Range = ______

8. A weather station recorded the following daily high temperatures for one month:

75, 76, 76, 78, 85, 82, 85, 88, 90, 90
88, 95, 96, 92, 88, 88, 80, 80, 78, 80
78, 76, 77, 75, 75, 74, 70, 70, 72, 73

(a) Mean = ______ (b) Median = ______ (c) Range = ______

NAME ______________ SECTION ______________ DATE ______________

ANSWERS

9a. ______
b. ______
c. ______

10a. ______
b. ______
c. ______

11a. ______
b. ______
c. ______

12a. ______
b. ______
c. ______

9. Police radar measured the following speeds of 35 cars on one street:

28, 24, 22, 38, 40, 25, 24, 35, 25, 23
22, 50, 31, 37, 45, 28, 30, 30, 30, 25
35, 32, 45, 52, 24, 26, 18, 20, 30, 32
33, 48, 58, 30, 25

(a) Mean = ______ (b) Median = ______ (c) Range = ______

10. The city planning department issued the following numbers of building permits over a three-week period:

17, 19, 18, 35, 30
29, 23, 14, 18, 16
29, 18, 18, 25, 30

(a) Mean = ______ (b) Median = ______ (c) Range = ______

11. On a one-day fishing trip, Mr. and Mrs. Johansen recorded the following lengths of fish they caught (measured in inches):

14.3, 13.6, 10.5, 15.5, 20.1
10.9, 12.4, 25.0, 30.2, 32.5

(a) Mean = ______ (b) Median = ______ (c) Range = ______

12. A machine puts out parts whose thickness is measured to the nearest hundredth of an inch. One hundred parts were measured, and the results are tallied in the following chart:

Thickness Measured	0.80	0.83	0.84	0.85	0.87
Number of Parts	22	41	14	20	3

(a) Mean = ______ (b) Median = ______ (c) Range = ______

13a. ______________

b. ______________

c. ______________

14. ______________

15. [Respond below exercise.] ______________

13. On a history exam, two students scored 95 each, six students scored 90 each, three students scored 80 each, and one student scored 50. What was (a) the class mean, (b) the class median, and (c) the range of the scores?

14. Nick bought shares in the stock market from two companies. He paid $4500 for 900 shares in one company and $6900 for 1100 shares in a second company. What did he pay as an average price per share for the 2000 shares?

Writing and Thinking about Mathematics

15. Suppose that you are to take four hourly exams and a final exam in your chemistry course. Each exam has a maximum of 100 points, and you must average between 75 and 82 points to receive a passing grade of C. If you have scores of 83, 65, 70, and 78 on the hourly exams, what is the minimum score you can make on the final exam to receive a grade of C? (First explain the strategy you are going to use in solving the problem, and then solve the problem.)

S.2 Circle Graphs and Pictographs

OBJECTIVES

1. *Understand circle graphs and interpret the data shown.*
2. *Learn to perform appropriate operations related to the data shown in a circle graph.*
3. *Understand pictographs and interpret the data shown.*
4. *Learn to perform appropriate operations related to the data shown in a pictograph.*

Introduction to Graphs

Graphs are **pictures** of numerical information. Graphs appear almost daily in newspapers and magazines and frequently in textbooks and corporate reports. Well-drawn graphs can organize and communicate information accurately, effectively, and fast. Most computers can be programmed to draw graphs, and anyone whose work involves a computer in any way will probably be expected to understand graphs and even to create graphs.

There are many different types of graphs, each particularly well suited to the display and clarification of certain types of information. In this section, we will discuss two types of graphs: circle graphs and pictographs. Their basic uses are listed below.

1. **Circle Graph:** Helps in understanding percents or parts of a whole.
2. **Pictograph:** Emphasizes the topic being related as well as quantities.

A common characteristic of all graphs is that they are intended to communicate information about numerical data quickly and easily. With this in mind, note the following three properties of all graphs:

1. They should be clearly labeled.
2. They should be easy to read.
3. They should have appropriate titles.

Reading Circle Graphs and Pictographs

The following examples illustrate how to read and calculate with the information given in the two types of graphs just mentioned: circle graphs and pictographs. The questions related to each graph are designed to develop your understanding by having you read information directly from the graph or perform some calculation related to the information given in the graph.

Even though some of the blanks in the completion examples are filled in, you should verify each answer by referring to the graph and making any related calculations. **You are to fill in any blanks that have been left empty.**

EXAMPLE 1 Circle Graph

The circle graph shown in Figure S.1 represents the various sources of income for a city government with a total income of \$100,000,000. (a) What is the city's largest source of income? (b) What percent of income comes from Goods and Services? (c) What is the ratio of income from Taxes to the total income?

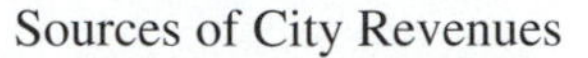

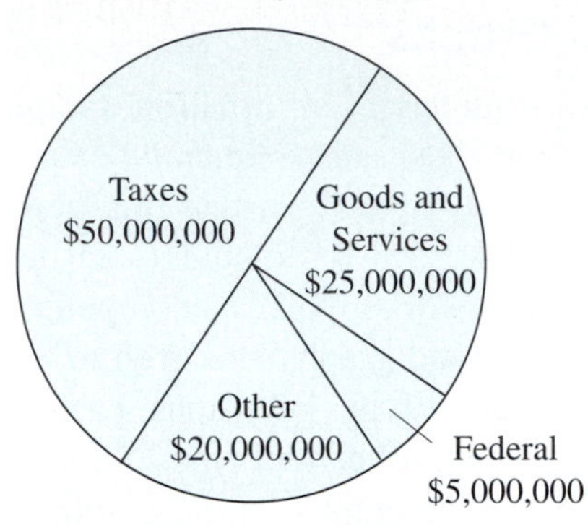

Figure S.1 Circle Graph

Solution

(a) The largest source of income is Taxes.

(b) To find percent, we set up a ratio, reduce, and then divide:

$$\frac{25{,}000{,}000 \text{ Goods and Services}}{100{,}000{,}000 \text{ Total Income}} = \frac{1}{4} \qquad \frac{1}{4} = 25\%$$

25% of the city's revenue comes from Goods and Services.

(c) The ratio of income from Taxes to the total revenue is:

$$\frac{50{,}000{,}000 \text{ Taxes}}{100{,}000{,}000 \text{ Total}} = \frac{1}{2}$$

✔ *COMPLETION EXAMPLE 2 Circle Graph*

Figure S.2 shows percents budgeted for various items in a home for one year. Suppose a single person has an annual income of \$15,000. Using Figure S.2, calculate how much will be allocated to each item indicated in the graph.

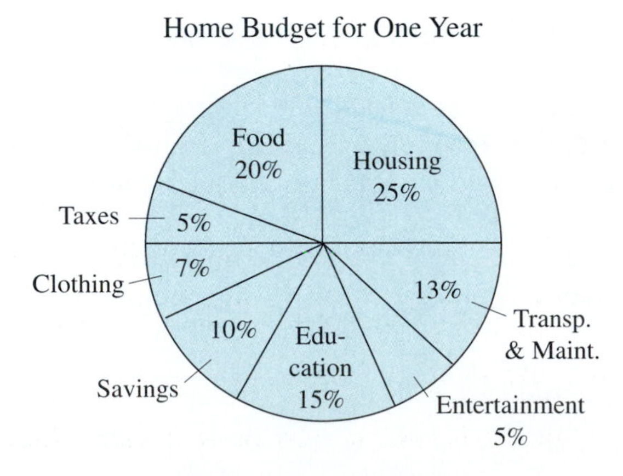

Figure S.2 Circle Graph

Solution

Item		**Amount**
Housing	0.25 × $15,000 =	$ 3750.00
Food	0.20 × $15,000 =	3000.00
Taxes	0.05 × $15,000 =	750.00
Clothing	0.07 × $15,000 =	1050.00
Savings	0.10 × $15,000 =	________
Education	0.15 × $15,000 =	________
Entertainment	____________ =	________
Transportation and Maintenance	____________ =	________

What is the total of all the amounts? ____________

☑ *COMPLETION EXAMPLE 3* *Pictograph*

Figure S.3 shows a pictograph of new home construction in five counties in 1994.

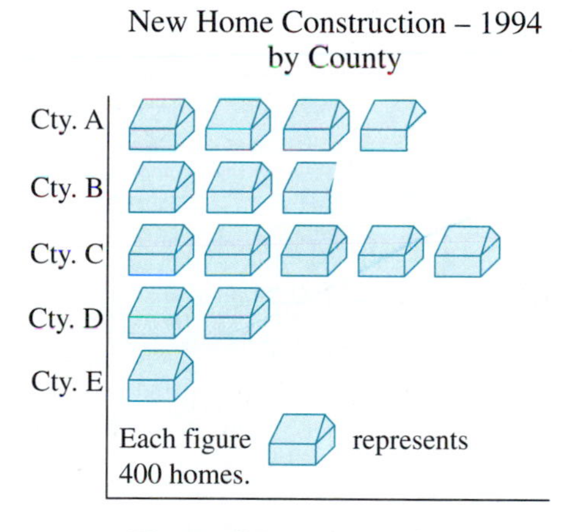

Figure S.3 Pictograph

(a) Which county had the greatest number of new homes built?
County C

(b) About how many homes were built in County B? 1100

(c) Which county had the smallest number of new homes built?
____________ How many? ____________

(d) What was the difference in new home construction in Counties C and D? ____________

Completion Example Answers

2. Savings **\$1500.00**
   Education **\$2250.00**
   Entertainment **0.05 × 15,000 = \$750.00**
   Transportation and Maintenance **0.13 × 15,000 = \$1950.00**
   Total of all amounts is 100% of \$15,000, which is **\$15,000.**

3. (c) **County E; 400 homes**
   (d) **1200 homes**

NAME ______________________ SECTION ______________ DATE ________

ANSWERS

1a. [Respond below exercise.] ______________

b. ______________

c. ______________

d. ______________

2. [Respond below exercise.] ______________

Exercises S.2

1. The school budget shown is based on a total budget of \$34,500,000. (a) What amount will be spent on each category? (b) How much more will be spent on teachers' salaries than on administrative salaries? (c) What percent will be spent on items other than salaries? (d) How much will be spent on items other than salaries?

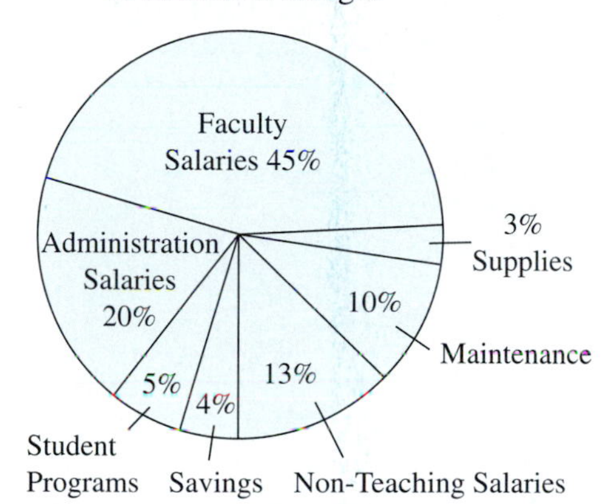

2. Station KCBA is off the air from 2 A.M. to 6 A.M., so there are only 20 hours of programming daily. Sports are not shown here because they are considered special events. In the 20-hour period shown, how much time (in minutes) is devoted to each category?

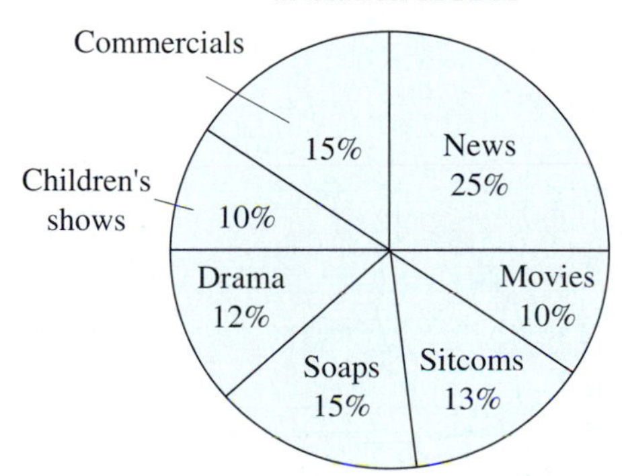

3a. ______________

[Respond below exercise.]

b. ______________

c. ______________

[Respond below exercise.]

4a. ______________

b. ______________

c. ______________

3. Sally's monthly car expenses are shown in the circle graph. (a) What were her total car expenses for the month? (b) What percent of her expenses did she spend on each category? (c) What was the ratio of her insurance expenses to her gas expenses?

Monthly Car Expenses

Repairs
$80
Gas
$50
Loan Payment
$300
Insurance
$70

4. Mike just graduated from college and decided that he should try to live within a budget. The circle graph shows the categories he chose and the percents he allowed. His beginning salary was $24,000. (a) How much did he budget for each category? (b) What category was smallest in his budget? (c) What total amount did he budget for personal items (savings, clothing, and fun)?

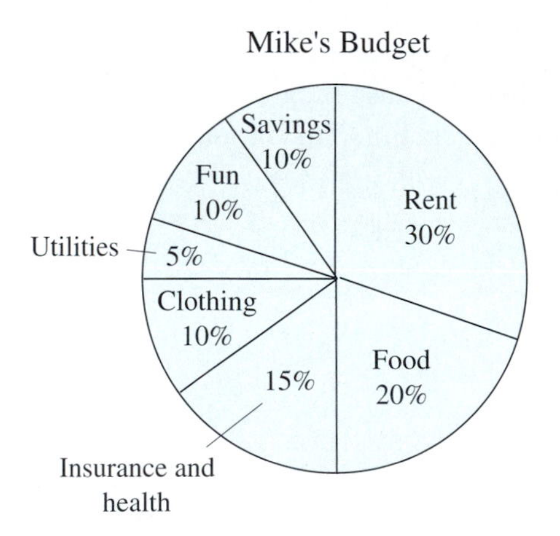

NAME ______________________ SECTION ______________ DATE __________

ANSWERS

5. The pictograph shows the numbers of men and women marathon runners in one particular race from three different cities.

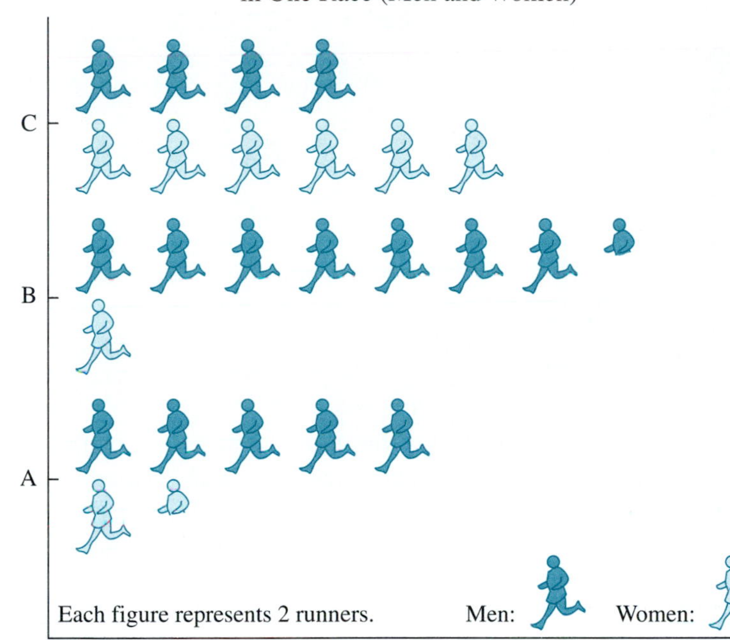

(a) Which city had the greatest number of marathon runners in the race? How many?
(b) What was the total number of runners in the race?
(c) Which city had the greatest number of women runners in the race? How many?
(d) What percent of the runners were men? What percent were women?

5a. ______________

b. ______________

c. ______________

d. ______________

6. The pictograph shows the profits earned by four different local banks during the last year.

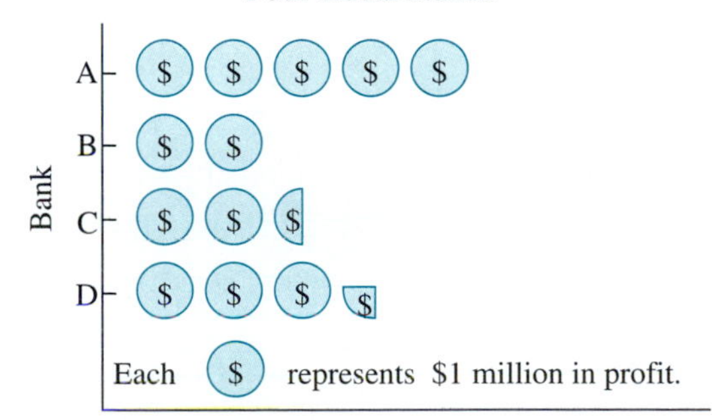

(a) Which bank earned the most money last year? How much?
(b) Which bank earned the least money last year? How much?
(c) What percent of the total amount earned by all the banks was made by Bank B?

6a. ______________

b. ______________

c. ______________

7a. ______________

b. ______________

c. ______________

7. The pictograph shows the annual sales of boats by one manufacturer for the years 1988–1992.

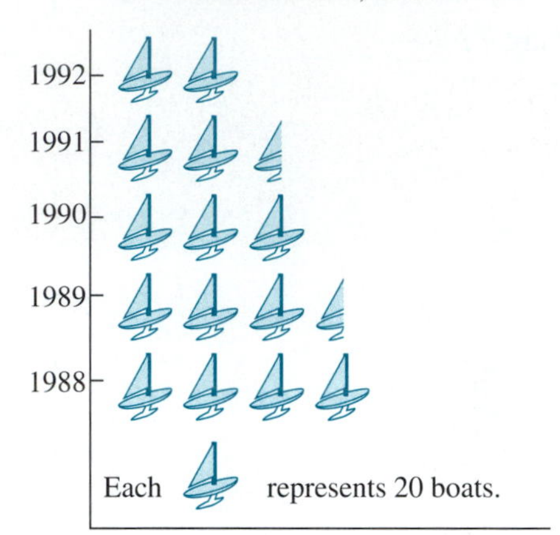

(a) During which year did the company sell the most boats? How many?
(b) What was the difference between the sales during the best year and the sales during the worst year?
(c) What was the average number of sales per year over the five-year period?

S.3 Bar Graphs and Line Graphs

OBJECTIVES

1. *Understand bar graphs and line graphs and interpret the data shown.*
2. *Learn to perform appropriate operations related to the data shown in a bar graph and in a line graph.*

Reading Bar Graphs and Line Graphs

In this section, we will discuss two more types of graphs: bar graphs and line graphs. The general purpose of each of these types of graphs is described below.

1. **Bar Graph:** Emphasizes comparative amounts.
2. **Line Graph:** Indicates tendencies (or trends) over a period of time.

Remember the following characteristics of all types of graphs.

1. They should be clearly labeled.
2. They should be easy to read.
3. They should have appropriate titles.

As in Section S.2, the following completion examples are related to specific graphs. **You are to complete blanks that are not filled in by referring to the corresponding graph.**

✓ COMPLETION EXAMPLE 1 *Bar Graph*

Figure S.4 shows a bar graph. Note that the scale on the left and the data on the bottom (months) are clearly labeled and the graph itself has a title. The following questions can be easily answered by looking at the graph.

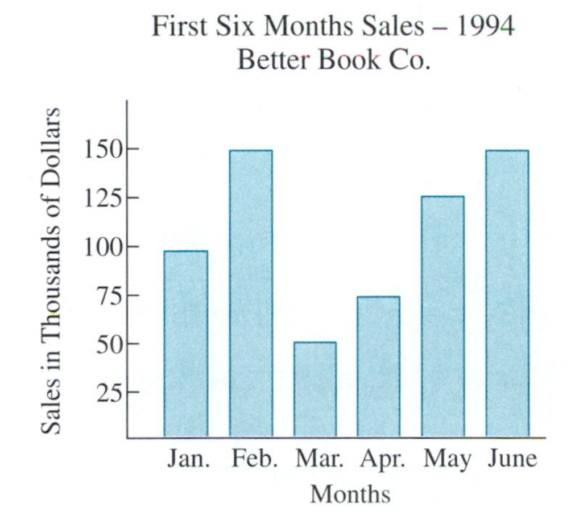

Figure S.4 Bar Graph

(a) What were the sales in January? $100,000

(b) During what month were sales lowest? March

(c) During what months were sales highest? February and June

(d) What were the sales during each of the highest sales months?

____

(e) What were the sales in April? ______________________

The following questions require some calculations after reading the graph.

(f) What was the amount of decrease in sales between February and March?

	$150,000	February sales
−	50,000	March sales
	$100,000	decrease in sales

(g) What was the percent of decrease?

$$\frac{\$100{,}000 \text{ decrease}}{\$150{,}000 \text{ February sales}} = \frac{2}{3} = 0.666$$

$$= 67\% \text{ decrease (approximately)}$$

☑ COMPLETION EXAMPLE 2 *Line Graph*

Figure S.5 is a line graph that shows the relationships between daily high and low temperatures. From the graph you can see that temperatures tended to rise during the week but fell sharply on Saturday.

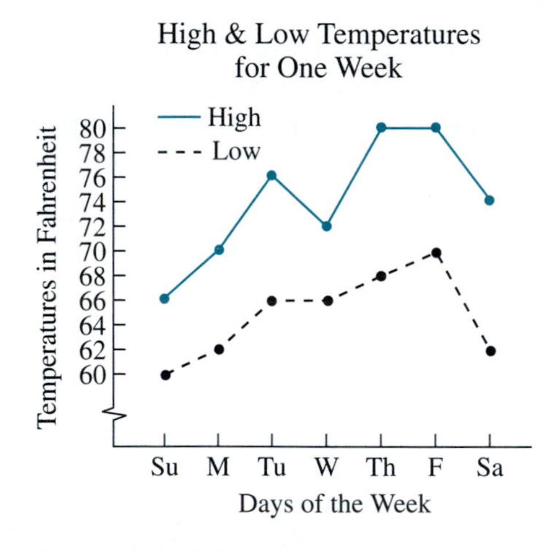

Figure S.5 Line Graph

What was the lowest high temperature? 66°

On what day did this occur? Sunday

What was the highest low temperature? 70°

On what day did this occur? Friday

Find the average difference between the daily high and low temperatures for the week shown.

Solution

First find the differences, then average these differences.

Sunday	66 − 60 =	6°
Monday	70 − 62 =	8°
Tuesday	76 − 66 =	10°
Wednesday	72 − 66 =	______
Thursday	80 − 68 =	______
Friday	80 − 70 =	______
Saturday	74 − 62 =	______
		______ total of differences

$7\overline{)\quad\quad}$ average difference (approximately)

Completion Example Answers

1. (d) **$150,000**
   (e) **$75,000**

2. | | |
   |---|---|
   | Sunday | 6° |
   | Monday | 8° |
   | Tuesday | 10° |
   | Wednesday | **6°** |
   | Thursday | **12°** |
   | Friday | **10°** |
   | Saturday | **12°** |
   | | **64°** |

$$\begin{array}{r} \mathbf{9.1^\circ} \\ 7\,\overline{)\,\mathbf{64.0}} \\ \underline{\mathbf{63}} \\ \mathbf{10} \\ \underline{\mathbf{7}} \\ \mathbf{3} \end{array}$$

average difference (approximately)

NAME ______________ SECTION ______________ DATE ______________

Exercises S.3

Answer the questions related to each of the graphs. Some questions can be answered directly from the graphs; others may require some calculations.

1. The bar graph shows the numbers of students in five fields of study at a university.

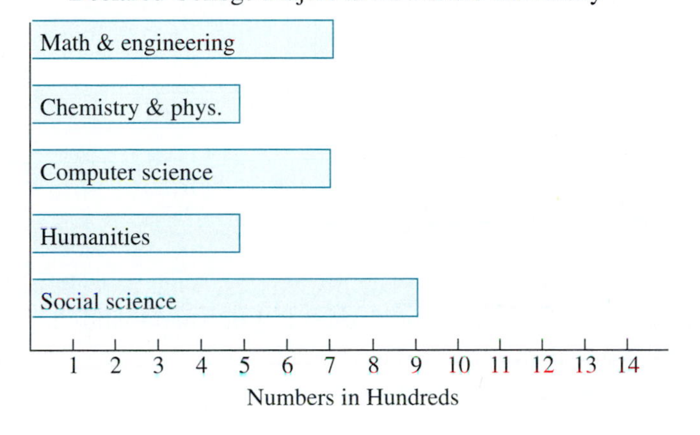

(a) Which field of study has the largest number of declared majors?
(b) Which field of study has the smallest number of declared majors?
(c) How many declared majors are indicated in the entire graph?
(d) What percent are computer science majors?

2. The bar graph shows the numbers of vehicles that crossed one intersection during a two-week period.

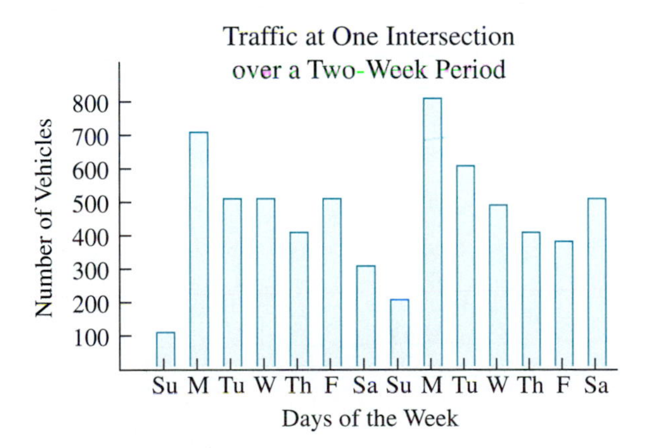

(a) On which day did the highest number of vehicles cross the intersection? How many crossed that day?
(b) What was the average numnber of vehicles that crossed the intersection on the two Sundays?
(c) What was the total number of vehicles that crossed the intersection during the two weeks?
(d) What percent of the total traffic was counted on Saturdays?

ANSWERS

1a. ______________
b. ______________
c. ______________
d. ______________

2a. ______________
b. ______________
c. ______________
d. ______________

3a. ______________

b. ______________

c. ______________

d. [Respond below exercise.] ______________

e. [Respond below exercise.] ______________

3. In comparing the following two graphs, assume that all five students graduated with comparable grades from the same high school.

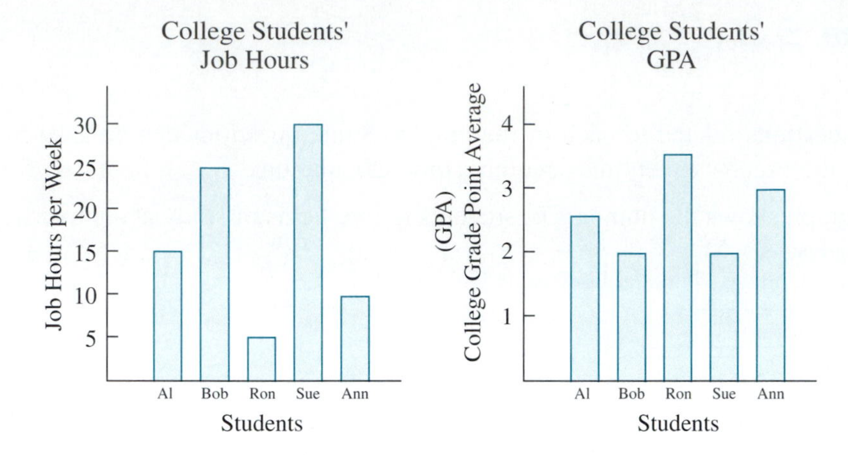

(a) Who worked the most hours per week?

(b) Who had the lowest GPA?

(c) If Ron spent 30 hours per week studying for his classes, what percent of his total work week (part-time work plus study time) did he spend studying?

(d) Which two students worked the most hours? Which two students had the lowest GPAs? Do you think that this is typical?

(e) Do you think that the two graphs shown here could be set as one graph? If so, show how you might do this.

NAME ______________________ SECTION ______________ DATE __________

ANSWERS

4a. ________

b. ________

c. ________

d. ________

5a. ________

b. ________

c. ________

4.

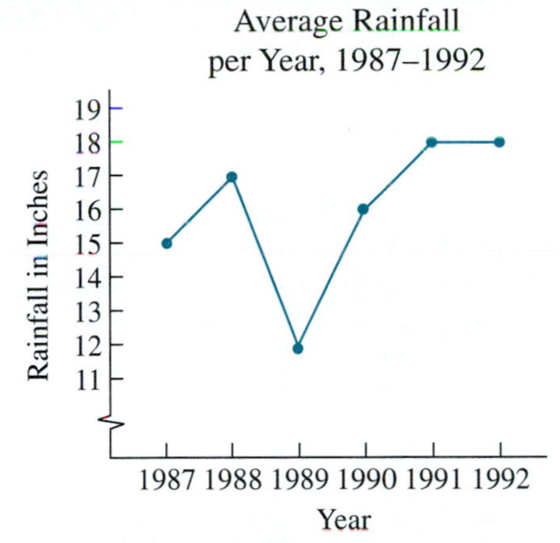

(a) What year had the least rainfall?
(b) What was the most rainfall in a year?
(c) What year had the most rainfall?
(d) What was the average rainfall over the 6-year period?

5.

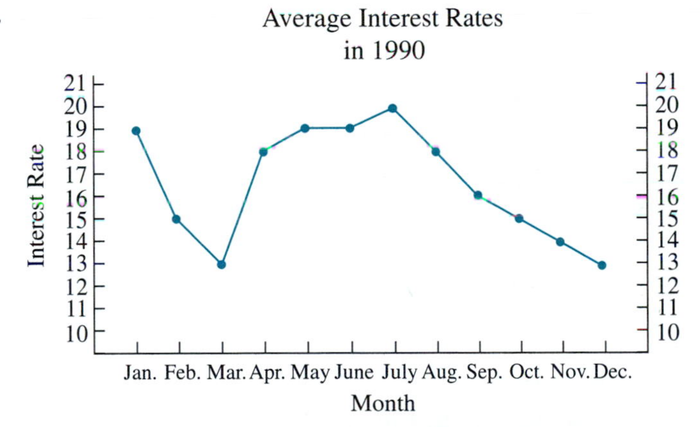

(a) During what month or months in 1990 were interest rates highest?
(b) Lowest?
(c) What was the average of the interest rates over the entire year?

6a. ________

b. ________

c. ________

d. ________

e. ________

f. ________

7a. ________

b. ________

c. ________

d. ________

e. ________

f. ________

6.

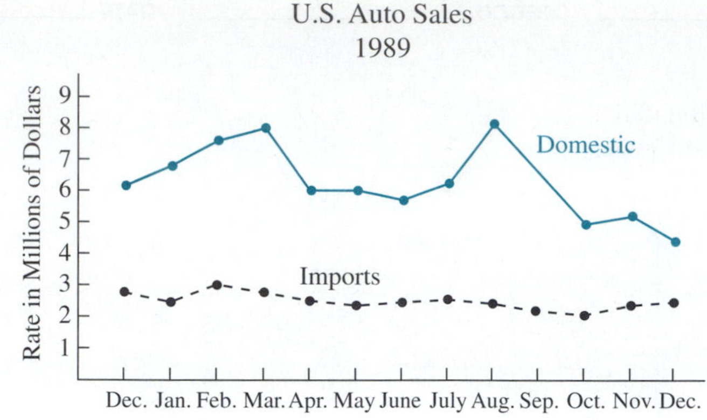

(a) During which month were domestic sales highest?
(b) How much higher were they than for the lowest month?
(c) What was the difference in import sales for the same two months?
(d) What was the difference between domestic and import sales in March?
(e) What percent of sales were imports in December 1988?
(f) In December 1989? (Answers will be approximate.)

7.

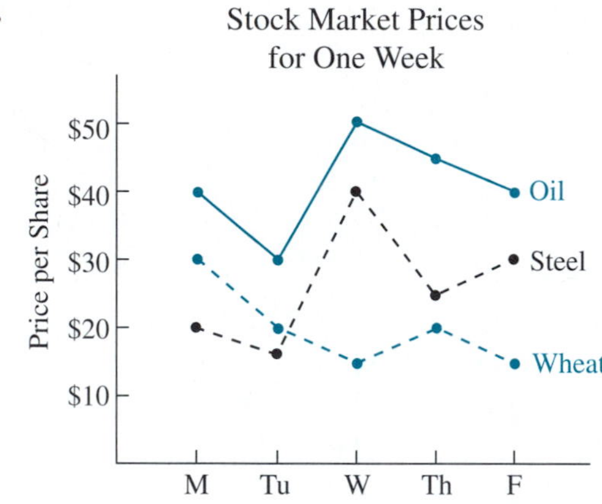

(a) If on Monday morning you had 100 shares of each of the three stocks shown (oil, steel, wheat), and you held the stock all week, on which stock would you have lost money?
(b) How much would you have lost?
(c) On which stock would you have gained money?
(d) How much would you have gained?
(e) On which stock could you have made the most money if you had sold at the best time?
(f) How much could you have made?

NAME ______________________ SECTION ______________________ DATE ______________

8. Growth of the Information Economy

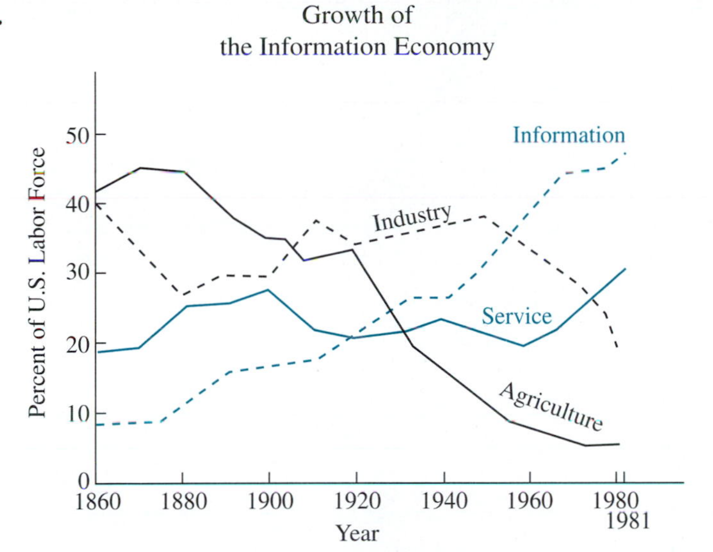

(a) What percent of workers were in each of the four areas in 1860?

(b) In 1980?

(c) Which area of work seems to have had the most stable percent of workers between 1860 and 1980?

(d) What is the difference between the highest and lowest percents for this area?

(e) Which area has had the most growth?

(f) What was its lowest percent and when?

(g) What was its highest percent and when?

(h) Which area has had the most decline?

ANSWERS

8a. [Respond below exercise.]

b. [Respond below exercise.]

c. ______________

d. ______________

e. ______________

f. ______________

g. ______________

h. ______________

S.4 Histograms and Frequency Polygons

OBJECTIVES

1. *Understand histograms and interpret the data shown.*
2. *Learn to perform appropriate operations related to the data shown in a histogram.*

Reading Histograms and Frequency Polygons

In this section, we will discuss two more types of graphs: histograms and frequency polygons. The general purpose or nature of each of these types of graphs is described below.

1. **Histogram:** A special type of bar graph with intervals of numbers indicated on the base line.
2. **Frequency Polygon:** A type of line graph formed by connecting the midpoints of the tops of the rectangles of a histogram.

A histogram looks very much like a bar graph, but there are some important distinctions that must be understood. In a histogram the base line is labeled with numbers that indicate the boundaries of a range of numbers called a **class.** The bars are placed next to each other with no spaces between them. The following terminology is related to histograms (and to frequency polygons).

Terms Related to Histograms

Class: A range (or interval) of numbers that contains data items.
Lower class limit: The smallest number that belongs to a class.
Upper class limit: The largest number that belongs to a class.
Class boundaries: Numbers that are halfway between the upper limit of one class and the lower limit of the next class.
Class width: The difference between the class boundaries of a class (the width of a bar).
Frequency: The number of data items in a class.

EXAMPLE 1 Histogram

The final statistics for the 1991 baseball season include the total number of runs scored over the course of the entire season by each of 26 teams in the major leagues that year. Those totals have been grouped as shown in Table S.1, and Figure S.6 shows the related histogram that can be drawn based on this data. (**Note:** Class boundaries, not class limits, are used in marking the baseline of the graph, and no data item is equal to any class boundary. This eliminates any ambiguity regarding the class to which a particular data item belongs.)

Table S.1 A Frequency Distribution of Total Runs Scored per Team in Major League Baseball, 1991

Class Number	Class Limits (Range of total runs)	Frequency (Number of teams in each class)
1	571–610	3
2	611–650	4
3	651–690	7
4	691–730	3
5	731–770	5
6	771–810	2
7	811–850	2

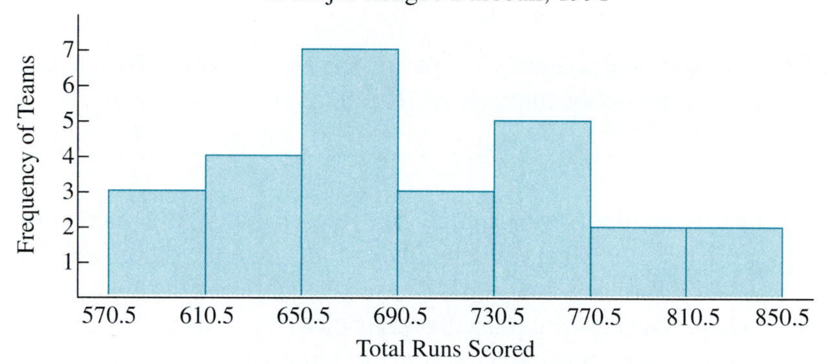

Figure S.6 Histogram for Table S.1

The following statements describe various characteristics and facts related to the histogram in Figure S.6.

(a) There are 7 classes represented.

(b) For the first class, 610 is the upper class limit and 571 is the lower class limit.

(c) There are 7 teams in the third class. That is, for these 7 teams the total number of runs that each scored in 1991 was between 651 and 690, inclusive.

(d) The class boundaries for the second class are 610.5 and 650.5.

(e) The width of each class is 40 $(650.5 - 610.5 = 40)$. (Note that this is **not** the difference in the class limits since $650 - 611 = 39$.)

(f) The percent of teams in the third class is found by dividing the frequency of the class (7) by the total number of teams (26):

$$\frac{7}{26} \approx 0.2692 \approx 26.9\%$$

Therefore, about 26.9% of the data are in the third class.

COMPLETION EXAMPLE 2

Figure S.7 shows a histogram that summarizes the scores of 50 students on an English placement test. Answer the following questions by referring to the graph.

(a) How many classes are represented? 6

(b) What are the class limits of the first class? 201 and 250

(c) What are the class boundaries of the second class? 250.5 and 300.5

(d) What is the width of each class? __________

(e) Which class has the greatest frequency? __________

(f) What is this frequency? __________

(g) What percent of the scores are between 200.5 and 250.5? __________

(h) What percent of the scores are above 400? __________

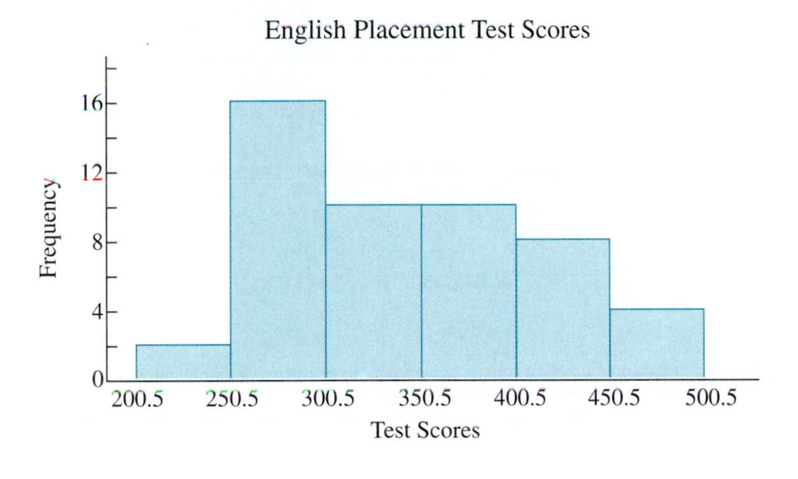

Figure S.7 Histogram of test scores

A variation on both the histogram and the line graph can be created by connecting the midpoints of the tops of the frequency rectangles with straight line segments. The resulting line graph is called a **frequency polygon.** Two additional line segments may be drawn to connect the points at each end to the base line. (See Figures S.8 and S.9.) This is done to indicate the fact that the frequency is 0 for any data outside the classes indicated by the histogram.

COMPLETION EXAMPLE 3

Answer the following questions by referring to the frequency polygon in Figure S.9.

(a) How many students scored in the highest class? 5

(b) What are the boundaries of the highest class? __________

(c) What are the limits of this class? __________

(d) What is the class width? __________

(e) How many students took the typing test? __________

(f) If a score of 51 or higher is passing, what percent of the scores were passing? __________

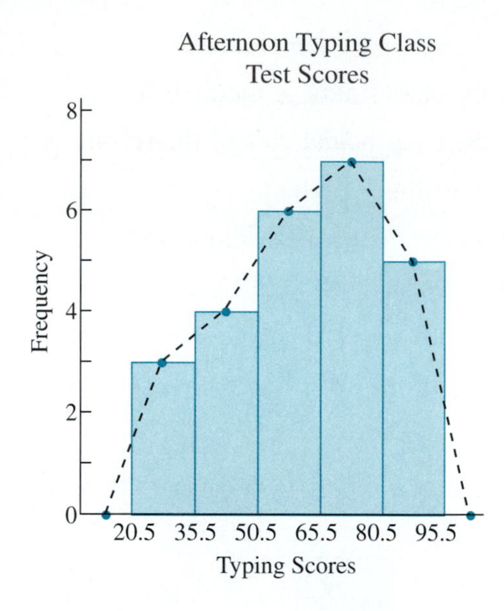

Figure S.8 Histogram of typing scores

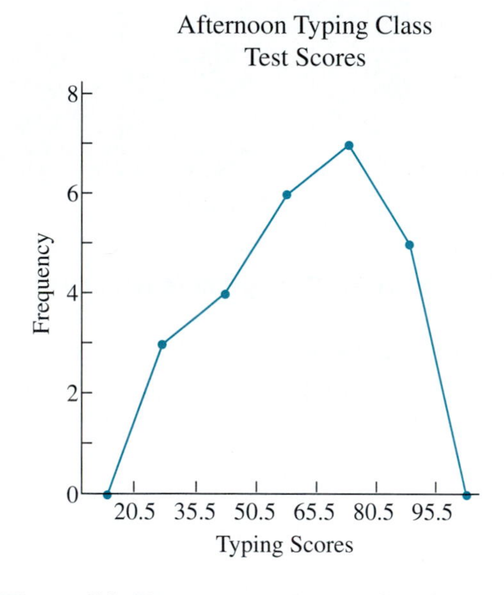

Figure S.9 Frequency polygon of typing scores

Completion Example Answers

2. (d) **50**
   (e) **The second class**
   (f) **16**
   (g) **4%**
   (h) **24%**

3. (b) **80.5 and 95.5**
   (c) **81 and 95**
   (d) **15**
   (e) **25**
   (f) **72%**

NAME ________________ SECTION ________________ DATE ________

ANSWERS

1a. ________

b. ________

c. ________

d. ________

e. ________

f. ________

g. ________

h. ________

Exercises S.4

Answer the questions related to each of the graphs.

1. Tread Life for New Tires

Number of Miles	Frequency

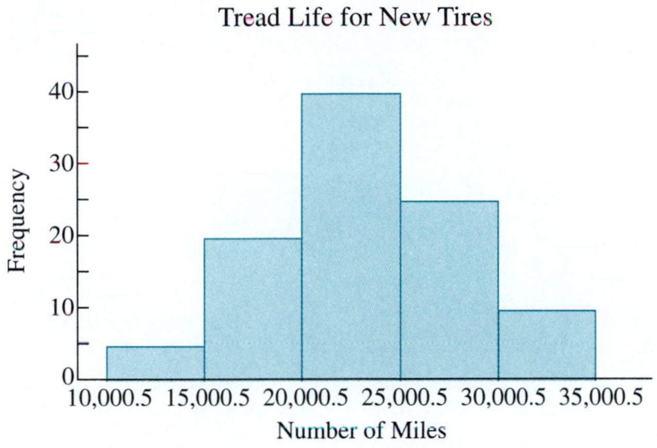

(a) How many classes are represented?

(b) What is the width of each class?

(c) Which class has the highest frequency?

(d) What is this frequency?

(e) What are the class boundaries of the second class?

(f) How many tires were tested?

(g) What percent of the tires were in the first class?

(h) What percent of the tires lasted more than 25,000 miles?

2a. ______________

b. ______________

c. ______________

d. ______________

e. ______________

f. ______________

g. ______________

h. ______________

2.

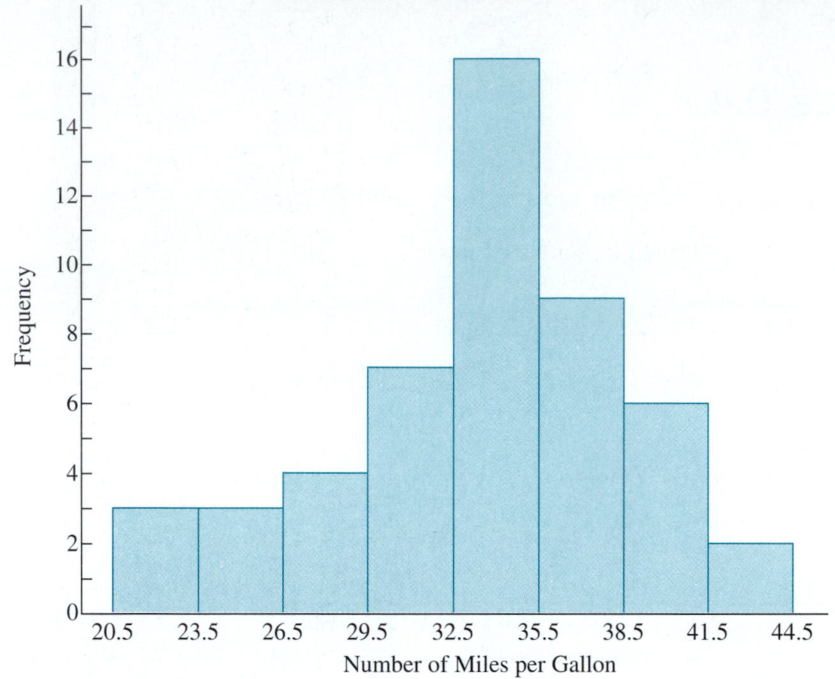

(a) How many classes are represented?

(b) What is the class width?

(c) Which class has the smallest frequency?

(d) What is this frequency?

(e) What are the class limits for the third class?

(f) How many cars were tested?

(g) How many cars tested below 30 miles per gallon?

(h) What percent of the cars tested above 38 miles per gallon?

3.

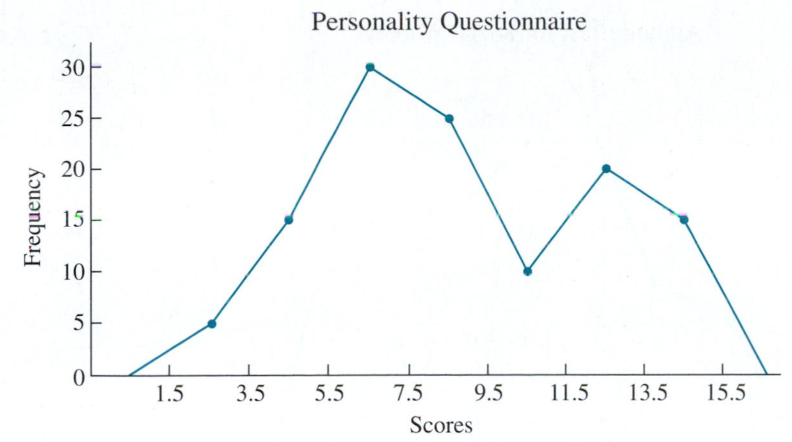

(a) Which class contains the highest number of test scores?

(b) How many scores are between 3.5 and 5.5?

(c) How many students took the test?

(d) How many students scored less than 7.5?

(e) How many students scored more than 11?

(f) Approximately what percent of the students scored between 5.5 and 11.5?

ANSWERS

3a. ______________

b. ______________

c. ______________

d. ______________

e. ______________

f. ______________

4a. ______________

b. ______________

c. ______________

d. ______________

e. ______________

f. ______________

4.

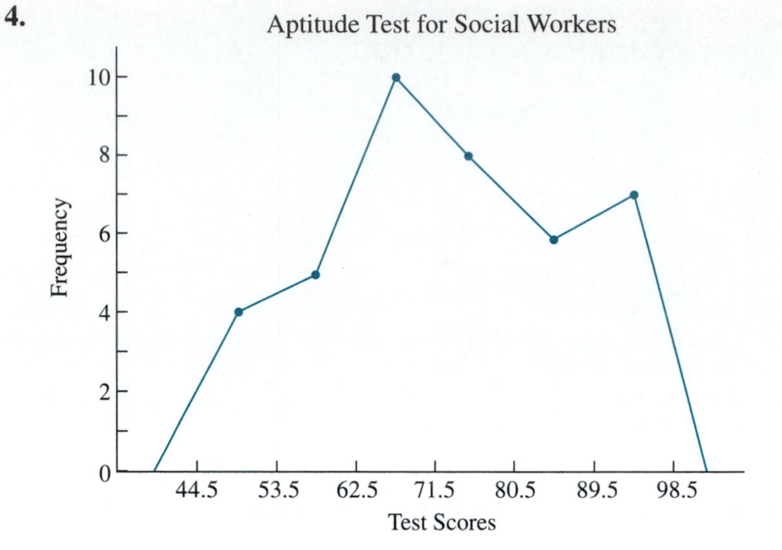

(a) How many people took the aptitude test?

(b) How many classes are represented?

(c) What is the class width?

(d) How many people scored below 63?

(e) What percent of the scores were in the first class?

(f) If a score of 72 or higher is considered acceptable for employment as a social worker, what percent of the test takers are employable?

Index of Key Ideas and Terms

NAME ______________________ SECTION ______________________ DATE ____________

ANSWERS

1a. ____________

b. ____________

c. ____________

2a. ____________

b. ____________

c. ____________

3. ____________

Chapter Test

Find (a) the mean, (b) the median, and (c) the range for the data given in Exercises 1 and 2.

1. The number of hours of television viewing per day for one person for eleven days:

3, 2, 2, 0, 1, 4, 1, 2, 3, 2, 2

2. The number of inches of rain per month for six months in a particular region:

2.54, 10.16, 7.62, 20.32, 12.70, 7.62

In Problems 3–5, use the circle graph below.

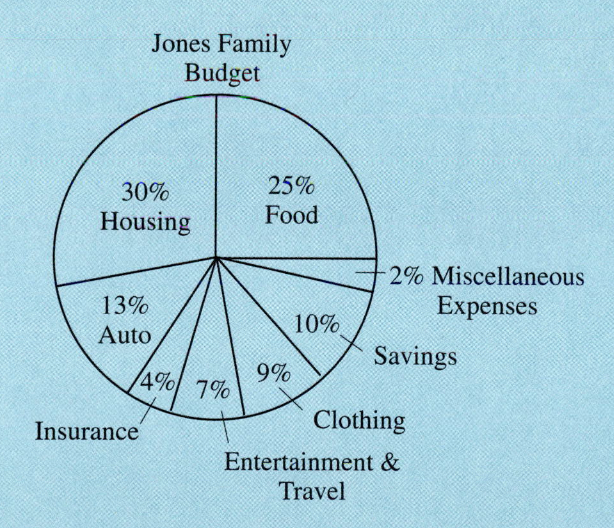

3. If the combined Jones family income is $3300 per month, how much do they spend for food in one year (to the nearest dollar)?

4. ______________

5. ______________

6. ______________

7. ______________

4. How much more do they save in one month than they spend for clothing (to the nearest dollar)?

5. The Jones family would like to take a $3000 tour of Europe. Will one year's entertainment and travel budget accommodate the trip? If so, what will they have left over? If not, how much will they be short?

In Problems 6–8, refer to the graph below.

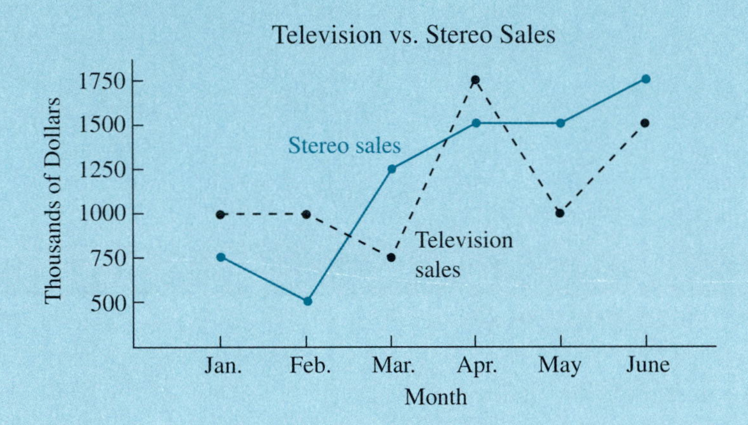

6. What were average monthly television sales during the six months (to the nearest dollar)?

7. What percent of the total March sales were stereos (to the nearest tenth of a percent)?

NAME ______________ SECTION ______________ DATE ________

ANSWERS

8. What percent of the total six months' sales were televisions (to the nearest tenth of a percent)?

8. ________

9a. ________

b. ________

c. ________

10. ________

11. ________

In Problems 9–11, refer to the histogram below.

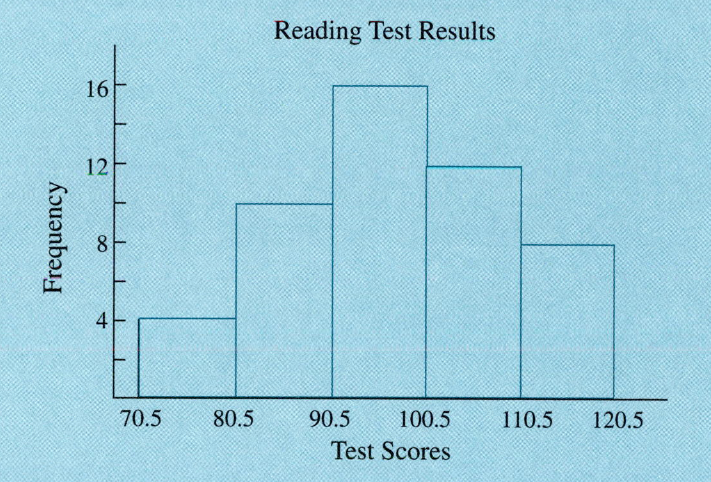

9. (a) How many classes are represented?
(b) What is the width of each class?
(c) What is the frequency of the fourth class?

10. What percent of the scores were between 80.5 and 100.5?

11. What percent of the scores were below 101?

12. [Respond below exercise.] ______

12. Draw a histogram that illustrates the data given in the following frequency distribution table. Give the graph a title and label the scales appropriately. After you have drawn the histogram, draw the corresponding frequency polygon over the top of the histogram.

A Frequency Distribution of Mathematics Exam Scores

Class Number	Class Limits (Range of exam scores)	Frequency (Number of students in each class)
1	50–59	1
2	60–69	4
3	70–79	8
4	80–89	5
5	90–99	2
		20

NAME ______________ SECTION ______________ DATE ______________

ANSWERS

Cumulative Review

All fractions should be reduced to lowest terms.

1. Write the number three hundred fifty-seven and forty-two thousandths in decimal notation.

2. Round off 28.4963 to the nearest hundredth.

3. Find the decimal equivalent of $\frac{7}{5}$.

4. Find the quotient of 77.35 and 8.2 to the nearest thousandth.

5. Simplify the expression: $\sqrt{75} + 2\sqrt{48}$.

6. Find the LCM of 45, 63, and 77.

7. Evaluate the expression: $6^2 + (12 \div 3 \cdot 2^2 + 2^3) \div 3 \cdot 2$.

8. Find (a) the mean, (b) the median, and (c) the range of the following set of numbers:

   36, 61, 59, 51, 51, 71, 56, 79

9. (a) If an investment pays 6.5% compounded daily, what will be the value of $5000 in five years? (b) In about how many years will the amount be doubled to $10,000? (**Hint:** This will involve some experimenting with your calculator.)

1. ______________
2. ______________
3. ______________
4. ______________
5. ______________
6. ______________
7. ______________

8a. ______________

b. ______________

c. ______________

9a. ______________

b. ______________

10. ____________

11. ____________

12. ____________

13. ____________

14. ____________

15. [Respond below exercise.]

16. ____________

17. ____________

10. A note for \$3500 at 7% interest was paid off with \$3622.50. What was the length of time of the note?

11. 75 is ________% of 200.

12. 80% of ________ is 232.

13. 26.8% of 16,240 is ________.

14. 125% of 300 is ________.

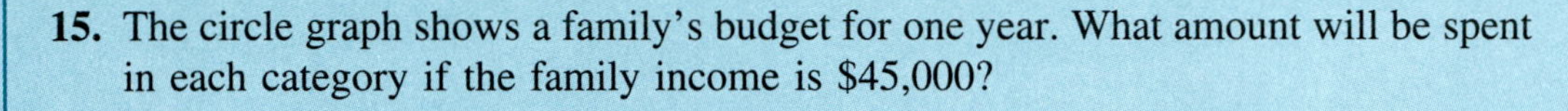

15. The circle graph shows a family's budget for one year. What amount will be spent in each category if the family income is \$45,000?

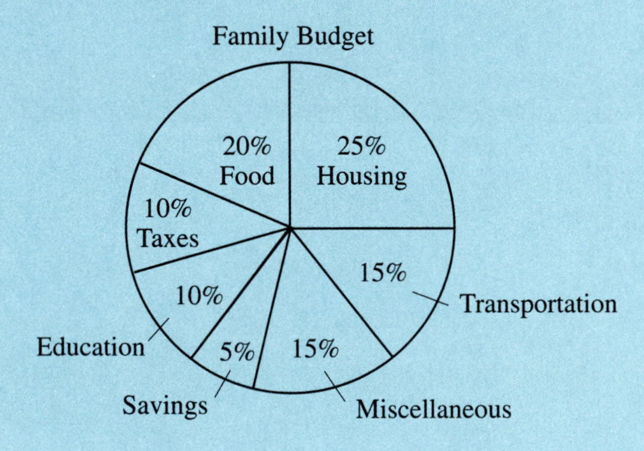

16. Solve for x:

$5x - 20 = 15x + 4.5$

17. Solve for y:

$1.2y + 2.4y = 26 + 10$

NAME ______________ SECTION ______________ DATE ______________

ANSWERS

18a. ______________
b. ______________
c. ______________

19a. ______________
b. ______________

20. ______________

21. ______________

22a. ______________
b. ______________
c. ______________
d. ______________

23a. ______________
b. ______________
c. ______________
d. ______________

18. A home sold for \$132,000. The buyer has to pay a down payment of 25% of the selling price, a loan fee of 1.5% of the mortgage, \$300 for fire insurance, \$75 for recording fees, \$700 for taxes, and \$550 for legal fees. (a) What is the amount of the down payment? (b) What is the amount of the mortgage? (c) How much cash does the buyer need to complete the purchase?

19. (a) Using radical notation, find the length of the hypotenuse of a right triangle with legs of length 8 feet and 4 feet. (b) Use a calculator to find this value to the nearest tenth of a foot.

20. Add and simplify:

$$\begin{array}{r} 12\text{ hr } 23\text{ min } 12\text{ sec} \\ +\ \ 6\text{ hr } 39\text{ min } 48\text{ sec} \\ \hline \end{array}$$

21. Find the difference:

$$\begin{array}{r} 4\text{ yd } 2\text{ ft } 5\text{ in.} \\ -\ 1\text{ yd } 2\text{ ft } 9\text{ in.} \\ \hline \end{array}$$

22. Convert the following metric units as indicated.

(a) 35 cm = __________ mm (b) 35 cm^2 = __________ mm^2

(c) 762 g = __________ kg (d) 16.3 L = __________ mL

23. Convert the following U.S. customary units as indicated.

(a) 10 ft = __________ in. (b) 10 ft^2 = __________ $in.^2$

(c) 3 yd^2 = __________ ft^2 (d) 3 acres = __________ ft^2

24a. ______________

b. ______________

c. ______________

d. ______________

24. The original price of a microwave oven was \$300 and it is marked on sale at a 30% discount. (a) Find the amount of the discount. (b) Find the profit if the store paid the manufacturer \$150 for the oven. (c) Find the percent of profit based on cost to the store. (d) Find the percent of profit based on selling price.

25a. ______________

b. ______________

c. ______________

d. ______________

25. Evaluate each of the following expressions.

(a) $-7 + 5 - 10$

(b) $(-32)(-5)(-1)$

(c) $\frac{-45}{-9}$

(d) $|-6| + |-2| - |-8|$

Geometry

What to Expect in this Chapter

This chapter introduces some of the basic concepts of plane geometry. Many of these ideas are considered fundamental knowledge for future courses in mathematics.

In Section G.1, we introduce three undefined terms: **point, line,** and **plane.** These concepts form the basis for the study of plane geometry as organized and developed by Euclid around the year 300 B.C. The concept of **length** is developed through a discussion of **perimeter (circumference** for a circle) and the formulas for the perimeters of six types of geometric figures: square, rectangle, parallelogram, triangle, circle, and trapezoid. In Section G.2, the concept of **area** is discussed and the formulas for the areas of these same six types of geometric figures are presented.

The concept of **volume** and formulas for finding the volume of five three-dimensional figures are presented in Section G.3. The figures discussed are rectangular solids, rectangular pyramids, right circular cylinders, right circular cones, and spheres.

Section G.4 begins with the definition of an **angle** and discusses how to use a protractor for measuring angles. We show that an angle is classified according to its measure as acute, right, obtuse, or straight. The chapter closes with a detailed discussion of **triangles** in Section G.5. A triangle can be classified in two ways: by the lengths of its sides and by the measures of its angles. Two similar triangles have the following two properties: corresponding angles have the same measure and corresponding sides are proportional.

Mathematics at Work!

Although most people are not aware of it, mathematics is part and parcel of our daily lives. Even in the arts, mathematics can play a central role. Geometry, in particular, is integral to much of art, employing concepts such as a vanishing point and ways of creating three-dimensional perspective on a two-dimensional canvas.

One famous artist who made use of geometry throughout most of his work was Maurits Escher, born in the Netherlands in 1898. Escher was encouraged by his father to become an architect because of his artistic abilities. However, his abilities lay more in the form of pure and decorative art, and he created many famous pieces in several mediums: woodcuts, lithographs, mezzotints, sketches, and watercolors.

Waterfall (pictured here) illustrates the use of geometry in creating an optical illusion in which water appears to fall and then flow back up.

G.1 Length and Perimeter

Introduction to Geometry

Plane geometry is the study of the properties of figures in a plane. You should understand that geometry is an important part of the development of mathematics and that basic mathematical concepts are geometric as well as arithmetic and algebraic.

The three most basic ideas in plane geometry are **point, line,** and **plane.** These terms are considered so fundamental that any attempt to define their meanings would use terms more complicated and more difficult to understand than the terms being defined. Thus, they are simply not defined and are called **undefined terms.** As we will see, these undefined terms provide the foundation for the study of geometry and the definitions of other geometric terms such as line segment, ray, angle, triangle, circle, and so on.

Undefined Term	Representation	Discussion
1. Point	A • point A	A dot represents a point. Points are labeled with capital letters.
2. Line	ℓ A B line ℓ or line $\overleftrightarrow{AB}$	A line has no beginning or end. Lines are labeled with small letters or by two points on the line.
3. Plane	P plane P	Flat surfaces, such as a tabletop or wall, represent planes. Planes are labeled with capital letters.

In a more formal approach to plane geometry than we will use in this text, these undefined terms, along with certain assumptions known as axioms, other statements called theorems, and a formal system of logic, are studied in detail. This approach to the study of geometry is credited to Euclid, which is why the plane geometry courses given in high school are generally known as courses in Euclidean geometry.

In this chapter, we will discuss formulas and properties related to several plane figures, including rectangles, triangles, and circles, and some three-dimensional figures such as rectangular solids and spheres. The formulas do not depend on any particular measurement system, and both the metric system and the U.S. customary system of measure will be used in the exercises. **Be sure to label all answers with the correct unit of measure indicating length, area, or volume.**

OBJECTIVES

1. *Recognize the terms* **point, line,** *and* **plane** *and know that they are undefined terms.*
2. *Understand the concepts of length and perimeter.*
3. *Learn how to apply the appropriate formula to find the perimeter of a given geometric figure.*
4. *Know how to label answers with the correct units.*

Length and Perimeter

A **formula** is a general statement (usually an equation) that relates two or more variables. The geometric figures and formulas discussed in this section illustrate applications of measures of length. From the metric system, some units of length are meter (m), decimeter (dm), centimeter (cm), millimeter (mm), and kilometer (km). From the U.S. customary system, some units of length are foot (ft), inch (in.), yard (yd), and mile (mi). Equivalent measures in both the metric system and the U.S. customary system were discussed in the chapter on measurements, which may or may not be included in your version of this textbook.

Important Terms

Perimeter: Total distance around a plane geometric figure.
Circumference: Perimeter of a circle.
Radius: Distance from the center of a circle to a point on the circle.
Diameter: Distance from one point on a circle to another point on the circle measured through the center.

For the formulas related to circles, we designate the length of a diameter with the letter d and the length of a radius with the letter r. Note that a diameter of a circle is twice as long as a radius. That is, $d = 2r$.

Six Geometric Figures and the Formulas for Finding Their Perimeters

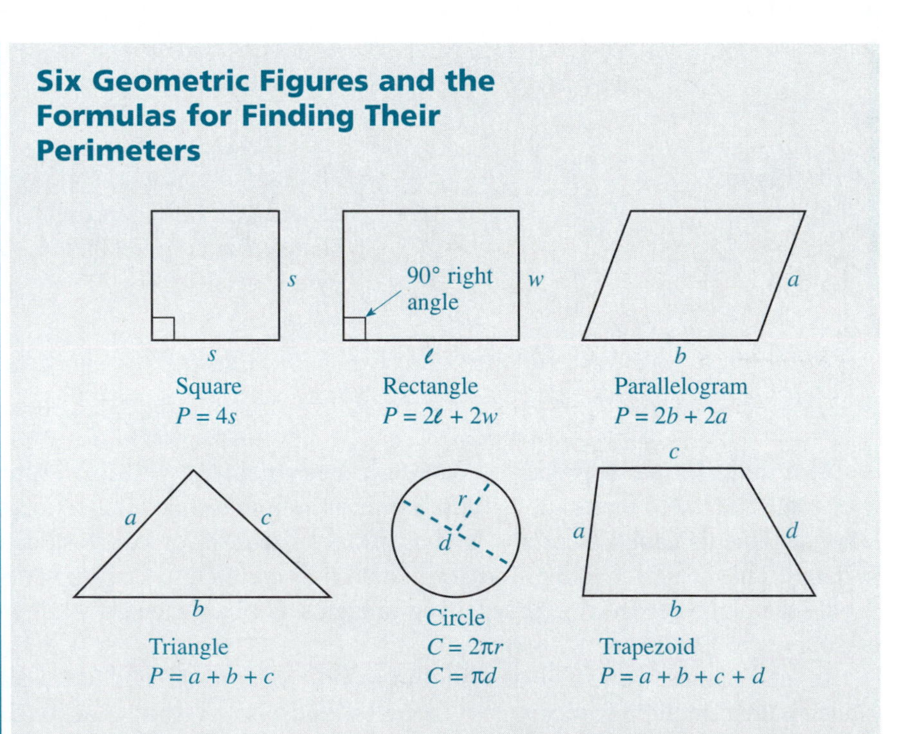

Note: π is the symbol that is used for the constant number 3.1415926535 . . . . This constant is an infinite decimal with no pattern to its digits. For our purposes, we will use $\pi = 3.14$, but you should be aware that 3.14 is only an approximation of π.

EXAMPLE 1

Find the perimeter of a rectangle with length 20 in. and width 15 in.

Solution

Sketch the figure first.

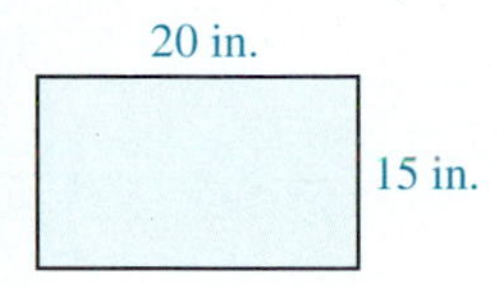

$$P = 2\ell + 2w$$

$$P = 2 \cdot 20 + 2 \cdot 15$$

$$= 40 + 30 = 70 \text{ in.}$$

The perimeter is 70 inches.

EXAMPLE 2

Find the circumference of a circle with a diameter 3 m.

Solution

Sketch the figure first.

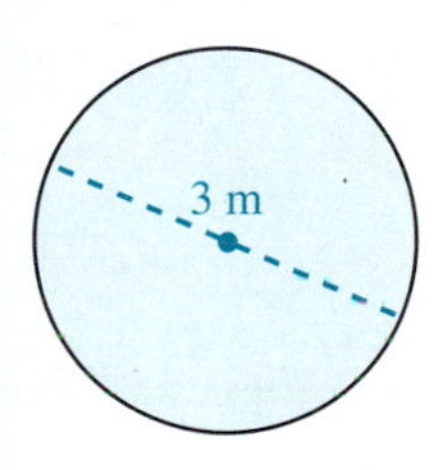

$$C = \pi d$$

$$C = 3.14(3) = 9.42 \text{ m}$$

The circumference is 9.42 m. (**Note:** Remember that 9.42 m is only an approximate answer because 3.14 is an approximation of π.)

EXAMPLE 3

Find the perimeter of a triangle with sides of 4 cm, 0.7 dm, and 80 mm. Write your answer in millimeters.

Solution

Sketch the figure and change all the units to millimeters.

4 cm = 40 mm
0.7 dm = 70 mm
80 mm

$$P = a + b + c$$

$$P = 40 + 70 + 80$$

$$= 190 \text{ mm}$$

The perimeter is 190 mm.

1. Find the circumference of a circle with radius 4 in.

2. Find the perimeter of a square with sides of 3.6 cm.

✓ Completion Example 4

Find the circumference of a circle with radius 6 ft.

Solution

Sketch the figure first.

$C = 2\pi r$

$C =$ ______ · ______ · ______

$=$ ______ ft

The circumference is ______ ft.

Now Work Exercises 1 and 2 in the Margin.

Example 5

Find the perimeter of a figure that is a semicircle (half of a circle) with a diameter of 20 centimeters.

Solution

A sketch of the figure will help.

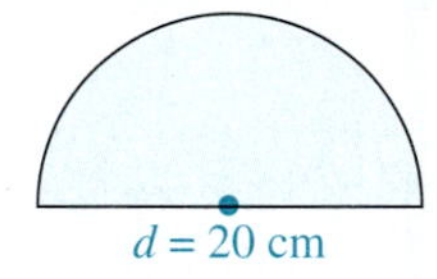

We find the perimeter of the figure by adding the length of the diameter (20 cm) to the length of the semicircle.

$$\text{Length of semicircle} = \frac{1}{2}\pi d = \frac{1}{2}(3.14) \cdot 20 = (1.57) \cdot 20$$

$$= 31.4 \text{ cm}$$

$$\text{Perimeter of figure} = 31.4 + 20 = 51.4 \text{ cm}$$

Completion Example Answer

4. $C = 2\pi r$

$C = \mathbf{2 \cdot 3.14 \cdot 6}$

$= \mathbf{37.68}$ ft

The circumference is **37.68** ft.

NAME ________________ SECTION ________________ DATE ________

ANSWERS

1. [Respond in exercise.]
2. ________
3. ________
4. ________
5. ________

Exercises G.1

1. Match each formula for perimeter to its corresponding geometric figure.

___ (a) square	A. $P = 2l + 2w$
___ (b) parallelogram	B. $P = 4s$
___ (c) circle	C. $P = 2b + 2a$
___ (d) rectangle	D. $C = 2\pi r$
___ (e) trapezoid	E. $P = a + b + c$
___ (f) triangle	F. $P = a + b + c + d$

2. Find the perimeter of a triangle with sides of 4 cm, 8.3 cm, and 6.1 cm.

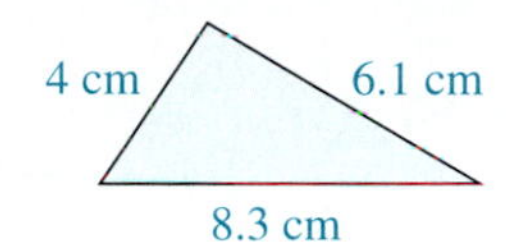

3. Find the perimeter of a rectangle with length 35 mm and width 17 mm.

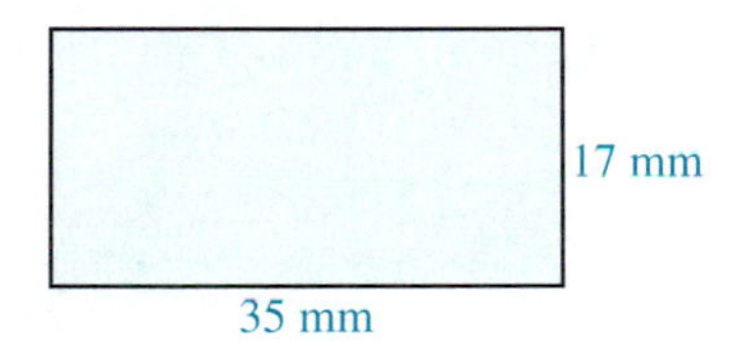

4. Find the perimeter of a square with sides of 13.3 m.

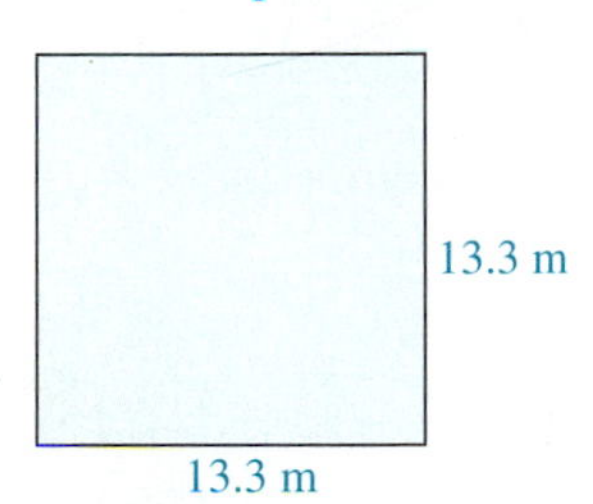

5. Find the circumference of a circle with radius 5 ft. (Use $\pi = 3.14$.)

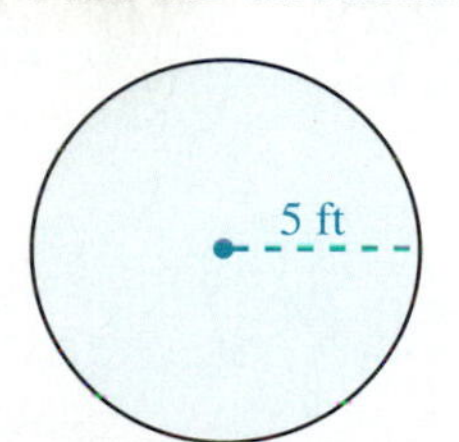

6. ______________

7. ______________

8. ______________

9. ______________

10. ______________

11. ______________

6. Find the perimeter of a parallelogram with one side 43 cm and another side 20 mm. Write your answer in millimeters.

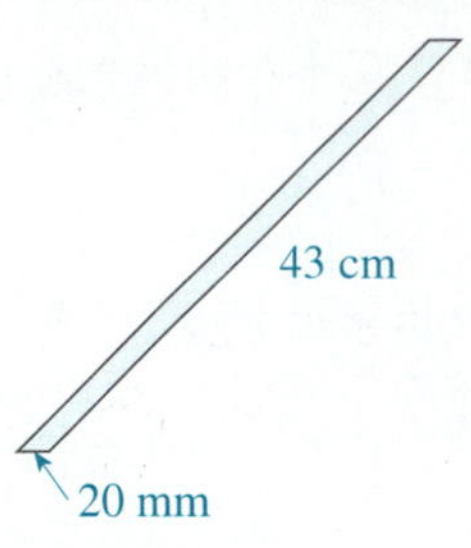

7. Find the circumference of a circle with diameter 6.2 yd. (Use $\pi = 3.14$.)

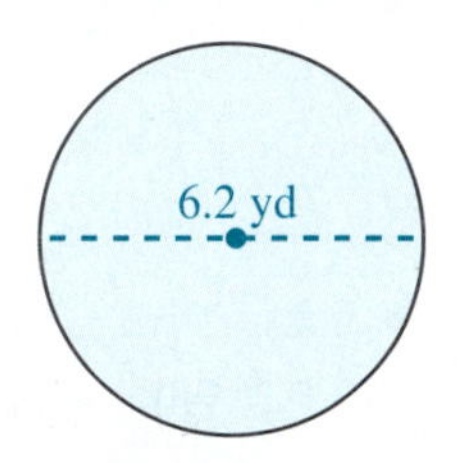

8. Find the perimeter of a rectangle with length 50 m and width 50 dm. Write your answer in meters.

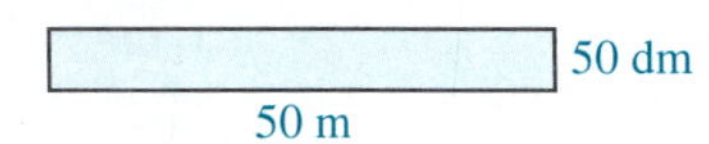

9. Find the perimeter of a triangle with sides of 5 cm, 55 mm, and 0.3 dm. Write your answers in centimeters.

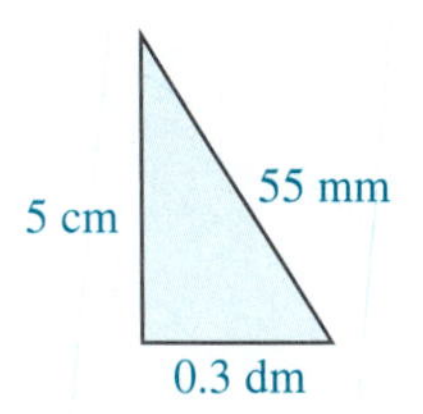

10. Find the circumference of a circle in meters if its radius is 70 cm. (Use $\pi = 3.14$.)

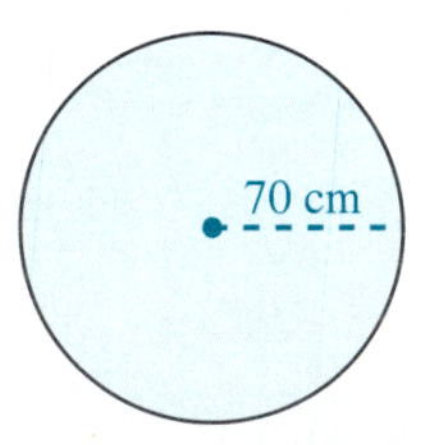

11. Find the perimeter of a square in meters if one side is 4 km long.

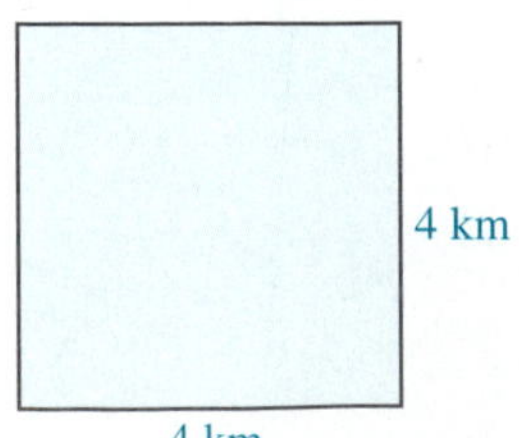

NAME ____________________ SECTION ______________ DATE __________

ANSWERS

12. ______________

13. ______________

14. ______________

15. ______________

16. ______________

17. ______________

18. ______________

19. ______________

12. Find the perimeter of a triangle with sides of 1 yd, 4 ft, and 5 ft. Write your answer in feet.

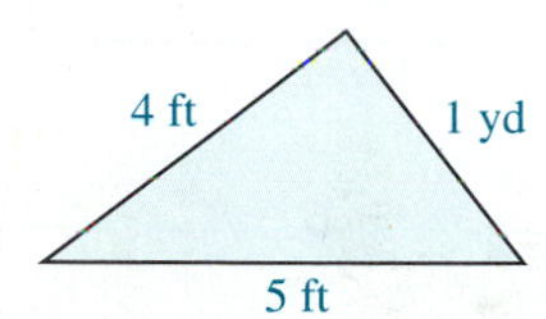

13. Find the perimeter of the trapezoid shown here.

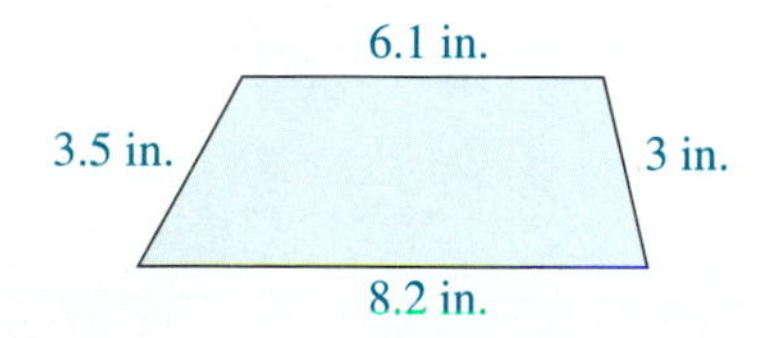

14. Find the perimeter of a parallelogram with sides of $3\frac{1}{2}$ ft and 14 inches. Write your answer in both inches and feet.

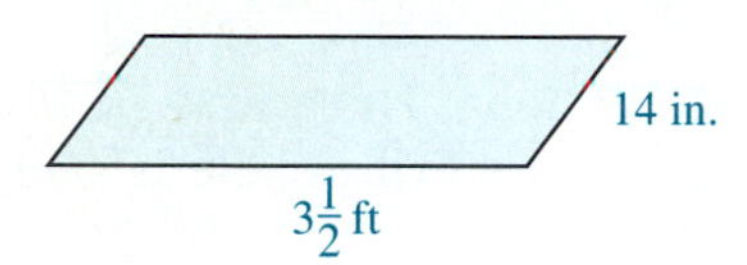

15. Find the circumference of a circle with diameter 1.5 in. Write your answer in centimeters.

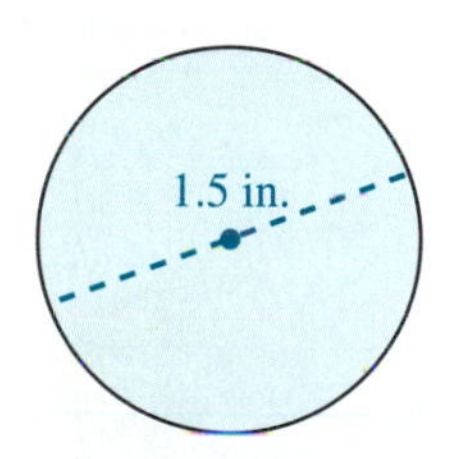

Find the perimeter of each of the following figures with the indicated dimensions.

16.

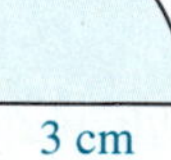

17.

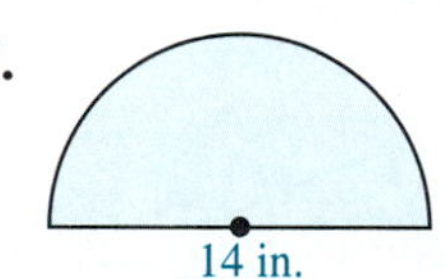

18.

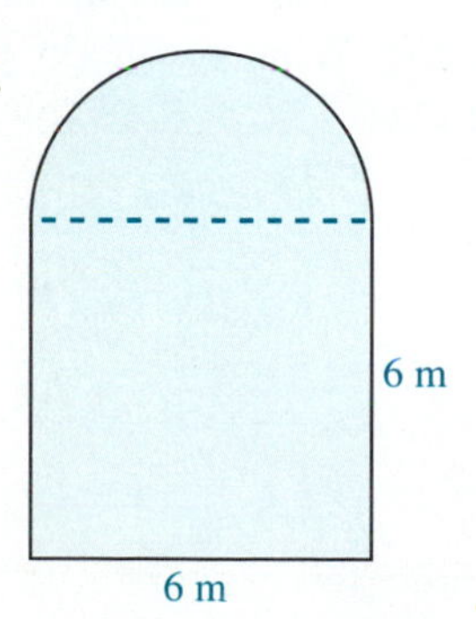

19.

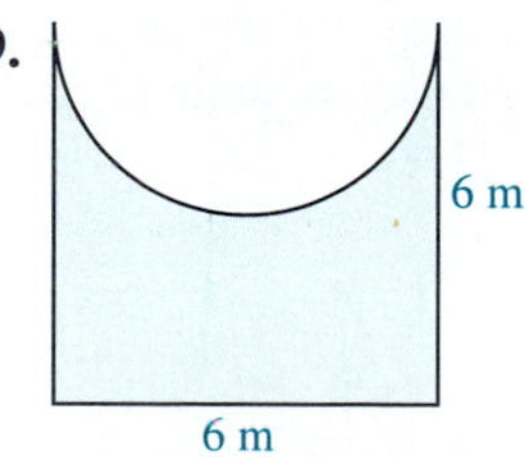

20. ______________

21. ______________

22. ______________

23. ______________

24. [Respond below exercise.]

25. [Respond below exercise.]

20.

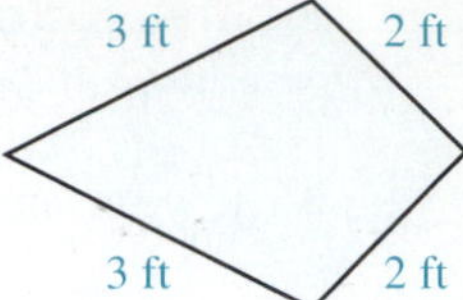

21.

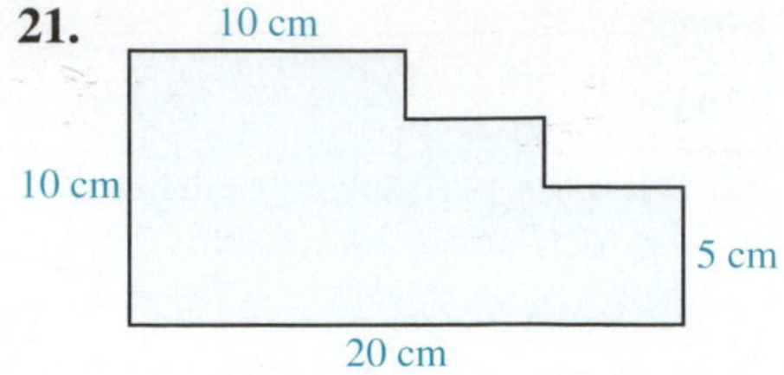

22.

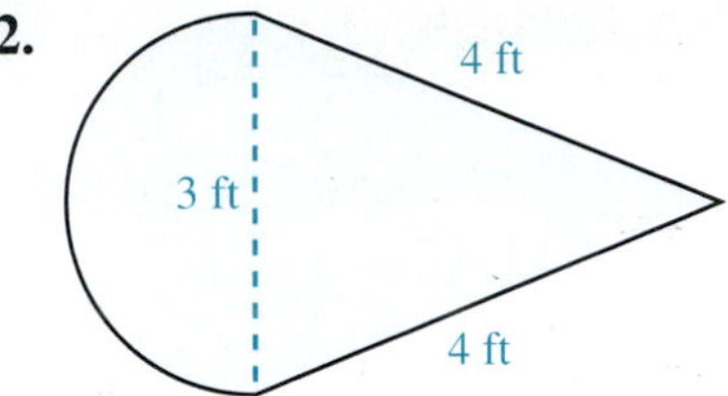

23.

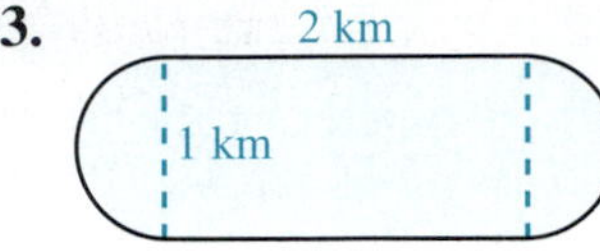

Writing and Thinking about Mathematics

24. First draw a square with sides that are each 2 inches long. Next, draw a circle inside the square that just touches each side of the square. (This circle is said to be **inscribed** in the square.)

(a) What is the length of a diameter of this circle?

(b) Do you think that the circumference is less than 8 inches or more than 8 inches? Why?

25. With a piece of string and a ruler, measure the circumference and the diameter of each of several circular figures in your home. With a calculator, divide each circumference by the length of the corresponding diameter. What result do you observe? Discuss your results with other students in the class.

G.2 Area

OBJECTIVES

1. *Understand the concept of area.*
2. *Learn how to apply the appropriate formula to find the area of a given geometric figure.*
3. *Know how to label answers with the correct units.*

The Concept of Area

Area is a measure of the interior of (or surface enclosed by) a figure in a plane and is measured in **square units.** The concept of area is illustrated in Figure G.1 in terms of square inches.

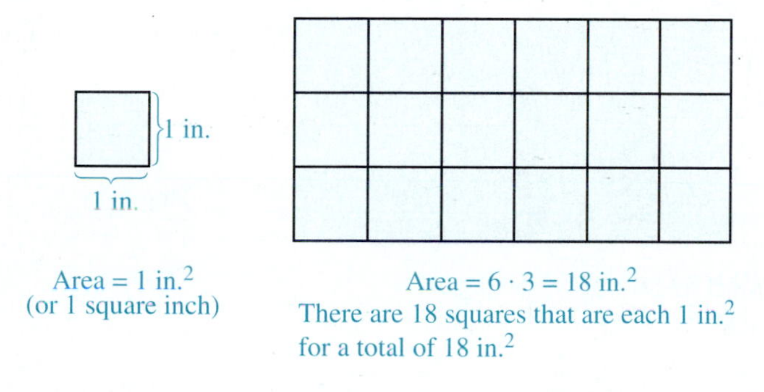

Area = 1 in.^{2}
(or 1 square inch)

Area = $6 \cdot 3 = 18$ in.^{2}
There are 18 squares that are each 1 in.^{2} for a total of 18 in.^{2}

Figure G.1

In the metric system, some units of area are square meters (m^2), square centimeters (cm^2), and square millimeters (mm^2). In the U.S. customary system, some units of area are square feet (ft^2), square inches (in.^{2}), and square yards (yd^2).

Formulas for Area

The following formulas for the areas of the geometric figures shown are valid regardless of the system of measurement used. **Be sure to label your answers with the correct units.**

Note: In triangles and other figures, we have used the letter h to represent the **height** of the figure. The height is also called the **altitude.**

Six Geometric Figures and the Formulas for Finding Their Areas

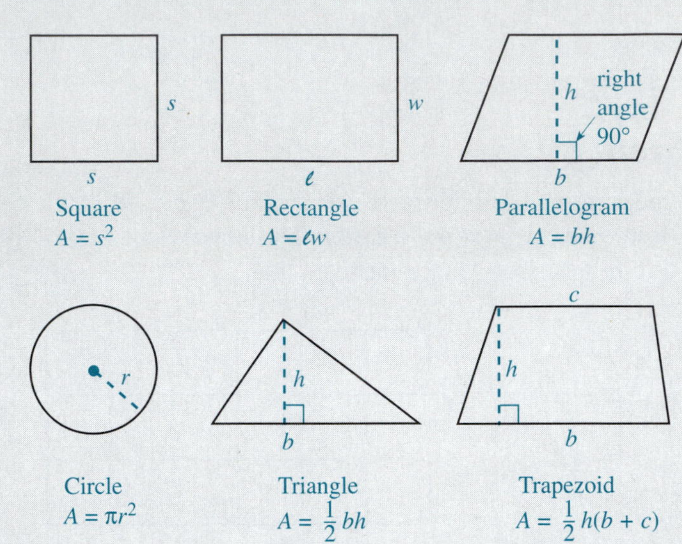

Example 1

Find the area of the figure shown here with the indicated dimensions.

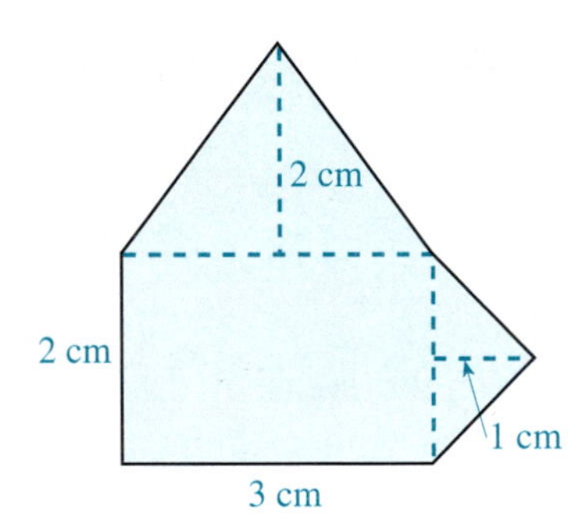

Solution

There are two triangles and one rectangle.

Rectangle	**Larger Triangle**	**Smaller Triangle**
$A = lw$	$A = \frac{1}{2}bh$	$A = \frac{1}{2}bh$
$A = 2 \cdot 3 = 6 \text{ cm}^2$	$A = \frac{1}{2} \cdot 3 \cdot 2 = 3 \text{ cm}^2$	$A = \frac{1}{2} \cdot 2 \cdot 1 = 1 \text{ cm}^2$

$$\begin{aligned} \text{Total area} &= 6 \text{ cm}^2 + 3 \text{ cm}^2 + 1 \text{ cm}^2 \\ &= 10 \text{ cm}^2 \end{aligned}$$

Example 2

Find the area of the washer (shaded portion) with dimensions as shown. (Use $\pi = 3.14$.)

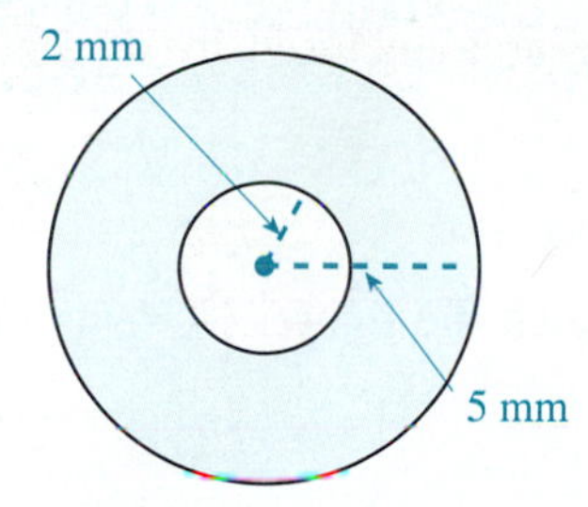

Solution

Subtract the area of the inside (smaller) circle from the area of the outside (larger) circle.

Larger Circle	**Smaller Circle**
$A = \pi r^2$	$A = \pi r^2$
$A = 3.14(5^2)$	$A = 3.14(2^2)$
$= 3.14(25)$	$= 3.14(4)$
$= 78.50\text{ mm}^2$	$= 12.56\text{ mm}^2$

Washer

$$\begin{array}{r} 78.50\text{ mm}^2 \\ -12.56\text{ mm}^2 \\ \hline 65.94\text{ mm}^2 \end{array}$$ area of washer

Example 3

Find the area of a trapezoid with altitude 6 in. and parallel sides of length 1 ft and 2 ft.

Solution

First draw a figure and label the lengths of the known parts. The units of length must be the same throughout, either all in inches or all in feet.

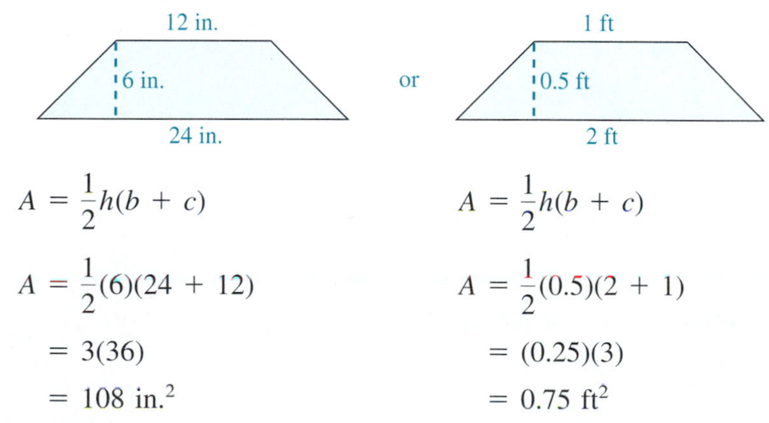

$$A = \frac{1}{2}h(b + c) \qquad A = \frac{1}{2}h(b + c)$$

$$A = \frac{1}{2}(6)(24 + 12) \qquad A = \frac{1}{2}(0.5)(2 + 1)$$

$$= 3(36) \qquad = (0.25)(3)$$

$$= 108\text{ in.}^2 \qquad = 0.75\text{ ft}^2$$

Two correct areas are 108 in.^{2} and 0.75 ft^2. (Note that, since

$$1\text{ ft}^2 = (12\text{ in.})(12\text{ in.}) = 144\text{ in.}^2,$$

we have $0.75\text{ ft}^2 = 0.75(144)\text{ in.}^2 = 108\text{ in.}^2$ The two answers are equivalent. They are simply in different units of area.)

1. Find the area of a circle with diameter 6 in.

2. Find the area of a triangle with base 10 cm and height 5 cm.

✓ *COMPLETION EXAMPLE 4*

Find the area enclosed by a semicircle and diameter if the diameter is 10 in.

Solution

First sketch the figure. (A semicircle is half of a circle.)

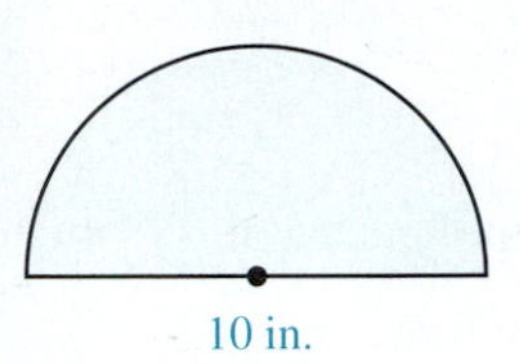

For a circle, $A = \pi r^2$.

Thus, for a semicircle, $A = \frac{1}{2}\pi r^2$.

For this semicircle, $d = 10$, so $r = 5$.

$$A = \frac{1}{2}(\_\_\_\_\_) \cdot (\_\_\_\_\_)^2$$

$$= \_\_\_\_\_ \cdot \_\_\_\_\_$$

$$= \_\_\_\_\_ \text{ in.}^2$$

The area enclosed by the semicircle and a diameter is _____ in.^{2}

Now Work Exercises 1 and 2 in the Margin.

Completion Example Answer

4. $A = \frac{1}{2}\mathbf{(3.14)(5)^2}$

$= \mathbf{1.57 \cdot 25}$

$= \mathbf{39.25}$ in.^{2}

The area enclosed by the semicircle and a diameter is **39.25** in.^{2}

NAME ______________ SECTION ______________ DATE __________

ANSWERS

1. [Respond in exercise.]

2. ______________

3. ______________

4. ______________

Exercises G.2

1. Match each formula for area to its corresponding geometric figure.

___ (a) square	A. $A = lw$
___ (b) parallelogram	B. $A = bh$
___ (c) circle	C. $A = s^2$
___ (d) rectangle	D. $A = \pi r^2$
___ (e) trapezoid	E. $A = \frac{1}{2}bh$
___ (f) triangle	F. $A = \frac{1}{2}h(b + c)$

First sketch the figure and label the given parts; then find the area of each of the following figures.

2. A rectangle 35 in. long and 25 in. wide.

3. A triangle with base 2 cm and altitude 6 cm.

4. A triangle with base 5 mm and altitude 8 mm.

5. ______________

6. ______________

7. ______________

8. ______________

9. ______________

5. A circle of radius 5 yd. (Use $\pi = 3.14$.)

6. A circle of radius 1.5 ft. (Use $\pi = 3.14$.)

7. A trapezoid with parallel sides of 3 in. and 10 in., and altitude of 35 in.

8. A trapezoid with parallel sides of 3.5 mm and 4.2 mm, and altitude of 1 cm.

9. A parallelogram with altitude 10 cm and a base of 5 mm.

NAME ______________________ SECTION ______________ DATE __________

ANSWERS

10. ________
11. ________
12. ________
13. ________
14. ________
15. ________
16. ________
17. ________
18. ________
19. ________

Find the area of each of the following figures with the indicated dimensions. (Use $\pi = 3.14$.)

10.

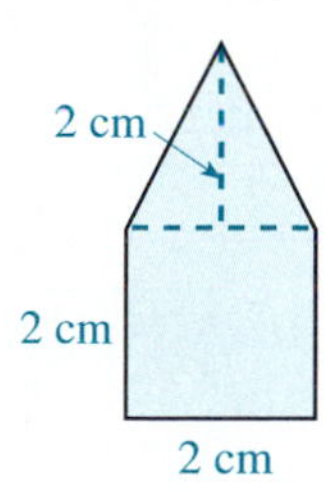

11.

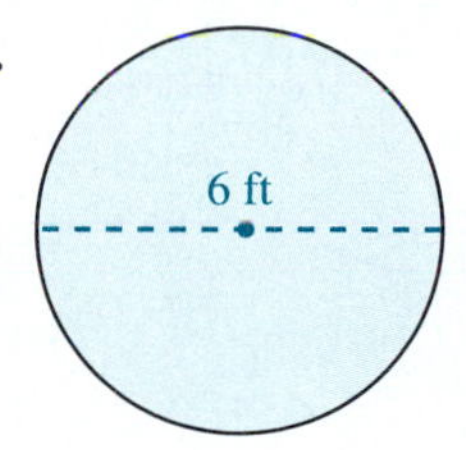

12.

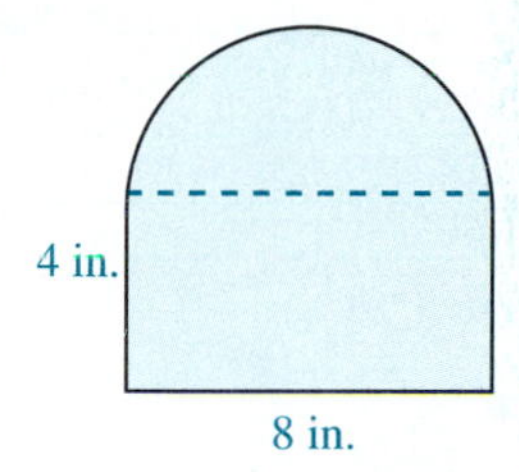

13.

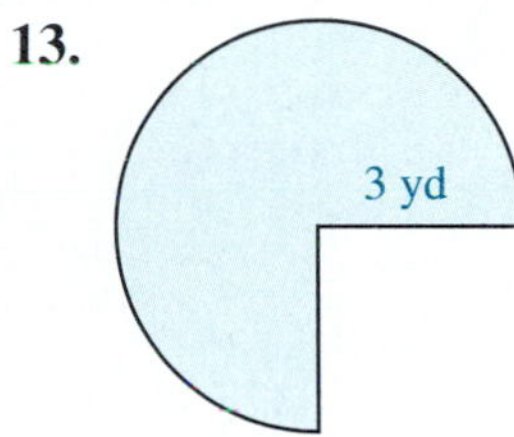

14.

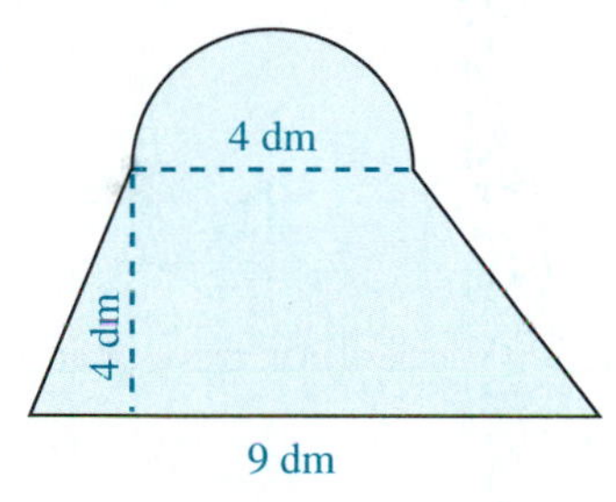

15.

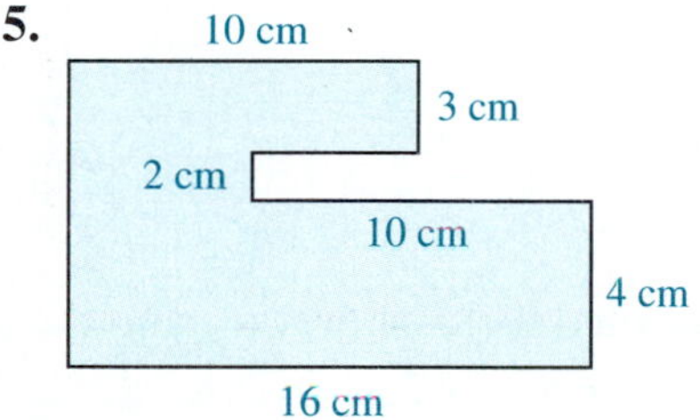

16.

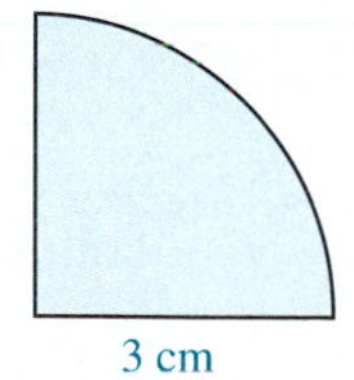

17.

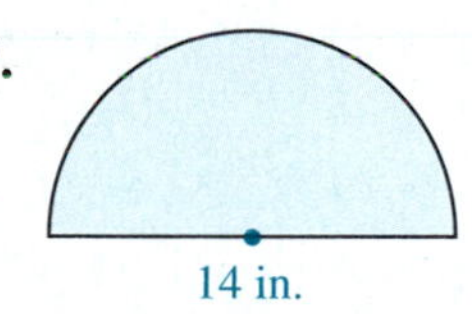

18.

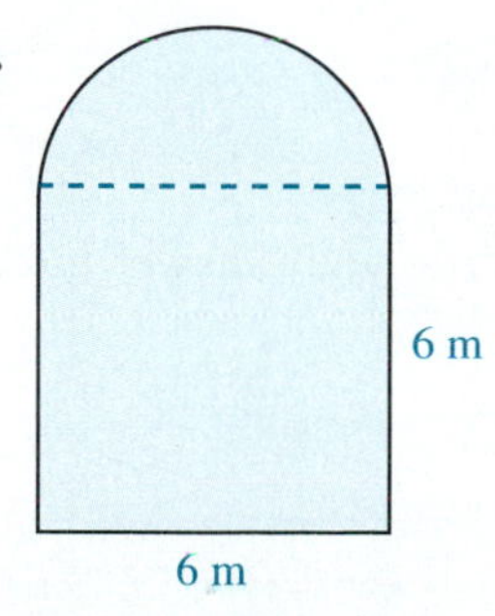

19.

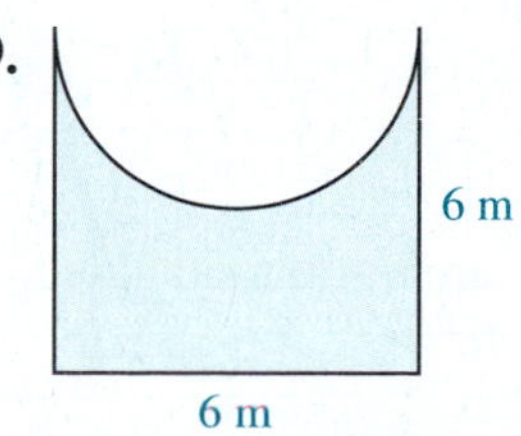

20. ________________

21. ________________

22. ________________

23. ________________

24. ________________

25. ________________

26. ________________

Find the areas of the shaded portions. (Use $\pi = 3.14$.)

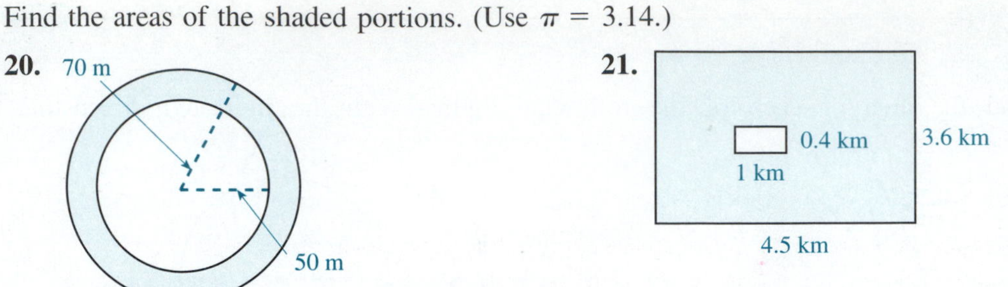
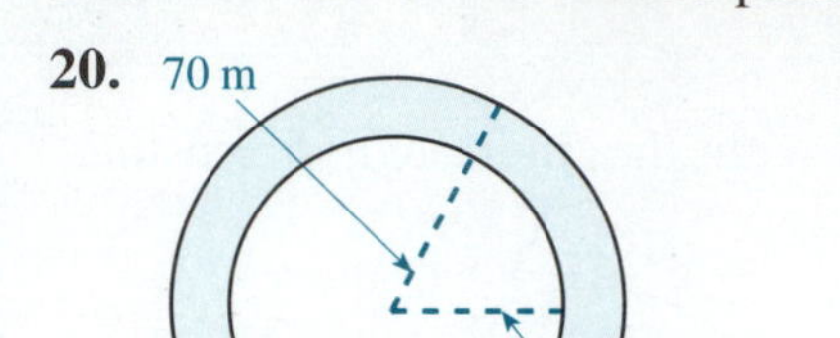

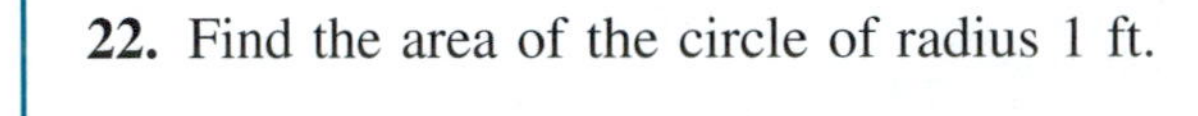

22. Find the area of the circle of radius 1 ft.

23. Find the area of a circle with diameter 1 ft.

24. Find the area of a rectangle 2 m long and 60 cm wide. Write your answer in both cm^2 and m^2.

25. Find the area of a square with sides of 50 cm. Write your answer in both cm^2 and m^2.

26. Find the area of a rectangle 0.5 ft long and 35 in. wide. Write your answer in both $in.^2$ and ft^2.

G.3 Volume

OBJECTIVES

1. Understand the concept of volume.
2. Learn how to apply the appropriate formula to find the volume of a given geometric figure.
3. Know how to label answers with the correct units.

The Concept of Volume

Volume is the measure of the space enclosed by a three-dimensional figure and is measured in **cubic units.** The concept of volume is illustrated in Figure G.2 in terms of cubic inches.

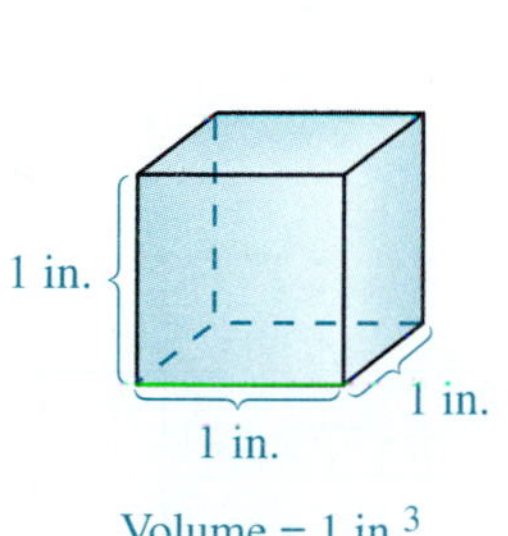

Volume = 1 in.^{3}
(or 1 cubic inch)

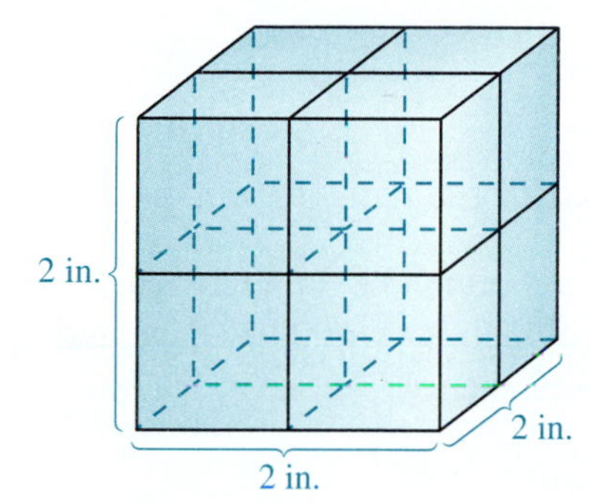

Volume = length x height x width = 2 x 2 x 2 = 8 in.^{3}
There are a total of 8 cubes that are each 1 in.^{3} for a total of 8 in.^{3}

Figure G.2

In the metric system, some of the units of volume are cubic meters (m^3), cubic decimeters (dm^3), cubic centimeters (cm^3), and cubic millimeters (mm^3). In the U.S. customary system, some of the units of volume are cubic feet (ft^3), cubic inches (in.^{3}), and cubic yards (yd^3).

Formulas for Volume

The following formulas for the volumes of the common geometric solids shown are valid regardless of the measurement system used. Always be sure to label your answers with the correct units.

Five Geometric Solids and the Formulas for Their Volumes

h
w
ℓ

Rectangular solid
$V = \ell wh$

h
w
ℓ

Rectangular pyramid
$V = \frac{1}{3}\ell wh$

h
r

Right circular cylinder
$V = \pi r^2 h$

1. Find the volume of a sphere with diameter 8 ft.

continued

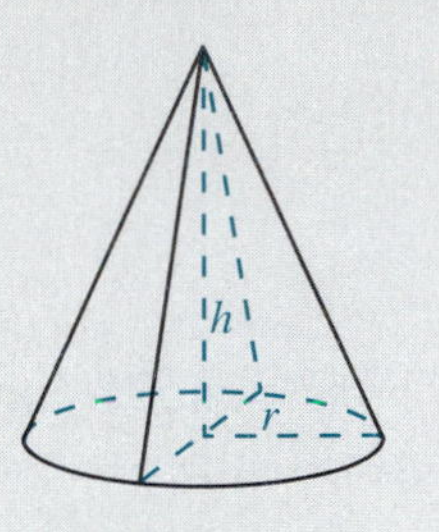

Right circular cone
$V = \frac{1}{3}\pi r^2 h$

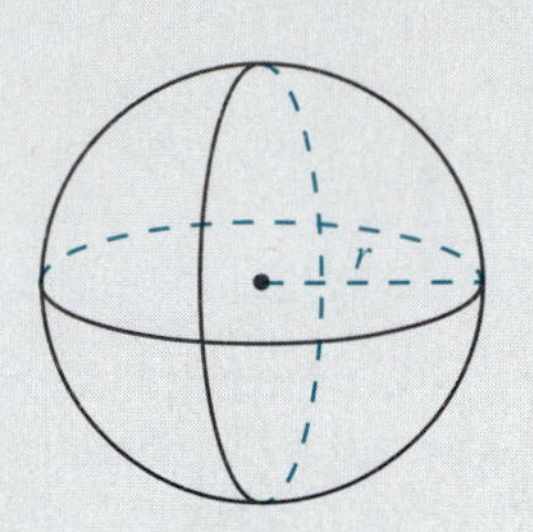

Sphere
$V = \frac{4}{3}\pi r^3$

EXAMPLE 1

Find the volume of the rectangular solid with length 8 inches, width 4 inches, and height 12 inches.

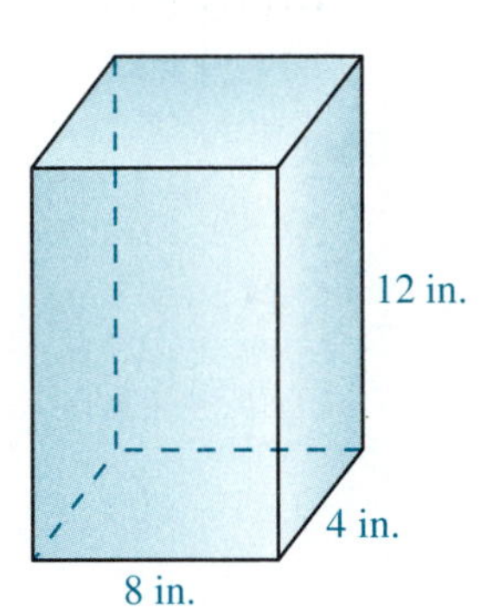

Solution

$V = \ell wh$

$V = 8 \cdot 4 \cdot 12$

$= 384 \text{ in.}^3$

EXAMPLE 2

Find the volume of the solid with the dimensions indicated.

Solution

On top of the cylinder is a hemisphere (half a sphere). Find the volume of the cylinder and the volume of the hemisphere and add the results.

Cylinder

$V = \pi r^2 h$

$V = 3.14(5^2)(3)$

$= 235.5 \text{ cm}^3$

Hemisphere

$V = \frac{1}{2} \cdot \frac{4}{3}\pi r^3$ one-half volume of a sphere

$V = \frac{2}{3}(3.14)(5^3)$

$= 261.67 \text{ cm}^3$

Total Volume

$$\begin{array}{r} 235.50 \text{ cm}^3 \\ +\ 261.67 \text{ cm}^3 \\ \hline 497.17 \text{ cm}^3 \end{array}$$ total volume

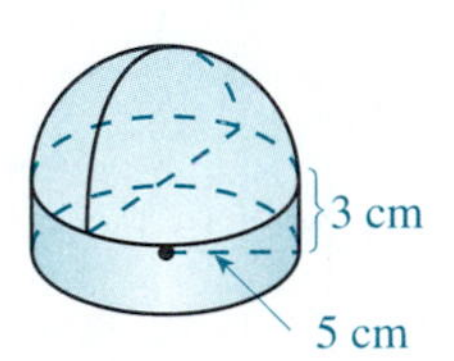

Now Work Exercise 1 in the Margin.

NAME ______________________ SECTION ______________ DATE ________

ANSWERS

1. [Respond in exercise.] ______________

2. ______________

3. ______________

4. ______________

5. ______________

6. ______________

7. ______________

Exercises G.3

1. Match each formula for volume to its corresponding geometric figure.

___ (a) rectangular solid	A. $V = \frac{4}{3}\pi r^3$
___ (b) rectangular pyramid	B. $V = \frac{1}{3}\pi r^2 h$
___ (c) right circular cylinder	C. $V = lwh$
___ (d) right circular cone	D. $V = \pi r^2 h$
___ (e) sphere	E. $V = \frac{1}{3}lwh$

Find the volume of each of the following solids in a convenient unit. (Use $\pi = 3.14$.)

2. A rectangular solid with length 5 in., width 2 in., and height 7 in.

3. A right circular cylinder 15 in. high and 1 ft in diameter.

4. A sphere with radius 4.5 cm.

5. A sphere with diameter 12 ft.

6. A right circular cone 3 dm high with a 2-dm radius.

7. A rectangular pyramid with length 8 cm, width 10 mm, and height 3 dm.

8. ______________

9. ______________

10. ______________

11. ______________

12. ______________

13. ______________

Find the volume of each of the following solids with the dimensions indicated.

8.

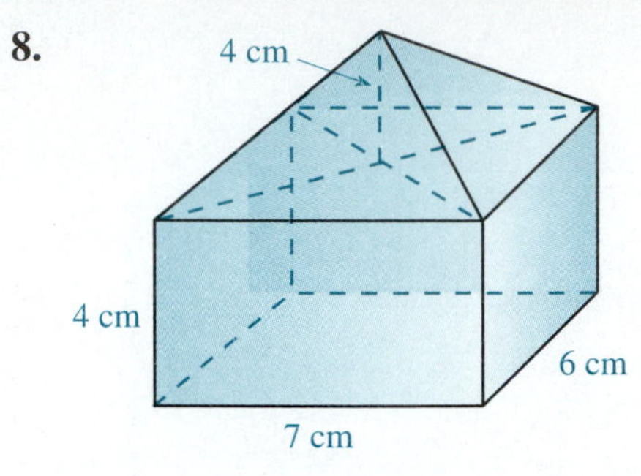

9.

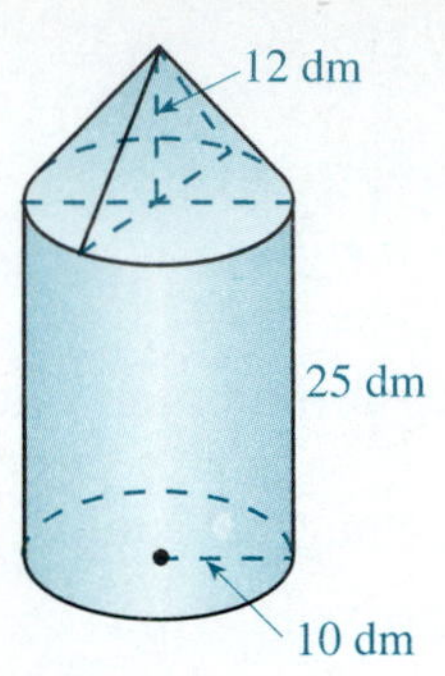

10.

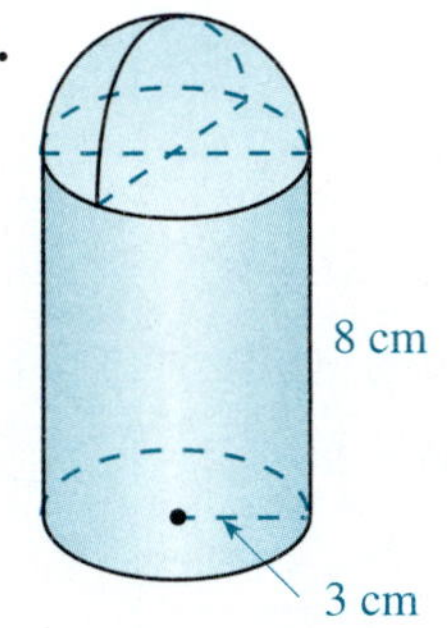

11.

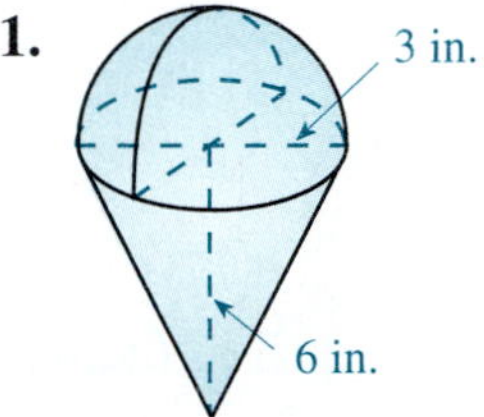

12.

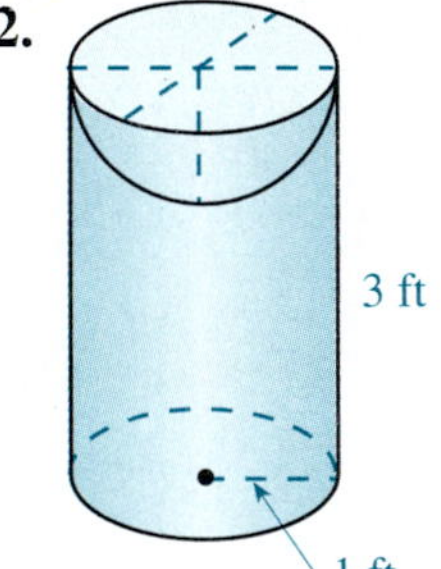

13.

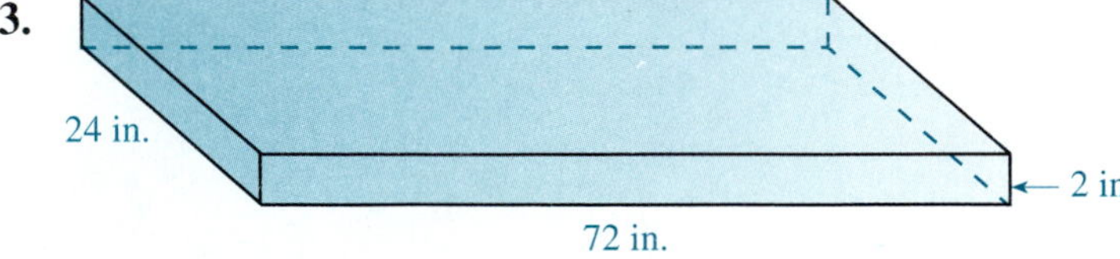

G.4 Angles

Objectives

1. Know the definition of an angle.
2. Learn how to use a protractor to measure angles.
3. Learn how to classify an angle by its measure as **acute, right, obtuse,** or **straight.**
4. Be able to determine when two angles are **complementary, supplementary,** or **equal** and know the meanings of these terms.
5. Know the meanings of the terms **vertical angles** and **adjacent angles.**

Measuring Angles

We begin the discussion of angles with two definitions using the undefined terms **point** and **line.**

Definitions

A **ray** consists of a point (called the **endpoint**) and all the points on a line on one side of that point.

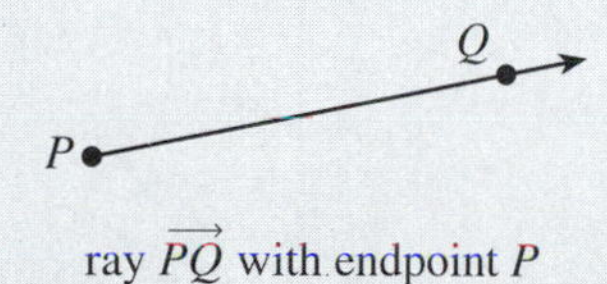

ray $\overrightarrow{PQ}$ with endpoint P

An **angle** consists of two rays with a common endpoint (called a **vertex**).

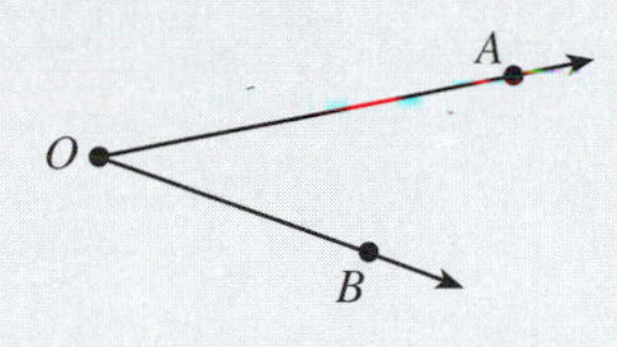

$\angle AOB$ with vertex O

In an angle, the two rays are called the **sides** of the angle.

Every angle has a **measurement,** or **measure,** associated with it. Suppose that a circle is divided into 360 equal arcs. If two rays are drawn from the center of the circle through two consecutive points of division on the circle, then that angle is said to **measure one degree** (symbolized 1°). For example, in Figure G.3, a device called a protractor shows that the measure of $\angle AOB$ is 60 degrees. (We write $m\angle AOB = 60°$.)

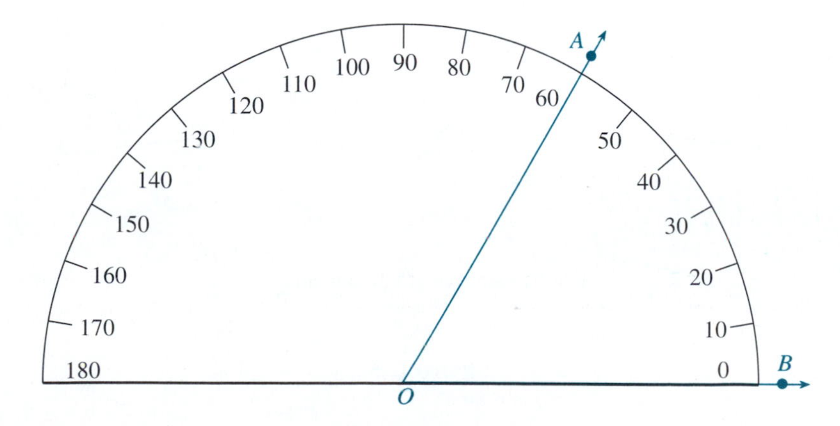

The protractor shows $m\angle AOB = 60°$

Figure G.3

To measure an angle with a protractor, lay the bottom edge of the protractor along one side of the angle, with the vertex at the marked center point. Then read the measure from the protractor where the other side of the angle crosses it. (See Figure G.4.) (**Note:** You may need to extend the side to be able to read where it crosses your protractor.)

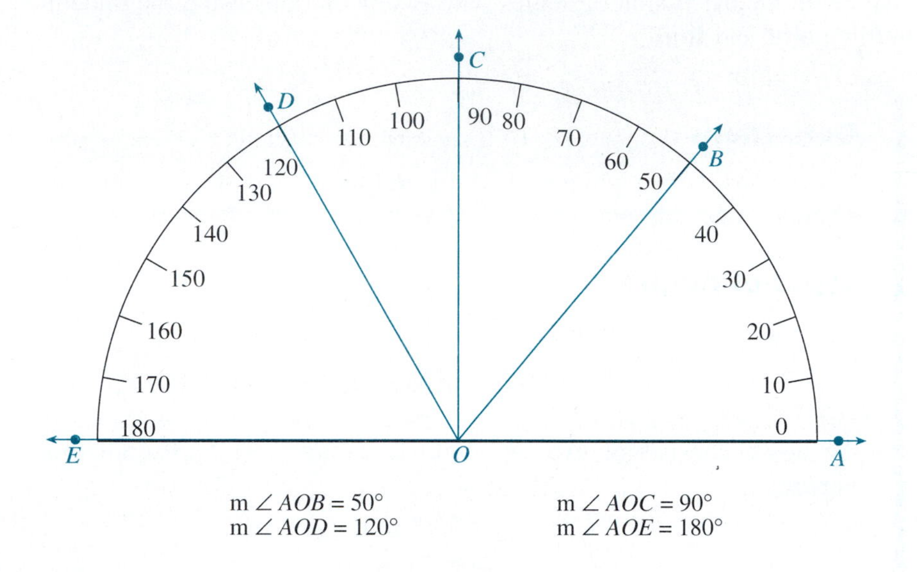

Figure G.4

Three common ways of labeling angles (Figure G.5) are

1. Using three capital letters with the vertex as the middle letter.
2. Using single numbers such as 1, 2, 3.
3. Using the single capital letter at the vertex when the meaning is clear.

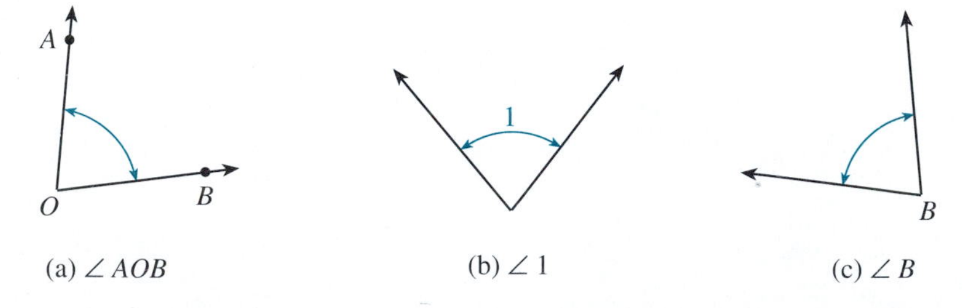

Three ways of labeling angles

Figure G.5

Classifying Angles

Angles can be classified (or named) according to their measures.

> **Note:** We will use the two inequality symbols
>
> $<$ (read "is less than")
>
> $>$ (read "is greater than")
>
> to indicate the relative sizes of the measures of angles.

Types of Angles

Name	Measure	Example
1. Acute	$0° < m\angle A < 90°$	$\angle A$ is an acute angle.

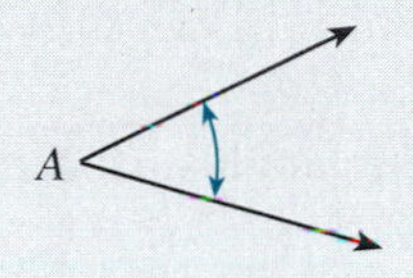

Name	Measure	Example
2. Right	$m\angle A = 90°$	$\angle A$ is a right angle.

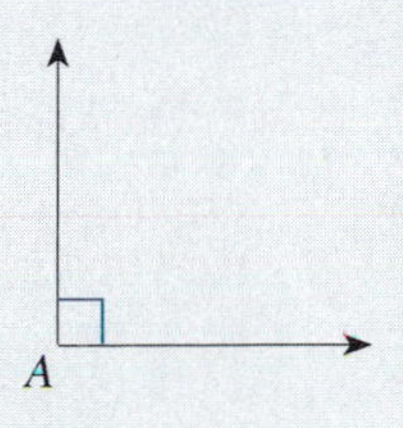

Name	Measure	Example
3. Obtuse	$90° < m\angle A < 180°$	$\angle A$ is an obtuse angle.

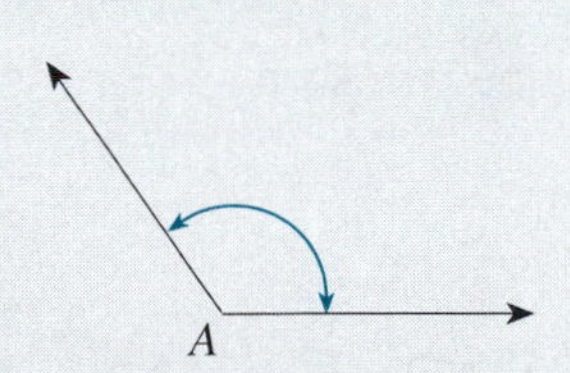

Name	Measure	Example
4. Straight	$m\angle A = 180°$	The rays are in opposite directions. $\angle A$ is a straight angle.

A

The following figure is used for Examples 1 and 2.

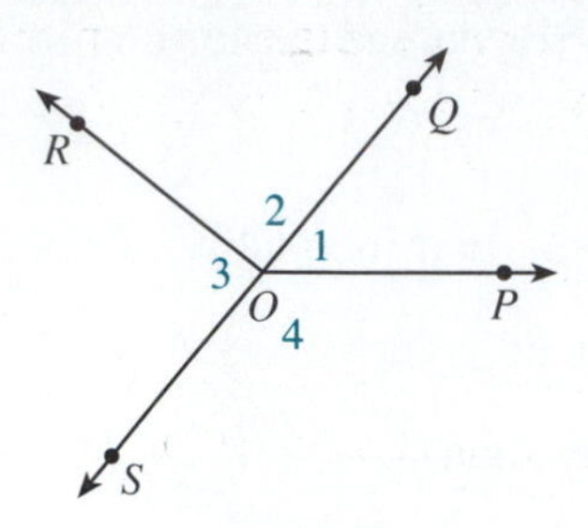

EXAMPLE 1

Use a protractor to check the measures of the following angles.

(a) $m\angle 1 = 45°$ (b) $m\angle 2 = 90°$

(c) $m\angle 3 = 90°$ (d) $m\angle 4 = 135°$

EXAMPLE 2

Tell whether each of the following angles is acute, right, obtuse, or straight. (a) $\angle 1$ (b) $\angle 2$ (c) $\angle POR$

Solution

(a) $\angle 1$ is acute since $0° < m\angle 1 < 90°$.

(b) $\angle 2$ is a right angle since $m\angle 2 = 90°$.

(c) $\angle POR$ is obtuse since $m\angle POR = 45° + 90° = 135° > 90°$.

Definition

Two angles are

1. **Complementary** if the sum of their measures is 90°.
2. **Supplementary** if the sum of their measures is 180°.
3. **Equal** if they have the same measure.

EXAMPLE 3

In the figure below, (a) $\angle 1$ and $\angle 2$ are complementary since $m\angle 1 + m\angle 2 = 90°$. (b) $\angle COD$ and $\angle COA$ are supplementary since $m\angle COD + m\angle COA = 70° + 110° = 180°$. (c) $\angle AOD$ is a straight angle since $m\angle AOD = 180°$. (d) $\angle BOA$ and $\angle BOD$ are supplementary, and in this case $m\angle BOA = m\angle BOD = 90°$.

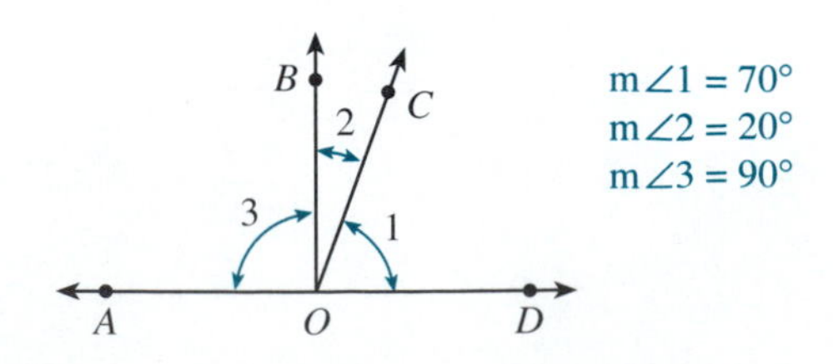

EXAMPLE 4

In the figure below, $\overleftrightarrow{PS}$ is a straight line and $m\angle QOP = 30°$. Find the measure of (a) $\angle QOS$ and (b) $\angle SOP$. (c) Are any pairs supplementary?

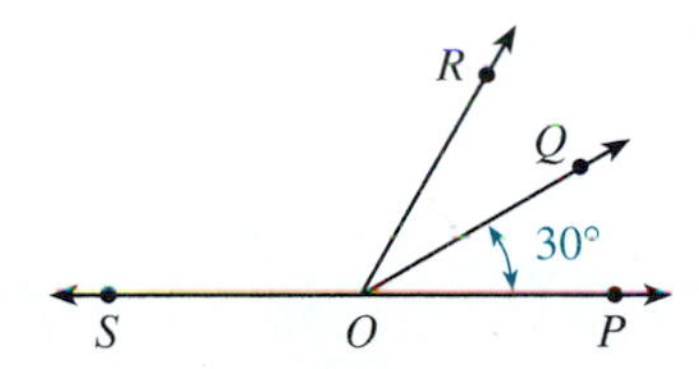

Solution

(a) $m\angle QOS = 150°$.

(b) $m\angle SOP = 180°$.

(c) Yes. $\angle QOP$ and $\angle QOS$ are supplementary and $\angle ROP$ and $\angle ROS$ are supplementary.

If two lines intersect, then two pairs of vertical angles are formed. **Vertical angles** are also called **opposite angles.** (See Figure G.6.)

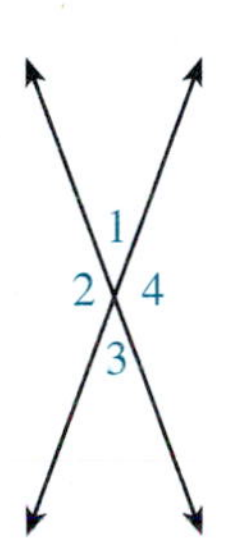

$\angle 1$ and $\angle 3$ are vertical angles.
$\angle 2$ and $\angle 4$ are vertical angles.

Figure G.6

Vertical angles are equal. That is, **vertical angles have the same measure.** (In Figure G.6, $m\angle 1 = m\angle 3$ and $m\angle 2 = m\angle 4$.)

Two angles are **adjacent** if they have a common side. (See Figure G.7.)

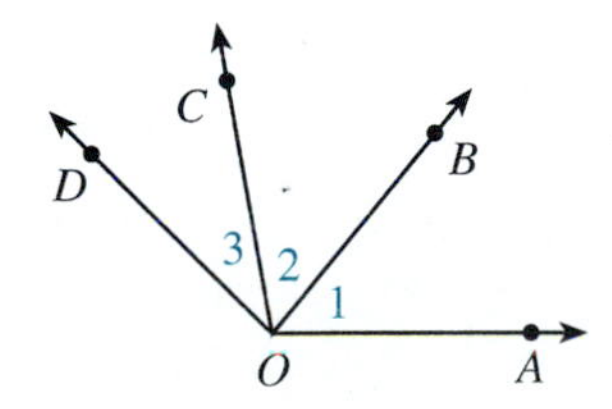

$\angle 1$ and $\angle 2$ are adjacent. They have the common side $\overrightarrow{OB}$.

$\angle 2$ and $\angle 3$ are adjacent. They have the common side $\overrightarrow{OC}$.

Figure G.7

EXAMPLE 5

In the figure below, $\overleftrightarrow{AC}$ and $\overleftrightarrow{BD}$ are straight lines. (a) Name an angle adjacent to $\angle EOD$. (b) What is $m\angle AOD$?

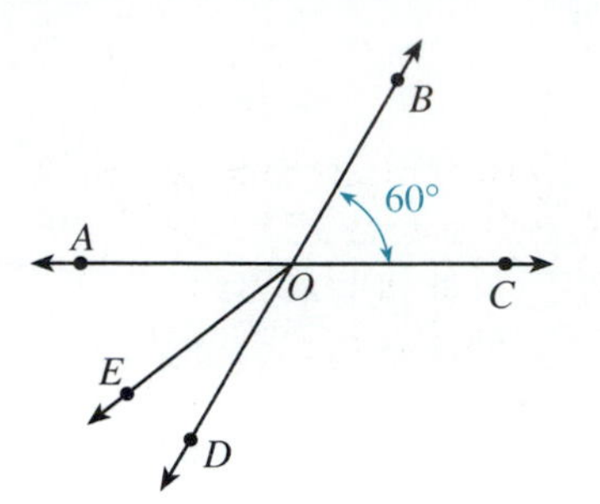

Solution

(a) Three angles adjacent to $\angle EOD$ are: $\angle AOE$ $\angle BOE$, and $\angle COD$.

(b) Since $\angle BOC$ and $\angle AOD$ are vertical angles, they have the same measure. So, $m\angle AOD = 60°$.

NAME ________________ SECTION ________________ DATE ____________

ANSWERS

Exercises G.4

1. (a) If $\angle 1$ and $\angle 2$ are complementary and $m\angle 1 = 15°$, what is $m\angle 2$?
(b) If $m\angle 1 = 3°$, what is $m\angle 2$?
(c) If $m\angle 1 = 45°$, what is $m\angle 2$?
(d) If $m\angle 1 = 75°$, what is $m\angle 2$?

2. (a) If $\angle 3$ and $\angle 4$ are supplementary and $m\angle 3 = 45°$, what is $m\angle 4$?
(b) If $m\angle 3 = 90°$, what is $m\angle 4$?
(c) If $m\angle 3 = 110°$, what is $m\angle 4$?
(d) If $m\angle 3 = 135°$, what is $m\angle 4$?

3. The supplement of an acute angle is an obtuse angle.
(a) What is the supplement of a right angle?
(b) What is the supplement of an obtuse angle?
(c) What is the complement of an acute angle?

4. In the figure shown below, $\overleftrightarrow{DC}$ is a straight line and $m\angle BOA = 90°$.
(a) What type of angle is $\angle AOC$?
(b) What type of angle is $\angle BOC$?
(c) What type of angle is $\angle BOA$?
(d) Name a pair of complementary angles.
(e) Name two pairs of supplementary angles.

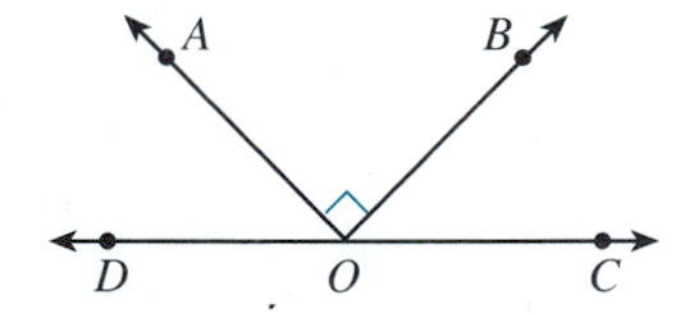

5. An **angle bisector** is a ray that divides an angle into two angles with equal measures. If $\overrightarrow{OX}$ bisects $\angle COD$ and $m\angle COD = 50°$, what is the measure of each of the equal angles formed?

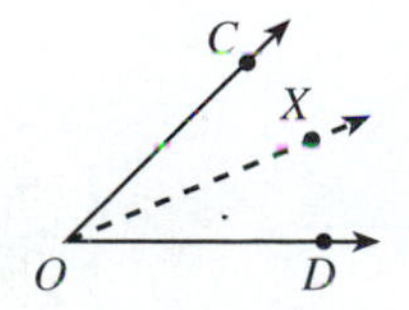

1a. ________
b. ________
c. ________
d. ________

2a. ________
b. ________
c. ________
d. ________

3a. ________
b. ________
c. ________

4a. ________
b. ________
c. ________
d. ________
e. ________

5. ________

6. ______________

7. ______________

8. ______________

9. ______________

10. ______________

11. ______________

[Respond below exercise.]
12. ______________

13. ______________

14. ______________

15. ______________

16. ______________

In the figure shown below, m$\angle AOB$ = 30°, and m$\angle BOC$ = 80° and $\overrightarrow{OX}$ and $\overrightarrow{OY}$ are angle bisectors. Find the measures of the following angles.

6. $\angle AOX$ **7.** $\angle BOY$

8. $\angle COX$ **9.** $\angle YOX$

10. $\angle AOY$ **11.** $\angle AOC$

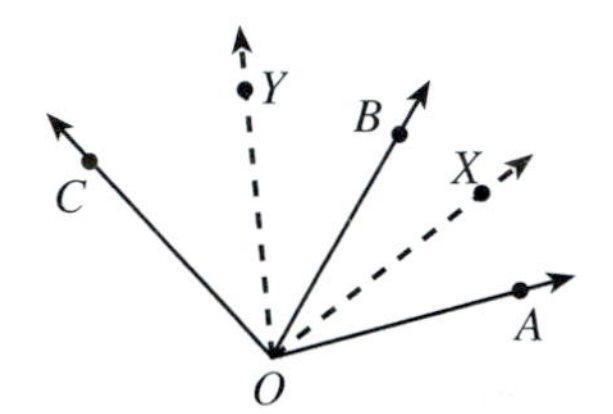

12. In the figure shown below:
(a) Name all the pairs of supplementary angles.
(b) Name all the pairs of complementary angles.

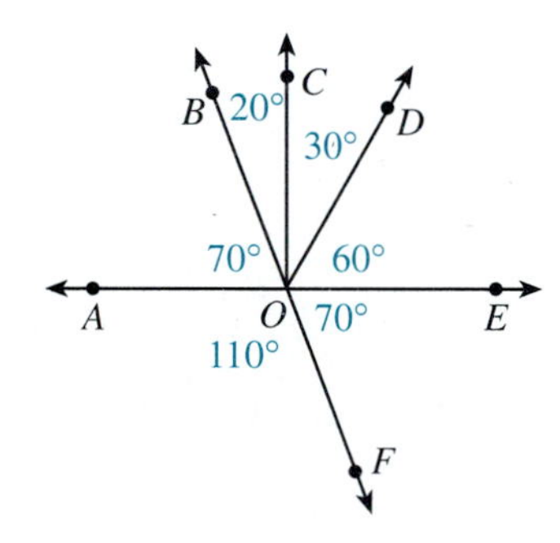

Use a protractor to measure all the angles in each figure. Each line segment may be extended as a ray to form the side of an angle.

13.

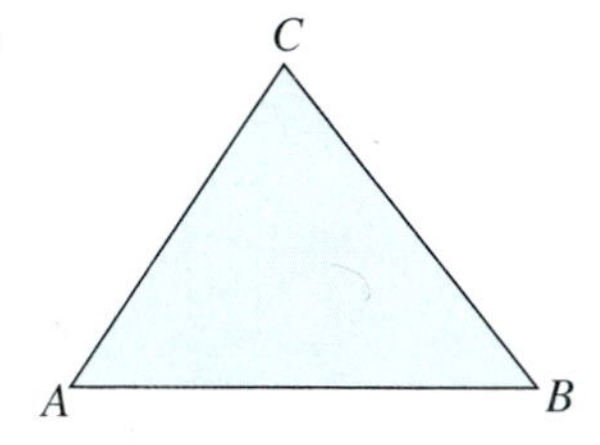

14.

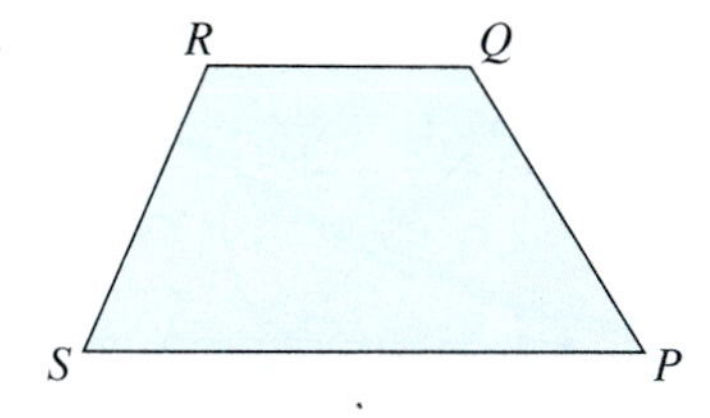

15.

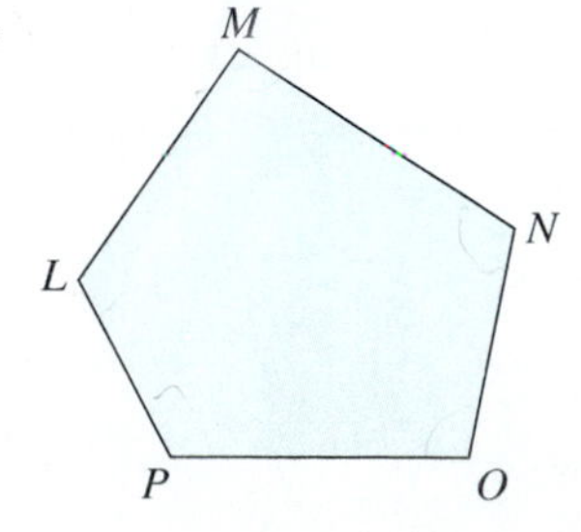

16.

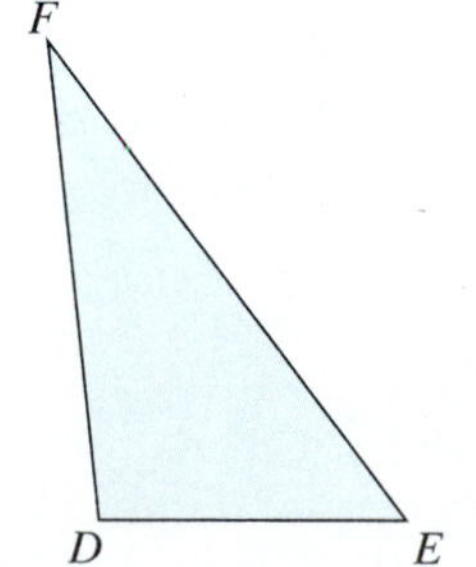

NAME ______________ SECTION ______________ DATE ______________

ANSWERS

17. Name the type of angle formed by the hands on a clock (a) at six o'clock; (b) at three o'clock; (c) at one o'clock; (d) at five o'clock.

17a. ______________

b. ______________

c. ______________

d. ______________

18. What is the measure of each angle formed by the hands of the clock in Exercise 17?

18a. ______________

b. ______________

c. ______________

d. ______________

19. The figure below shows two interesecting lines.
(a) If m∠1 = 30°, what is m∠2?
(b) Is m∠3 = 30°? Give a reason for your answer other than the fact that ∠1 and ∠3 are vertical angles.
(c) Name four pairs of adjacent angles.

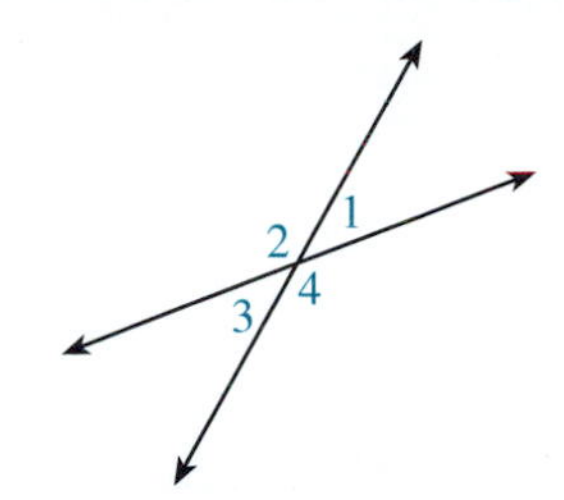

19. [Respond below exercise.]

20. In the figure shown below, $\overleftrightarrow{AB}$ is a straight line.
(a) Name two pairs of adjacent angles.
(b) Name two vertical angles if there are any.

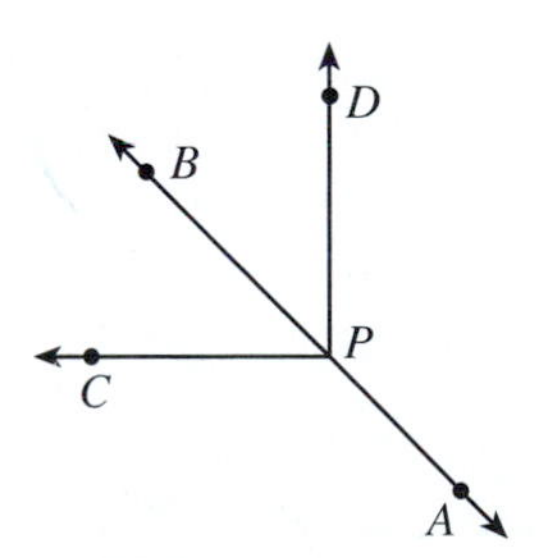

20. [Respond below exercise.]

21. Given that m∠1 = 30° in the figure shown below, find the measures of the other three angles.

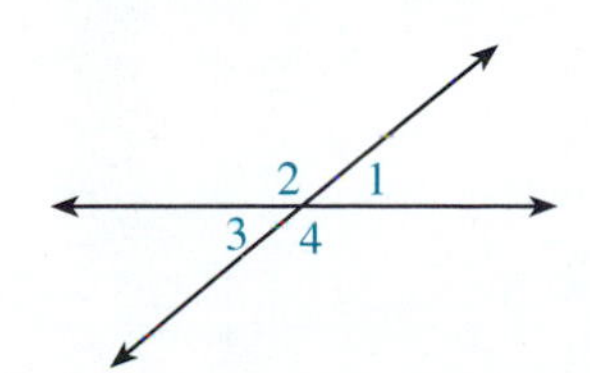

21. [Respond below exercise.]

[Respond below
22. exercise.]

[Respond below
23. exercise.]

22. In the figure shown below, ℓ, m, and n are straight lines with $m\angle 1 = 20°$ and $m\angle 6 = 90°$.
(a) Find the measures of the other four angles.
(b) Which angle is supplementary to $\angle 6$?
(c) Which angles are complementary to $\angle 1$?

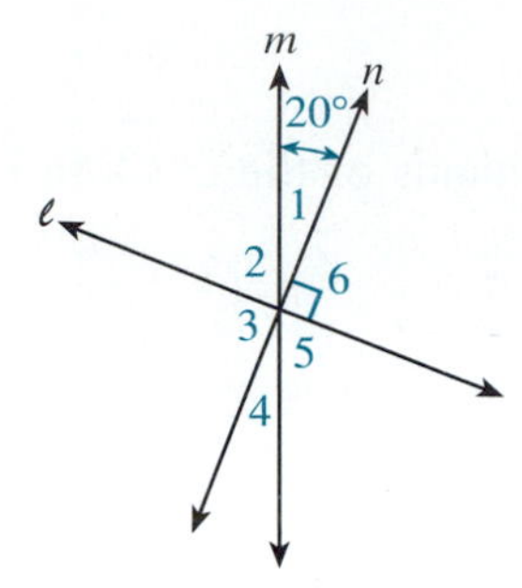

23. In the figure shown below, $m\angle 2 = m\angle 3 = 40°$. Find all other pairs of angles that have equal measures.

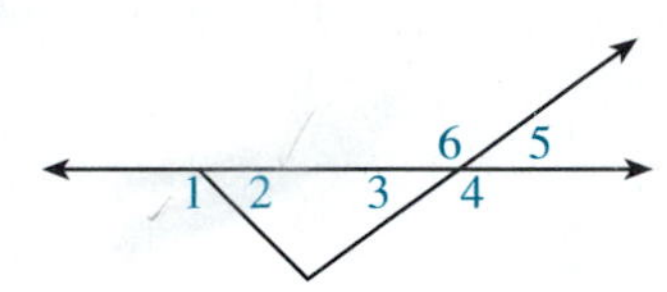

G.5 Triangles

OBJECTIVES

1. *Learn how to classify a triangle by its sides:* ***scalene, isosceles,*** *or* ***equilateral.***
2. *Learn how to classify a triangle by its angles:* ***acute, right,*** *or* ***obtuse.***
3. *Know the following two important statements about any triangle:*
   (a) *The sum of the measures of the angles is 180°.*
   (b) *The sum of the lengths of any two sides must be greater than the length of the third side.*
4. *Know that in similar triangles:*
   (a) *Corresponding angles have the same measure.*
   (b) *Corresponding sides are proportional.*

Triangles

A **triangle** consists of the three line segments that join three points that do not lie on a straight line. The line segments are called the **sides** of the triangle, and the points are called the **vertices** of the triangle. If the points are labeled A, B, and C, the triangle is symbolized $\triangle ABC$. (See Figure G.8.)

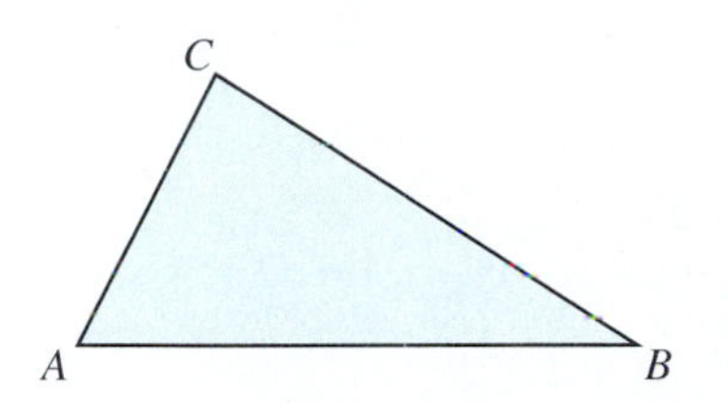

$\triangle ABC$ with vertices A, B, and C and sides $\overline{AB}$, $\overline{BC}$, and $\overline{AC}$.

Figure G.8

The sides of a triangle are said to determine three angles, and these angles are labeled by the vertices. Thus, the angles of $\triangle ABC$ are $\angle A$, $\angle B$, and $\angle C$. (Since the definition of angle involves rays, we can think of the sides of the triangle extended as rays that form these angles.)

Triangles are classified in two ways:

1. According to the lengths of their sides, and
2. According to the measures of their angles.

The corresponding names and properties are listed in the following tables.

Special Note: The line segment with endpoints A and B is indicated by placing a bar over the letters, as in $\overline{AB}$. The length of the segment is indicated by writing only the letters, as in AB.

Triangles Classified by Sides

Name	Property	Example
1. Scalene	No two sides are equal.	$\triangle ABC$ is scalene since no two sides are equal.

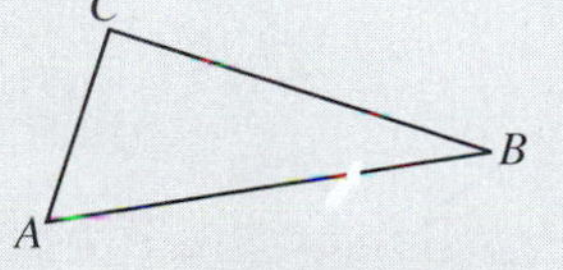

2. Isosceles — At least two sides are equal. — $\triangle PQR$ is isosceles since $PR = QR$.

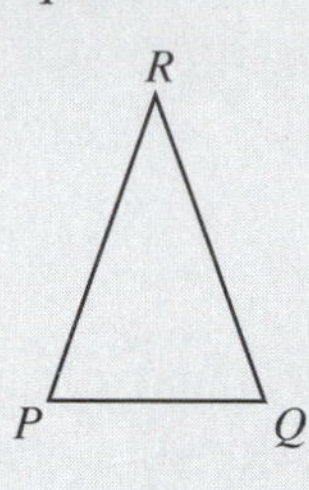

3. Equilateral — All three sides are equal. — $\triangle XYZ$ is equilateral since $XY = XZ = YZ$.

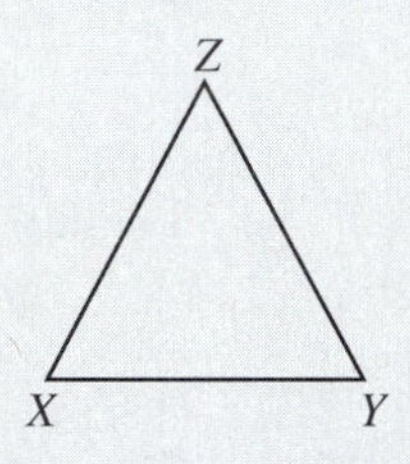

Triangles Classified by Angles

Name	Property	Example
1. Acute	All three angles are acute.	$\angle A$, $\angle B$, $\angle C$ are all acute so $\triangle ABC$ is acute.

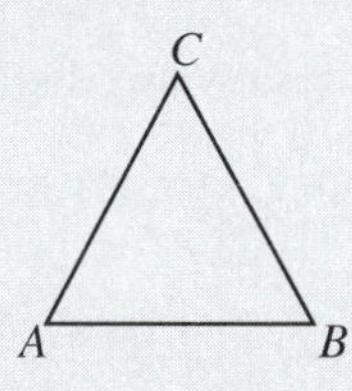

Name	Property	Example
2. Right	One angle is a right angle.	$m\angle P = 90°$ so $\triangle PQR$ is a right triangle.

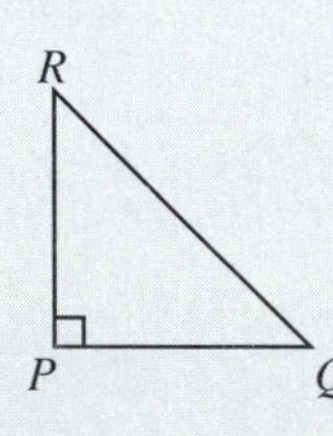

Name	Property	Example
3. Obtuse	One angle is an obtuse angle.	$\angle X$ is obtuse so $\triangle XYZ$ is an obtuse triangle.

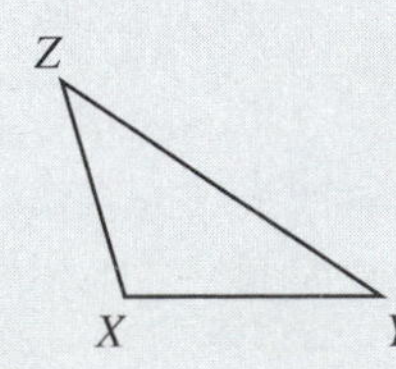

Every triangle is said to have six parts—namely, three angles and three sides. Two sides of a triangle are said to **include** the angle at their common endpoint or vertex. The third side is said to be **opposite** this angle.

The sides in a right triangle have special names. The longest side, opposite the right angle, is called the **hypotenuse,** and the other two sides are called **legs.** (See Figure G.9.)

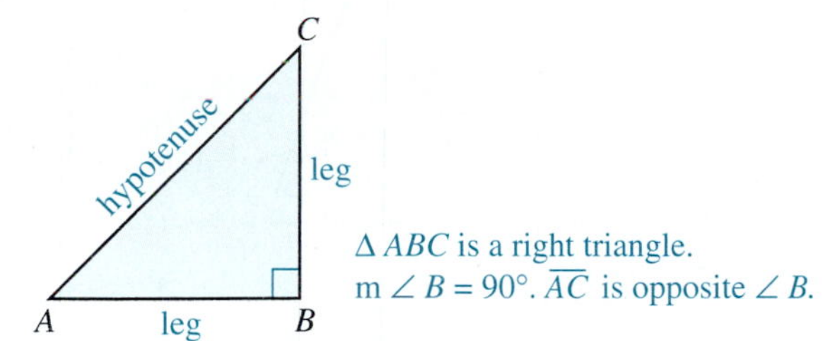

Figure G.9

Two Important Statements about any Triangle

1. The sum of the measures of the angles is 180°.
2. The sum of the lengths of any two sides must be greater than the length of the third side.

EXAMPLE 1

In $\triangle ABC$ below, $AB = AC$. What kind of triangle is $\triangle ABC$?

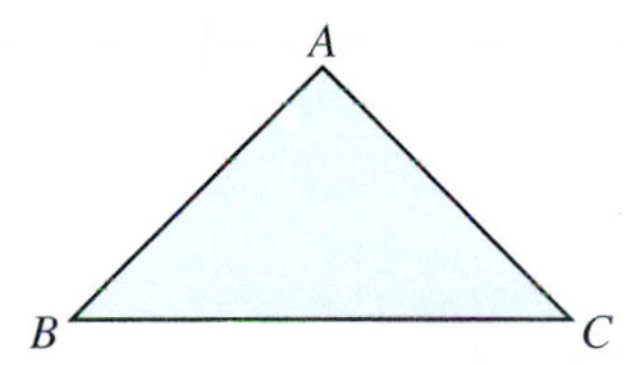

Solution

$\triangle ABC$ is isosceles because two sides are equal.

EXAMPLE 2

Suppose the lengths of the sides of $\triangle PQR$ are as shown in the figure below. Is this possible?

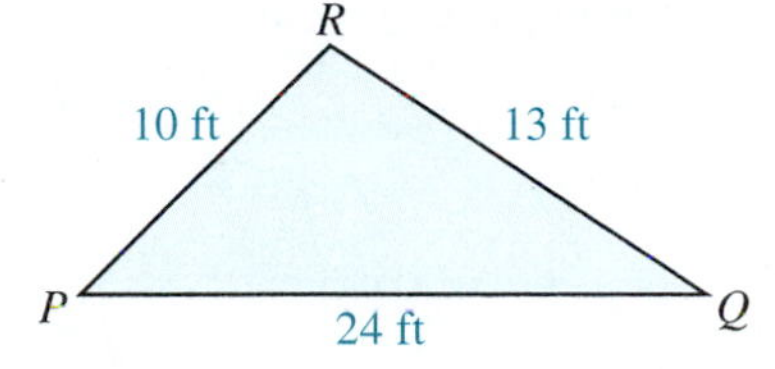

Solution

This is not possible because $PR + QR = 10 \text{ ft} + 13 \text{ ft} = 23 \text{ ft}$ and $PQ = 24 \text{ ft}$, which is longer than the sum of the other two sides. In a triangle, the sum of the lengths of any two sides must be greater than the length of the third side.

EXAMPLE 3

In $\triangle BOR$ below, $m\angle B = 50°$ and $m\angle O = 70°$. (a) What is $m\angle R$? (b) What kind of triangle is $\triangle BOR$? (c) Which side is opposite $\angle R$? (d) Which sides include $\angle R$? (e) Is $\triangle BOR$ a right triangle? Why or why not?

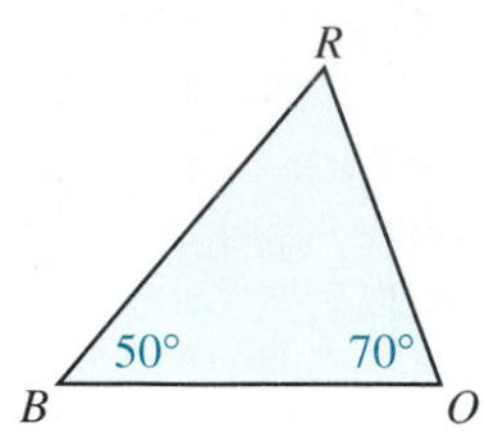

Solution

(a) The sum of the measures of the angles must be 180°. Since $50° + 70° = 120°$,

$$m\angle R = 180° - 120° = 60°$$

(b) $\triangle BOR$ is an acute triangle since all the angles are acute. Also, $\triangle BOR$ is scalene because no two sides are equal.

(c) $\overline{BO}$ is opposite $\angle R$.

(d) $\overline{RB}$ and $\overline{RO}$ include $\angle R$.

(e) $\triangle BOR$ is not a right triangle because none of the angles is a right angle.

Similar Triangles

Two triangles are said to be **similar triangles** if they have the same "shape." They may or may not have the same "size." More formally, two triangles are **similar** if they have the following two properties.

Two Triangles Are Similar If:

1. The **corresponding angles have the same measure.** (We say that the corresponding angles are equal.)
2. Their **corresponding sides are proportional.** (See Figure G.10.)

In similar triangles, **corresponding sides** are those sides opposite the equal angles (angles with the same measure) in the respective triangles. (See Figure G.10 and the following discussion.)

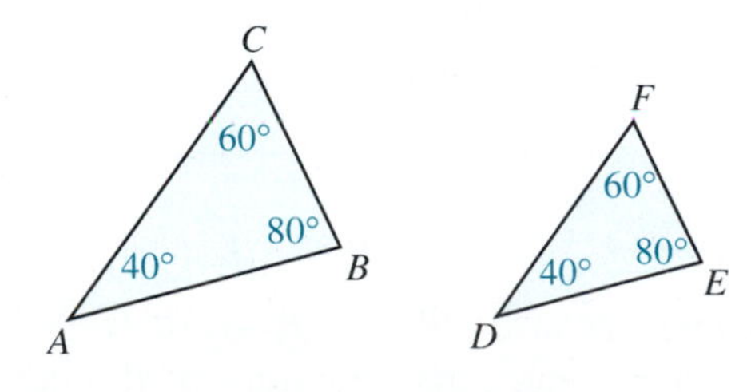

Figure G.10

For the similar triangles in Figure G.10, the following relationships are true.

$\angle A$ corresponds to $\angle D$; $m\angle A = m\angle D$.	$\overline{AB}$ corresponds to $\overline{DE}$.
$\angle B$ corresponds to $\angle E$; $m\angle B = m\angle E$.	$\overline{BC}$ corresponds to $\overline{EF}$.
$\angle C$ corresponds to $\angle F$; $m\angle C = m\angle F$.	$\overline{AC}$ corresponds to $\overline{DF}$.
The corresponding angles are equal.	$\frac{AB}{DE} = \frac{BC}{EF} = \frac{AC}{DF}$ **The corresponding sides are proportional.**

We write $\triangle ABC \sim \triangle DEF$. ($\sim$ is read "is similar to.")

Note: The notation for similar triangles indicates the respective correspondences of angles and sides by the order in which the vertices of each triangle are identified. For example, we could have written $\triangle BCA \sim \triangle EFD$ or $\triangle CAB \sim \triangle FDE$. Be sure to follow this pattern when indicating similar triangles.

EXAMPLE 4

Given the two triangles $\triangle ABC$ and $\triangle AXY$ with $m\angle ABC = m\angle AXY = 90°$, as shown in the figure, determine whether or not $\triangle ABC \sim \triangle AXY$.

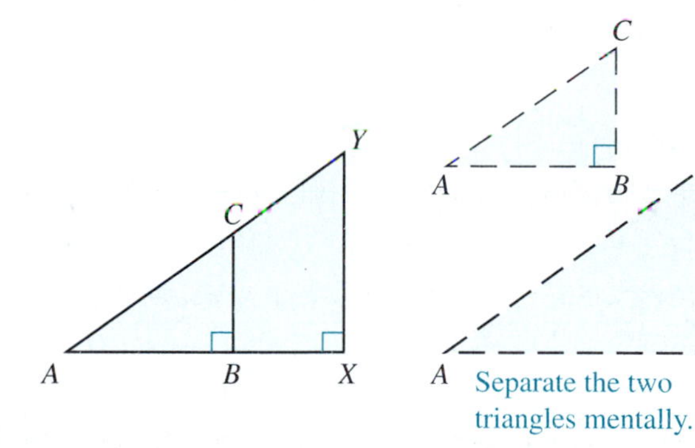

Solution

We can show that the corresponding angles are equal as follows:

$m\angle CAB = m\angle YAX$ because they are the same angle.

$m\angle CBA = m\angle YXA$ because both are right angles (90°).

$m\angle BCA = m\angle XYA$ because the sum of the measures of the angles in each triangle must be 180°.

Therefore, the corresponding angles are equal, and the triangles are similar.

Example 5

Refer to the figure used in Example 4. If $AB = 4$ centimeters, $BX = 2$ centimeters, and $BC = 3$ centimeters, find XY.

Solution

From Example 4 we know that the two triangles are similar; therefore, their corresponding sides are proportional. Since $\overline{AB}$ and $\overline{AX}$ are corresponding sides (they are opposite equal angles) and $\overline{BC}$ and $\overline{XY}$ are corresponding sides (they are opposite equal angles), the following proportion is true:

$$\frac{AB}{AX} = \frac{BC}{XY}$$

But, $AX = AB + BX = 4 + 2 = 6$ cm.
Thus,

$$\frac{4\text{ cm}}{6\text{ cm}} = \frac{3\text{ cm}}{XY}$$

$$4 \cdot XY = 3 \cdot 6$$

$$\frac{\cancel{4} \cdot XY}{\cancel{4}} = \frac{18}{4}$$

$$XY = 4.5\text{ cm}$$

NAME ______________ SECTION ______________ DATE ______________

ANSWERS

1. ______________
2. ______________
3. ______________
4. ______________
5. ______________
6. ______________
7. ______________
8. ______________
9. ______________
10. ______________

Exercises G.5

Name each of the following triangles, given the indicated measures of angles and lengths of sides.

1.

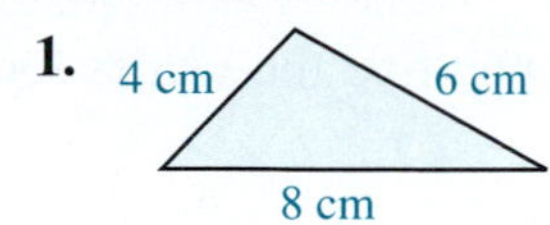

2.

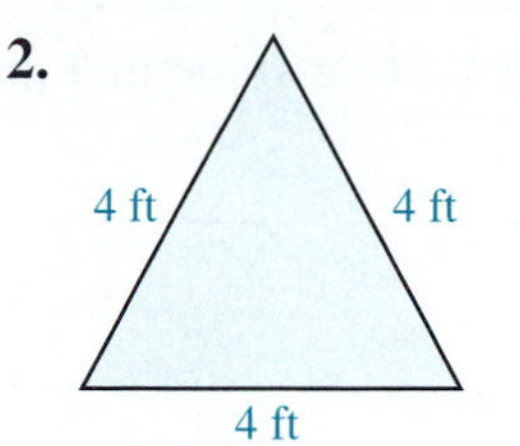

3.

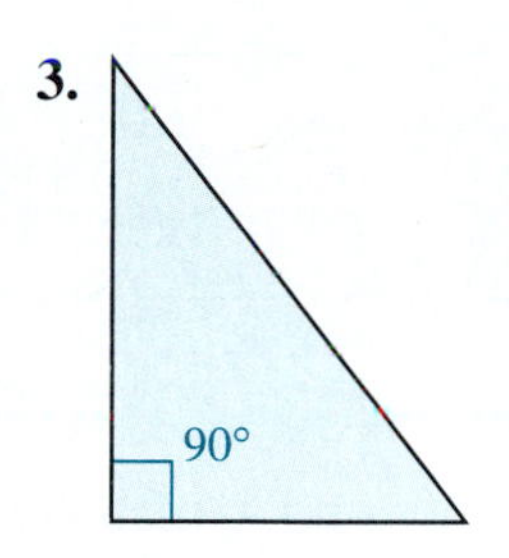

4.

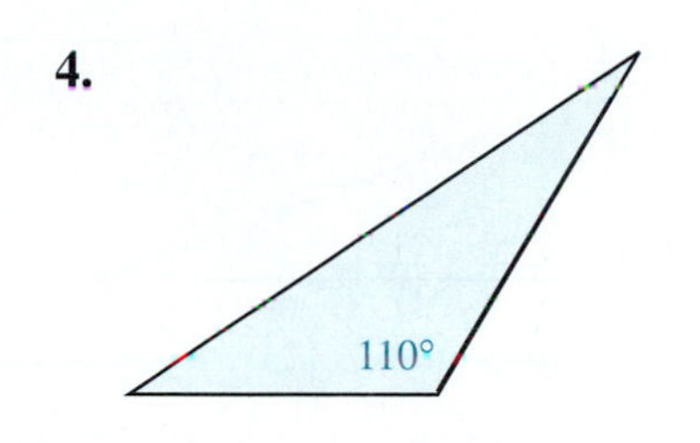

5.

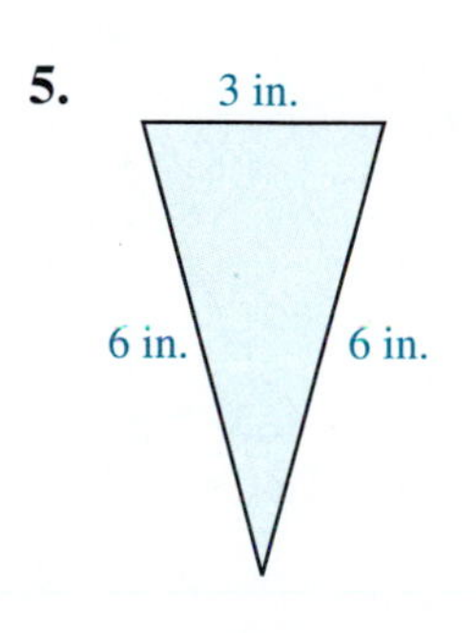

6.

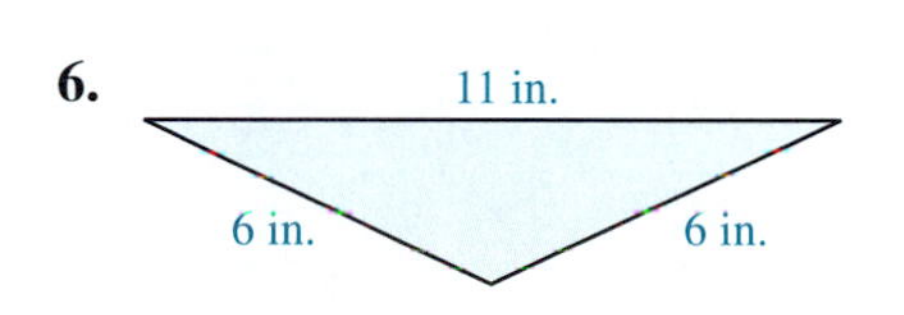

7.

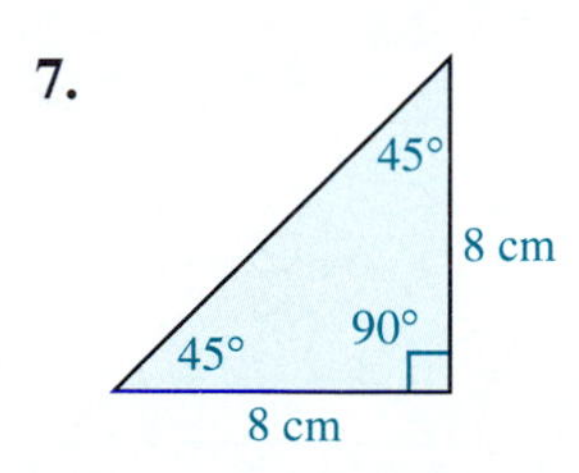

8.

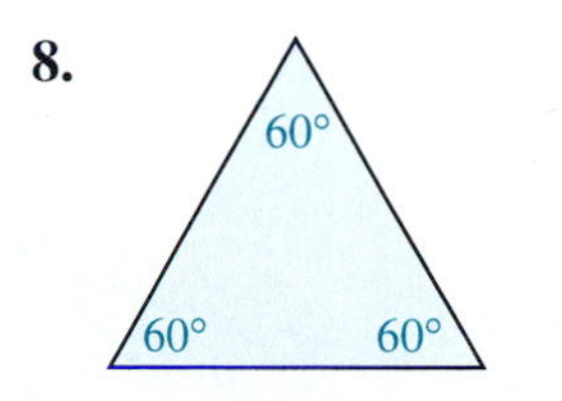

9.

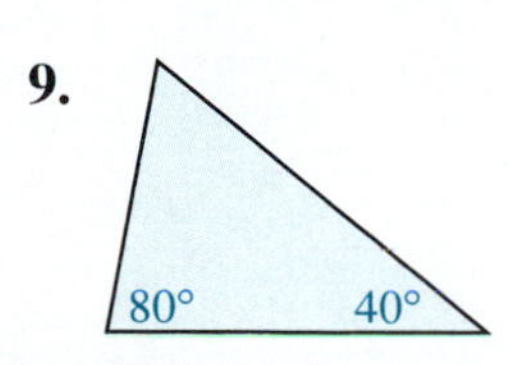

10.

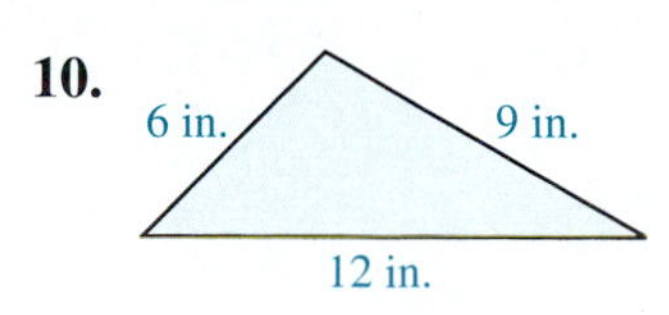

11. ______________

12. ______________

13. ______________

14. ______________

15. ______________

16. ______________

17. [Respond below exercise.]

18. [Respond below exercise.]

19. [Respond below exercise.]

20. [Respond below exercise.]

11.

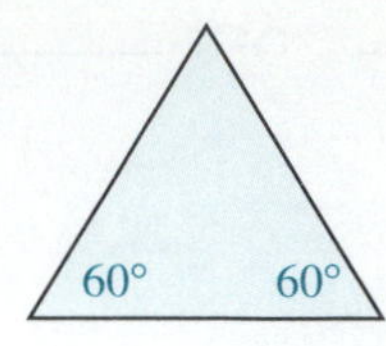

12.

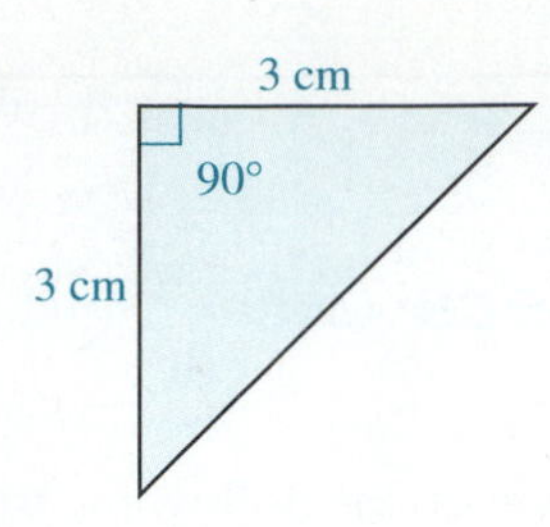

In Exercises 13–16, assume that the given triangles are similar and find the values for x and y.

13.

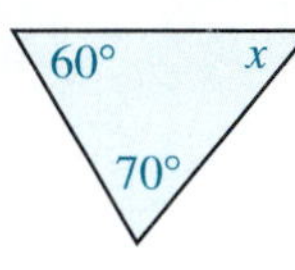

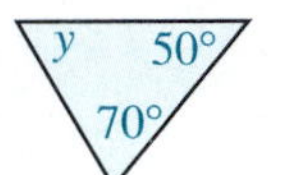

14.

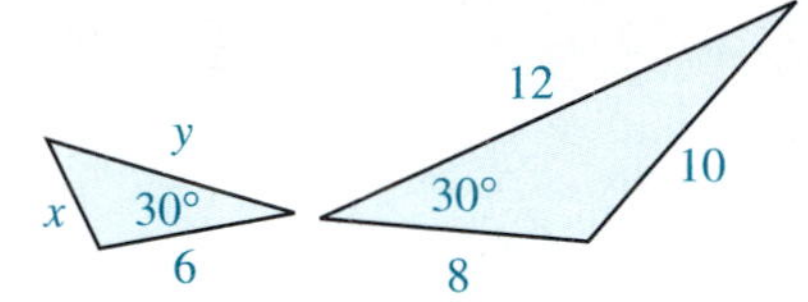

15.

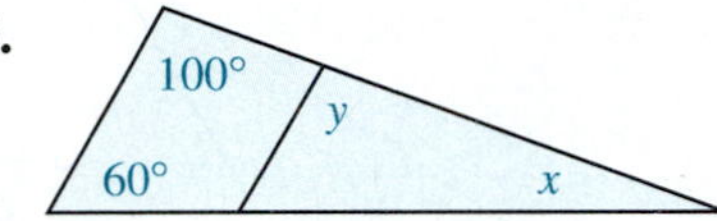

16.

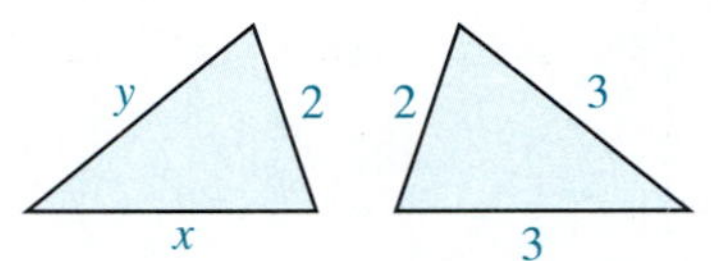

In Exercises 17–20, determine whether each pair of triangles is similar. If the triangles are similar, explain why and indicate the similarity by using the $\sim$ symbol.

17.

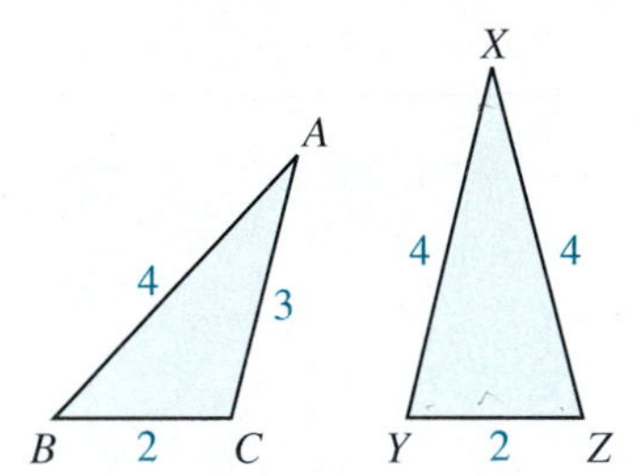

18.

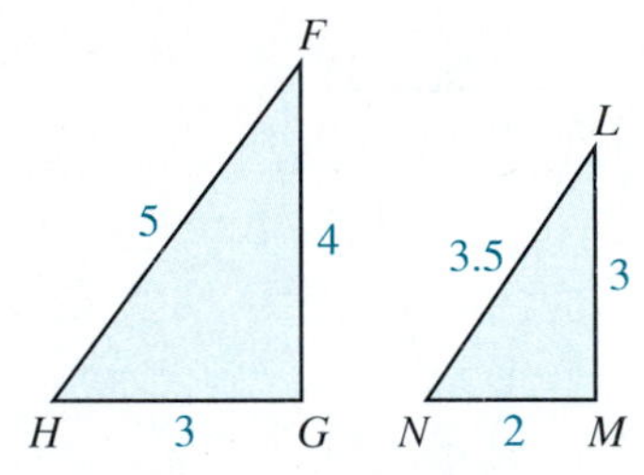

19.

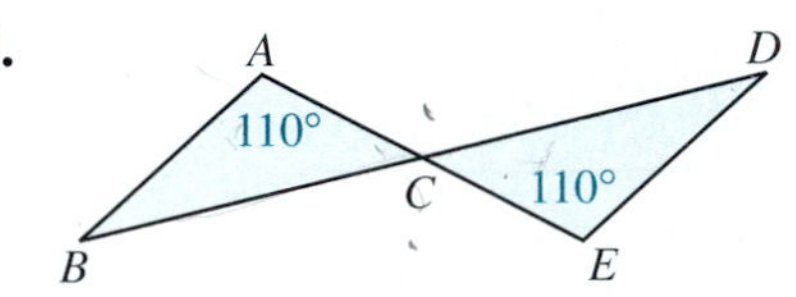

20.

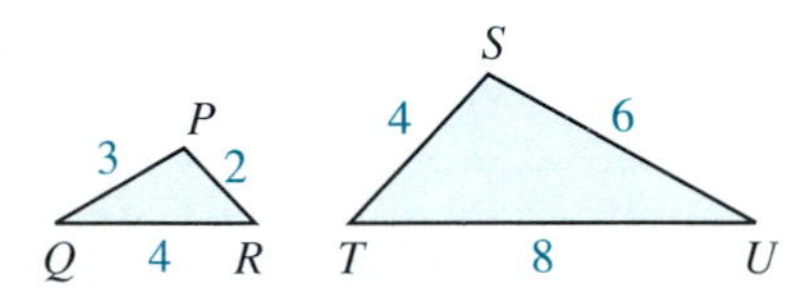

NAME ______________________ SECTION ______________ DATE __________

ANSWERS

21. [Respond below exercise.] ____________

22. [Respond below exercise.] ____________

23. [Respond below exercise.] ____________

21. In the figure below, $m\angle Q = m\angle S$. Is $\triangle PQR \sim \triangle PST$? Explain your reasoning.

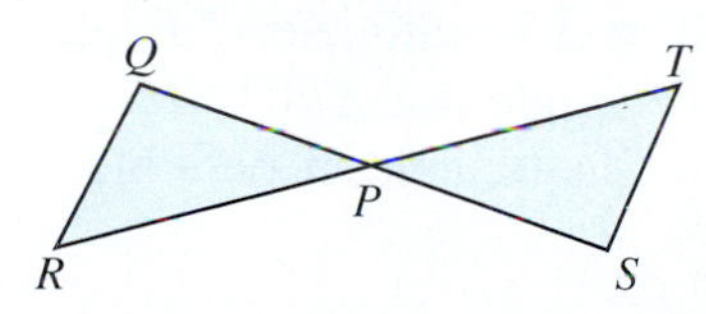

22. In $\triangle XYZ$ and $\triangle UVW$ below, $m\angle Z = 30°$ and $m\angle W = 30°$.

(a) If both triangles are isosceles, what are the measures of the other four angles? (In an isosceles triangle, the angles opposite the equal sides must be equal.)

(b) Are the triangles similar? Explain.

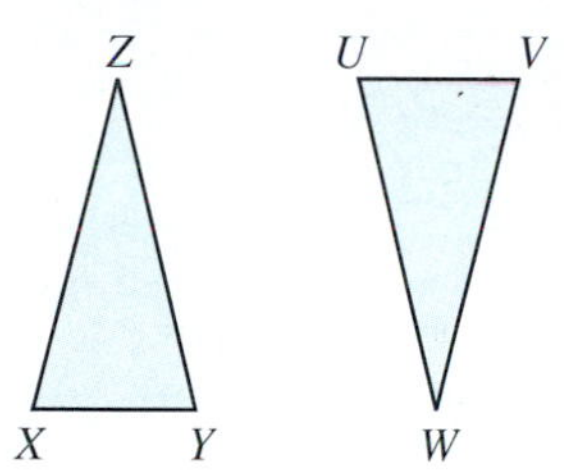

23. The Pythagorean Theorem (see Section 6.7) states that in a right triangle, the square of the hypotenuse is equal to the sum of the squares of the two legs.

(a) Are the triangles $\triangle ABC$ and $\triangle DEF$ right triangles? Explain.

(b) If so, which sides are the hypotenuses?

(c) Are the triangles similar? Explain.

(d) If so, which angles are equal?

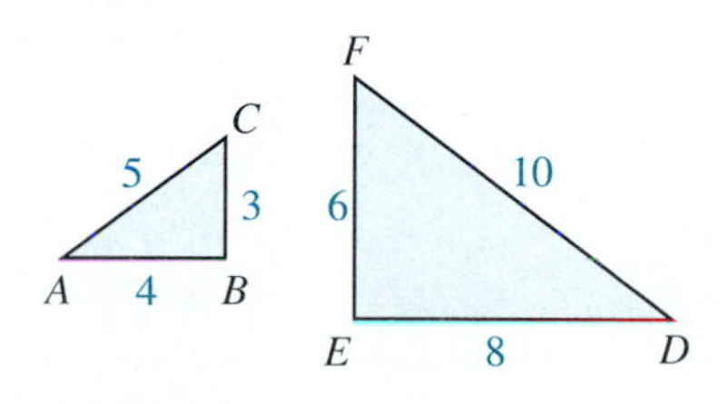

[Respond below

24. exercise.]

Writing and Thinking about Mathematics

24. (a) Is it possible to form a triangle, $\triangle STV$, if $ST = 12$ centimeters, $TV = 9$ centimeters, and $SV = 15$ centimeters? Explain your reasoning.

(b) If so, what kind of triangle is $\triangle STV$?

(c) Use a ruler to draw this triangle, if possible.

Index of Key Ideas and Terms

NAME ______________________ SECTION ______________________ DATE ______________

Chapter Test

1. In the figure shown below, $\overleftrightarrow{AD}$ and $\overleftrightarrow{BE}$ are straight lines.
(a) What type of angle is $\angle BOC$?
(b) What type of angle is $\angle AOE$?
(c) Name a pair of vertical angles.
(d) Name two pairs of supplementary angles.

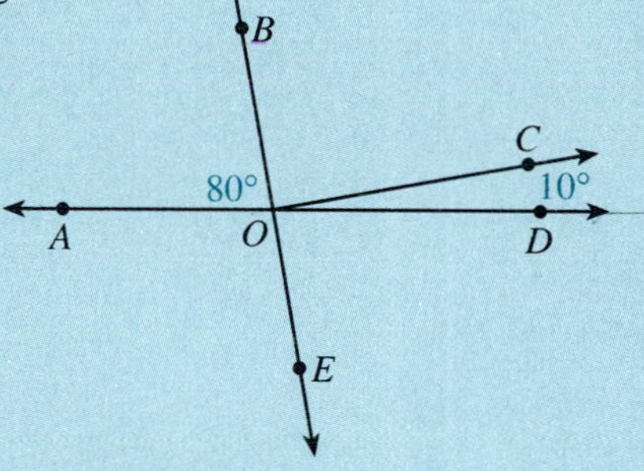

2. (a) If $\angle 1$ and $\angle 2$ are complementary and m $\angle 1 = 35°$, what is m $\angle 2$?
(b) If $\angle 3$ and $\angle 4$ are supplementary and m $\angle 3 = 15°$, what is m $\angle 4$?

3. Find (a) the circumference and (b) the area of the following circle. (Use $\pi = 3.14$.)

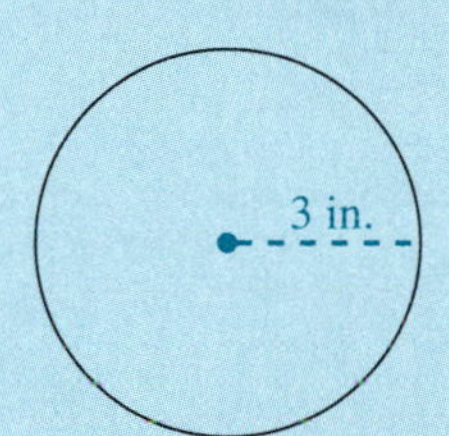

4. Find (a) the perimeter and (b) the area of the trapezoid with dimensions shown here.

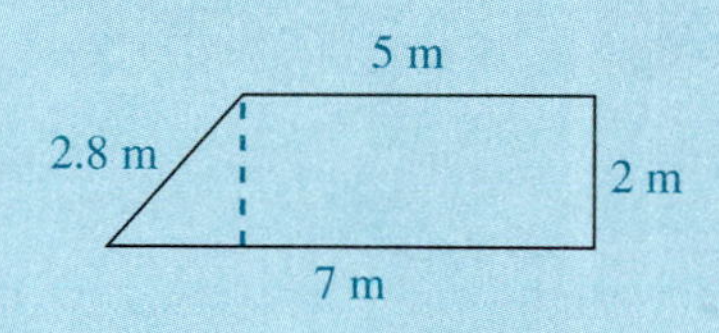

5. (a) Find the perimeter of a rectangle that is $8\frac{1}{2}$ inches wide and $11\frac{2}{3}$ inches long.
(b) Find the area of the rectangle.

6. Find the area of a circle that has a diameter of 14 inches.

7. Find the area of a right triangle if its legs are 4 centimeters and 5 centimeters long.

8. Find the volume of the cylinder shown here with the given dimensions. (Use $\pi = 3.14$.)

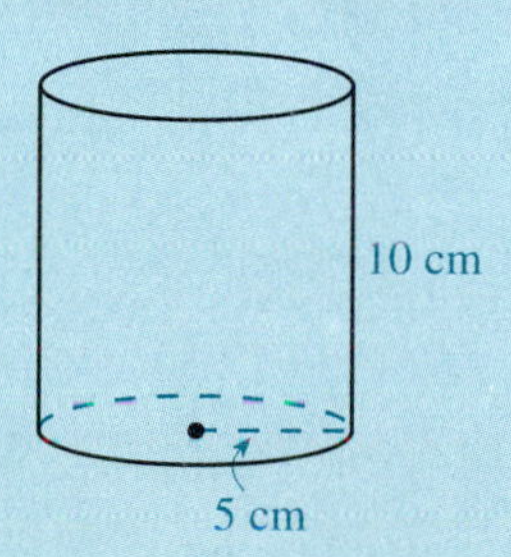

9. Find the volume of the rectangular solid shown here with the given dimensions.

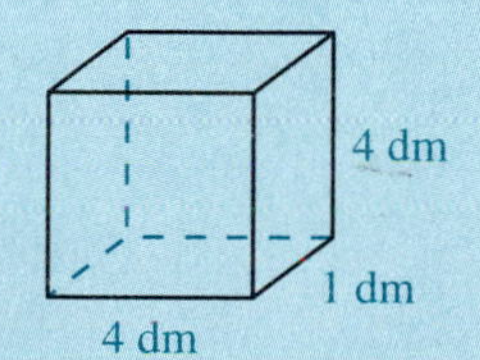

ANSWERS

1. [Respond below exercise.]
2a. ______
b. ______
3a. ______
b. ______
4a. ______
b. ______
5a. ______
b. ______
6. ______
7. ______
8. ______
9. ______

10a. ____________

b. ____________

c. ____________

d. ____________

11a. ____________

b. ____________

c. ____________

12. ____________

13. [Respond below exercise.] ____________

14. ____________

10. Name the type of each of the following triangles based on the measures and shapes shown.

(a)

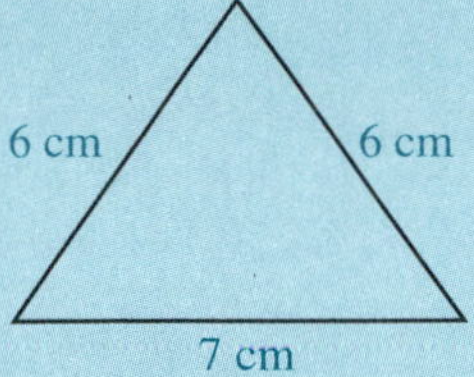
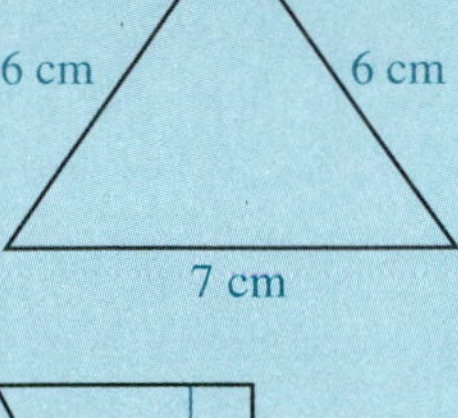

(b)

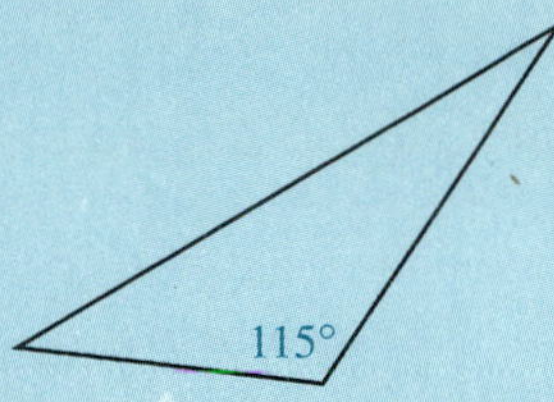

(c)

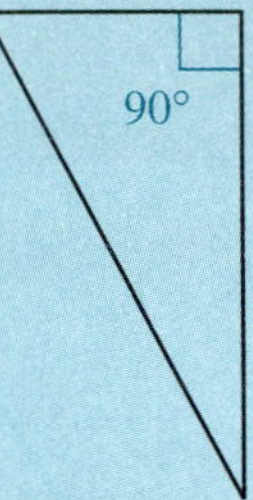

(d)

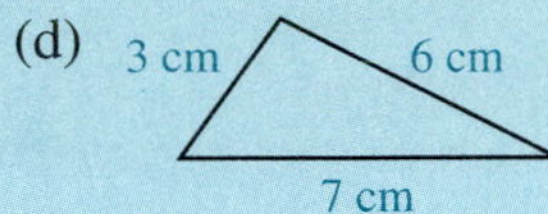

11. (a) Find the measure of $\angle x$.
(b) What kind of triangle is $\triangle RST$?
(c) Which side is opposite $\angle S$?

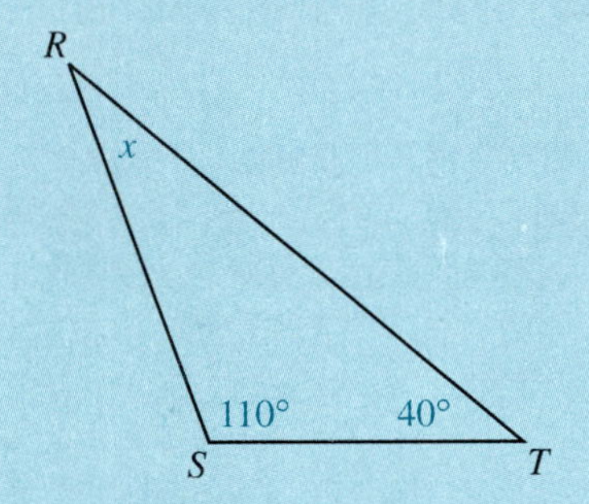

12. Find the value of x given that $\triangle ABC \sim \triangle ADE$.

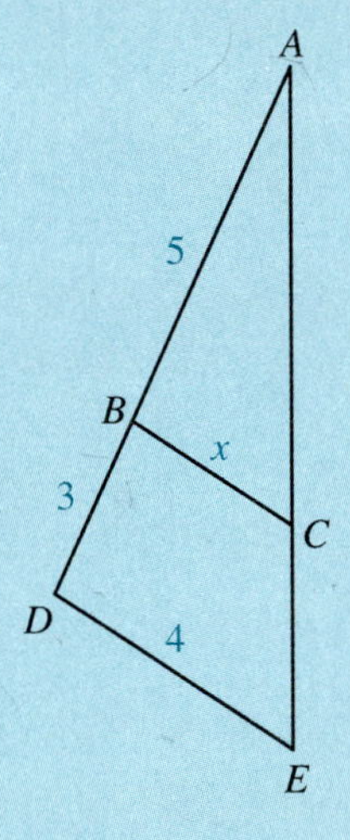

13. Determine whether or not $\triangle AOB \sim \triangle COD$. Explain your reasoning.

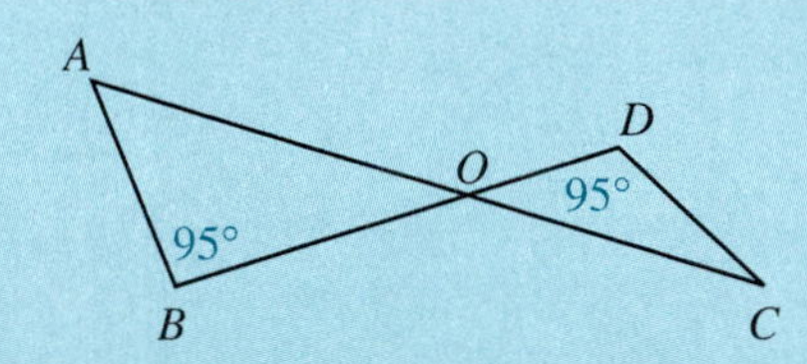

14. Given that $\triangle ABC \sim \triangle ADE$, find the value of x and the value of y.

NAME ______________ SECTION ______________ DATE ______________

Cumulative Review

In Exercises 1–4, choose the correct answer by performing the related operations mentally.

1. Write $\frac{11}{5}$ as a percent.

(a) 200% (b) 210% (c) 220% (d) 240%

2. Write 150% as a mixed number with the fraction part reduced.

(a) $1\frac{1}{2}$ (b) $1\frac{5}{8}$ (c) $1\frac{3}{4}$ (d) $2\frac{1}{2}$

3. Write $2\frac{1}{3}$ as a percent.

(a) 213% (b) 230% (c) 233% (d) $233\frac{1}{3}\%$

4. Find 100% of 75.

(a) 7.5 (b) 75 (c) 750 (d) 7500

5. Write the number 423.85 in its English word form.

6. Round off 166.075 to the nearest hundred.

7. Find the average (mean) of the numbers 5000, 6250, 9475, and 8672.

In Exercises 8–10, (a) estimate the answer, then (b) find the actual answer.

8. $\begin{array}{r} 75.63 \\ 81.45 \\ +\ 146.98 \\ \hline \end{array}$

9. $\begin{array}{r} 8000.0 \\ -\ 6476.9 \\ \hline \end{array}$

10. $\begin{array}{r} 43.8 \\ \times\ 2.7 \\ \hline \end{array}$

Perform the indicated operation and reduce all fractions.

11. $\frac{16}{0}$

12. $\frac{0}{-3}$

13. $(-500)(7000)$

14. $(-2.4)(-6.1)$

15. $(-27) + (-10) + (-5)$

16. $(15 - 17)(21 - 32)$

17. $(-7)^2 \cdot 5 + 2.1(-6) \div \frac{1}{2}$

18. $\frac{1}{2} + \frac{2}{3} \cdot \left(\frac{1}{2}\right)^2 - \frac{9}{10}$

ANSWERS

1. ______________
2. ______________
3. ______________
4. ______________
5. [Respond below exercise.]
6. ______________
7. ______________
8. ______________
9. ______________
10. ______________
11. ______________
12. ______________
13. ______________
14. ______________
15. ______________
16. ______________
17. ______________
18. ______________

19. ____________

20. ____________

21a. ____________

b. ____________

22. ____________

23. ____________

24. ____________

25. ____________

26a. ____________

b. ____________

19. Find the area of a triangle with base 20 cm and height 14 cm.

20. What is the circumference of a circle with diameter 2 ft 4 in.? (Use $\pi = 3.14$.)

21. Find (a) the perimeter and (b) the area of the trapezoid shown.

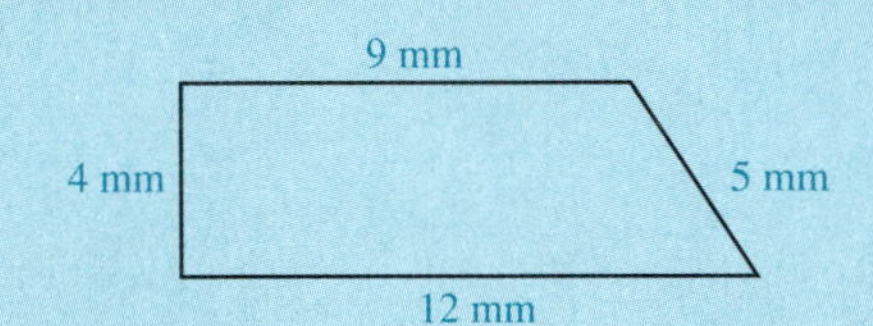

22. What is the rate of interest if the simple interest earned on \$5000 for 90 days is \$156.25?

23. Find the interest earned if a savings account of \$10,000 is compounded daily at 8% for 7 years.

24. A student decided to buy a new typewriter. She paid \$630.50. This was not the original price. She received a 3% discount for paying cash. What was the original price?

25. The scale on a map indicates that $1\frac{3}{4}$ inches represents 50 miles. How many miles apart are two cities marked 2 inches apart on the map?

26. Find the length of the hypotenuse of a right triangle with both legs of length 15 feet. (Write the answer in both radical form and decimal form.)

Additional Topics from Algebra

What to Expect in this Chapter

This chapter is designed to provide a first look at a few slightly more advanced topics from a beginning algebra course. That is, it is meant to provide a "jump start" into algebra for those students who plan to continue their studies in mathematics.

Section T.1 shows how to solve first-degree inequalities and how to graph the solutions on number lines. Sections T.2 and T.3 deal with concepts related to graphing in two dimensions.

Section T.4 provides an introduction to polynomials: classification of polynomials and addition and subtraction with polynomials. Multiplication with polynomials is presented in Section T.5.

Mathematics at Work!

In many real-life applications, the relationship between two variables can be expressed in terms of a formula in which both variables are first-degree. For example, the distance you travel on your bicycle at an average speed of 15 miles per hour can be represented by the formula $d = 15t$, where t is the time (in hours) that you ride.

Relationships indicated by first-degree equations and formulas can be "pictured" as straight-line graphs. Once the correct graph has been drawn, this graph can serve in place of the formula in finding a value (or an approximate value) of one of the variables corresponding to a known value of the other variable.

The formula $C = \frac{5}{9}(F - 32)$ represents the relationship between degrees Fahrenheit (F), and degrees Celsius (C). A portion of the graph of this relationship is shown below. (See Section T.2, Exercise 33.)

(a) From the graph, estimate the Celsius temperature corresponding to 32° F. (That is, estimate the value of C given that $F = 32$.)

(b) From the graph, estimate the Celsius temperature corresponding to 59° F. (That is, estimate the value of C given that $F = 59$.)

An Eskimo ice fishes along the Arctic coast, where mean winter temperatures are about −30° F (−34° C).

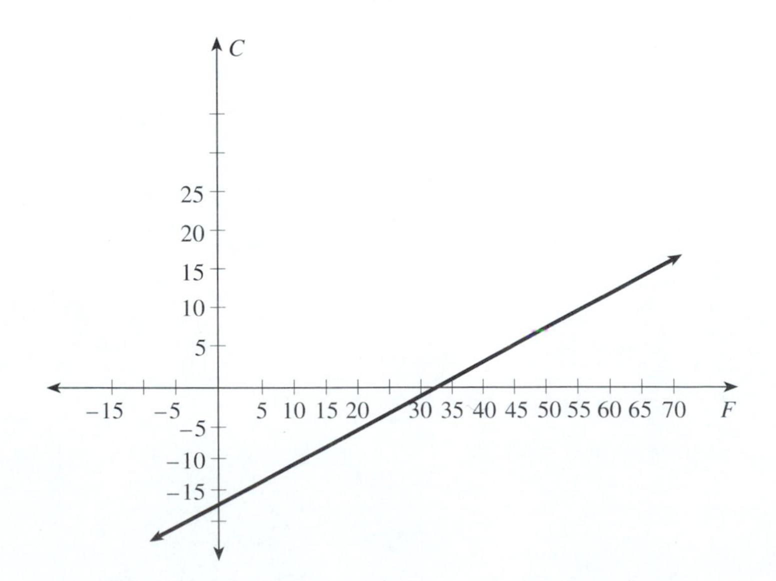

T.1 First-Degree Inequalities ($ax + b < c$)

OBJECTIVES

1. *Be able to read inequalities from left to right and from right to left.*
2. *Be able to determine whether a given inequality is true or false.*
3. *Know the different types of intervals of real numbers.*
4. *Learn how to solve and graph the solutions to inequalities of the form* $ax + b < c$.

Real Numbers

Real numbers include radicals such as $\sqrt{2}$, $\sqrt{3}$, $\sqrt[3]{15}$, and $\sqrt[4]{10}$, integers such as 5, 0, and -3, fractions such as $\frac{1}{5}$, $\frac{3}{100}$, and $-\frac{5}{8}$, and decimals such as 0.007, 4.56, and 0.3333. . . . In fact, real numbers include all those numbers that can be classified as either **rational numbers** or **irrational numbers.** (See Sections 5.6 and 5.7.) All of the numbers that we have discussed in this text are real numbers.

Rational numbers are numbers that can be represented as **terminating decimals** or as **infinite repeating decimals.** The whole numbers, integers, and fractions that we have studied are **all** rational numbers. Examples of rational numbers are

$$0, \quad -6, \quad \frac{2}{3}, \quad 27.1, \quad -90, \quad \text{and} \quad 19.3$$

Irrational numbers are numbers that can be represented as **infinite nonrepeating decimals.** Examples of irrational numbers are

$$\pi = 3.1415926535...$$

$$\sqrt{2} = 1.414213562...$$

$$-\sqrt{5} = -2.236067977...$$

Most calculators give irrational numbers to eight- or nine-digit accuracy.

The diagram in Figure T.1 illustrates the relationships among various types of real numbers.

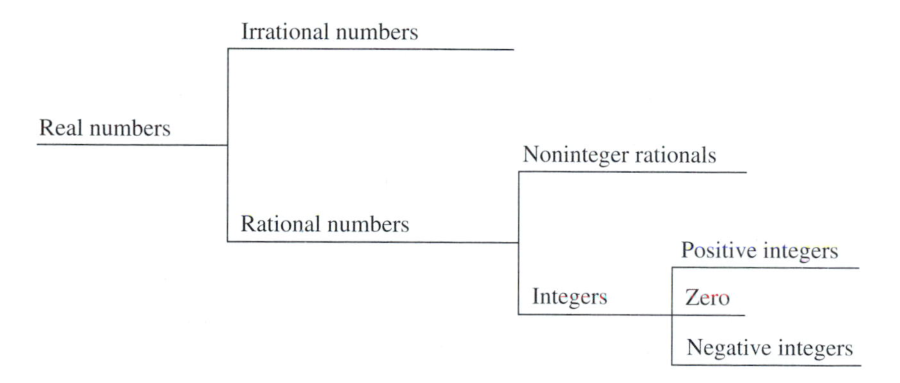

Figure T.1

Example 1

Given the set of real numbers

$$A = \left\{-5, -\pi, -1.3, 0, \sqrt{2}, \frac{4}{5}, 8\right\}$$

determine which numbers in A are (a) integers, (b) rational numbers, (c) irrational numbers.

Solution

(a) Integers: $-5, 0, 8$

(b) Rational numbers: $-5, 0, 8, -1.3, \frac{4}{5}$

Note that each integer is also a rational number.

(c) Irrational numbers: $-\pi, \sqrt{2}$

Real Number Lines and Inequalities

Numbers lines are called **real number lines** because of the following important relationship between real numbers and points on a line.

> **There is a one-to-one correspondence between the real numbers and the points on a line.** That is, each point on a number line corresponds to one real number, and each real number corresponds to one point on a number line.

The locations of some real numbers, including integers, are shown in Figure T.2. Sometimes a calculator may be used to find the approximate value (and therefore the approximate location) of an unfamiliar real number. For example, a calculator will show that

$$\sqrt{6} = 2.449489743\ldots$$

Therefore, $\sqrt{6}$ is located between 2.4 and 2.5 on a real number line.

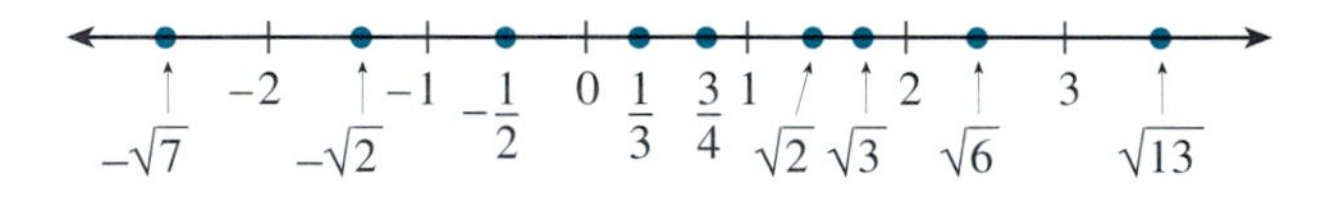

Figure T.2

To compare two real numbers, we use the following symbols of equality and inequality (reading from **left to right**):

$a = b$ a is equal to b

$a < b$ a is less than b

$a \leq b$ a is less than or equal to b

$a > b$ a is greater than b

$a \geq b$ a is greater than or equal to b

These symbols can also be read from **right to left.** For example, we can read

$a < b$ as "b is greater than a"

and

$a > b$ as "b is less than a"

A slash, /, through a symbol negates that symbol. Thus, for example, $\neq$ is read "is not equal to" and $\not<$ is read "is not less than."

EXAMPLE 2

Write the meaning of each of the following inequalities.

(a) $6 < 7.5$

(b) $-3 > -10$

(c) $-14 \neq |-14|$

Solution

(a) $6 < 7.5$ "6 is less than 7.5" or "7.5 is greater than 6"

(b) $-3 > -10$ "-3 is greater than -10" or "-10 is less than -3"

(c) $-14 \neq |-14|$ "-14 is not equal to the absolute value of -14"

On a number line, smaller numbers are to the left of larger numbers. (Or, larger numbers are to the right of smaller numbers.) Thus, as shown in Figure T.3, we have

$-1 < 2 \qquad -2 < -1 \qquad c > a \qquad b > 0$

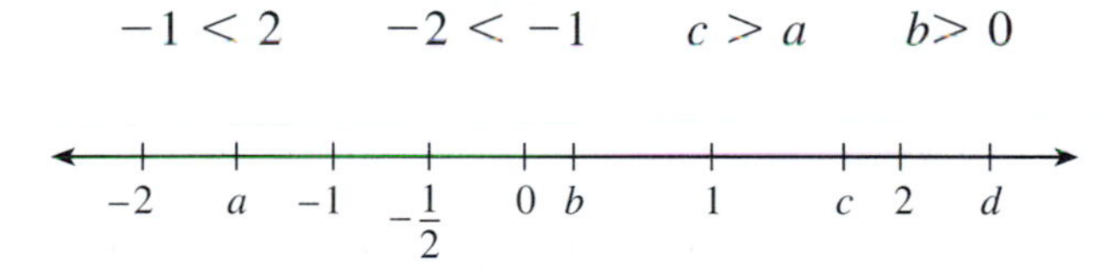

Figure T.3

Suppose that a and b are two real numbers and that $a < b$. The set of all real numbers between a and b is called an **interval of real numbers.** Intervals are classified and graphed as indicated in the box that follows. You should keep in mind the following facts as you consider the boxed information.

1. x is understood to represent real numbers.
2. Open dots at endpoints a and b indicate that these points are **not** included in the graph.
3. Solid dots at endpoints a and b indicate that these points **are** included in the graph.

Intervals of Real Numbers

Name	Symbolic Representation	Graph
Open Interval	$a < x < b$	
Closed Interval	$a \leq x \leq b$	
Half-open Interval	$a \leq x < b$	
	$a < x \leq b$	
Open Interval	$x > a$	
	$x < a$	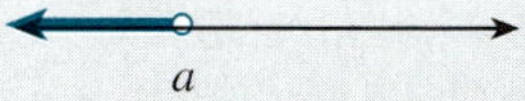
Half-open Interval	$x \geq a$	
	$x \leq a$	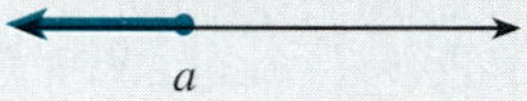

EXAMPLE 3

Graph the closed interval $4 \leq x \leq 6$.

Solution

Note that 4 and 6 are in the interval, as are all the real numbers between 4 and 6. For example, $4\frac{1}{2}$ and 5.99 are in the interval since

$$4 \leq 4\frac{1}{2} \leq 6 \quad \text{and} \quad 4 \leq 5.99 \leq 6$$

EXAMPLE 4

Represent the following graph using algebraic notation and tell what kind of interval it is.

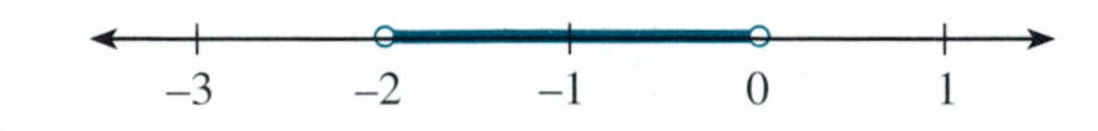

Solution

$-2 < x < 0$ is an open interval.

Now Work Exercises 1–3 in the Margin.

Solving Linear Inequalities

Inequalities of the forms

$$ax + b < c \quad \text{and} \quad ax + b \leq c$$
$$ax + b > c \quad \text{and} \quad ax + b \geq c$$
$$c < ax + b < d \quad \text{and} \quad c \leq ax + b \leq d$$

are all called **linear inequalities** or **first-degree inequalities.**

The solutions to linear inequalities are intervals of real numbers, and the methods for solving linear inequalities are similar to those used in solving linear (or first-degree) equations. The rules are the same with one important exception:

Multiplying (or dividing) both sides of an inequality by a negative number "reverses the sense" of the inequality.

Consider the following examples.

We know that $4 < 10$:

Add	**Multiply by 2**	**Add −5**
$4 < 10$	$4 < 10$	$4 < 10$
$4 + 3 \; ? \; 10 + 3$	$2 \cdot 4 \; ? \; 2 \cdot 10$	$4 + (-5) \; ? \; 10 + (-5)$
$7 < 13$	$8 < 20$	$-1 < 5$

In each instance, the sense of the inequality stayed the same after the operation, namely $<$.

Now we see that multiplying both sides by a **negative** number reverses the sense (in these examples, from $<$ to $>$):

Multiply by −6		**Divide by −2**
$4 < 10$		$4 < 10$
$-6 \cdot 4 \; ? \; -6 \cdot 10$		$\frac{4}{-2} \; ? \; \frac{10}{-2}$
$-24 > -60$	In both cases the sense is reversed from $<$ to $>$.	$-2 > -5$

Rules for Solving Linear (or First-Degree) Inequalities

1. The same number (positive or negative) may be added to both sides, and the sense of the inequality will remain the same.
2. Both sides may be multiplied by (or divided by) the same **positive** number, and the sense of the inequality will remain the same.
3. Both sides may be multiplied by (or divided by) the same **negative** number, but the sense of the inequality must be **reversed.**

1. Graph the closed interval $-2 \leq x \leq 1$.

2. Graph the open interval $x > 3$.

3

3. Represent the graph below using algebraic notation and tell what type of interval it is.

As with solving an equation, the object of solving an inequality is to find equivalent inequalities that are simpler than the original and to isolate the variables on one side of the inequality. The difference between the solutions of inequalities and equations is that the solution to an inequality consists of an interval of numbers (an infinite set of numbers) whereas the solution to a first-degree equation is a single number.

EXAMPLE 5

Solve the inequality $x - 3 < 2$ and graph the solution on a number line.

Solution

$$x - 3 < 2$$

$$x - 3 + 3 < 2 + 3 \quad \text{Add 3 to both sides.}$$

$$x < 5$$

5

EXAMPLE 6

Solve the inequality $-2x + 7 \geq 4$ and graph the solution on a number line.

Solution

$$-2x + 7 \geq 4$$

$$-2x + 7 - 7 \geq 4 - 7 \quad \text{Add } -7 \text{ to both sides.}$$

$$-2x \geq -3$$

$$\frac{-2x}{-2} \leq \frac{-3}{-2} \quad \text{Divide both sides by } -2 \text{ and } \textbf{reverse the sense.}$$

$$x \leq \frac{3}{2}$$

$\frac{3}{2}$

✓ COMPLETION EXAMPLE 7

Solve the inequality $7y - 8 > y + 10$ and graph the solution on a number line.

Solution

$7y - 8 > y + 10$

$7y - 8 - y > y + 10 - y$ Add _______ to both sides.

$6y - 8 >$ ______ Simplify.

$6y - 8 +$ ______ $>$ ______ $+$ ______ Add _______ to both sides.

$6y >$ ______ Simplify.

$\frac{6y}{\_\_\_\_} > \frac{\_\_\_\_}{\_\_\_\_}$ Divide both sides by _______.

$y >$ ______ Simplify.

Graph.

EXAMPLE 8

Find the values of x that satisfy **both** of the inequalities:

$5 < 2x + 3$ **and** $2x + 3 < 10$

Graph the solution on a number line.

Solution

Since the variable expression $2x + 3$ is the same in both inequalities and $5 < 10$, we can write the two inequalities in one expression and solve both inequalities at the same time.

$5 < 2x + 3 < 10$

$5 - 3 < 2x + 3 - 3 < 10 - 3$ Add -3 to each part.

$2 < 2x < 7$ Simplify.

$\frac{2}{2} < \frac{2x}{2} < \frac{7}{2}$ Divide each part by 2.

$1 < x < 3.5$ Thus, x is greater than 1 **and** less than 3.5.

0 1 2 3 4 5
3.5

Now Work Exercises 4–6 in the Margin.

Solve each inequality and graph the solution on a number line.

4. $2x - 1 > 7$

5. $3x + 2 \le 5x + 1$

6. $-4 < 5y + 1 \le 11$

Completion Example Answer

7. $7y - 8 > y + 10$

$7y - 8 - y > y + 10 - y$	Add $-y$ to both sides.
$6y - 8 > \mathbf{10}$	Simplify.
$6y - 8 + \mathbf{8} > \mathbf{10} + \mathbf{8}$	Add **8** to both sides.
$6y > \mathbf{18}$	Simplify.
$\frac{6y}{\mathbf{6}} > \frac{\mathbf{18}}{\mathbf{6}}$	Divide both sides by **6.**
$y > \mathbf{3}$	Simplify.
	Graph.

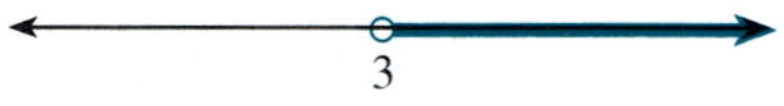

3

NAME ______________________ SECTION ______________ DATE ________

ANSWERS

11. ____________

12. ____________

13. ____________

14. ____________

15. ____________

Exercises T.1

Below each exercise, write each of the following inequalities in words and state whether it is true or false. If an inequality is false, rewrite it in a correct form. (There may be more than one way to correct a false statement.)

1. $3 \neq -3$

2. $-5 < -2$

3. $-13 > -1$

4. $|-7| \neq |+7|$

5. $|-7| < +5$

6. $|-4| > |+3|$

7. $-\frac{1}{2} < -\frac{3}{4}$

8. $\sqrt{2} > \sqrt{3}$

9. $-4 < -6$

10. $|-6| > 0$

Represent each of the following graphs with algebraic notation and tell what kind of interval it is.

11. 

12. 

13. 

14. 

15. 

16. ____________

17. ____________

18. ____________

19. ____________

20. ____________

21. ____________

22. ____________

23. ____________

24. ____________

25. ____________

26. ____________

27. ____________

28. ____________

29. ____________

30. ____________

31. ____________

In Exercises 16–27, graph the interval on the number line provided. Use the answer blank in the margin to tell what kind of interval each is.

16. $5 < x < 8$

17. $-2 < x < 0$

18. $3 \le y \le 6$

19. $-3 \le y \le 5$

20. $-2 < y \le 1$

21. $4 \le x < 7$

22. $47 \le x \le 52$

23. $-12 < z < -7$

24. $x > -5$

25. $x \ge 0$

26. $x < -\sqrt{3}$

27. $x \le \frac{2}{3}$

In Exercises 28–55, solve each inequality and write the solution in the answer blank in the margin. Then graph the solution on the number line provided.

28. $x + 5 < 6$

29. $x - 6 > -2$

30. $y - 4 \ge -1$

31. $y + 5 \le 2$

NAME ______________________ SECTION ________________ DATE __________

32. $3y \leq 4$

33. $5y > -6$

34. $10y + 1 > 5$

35. $7x - 2 < 9$

36. $x + 3 < 4x + 3$

37. $x - 4 > 2x + 1$

38. $3x - 5 \geq x + 5$

39. $3x - 8 \leq x + 2$

40. $\frac{1}{2}x - 2 < 6$

41. $\frac{1}{3}x + 1 > -1$

42. $-x + 4 < -1$

43. $-x - 5 \geq -4$

44. $8y - 2 \geq 5y + 1$

45. $6x + 3 > x - 2$

ANSWERS

32. ____________

33. ____________

34. ____________

35. ____________

36. ____________

37. ____________

38. ____________

39. ____________

40. ____________

41. ____________

42. ____________

43. ____________

44. ____________

45. ____________

46. ______________

47. ______________

48. ______________

49. ______________

50. ______________

51. ______________

52. ______________

53. ______________

54. ______________

55. ______________

46. $-5x - 7 \geq -7 + 2x$

47. $3x + 15 < x + 5$

48. $6 \leq x + 9 \leq 7$

49. $-2 \leq x - 3 \leq 1$

50. $-5 \leq 4y + 3 \leq 0$

51. $0 \leq 3x - 1 \leq 5$

52. $-2 < 5x + 3 \leq 8$

53. $-5 \leq y - 2 \leq -1$

54. $5 \leq -2x - 5 < 7$

55. $14 < -5x - 1 \leq 24$

T.2 Graphing Ordered Pairs of Real Numbers

OBJECTIVES

1. *Learn the terminology related to graphs in two dimensions, such as* ***ordered pairs, first coordinate, second coordinate, axis,*** *and* ***quadrant.***
2. *Be able to list the set of ordered pairs represented on a graph.*
3. *Know how to graph a set of ordered pairs of real numbers.*

Equations in Two Variables

Equations such as

$$d = 40t, \qquad I = 0.18P, \qquad \text{and} \qquad y = 2x - 5$$

represent relationships between pairs of variables. These equations are said to be **equations in two variables.** The first equation, $d = 40t$, can be interpreted as follows: The distance d traveled in time t at a rate of 40 miles per hour is found by multiplying 40 by t (where t is measured in hours). Thus, if $t = 3$ hours, then $d = 40(3) = 120$ miles. The pair (3, 120) is called an **ordered pair** and is in the form (t, d).

We say that the ordered pair (3, 120) **is a solution of** (or **satisfies**) the equation $d = 40t$. Similarly, (5, 200) represents $t = 5$ and $d = 200$ and satisfies the equation $d = 40t$. In the same way, (100, 18) satisfies the equation $I = 0.18P$, where $P = 100$ and $I = 0.18(100) = 18$. In this equation, the interest I is equal to 18% (or 0.18) times the principal P and the ordered pair (100, 18) is in the form (P, I).

For the equation $y = 2x - 5$, ordered pairs are in the form (x, y) and (3, 1) satisfies the equation: if $x = 3$, then $y = 2(3) - 5 = 1$. In the ordered pair (x, y), x is called the **first coordinate** (or **first component**) and y is called the **second coordinate** (or **second component**). To find ordered pairs that satisfy an equation in two variables, we can **choose any value** for one variable and find the corresponding value for the other variable by substituting into the equation. For example,

if $y = 2x - 5$, then

for $x = 2$ we have $y = 2(2) - 5 = 4 - 5 = -1$

for $x = 0$ we have $y = 2(0) - 5 = 0 - 5 = -5$

for $x = 6$ we have $y = 2(6) - 5 = 12 - 5 = 7$

All the ordered pairs $(2, -1)$, $(0, -5)$, and $(6, 7)$ satisfy the equation $y = 2x - 5$. There are an infinite number of such ordered pairs.

Since the equation $y = 2x - 5$ is solved for y, we say that the value of y "depends" on the choice of x. Thus, in an ordered pair of the form (x, y), the second coordinate y is called the **dependent variable** and the first coordinate x is called the **independent variable.**

Some examples of ordered pairs for each of the three equations discussed are shown below in table form. **Remember that the choices for the value of the independent variable are arbitrary; other values could have been chosen.**

$d = 40t$

t	d
1	40
2	80
3	120
4	160

$I = 0.18P$

P	I
100	18
200	36
1000	180
5000	900

$y = 2x - 5$

x	y
−2	−9
0	−5
1	−3
5	5

Graphing Ordered Pairs

Ordered pairs of real numbers can be graphed as points in a plane by using the **Cartesian coordinate system** [named after the famous French mathematician Rene Descartes (1596–1650)]. In this system, the plane is separated into four **quadrants** by two number lines that are perpendicular to each other. The lines intersect at a point called the **origin,** represented by the ordered pair (0, 0). The horizontal number line represents the independent variable and is called the **horizontal axis** (or the ***x*-axis**). The vertical number line represents the dependent variable and is called the **vertical axis** (or the ***y*-axis**). (See Figure T.4.)

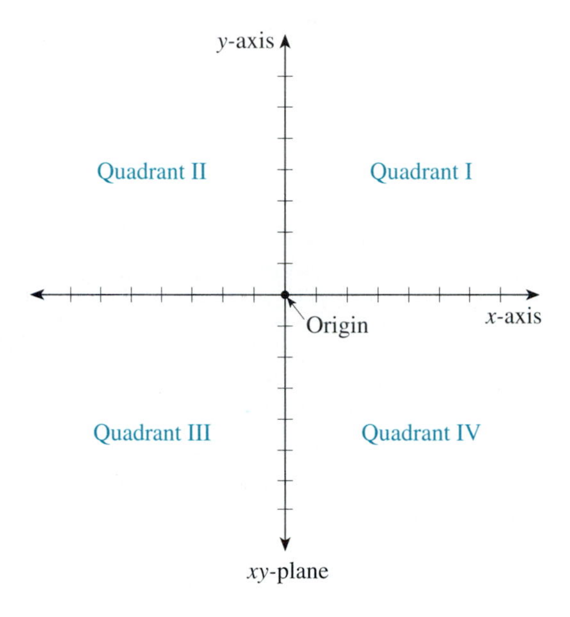

Figure T.4

On the x-axis, positive values of x are indicated to the right of the origin and negative values of x are to the left of the origin. On the y-axis, positive values of y are indicated above the origin and negative values of y are below the origin. The relationship between ordered pairs of real numbers and the points in a plane is similar to the correspondence between real numbers and points on a number line and is expressed in the following important statement.

There is a one-to-one correspondence between the points in a plane and ordered pairs of real numbers. That is, each point in a plane corresponds to one ordered pair of real numbers, and each ordered pair of real numbers corresponds to one point in a plane. (See Figure T.5.)

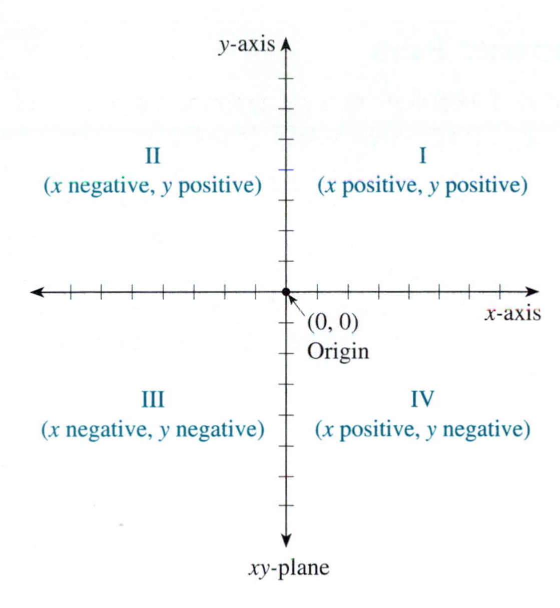

Figure T.5

The graphs of the points A (3, 1), B (−2, 3), C (−3, −1), D (1, −2), and E (2, 0) are shown in Figure T.6. The point E (2, 0) is on an axis and not in any quadrant. Each ordered pair is called the **coordinates** of the corresponding point. For example, the coordinates of point A are given by the ordered pair (3, 1).

POINT	QUADRANT
$A(3, 1)$	I
$B(-2, 3)$	II
$C(-3, -1)$	III
$D(1, -2)$	IV
$E(2, 0)$	x-axis

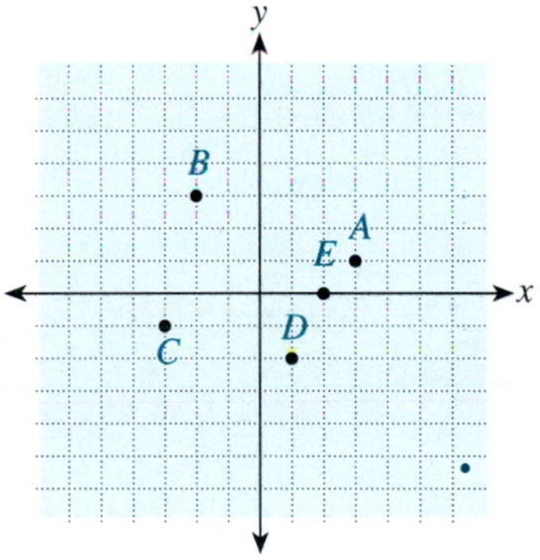

Figure T.6

Note: Unless a scale is labeled on the x-axis or on the y-axis, the grid lines are assumed to be one unit apart in both the horizontal and vertical directions.

EXAMPLE 1

Graph the set of ordered pairs:

$\{(-2, 1), (0, 3), (1, 2), (2, -2)\}$

Solution

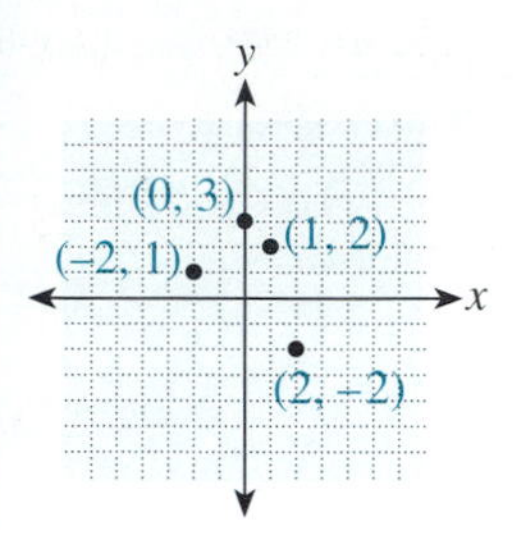

EXAMPLE 2

Graph the set of ordered pairs:

$\{(-3, -5), (-2, -3), (-1, -1), (0, 1), (1, 3)\}$

Solution

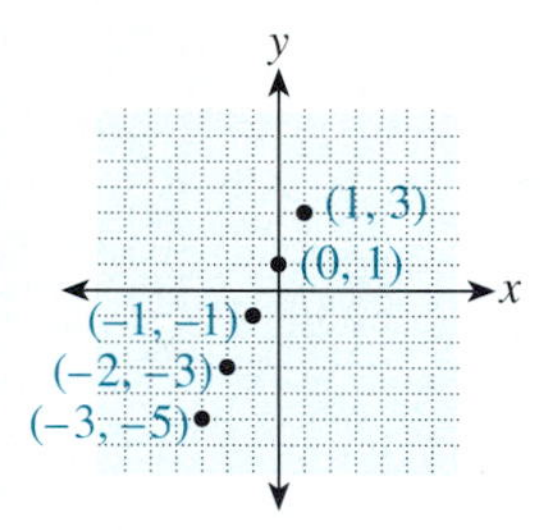

EXAMPLE 3

The graph of a set of points is given. List the set of ordered pairs that correspond to the points in the graph.

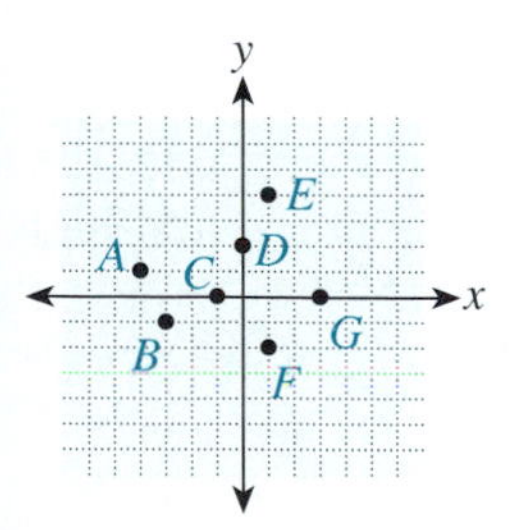

Solution

$\{A\ (-4, 1), B\ (-3, -1), C\ (-1, 0), D\ (0, 2), E\ (1, 4), F\ (1, -2), G\ (3, 0)\}$

NAME ______________________ SECTION ______________ DATE ________

Exercises T.2

Below each graph in Exercises 1–10, list the set of ordered pairs that correspond to the points in the graph.

1.

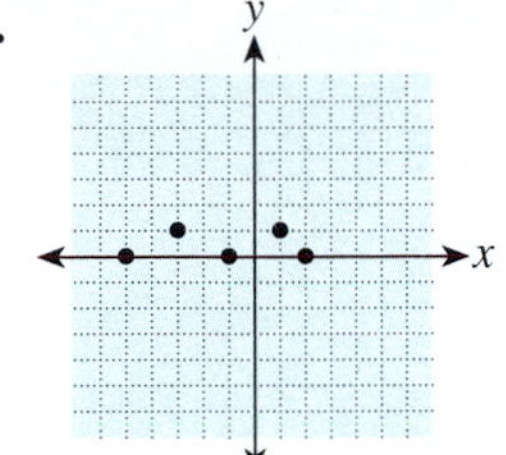

2.

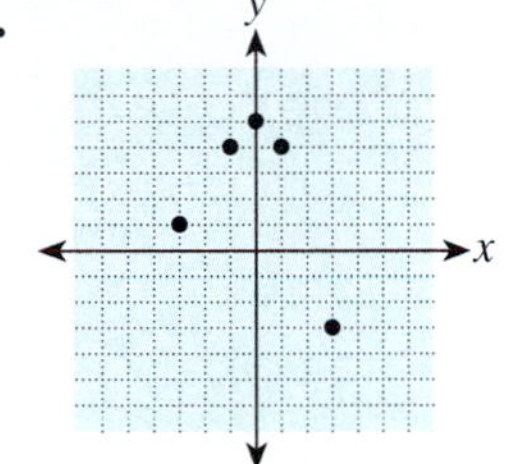

3.

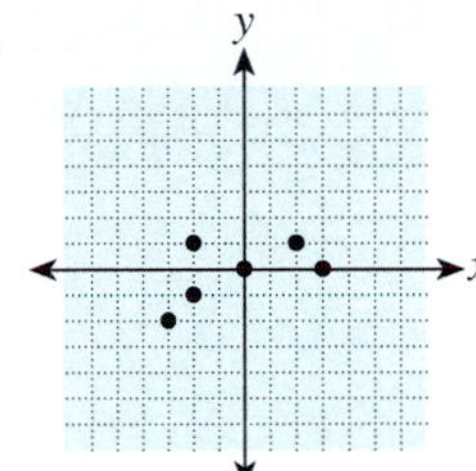

4.

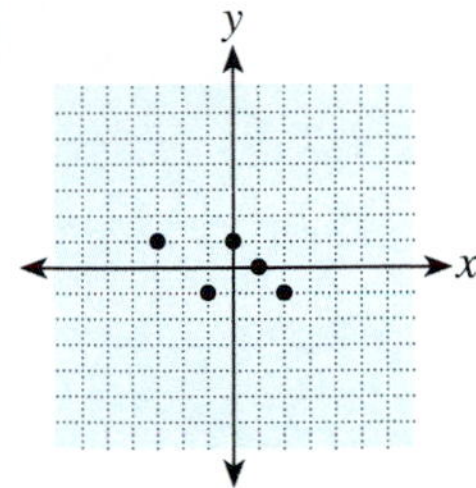

5.

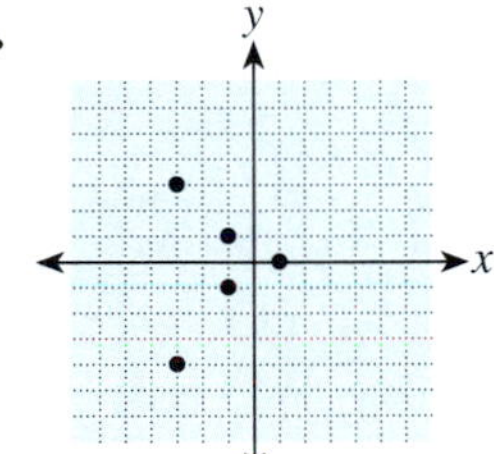

6.

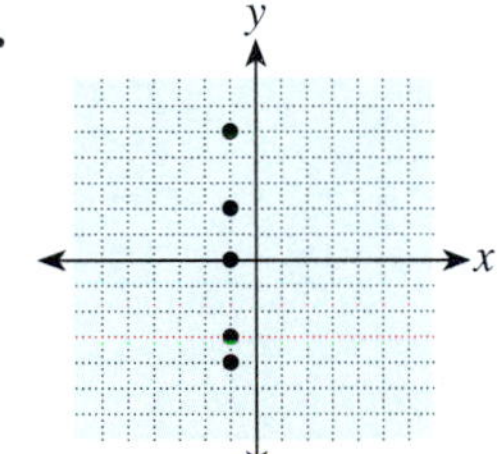

7.

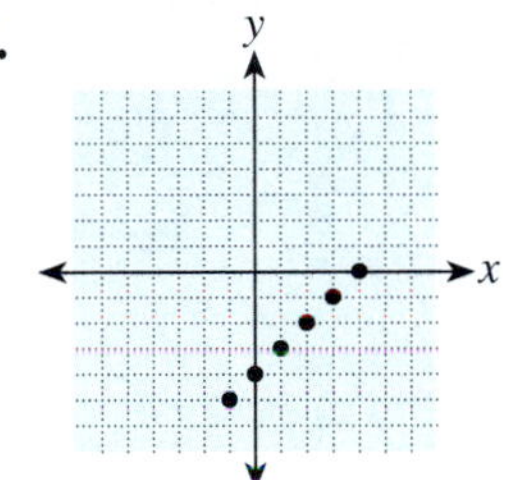

8.

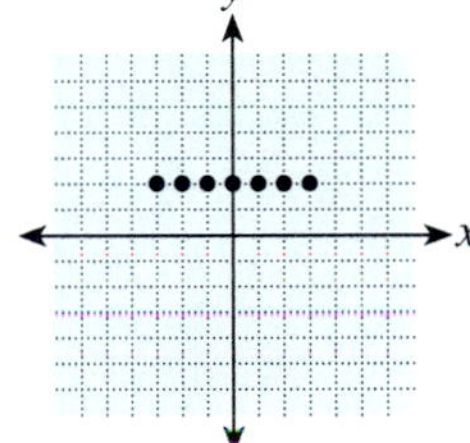

9.

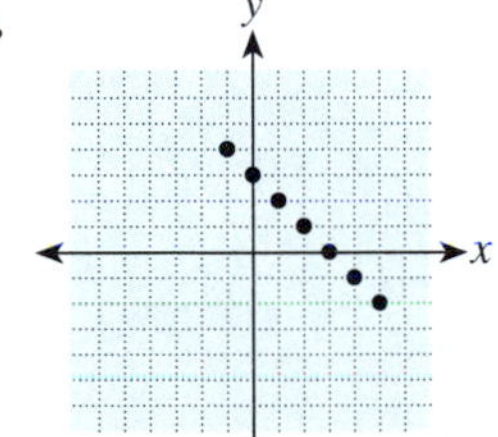

10.

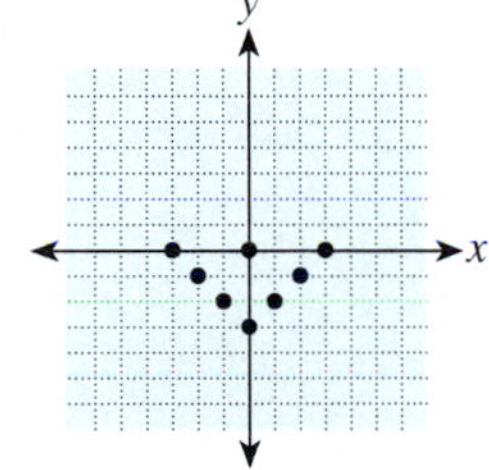

Graph each of the following sets of ordered pairs.

11. $\{(-2, 4), (-1, 3), (0, 1), (1, -2), (1, 3)\}$

12. $\{(-5, 1), (-3, 2), (-2, -1), (0, 2), (2, -1)\}$

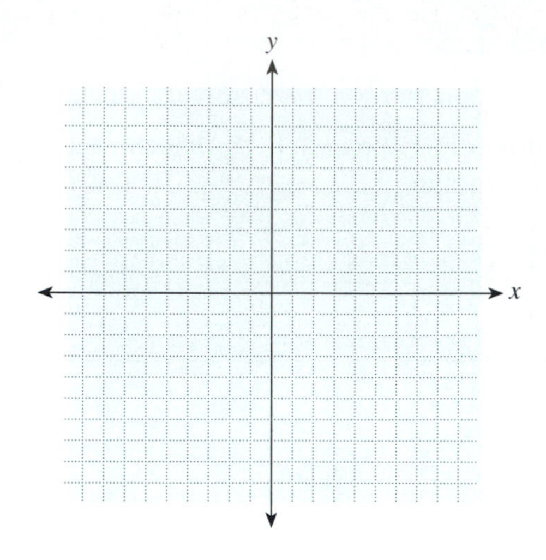

13. $\{(-1, 2), (1, 3), (2, -2), (3, 4), (4, -2)\}$

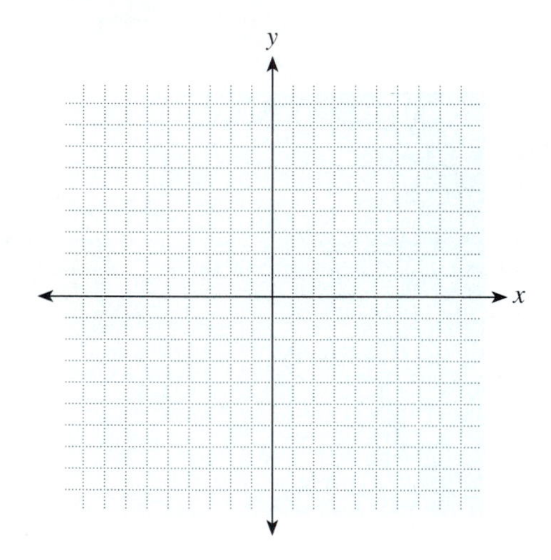

14. $\{(-2, 3), (-1, 0), (0, -3), (2, 3), (4, -1)\}$

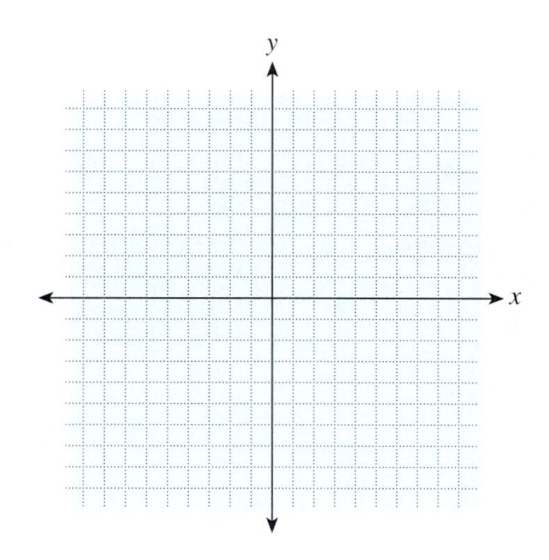

15. $\{(0, -3), (1, -1), (2, 1), (3, 3), (4, 5)\}$

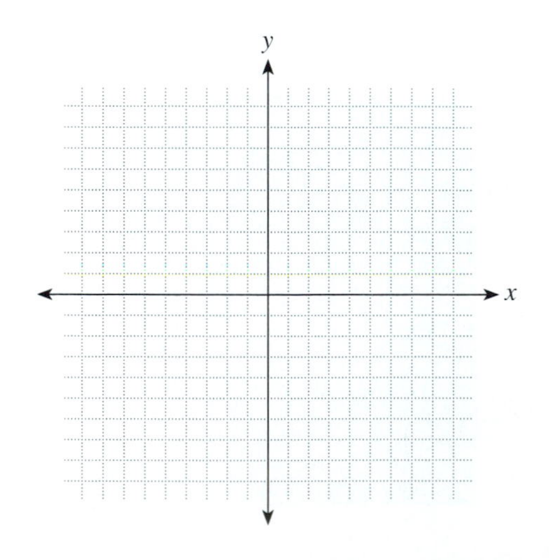

16. $\{(-3, 3), (-2, 2), (-1, 1), (0, 0), (1, -1)\}$

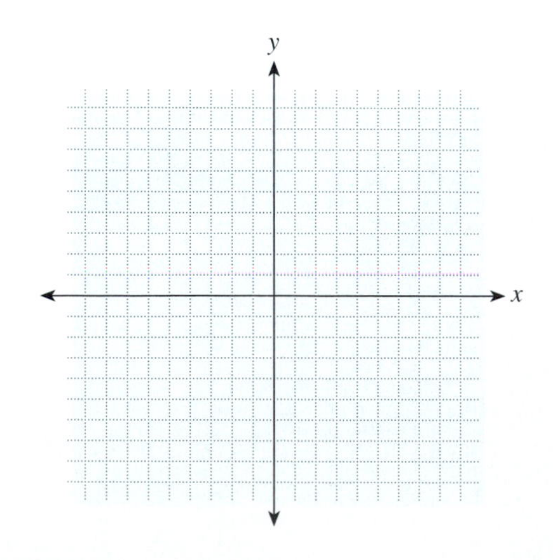

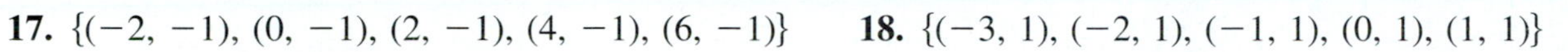
NAME ______________ SECTION ______________ DATE ______________

17. $\{(-2, -1), (0, -1), (2, -1), (4, -1), (6, -1)\}$

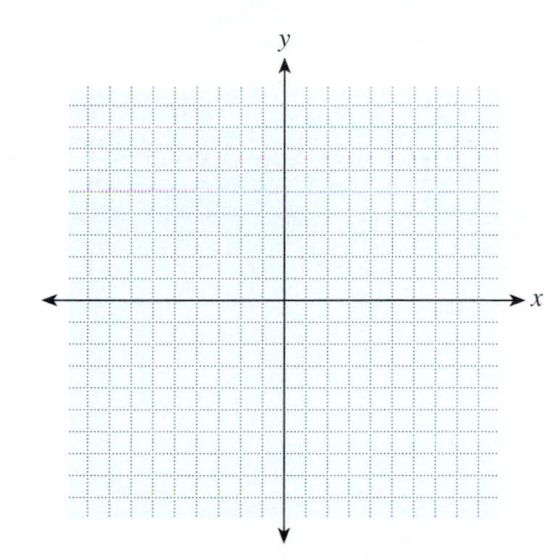

18. $\{(-3, 1), (-2, 1), (-1, 1), (0, 1), (1, 1)\}$

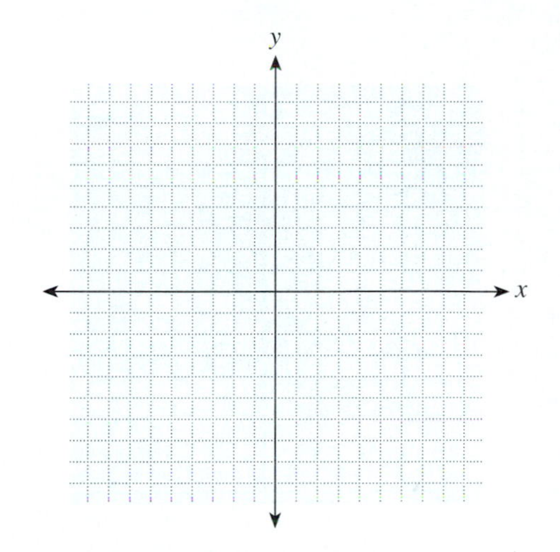

19. $\{(-3, 3), (-1, 1), (0, 0), (1, 1), (3, 3)\}$

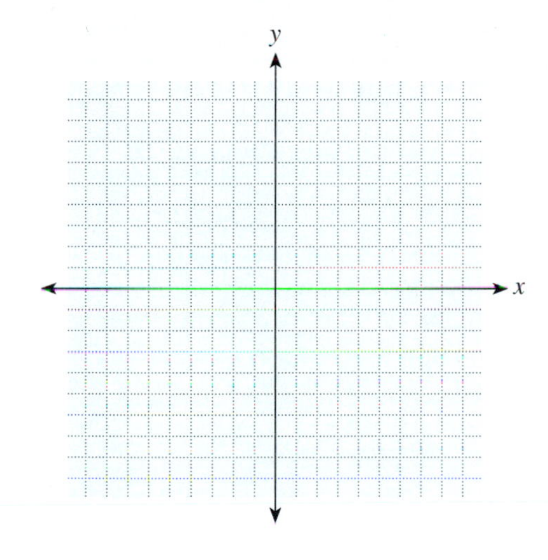

20. $\{(-2, -2), (-1, -1), (0, 0), (1, -1), (2, -2)\}$

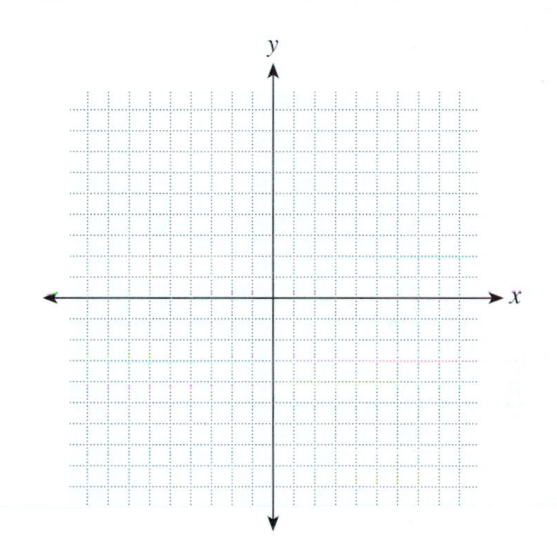

21. $\{(-3, 9), (-2, 4), (-1, 1), (1, 1), (2, 4), (3, 9)\}$

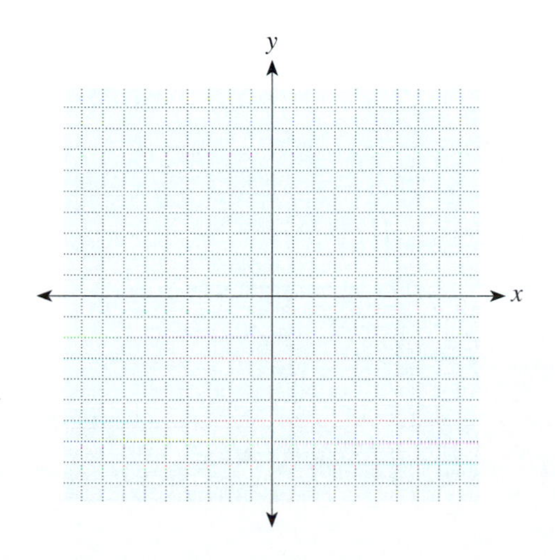

22. $\{(-3, -9), (-2, -4), (-1, -1), (1, -1), (2, -4), (3, -9)\}$

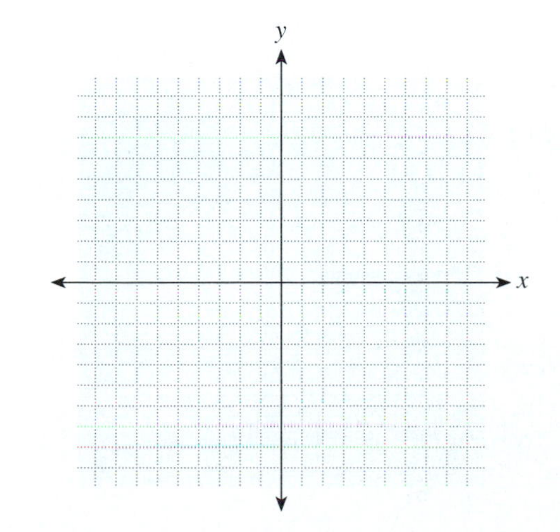

23. $\{(-4, 0), (-2, 0), (0, 0), (2, 0), (4, 0)\}$

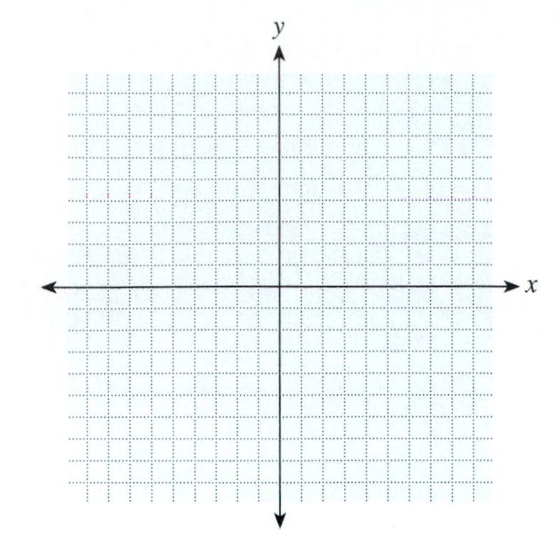

24. $\{(0, -3,), (0, -1), (0, 0), (0, 1), (0, 3)\}$

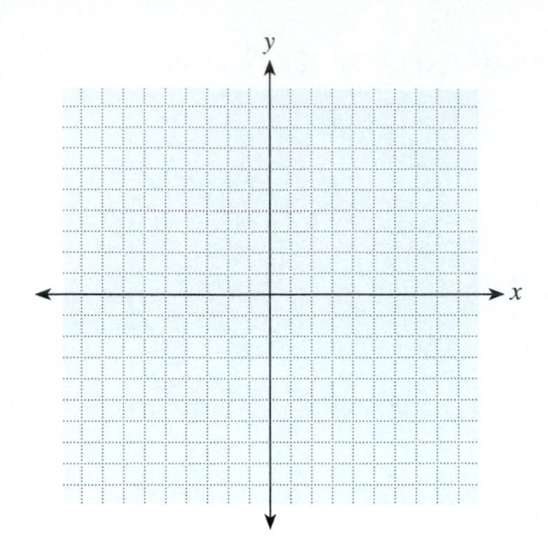

25. $\{(-2, 1), (1, 4), (2, 5), (3, 6), (4, 7)\}$

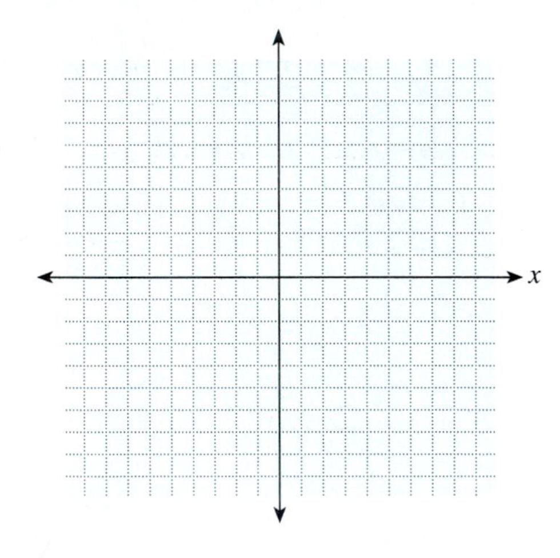

26. $\{(-1, -5), (0, -2), (1, 1), (2, 4), (3, 7)\}$

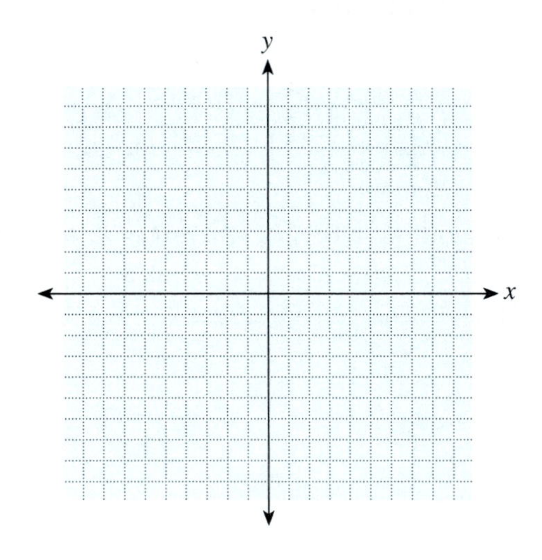

27. $\{(-2, -7), (-1, -5), (2, 1), (3, 3), (4, 5)\}$

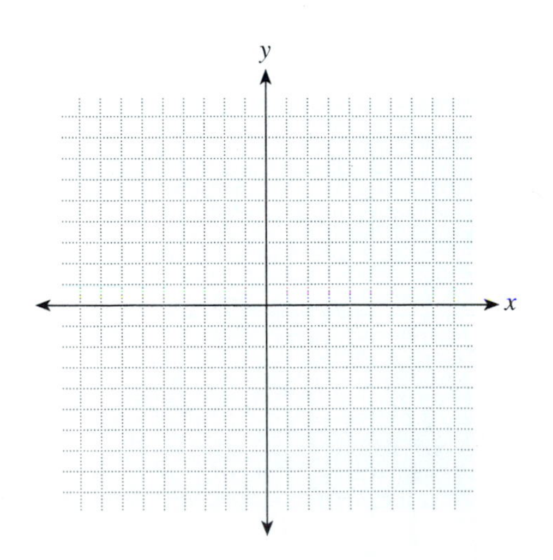

28. $\{(0, 1), (1, -1), (2, -3), (3, -5), (5, -9)\}$

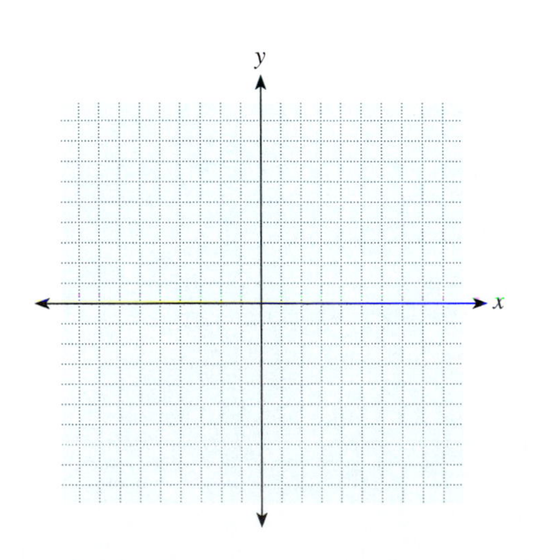

NAME ______________________ SECTION ______________________ DATE ____________

29. $\{(-2, 8), (0, 2), (1, -1), (2, -4), (3, -7)\}$

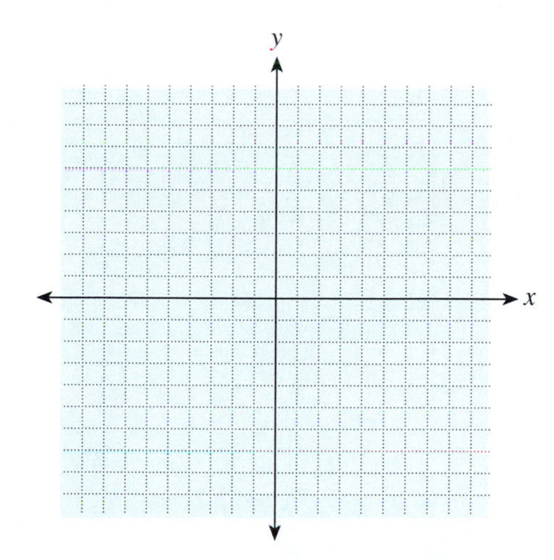

30. $\{(-2, -1), (-1, 1), (1, 5), (2, 7), (3, 9)\}$

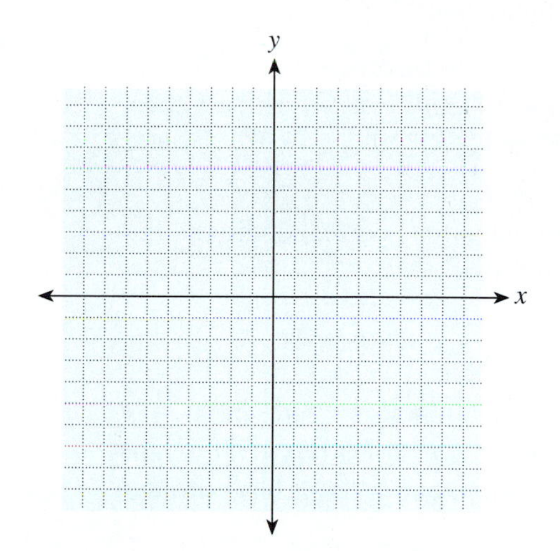

For the graphs in Exercises 31–34, mark an appropriate scale on both the horizontal and vertical axes. These scales need not be the same.

31. Given the equation $I = 0.12P$, where I is the interest earned on a principal P at a rate of 12%:

(a) Make a table of ordered pairs for the values of P and I if P has the values \$100, \$200, \$300, \$400, and \$500.

(b) Graph the points corresponding to the ordered pairs.

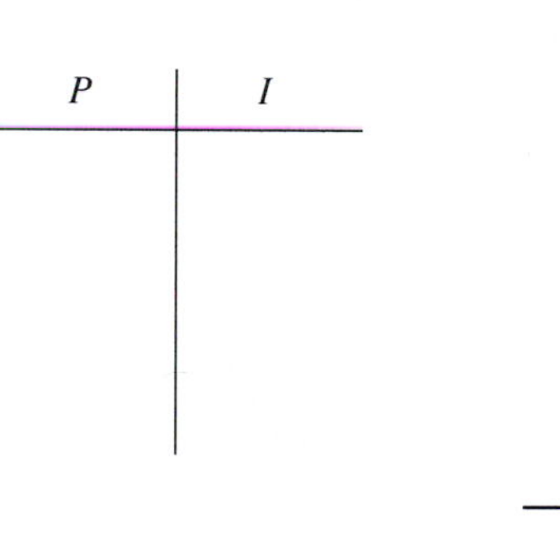

32. Given the equation $d = 16t^2$, where d is the distance an object falls in feet and t is the time in seconds that the object falls:

(a) Make a table of ordered pairs for the values of t and d if t has the values 1, 2, 3, 4, and 5 seconds.

(b) Graph the points corresponding to the ordered pairs.

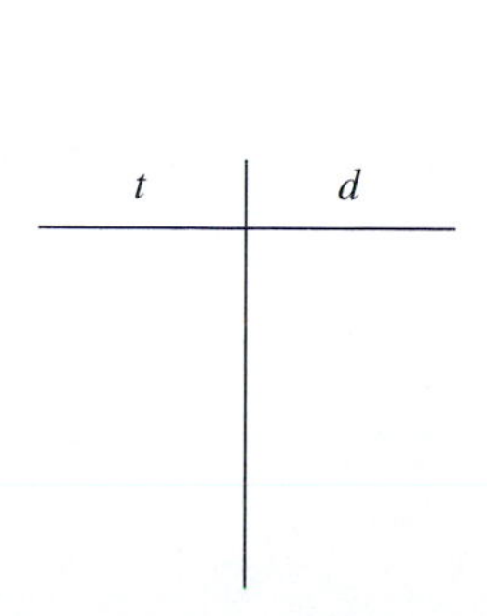

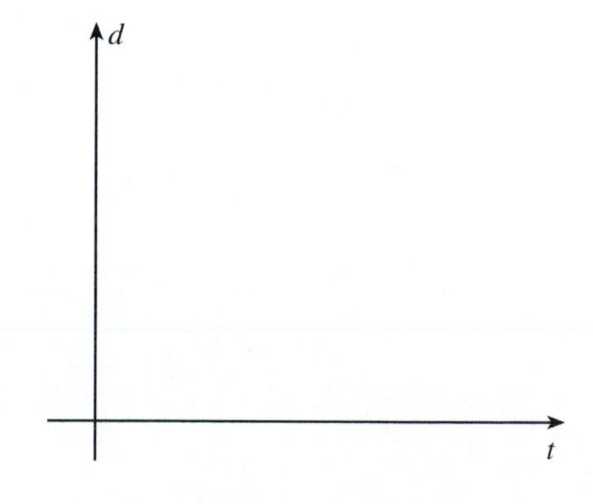

33. Given the equation $F = \frac{9}{5}C + 32$, where C is the temperature in degrees Celsius and F is the corresponding temperature in degrees Fahrenheit:

(a) Make a table of ordered pairs for the values of C and F if C has the values $-20°$, $-15°$, $-10°$, $-5°$, $0°$, $5°$, and $10°$.

(b) Graph the points corresponding to the ordered pairs.

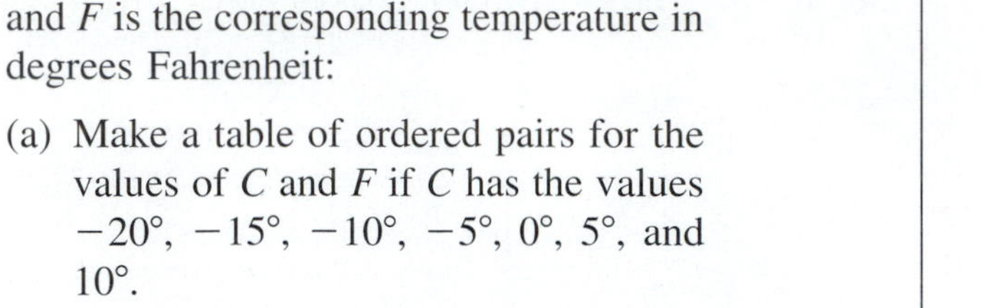

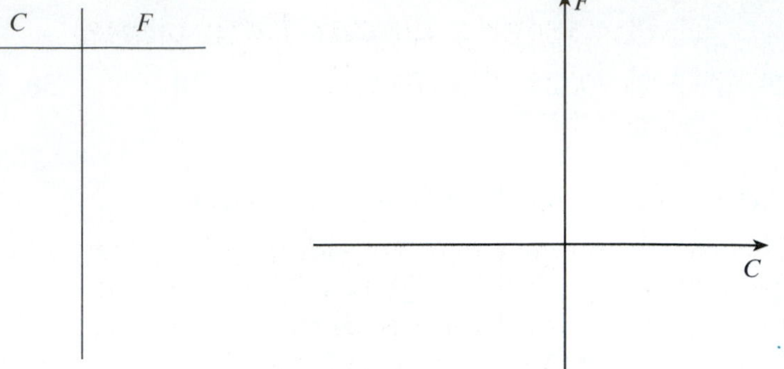

34. Given the equation $V = 25h$, where V is the volume (in cm^3) of a box with height h in centimeters and a fixed base of area $25\ cm^2$:

(a) Make a table of ordered pairs for the values of h and V if h has the values of 3 cm, 5 cm, 6 cm, 8 cm, and 10 cm.

(b) Graph the points corresponding to the ordered pairs.

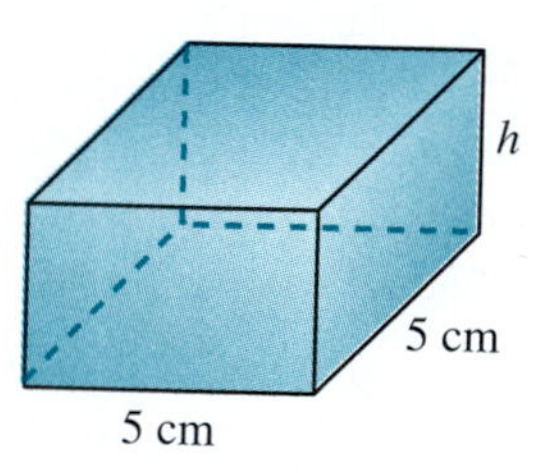

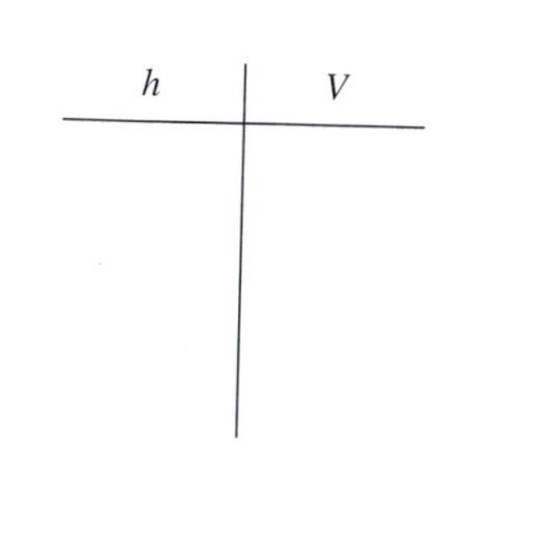

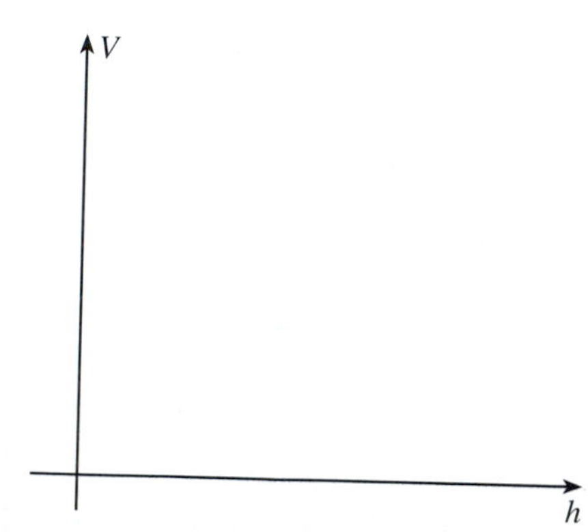

T.3 Graphing Linear Equations ($Ax + By = C$)

OBJECTIVES

1. *Be familiar with the standard form for the equation of a line:* $Ax + By = C$.
2. *Know how to find the y-intercept and the x-intercept of a line.*
3. *Know how to plot points that satisfy a linear equation and draw the graph of the corresponding line.*

Linear Equations in Standard Form

There are an **infinite number** of ordered pairs of real numbers that satisfy the equation $y = 3x + 1$. In Section T.2, we substituted only integer values for x in introducing the concepts of ordered pairs and graphing ordered pairs. However, the discussion was based on ordered pairs of real numbers, and this means that we can also substitute values for x that are rational numbers or irrational numbers. That is, fractions and radicals can be substituted just as well as integers. The corresponding y values may also be expressions with fractions and radicals. For example,

$$\text{if } x = \frac{1}{3}, \quad \text{then} \quad y = 3 \cdot \frac{1}{3} + 1 = 1 + 1 = 2$$

$$\text{if } x = -\frac{3}{4}, \quad \text{then} \quad y = 3\left(-\frac{3}{4}\right) + 1 = -\frac{9}{4} + \frac{4}{4} = -\frac{5}{4}$$

$$\text{if } x = \sqrt{2}, \quad \text{then} \quad y = 3\sqrt{2} + 1$$

The important idea here is that even though we cannot actually substitute all real numbers for x (we do not have enough time or paper), there is a corresponding real value for y for any real value of x we choose. In Figure T.7 a few points that satisfy the equation $y = 3x + 1$ have been graphed so that you can observe a pattern.

x	$y = 3x + 1$
-2	$y = 3(-2) + 1 = -5$
-1	$y = 3(-1) + 1 = -2$
0	$y = 3(0) + 1 = 1$
$\frac{2}{3}$	$y = 3\left(\frac{2}{3}\right) + 1 = 3$
2	$y = 3(2) + 1 = 7$

Figure T.7

The points in Figure T.7 appear to lie on a straight line and, in fact, they do. We can draw a straight line through all the points, as shown in Figure T.8, and **any point that lies on the line will satisfy the equation.**

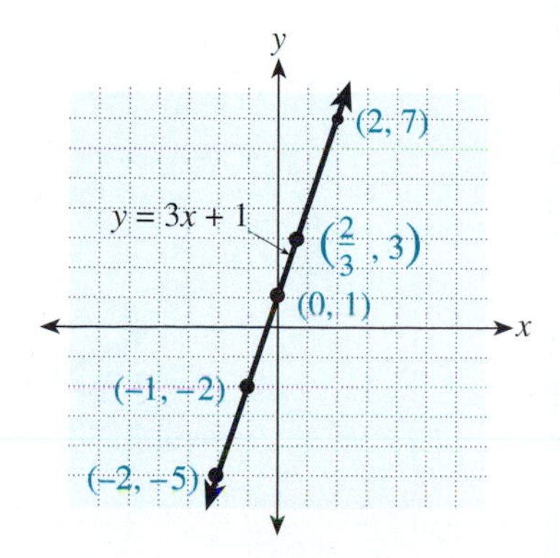

Figure T.8

The points (ordered pairs of real numbers) that satisfy any equation of the form

$$Ax + By = C \quad \text{where } A \text{ and } B \text{ are not both } 0$$

will lie on a straight line. The equation is a **linear equation in two variables** and is in the **standard form** for the equation of a line.

The linear equation $y = 3x + 1$ is solved for y and is not in standard form. However, it can be written in the standard form as

$$-3x + y = 1 \quad \text{or} \quad 3x - y = -1$$

All three forms are acceptable and correct.

EXAMPLE 1 Write the linear equation $y = -2x + 5$ in standard form.

Solution

By adding $2x$ to both sides we get the standard form:

$$2x + y = 5$$

Graphing Linear Equations

Since we know that the graph of a linear equation is a straight line, we need only graph two points (because two points determine a line) and draw the line through these two points. A good check against possible error is to locate three points instead of only two. Also, the values chosen for x should be such that the points are not too close together. You may choose any values for x (or y) that you like. Regardless of your choices, the points will all lie on the same line.

EXAMPLE 2 Draw the graph of the linear equation $x + 2y = 6$.

Solution

We find and plot three ordered pairs that satisfy the equation and then sketch the graph. (*Note:* These choices for x were arbitrary. Other choices will give points on the same line.)

$$x = -2 \qquad -2 + 2y = 6 \qquad 2y = 8 \qquad y = 4$$

$$x = 0 \qquad 0 + 2y = 6 \qquad 2y = 6 \qquad y = 3$$

$$x = 2 \qquad 2 + 2y = 6 \qquad 2y = 4 \qquad y = 2$$

x	y
−2	4
0	3
2	2

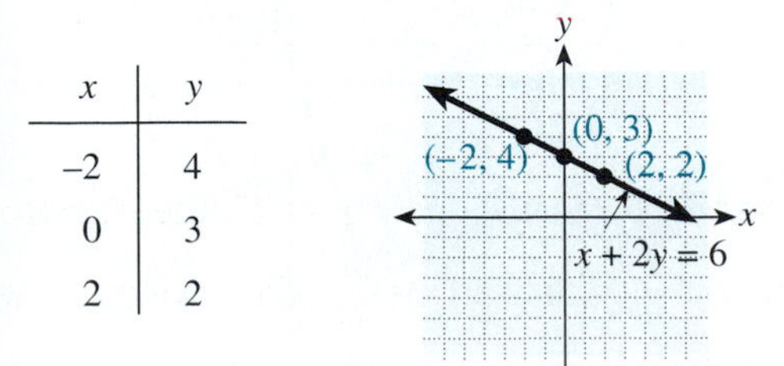

(Locating three points helps in avoiding errors. Avoid choosing points close together.)

The **y-intercept** of a line is the point where the line crosses the y-axis. This point can be located by letting $x = 0$. Similarly, the **x-intercept** is the point where the line crosses the x-axis and is found by letting $y = 0$. When the line is not vertical or horizontal, the y-intercept and the x-intercept are generally easy to locate and are frequently used for drawing the graph of a linear equation.

EXAMPLE 3

Draw the graph of the linear equation $x - 2y = 8$ by locating the y-intercept and the x-intercept.

Solution

Find the y-intercept:

$$x = 0$$
$$0 - 2y = 8$$
$$y = -4$$

Find the x-intercept:

$$y = 0$$
$$x - 2 \cdot 0 = 8$$
$$x = 8$$

x	y
0	−4
8	0

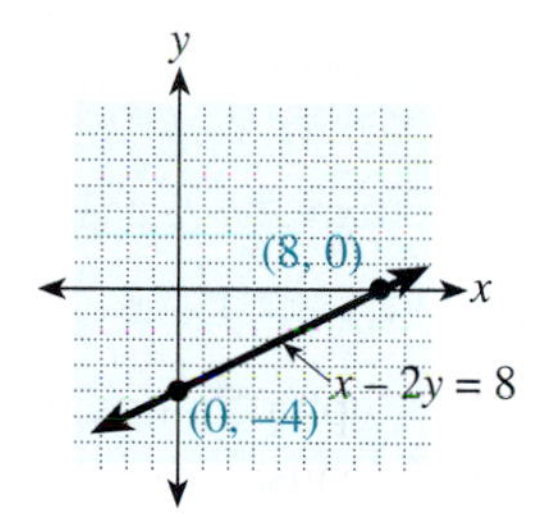

EXAMPLE 4

Locate the y-intercept and the x-intercept and draw the graph of the linear equation $3x - y = 3$.

Solution

Find the y-intercept:

$$x = 0$$
$$3 \cdot 0 - y = 3$$
$$y = -3$$

Find the x-intercept:

$$y = 0$$
$$3x - 0 = 3$$
$$x = 1$$

x	y
0	−3
1	0

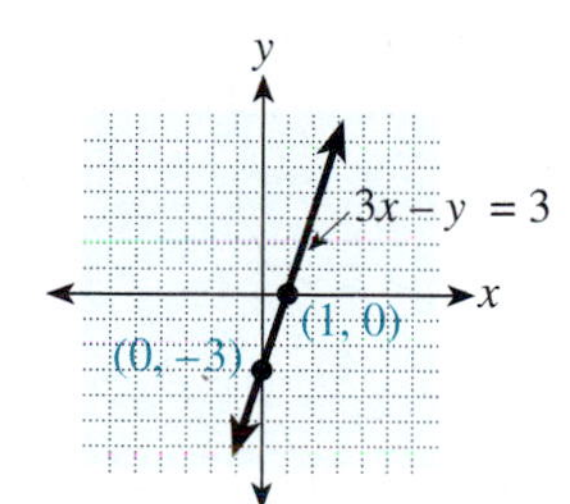

If $A = 0$ and $B \neq 0$ in the standard form $Ax + By = C$, the equation takes the form $By = C$ and can be solved for y as $y = \frac{C}{B}$. For example, we can write $0x + 3y = 6$ as $y = 2$. Thus, no matter what value x has, the value of y is 2. The graph of the equation $y = 2$ is a **horizontal line,** as shown in Figure T.9. For horizontal lines (other than the x-axis itself) there is no x-intercept.

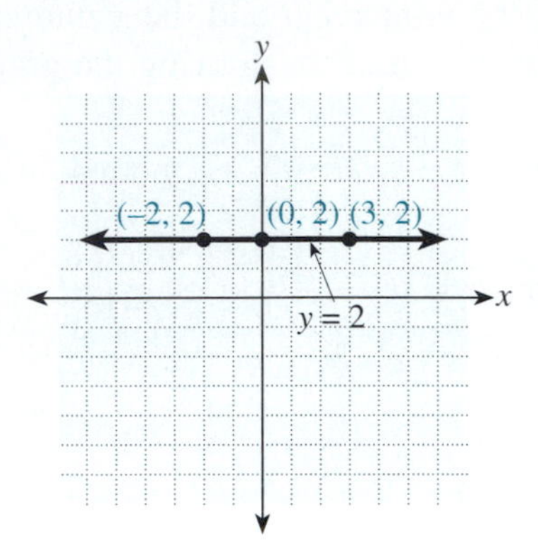

The y-coordinate is 2 for every point on the line $y = 2$.

Figure T.9

If $B = 0$ and $A \neq 0$ in the standard form $Ax + By = C$, the equation takes the form $Ax = C$ or $x = \frac{C}{A}$. So, we can write $5x + 0y = -5$ as $x = -1$. Thus, no matter what value y has, the value of x is -1. The graph of the equation $x = -1$ is a **vertical line,** as shown in Figure T.10. For vertical lines (other than the y-axis itself) there is no y-intercept.

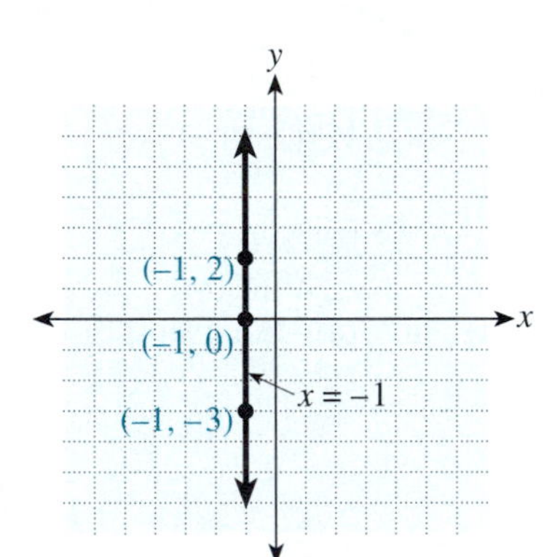

The x-coordinate is −1 for every point on the line $x = -1$.

Figure T.10

EXAMPLE 5

Graph the horizontal line $y = -1$ and the vertical line $x = 3$ on the same coordinate system.

Solution

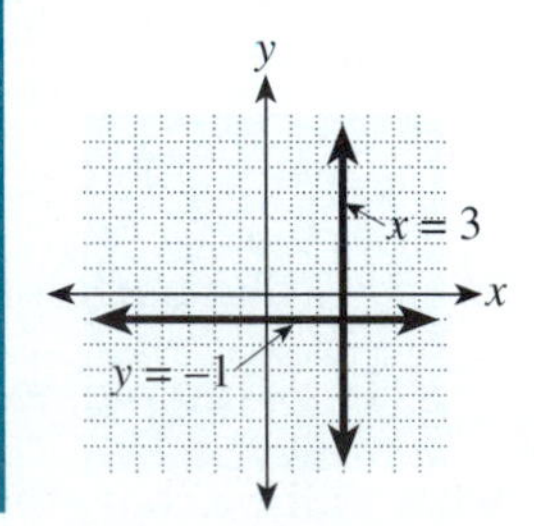

NAME ______________________ SECTION ______________ DATE ________

Exercises T.3

Graph the following linear equations. Label at least three points on each line.

1. $y = x + 1$

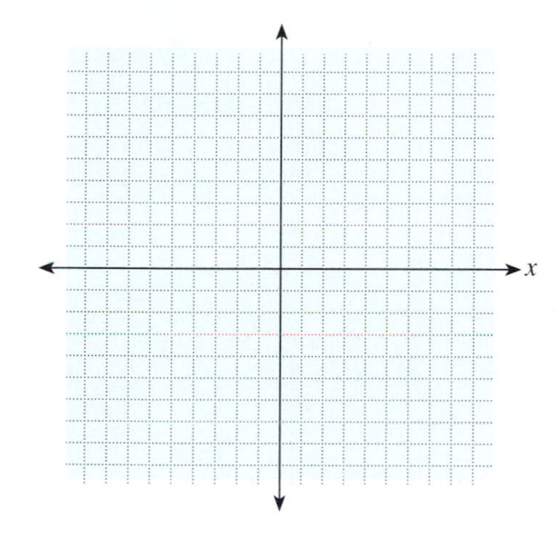

2. $y = x + 3$

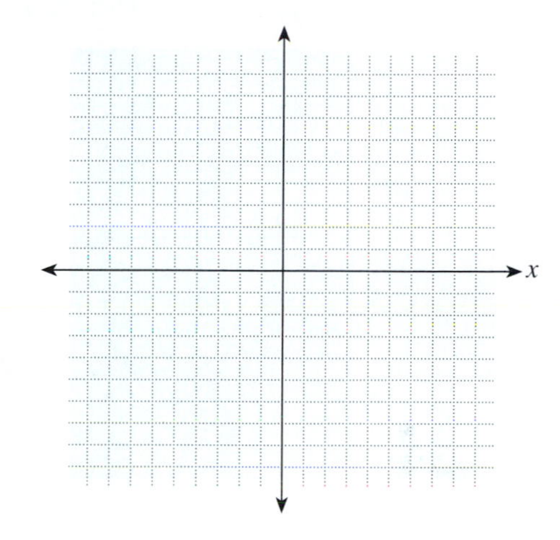

3. $y = x - 6$

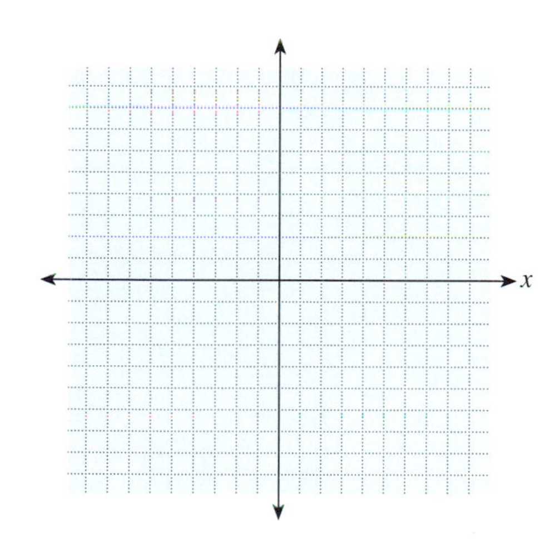

4. $y = x - 2$

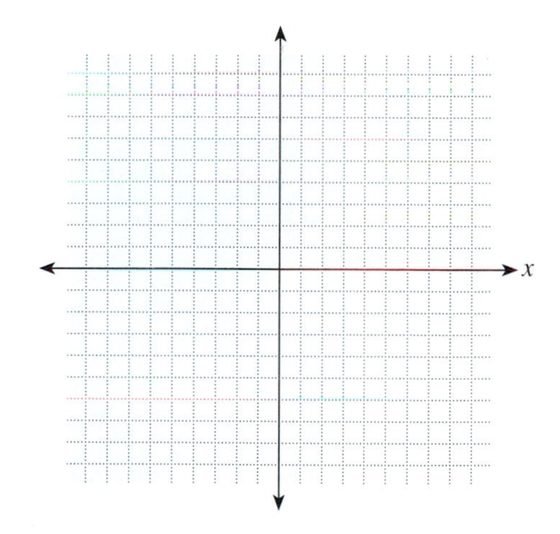

5. $y = 2x$

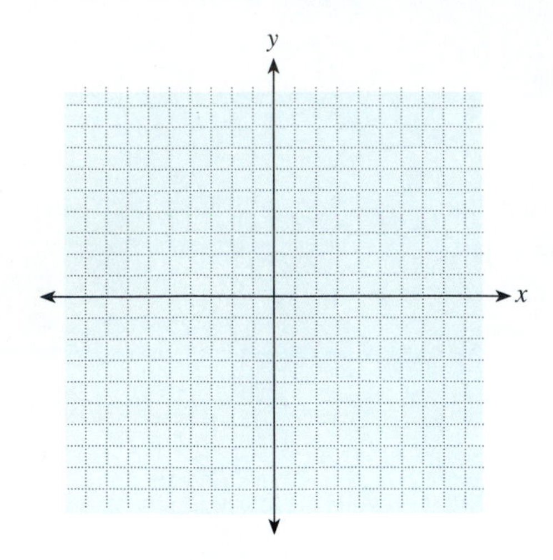

6. $y = 4x$

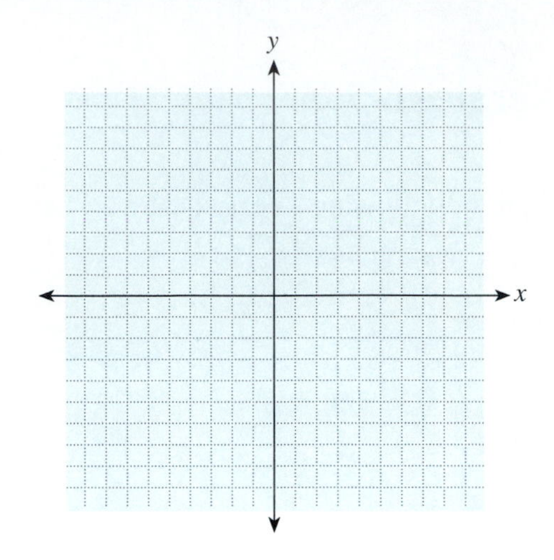

7. $y = -x$

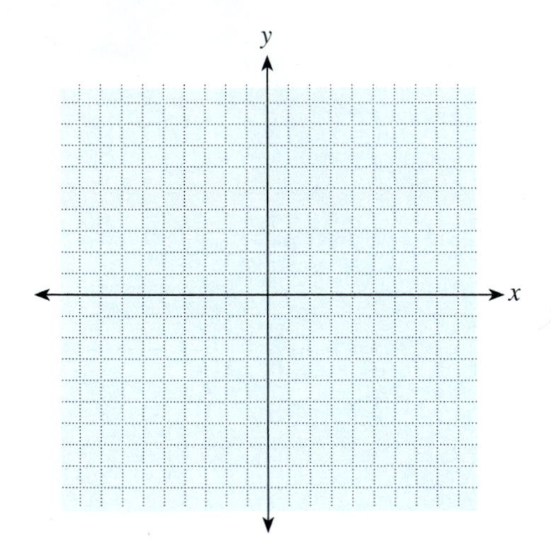

8. $y = -3x$

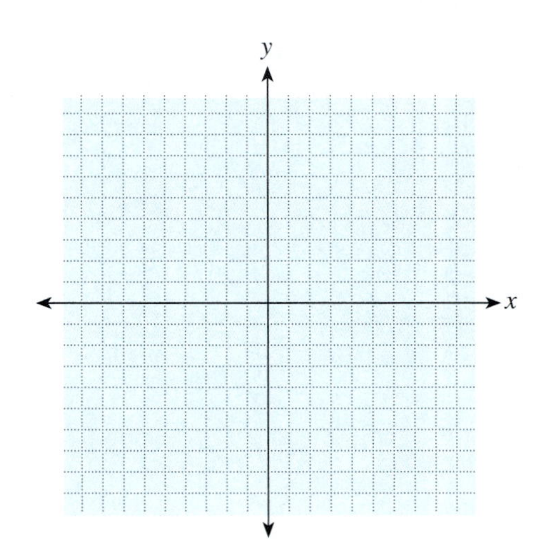

9. $y = x$

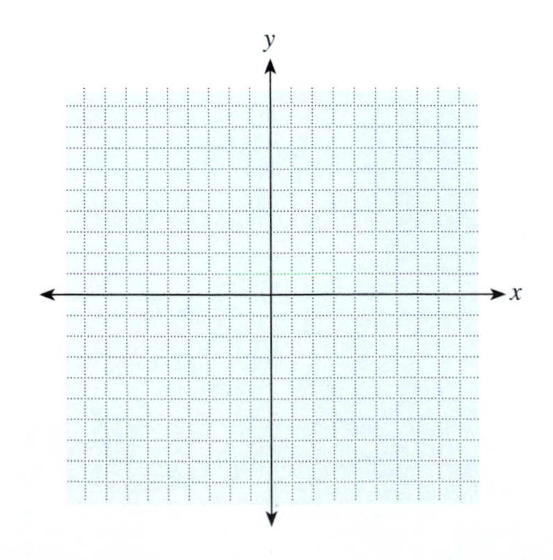

10. $y = 2 - x$

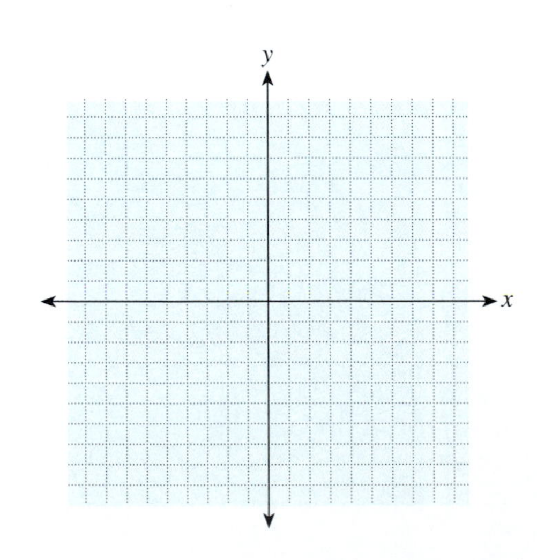

NAME ______________________ SECTION ______________ DATE ________

11. $y = 4 - x$

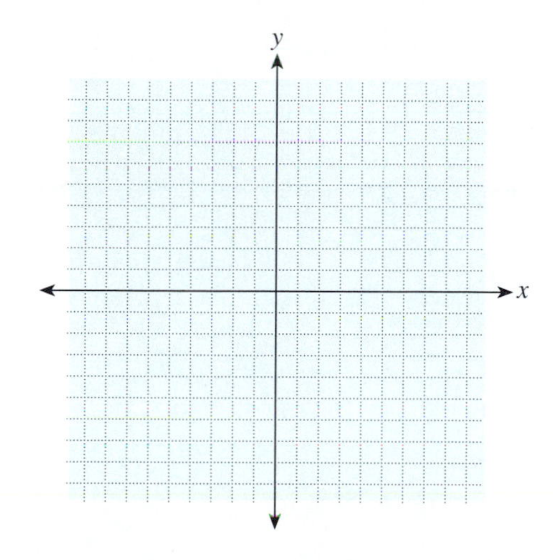

12. $y = 7 - x$

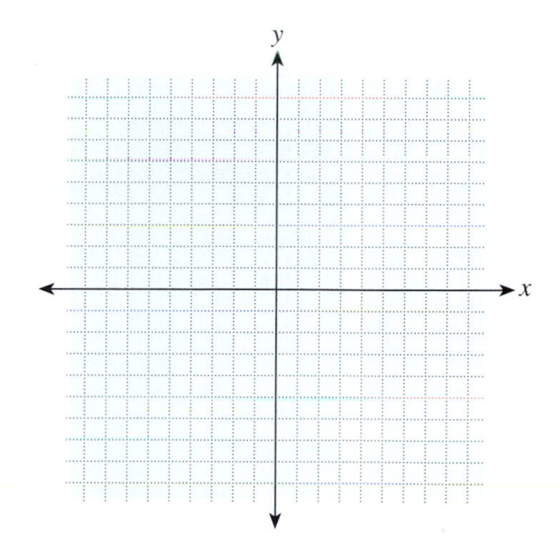

13. $y = 2x + 1$

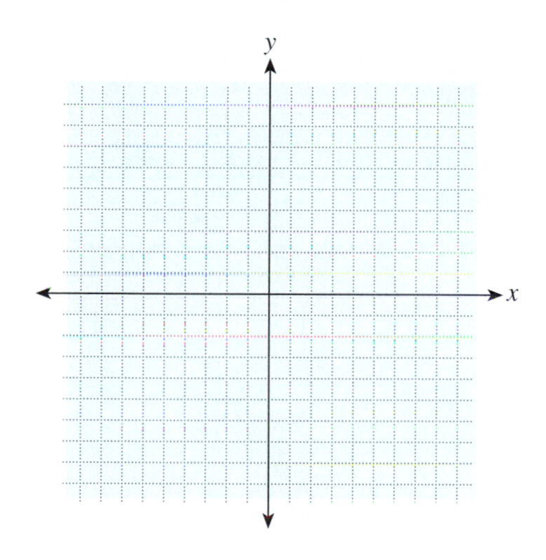

14. $y = 2x - 2$

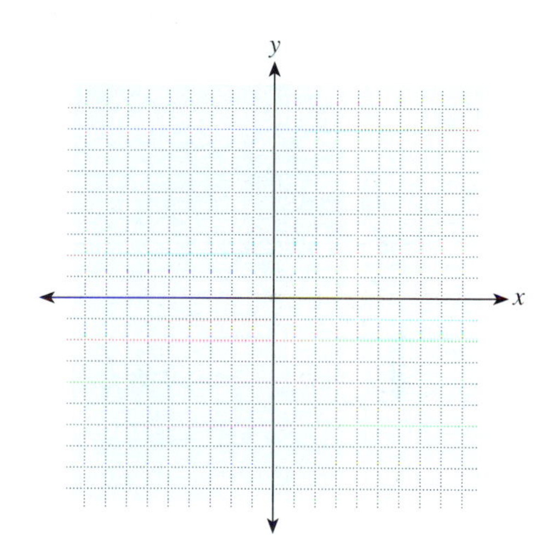

15. $y = 2x - 5$

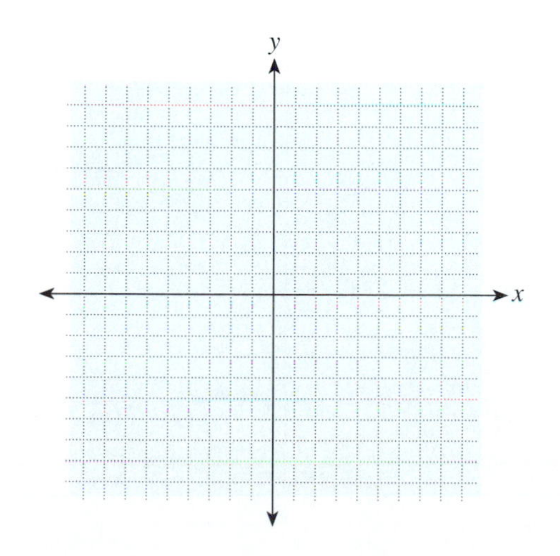

16. $y = 2x + 3$

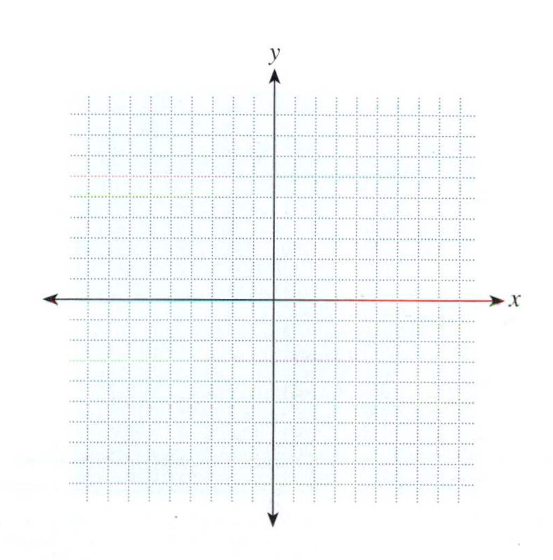

17. $y = -2x + 4$

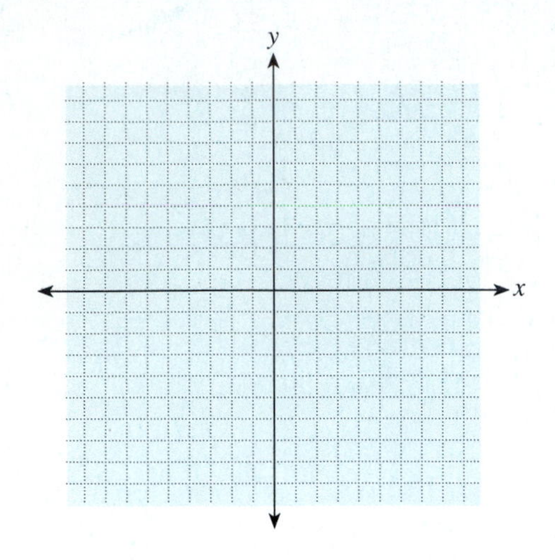

18. $y = -2x - 1$

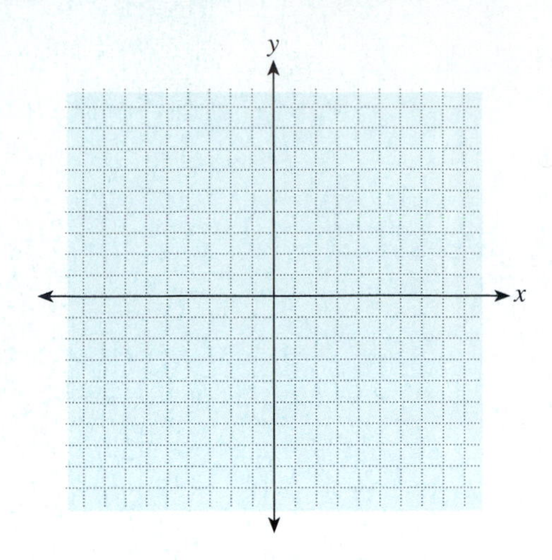

19. $y = -3x + 3$

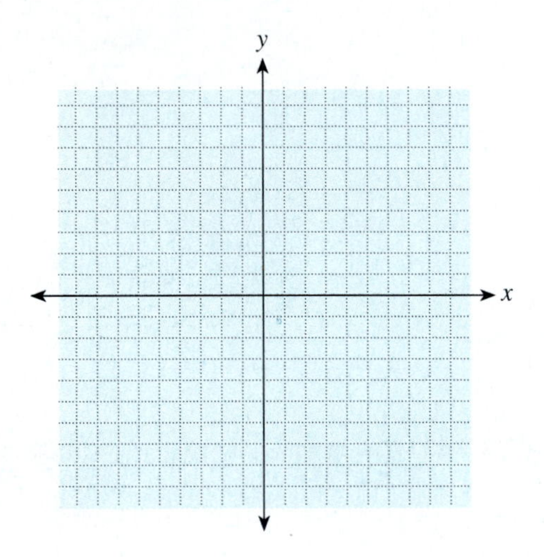

20. $y = -3x - 5$

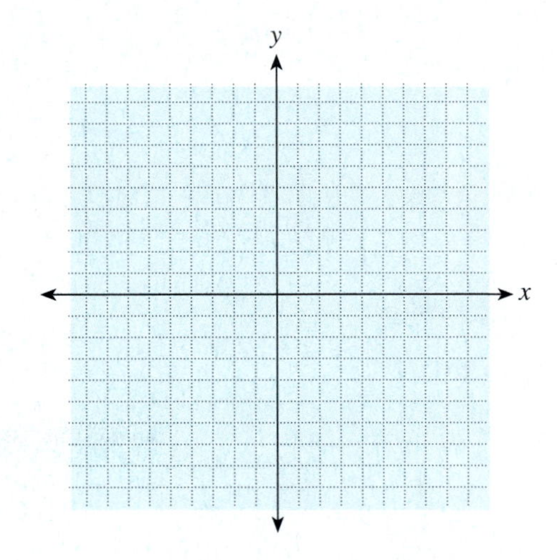

21. $x - 2y = 4$

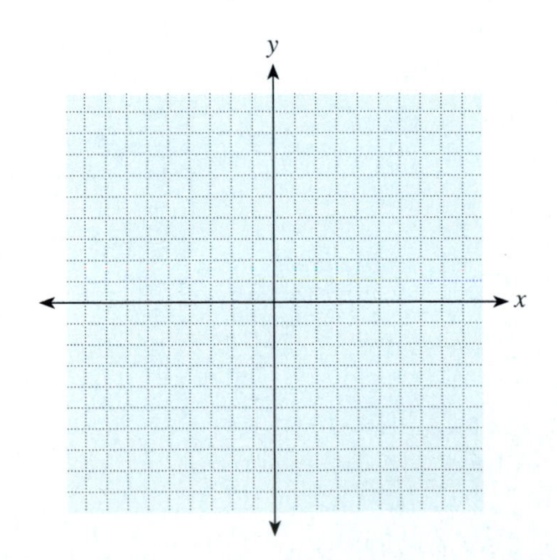

22. $x - 4y = 12$

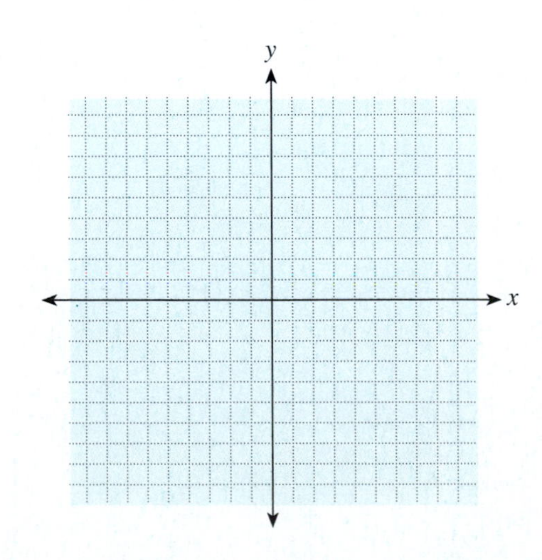

NAME ____________________ SECTION ____________________ DATE ____________

23. $-2x + 3y = 6$

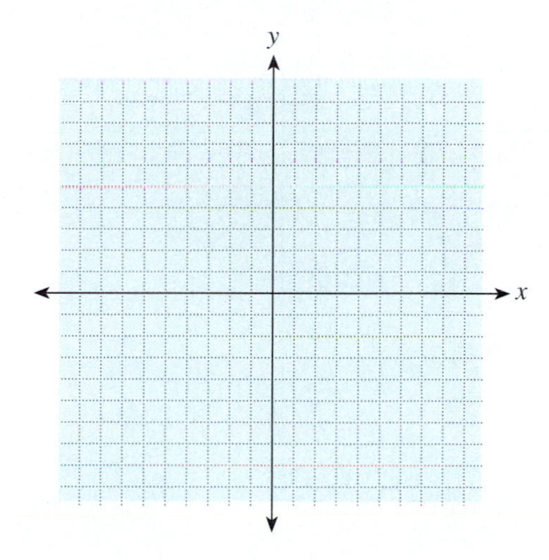

24. $2x - 5y = 10$

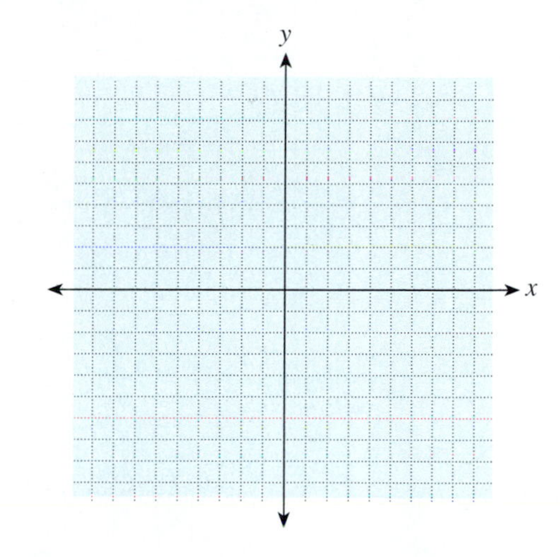

25. $-2x + y = 4$

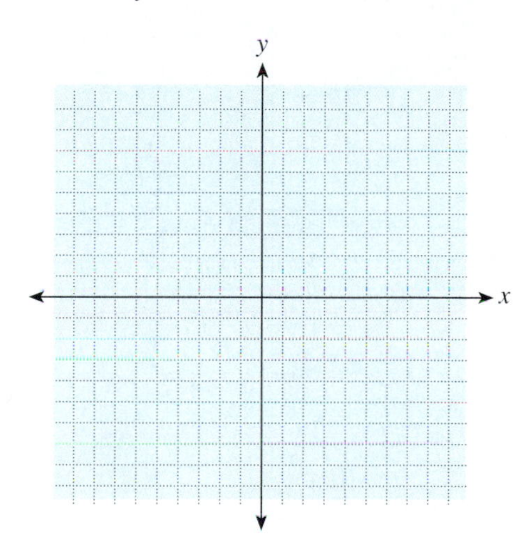

26. $3x + 4y = 12$

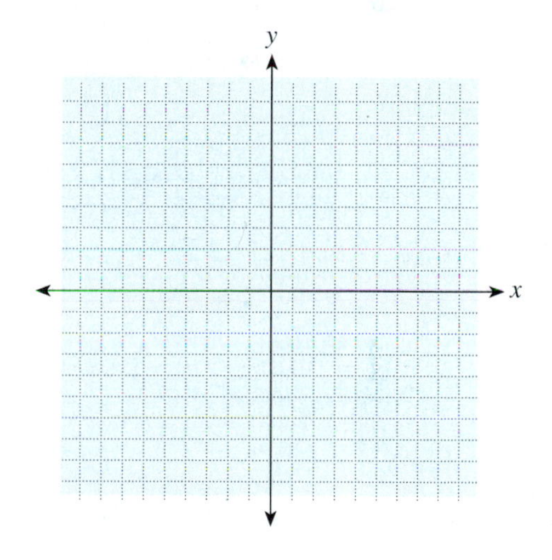

27. $3x + 5y = 15$

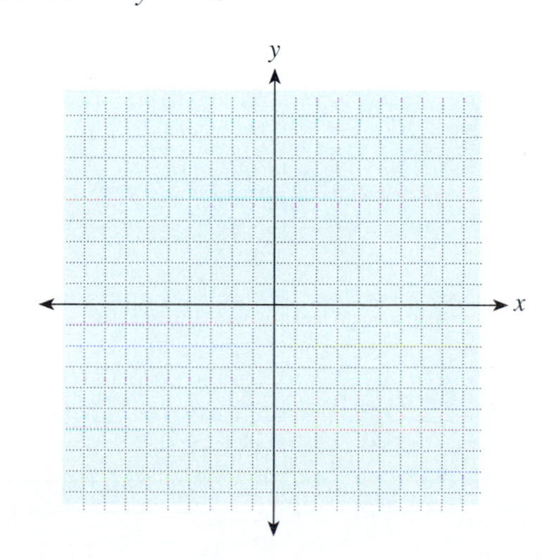

28. $4x + y = 8$

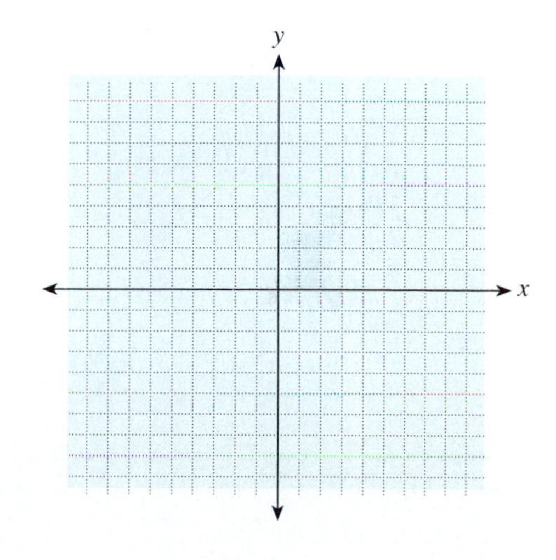

29. $x + 4y = 8$

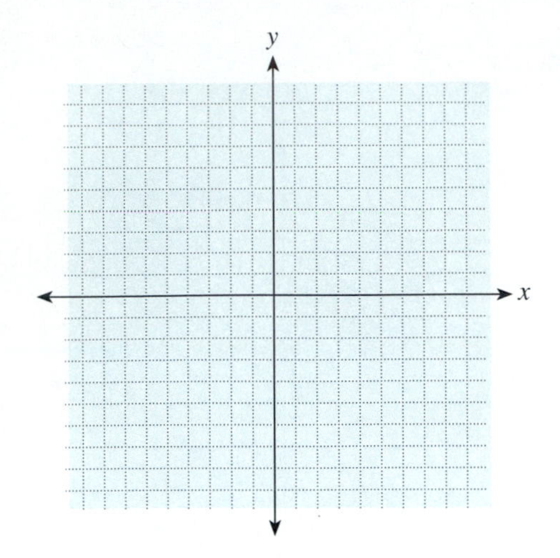

30. $2x - 3y = 6$

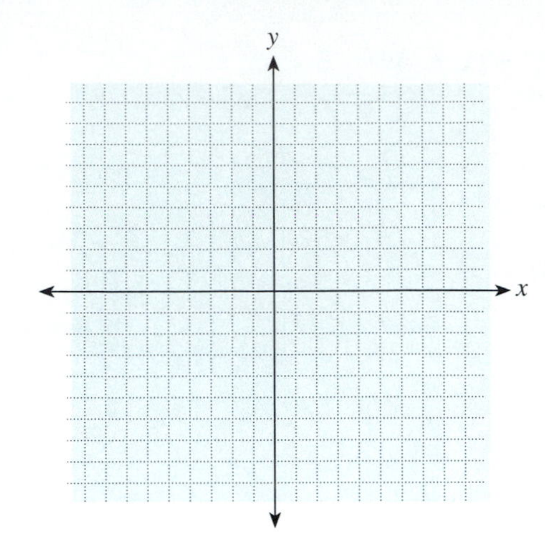

31. $x = 3$

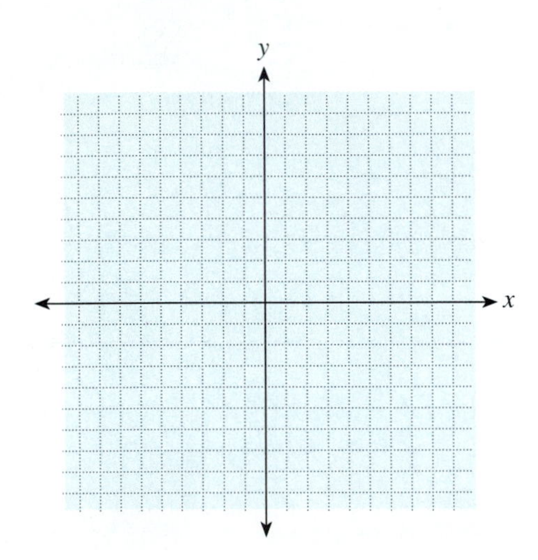

32. $y = 7$

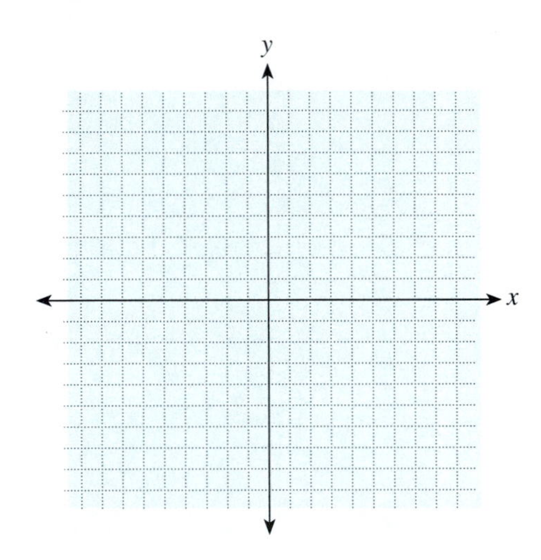

33. $y = -3$

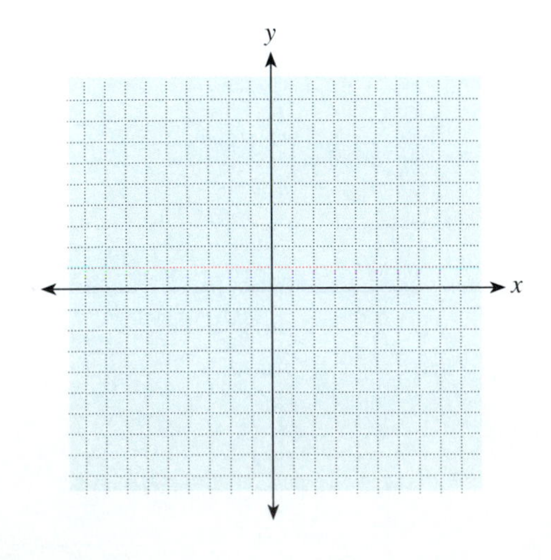

34. $x = -6$

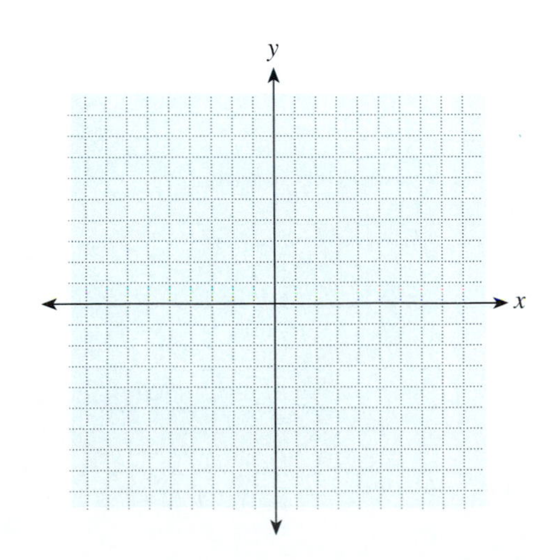

NAME ______________________ SECTION ________________ DATE __________

35. $x - 4 = 0$

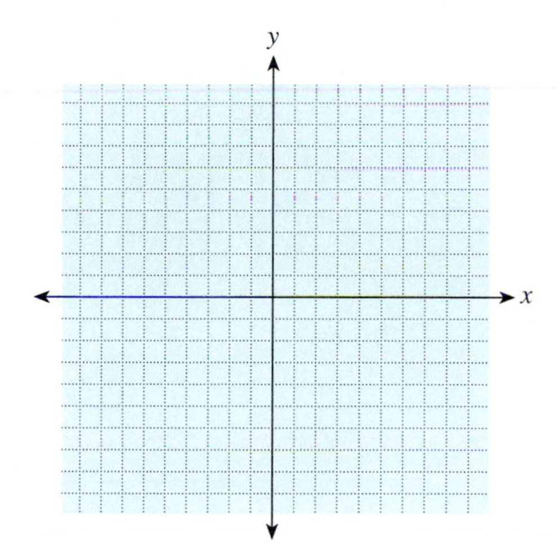

36. $y + 2 = 0$

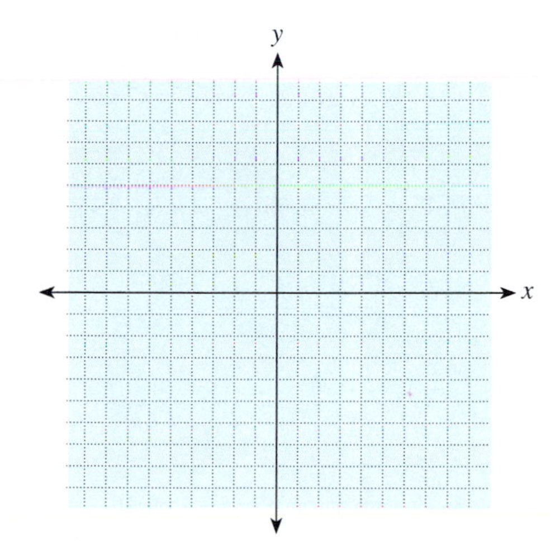

37. $2y - 5 = 0$

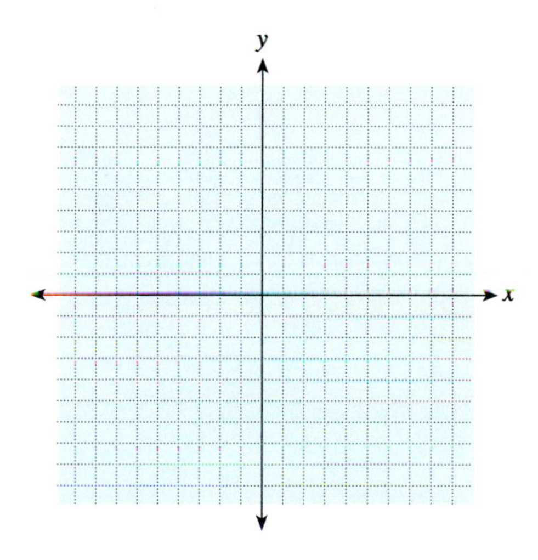

38. $3x = -8$

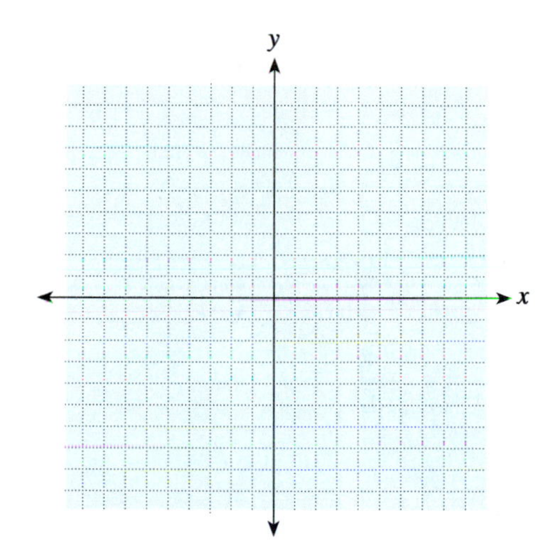

39. $4x - 7 = 0$

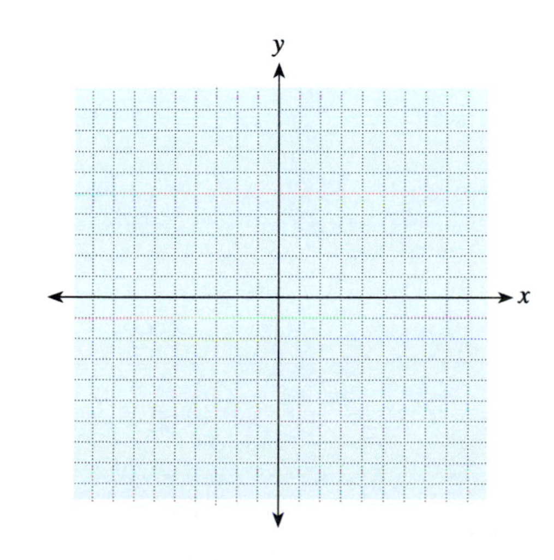

40. $3y + 8 = 0$

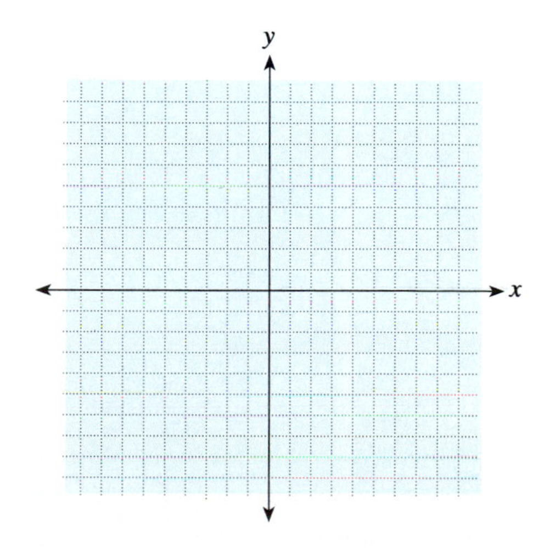

T.4 Addition and Subtraction with Polynomials

OBJECTIVES

1. *Understand how to classify polynomials.*
2. *Learn how to add polynomials.*
3. *Learn how to subtract polynomials.*

Classification of Polynomials

An **algebraic term** (discussed previously in Section 9.5) is an expression that involves only multiplication and/or division with constants and/or variables. **Like terms** (or **similar terms**) are terms that contain the same variable(s) to the same power or are terms that are constants. The **distributive property** can be used to combine like terms. For example,

$$7x + 5x = (7 + 5)x = 12x$$

and

$$15x^2 - 20x^2 = (15 - 20)x^2 = -5x^2$$

The distributive property is restated here for your reference and convenience.

Distributive Property of Multiplication over Addition

For real numbers a, b, and c,

$$a(b + c) = ab + ac$$

A **monomial** is a single term with no variable in a denominator and with variables that have only whole number exponents.

The general form of a **monomial in x** is

kx^n, where n is a whole number and k is any real number.

n is called the **degree** of the monomial, and k is the **coefficient.**

Consider a constant monomial such as 7. We can write

$$7 = 7 \cdot 1 = 7x^0.$$

With this reasoning, we say that a nonzero constant is a **monomial of 0 degree.** The constant 0 is a special case. Since we can write

$$0 = 0x^2 = 0x^4 = 0x^{63}$$

(in effect, any exponent can be used on x), we say that **0 is a monomial of no degree.**

A **polynomial** is a monomial or the sum of monomials. The **degree of a polynomial** is the largest of the degrees of its terms after all like terms have been combined. Generally, for easy reading, a polynomial is written so that the degrees of its terms either decrease from left to right or increase from left to right. If the degrees decrease, we say that the terms are written in **descending order.** If the degrees increase, we say that the terms are written in **ascending order.** For example,

$3x^4 + 5x^2 - 8x + 34$ is a fourth-degree polynomial written in descending order.

$15 - 3x + 4x^2 + x^3$ is a third-degree polynomial written in ascending order.

For consistency and because of the style used in multiplying polynomials, the polynomials in this text will be written in descending order.

Note: Although polynomials may have more than one variable, in this text we will limit our discussion to polynomials in only one variable.

Some forms of polynomials are used so frequently that they have been given special names, as indicated in the following box.

Classification of Polynomials		Example
Monomial:	polynomial with one term	$-2x^3$
Binomial:	polynomial with two terms	$7x + 23$
Trinomial:	polynomial with three terms	$a^2 + 5a + 6$

Example 1

Simplify each of the following polynomials and tell its degree and what type of polynomial it is.

(a) $4x^2 - 7x^2$

(b) $x^3 + 8x - 9 - x^3$

Solution

(a) $4x^2 - 7x^2 = (4 - 7)x^2 = -3x^2$ Second-degree monomial

(b) $x^3 + 8x - 9 - x^3 = (1 - 1)x^3 + 8x - 9$

$= 0x^3 + 8x - 9$

$= 8x - 9$ First-degree binomial

Addition with Polynomials

The **sum** of two or more polynomials is found by combining like terms. The polynomials may be written horizontally (as in Example 2) or vertically with like terms aligned (as in Example 3).

EXAMPLE 2

Add the polynomials as indicated.

$$\begin{aligned}(4x^3 - 5x^2 + 15x - 10) &+ (-7x^2 - 2x + 3) + (4x^2 + 9)\\ &= 4x^3 + (-5x^2 - 7x^2 + 4x^2) + (15x - 2x)\\ &\quad + (-10 + 3 + 9)\\ &= 4x^3 - 8x^2 + 13x + 2\end{aligned}$$

EXAMPLE 3

Find the sum.

$$\begin{array}{r} x^3 - 4x^2 + 2x + 13 \\ \underline{2x^3 + x^2 - 5x + 6} \\ 3x^3 - 3x^2 - 3x + 19 \end{array}$$

COMPLETION EXAMPLE 4

Write the sum

$$(5x^3 - 8x^2 - 10x + 2) + (3x^3 + 4x^2 - x - 7)$$

in the vertical format and find the sum.

Solution

$$5x^3 - 8x^2 - 10x + 2$$

____

____

Now Work Exercises 1 and 2 in the Margin.

Subtraction with Polynomials

The opposite of a polynomial can be indicated by writing a negative sign in front of the polynomial. In this case the sign of every term in the polynomial is changed. For example,

$$-(5x^2 - 6x - 1) = -5x^2 + 6x + 1$$

We can also think of this type of expression as indicating multiplication by -1 and use the distributive property as follows:

$$\begin{aligned}-(5x^2 - 6x - 1) &= -1(5x^2 - 6x - 1)\\ &= -1(5x^2) - 1(-6x) - 1(-1)\\ &= -5x^2 + 6x + 1\end{aligned}$$

Find the indicated sums.

1. $(2x^2 - 5x + 3) + (x^2 - 7) + (2x + 10)$

2. $$\begin{array}{r} x^4 - 5x^3 + 7x^2 + 12 \\ \underline{3x^4 - 6x^3 - 9x^2 - 2x - 14} \end{array}$$

Find the indicated differences.

3. $$\begin{array}{r} (-2x^3 + 5x^2 + 8x - 1) \\ -(2x^3 - x^2 - 6x + 13) \\ \hline \end{array}$$

4. $$\begin{array}{r} 15x^3 + 10x^2 - 17x - 25 \\ -(12x^3 + 10x^2 + 3x - 16) \\ \hline \end{array}$$

The answer is the same in either case. Therefore, the **difference** of two polynomials can be found by **adding the opposite** of the polynomial being subtracted. The polynomials can be written horizontally (as in Example 5) or vertically with like terms aligned (as in Example 6).

Example 5

Subtract the polynomials as indicated.

$$\begin{aligned} &(9x^3 + 4x^2 - 15) - (-2x^3 + x^2 - 5x - 6) \\ &= 9x^3 + 4x^2 - 15 + 2x^3 - x^2 + 5x + 6 \\ &= 11x^3 + 3x^2 + 5x - 9 \end{aligned}$$

When using the vertical alignment format, write a 0 as a placeholder for any missing powers of the variable in order to maintain proper alignment and ensure that like terms will be subtracted (or added).

Example 6

Find the difference.

$$\begin{array}{r} 8x^4 + 2x^3 - 5x^2 + 0x - 7 \\ -(3x^4 + 5x^3 - x^2 + 6x - 11) \\ \hline \end{array}$$

Be sure to change the sign of each term in the polynomial being subtracted and then combine like terms.

$$\begin{array}{rl} 8x^4 + 2x^3 - 5x^2 + 0x - 7 & \\ -3x^4 - 5x^3 + x^2 - 6x + 11 & \leftarrow \text{signs changed} \\ \hline 5x^4 - 3x^3 - 4x^2 - 6x + 4 & \leftarrow \text{difference} \end{array}$$

Now Work Exercises 3 and 4 in the Margin.

Completion Example Answer

4. $$\begin{array}{r} 5x^3 - 8x^2 - 10x + 2 \\ \mathbf{3x^3 + 4x^2 - x - 7} \\ \hline \mathbf{8x^3 - 4x^2 - 11x - 5} \end{array}$$

NAME ______________ SECTION ______________ DATE ______________

ANSWERS

Exercises T.4

Simplify each polynomial and write the simplified expression in the answer blank. Then, below each exercise tell what type of polynomial it is and its degree.

1. $5x + 6x - 10 + 3$

2. $8x - 9x + 14 - 5$

3. $3x^2 - x^2 + 7x - x + 2$

4. $4x^2 + 3x^2 - x + 2x + 18$

5. $a^3 + 4a^2 + a^2 - a^3$

6. $y^4 - 2y^3 + 3y^3 - y^4$

7. $-2y^2 - y^2 + 10y - 3y + 2 + 5$

8. $-5a^2 + 2a^2 - 4a - 2a + 4 + 1$

9. $5x^3 + 2x - 8x + 17 + 3x$

10. $-4x^3 + 5x^2 - 3x^2 + 12x - x$

Add or subtract as indicated and simplify if possible.

11. $(3x - 5) + (2x - 5)$

12. $(7x + 8) + (-3x + 8)$

13. $(x^2 + 4x - 6) + (x^2 - 4x + 2)$

14. $(x^2 - 3x - 10) + (2x^2 - 3x - 10)$

1. ______________
2. ______________
3. ______________
4. ______________
5. ______________
6. ______________
7. ______________
8. ______________
9. ______________
10. ______________
11. ______________
12. ______________
13. ______________
14. ______________

15. ________

16. ________

17. ________

18. ________

19. ________

20. ________

21. ________

22. ________

23. ________

24. ________

25. ________

26. ________

15. $(4y^3 + 2y - 7) + (3y^2 - 2)$

16. $(-2y^3 + y^2 - 4) + (3y^2 - 6)$

17. $(8x + 3) - (7x + 5)$

18. $(4x - 9) - (5x + 2)$

19. $(2a^2 + 3a - 1) - (a^2 - 2a - 1)$

20. $(a^2 - 5a + 3) - (2a^2 + 5a - 3)$

21. $(9x^3 + x^2 - x) - (-3x^3 + 5x)$

22. $(4x^3 - 9x + 11) - (-x^3 + 2x + 1)$

Add in Exercises 22–26.

23. $$\begin{array}{r} x^2 + 5x - 7 \\ \underline{-3x^2 + 2x - 1} \end{array}$$

24. $$\begin{array}{r} 2x^2 + 4x - 6 \\ \underline{3x^2 - 9x + 2} \end{array}$$

25. $$\begin{array}{r} x^3 + 2x^2 + x \\ \underline{2x^3 - 2x^2 - 2x + 6} \end{array}$$

26. $$\begin{array}{r} x^3 + 6x^2 + 7x - 8 \\ \underline{7x^2 + 2x + 1} \end{array}$$

NAME ______________ SECTION ______________ DATE ______________

ANSWERS

Subtract in Exercises 27–30.

27. $\begin{array}{r} 9x^2 + 3x - 2 \\ -(4x^2 + 5x + 3) \\ \hline \end{array}$

28. $\begin{array}{r} -3x^2 + 6x - 7 \\ -(\ 2x^2 - \ x + 7) \\ \hline \end{array}$

29. $\begin{array}{r} 4x^3 \qquad\quad + 10x - 15 \\ -(\ x^3 - 5x^2 - \ 3x - \ 9) \\ \hline \end{array}$

30. $\begin{array}{r} x^3 - 8x^2 + 11x + 6 \\ -(-3x^3 + 8x^2 - \ 2x + 6) \\ \hline \end{array}$

27. ______________

28. ______________

29. ______________

30. ______________

31. [Respond below exercise.] ______________

Writing and Thinking about Mathematics

31. (a) Find the sum of these two polynomials:

$\begin{array}{r} 2x^3 - 4x^2 + 3x + 20 \\ 3x^3 + \ x^2 - 5x + 10 \\ \hline \end{array}$

(b) Now substitute 1 for x in each of the polynomials in part (a), including the sum. Does the sum of the first two values equal the value of the sum?

(c) Repeat the process in part (b) using $x = 3$.

(d) Substituting a value of x in each polynomial seems to provide a method for checking answers. Do you think that this method will catch all errors? Would substituting $x = 0$ be a good idea? Briefly discuss your reasoning.

[Respond below
32. exercise.] ______

32. (a) Find the difference of these two polynomials:

$$\begin{array}{r} 2x^3 - 4x^2 + 3x + 20 \\ -(3x^3 + x^2 - 5x + 10) \\ \hline \end{array}$$

(b) Now, substitute 2 for x in each of the polynomials in part (a), including the difference. Does the difference of the first two values equal the value of the difference?

(c) Repeat the process in part (b) using $x = 3$.

(d) Substituting a value for x in each polynomial seems to provide a method for checking answers. Do you think that this method will catch all errors? Would substituting $x = 0$ be a good idea? Briefly discuss your reasoning.

T.5 Multiplication with Polynomials

OBJECTIVES

1. *Learn how to use the distributive property to multiply a monomial and a polynomial with two or more terms.*
2. *Learn how to multiply two binomials.*
3. *Know how to use a vertical format to multiply two polynomials.*

Review of Exponents and the Product Rule for Exponents

Exponents have been discussed in several sections and used in conjunction with a variety of topics throughout the text. Whole-number exponents were introduced in Chapter 2, where we defined

$$\underbrace{a \cdot a \cdot a \cdot \ldots \cdot a}_{n \text{ factors}} = a^n \quad \text{and} \quad a = a^1$$

We also stated that for $a \neq 0$, $a^0 = 1$. (0^0 is undefined.)

Expressions with exponents can be evaluated with a calculator that has a key marked $\boxed{x^y}$ (or a similar symbol). In Chapter 5, we noted that calculators use scientific notation for very large and very small decimal numbers, and scientific notation involves both positive and negative integer exponents. For example,

$$3{,}200{,}000 = 3.2 \times 10^6 \quad \text{and} \quad 0.000047 = 4.7 \times 10^{-5}.$$

In Section 8.2, we again used exponents to find compound interest with the formula $A = P\left(1 + \frac{r}{n}\right)^{nt}$.

One of the basic rules (or properties) of exponents deals with multiplying powers that have the same base. For example,

$$a^2 \cdot a^4 = \underbrace{(a \cdot a)}_{2 \text{ factors}} \cdot \underbrace{(a \cdot a \cdot a \cdot a)}_{4 \text{ factors}} = \underbrace{a \cdot a \cdot a \cdot a \cdot a \cdot a}_{6 \text{ factors}} = a^6$$

or

$$a^2 \cdot a^4 = a^{2+4} = a^6$$

The property of exponents illustrated by this example is called the **Product Rule for Exponents.**

Product Rule for Exponents

For any real number a and whole number m and n,

$$a^m \cdot a^n = a^{m+n}$$

(To multiply two powers with the same base, keep the base and add the exponents.)

Note: The Product Rule for Exponents is stated here only for whole number exponents because those are the only exponents we use with polynomials. However, this rule and many other rules for exponents are valid for negative integer exponents and fractional exponents as well. You will study these ideas in detail in later courses in mathematics.

Use the Product Rule for Exponents to simplify each of the following expressions.

1. $8^2 \cdot 8$

2. $(-2)^3 \cdot (-2)^4$

3. $-5x^3 \cdot 2x^3$

Example 1

Use the Product Rule for Exponents to simplify each of the following expressions.

(a) $7^2 \cdot 7^3$ (b) $x^3 \cdot x^5$ (c) $3y \cdot 5y^9$

Solution

Use the Product Rule in each case.

(a) $7^2 \cdot 7^3 = 7^{2+3} = 7^5$ (or 16,807)

(b) $x^3 \cdot x^5 = x^{3+5} = x^8$

(c) $3y \cdot 5y^9 = (3 \cdot 5)(y^1 \cdot y^9) = 15 \cdot y^{1+9} = 15y^{10}$

Now Work Exercises 1–3 in the Margin.

Multiplication with Polynomials

We can find **the product of a monomial and a polynomial of two or more terms** by using the distributive property and the Product Rule for Exponents as follows:

$$6x(3x + 5) = 6x \cdot 3x + 6x \cdot 5 = 18x^2 + 30x$$

$$\begin{aligned} 4x^2(3x^2 - 5x + 2) &= 4x^2(3x^2) + 4x^2(-5x) + 4x^2(+2) \\ &= 12x^4 - 20x^3 + 8x^2 \end{aligned}$$

$$\begin{aligned} -2a^3(5a^2 + 3a - 7) &= (-2a^3)(5a^2) + (-2a^3)(+3a) + (-2a^3)(-7) \\ &= -10a^5 - 6a^4 + 14a^3 \end{aligned}$$

Now, to **multiply two binomials,** such as $(x + 5)(x + 8)$, we can apply the distributive property three times. For the first step we treat the binomial $(x + 8)$ as a single term and multiply on the right as follows:

$$(b + c)a = ba + ca$$

$$(x + 5)(x + 8) = x(x + 8) + 5(x + 8)$$

The product is completed by applying the distributive property twice more and combining like terms:

$$\begin{aligned} (x + 5)(x + 8) &= x(x + 8) + 5(x + 8) \\ &= x \cdot x + x \cdot 8 + 5 \cdot x + 5 \cdot 8 \\ &= x^2 + 8x + 5x + 40 \\ &= x^2 + 13x + 40 \end{aligned}$$

Example 2

Find each product.

(a) $3a(2a^2 + 3a - 4)$ (b) $-2x^3(7x^2 + 4x - 9)$

Solution

(a) $\begin{aligned} 3a(2a^2 + 3a - 4) &= 3a(2a^2) + 3a(3a) + 3a(-4) \\ &= 6a^3 + 9a^2 - 12a \end{aligned}$

(b) $\begin{aligned} -2x^3(7x^2 + 4x - 9) &= -2x^3(7x^2) + (-2x^3)(+4x) + (-2x^3)(-9) \\ &= -14x^5 - 8x^4 + 18x^3 \end{aligned}$

Example 3

Find each product.

(a) $(x + 5)(x - 10)$

(b) $(2x - 3)(2x - 7)$

Solution

(a) $(x + 5)(x - 10) = x(x - 10) + 5(x - 10)$
$= x^2 - 10x + 5x - 50$
$= x^2 - 5x - 50$

(b) $(2x - 3)(2x - 7) = 2x(2x - 7) - 3(2x - 7)$
$= 4x^2 - 14x - 6x + 21$
$= 4x^2 - 20x + 21$

Now Work Exercises 4–6 in the Margin.

Another technique for finding the product of two polynomials is outlined as follows:

Multiplying Polynomials Vertically

1. Write the polynomials in a vertical format with one polynomial directly below the other.
2. Multiply each term of the top polynomial by each term in the bottom polynomial and align like terms.
3. Combine like terms.

This technique is illustrated in Examples 4 and 5.

Example 4

Find the product $(2x + 3)(3x^2 + 4x - 5)$.

Solution

Arrange the polynomials in a vertical format and multiply each term in the top polynomial by $2x$.

$$\begin{array}{r} 3x^2 + 4x - 5 \\ 2x + 3 \\ \hline 6x^3 + 8x^2 - 10x \end{array}$$

Next multiply each term in the top polynomial by 3 and align like terms.

$$\begin{array}{rrrr} & 3x^2 & + 4x & - 5 \\ & & 2x & + 3 \\ \hline 6x^3 & + 8x^2 & - 10x & \\ & 9x^2 & + 12x & - 15 \\ \hline \end{array}$$

Find each indicated product and simplify.

4. $5a(3a^2 + 9a - 2)$

5. $7x^2(-2x^3 + 8x^2 + 3x - 5)$

6. $(x - 9)(2x + 1)$

Multiply as indicated and simplify.

7. $(2x + 5)(x^2 - 8x + 1)$

8. (**Note:** Be sure to align like terms as you multiply.)

$$\begin{array}{l} 2x^3 + 4x^2 - 7 \\ \underline{5x + 6} \end{array}$$

Now combine like terms to find the product.

$$\begin{array}{rrrrl} 3x^2 & + 4x & - 5 & & \\ 2x & + 3 & & & \\ \hline 6x^3 + & 8x^2 & - 10x & & \\ & 9x^2 & + 12x & - 15 & \\ \hline 6x^3 + & 17x^2 & + 2x & - 15 & \text{product} \end{array}$$

EXAMPLE 5

Multiply: $(3x - 1)(x^3 - 5x^2 + 2x + 6)$.

Solution

$$\begin{array}{rrrrr} x^3 & - 5x^2 & + 2x & + 6 & \\ 3x & - 1 & & & \\ \hline 3x^4 - 15x^3 & + 6x^2 & + 18x & & \\ - x^3 & + 5x^2 & - 2x & - 6 & \\ \hline 3x^4 - 16x^3 & + 11x^2 & + 16x & - 6 & \end{array}$$

✓ COMPLETION EXAMPLE 6

Find the product: $(x^2 + x + 2)(x^2 + 3x - 4)$.

Solution

$$\begin{array}{ll} x^2 + 3x - 4 & \\ x^2 + x + 2 & \\ \hline x^4 + 3x^3 - 4x^2 & \text{Multiply by } x^2. \\ \_\_\_\_\_\_\_\_\_\_ & \text{Multiply by } x. \\ \_\_\_\_\_\_\_\_\_\_ & \text{Multiply by } 2. \\ \hline \_\_\_\_\_\_\_\_\_\_ & \text{Combine like terms.} \end{array}$$

Now Work Exercises 7 and 8 in the Margin.

Completion Example Answer

6.
$$\begin{array}{ll} x^2 + 3x - 4 & \\ x^2 + x + 2 & \\ \hline x^4 + 3x^3 - 4x^2 & \text{Multiply by } x^2. \\ \mathbf{x^3 + 3x^2 - 4x} & \text{Multiply by } x. \\ \mathbf{2x^2 + 6x - 8} & \text{Multiply by } 2. \\ \hline \mathbf{x^4 + 4x^3 + x^2 + 2x - 8} & \text{Combine like terms.} \end{array}$$

NAME ______________ SECTION ______________ DATE ______________

ANSWERS

1. ______
2. ______
3. ______
4. ______
5. ______
6. ______
7. ______
8. ______
9. ______
10. ______
11. ______
12. ______
13. ______
14. ______
15. ______
16. ______

Exercises T.5

Find each indicated product and simplify if possible.

1. $-4x^2(-3x^2)$

2. $(-5x^3)(-2x^2)$

3. $9a^2(2a)$

4. $-7a^2(2a^4)$

5. $4y(2y^2 + y + 2)$

6. $5y(3y^2 - 2y + 1)$

7. $-1(4x^3 - 2x^2 + 3x - 5)$

8. $-1(7x^3 + 3x^2 - 4x - 2)$

9. $7x^2(-2x^2 + 3x - 12)$

10. $-3x^2(x^2 - x + 13)$

11. $(x + 2)(x + 5)$

12. $(x + 3)(x + 10)$

13. $(x - 4)(x - 3)$

14. $(x - 7)(x - 2)$

15. $(a + 6)(a - 2)$

16. $(a - 7)(a + 5)$

17. ______________
18. ______________
19. ______________
20. ______________
21. ______________
22. ______________
23. ______________
24. ______________
25. ______________
26. ______________
27. ______________
28. ______________
29. ______________
30. ______________
31. ______________
32. ______________

17. $(7x + 1)(x - 3)$

18. $(5x - 6)(3x + 2)$

19. $(2x + 3)(2x - 3)$

20. $(6x + 5)(6x - 5)$

21. $(x + 3)(x - 3)$

22. $(x + 5)(x - 5)$

23. $(x + 1.5)(x - 1.5)$

24. $(x + 2.5)(x - 2.5)$

25. $(3x + 7)(3x - 7)$

26. $(5x + 1)(5x - 1)$

Find the indicated products. Be sure to align like terms, particularly if some degrees are missing.

27. $$\begin{array}{l} x^3 + 5x + 3 \\ \underline{x + 4} \end{array}$$

28. $$\begin{array}{l} x^3 - 2x + 1 \\ \underline{x + 5} \end{array}$$

29. $$\begin{array}{l} 2x^3 + x - 7 \\ \underline{x - 2} \end{array}$$

30. $$\begin{array}{l} 4x^3 - x - 8 \\ \underline{x - 6} \end{array}$$

31. $$\begin{array}{l} x^3 - 6x - 10 \\ \underline{3x + 1} \end{array}$$

32. $$\begin{array}{l} x^3 - 5x - 11 \\ \underline{2x + 9} \end{array}$$

NAME ______________________ SECTION ______________ DATE ________

ANSWERS

33. $\begin{array}{l} x^4 + 2x^3 - 5x^2 + 3 \\ \underline{x^2 + x - 1} \end{array}$

34. $\begin{array}{l} x^4 - 3x^2 + 4x - 1 \\ \underline{x^2 - 2x + 1} \end{array}$

35. $\begin{array}{l} x^3 - 7x + 2 \\ \underline{x^2 + 2x + 3} \end{array}$

36. $\begin{array}{l} x^3 + 3x - 4 \\ \underline{2x^2 - x + 3} \end{array}$

37. $\begin{array}{l} x^4 + 3x^2 - 5 \\ \underline{x^4 - x^2 + 1} \end{array}$

38. $\begin{array}{l} x^4 - 4x^2 + 7 \\ \underline{x^4 + x^2 - 1} \end{array}$

39. $\begin{array}{l} x^3 + x^2 + x + 1 \\ \underline{x^3 + x^2 + x + 1} \end{array}$

40. $\begin{array}{l} x^3 - x^2 + x - 1 \\ \underline{x^3 - x^2 + x - 1} \end{array}$

33. ________

34. ________

35. ________

36. ________

37. ________

38. ________

39. ________

40. ________

41. [Respond below exercise.]

Writing and Thinking about Mathematics

41. The sides of the square shown in the figure are of length $(a + b)$. The area of the square is $(a + b)^2$. Show that $(a + b)^2$ is equal to the sum of the indicated areas in the figure.

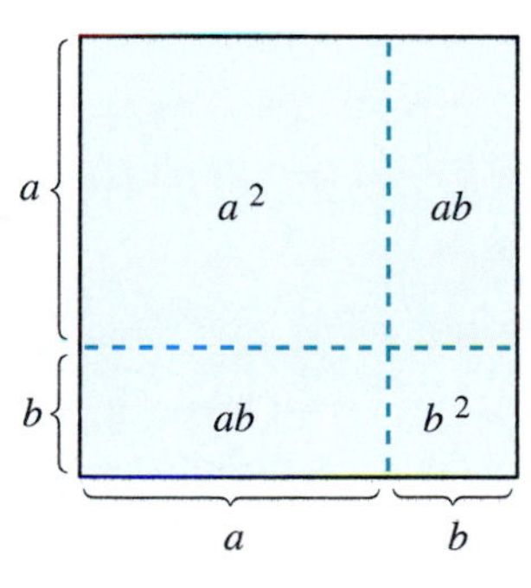

42. [Respond below exercise.] ____________

42. (a) Find the product of these two polynomials:

$$\begin{array}{l} 2x^3 - 4x^2 - 5x + 6 \\ \underline{5x - 2 \qquad\qquad\quad} \end{array}$$

(b) Now substitute 2 for x in each of the polynomials in part (a), including the product. Does the product of the first two values equal the value of the product?

(c) Repeat the process in part (b) using $x = -2$.

(d) Substituting a value for x in each polynomial seems to provide a method for checking answers. Do you think that this method will catch all errors? Would substituting $x = 0$ be a good idea? Briefly discuss your reasoning.

Index of Key Ideas and Terms

NAME ______________ SECTION ______________ DATE ______________

Chapter Test

1. Represent the following graph by using algebraic notation and tell what type of interval it is.

2. Graph the interval $\frac{1}{2} \leq x \leq 3$ and tell what type of interval it is.

Solve each inequality and graph the solution on a number line.

3. $2x - 5 \leq 5x - 2$

4. $7x + 4 < 16 + x$

List the set of ordered pairs that correspond to the points in each graph.

5.

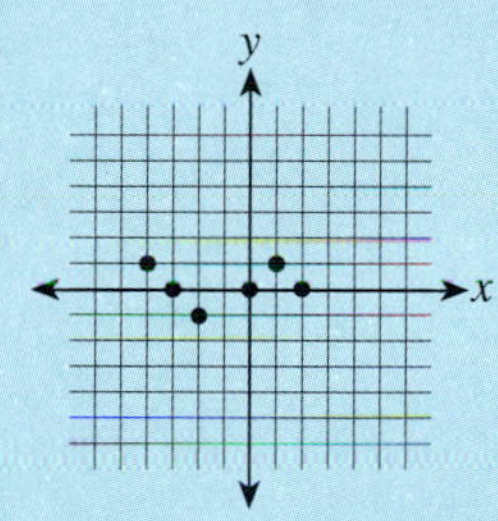

6.

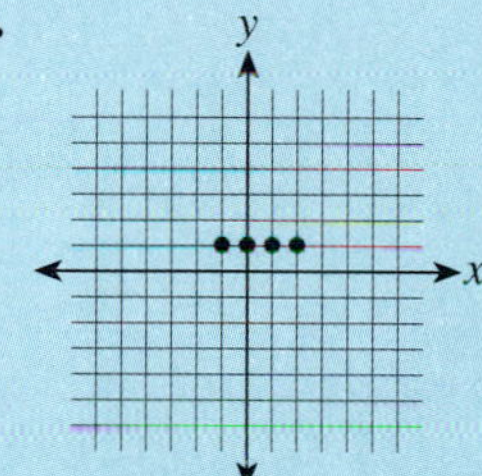

7. Graph the following set of ordered pairs:
$\{(-2, 5), (-1, 3), (0, 1), (1, -1), (2, 0)\}$

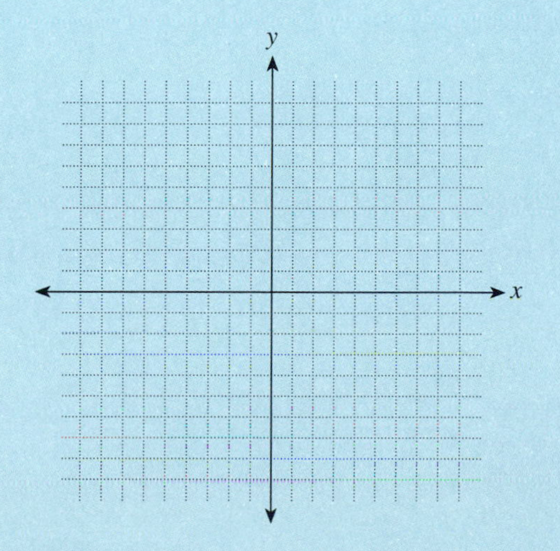

ANSWERS

1. ______________

2. ______________

3. ______________

4. ______________

5. ______________

6. ______________

7. [Use graph provided.]

8. [Use graph provided.] ______

9. [Use graph provided.] ______

10. [Use graph provided.] ______

11. ______

12. ______

13. ______

14. ______

15. ______

Graph the following linear equations.

8. $y = -2x + 1$

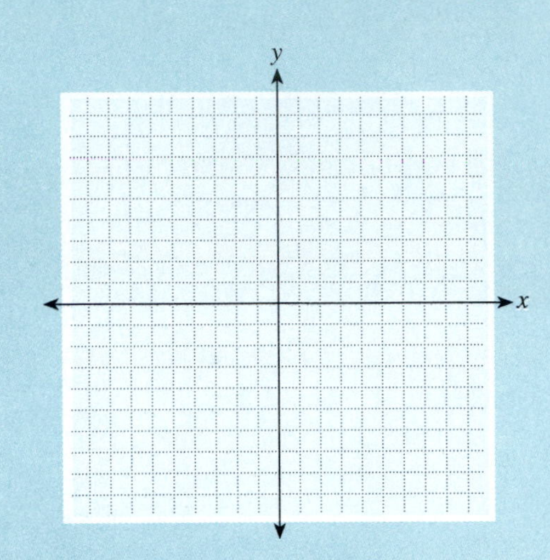

9. $2x - 3y = 6$

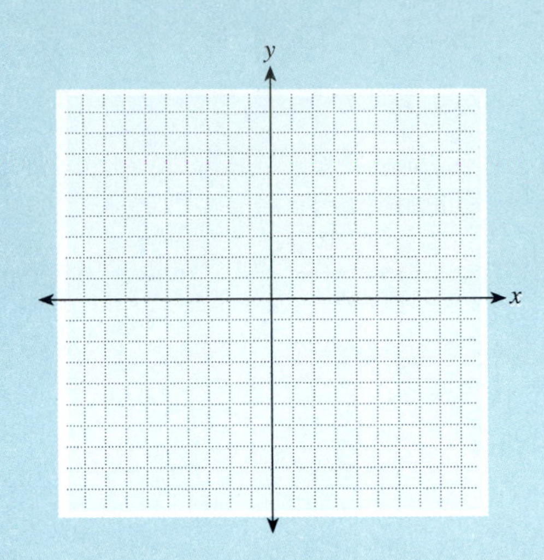

10. $x = 2.5$

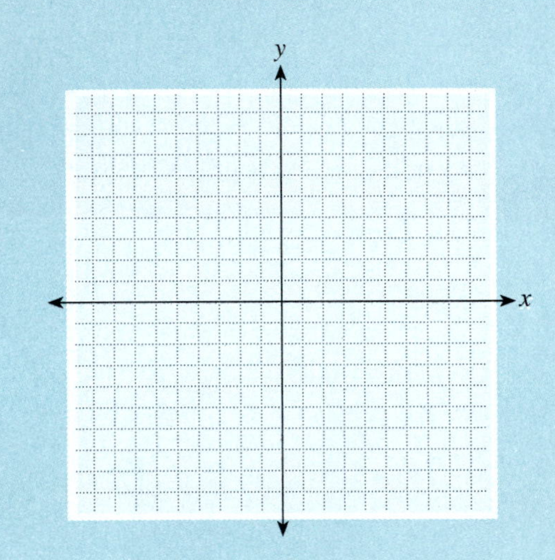

Use the Product Rule for Exponents to simplify each of the following expressions.

11. $3^3 \cdot 3^5$

12. $y^4 \cdot y^4$

13. $2x^2 \cdot 15x^3$

14. Simplify the following polynomial and state its degree:

$5x^3 - x^3 + 4x^2 + 5x - 3x^2 + 7 - 12x$

15. Evaluate the polynomial in Problem 14 for $x = -2$.

NAME ______________ SECTION ______________ DATE ______________

ANSWERS

16. ______________

17. ______________

18. ______________

19. ______________

20. ______________

21. ______________

22. ______________

23. ______________

24. ______________

25. ______________

Add or subtract as indicated and simplify if possible.

16. $(4y^3 + 7y^2 - 8y - 14) + (-5y^3 - 2y^2 + 6y - 4)$

17. $(8x + 17) - (4x - 10)$

18. $(2a^2 + 5a - 3) - (4a^2 + 9a - 3)$

19. $(x^3 + 3x^2 - 7) - (x^3 - 5x^2 + 3x + 7)$

Find each of the following products.

20. $5x(3x^2 + 2x - 4.1)$

21. $2x^3(4x^2 - 8x + 5)$

22. $(a + 6)(a + 6)$

23. $(4y - 11)(2y + 3)$

24. $$\begin{array}{r} x^3 + 2x^2 - 3x + 2 \\ 5x - 4 \\ \hline \end{array}$$

25. $$\begin{array}{r} 2x^2 - 6x + 7 \\ x^2 + 3x + 1 \\ \hline \end{array}$$

NAME ______________ SECTION ______________ DATE ______________

Cumulative Review

All answers should be in simplest form. Reduce all fractions to lowest terms and express all improper fractions as mixed numbers.

Find the value of each variable.

1. $x = \frac{2}{3} + \frac{5}{6} + \frac{2}{9}$

2. $y = 17\frac{5}{8} + 12\frac{7}{10}$

3. $6.09 + 10.6 + 7 = t$

4. $(-2.03)+(16.7)+(-5.602)=w$

Find each difference.

5. $\frac{3}{4} - \frac{5}{9}$

6. $4\frac{1}{6} - 2\frac{2}{3}$

7. $142.01 - 67.135$

8. Subtract -4.5 from -0.03.

Find the value of each variable.

9. $x = 2400 \cdot 30{,}000$

10. $y = \frac{5}{12} \cdot \frac{3}{5} \cdot \frac{4}{7}$

11. $\left(3\frac{2}{3}\right)\left(4\frac{4}{11}\right) = q$

12. $(-7.03)(0.28) = p$

Find each quotient.

13. $\frac{4}{9} \div \frac{8}{15}$

14. $5\frac{1}{4} \div \frac{7}{10}$

15. $1.0336 \div 1.7$

16. $-3.7 \div 0$

ANSWERS

1. ______________
2. ______________
3. ______________
4. ______________
5. ______________
6. ______________
7. ______________
8. ______________
9. ______________
10. ______________
11. ______________
12. ______________
13. ______________
14. ______________
15. ______________
16. ______________

17. ______________

18. [Respond below exercise.]

19. ______________

20. ______________

21. ______________

22. ______________

23. ______________

24. ______________

25. ______________

26. ______________

27. ______________

28. ______________

29. ______________

17. Find the prime factorization of 792.

18. Is 45 a factor (or divisor) of the product $2 \cdot 3 \cdot 6 \cdot 5 \cdot 7$? Explain.

Round off each of the following numbers as indicated.

19. 5987.04 (nearest ten)

20. 723.149 (nearest tenth)

Evaluate each of the following expressions.

21. $(2^2 \cdot 3 \div 4 + 2) \cdot 5 - 3$

22. $\left(\frac{1}{2}\right)^2 - \left(\frac{1}{3}\right)^2$

23. Simplify: $3\sqrt{20} + 4\sqrt{75} - \sqrt{45}$.

24. Fill in the blank with the appropriate symbol: $<$, $>$, or $=$.

$-|-7|$ ______ $-(-7)$

25. Write an algebraic expression for the following English phrase:
5 less than twice the sum of a number and 6.

Solve each of the following equations.

26. $3x + 4 = x - 6$

27. $-2(3 - x) = 3(x + 2)$

28. $\frac{5}{6}x - 1 = \frac{2}{3}x + 2$

29. Solve $2 < 3 - 4x \leq 5$ and graph the solution on the number line.

NAME ______________________ SECTION ______________ DATE ________

ANSWERS

30. Find the length of the hypotenuse of a right triangle if one of its legs is 5 inches long and the other leg is 10 inches long.

31. List the set of ordered pairs that correspond to the points in the graph.

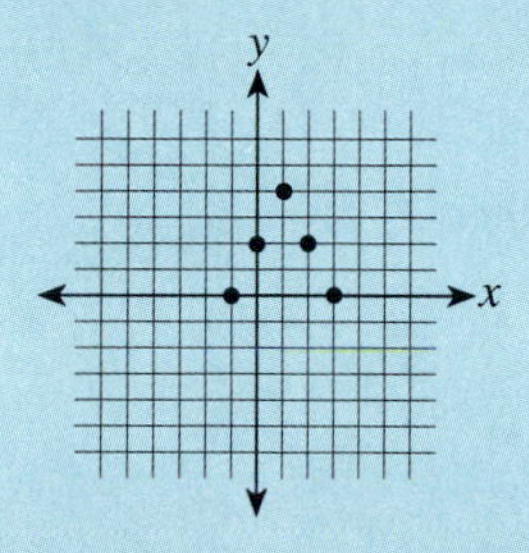

Graph each of the following linear equations.

32. $3x + 4y = 0$

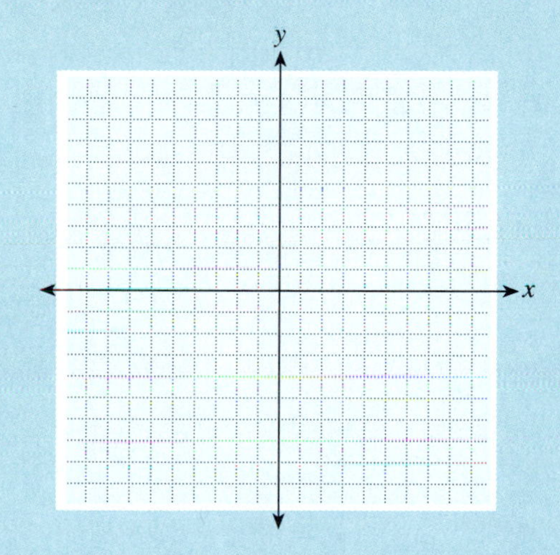

33. $y = \frac{1}{3}x + 5$

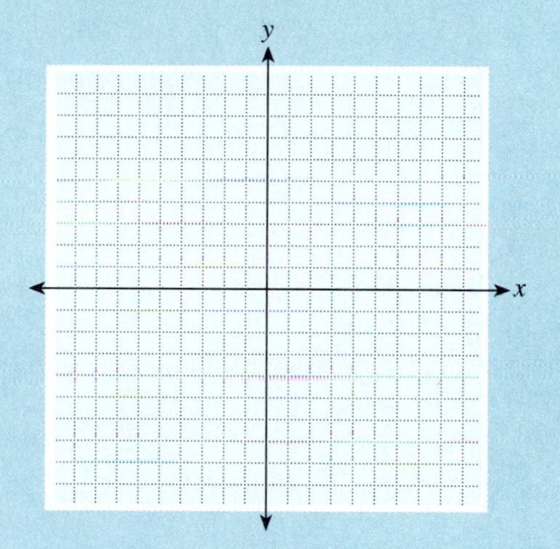

34. 16% of what number is 5.76?

35. What percent of 2.1 is 4.41?

36. Find (a) the perimeter and (b) the area of a rectangle that is 6.5 feet long and 2.4 feet wide.

30. ____________

31. ____________

32. [Use graph provided.]

33. [Use graph provided.]

34. ____________

35. ____________

36a. ____________

b. ____________

37a. ____________

b. ____________

38. ____________

39a. ____________

b. ____________

40. ____________

41. ____________

42. ____________

43a. ____________

b. ____________

c. ____________

37. Find (a) the circumference and (b) the area of a circle that has a radius of 20 inches. (Use $\pi = 3.14$.)

38. Two towns are shown 13.5 centimeters apart on a map that has a scale of 3 centimeters to 50 miles. What is the actual distance between the towns (in miles)?

39. (a) If \$1550 is deposited in an account paying 8% compounded monthly, what will be the total amount in the account at the end of three years (to the nearest dollar)? (b) How much interest will be earned?

40. A customer received a 2% discount for paying cash for her new computer. If she paid \$2009 in cash, what was the amount of the discount?

41. Evaluate the polynomial $4x^3 - 5x^2 - 6x + 22$ for $x = -3$.

42. Write the number 3,450,000 in scientific notation.

43. Find (a) the mean, (b) the median, and (c) the range of the following set of numbers: 27, 36, 45, 72, 63, 36, 27, 18, 36, 90

NAME ____________________ SECTION ____________________ DATE ____________

ANSWERS

44. Add and simplify:

8 hr 25 min 15 sec
+3 hr 36 min 55 sec

45. Find the difference:

5 yd 2 ft 3 in.
−1 yd 1 ft 7 in.

46. Use the Product Rule for Exponents to simplify the expression $6x^3 \cdot 7x^9$.

47. Find the sum: $(-3x^3 + 4x^2 - 10x + 33) + (2x^3 + 5x^2 - x - 10)$.

48. Find the difference: $(8x^2 - 14x - 12) - (3x^2 - 10x - 30)$.

49. Find the product: $(5x + 3)(2x - 3)$.

50. Find the product:

$x^2 - 3x - 4$
$x^2 + x + 2$

44. ____________

45. ____________

46. ____________

47. ____________

48. ____________

49. ____________

50. ____________

Tables

Table 1 U.S. Customary and Metric Equivalents

In the following tables, the equivalents are rounded off.

U.S. to Metric	Metric to U.S.
Length Equivalents	
1 in. = 2.54 cm (exact)	1 cm = .394 in.
1 ft = 0.305 m	1 m = 3.28 ft
1 yd = 0.914 m	1 m = 1.09 yd
1 mi = 1.61 km	1 km = 0.62 mi
Area Equivalents	
1 in.^{2} = 6.45 cm^2	1 cm^2 = 1.55 in.^{2}
1 ft^2 = 0.093 m^2	1 m^2 = 10.764 ft^2
1 yd^2 = 0.836 m^2	1 m^2 = 1.196 yd^2
1 acre = 0.405 ha	1 ha = 2.47 acres
Volume Equivalents	
1 in.^{3} = 16.387 cm^3	1 cm^3 = 0.06 in.^{3}
1 ft^3 = 0.028 m^3	1 m^3 = 35.315 ft^3
1 qt = 0.946 L	1 L = 1.06 qt
1 gal = 3.785 L	1 L = 0.264 gal
Mass Equivalents	
1 oz = 28.35 g	1 g = 0.035 oz
1 lb = 0.454 kg	1 kg = 2.205 lb

Table 2 Centigrade and Fahrenheit Equivalents

Conversion Formulas: $F = \frac{9}{5}C + 32 \quad C = \frac{5}{9}F - 32)$

Centigrade	Fahrenheit	
100°	212°	← water boils at sea level
95°	203°	
90°	194°	
85°	185°	
80°	176°	
75°	167°	
70°	158°	
65°	149°	
60°	140°	
55°	131°	
50°	122°	
45°	113°	
40°	104°	
35°	95°	
30° (comfort range)	86° (comfort range)	
25° (comfort range)	77° (comfort range)	
20° (comfort range)	68° (comfort range)	
15° (comfort range)	59° (comfort range)	
10°	50°	
5°	41°	
0°	32°	← water freezes at sea level

About Table 3 and the Value of π

As discussed in the text on page 577, π is an irrational number, and so the decimal form of π is an infinite non-repeating decimal. Mathematicians even in ancient times realized that π is a constant value obtained from the ratio of a circle's circumference to its diameter, but they had no sense that it might be an irrational number. As early as about 1800 B.C. the Babylonians gave π a value of 3, and around 1600 B.C. the ancient Egyptians were using the approximation of 256/81, what would be a decimal value of about 3.1605. In the third century B.C., the Greek mathematician Archimedes used polygons approximating a circle to determine that the value of π must lie between 223/71(= 3.1408) and 22/7(= 3.1429). He was thus accurate to two decimal places. About seven hundred years later, in the fourth century A.D., Chinese mathematician Tsu Chung-Chi refined Archimedes' method and expressed the constant as 355/113, which was correct to six decimal places. By 1610, Ludolph van Ceulen of Germany had also used a polygon method to find π accurate to 35 decimal places.

Knowing that the decimal expression of π would not terminate, mathematicians still sought a repeating pattern in its digits. Such a pattern would mean that π was a rational number and that there would be some ratio of two whole numbers that would produce the correct decimal representation. Finally, in 1767, Johann Heinrich Lambert provided a proof to show that π is indeed irrational, and thus is non-repeating as well as non-terminating.

Since Lambert's proof, mathematicians have still made an exercise of calculating π to more and more decimal places. The advent of the computer age in this century has made that work immeasurably easier, and on occasion you will still see newspaper articles pronouncing that mathematics researchers have reached a new high in the number of decimal places in their approximations. In 1988 that number was 201,326,000 decimal places. Within one year that record was more than doubled, and most recent approximations of π now reach beyond one billion decimal places! For your understanding, appreciation, and interest, the value of π is given in Table 3 to a mere 3742 decimal places as calculated by a computer program. To show π calculated to one billion decimal places would take every page of nearly 400 copies of this text!

Table 3 The Value of π

π =

3.1415926535897932384626433832795028841971693993751058209749445923
0781640628620899862803482534211706798214808651328230664709384460 95
5058223172535940812848111745028410270193852110555964462294895493 03
8196442881097566593344612847564823378678316527120190914564856692 34
6034861045432664821339360726024914127372458700660631558817488152 09
2096282925409171536436789259036001133053054882046652138414695194 15
1160943305727036575959195309218611738193261179310511854807446237 99
6274956735188575272489122793818301194912983367336244065664308602 13
9494639522473719070217986094370277053921717629317675238467481846 76
6940513200056812714526356082778577134275778960917363717872146844 09
0122495343014654958537105079227968925892354201995611212902196086 40
3441815981362977477130996051870721134999999837297804995105973173 28
1609631859502445945534690830264252230825334468503526193118817101 00
0313783875288658753320838142061717766914730359825349042875546873 11
5956286388235378759375195778185778053217122680661300192787661119 59
0921642019893173507587940712544731117180978047272317724064216006 38
9377278375236525307175633161841241921971450150701067309483743038 59
6107212000873347515676002652778929047731302426815346962841501296 96
7149153770865497958835174652289388524151403102528433113942910586 44
7109152700677848924200720807339737603843357971117145532213422958 36
6204314895098162384143929756056081970942213639299200826062291594 66
4777545883055143213228365671403858649396558636890790649697410466 73
0752085836060421715296142027349946507783597117173631648947718571 14
5707433080042225077388362337053066816120701237140505070262124113 00
6475595573605985237228154083535666110093396951576034023178723400 19
7903862246839077784495488003826979702310069628445147047512091778 28
2586365420964030128197646573886650226384627183308704508541540196 18
2983401832216552790654382377292917044216405342359254109248408909 30
1521626930313974946993462596074378786192675563872651615680600411 43
1408991739953266556307996722826428009406044939515466071524746165 34
7930533696976479739765562615755864901081820059891517532404513440 39
3172323462947355195492311022634818250130360202162540464228269037 89
2821620948188787506276307819506946613872673520891591046775285039 87
3018784580883194426986016953458676427241746672722347209015942845 90
5920875245380769966194274846705687861752240243662392864764276375 04
6739947348234253454079685236817711272583883473113579206721044307 35
2880542424950165232933275511013326277993525979613388163804617128 15
9338408230697325013514612614930586533610412663736836281623393781 99
0136329784635436278766739495587394488795635001639455813075811257 36
7575119112003664022890382906680608511677828978463483479665115515 82
3437372754031302693933647933155652090445499032805430672429141363 17
9966536861999193366332640659366926394230517181700254872542943297 13
7699010195993591671882970531088439848643326720099910089541798418 77
1428584586401048581003046178651439157666528746685960809259799616 35
7859163546988513283836284682832965411704855475274864957247192469 71
0310237595356206021980631551215815303427029482390497850838959700 07
9737143923901518290675611595974624355803572628454843929317302355 23
7279756190034940650829785564455235314661862934895982703648185859 33
2800598936215112267099531557512033918835953248898738989312902621 40
3982479724125133319915432627152932188711037440670042990200494965 67
7555396168040711287181677406532621646942907409556695005140066643 63
4285900913660355785104356290341891513505082621315326392403652173 07
6077943027726311565001089886572280029598027701808282028401345492 73
9102272751804013465732315846999322733293532052271112233471198170 08
5095373427609188255889100141736202458248800978143099741280273175 39
0768213563373855053890490536898890516077937973190669787062683955 20
7857573502722867787428244462059949689243235966 28. . .

Answer Key

CHAPTER 1

Margin Exercises 1.1

1. 4000 + 600 + 30 +5 **2.** 70,000 + 3000 + 400 + 20 + 8 **3.** 521
4. 9000 + 500 + 60 + 7; nine thousand five hundred sixty-seven
5. 20,000 + 5000 + 400 + 0 + 0; twenty-five thousand, four hundred
6. 6792; six thousand seven hundred ninety-two **7.** 23,642 **8.** 363,975
9. six million, three hundred thousand, five hundred **10.** four million, eight hundred seventy-five thousand

Exercises 1.1, p. 9

1. 0, 1, 2, 3, 4, 5, 6, 7, 8, 9 **3.** 1, 10, 100, 1000, 10,000, 100,000, and so on. **5.** four **7.** 30 + 7
9. 50 + 6 **11.** 1000 + 800 + 90 + 2 **13.** 20,000 + 5000 + 600 + 50 + 8 **15.** eighty-three
17. ten thousand, five hundred **19.** five hundred ninety-two thousand, three hundred
21. seventy-one million, five hundred thousand **23.** 98 **25.** 573 **27.** 10,011 **29.** 537,082
31. 2 billion, 4 hundred million, 3 million, 1 hundred thousand, 8 ten thousand, 9 thousand, 5 hundred
33. ninety-three million; one hundred forty-nine million, seven hundred thirty thousand
35. three hundred fifty-two thousand, one hundred forty-three; nine hundred twelve thousand, fifty

Margin Exercises 1.2

1. 13; commutative property of addition **2.** 15; commutative property of addition
3. 15; associative property of addition **4.** 17; associative property of addition **5.** 10 **6.** 15 **7.** 3 **8.** 12
9. 6889 **10.** 7797 **11.** 6994 **12.** 5708 **13.** 1006 **14.** 13,388 **15.** 123,041 **16.** 160
17. 2470 **18.** 494,435 **19.** 233 **20.** 1511 **21.** 1344 **22.** 539 **23.** 38 **24.** 288 **25.** 3858
26. 54

Exercises 1.2, p. 23

1. commutative property of addition **3.** associative property of addition **5.** additive identity
7. associative property of addition **9.** commutative property of addition **11.** 6 **13.** 9 **15.** 7 **17.** 1
19. 6 **21.**

+	5	8	7	9
3	8	11	10	12
6	11	14	13	15
5	10	13	12	14
2	7	10	9	11

23. 125 **25.** 1552 **27.** 13,324 **29.** 197 **31.** 835 **33.** 553 **35.** 4168 **37.** 9064
39. 1,463,930 **41.** 2,817,126 **43.** 11 **45.** 20 **47.** 5 **49.** 126 **51.** 376 **53.** 395 **55.** 966
57. 1424 **59.** 5871 **61.** 3,800,559 **63.** 5,671,011 **65.** \$6,313,323 **67.** \$493,154 **69.** 790
71. 140 **73.** \$1445 **75.** \$45,250

Margin Exercises 1.3

1. 60 **2.** 300 **3.** 2000 **4.** 580 **5.** 800 **6.** 2000 **7.** 2600 **8.** 44,000 **9.** 2000 **10.** 1900
11. 4400 (estimate); 4876 (sum) **12.** 5000 (estimate); 4805 (difference)

Exercises 1.3, p. 37

1. 760 **3.** 80 **5.** 300 **7.** 990 **9.** 4200 **11.** 500 **13.** 600 **15.** 76,500 **17.** 7000 **19.** 8000
21. 13,000 **23.** 62,000 **25.** 80,000 **27.** 260,000 **29.** 120,000 **31.** 180,000 **33.** 100 **35.** 1000

37. 180 (estimate); 168 (sum) **39.** 1200 (estimate); 1173 (sum) **41.** 1900 (estimate); 1881 (sum)
43. 9500 (estimate); 9224 (sum) **45.** 6000 (estimate); 5467 (difference) **47.** 5000 (estimate); 5931 (difference)
49. 20,000 (estimate); 20,804 (difference) **51.** about 700,000,000 **53.** 60 centimeters **55.** $200

Margin Exercises 1.4

1. 56; commutative property of multiplication **2.** 5; multiplicative identity
3. 42; associative property of multiplication **4.** 32; associative property of multiplication **5.** 0; zero factor law
6. 8000 **7.** 1500 **8.** 2500 **9.** 3000 **10.** 270,000 **11.** 1792 **12.** 4067 **13.** 23,808 **14.** 852
15. 3145 **16.** 6552 **17.** 49,600 **18.** 61,100 **19.** 8,496,000

Exercises 1.4, p. 51

1. 72 **3.** 24 **5.** 7 **7.** 0 **9.** 28; commutative property of multiplication **11.** 5; multiplicative identity
13. 250 **15.** 4000 **17.** 16,000 **19.** 900 **21.** 16,000,000 **23.** 5000 **25.** 36,000 **27.** 150,000
29. 240 (estimate); 224 (product) **31.** 450 (estimate); 432 (product) **33.** 240 (estimate); 252 (product)
35. 2400 (estimate); 2352 (product) **37.** 1000 (estimate); 960 (product) **39.** 7200 (estimate); 7055 (product)
41. 600 (estimate); 544 (product) **43.** 800 (estimate); 880 (product) **45.** 600 (estimate); 375 (product)
47. 1800 (estimate); 2064 (product) **49.** 100 (estimate); 156 (product) **51.** 4000 (estimate); 5166 (product)
53. 3000 (estimate); 2850 (product) **55.** 20,000 (estimate); 29,601 (product)
57. 10,000 (estimate); 9800 (product) **59.** 140,000 (estimate); 125,178 (product)
61. 231 (sum); 58 (difference); 13,398 (product) **63.** $2760 increase per year; $165,600 over five years
65. $230,400 **67.** 16,000 square meters (estimate); 17,630 square meters in area
69. $16,000,000 (estimate); $15,200,000 total value
Check Your Number Sense **71.** $200 **73.** (a) C (b) B (c) A (d) D

Margin Exercises 1.5

1. 94 R 3 **2.** 548 R 10 **3.** 46 R 3 **4.** 20 R 4 **5.** 602 R 0 **6.** 200 **7.** 2000

Exercises 1.5, p. 65

1. 4 **3.** 9 **5.** 8 **7.** 5 **9.** 5 **11.** 5 **13.** 8 **15.** 9 **17.** 9 **19.** 8 **21.** 0 **23.** 1
25. 30 R 0; 7 and 30 are factors of 210. **27.** 12 R 0; 11 and 12 are factors of 132.
29. 17 R 0; 3 and 17 are factors of 51. **31.** 20 R 13; 20 is not a factor of 413.
33. 14 R 0; 13 and 14 are factors of 182. **35.** 3 R 10; 14 and 3 are not factors of 52.
37. 38 R 6; 11 and 38 are not factors of 424. **39.** 15 R 5 **41.** 42 R 3 **43.** 20 R 0 **45.** 400 R 3
47. 301 R 4 **49.** 3 R 3 **51.** 54 R 3 **53.** 22 R 74 **55.** 7 R 358 **57.** (a) about $40 (b) $38
59. $31{,}590 \div 45 = 702$ (or $31{,}590 \div 702 = 45$)
Check Your Number Sense **61.** (a) C (b) A (c) E (d) D (e) B

Margin Exercises 1.6

1. 101 **2.** 150 centimeters (perimeter); 750 square centimeters (area)

Exercises 1.6, p. 75

1. 16
3. 18,817; 3 is not a factor of the product because the sum of the digits is $1 + 8 + 8 + 1 + 7 = 25$ and 25 is not divisible by 3.
5. $160 **7.** $786 **9.** $874 **11.** $485 **13.** $316 **15.** (a) 104 meters (b) 555 square meters
17. 120 in. **19.** 103 **21.** 6 **23.** 485 **25.** 85 **27.** $18 per share; $1200 profit
29. (a) $665 (b) $2495 **31.** 52,050 **33.** (a) 12,730 (b) $10,624 (c) $12,626 (d) $4872

Chapter 1 Test, p. 85

1. 600 + 50 + 3; six hundred fifty-three **2.** 8000 + 900 + 50 + 2; eight thousand nine hundred fifty-two **3.** any example of the form a + b = b + a **4.** any example of the form a · 1 = a **5.** 1340 **6.** 16,000 **7.** 250,000 **8.** 13,300 (estimate); 12,909 (sum) **9.** 5,000,000 (estimate); 5,123,250 (sum) **10.** 800 (estimate); 712 (difference) **11.** 5000 (estimate); 5217 (difference) **12.** 2400 (estimate); 2584 (product) **13.** 270,000 (estimate); 220,405 (product) **14.** 403 R 0 **15.** 173 R 26 **16.** 54 **17.** 74 **18.** 83 **19.** (a) about \$540 (b) \$552 (c) \$278 **20.** (a) 240 meters (b) 3500 square meters

CHAPTER 2

Margin Exercises 2.1

1. exponent 2; base 8; power 64 **2.** exponent 3, base 6; power 216 **3.** exponent 2; base 9; power 81 **4.** exponent 6; base 2; power 64 **5.** exponent 4; base 1; power 1 **6.** 8^3 **7.** $2^2 \cdot 3 \cdot 5^2$ **8.** $3^2 \cdot 4^3$ **9.** 256 **10.** 169 **11.** 23 **12.** 13 **13.** 5 **14.** 9 **15.** 12

Exercises 2.1, p. 95

1. (a) 2 (b) 3 (c) 8 **3.** (a) 4 (b) 2 (c) 16 **5.** (a) 9 (b) 2 (c) 81 **7.** (a) 10 (b) 4 (c) 10,000 **9.** (a) 3 (b) 6 (c) 729 **11.** 5^2 **13.** 3^3 **15.** 2^3 **17.** 10^2 **19.** 6^5 **21.** $2^3 \cdot 3^2$ **23.** $2 \cdot 3^2 \cdot 11^2$ **25.** 64 **27.** 196 **29.** 144 **31.** 1600 **33.** (a) multiply (b) divide (c) add **35.** (a) divide within parentheses (b) add within parentheses (c) divide **37.** 3 **39.** 7 **41.** 22 **43.** 3 **45.** 5 **47.** 5 **49.** 5 **51.** 3 **53.** 7 **55.** 0 **57.** 6 **59.** 3 **61.** 27 **63.** 26 **65.** 0 **67.** 69 **69.** 68 **71.** 140 **73.** 5 **75.** 0 **77.** 9 **79.** 12 **81.** 24 **83.** 118 **85.** 0

Margin Exercises 2.2

1. Yes, 4 divides 9044 because 4 divides the two digit number 44.
2. No, 3 does not divide 106 because 3 does not divide the sum of the digits: 1 + 0 + 6 = 7.
3. Both 3 and 9 divide 306 because both divide the sum of the digits: 3 + 0 + 6 = 9.
4. 3165 is divisible by 5 because the units digit is 5. 3165 is not divisible by 10 because the units digit is not 0.
5. 463 is not divisible by 2 because the units digit is not even. Therefore, the number is not an even number. 463 is not divisible by 4 because the two digit number 63 is not divisible by 4. (Also, if a number is not divisible by 2, it cannot be divisible by 4.) 463 is not divisible by 5 because the units digit is not 0 or 5.
6. 842 is divisible by 2. **7.** 9030 is divisible by 2, 3, 5, and 10. **8.** 4031 is not divisible by any of these numbers.

Exercises 2.2, p. 107

1. 2, 3, 4, 9 **3.** 3, 5 **5.** 2, 3, 5, 10 **7.** 2, 4 **9.** 2, 3, 4 **11.** none **13.** 3, 9 **15.** none **17.** none **19.** 3 **21.** 3, 5 **23.** 2, 3, 4, 5, 9, 10 **25.** 2, 3 **27.** none **29.** 2 **31.** 3, 5 **33.** 3, 9 **35.** 2, 3, 9 **37.** 2, 4, 5, 10 **39.** 3, 9 **41.** 2, 3, 4, 5, 10 **43.** 2, 4 **45.** 2, 3, 4, 5, 10 **47.** 3, 5 **49.** none **51.** 2, 3, 9 **53.** 2, 3, 4 **55.** 2, 3, 4, 9 **57.** 3, 9 **59.** 2, 3, 4
61. $2 \cdot 3 \cdot 3 \cdot 5 = 6 \cdot 15$; The product is 90, and 6 divides the product 15 times.
63. $2 \cdot 3 \cdot 5 \cdot 7 = 14 \cdot 15$; The product is 210, and 14 divides the product 15 times.
65. The product $3 \cdot 3 \cdot 5 \cdot 7 = 315$, and since this product does not have 2 as a factor, it is not divisible by 10.
67. $2 \cdot 2 \cdot 3 \cdot 5 \cdot 5 = 25 \cdot 12$; The product is 300, and 25 divides the product 12 times.
69. $3 \cdot 3 \cdot 5 \cdot 7 \cdot 11 = 21 \cdot 165$; The product is 3465, and 21 divides the product 165 times.

The Recycle Bin, p. 110

1. 850 **2.** 1930 **3.** 400 **4.** 2600 **5.** 14,000 **6.** 21,000 **7.** 690 (estimate); 693 (sum) **8.** 4700 (estimate); 4450 (sum) **9.** 5000 (estimate); 5225 (difference) **10.** 20,000 (estimate); 28,522 (difference)

Margin Exercises 2.3

1. 17 is prime because it has only two factors, 1 and 17.
2. 28 is not prime. It is composite because it has more than two factors; namely, 1, 18, 2, 4, 7, and 14.
3. 25 is not prime. It is composite because it has more than two divisors; namely, 1, 25, and 5.
4. 2 is prime because it has only two divisors, 1 and 2. **5.** 31 is prime because it has only two divisors, 1 and 31.
6. Division by prime numbers shows that $187 = 11 \cdot 17$. Therefore, 187 is composite.
7. Division by prime numbers up to 17 shows no divisors. When 233 is divided by 17 the quotient is less that 17. Therefore, 233 is prime and its only divisors are 1 and 233.

Exercises 2.3, p. 117

1. 5, 10, 15, 20, 25, 30, 35, . . . **3.** 11, 22, 33, 44, 55, 66, 77, . . . **5.** 12, 24, 36, 48, 60, 72, 84, . . .
7. 16, 32, 48, 64, 80, 96, 112, . . .
9. 2, 3, 5, 7, 11, 13, 17, 19, 23, 29, 31, 37, 41, 43, 47, 53, 59, 61, 67, 71, 73, 79, 83, 89, 97

1	**2**	**3**	4	**5**	6	**7**	8	9	10
11	12	**13**	14	15	16	**17**	18	**19**	20
21	22	**23**	24	25	26	27	28	**29**	30
31	32	33	34	35	36	**37**	38	39	40
41	42	**43**	44	45	46	**47**	48	49	50
51	52	**53**	54	55	56	57	58	**59**	60
61	62	63	64	65	66	**67**	68	69	70
71	72	**73**	74	75	76	77	78	**79**	80
81	82	**83**	84	85	86	87	88	**89**	90
91	92	93	94	95	96	**97**	98	99	100

11. composite; 1 and 39; 3 and 13 **13.** composite; 1 and 51; 3 and 17 **15.** prime
17. composite; 3 and 35; 5 and 21 **19.** prime **21.** prime **23.** composite; 1 and 713; 23 and 31
25. composite; 1 and 527; 17 and 31 **27.** 4 and 6 **29.** 2 and 8 **31.** 2 and 7 **33.** 4 and 5 **35.** 3 and 16
37. 1 and 7 **39.** 3 and 17 **41.** 4 and 4 **43.** 4 and 13 **45.** 4 and 18

The Recycle Bin, p. 120

1. commutative property of multiplication **2.** associative property of multiplication **3.** 16,000 **4.** 150,000
5. 840,000 **6.** 22,922 **7.** 84,812 **8.** 178,500

Margin Exercises 2.4

1. $42 = 2 \cdot 3 \cdot 7$ **2.** $56 = 2 \cdot 2 \cdot 2 \cdot 7$ **3.** $230 = 2 \cdot 5 \cdot 23$ **4.** $165 = 3 \cdot 5 \cdot 11$
5. 1, 18, 2, 3, 6, and 9 are all factors of 18. **6.** 1, 63, 3, 7, 9, and 21 are all factors of 63.

Exercises 2.4, p. 125

1. $24 = 2^3 \cdot 3$ **3.** $27 = 3^3$ **5.** $36 = 2^2 \cdot 3^2$ **7.** $72 = 2^3 \cdot 3^2$ **9.** $81 = 3^4$ **11.** $125 = 5^3$
13. $75 = 3 \cdot 5^2$ **15.** $210 = 2 \cdot 3 \cdot 5 \cdot 7$ **17.** $250 = 2 \cdot 5^3$ **19.** $168 = 2^3 \cdot 3 \cdot 7$ **21.** $126 = 2 \cdot 3^2 \cdot 7$
23. prime **25.** $51 = 3 \cdot 17$ **27.** $121 = 11^2$ **29.** $225 = 3^2 \cdot 5^2$ **31.** $32 = 2^5$ **33.** $108 = 2^2 \cdot 3^3$
35. prime **37.** $78 = 2 \cdot 3 \cdot 13$ **39.** $10{,}000 = 2^4 \cdot 5^4$ **41.** 1, 12, 2, 3, 4, 6 **43.** 1, 28, 2, 4, 7, 14
45. 1, 121, 11 **47.** 1, 105, 3, 5, 7, 15, 21, 35 **49.** 1, 97

The Recycle Bin, p. 128

1. 1525 **2.** $12,942 **3.** 104 **4.** (a) 30 meters (b) 30 square meters

---

Margin Exercises 2.5

1. $\text{LCM} = 2 \cdot 2 \cdot 5 \cdot 7 = 140$ **2.** (a) $\text{LCM} = 2 \cdot 3 \cdot 5 \cdot 5 = 150$ (b) $150 = 30 \cdot 5$ and $150 = 50 \cdot 3$
3. (a) $\text{LCM} = 2 \cdot 2 \cdot 3 \cdot 3 \cdot 5 \cdot 5 = 900$ (b) $900 = 18 \cdot 50$, $900 = 20 \cdot 45$, and $900 = 25 \cdot 36$

Exercises 2.5, p. 137

1. LCM $= 2 \cdot 2 \cdot 2 \cdot 3 = 2^3 \cdot 3 = 24$ **3.** LCM $= 2 \cdot 2 \cdot 3 \cdot 3 = 2^2 \cdot 3^2 = 36$ **5.** LCM $= 2 \cdot 5 \cdot 11 = 110$
7. LCM $= 2 \cdot 2 \cdot 3 \cdot 5 = 2^2 \cdot 3 \cdot 5 = 60$ **9.** LCM $= 2 \cdot 2 \cdot 2 \cdot 5 \cdot 5 = 2^3 \cdot 5^2 = 200$
11. LCM $= 2 \cdot 2 \cdot 7 \cdot 7 = 2^2 \cdot 7^2 = 196$ **13.** LCM $= 2 \cdot 2 \cdot 2 \cdot 2 \cdot 3 \cdot 5 = 2^4 \cdot 3 \cdot 5 = 240$
15. LCM $= 2 \cdot 2 \cdot 5 \cdot 5 = 2^2 \cdot 5^2 = 100$ **17.** LCM $= 2 \cdot 2 \cdot 5 \cdot 5 \cdot 7 = 2^2 \cdot 5^2 \cdot 7 = 700$
19. LCM $= 2 \cdot 2 \cdot 3 \cdot 3 \cdot 7 = 2^2 \cdot 3^2 \cdot 7 = 252$ **21.** LCM $= 2 \cdot 2 \cdot 2 = 2^3 = 8$
23. LCM $= 2 \cdot 2 \cdot 2 \cdot 3 \cdot 5 \cdot 13 = 2^3 \cdot 3 \cdot 5 \cdot 13 = 1560$ **25.** LCM $= 2 \cdot 2 \cdot 3 \cdot 5 = 2^2 \cdot 3 \cdot 5 = 60$
27. LCM $= 2 \cdot 2 \cdot 2 \cdot 2 \cdot 3 \cdot 5 = 2^4 \cdot 3 \cdot 5 = 240$
29. LCM $= 2 \cdot 3 \cdot 3 \cdot 5 \cdot 5 \cdot 5 = 2 \cdot 3^2 \cdot 5^3 = 2250$ **31.** LCM $= 2 \cdot 3 \cdot 11 \cdot 11 = 2 \cdot 3 \cdot 11^2 = 726$
33. LCM $= 2 \cdot 3 \cdot 3 \cdot 5 \cdot 29 = 2 \cdot 3^2 \cdot 5 \cdot 29 = 2610$ **35.** LCM $= 3 \cdot 3 \cdot 3 \cdot 5 \cdot 5 = 3^3 \cdot 5^2 = 675$
37. LCM $= 2 \cdot 2 \cdot 2 \cdot 3 \cdot 5 = 2^3 \cdot 3 \cdot 5 = 120$ **39.** LCM $= 2 \cdot 2 \cdot 2 \cdot 2 \cdot 3 \cdot 5 \cdot 7 = 2^4 \cdot 3 \cdot 5 \cdot 7 = 1680$
41. (a) LCM $= 2^3 \cdot 3 \cdot 5 = 120$ (b) $120 = 8 \cdot 15$; $120 = 10 \cdot 12$
43. (a) LCM $= 2^3 \cdot 3 \cdot 5 = 120$ (b) $120 = 10 \cdot 12$; $120 = 15 \cdot 8$; $120 = 24 \cdot 5$
45. (a) LCM $= 2 \cdot 3^3 \cdot 5 = 270$ (b) $270 = 6 \cdot 45$; $270 = 18 \cdot 15$; $270 = 27 \cdot 10$
47. (a) LCM $= 2 \cdot 3^2 \cdot 5 \cdot 7^2 = 4410$ (b) $4410 = 45 \cdot 98$; $4410 = 63 \cdot 70$
49. (a) LCM $= 3^2 \cdot 11^2 \cdot 13 = 14{,}157$ (b) $14{,}157 = 99 \cdot 143$; $14{,}157 = 363 \cdot 39$
51. (a) 70 min (b) 7 laps and 5 laps **53.** (a) 48 hr (b) 4 orbits and 3 orbits
55. (a) 180 days (b) 18, 15, 12, and 10 trips

Chapter 2, Test, p. 143

1. 3 is the exponent, 7 is the base, and 343 is the power **2.** (a) 7, 11, 13, 17, 19 (b) 49, 121, 169, 289, 361
3. 13 **4.** 48 **5.** 0 **6.** 72 **7.** 2, 3, 5, 9, 10 **8.** none **9.** 2, 3, 4, 9 **10.** 2, 4, 5, 10
11. 13, 26, 39, 52, 65, 78, 91 **12.** $7452 \div 81 = 92$ so 81 is a factor of 7452 and $81 \cdot 92 = 7452$.
13. Yes. $2 \cdot 5 \cdot 6 \cdot 7 \cdot 9 = 42 \cdot 90$. So, 42 divides the product 90 times. **14.** 1, 60, 2, 3, 4, 5, 6, 10, 12, 15, 20, 30
15. $124 = 2 \cdot 2 \cdot 31 = 2^2 \cdot 31$ **16.** $165 = 3 \cdot 5 \cdot 11$ **17.** 107 is prime
18. (a) LCM $= 2 \cdot 2 \cdot 3 \cdot 7 = 2^2 \cdot 3 \cdot 7 = 84$ (b) $84 = 4 \cdot 21$; $84 = 14 \cdot 6$; $84 = 21 \cdot 4$
19. (a) LCM $= 2 \cdot 2 \cdot 3 \cdot 5 = 2^2 \cdot 3 \cdot 5 = 60$ (b) $60 = 6 \cdot 10$; $60 = 15 \cdot 4$; $60 = 60 \cdot 1$
20. (a) LCM $= 2 \cdot 2 \cdot 2 \cdot 3 \cdot 5 \cdot 7 = 2^3 \cdot 3 \cdot 5 \cdot 7 = 840$ (b) $840 = 8 \cdot 105$; $840 = 10 \cdot 84$; $840 = 15 \cdot 56$; $840 = 28 \cdot 30$ **21.** (a) every 36 days (b) every 180 days

---

Cumulative Review, p. 145

1. (a) $50{,}000 + 0 + 700 + 30 + 2$ (b) fifty thousand, seven hundred thirty-two
2. commutative property of multiplication **3.** associative property of multiplication **4.** additive identity
5. commutative property of addition **6.** 42,000 **7.** (a) 1980 (b) 2065 **8.** (a) 5200 (b) 5244
9. (a) 6300 (b) 6364 **10.** (a) 100 (b) 130 **11.** 27,000 **12.** 3 **13.** 57 **14.** 2, 4, 5, 10
15. 307 is prime **16.** 2 is the only even prime number. **17.** 1, 65, 5, 13 **18.** $475 = 5 \cdot 5 \cdot 19 = 5^2 \cdot 19$
19. $2 \cdot 3 \cdot 4 \cdot 6 \cdot 7 \cdot 9 = (2 \cdot 4 \cdot 7) \cdot (3 \cdot 6 \cdot 9) = 56 \cdot 162$; So, 56 divides the product 162 times.
20. (a) LCM $= 3 \cdot 3 \cdot 3 \cdot 5 \cdot 7 = 3^3 \cdot 5 \cdot 7 = 945$ (b) $945 = 15 \cdot 63$; $945 = 27 \cdot 35$
21. (a) LCM $= 2 \cdot 2 \cdot 3 \cdot 5 \cdot 11 \cdot 11 = 2^2 \cdot 3 \cdot 5 \cdot 11^2 = 7260$ (b) $7260 = 33 \cdot 220$; $7260 = 44 \cdot 165$; $7260 = 55 \cdot 132$; $7260 = 121 \cdot 60$ **22.** 2333 miles **23.** 1728 square inches **24.** 85

CHAPTER 3

Margin Exercises 3.1

1. $\frac{3}{4} \cdot \frac{3}{5} = \frac{9}{20}$

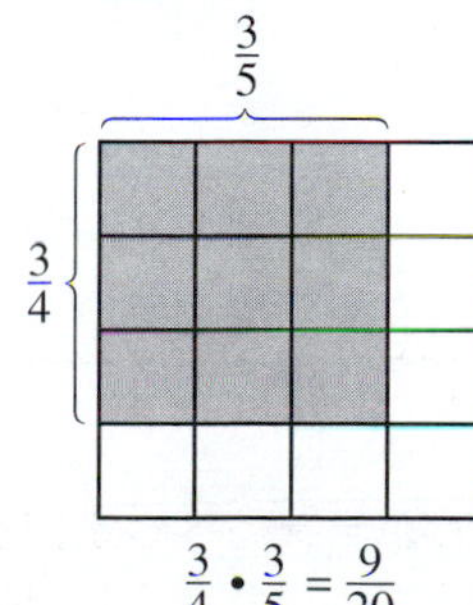

2. $\frac{3}{35}$ **3.** 0 **4.** $\frac{15}{14}$ **5.** $\frac{21}{40}$ **6.** $\frac{21}{32}$
7. commutative property of multiplication

8. commutative property of multiplication **9.** associative property of multiplication

10. associative property of multiplication **11.** $\frac{3}{5} = \frac{3}{5} \cdot \frac{4}{4} = \frac{12}{20}$ **12.** $\frac{1}{9} = \frac{1}{9} \cdot \frac{8}{8} = \frac{8}{72}$ **13.** $\frac{5}{8} = \frac{5}{8} \cdot \frac{5}{5} = \frac{25}{40}$

14. $\frac{3}{11} = \frac{3}{11} \cdot \frac{4}{4} = \frac{12}{44}$

Exercises 3.1, p. 161

1. 0 **3.** 0 **5.** undefined **7.** undefined **9.** Answers will vary. See the discussion in the text on page 153.

11. $\frac{1}{3}$ **13.** $\frac{1}{4}$ **15.** $\frac{1}{4}$ **17.** $\frac{1}{5} \cdot \frac{3}{4} = \frac{3}{20}$ **19.** $\frac{2}{3} \cdot \frac{4}{7} = \frac{8}{21}$

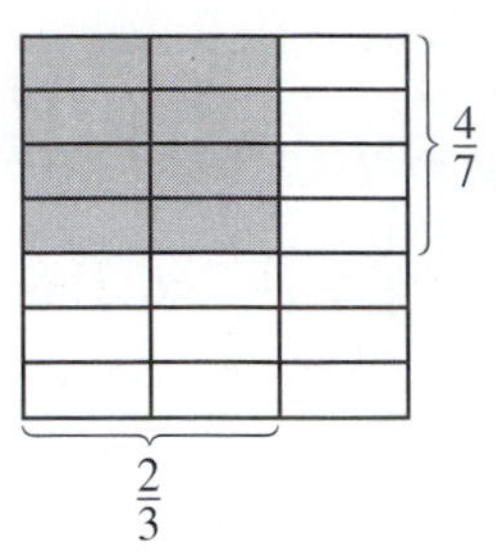

21. $\frac{3}{32}$ **23.** $\frac{9}{49}$ **5.** $\frac{15}{32}$ **27.** 0 **29.** $\frac{35}{12}$ **31.** $\frac{10}{1}$ or 10 **33.** $\frac{45}{2}$ **35.** $\frac{32}{15}$ **37.** $\frac{99}{20}$ **39.** $\frac{48}{455}$

41. $\frac{1}{10}$ **43.** $\frac{1}{8}$ **45.** $\frac{12}{35}$ **47.** commutative property of multiplication

49. associative property of multiplication **51.** $\frac{7}{20}; \frac{13}{20}$ **53.** $\frac{3}{8}$ **55.** $\frac{5}{8} = \frac{5}{8} \cdot \frac{3}{3} = \frac{15}{24}$ **57.** $\frac{2}{5} = \frac{2}{5} \cdot \frac{5}{5} = \frac{10}{25}$

59. $\frac{1}{9} = \frac{1}{9} \cdot \frac{5}{5} = \frac{5}{45}$ **61.** $\frac{5}{8} = \frac{5}{8} \cdot \frac{2}{2} = \frac{10}{16}$ **63.** $\frac{14}{3} = \frac{14}{3} \cdot \frac{3}{3} = \frac{42}{9}$ **65.** $\frac{9}{16} = \frac{9}{16} \cdot \frac{6}{6} = \frac{54}{96}$

67. $\frac{10}{11} = \frac{10}{11} \cdot \frac{4}{4} = \frac{40}{44}$ **69.** $\frac{11}{12} = \frac{11}{12} \cdot \frac{4}{4} = \frac{44}{48}$ **71.** $\frac{2}{3} = \frac{2}{3} \cdot \frac{16}{16} = \frac{32}{48}$ **73.** $\frac{9}{10} = \frac{9}{10} \cdot \frac{10}{10} = \frac{90}{100}$

75. $\frac{7}{10} = \frac{7}{10} \cdot \frac{7}{7} = \frac{49}{70}$ **77.** $\frac{5}{12} = \frac{5}{12} \cdot \frac{9}{9} = \frac{45}{108}$ **79.** $\frac{7}{6} = \frac{7}{6} \cdot \frac{6}{6} = \frac{42}{36}$

Margin Exercises 3.2

1. $\frac{2}{9}$ **2.** $\frac{7}{8}$ **3.** $\frac{3}{2}$ **4.** $\frac{12}{25}$ **5.** $\frac{1}{6}$ **6.** $\frac{3}{4}$ **7.** $\frac{16}{25}$ **8.** $\frac{2}{3}$ **9.** $\frac{4}{3}$ **10.** $\frac{1}{24}$ **11.** $\frac{4}{3}$ **12.** $\frac{1}{3}$ **13.** $\frac{9}{4}$

Exercises 3.2, p. 173

1. $\frac{1}{3}$ **3.** $\frac{3}{4}$ **5.** $\frac{2}{5}$ **7.** $\frac{7}{18}$ **9.** $\frac{0}{25}$ or 0 **11.** $\frac{2}{5}$ **13.** $\frac{5}{6}$ **15.** $\frac{2}{3}$ **17.** $\frac{2}{3}$ **19.** $\frac{6}{25}$ **21.** $\frac{2}{3}$

23. $\frac{12}{35}$ **25.** $\frac{2}{9}$ **27.** $\frac{25}{76}$ **29.** $\frac{1}{2}$ **31.** $\frac{3}{8}$ **33.** $\frac{12}{13}$ **35.** $\frac{11}{15}$ **37.** $\frac{8}{9}$ **39.** $\frac{2}{1}$ or 2 **41.** $\frac{8}{9}$ **43.** $\frac{5}{7}$

45. $\frac{1}{3}$ **47.** 1 **49.** $\frac{1}{6}$ **51.** $\frac{1}{12}$ **53.** $\frac{2}{25}$ **55.** $\frac{10}{3}$ **57.** $\frac{4}{15}$ **59.** $\frac{5}{18}$ **61.** $\frac{1}{4}$ **63.** $\frac{21}{4}$ **65.** $\frac{77}{4}$

67. $\frac{8}{5}$ **69.** $\frac{2}{7}$ **71.** $\frac{625}{128}$ feet **73.** 10 feet, 5 feet, $\frac{5}{2}$ feet, and $\frac{5}{2}$ feet **75.** $\frac{1}{12}$ **77.** 225 "earth-days"

The Recycle Bin, p. 178

1. 3, 5, 9 **2.** 2, 3 **3.** 2, 5, 10 **4.** 2, 4, 5, 10 **5.** 2, 4, 5, 10 **6.** 1, 30, 2, 3, 5, 6, 10, 15
7. 1, 48, 2, 3, 4, 6, 8, 12, 16, 24 **8.** 1, 63, 3, 7, 9, 21

Margin Exercises 3.3

1. $\frac{8}{7}$ **2.** $\frac{10}{1}$ or 10 **3.** $\frac{1}{16}$ **4.** $\frac{6}{5}$ **5.** $\frac{1}{4}$ **6.** $\frac{1}{1}$ or 1 **7.** $\frac{6}{5}$

Exercises 3.3, p. 183

1. $\frac{13}{12}$ **3.** 0 **5.** $\frac{8}{9}$ **7.** $\frac{5}{7}$ **9.** $\frac{7}{5}$ **11.** $\frac{1}{3}$ **13.** $\frac{3}{4}$ **15.** $\frac{4}{9}$ **17.** $\frac{5}{8}$ **19.** $\frac{39}{32}$ **21.** $\frac{9}{10}$ **23.** 1 **25.** $\frac{32}{21}$ **27.** $\frac{10}{3}$ **29.** $\frac{16}{21}$ **31.** $\frac{7}{60}$ **33.** $\frac{63}{50}$ **35.** $\frac{7}{5}$ **37.** $\frac{5}{7}$ **39.** $\frac{8}{3}$ **41.** $\frac{3}{8}$ **43.** $\frac{3}{2}$ **45.** $\frac{1}{4}$ **47.** $\frac{22}{7}$ **49.** $\frac{448}{1}$ or 448 **51.** $\frac{50}{27}$ **53.** 1250 freshmen

Check Your Number Sense **55.** more than 45; 60 passengers

57. More because you will be getting $\frac{1}{10}$ of a number larger than \$1200 as your raise. This number will be more than \$120 and it will be added to a number larger than \$1200.

The Recycle Bin, p. 188

1. LCM $= 2 \cdot 3 \cdot 5 \cdot 13 = 390$ **2.** LCM $= 2 \cdot 2 \cdot 3 \cdot 3 \cdot 7 = 2^2 \cdot 3^2 \cdot 7 = 252$
3. LCM $= 2 \cdot 2 \cdot 5 \cdot 5 = 2^2 \cdot 5^2 = 100$ **4.** LCM $= 3 \cdot 13 \cdot 17 = 663$
5. LCM $= 2 \cdot 2 \cdot 3 \cdot 5 \cdot 5 = 2^2 \cdot 3 \cdot 5^2 = 300$ **6.** LCM $= 2 \cdot 2 \cdot 2 \cdot 11 \cdot 11 = 2^3 \cdot 11^2 = 968$
7. (a) LCM $= 2 \cdot 5 \cdot 5 \cdot 5 = 2 \cdot 5^3 = 250$ (b) $250 = 50 \cdot 5$; $250 = 125 \cdot 2$
8. (a) LCM $= 2 \cdot 2 \cdot 2 \cdot 3 \cdot 5 = 2^3 \cdot 3 \cdot 5 = 120$ (b) $120 = 20 \cdot 6$; $120 = 24 \cdot 5$; $120 = 30 \cdot 4$
9. (a) LCM $= 2 \cdot 2 \cdot 3 \cdot 5 \cdot 7 = 2^2 \cdot 3 \cdot 5 \cdot 7 = 420$ (b) $420 = 12 \cdot 35$; $420 = 70 \cdot 6$
10. (a) LCM $= 3 \cdot 3 \cdot 5 \cdot 7 \cdot 11 = 3^2 \cdot 5 \cdot 7 \cdot 11 = 3465$ (b) $3465 = 45 \cdot 77$; $3465 = 63 \cdot 55$; $3465 = 99 \cdot 35$
11. 90 days

Margin Exercises 3.4

1. $\frac{5}{7}$ **2.** $\frac{1}{2}$ **3.** $\frac{11}{16}$ **4.** $\frac{5}{4}$ **5.** $\frac{53}{40}$ **6.** $\frac{23}{72}$ **7.** $\frac{3}{10}$ **8.** $\frac{11}{15}$ **9.** $\frac{1}{2}$ **10.** $\frac{2}{3}$ **11.** $\frac{17}{30}$ **12.** $\frac{25}{48}$

Exercises 3.4, p. 197

1. 1 **3.** $\frac{1}{5}$ **5.** $\frac{3}{2}$ **7.** $\frac{6}{5}$ **9.** $\frac{3}{5}$ **11.** $\frac{17}{15}$ **13.** $\frac{23}{20}$ **15.** $\frac{43}{200}$ **17.** $\frac{11}{16}$ **19.** 1 **21.** $\frac{3}{4}$ **23.** $\frac{8}{5}$ **25.** $\frac{151}{140}$ **27.** $\frac{377}{280}$ **29.** $\frac{3}{5}$ **31.** $\frac{5}{3}$ **33.** $\frac{119}{72}$ **35.** $\frac{19}{100}$ **37.** $\frac{49}{36}$ **39.** $\frac{73}{96}$ **41.** $\frac{3}{7}$ **43.** $\frac{3}{5}$ **45.** $\frac{1}{2}$ **47.** $\frac{1}{3}$ **49.** $\frac{3}{5}$ **51.** $\frac{1}{2}$ **53.** $\frac{13}{30}$ **55.** $\frac{1}{12}$ **57.** $\frac{9}{32}$ **59.** $\frac{13}{20}$ **61.** $\frac{7}{54}$ **63.** $\frac{1}{40}$ **65.** $\frac{7}{60}$ **67.** $\frac{27}{8}$ **69.** $\frac{3}{20}$ **71.** 0 **73.** $\frac{1}{10}$ **75.** $\frac{1}{3}$ **77.** $\frac{17}{20}$ **79.** $\frac{5}{24}$ **81.** $\frac{1}{50}$ **83.** 1 oz **85.** $\frac{23}{20}$ in. **87.** $\frac{5}{16}$ **89.** $\frac{11}{12}$ **91.** $\frac{1}{4}$

The Recycle Bin, p. 204

1. 81 **2.** 169 **3.** 225 **4.** 324 **5.** 0 **6.** 29 **7.** 74 **8.** 0 **9.** 70 **10.** 4

Margin Exercises 3.5

1. $\frac{9}{22} = \frac{27}{66}$ and $\frac{10}{33} = \frac{20}{66}$; so, $\frac{9}{22}$ is larger by $\frac{7}{66}$.

2. $\frac{7}{12} = \frac{21}{36}$, $\frac{5}{9} = \frac{20}{36}$, and $\frac{2}{3} = \frac{24}{36}$; so, the numbers in order are $\frac{5}{9}$, $\frac{7}{12}$, and $\frac{2}{3}$. **3.** $\frac{15}{8}$ **4.** $\frac{25}{84}$

Exercises 3.5, p. 211

1. $\frac{3}{4}$ is larger by $\frac{1}{12}$. **3.** $\frac{17}{20}$ is larger by $\frac{1}{20}$. **5.** $\frac{13}{20}$ is larger by $\frac{1}{40}$. **7.** $\frac{14}{35}$ and $\frac{12}{30}$ are equal. Both reduce to $\frac{2}{5}$.

9. $\frac{11}{48}$ is larger by $\frac{1}{60}$. **11.** $\frac{3}{5}, \frac{2}{3}, \frac{7}{10}; \frac{7}{10} - \frac{3}{5} = \frac{7}{10} - \frac{6}{10} = \frac{1}{10}$ **13.** $\frac{11}{12}, \frac{19}{20}, \frac{7}{6}; \frac{7}{6} - \frac{11}{12} = \frac{14}{12} - \frac{11}{12} = \frac{3}{12} = \frac{1}{4}$

15. $\frac{1}{4}, \frac{1}{3}, \frac{1}{2}; \frac{1}{2} - \frac{1}{4} = \frac{2}{4} - \frac{1}{4} = \frac{1}{4}$ **17.** $\frac{13}{18}, \frac{7}{9}, \frac{31}{36}; \frac{31}{36} - \frac{13}{18} = \frac{31}{36} - \frac{26}{36} = \frac{5}{36}$

19. $\frac{20}{10{,}000}, \frac{3}{1000}, \frac{1}{100}; \frac{1}{100} - \frac{20}{10{,}000} = \frac{100}{10{,}000} - \frac{20}{10{,}000} = \frac{80}{10{,}000} = \frac{1}{125}$ **21.** $\frac{2}{3}$ **23.** $\frac{13}{9}$ **25.** $\frac{187}{32}$ **27.** $\frac{1}{4}$

29. $\frac{8}{39}$ **31.** $\frac{15}{64}$ **33.** $\frac{29}{36}$ **35.** $\frac{3}{16}$ **37.** $\frac{11}{70}$ **39.** $\frac{25}{21}$ **41.** $\frac{33}{4}$

Check Your Number Sense **43.** (a) Yes. We can be guaranteed that their sum will be more than 1 if each of the fractions is greater than $\frac{1}{2}$. (b) No. If a fraction is less than 1, then that fraction *of* another number will be less than the other number. If the other number is less than 1, then the product must be less than 1.

45. Yes, the quotient will get smaller and smaller. The number is being divided into more parts. (See Exercise 44.) When 0 is divided by larger and larger numbers, the quotient remains 0.

Chapter 3 Test, p. 217

1. $\frac{8}{5}$ **2.** $\frac{35}{80}$ **3.** commutative property of multiplication **4.** $\frac{5}{6}$ **5.** $\frac{7}{5}$ **6.** $\frac{3}{4}$ **7.** $\frac{4}{15}$ **8.** $\frac{5}{72}$ **9.** 9

10. $\frac{2}{9}$ **11.** 25 **12.** $\frac{9}{80}$ **13.** $\frac{25}{44}$ **14.** $\frac{8}{11}$ **15.** $\frac{1}{2}$ **16.** $\frac{3}{8}$ **17.** $\frac{1}{2}$ **18.** $\frac{12}{5}$ **19.** 2 **20.** $\frac{1}{6}$

21. $\frac{1}{16}$ **22.** $\frac{13}{20}$ **23.** $\frac{11}{30}$ **24.** $\frac{1}{3}$

25. $\frac{7}{8} = \frac{21}{24}; \frac{3}{4} = \frac{18}{24}; \frac{2}{3} = \frac{16}{24}$; so, comparing the numerators of the fractions with the denominator 24 gives the following order to the original fractions: $\frac{2}{3}, \frac{3}{4}$, and $\frac{7}{8}$. **26.** $\frac{2}{3}$

Cumulative Review, p. 219

1. 2,500,000

2. (a) as examples: $3 + (4 + 5) = (3 + 4) + 5$ and $1 + (7 + 3) = (1 + 7) + 3$ (b) as examples: $8 \cdot 1 = 8$ and $29 \cdot 1 = 29$

3. (a) 3690 (estimate) (b) 3413 (sum) **4.** (a) 20,000 (estimate) (b) 24,192 (product)

5. (a) 1000 (estimate) (b) 996 (difference) **6.** (a) 250 (estimate) (b) 316 R10 (quotient and remainder)

7. 350,000

8. 2: because the number is even; 3: because the sum of the digits is 18 and 18 is divisible by 3; 5: because the units digit is 0; 9: because the sum of the digits is 18 and 18 is divisible by 9; 10: because the units digit is 0

9. 431 is prime. **10.** $780 = 2 \cdot 2 \cdot 3 \cdot 5 \cdot 13 = 2^2 \cdot 3 \cdot 5 \cdot 13$

11. (a) $\text{LCM} = 2 \cdot 3 \cdot 3 \cdot 5 \cdot 7 = 630$ (b) $630 = 18 \cdot 35$; $630 = 42 \cdot 15$; $630 = 90 \cdot 7$

12. (a) $\text{LCM} = 2 \cdot 2 \cdot 2 \cdot 2 \cdot 2 \cdot 3 \cdot 3 \cdot 5 \cdot 7 = 2^5 \cdot 3^2 \cdot 5 \cdot 7 = 10{,}080$ (b) $10{,}080 = 36 \cdot 280$; $10{,}080 = 60 \cdot 168$; $10{,}080 = 84 \cdot 120$; $10{,}080 = 96 \cdot 105$

13. 32 **14.** $\frac{1}{24}$ **15.** $\frac{139}{120}$ **16.** $\frac{79}{102}$ **17.** $\frac{9}{8}$ **18.** $\frac{9}{7}$ **19.** 95 **20.** \$3680 **21.** 775

22. 800 square centimeters **23.** about 4000 miles **24.** (a) $\frac{8}{105}$ (b) $\frac{4}{35}$ (c) $\frac{2}{3}$

CHAPTER 4

Margin Exercises 4.1

1. $\frac{57}{8}$ **2.** $\frac{47}{4}$ **3.** $\frac{73}{9}$ **4.** $\frac{11}{2}$ **5.** $2\frac{4}{7}$ **6.** $1\frac{1}{4}$ **7.** $12\frac{1}{2}$ **8.** $2\frac{1}{3}$

Exercises 4.1, p. 229

1. $\frac{4}{3}$ **3.** $\frac{4}{3}$ **5.** $\frac{3}{2}$ **7.** $\frac{7}{5}$ **9.** $\frac{5}{4}$ **11.** $4\frac{1}{6}$ **13.** $1\frac{1}{3}$ **15.** $1\frac{1}{2}$ **17.** $6\frac{1}{7}$ **19.** $7\frac{1}{2}$ **21.** $3\frac{1}{9}$ **23.** 3 **25.** $4\frac{1}{2}$ **27.** 3 **29.** $1\frac{3}{4}$ **31.** $\frac{37}{8}$ **33.** $\frac{76}{15}$ **35.** $\frac{46}{11}$ **37.** $\frac{7}{3}$ **39.** $\frac{32}{3}$ **41.** $\frac{34}{5}$ **43.** $\frac{50}{3}$ **45.** $\frac{101}{5}$ **47.** $\frac{92}{7}$ **49.** $\frac{51}{3}$ or 17 **51.** $2\frac{19}{24}$ feet **53.** net gain of $1\frac{3}{4}$ dollars

The Recycle Bin, p. 232

1. $\frac{7}{9}$ **2.** $\frac{2}{5}$ **3.** $\frac{3}{1}$ or 3 **4.** $\frac{6}{13}$ **5.** $\frac{1}{5}$ **6.** $\frac{3}{16}$ **7.** $\frac{5}{9}$ **8.** $\frac{15}{14}$

Margin Exercises 4.2

1. 15 **2.** $3\frac{1}{2}$ **3.** $189\frac{1}{16}$ **4.** 18 **5.** $\frac{8}{5}$ or $1\frac{3}{5}$ **6.** $1\frac{9}{16}$ **7.** 1 **8.** 6

Exercises 4.2, p. 239

1. $\frac{91}{12}$ or $7\frac{7}{12}$ **3.** $\frac{21}{2}$ or $10\frac{1}{2}$ **5.** $\frac{45}{2}$ or $22\frac{1}{2}$ **7.** $\frac{187}{6}$ or $31\frac{1}{6}$ **9.** 8 **11.** 30 **13.** $\frac{87}{4}$ or $21\frac{3}{4}$ **15.** 1 **17.** $\frac{55}{2}$ or $27\frac{1}{2}$ **19.** $\frac{77}{4}$ or $19\frac{1}{4}$ **21.** $\frac{1}{7}$ **23.** $\frac{1}{10}$ **25.** 17 **27.** $\frac{21}{10}$ or $2\frac{1}{10}$ **29.** $\frac{18,271}{10}$ or $1827\frac{1}{10}$ **31.** $\frac{226,226}{75}$ or $3016\frac{26}{75}$ **33.** 40 **35.** 20 **37.** $\frac{21}{16}$ or $1\frac{5}{16}$ **39.** $\frac{117}{35}$ or $3\frac{12}{35}$ **41.** $\frac{1}{3}$ **43.** $\frac{5}{9}$ **45.** $\frac{10}{39}$ **47.** $\frac{29}{155}$ **49.** $\frac{28}{17}$ or $1\frac{11}{17}$ **51.** $\frac{200}{301}$ **53.** $\frac{7}{5}$ or $1\frac{2}{5}$ **55.** $\frac{41}{3}$ or $13\frac{2}{3}$ **57.** $\frac{63}{5}$ or $12\frac{2}{3}$ **59.** $\frac{20}{11}$ or $1\frac{9}{11}$ **61.** 177 miles **63.** 90 pages; 18 hours **65.** 10 feet; 22 feet **67.** $\frac{50}{27}$ or $1\frac{23}{27}$ **69.** 175 passengers **71.** $3200 **73.** (a) $460\frac{1}{4}$ miles (b) $2325\frac{3}{4}$¢ (or 2326¢ or $23\frac{26}{100}$ dollars) **75.** $162\frac{1}{2}$ miles

Check Your Number Sense **77.** (a) More than 1 since the product is greater than the first factor. (b) Less than 2 since the product is less than twice the first factor. The number is $\frac{35}{19}$ or $1\frac{16}{19}$.

The Recycle Bin, p. 246

1. $\frac{17}{21}$ **2.** $\frac{61}{40}$ **3.** $\frac{31}{30}$ **4.** $\frac{59}{45}$ **5.** $\frac{17}{40}$ **6.** $\frac{67}{84}$ **7.** $\frac{17}{64}$ **8.** $\frac{1}{6}$

Margin Exercises 4.3

1. $5\frac{5}{6}$ **2.** $36\frac{9}{10}$ **3.** $15\frac{1}{2}$ **4.** $12\frac{3}{10}$

Exercises 4.3, p. 251

1. (a) 14 (estimate) (b) $14\frac{5}{7}$ (sum) **3.** (a) 7 (estimate) (b) 8 (sum) **5.** (a) 12 (estimate) (b) $12\frac{3}{4}$ (sum) **7.** (a) 8 (estimate) (b) $8\frac{2}{3}$ (sum) **9.** (a) 34 (estimate) (b) $34\frac{7}{18}$ (sum) **11.** (a) 7 (estimate) (b) $7\frac{2}{3}$ (sum) **13.** (a) 42 (estimate) (b) $42\frac{7}{20}$ (sum) **15.** (a) 10 (estimate) (b) $10\frac{3}{4}$ (sum) **17.** (a) 12 (estimate) (b) $12\frac{7}{27}$ (sum) **19.** (a) 18 (estimate) (b) $18\frac{23}{45}$ (sum)

21. (a) 27 (estimate) (b) $28\frac{1}{2}$ (sum) **23.** (a) 9 (estimate) (b) $9\frac{29}{40}$ (sum) **25.** (a) 12 (estimate) (b) $12\frac{47}{60}$ (sum)

27. (a) 62 (estimate) (b) $63\frac{11}{24}$ (sum) **29.** (a) 53 (estimate) (b) $53\frac{83}{192}$ (sum)

31. (a) 18 (estimate) (b) $18\frac{7}{8}$ (sum) **33.** (a) 36 (estimate) (b) $36\frac{23}{24}$ (sum)

35. (a) 88 (estimate) (b) $90\frac{1}{8}$ (sum) **37.** $35\frac{7}{10}$ kilometers **39.** $12\frac{1}{8}$ **41.** $\$142\frac{4}{5}$ million

Check Your Number Sense **43.** (a) 6 (b) 600 (c) 60,000

Margin Exercises 4.4

1. $6\frac{1}{2}$ **2.** $8\frac{17}{30}$ **3.** $4\frac{15}{28}$ **4.** $7\frac{5}{11}$ **5.** $4\frac{2}{3}$

Exercises 4.4, p. 259

1. $4\frac{1}{2}$ **3.** $1\frac{5}{12}$ **5.** 4 **7.** $3\frac{1}{2}$ **9.** $3\frac{1}{4}$ **11.** $3\frac{5}{8}$ **13.** $4\frac{2}{3}$ **15.** $6\frac{7}{12}$ **17.** $10\frac{4}{5}$ **19.** $3\frac{9}{10}$ **21.** $7\frac{4}{5}$

23. $1\frac{11}{16}$ **25.** 17 **27.** $13\frac{20}{21}$ **29.** $16\frac{5}{24}$ **31.** (a) $\frac{3}{5}$ hours (b) 36 minutes **33.** $\frac{5}{6}$ hours **35.** $\$4\frac{5}{8}$

37. $219\frac{13}{16}$ pounds **39.** $4\frac{13}{20}$ parts

The Recycle Bin, p. 262

1. 18 **2.** 101 **3.** $\frac{47}{18}$ or $2\frac{11}{18}$ **4.** 0

Margin Exercises 4.5

1. 2 **2.** 30 **3.** $\frac{1}{4}$ **4.** $\frac{2}{91}$

Exercises 4.5, p. 267

1. $\frac{44}{25}$ or $1\frac{19}{25}$ **3.** $\frac{52}{75}$ **5.** $\frac{25}{21}$ or $\frac{4}{21}$ **7.** 4 **9.** $\frac{7}{45}$ **11.** $\frac{429}{365}$ or $1\frac{64}{365}$ **13.** $\frac{69}{35}$ or $1\frac{34}{35}$ **15.** $\frac{1}{5}$

17. $\frac{172}{9}$ or $19\frac{1}{9}$ **19.** $\frac{791}{120}$ or $6\frac{71}{120}$ **21.** $\frac{3}{5}$ **23.** $\frac{15}{4}$ or $3\frac{3}{4}$ **25.** $\frac{543}{40}$ or $13\frac{23}{40}$ **27.** $\frac{42}{5}$ or $8\frac{2}{5}$

29. $\frac{47}{40}$ or $1\frac{7}{40}$ **31.** $6\frac{31}{32}$

Margin Exercises 4.6

1. 4 hr 15 min **2.** 11 ft 1 in. **3.** 34 min 10 sec **4.** 2 gal 1 qt 19 fl oz

Exercises 4.6, p. 275

1. 4 ft 8 in. **3.** 7 lb 4 oz **5.** 6 min 20 sec **7.** 3 days 6 hr **9.** 9 gal 1 qt
11. 5 pt 4 fl oz (or 2 qt 1 pt 4 fl oz) **13.** 9 ft 7 in. (or 3 yd 7 in.) **15.** 18 lb 2 oz **17.** 18 min 10 sec
19. 4 hr 11 min 15 sec **21.** 6 days 2 hr 35 min **23.** 13 gal 3 qt
25. 7 gal 2 qt 20 fl oz (or 7 gal 2 qt 1 pt 4 fl oz) **27.** 12 yd 2 ft 6 in. **29.** 2 yd 2 ft 2 in.
31. 3 gal 2 qt 30 fl oz (or 3 gal 2 qt 1 pt 14 fl oz) **33.** 2 hr 45 min **35.** 4 min 50 sec **37.** 4 lb 8 oz
39. 3 ft 8 in. (or 1 yd 8 in.) **41.** 7 ft 1 in. (or 2 yd 1 ft 1 in.) **43.** 3 hr 40 min **45.** 7 gal 1 qt
47. 16 ft 8 in. (or 5 yd 1 ft 8 in.)

Chapter 4 Test, p. 281

1. 0 **2.** (a) $3\frac{2}{5}$ (b) $3\frac{1}{33}$ **3.** (a) $\frac{53}{8}$ (b) $\frac{43}{10}$ **4.** $6\frac{7}{12}$ **5.** $2\frac{5}{6}$ **6.** $24\frac{1}{6}$ **7.** $10\frac{5}{24}$ **8.** $1\frac{2}{5}$ **9.** $7\frac{1}{20}$
10. $1\frac{37}{42}$ **11.** $4\frac{17}{40}$ **12.** 20 **13.** $\frac{15}{4}$ or $3\frac{3}{4}$ **14.** $\frac{1}{3}$ **15.** $\frac{3}{2}$ or $1\frac{1}{2}$ **16.** $\frac{35}{3}$ or $11\frac{2}{3}$ **17.** $\frac{83}{105}$
18. $\frac{99}{8}$ or $12\frac{3}{8}$ **19.** 5 hr 40 min. **20.** 19 ft 3 in. (or 6 yd 1 ft 3 in.) **21.** $16\frac{9}{10}$ centimeters
22. 561 square inches

Cumulative Review, p. 283

1. 50,000 + 3000 + 400 + 60 + 0; fifty-three thousand four hundred sixty **2.** 270,000
3. commutative property of addition **4.** (a) B (b) A (c) D (d) C **5.** 5,600,000 **6.** 158
7. 2, 3, 5, 7, 11, 13, 17, 19, 23, 29 **8.** (a) $170 = 2 \cdot 5 \cdot 17$ (b) $305 = 5 \cdot 61$ **9.** 0, undefined **10.** 43
11. 13,250 **12.** 2527 **13.** 26,352 **14.** 203 **15.** $\frac{7}{48}$ **16.** $\frac{1}{20}$ **17.** $\frac{7}{24}$ **18.** $\frac{28}{15}$ or $1\frac{13}{15}$ **19.** $5\frac{8}{35}$
20. $\frac{7}{4}$ or $1\frac{3}{4}$ **21.** $\frac{315}{2}$ or $157\frac{1}{2}$ **22.** 3 **23.** $\frac{241}{90}$ or $2\frac{61}{90}$ **24.** $\frac{7}{10} = \frac{42}{60}$; $\frac{3}{4} = \frac{45}{60}$; $\frac{5}{6} = \frac{50}{60}$; so, the fractions in order are $\frac{7}{10}$, $\frac{3}{4}$, and $\frac{5}{6}$. **25.** (a) 5 hr 11 min 15 sec (b) 28 lb 6 oz **26.** \$941 **27.** $16\frac{1}{2}$ gal
28. \$44 **29.** \$225 **30.** 3 hr 50 min; 15 hr 20 min

CHAPTER 5

Margin Exercises 5.1

1. twenty and seven tenths **2.** nine and four hundredths **3.** eighteen and six hundred fifty-one thousandths
4. two and eight ten-thousandths **5.** 10.11 **6.** 4.005 **7.** 800.3 **8.** 1600.264 **9.** 3.3 **10.** 0.079
11. 8000 **12.** 0.00069

Exercises 5.1, p. 295

1. 37.498 **2.** 4.11 **5.** 95.2 **7.** 62.7 **9.** $82\frac{56}{100}$ **11.** $10\frac{576}{1000}$ **13.** $65\frac{3}{1000}$ **15.** 0.3 **17.** 0.17
19. 60.028 **21.** 850.0036 **23.** five tenths **25.** five and six hundredths **27.** seven and three thousandths
29. ten and four thousand six hundred thirty-eight ten-thousandths
31. (a) 2 (b) 4 (c) 7 (d) 7, 4, 5, 7, 2 (e) 8500 **33.** 4.8 **35.** 76.3 **37.** 89.0 **39.** 18.0 **41.** 0.39
43. 5.72 **45.** 7.00 **47.** 0.08 **49.** 0.067 **51.** 0.634 **53.** 32.479 **55.** 0.002 **57.** 479 **59.** 20
61. 650 **63.** 6333 **65.** 5160 **67.** 500 **69.** 1000 **71.** 92,540 **73.** 7000 **75.** 48,000
77. 217,000 **79.** 4,501,000 **81.** 0.00058 **83.** 470 **85.** 500 **87.** 3.230 **89.** 80,000
91. two and fifty-four hundredths **93.** one hundred seventy-six and four hundred fifty-seven thousandths
95. three and fourteen thousand, one hundred fifty-nine hundred-thousandths **97.** four hundred five thousandths

The Recycle Bin, p. 300

1. (a) 9000 (estimate) (b) 9107 (sum) **2.** (a) 1600 (estimate) (b) 1555 (sum)
3. (a) 550,000 (estimate) (b) 562,712 (sum) **4.** (a) 200 (estimate) (b) 188 (difference)
5. (a) 4000 (estimate) (b) 4001 (difference) **6.** (a) 85,000 (estimate) (b) 84,891 (difference) **7.** 350

Margin Exercises 5.2

1. 50.78 **2.** 10.33 **3.** 59.804 **4.** 2.456 **5.** 9.824

Exercises 5.2, p. 305

1. 2.3 **3.** 7.55 **5.** 72.31 **7.** 276.096 **9.** 44.6516 **11.** 118.333 **13.** 7.148 **15.** 93.877 **17.** 103.429 **19.** 137.150 **21.** 1.44 **23.** 15.89 **25.** 64.947 **27.** 4.7974 **29.** 2.9434 **31.** $4.50 **33.** (a) $94.85 (b) $5.15 **35.** (a) $95.50 (b) $4.50 **37.** $1300 **39.** 2687.725 million bushels of grain **41.** 0.188919

Check Your Number Sense **43.** (a) no (b) yes (c) yes; The sum cannot be more than 2 because each number is less than 1. The sum can be more than 1 if at least one of the numbers is more than 0.5. The sum can be less than 1 if both numbers are less than 0.5.

The Recycle Bin, p. 310

1. 350,000 **2.** 525 **3.** 7020 **4.** 2,000,000 (estimate); 1,714,125 (product) **5.** 168; 79; 13,272

Margin Exercises 5.3

1. 0.16 **2.** 0.224 **3.** 11.38081 **4.** 87.5 **5.** 630 **6.** 189.4 **7.** 560 cm **8.** 352.5 mm **9.** 16 430 m

Exercises 5.3, p. 317

1. (a) B (b) C (c) E (d) A (e) D **3.** 0.42 **5.** 0.04 **7.** 21.6 **9.** 0.42 **11.** 0.004 **13.** 0.108 **15.** 0.0276 **17.** 0.0486 **19.** 0.0006 **21.** 0.375 **23.** 1.4 **25.** 1.725 **27.** 5.063 **29.** 0.08 **31.** 346 **33.** 782 **35.** 1610 **37.** 4.35 **39.** 18.6 **41.** 380 **43.** 50 **45.** 74,000 **47.** 130 mm **49.** 1500 cm **51.** 6170 mm **53.** 16 000 m **55.** 600 m = 60 000 cm = 600 000 mm **57.** 20 m = 2000 cm = 20 000 mm **59.** 0.009 (estimate); 0.00954 (product) **61.** 2 (estimate); 2.032 (product) **63.** 0.0014 (estimate); 0.0013845 (product) **65.** 2 (estimate); 1.717202 (product) **67.** 1.6 (estimate); 1.4094 (product) **69.** 3.6 (estimate); 3.4314 (product) **71.** $240.90 **73.** (a) $636 (b) $617.98 **75.** 12 ft (perimeter); 9 square feet (area) **77.** $30,855

The Recycle Bin, p. 322

1. 8 R 0 **2.** 15 R 15 **3.** 81 R 7 **4.** 203 R 100 **5.** (a) about 1 square mile per person (b) about 2.5 square kilometers per person

---

Margin Exercises 5.4

1. 4.7 **2.** 1.01 **3.** 3.5 **4.** 0.26 **5.** 728.68 **6.** 0.4231 **7.** 32.8 **8.** 0.00576 **9.** 3.5 m **10.** 6.8 cm **11.** 3.952 m

Exercises 5.4, p. 331

1. (a) B (b) E (c) D (d) C (e) A **3.** 2.34 **5.** 0.99 **7.** 0.08 **9.** 2056 **11.** 20 **13.** 56.9 **15.** 0.7 **17.** 0.1 **19.** 21.0 **21.** 0.01 **23.** 5.70 **25.** 2.74 **27.** 5.04 **29.** 0.784 **31.** 0.5036 **33.** 0.07385 **35.** 16.7 **37.** 0.785 **39.** 0.01699 **41.** 0.5 cm **43.** 0.83 m **45.** 0.344 m **47.** 1.5 km **49.** 9.72 cm **51.** 3.2 cm = 0.32 dm = 0.032 m **53.** 3.5 (estimate); 3.087 (quotient) **55.** 0.222 (estimate); 0.285 (quotient) **57.** 0.005 (estimate); 0.007 (quotient) **59.** (a) about 400 miles (b) 442.8 miles **61.** (a) about $1 per pound (b) $1.25 per pound **63.** $239.56 **65.** $295 **67.** 5620 at bats **69.** 285 free throws **71.** 42.2 kilometers

Check Your Number Sense **73.** (a) $5000 (b) $6500

Margin Exercises 5.5

1. 7.832×10^8 **2.** 3.9×10^{-4} **3.** 237,000 **4.** 0.00065

Exercises 5.5, p. 341

1. 5.0×10^6 **3.** 7.5×10^5 **5.** 6.7×10^7 **7.** 1.75×10^8 **9.** 2.137×10^{11} **11.** 6.2×10^{-4}
13. 2.5×10^{-5} **15.** 8.0×10^{-7} **17.** 6.71×10^{-9} **19.** 3.21×10^{-12} **21.** 5700 **23.** 75,400
25. 4,720,000 **27.** 0.0057 **29.** 0.000 018 4 **31.** 0.000 000 052 4 **33.** no; 5.671×10^4 **35.** yes
37. no; 4.743×10^{-3} **39.** no; 1.786×10^{-5} **41.** (a) 4.0×10^4 (b) 40,000
43. (a) 5.358×10^9 (b) 5,358,000,000 **45.** (a) 3.08×10^8 (b) 308,000,000 **47.** (a) 2×10^{-7} (b) 0.000 000 2
49. (a) 3.39×10^8 (b) 339,000,000 **51.** (a) 3×10^{10} (b) 300,000,000 m per sec, 30,000,000,000 cm per sec
53. 0.000 000 000 000 000 000 000 0199 26 **55.** 250,000,000
57. 81,170,000,000,000,000; 67,470,000,000,000,000

Exercises 5.6, p. 351

1. $\frac{9}{10}$ **3.** $\frac{5}{10}$ **5.** $\frac{62}{100}$ **7.** $\frac{57}{100}$ **9.** $\frac{526}{1000}$ **11.** $\frac{16}{1000}$ **13.** $\frac{51}{10}$ **15.** $\frac{815}{100}$ **17.** $\frac{1}{8}$ **19.** $\frac{9}{50}$
21. $\frac{9}{40}$ **23.** $\frac{17}{100}$ **25.** $\frac{16}{5}$ or $3\frac{1}{5}$ **27.** $\frac{25}{4}$ or $6\frac{1}{4}$ **29.** $0.\overline{6}$ **31.** $0.\overline{45}$ **33.** 0.6875 **35.** $0.\overline{428571}$
37. $0.1\overline{6}$ **39.** $0.\overline{5}$ **41.** 0.292 **43.** 0.417 **45.** 0.031 **47.** 1.231 **49.** 0.938 **51.** 0.70 **53.** 1.635
55. 14.98 **57.** 1.125 **59.** 13.51 **61.** 0.2555 **63.** 11.0825 **65.** 2.638 **67.** 27.3
69. 2.3 is larger; 0.05 **71.** 0.28 is larger; $0.00\overline{72}$ **73.** $70\frac{3}{10}$ **75.** $26\frac{3}{10}$; $24\frac{1}{10}$ **77.** $22\frac{8}{10}$ (or $22\frac{4}{5}$)
79. 0.06; 20 **81.** 5.72

Margin Exercises 5.7

1. 2.6458 **2.** 8.3666 **3.** 24.4949 **4.** 94.8683 **5.** $4\sqrt{5}$ **6.** $3\sqrt{10}$ **7.** $10\sqrt{30}$
8. Yes. $16^2 + 30^2 = 256 + 900 = 1156 = 34^2$ **9.** No. $4^2 + 5^2 = 16 + 25 = 41 \neq 6^2$

Exercises 5.7, p. 363

1. yes ($144 = 12^2$) **3.** yes ($81 = 9^2$) **5.** yes ($400 = 20^2$) **7.** no **9.** no
11. $(1.732)^2 = 2.999824$ and $(1.733)^2 = 3.003289$ **13.** $2\sqrt{3}$ **15.** $2\sqrt{6}$ **17.** $4\sqrt{3}$ **19.** $11\sqrt{3}$
21. $10\sqrt{5}$ **23.** $8\sqrt{2}$ **25.** $6\sqrt{2}$ **27.** $11\sqrt{5}$ **29.** 13 **31.** $20\sqrt{2}$ **33.** $3\sqrt{10}$ **35.** 16
37. yes; $6^2 + 8^2 = 10^2$ **39.** $c = \sqrt{5}$ **41.** $x = 5\sqrt{2}$ **43.** $c = 2\sqrt{29}$ **45.** 2.8284 **47.** 6.7082
49. 6.3246 **51.** 8.6603 **53.** 17.3205 **55.** 22.3607 **57.** 63.2 feet **59.** 522.0 feet
61. (a) 127.3 feet (b) closer to home plate **63.** 44.9 inches **65.** 7.1 centimeters

Chapter 5 Test, p. 371

1. thirty and six hundred fifty-seven thousandths **2.** $\frac{375}{1000} = \frac{3}{8}$ **3.** 0.3125 **4.** 203.02 **5.** 2000 **6.** 0.1
7. 103.758 **8.** 1.888 **9.** 122.11 **10.** 18.145 **11.** 37.313 **12.** 0.90395 **13.** 0.294 **14.** 12.80335
15. 1920 **16.** 0.03614 **17.** 17.83 **18.** 66.28 **19.** 681 m = 68 100 cm = 681 000 mm
20. 35.5 cm = 0.355 m **21.** 8.65×10^7 **22.** 3.72×10^{-4} **23.** $20\sqrt{2} \approx 28.2843$ **24.** $5\sqrt{5} \approx 11.1803$
25. (a) approximately 400 miles (b) 391.62 miles **26.** $20\sqrt{29} \approx 107.7$ yards

Cumulative Review, p. 373

1. (a) commutative property of addition (b) associative property of multiplication (c) additive identity **2.** 13
3. 28 **4.** 52 **5.** 45 **6.** $2^2 \cdot 3 \cdot 5 \cdot 7$ **7.** (a) 140 (b) 64 **8.** $\frac{1}{30}$ **9.** $8\frac{21}{40}$ **10.** $55\frac{29}{40}$ **11.** $32\frac{7}{20}$
12. 2 **13.** $\frac{11}{30}$ **14.** 6 **15.** $\frac{32}{9} = 3\frac{5}{9}$ **16.** 72 **17.** 18.9 **18.** 2.185
19. (a) 1.87×10^7 (b) 6.32×10^{-7} **20.** $12^2 + 16^2 = 144 + 256 = 400 = 20^2$
21. (a) approximately \$100,000 (b) \$107,700 **22.** \$300

CHAPTER 6

Margin Exercises 6.1

1. $\frac{3}{4}$ **2.** $\frac{3}{5}$ **3.** $\frac{\text{5 washers}}{\text{4 bolts}}$

Exercises 6.1, p. 379

1. $\frac{1}{2}$ **3.** $\frac{4}{1}$ **5.** $\frac{\text{50 miles}}{\text{1 hour}}$ **7.** $\frac{\text{25 miles}}{\text{1 gallon}}$ **9.** $\frac{\text{6 chairs}}{\text{5 people}}$ **11.** $\frac{3}{4}$ **13.** $\frac{8}{7}$ **15.** $\frac{\text{\$2 profit}}{\text{\$5 invested}}$ **17.** $\frac{1}{1}$ or 1
19. $\frac{\text{1 hit}}{\text{4 times at bat}}$ **21.** $\frac{7}{25}$ **23.** $\frac{9}{41}$

The Recycle Bin, p. 382

1. 11 **2.** $\frac{185}{304}$ **3.** $\frac{1}{20}$ **4.** 80 **5.** 149.76 **6.** 0.02118 **7.** 5.5 **8.** 400

Exercises 6.2, p. 387

1. 5.6¢/oz; 5.3¢/oz; 32 oz box **3.** 8.7¢/oz; 11.8¢/oz; 8 oz container **5.** 43.7¢/oz; 64.5¢/oz; 8 oz jar
7. 10.6¢/fl oz; 8.3¢/fl oz; 11.5¢/fl oz; 12 fl oz can **9.** 15.6¢/oz; 14.1¢/oz; 25.6 oz box
11. 12.4¢/oz; 7.5¢/oz; 16 oz box **13.** 2.1¢/sq ft; 2.2¢/sq ft; 2.8¢/sq ft; 200 sq ft box
15. 13.7¢/oz; 13.3¢/oz; 12 oz pkg. **17.** 4.1¢/oz; 5.6¢/oz; 5.7¢/oz; 7.3¢/oz; 9 lb 3 oz box
19. 0.9¢/fl oz; 1.4¢/fl oz; 1.9¢/fl oz; 1 gal bottle **21.** 10.7¢/oz; 12.2¢/oz; 13.8¢/oz; 14 oz box
23. 7.7¢/oz; 5.3¢/oz; 32 oz jar **25.** 8.7¢/oz; 6.9¢/oz; 16 oz jar **27.** 5.8¢/oz; 6.0¢/oz; 40 oz can
29. 3.0¢/oz; 3.9¢/oz; 30 oz container

Margin Exercises 6.3

1. True. $15 \cdot 28 = 20 \cdot 21$ **2.** False. $5\frac{1}{2} \cdot 3 \neq 2 \cdot 6\frac{1}{2}$ **3.** True. $(1.3)(2.1) = (1.5)(1.82)$

Exercises 6.3, p. 393

1. (a) 7 and 544 (b) 8 and 476 **3.** true **5.** true **7.** true **9.** true **11.** true **13.** false **15.** true
17. true **19.** true **21.** false **23.** true

Margin Exercises 6.4

1. $R = 30$ **2.** $x = 60$ **3.** $x = 6$

Exercises 6.4, p. 399

1. $x = 12$ **3.** $x = 20$ **5.** $B = 40$ **7.** $x = 50$ **9.** $A = \frac{21}{2}$ (or 10.5) **11.** $D = 100$ **13.** $x = 1$
15. $x = \frac{3}{5}$ (or 0.6) **17.** $w = 24$ **19.** $y = 6$ **21.** $x = \frac{3}{2}$ (or $1\frac{1}{2}$ or 1.5) **23.** $R = 60$ **25.** $A = 2$
27. $B = 120$ **29.** $R = \frac{100}{3}$ (or $33\frac{1}{3}$) **31.** $x = 22$ **33.** $x = 1$ **35.** $x = 15$ **37.** $B = 7.8$
39. $x = 1.56$ **41.** $R = 50$ **43.** $B = 48$ **45.** $A = 27.3$ **47.** $A = 255$ **49.** $B = 5800$
Check Your Number Sense **51.** (b) 25 **53.** (c) 100 **55.** (c) 3.0

The Recycle Bin, p. 402

1. 8.51 **2.** 42.6 **3.** 13.7 **4.** 18.2925 **5.** 21.95 **6.** 2.98

Margin Exercises 6.5

1. 6 pounds **2.** 126

Exercises 6.5, p. 407

1. 45 miles **3.** \$3.27 **5.** \$120 **7.** \$400 **9.** \$160,000 **11.** 200 feet **13.** 7.5 **15.** $11\frac{1}{4}$ hours
17. \$26.25 **19.** 259,200 revolutions **21.** 2100 minutes (or 35 hours) **23.** 30.48 centimeters
25. 180 minutes (or 3 hours) **27.** 18 biscuits **29.** 2.7 kilograms **31.** 500 milligrams **33.** 4 drams
35. 1920 minims **37.** 60 grams **39.** $\frac{1}{2}$ ounce
41. (a) 56 pounds (b) 4 bags (because you cannot buy part of a bag) (c) \$56
43. (a) 4 gallons (assuming you cannot buy a part of a gallon); 3 gallons 1 quart (if you can buy this paint in quarts) (b) yes (c) 0.88 gallon (or 0.13 quart if you can buy quarts)
Check Your Number Sense **45.** (c) 150 **47.** (c) 20,000

The Recycle Bin, p. 414

1. close to 6; $\frac{63}{10}$ (or 6.3 or $6\frac{3}{10}$) **2.** close to 20; $19\frac{7}{10}$ **3.** less than 6; 5 **4.** $1\frac{1}{3}$

Chapter 6 Test, p. 417

1. $\frac{7}{6}$ **2.** extremes **3.** false; $4 \cdot 14 \neq 6 \cdot 9$ **4.** $\frac{3}{5}$ **5.** $\frac{2}{5}$ **6.** $\frac{55 \text{ miles}}{1 \text{ hour}}$
7. 6.75¢/oz; 6.4¢/oz; 32 oz carton **8.** $x = 27$ **9.** $y = \frac{50}{3}$ **10.** $x = 75$ **11.** $y = 19.5$ **12.** $x = \frac{5}{6}$
13. $x = \frac{1}{90}$ **14.** $y = 4.5$ **15.** $x = 36$ **16.** 171.4 miles **17.** 24 miles **18.** \$295
19. 45 miles per hour **20.** 25 weeks **21.** 312 pounds **22.** \$7.50

Cumulative Review, p. 419

1. 5740 **2.** three and seventy-five thousandths **3.** $2 \cdot 5 \cdot 11 \cdot 41$ **4.** 79 **5.** 26 **6.** 210 **7.** 60
8. $\frac{1}{16}$ **9.** $2\frac{5}{16}$ **10.** $14\frac{2}{3}$ **11.** $3\frac{3}{5}$ **12.** $3\frac{1}{4}$ **13.** $38\frac{1}{4}$ **14.** 57 **15.** 0.036
16. (a) six hundred and six thousandths (b) six hundred six thousandths **17.** 1639.438 **18.** 16.172
19. 4077.36 **20.** 0.79 **21.** (a) 9.35×10^7 (b) 4.82×10^{-7} **22.** $x = 2\frac{1}{2}$ **23.** $A = 0.32$
24. (a) 45.5 miles per hour (b) 318.5 miles **25.** \$2600

CHAPTER 7

Margin Exercises 7.1

1. 9% **2.** 1.25% **3.** 125% **4.** $6\frac{1}{4}\%$ **5.** 4% **6.** 135% **7.** 93.6% **8.** 0.12 **9.** 0.182
10. 0.007

Exercises 7.1, p. 427

1. 60% **3.** 65% **5.** 100% **7.** 20% **9.** 15% **11.** 53% **13.** 125% **15.** 336% **17.** 0.48%
19. 2.14%

21. (a) $\frac{\$36 \text{ profit}}{\$200 \text{ invested}} = \frac{2 \cdot 18}{2 \cdot 100} = \frac{18}{100} = 18\%$ (b) $\frac{\$51 \text{ profit}}{\$300 \text{ invested}} = \frac{3 \cdot 17}{3 \cdot 100} = \frac{17}{100} = 17\%$; Investment (a) is better.
23. (a) $\frac{\$150 \text{ profit}}{\$1500 \text{ invested}} = \frac{15 \cdot 10}{15 \cdot 100} = \frac{10}{100} = 10\%$ (b) $\frac{\$200 \text{ profit}}{\$2000 \text{ invested}} = \frac{20 \cdot 10}{20 \cdot 100} = \frac{10}{100} = 10\%$; The investments give the same percent of profit.
25. 2% **27.** 10% **29.** 36% **31.** 40% **33.** 2.5% **35.** 5.5% **37.** 110% **39.** 200% **41.** 0.02 **43.** 0.18 **45.** 0.3 **47.** 0.0026 **49.** 1.25 **51.** 2.32 **53.** 0.173 **55.** 0.132 **57.** 0.0725 **59.** 0.085 **61.** 28% **63.** 0.45 **65.** 6.5% **67.** 1.33

The Recycle Bin, p. 430

1. $R = 80$ **2.** $R = 66\frac{2}{3}$ (or 66.7 to the nearest tenth) **3.** $A = 115$ **4.** $A = 45$ **5.** $B = 40$ **6.** $B = 74$

Margin Exercises 7.2

1. 65% **2.** 150% **3.** 25% **4.** 37.5% **5.** 7.5% **6.** 77.8% (or $77\frac{7}{9}\%$ or $77.\bar{7}\%$) **7.** $\frac{4}{5}$ **8.** $\frac{4}{25}$ **9.** $\frac{11}{200}$ **10.** $2\frac{7}{20}$

Exercises 7.2, p. 437

1. 3% **3.** 7% **5.** 50% **7.** 25% **9.** 55% **11.** 30% **13.** 20% **15.** 80% **17.** 26% **19.** 48% **21.** 12.5% **23.** 87.5% **25.** $55\frac{5}{9}\%$ (or 55.6%) **27.** $42\frac{6}{7}\%$ (or 42.9%) **29.** $63\frac{7}{11}\%$ (or 63.6%) **31.** $107\frac{1}{7}\%$ (or 107.1%) **33.** 105% **35.** 175% **37.** 137.5% **39.** 210% **41.** $\frac{1}{10}$ **43.** $\frac{3}{20}$ **45.** $\frac{1}{4}$ **47.** $\frac{1}{2}$ **49.** $\frac{3}{8}$ **51.** $\frac{1}{3}$ **53.** $\frac{33}{100}$ **55.** $\frac{1}{400}$ **57.** 1 **59.** $1\frac{1}{5}$ **61.** $\frac{3}{1000}$ **63.** $\frac{5}{8}$ **65.** $\frac{3}{400}$ **67.** (a) 0.55 (b) 55% **69.** (a) $\frac{7}{4}$ (or $1\frac{3}{4}$) (b) 175% **71.** (a) $\frac{21}{200}$ (b) 0.105 **73.** 4% **75.** (a) 8.3% (b) 25% (c) 6.25% **77.** 0.143%

The Recycle Bin, p. 440

1. $7200 **2.** 216 feet **3.** 15 gallons **4.** (a) 24 pounds (b) 3 bags

---

Margin Exercises 7.3

1. 12 **2.** 500 **3.** 22%

Exercises 7.3, p. 445

1. 9 **3.** 7.5 **5.** 24 **7.** 47 **9.** 250 **11.** 900 **13.** 62 **15.** 46 **17.** 150% **19.** 30% **21.** 150% **23.** $33\frac{1}{3}\%$ **25.** 17.5 **27.** 160 **29.** 230% **31.** 120 **33.** 620 **35.** 200% **37.** 33.6 **39.** 76.5 **41.** 62.1 **43.** 105 **45.** 160 **47.** 66.5% **49.** $16\frac{2}{3}\%$ **51.** 152,429 **53.** 23,220

Margin Exercises 7.4

1. 13.7 **2.** 146 **3.** 120% **4.** 59.4 **5.** 90

Exercises 7.4, p. 455

1. 7 **3.** 9 **5.** 9 **7.** 36 **9.** 150 **11.** 700 **13.** 75 **15.** 42 **17.** 150% **19.** 20% **21.** 50%

23. $33\frac{1}{3}\%$ **25.** 12.5 **27.** 110 **29.** 180% **31.** 72 **33.** 520 **35.** 200% **37.** 16.32 **39.** 58.5
41. 43.92 **43.** 72 **45.** 80 **47.** 40 **49.** 10 **51.** 37.5 **53.** 70 **55.** 76.3 **57.** 6800 **59.** 44%
Check Your Number Sense **61.** (c) **63.** (d) **65.** (c) **67.** (c) **69.** (c)

Exercises 7.5, p. 463

1. (a) $465 (b) $15,035 **3.** (a) $6.93 (b) $122.38 **5.** (a) $750 (b) $600 (c) $636
7. (a) $10.53 (b) 60% **9.** $4.30 **11.** $1.95 **13.** $5.70; $44.10 **15.** $29.90
17. (a) the paperback (b) $4.12 **19.** $32,800 **21.** (a) $161 (b) $5474

Exercises 7.6, p. 473

1. $5700 **3.** $1310 **5.** $28,000 **7.** 153 **9.** (a) $50 (b) 20% (c) 16.7% (or $16\frac{2}{3}\%$)
11. (a) $1875 (b) $2025 **13.** $120,000 **15.** (a) 88% (b) 250 (c) 30 **17.** 770,000 **19.** 423%

Chapter 7 Test, p. 481

1. 101% **2.** 0.3% **3.** 17.3% **4.** 32% **5.** 237.5% **6.** 0.7 **7.** 1.8 **8.** 0.093 **9.** $1\frac{3}{10}$
10. $\frac{71}{200}$ **11.** $\frac{43}{500}$ **12.** 5.6 **13.** $33\frac{1}{3}\%$ **14.** 30 **15.** 294.5 **16.** 20.5 **17.** 33.1% **18.** (c)
19. (d) **20.** (b) **21.** (c) **22.** (a) $1.05 (b) $3.75 (c) $2.70 **23.** (a) $62.50 (b) 25% (c) 20%
24. (a) $1200 (b) $24 **25.** $3300

Cumulative Review, p. 483

1. identity **2.** commutative property of addition **3.** (a) 13,000 (b) 3.097 **4.** 10,500 (estimate); 10,358 (sum)
5. 4000 (estimate); 3825 (difference) **6.** 100 (estimate); 93 (quotient) **7.** 8000 (estimate); 6552 (product) **8.** 3
9. 4, 9, 25, 49, 121, 169, 289, 361, 529, 841 **10.** (a) LCM = 4200 (b) $56 \cdot 75$, $60 \cdot 70$ **11.** $\frac{3}{20}$
12. $\frac{5}{2}$ (or $2\frac{1}{2}$) **13.** 9.475 (or $9\frac{19}{40}$) **14.** $\frac{11}{18}$ **15.** $x = 22$ **16.** 78 **17.** 400
18. (a) 5.63×10^8 (b) 9.42×10^{-5} **19.** (a) 3450 mm (b) 160 cm (c) 83 cm (d) 5.2 km
20. (a) $5\sqrt{7}$ (b) $6\sqrt{3}$ **21.** 25 feet **22.** (a) $100 (b) $33\frac{1}{3}\%$ (c) 25% **23.** (a) $4.20 (b) $31.75
24. (a) $25,000 (b) $21,300 (c) $4260 (d) 25%

CHAPTER 8

Margin Exercises 8.1

1. $210 **2.** $112.50 **3.** 12%

Exercises 8.1, p. 491

1. $30 **3.** $90 **5.** $26.67 **7.** $2500 **9.** 1 year **11.** $32 **13.** $20.25 **15.** $11.25 **17.** $32.67
19. $730 **21.** (a) $463.50 (b) $36.50 **23.** $37,500 **25.** 72 days
27. (a) $7.50 (b) 60 days (c) 18% (d) $100 **29.** 10% **31.** $460,000 **33.** 1 year
35. (a) $17.50 (b) 60 days (or 2 months) (c) 11.5% (d) $1133.33

Margin Exercises 8.2

1. $22.84 **2.** $6594.66

Exercises 8.2, p. 503

1. (a) \$13,261.30 (b) \$13,527.86 **3.** (a) \$306.82 (b) \$5306.82 (c) \$6.82 **5.** (a) \$339.08 (b) \$343.87
7. \$406.73 **9.** \$600
11. (a) \$634.13 (b) No. It will be less. (c) More frequent compounding gives higher interest.
13. (a) \$6863.93 (b) about \$20 (c) \$21.47 **15.** (a) \$3153.49 (b) definitely not the same. \$2234.08
17. (a) \$1638.62 (b) \$638.62 **19.** (a) \$3297.33 (b) \$1297.33

Exercises 8.3, p. 509

1. (a) \$40,625 (b) \$121,875 (c) \$44,462.50 **3.** (a) 2% (b) \$19,010
5. (a) \$24,000 (b) \$72,000 (c) \$1080 (d) \$25,915 **7.** (a) \$1405 (b) about 30.4% **9.** (a) \$738 (b) 41%

Exercises 8.4, p. 513

1. \$1099.20 **3.** \$4820 **5.** \$2101.40 **7.** \$1938
9. No. Her total expenses would be \$493 which would be about 23.5% of her income.

Chapter 8 Test, p.517

1. \$67.50 **2.** \$800 **3.** $\frac{1}{2}$ year (or 6 months) **4.** 14% **5.** (a) \$1061.37 (b) \$1346.86 **6.** \$2.96
7. \$13,863 **8.** \$518.90 **9.** (a) \$24,000 (b) \$960 **10.** \$25,810 **11.** (a) \$1090 (b) 36.3% (or $36\frac{1}{3}$%)
12. \$94,007.50

Cumulative Review, p. 519

1. 261,075 **2.** 300.004 **3.** (a) 16.94 (b) 500 **4.** 0.4 **5.** 0.525 **6.** 180% **7.** 0.015 **8.** $\frac{4}{3}$ (or $1\frac{1}{3}$)
9. $\frac{41}{11}$ (or $3\frac{8}{11}$) **10.** $\frac{8}{105}$ **11.** $10\frac{2}{15}$ **12.** $46\frac{5}{12}$ **13.** 12 **14.** $\frac{9}{5}$ (or $1\frac{4}{5}$) **15.** 5,600,000 **16.** 398.988
17. 75.744 **18.** 0.01161 **19.** 80.6 **20.** 65 **21.** 2, 3, 4 **22.** $2^2 \cdot 3^2 \cdot 11$ **23.** 210 **24.** 0
25. 7 and 24; 8 and 21 **26.** 50 **27.** 18.5 **28.** 250% **29.** $x = \frac{3}{8}$ **30.** 325 **31.** \$150 **32.** \$4536
33. (a) \$10,199.44 (b) \$10,271.35 **34.** 22.4 centimeters **35.** (a) 7.38×10^7 (b) 4.5×10^{-4}

CHAPTER 9

Margin Exercises 9.1

1. (a) −5 (b) +12 **2.** −4 −3 −2 −1 0 1 2 3 **3.** ••• −5 −4 −3 −2 −1 0 1 2 3 4 5 ••• **4.** true
5. false: $-4 < 4$ (or $-4 \leq 4$) **6.** true **7.** (a) 15 (b) 9 (c) 4 **8.** true **9.** false: $|-38| < |-39|$

Exercises 9.1, p. 531

1. 0 1 2 3 4
3. −4 −3 −2 −1 0 1 2
5. −11 −10 −9 −8 −7 −6
7. −6 −5 −4 −3 −2 −1 0 1 2 3 4 5 6
9. 0 1 2 3 4 5 6 7 8 9 10 11
11. 0 1 2 3 4 5 6 7 8 9 •••
13. −1 0 1 2 3 4
15. ••• −5 −4 −3 −2 −1 0 1

17. 10 **19.** -14 **21.** 0 **23.** 6 **25.** 24 **27.** 20 **29.** 0 **31.** $>$ **33.** $<$ **35.** $=$ **37.** $>$
39. true **41.** false: $21 = |-21|$ **43.** false: $-6 < |-6|$ **45.** false: $-4 \leq 4$ **47.** true **49.** false: $0 = |0|$

Margin Exercises 9.2

1. -16 **2.** 10 **3.** -13 **4.** -35 **5.** -7 **6.** 4 **7.** -20 **8.** -15

Exercises 9.2, p. 537

1. 2 **3.** 10 **5.** 19 **7.** -9 **9.** 0 **11.** -4 **13.** 2 **15.** -16 **17.** -4 **19.** 0 **21.** 4
23. 14 **25.** -1 **27.** -15 **29.** 41 **31.** 13 **33.** 10 **35.** -12 **37.** -10 **39.** 0
41. sometimes **43.** sometimes **45.** never **47.** $-14{,}625$ **49.** 11,700

Margin Exercises 9.3

1. $+5$ **2.** -7 **3.** -4 **4.** $+1$ **5.** -6 **6.** 5 **7.** -20 **8.** -12

Exercises 9.3, p. 545

1. -16 **3.** $+45$ **5.** $+8$ **7.** 3 **9.** 11 **11.** -7 **13.** -9 **15.** 4 **17.** -10 **19.** 1 **21.** 18
23. 17 **25.** 0 **27.** 0 **29.** -26 **31.** 30 **33.** 4 **35.** -9 **37.** 24 **39.** -29 **41.** 8 **43.** 6
45. 10 **47.** -4 **49.** 6 **51.** -6 **53.** -8 **55.** -18 **57.** 0 **59.** 4 **61.** -21 **63.** -5
65. 0 **67.** $<$ **69.** $=$ **71.** 5° **73.** $+54$ yards **75.** -2418 **77.** $-838{,}350$

Margin Exercises 9.4

1. -18 **2.** 16 **3.** 0 **4.** -8 **5.** 0 **6.** -72 **7.** -2 **8.** 20 **9.** 0 **10.** -18 **11.** undefined
12. 2

Exercises 9.4, p. 553

1. -15 **3.** 24 **5.** -20 **7.** 28 **9.** -50 **11.** -21 **13.** -48 **15.** 63 **17.** 0 **19.** 90
21. 24 **23.** 60 **25.** -42 **27.** -45 **29.** -60 **31.** 0 **33.** -1 **35.** 16 **37.** -84 **39.** 300
41. -3 **43.** -2 **45.** 4 **47.** 5 **49.** -5 **51.** -3 **53.** 2 **55.** 4 **57.** -3 **59.** -8 **61.** 2
63. 1 **65.** undefined **67.** 0 **69.** undefined **71.** -14 **73.** -14 **75.** -23

Margin Exercises 9.5

1. $11x$ **2.** $-7y$ **3.** $7x - 2$ **4.** $5n - 1$ **5.** 0 **6.** -1 **7.** 21 **8.** -26

Exercises 9.5, p. 563

1. $8x$ **3.** $6x$ **5.** $-7a$ **7.** $-12y$ **9.** $-9x$ **11.** $-8x$ **13.** $10x$ **15.** $-2c$ **17.** $-2n$
19. $2x^2 + 2x$ **21.** $3x^2 + 20$ **23.** $4x + 7$ **25.** $5x - 1$ **27.** $3ab + 6a + 2b$ **29.** $x^2 - 5x + 6$
31. $-3n - 2$ **33.** $-x^2 + 2x - 1$ **35.** $10x^2 + 28x - 76$ **37.** 0 **39.** -22 **41.** -10 **43.** 4
45. -32 **47.** 11 **49.** 39

Margin Exercises 9.6

1. $13x$ **2.** $n - 45$ **3.** a number decreased by 45 **4.** the quotient of 50 and a number **5.** $x = -4$
6. $y = -3$ **7.** $x = 12$ **8.** $x = -7$ **9.** $n = 1$ **10.** $x = -9$ **11.** $x = 5$ **12.** $-6 = y$

Exercises 9.6, p. 571

1. $n + 5$ **3.** $n + 9$ **5.** $\frac{x}{9}$ **7.** $x - 13$ **9.** $13 - x$ **11.** $2y + 4$ **13.** $8y + 6$ **15.** $\frac{n}{2} - 18$
17. $3 - 2n$ **19.** $6x + 4x$ **21.** the product of 5 and a number **23.** 6 less than a number

25. the sum of a number and 491 **27.** the quotient of a number and 17 **29.** one more than twice a number **31.** the quotient of a number and 3 increased by 20 **33.** 15 times a number decreased by 15 **35.** a number subtracted from 13 **37.** 5 **39.** 6 **41.** Write the equation. Subtract 10 from both sides. (or add −10 to both sides.) Simplify. Divide both sides by 3. Simplify. **43.** $x = 18$ **45.** $n = -16$ **47.** $y = 6$ **49.** $n = 224$ **51.** $x = -6$ **53.** $-6 = y$ **55.** $n = 24$ **57.** $x = 24$ **59.** $38 = x$ **61.** $x = -10$ **63.** $-2 = n$ **65.** $x = -7$ **67.** $n = 19$ **69.** $x = -10$

Exercises 9.7, p. 579

1. The sum of a number and 5 is equal to 16. **3.** Four times a number decreased by 8 is equal to 20. **5.** The quotient of a number and 3 increased by 1 is equal to 13. **7.** Translation of (a): $x - 12$; Equation: $x - 12 = 16$; Solution: $x = 28$ **9.** Translation of (a): $3y - 4$; Equation: $3y - 4 = 26$; Solution: $y = 10$ **11.** 11 **13.** 17 **15.** 5 **17.** 100 **19.** −4 **21.** 15 **23.** 1 **25.** 6 inches **27.** 13 millimeters **29.** 100° **31.** 6 feet **33.** 21.2 inches **35.** 3.5 centimeters **37.** $60,000 **39.** $210,069 **41.** 135,618 acres **43.** $x = 312.5$ miles; This is not a "better deal."

Chapter 9 Test; p. 589

1. 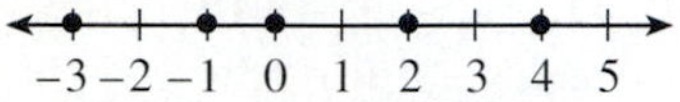

2. 11 **3.** (a) true (b) false: $|-15| > |-12|$ **4.** 8 **5.** −12 **6.** −9 **7.** −16 **8.** 16 **9.** 28 **10.** −90 **11.** 8 **12.** undefined **13.** −5 **14.** (a) $3x + 5$ (b) $6x^2 - 3x + 5$ **15.** (a) $5x - y - 7$ (b) 4 **16.** $x = 7$ **17.** $x = -5$ **18.** $6 = y$ **19.** $n = 1$ **20.** 15 **21.** 10 **22.** −3 **23.** 10 feet **24.** 14 centimeters

Cumulative Review, p. 591

1. $\frac{15}{16}$ **2.** $\frac{1}{6}$ **3.** $336\frac{29}{48}$ **4.** $189\frac{1}{2}$ **5.** $16\frac{4}{5}$ **6.** 766.234 **7.** 5.48 **8.** undefined **9.** $\frac{5}{8}$ **10.** 219.8 **11.** (c) **12.** (d) **13.** (c) **14.** (b) **15.** 200 **16.** 50 **17.** 7349.25 **18.** 3 km **19.** 86 mm **20.** $10\sqrt{6}$ **21.** $5\sqrt{10}$ **22.** 9.35×10^7 **23.** 2.47×10^{-3} **24.** $13,498.26 **25.** 354 miles **26.** (a) $500 (b) $66\frac{2}{3}\%$ (c) 40% **27.** $57\frac{1}{7}$ miles **28.** $x = -6$ **29.** $n = 10$ **30.** −5

MEASUREMENT

Margin Exercises M.1

1. 5000 mm **2.** 160 cm **3.** 83 000 mm **4.** 0.07 m **5.** 0.032 cm **6.** 3500 cm **7.** 35 000 mm **8.** 64 mm **9.** 0.0059 km **10.** 0.008 km **11.** 2200 mm^2 **12.** 5 cm^2 **13.** 370 cm^2 = 37 000 mm^2 **14.** 52 cm^2 = 0.52 dm^2 **15.** 360 m^2 **16.** 73 a = 7300 m^2 **17.** 540 a = 54 000 m^2 **18.** 495 a = 4.95 ha

Exercises M.1, p. M13

1. kilo-, hecto-, deka-, deci-, centi-, milli- **3.** 500 cm **5.** 600 cm **7.** 300 mm **9.** 1400 mm **11.** 18 mm **13.** 350 mm **15.** 160 dm **17.** 210 cm **19.** 5000 m **21.** 6400 m **23.** 2.6 cm **25.** 4.8 cm **27.** 1.2 cm **29.** 0.03 m **31.** 0.256 m **33.** 0.32 m **35.** 1.7 m **37.** 2.4 km **39.** 0.4 km **41.** 462 cm **43.** 0.052 m **45.** 6410 mm **47.** 0.5 cm **49.** 0.057 m **51.** 35 km **53.** 2300 m **55.** 0.0015 km **57.** 560 mm^2 **59.** 361 mm^2 **61.** 0.28 cm^2 **63.** 200 cm^2 **65.** 730 cm^2 = 73 000 mm^2 **67.** 60 cm^2 = 6000 mm^2 **69.** 290 dm^2 = 29 000 cm^2 = 2 900 000 mm^2 **71.** 780 m^2 **73.** 4 m^2 **75.** 869 a = 86 900 m^2 **77.** 16 a = 1600 m^2 **79.** 0.01 ha **81.** 500 ha **83.** 30 ha

Margin Exercises M.2

1. 43 000 mg **2.** 7800 mg **3.** 0.35 g **4.** 75 000 g **5.** 3.94 t **6.** 0.002 L **7.** 3600 L **8.** 0.5 L **9.** 500 cm^3 **10.** 4.2 kL

Exercises M.2, p. M25

1. 2000 mg **3.** 3.7 t **5.** 5.6 kg **7.** 0.091 t **9.** 700 mg **11.** 5000 kg **13.** 2000 kg
15. 896 000 mg **17.** 75 kg **19.** 7 000 000 g **21.** 0.000 34 kg **23.** 0.016 g **25.** 0.0923 kg
27. 7580 kg **29.** 2.963 t
31. 1 cm^3 = 1000 mm^3; 1 dm^3 = 1000 cm^3; 1 m^3 = 1000 dm^3; 1 km^3 = 1 000 000 000 m^3
33. 1 m = 10 dm; 1 m = 100 cm; 1m^2 = 100 dm^2; 1 m^2 = 10 000 cm^2; 1 m^3 = 1000 dm^3; 1 m^3 = 1 000 000 cm^3
35. 73 000 dm^3 **37.** 0.000 525 m^3 **39.** 8 700 000 cm^3 **41.** 0.045 cm^3 **43.** 0.000 019 dm^3
45. 2000 cm^3 **47.** 5300 mL **49.** 0.03 L **51.** 48 000 L **53.** 0.29 kL **55.** 80 000 mL = 80 000 cm^3

Margin Exercises M.3

1. 24 in. **2.** $\frac{2}{3}$ ft **3.** 8 qt **4.** 6000 lb **5.** $\frac{3}{4}$ hr **6.** 50.8 cm **7.** 91.4 m **8.** 32.8 ft **9.** 12.4 mi
10. 38.7 cm^2 **11.** 58.125 m^2 **12.** 538.2 ft^2 **13.** 7.41 acres **14.** 2.838 L **15.** 18.925 L **16.** 6 in.^{3}
17. 176.575 ft^3

Exercises M.3, p. M35

1. 60 in. **3.** 18 in. **5.** $2\frac{1}{2}$ ft **7.** 7 ft. **9.** 2 mi **11.** 2 yd **13.** 32 fl oz **15.** 6 pt **17.** 20 qt
19. 5 gal **21.** 48 oz **23.** 5000 lb **25.** 300 min **27.** 90 min **29.** 72 hr **31.** 300 sec **33.** 4 min
35. 2 days **37.** 77° F **39.** 10° C **41.** 0° C **43.** 2.742 m **45.** 96.6 km **47.** 124 mi **49.** 19.7 in.
51. 19.35 cm^2 **53.** 55.8 m^2 **55.** 83.6 m^2 **57.** 405 ha **59.** 741 acres **61.** 53.82 ft^2 **63.** 4.65 in.^{2}
65. 9.46 L **67.** 10.6 qt **69.** 4.54 kg **71.** 453.6 g **73.** $\frac{3}{4}$ acre **75.** (a) 1728 in.^{2} (b) $1\frac{1}{3}$ yd^2

Exercises M.4, p. M41

1. $\frac{3}{10}$ mL **3.** 60 mL **5.** 4.8 tsp **7.** 2.7 gtt **9.** 20 fluidrams **11.** $1\frac{1}{2}$ glassfuls **13.** 5 gtt **15.** 6 tsp
17. 3 tbsp **19.** 150 gtt **21.** 15 mL ampicillin **23.** 2 mL castor oil **25.** 4 mL Robitussin®
27. $\frac{2}{3}$ mL Polaramine® **29.** $7\frac{1}{2}$ mL castor oil

---

Chapter Test, p. M45

1. 0.07 **2.** 1500 **3.** 0.173 **4.** 300 **5.** 0.017 **6.** 103 000 **7.** 0.042 **8.** 6000 **9.** 40000
10. 0.0016 **11.** 2.7 **12.** 2 **13.** 2000 **14.** 0.175 **15.** 33 **16.** $\frac{39}{2} = 19\frac{1}{2}$ **17.** $\frac{37}{6} = 6\frac{1}{6}$ **18.** $\frac{2}{3}$
19. 76 **20.** 2600 **21.** $\frac{27}{8} = 3\frac{3}{8}$ **22.** $0.4 = \frac{2}{5}$ **23.** 192 **24.** 58 **25.** $2.8 = 2\frac{4}{5}$ **26.** $3.5 = 3\frac{1}{2}$
27. $25.6 = 25\frac{3}{5}$ **28.** $\frac{19}{2} = 9\frac{1}{2}$ **29.** $\frac{13}{16}$ **30.** $\frac{11}{8} = 1\frac{3}{8}$ **31.** 75 mL **32.** 20° C

Cumulative Review, p. M47

1. 7 **2.** 1.15 **3.** 0.81 **4.** 46 **5.** $2000 **6.** $10,798.61 **7.** $2907.04 **8.** 20% **9.** 130
10. 38.775 **11.** 12,040 **12.** (a) $630 (b) 25% **13.** 405 miles **14.** 12 fl oz for $1.09 is better.
15. (a) about $18 (b) $45 **16.** 19 hr 3 min **17.** 2 yd 2 ft 8 in.
18. (a) 356 mm (b) 0.1872 m (c) 19 500 mm^2 (d) 540 mL (e) 0.078 m^3 **19.** $16\sqrt{2}$
20. (a) $6\sqrt{10}$ in. (b) 18.97 in. **21.** (a) 8.9×10^8 (b) 6.32×10^{-5} **22.** (a) −21 (b) −16
23. (a) $x = -7$ (b) $y = 9.7$ **24.** 11

STATISTICS AND READING GRAPHS

Exercises S.1, p. S9

1. (a) 79 (b) 84 (c) 38 **3.** (a) 74 (b) 74 (c) 43 **5.** (a) 83.8 (b) 83.5 (c) 12
7. (a) 12,000 (b) 12,000 (c) 6000 **9.** (a) 32 (b) 30 (c) 40 **11.** (a) 18.5 (b) 14.9 (c) 22.0
13. (a) 85 (b) 90 (c) 45

Exercises S.2, p. S17

1. (a) Faculty salaries: \$15,525,000; Administration salaries: \$6,900,000; Student programs: \$1,725,000; Savings: \$1,380,000; Non-teaching salaries: \$4,485,000; Maintenance: \$3,450,000; Supplies: \$1,035,000 (b) \$8,625,000 (25%) (c) 22% (d) \$7,590,000
3. (a) \$500 (b) Gas: 10%; Repairs: 16%; Loan payment: 60%; Insurance: 14% (c) 7:5 (or 7 to 5)
5. (a) city C; 20 (b) 50 (c) city C; 12 (d) 66% men; 34% women
7. (a) 1988; 80 boats (b) 40 boats (c) 60 boats

Exercises S.3, p. S25

1. (a) Social science (b) Chemistry & phys. and Humanities (c) about 3280 (d) about 21%
3. (a) Sue (b) Bob and Sue (c) about 86% (d) Bob and Sue: Bob and Sue: yes, in most cases (e) no, since their vertical scales represent two different types of quantities.
5. (a) July (b) March and December (c) about 16.6%
7. (a) wheat (b) \$1400 (c) steel (d) \$1000 (e) steel (f) \$2000

Exercises S.4, p. S35

1. (a) 5 (b) 5000 (c) 3rd class (d) 40 (e) 15,000.5 and 20,000.5 (f) 100 (g) 5% (h) 35%
3. (a) 3rd class (b) 15 (c) 120 (d) 50 (e) 35 (f) 54%

Chapter Test, p. S41

1. (a) 2 (b) 2 (c) 4 **2.** (a) 10.16 (b) 8.89 (c) 17.78 **3.** \$9900 **4.** \$33
5. No; they will be short by \$228. **6.** \$1,666,667 **7.** 62.5% **8.** 49.1% **9.** (a) 5 (b) 10 (c) 12
10. 52% **11.** 60% **12.**

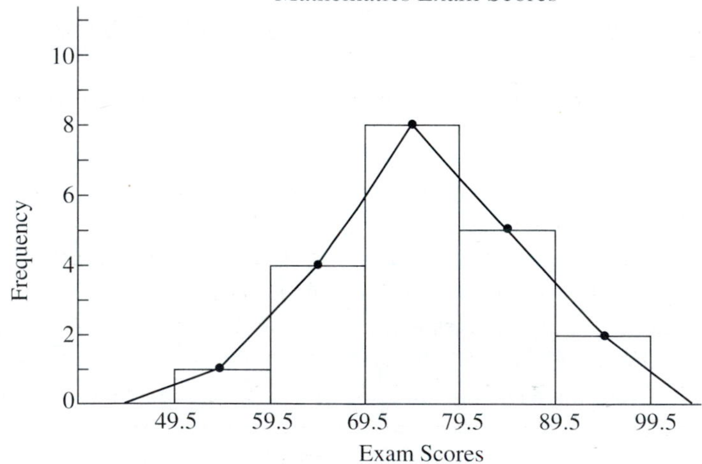

Cumulative Review, p. S45

1. 357.042 **2.** 28.50 **3.** 1.4 **4.** 9.433 **5.** $13\sqrt{3}$ **6.** 3465 **7.** 52 **8.** (a) 58 (b) 58.5 (c) 43
9. (a) \$6919.96 (b) about 11.5 years **10.** 6 months (0.5 year) **11.** 37.5% **12.** 290 **13.** 4352.32
14. 375

15. Food: \$9000; Housing: \$11,250; Transportation: \$6750; Miscellaneous: \$6750; Savings: \$2250; Education: \$4500; Taxes: \$4500
16. $x = -2.45$ **17.** $y = 10$ **18.** (a) \$33,000 (b) \$99,000 (c) \$36,110 **19.** (a) $4\sqrt{5}$ ft (b) 8.9 ft
20. 19 hr 3 min **21.** 2 yd 2 ft 8 in. **22.** (a) 350 mm (b) 3500 mm^2 (c) 0.762 kg (d) 16 300 mL
23. (a) 120 in. (b) 1440 in.^{2} (c) 27 ft^2 (d) 130,680 ft^2 **24.** (a) \$90 (b) \$60 (c) 40% (d) $28\frac{4}{7}$% (or 28.6%)
25. (a) -12 (b) -160 (c) 5 (d) 0

GEOMETRY

Margin Exercises G.1

1. 25.12 in. **2.** 14.4 cm

Exercises G.1, p. G7

1. (a) B (b) C (c) D (d) A (e) F (f) E **3.** 104 mm **5.** 31.4 ft **7.** 19.468 yd **9.** 13.5 cm
11. 16 000 m **13.** 20.8 in. **15.** 11.9634 cm **17.** 35.98 in. **19.** 27.42 m **21.** 60 cm **23.** 7.14 km

Margin Exercises G.2

1. 28.26 in.^{2} **2.** 25 cm^2

Exercises G.2, p. G15

1. (a) C (b) B (c) D (d) A (e) F (f) E **3.** 6 cm^2 **5.** 78.5 yd^2 **7.** 227.5 in.^{2} **9.** 500 mm^2 (or 5 cm^2)
11. 28.26 ft^2 **13.** 21.195 yd^2 **15.** 106 cm^2 **17.** 76.93 in.^{2} **19.** 21.87 m^2 **21.** 15.8 km^2
23. 0.785 ft^2 **25.** 2500 cm^2 and 0.25 m^2

Margin Exercise G.3

1. 267.95 ft^3

Exercises G.3, p. G21

1. (a) C (b) E (c) D (d) B (e) A **3.** 1695.6 in.^{3} **5.** 904.32 ft^3 **7.** 80 cm^3 **9.** 9106 dm^3
11. 113.04 in.^{3} **13.** 3456 in.^{3}

Exercises G.4, p. G29

1. (a) 75° (b) 87° (c) 45° (d) 15° **3.** (a) a right angle (b) an acute angle (c) an acute angle **5.** 25°
7. 40° **9.** 55° **11.** 110° **13.** m $\angle A = 55°$; m $\angle B = 50°$; m $\angle C = 75°$
15. m $\angle L = 118°$; m $\angle M = 92°$; m $\angle N = 112°$; m $\angle O = 102°$; m $\angle P = 116°$;
17. (a) straight angle (b) right angle (c) acute angle (d) obtuse angle
19. (a) 150° (b) yes: $\angle$ 3 and $\angle$ 2 are supplementary (c) $\angle$ 1 and $\angle$ 2; $\angle$ 2 and $\angle$ 3; $\angle$ 3 and $\angle$ 4; $\angle$ 4 and $\angle$ 1
21. m $\angle 2 = 150°$; m $\angle 3 = 30°$; m $\angle 4 = 150°$
23. m $\angle 3$ = m $\angle 5$; m $\angle 2$ = m $\angle 5$; m $\angle 1$ = m $\angle 4$; m $\angle 1$ = m $\angle 6$; m $\angle 4$ = m $\angle 6$

Exercises G.5, p. G39

1. scalene (and obtuse) **3.** right (and scalene) **5.** isosceles (and acute) **7.** isosceles (and right)
9. acute (and scalene) **11.** equilateral (and acute) **13.** $x = 50°$; $y = 60°$ **15.** $x = 20°$; $y = 100°$
17. The triangles are not similar. The corresponding sides are not proportional.
19. The triangles are similar. m $\angle ACB$ = m $\angle ECD$ because the angles are vertical angles. Since the sum of the measures of the angles of each triangle must be 180°, m $\angle B$ = m $\angle D$ and the corresponding angles have the same measures. Therefore, $\triangle ABC \sim \triangle EDC$.

21. Yes, $\triangle PQR \sim \triangle PST$. m $\angle QPR$ = m $\angle SPT$ because the angles are vertical angles. Since we are given that m $\angle Q$ = m $\angle S$ and the sum of the angles of each triangle must be 180°, the corresponding angles have the same measures and the triangles are similar.
23. (a) Yes: $3^2 + 4^2 = 5^2$ and $6^2 + 8^2 = 10^2$. (b) $\overline{AC}$ and $\overline{DF}$ are the hypotenuses. (c) The triangles are similar because the corresponding sides are proportional. (d) m $\angle A$ = m $\angle D$, m $\angle B$ = m $\angle E$, m $\angle C$ = m $\angle F$

Chapter Test, p. G45

1. (a) right angle (b) obtuse angle (c) $\angle AOB$ and $\angle DOE$ (or $\angle BOD$ and $\angle AOE$) (d) $\angle AOB$ and $\angle AOE$; $\angle AOB$ and $\angle BOD$ (or $\angle AOE$ and $\angle DOE$; $\angle DOE$ and $\angle DOB$)

2. (a) 55° (b) 165° **3.** (a) 18.84 in. (b) 28.26 in.^{2} **4.** (a) 16.8 m (b) 12 m^2 **5.** (a) $40\frac{1}{3}$ in. (b) $99\frac{1}{6}$ in.^{2}

6. 153.86 in.^{2} **7.** 10 cm^2 **8.** 785 cm^3 **9.** 16 dm^3
10. 10. (a) isosceles (and acute) (b) obtuse (and scalene) (c) right (and scalene) (d) scalene (and acute)
11. (a) $x = 30°$ (b) obtuse (and scalene) (c) $\overline{RT}$ **12.** $x = 2.5$
13. m $\angle BOA$ = m $\angle DOC$ because they are vertical angles and we are given that m $\angle B$ = m $\angle D$. Since the sum of the measures of the angles of each triangle must equal 180°, it follows that m $\angle A$ = m $\angle C$. Therefore, the corresponding angles are equal and the triangles are similar.
14. $x = 3$; $y = 4$

Cumulative Review, p. G47

1. (c) **2.** (a) **3.** (d) **4.** (b) **5.** four hundred twenty-three and eighty-five hundredths **6.** 200
7. 7349.25 **8.** 260 (estimate); 304.06 (sum) **9.** 2000 (estimate); 1523.1 (difference)
10. 120 (estimate); 118.26 (product) **11.** undefined **12.** 0 **13.** −3,500,000 **14.** 14.64 **15.** −42
16. 22 **17.** 219.8 **18.** $-\frac{7}{30}$ **19.** 140 cm^2 **20.** 87.92 in. (or 7.37 ft) **21.** (a) 30 mm (b) 42 mm^2
22. 12.5% **23.** $7505.64 **24.** $650.00 **25.** $57\frac{1}{7}$ (or 57.14 miles) **26.** (a) $15\sqrt{2}$ ft (b) 21.21320344 ft

ADDITIONAL TOPICS FROM ALGEBRA

Margin Exercises T.1

1. −2 1 **2.** 3

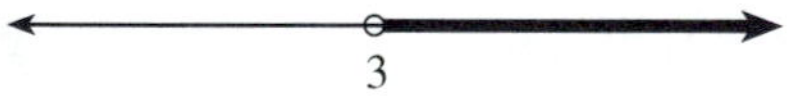

3. $-3 \le x < 2$; half-open interval **4.** $x > 4$ 4

5. $x \ge \frac{1}{2}$ $\frac{1}{2}$ 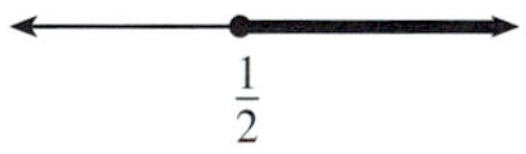 **6.** $-1 < y \le 2$ −1 2

Exercises T.1, p. T11

1. In words: Three is not equal to negative three. True
3. In words: Negative thirteen is greater than negative one. False. Correction: $-13 < -1$.
5. In words. The absolute value of negative seven is less than positive five. False. Correction: $|-7| > +5$.
7. In words: Negative one half is less than negative three fourths. False. Correction: $-\frac{1}{2} > -\frac{3}{4}$.
9. In words: Negative four is less than negative six. False. Correction: $-4 > -6$.
11. $-1 < x < 2$; open interval **13.** $x < 0$; open interval **15.** $x \ge -1$; half-open interval
17. open interval −2 −1 0 **19.** closed interval −3 0 5
21. half-open interval 4 5 6 7 **23.** open interval −12 −7

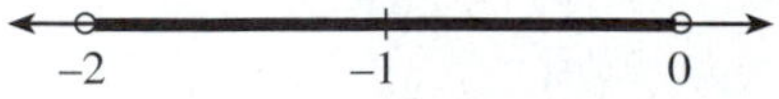

25. half-open interval (0)

27. half-open interval ($\frac{2}{3}$)

29. $x > 4$ (4)

31. $y \leq -3$ (−3)

33. $y > -\frac{6}{5}$ ($-\frac{6}{5}$)

35. $x < \frac{11}{7}$ ($\frac{11}{7}$)

37. $-5 > x$ (−5)

39. $x \leq 5$ (5)

41. $x > -6$ (−6)

43. $x \leq -1$ (−1)

45. $x > -1$ (−1)

47. $x < -5$ (−5)

49. $1 \leq x \leq 4$ (1, 4)

51. $\frac{1}{3} \leq x \leq 2$ ($\frac{1}{3}$, 2)

53. $-3 \leq y \leq 1$ (−3, 1)

55. $-5 \leq x < -3$ (−5, −3)

Exercises T.2, p. T19

1. $\{(-5, 0), (-3, 1), (-1, 0), (1, 1)\ (2, 0)\}$ **3.** $\{(-3, -2), (-2, -1), (-2, 1), (0, 0), (2, 1), (3, 0)\}$

5. $\{(-3, -4), (-3, 3,), (-1, -1), (-1, 1), (1, 0)\}$ **7.** $\{(-1, -5), (0, -4), (1, -3), (2, -2), (3, -1), (4, 0)\}$

9. $\{(-1, 4), (0, 3), (1, 2), (2, 1), (3, 0), (4, -1), (5, -2)\}$

11. **13.** **15.**

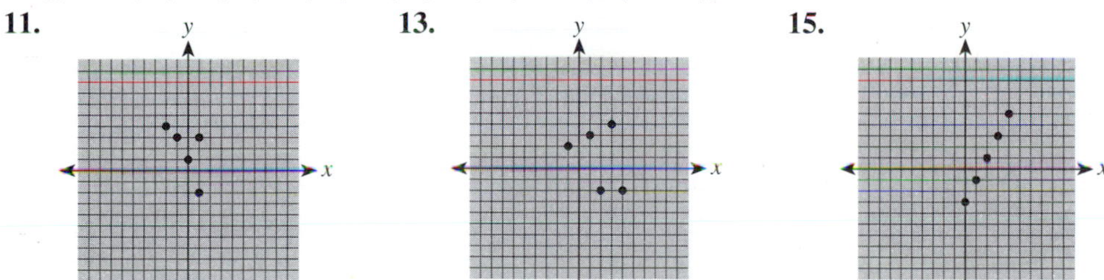

17. **19.** **21.**

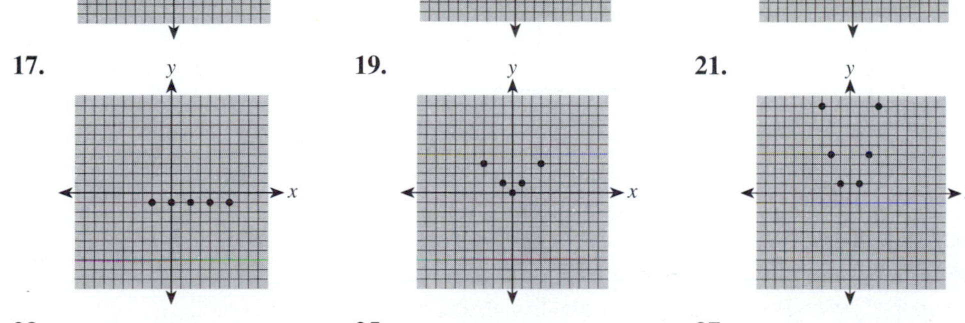

23. **25.** **27.**

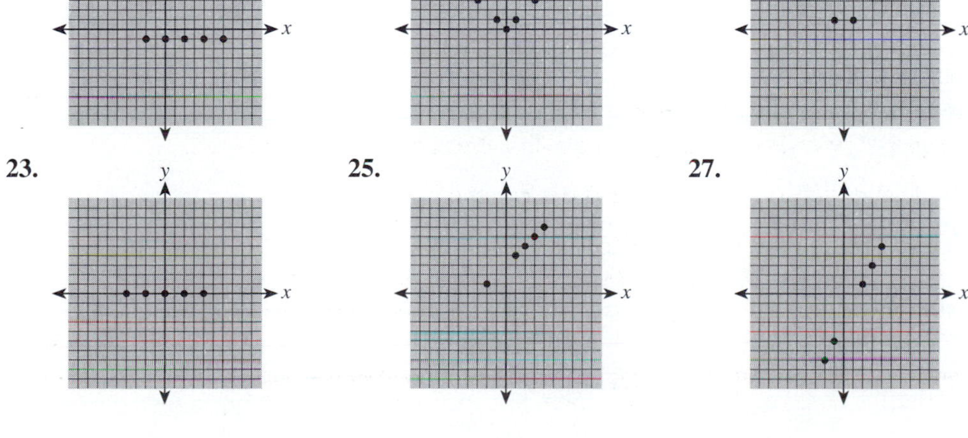

29.

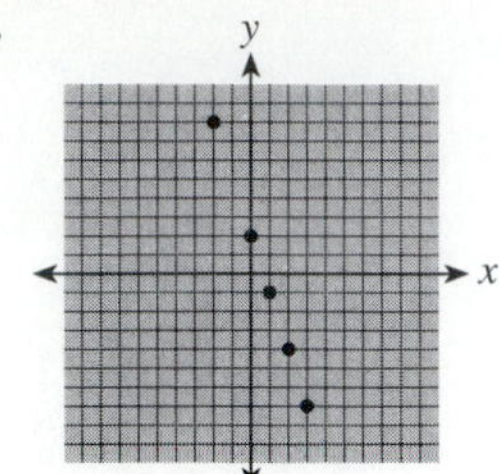

31.

P	I
100	12
200	24
300	36
400	48
500	60

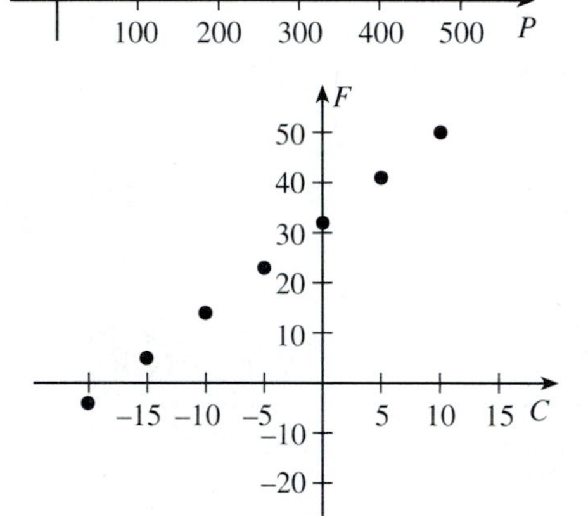

33.

C	F
−20	−4
−15	5
−10	14
−5	23
0	32
5	41
10	50

Exercises T.3, p. T29

1.

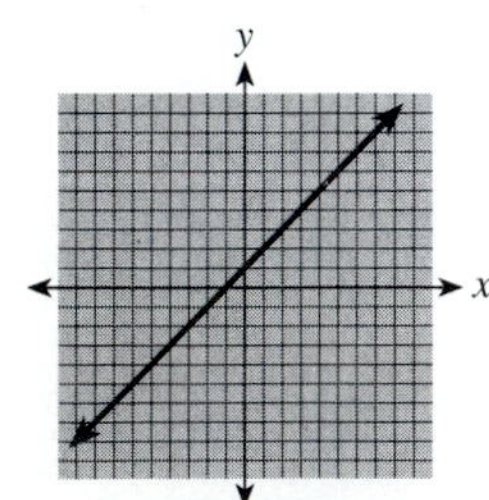

3.

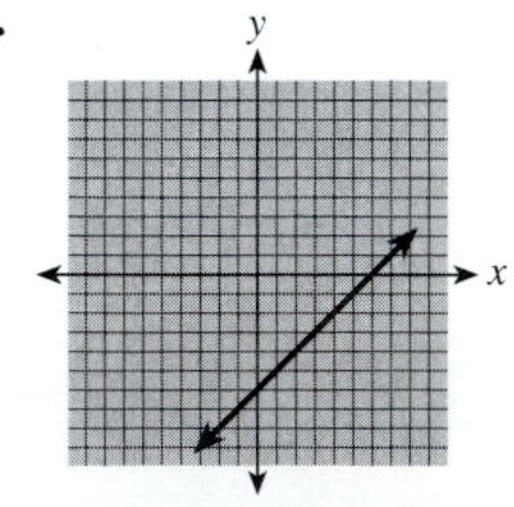

5.

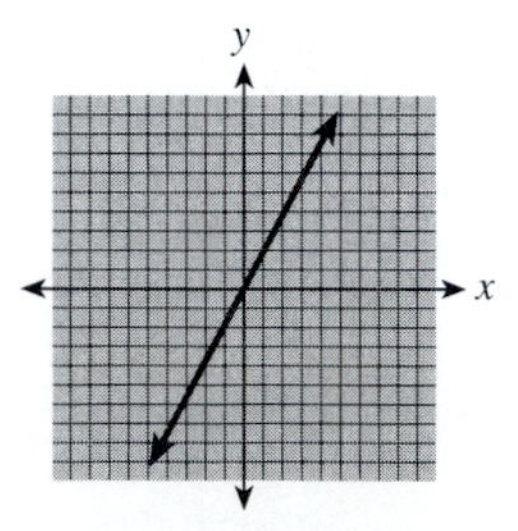

7.

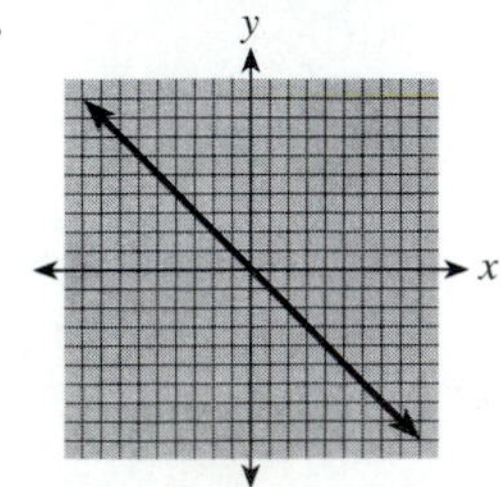

9.

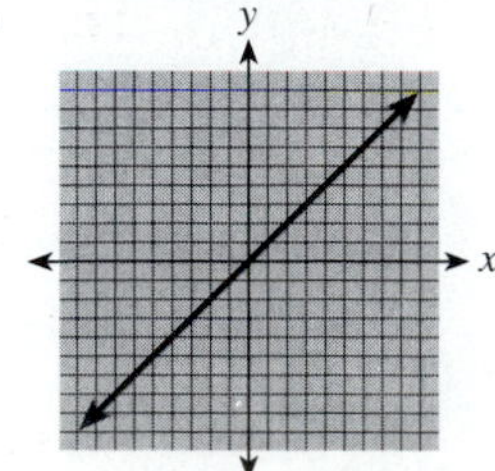

11.

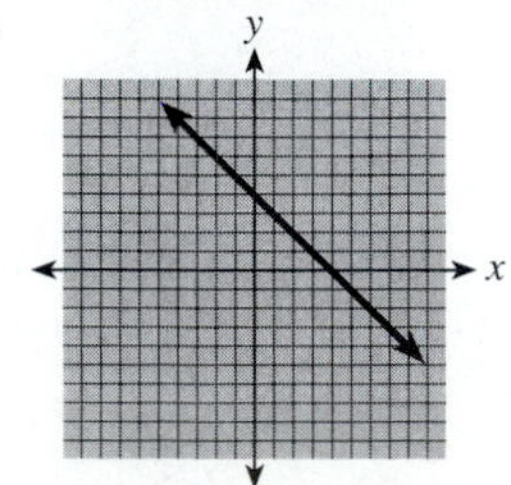

13.
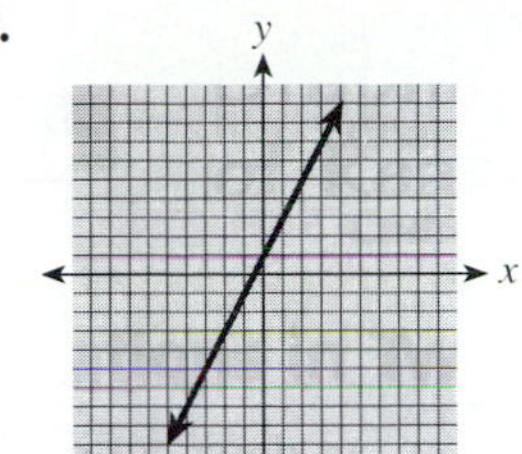

15.
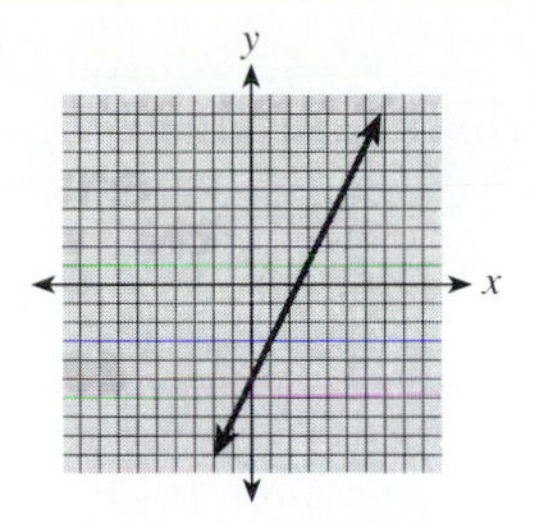

17.
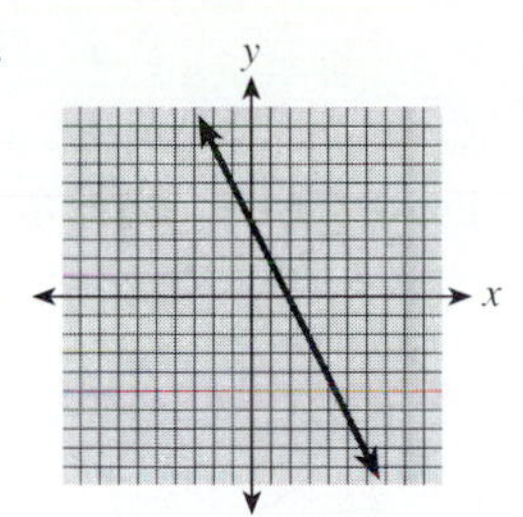

19.
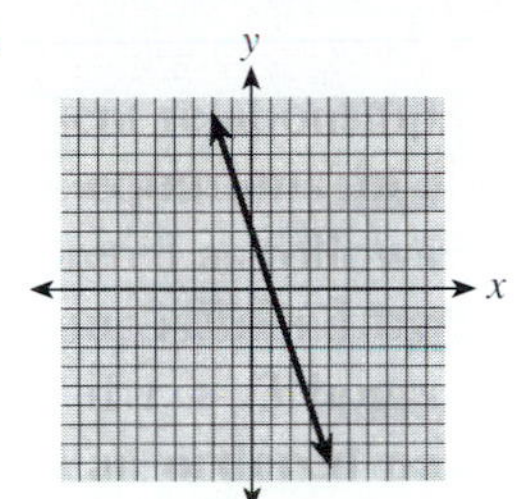

21.
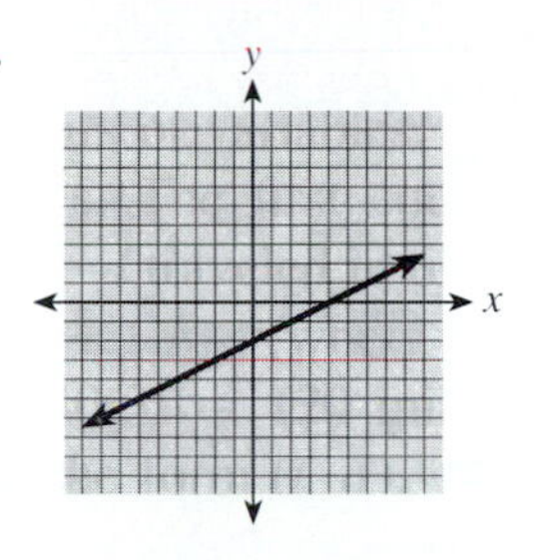

23.
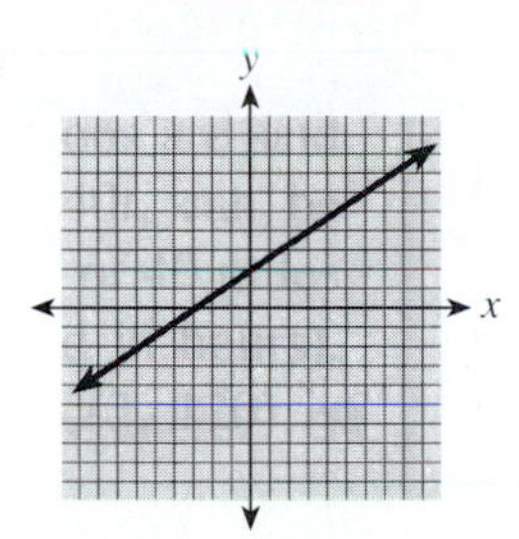

25.
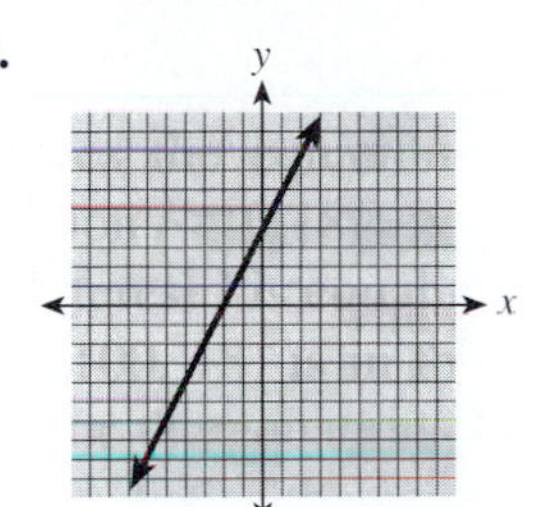

27.
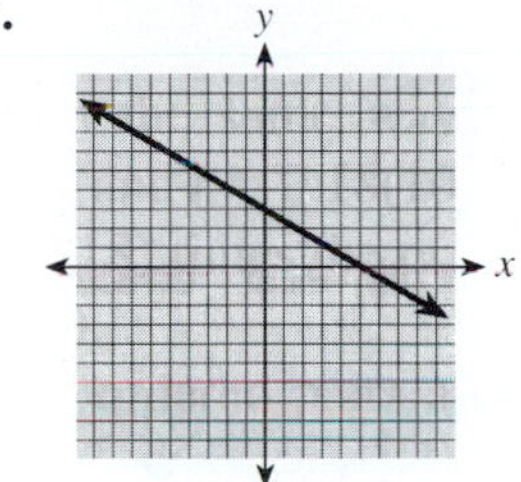

29.
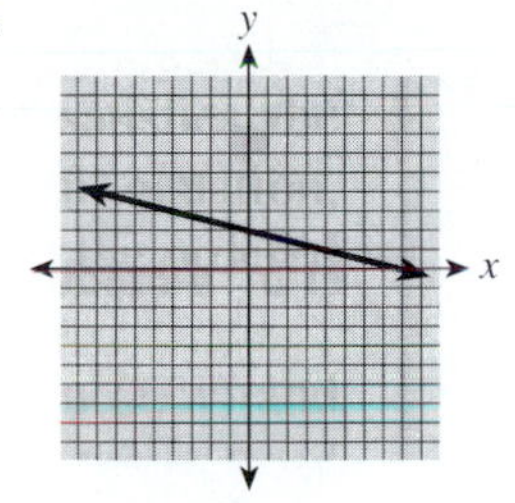

31.
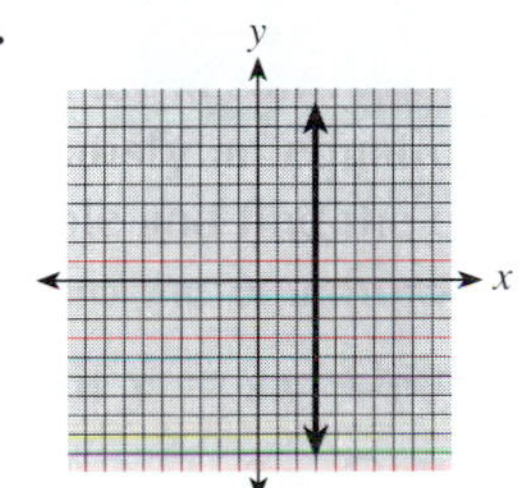

33.
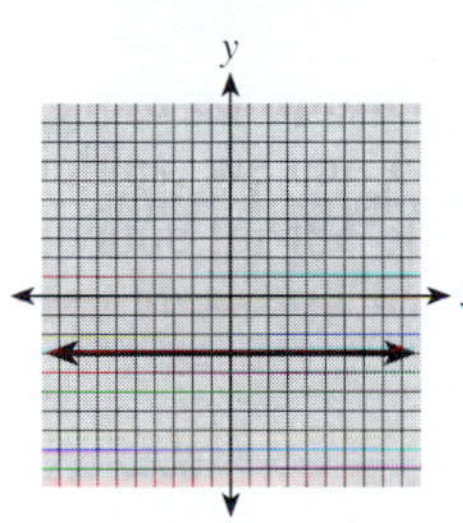

35.
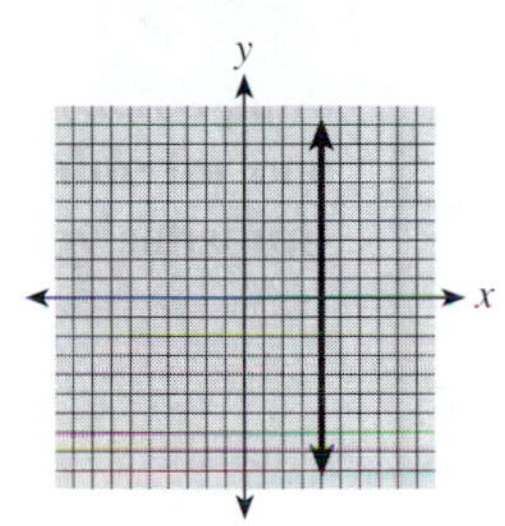

37.
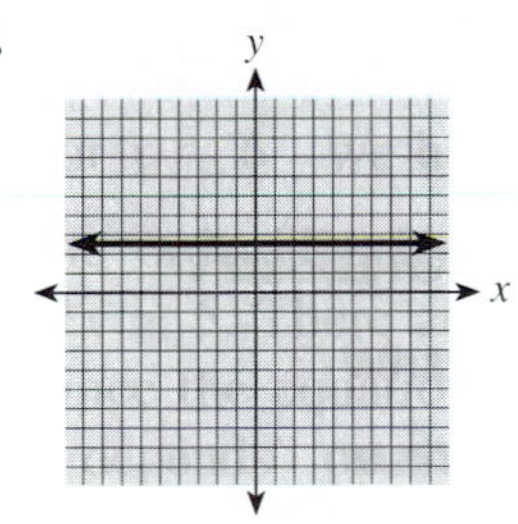

39.
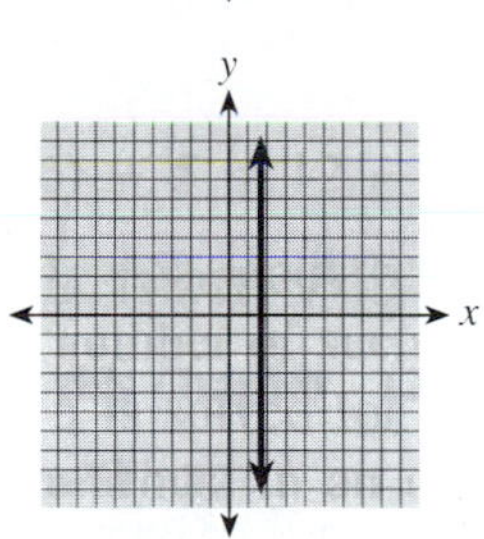

Margin Exercises T.4

1. $3x^2 - 3x + 6$ **2.** $4x^4 - 11x^3 - 2x^2 - 2x - 2$ **3.** $-4x^2 + 6x^2 + 14x - 14$ **4.** $3x^3 - 20x - 9$

Exercises T.4, p. T41

1. $11x - 7$; binomial; first-degree **3.** $2x^2 + 6x + 2$; trinomial; second-degree **5.** $5a^2$; monomial; second-degree **7.** $-3y^2 + 7y + 7$; trinomial; second-degree **9.** $5x^3 - 3x + 17$; trinomial; third-degree **11.** $5x - 10$ **13.** $2x^2 - 4$ **15.** $4y^3 + 3y^2 + 2y - 9$ **17.** $x - 2$ **19.** $a^2 + 5a$ **21.** $12x^3 + x^2 - 6x$ **23.** $-2x^2 + 7x - 8$ **25.** $3x^3 - x + 6$ **27.** $5x^2 - 2x - 5$ **29.** $3x^3 + 5x^2 + 13x - 6$

Margin Exercises T.5

1. 8^3 (or 512) **2.** $(-2)^7$ (or -128) **3.** $-10x^6$ **4.** $15a^3 + 45a^2 + 10a$ **5.** $-14x^5 + 56x^4 + 21x^3 - 35x^2$
6. $2x^2 - 17x - 9$ **7.** $2x^3 - 11x^2 - 38x + 5$ **8.** $10x^4 + 32x^3 + 24x^2 - 35x - 42$

Exercises T.5, p. T49

1. $12x^4$ **3.** $18a^3$ **5.** $8y^3 + 4y^2 + 8y$ **7.** $-4x^3 + 2x^2 - 3x + 5$ **9.** $-14x^4 + 21x^3 - 84x^2$
11. $x^2 + 7x + 10$ **13.** $x^2 - 7x + 12$ **15.** $a^2 + 4a - 12$ **17.** $7x^2 - 20x - 3$ **19.** $4x^2 - 9$ **21.** $x^2 - 9$
23. $x^2 - 2.25$ **25.** $9x^2 - 49$ **27.** $x^4 + 4x^3 + 5x^2 + 23x + 12$ **29.** $2x^4 - 4x^3 + x^2 - 9x + 14$
31. $3x^4 + x^3 - 18x^2 - 36x - 10$ **33.** $x^6 + 3x^5 - 4x^4 - 7x^3 + 8x^2 + 3x - 3$
35. $x^5 + 2x^4 - 4x^3 - 12x^2 - 17x + 6$ **37.** $x^8 + 2x^6 - 7x^4 + 8x^2 - 5$
39. $x^6 + 2x^5 + 3x^4 + 4x^3 + 3x^2 + 2x + 1$

Chapter Test, p. T55

1. $x > 4$, open interval **2.** closed interval

3. $-1 \le x$ **4.** $x < 2$

5. $\{(-4, 1), (-3, 0), (-2, -1), (0, 0), (1, 1), (2, 0)\}$ **6.** $\{(-1, 1), (0, 1), (1, 1), (2, 1)\}$

7. **8.** **9.** **10.**

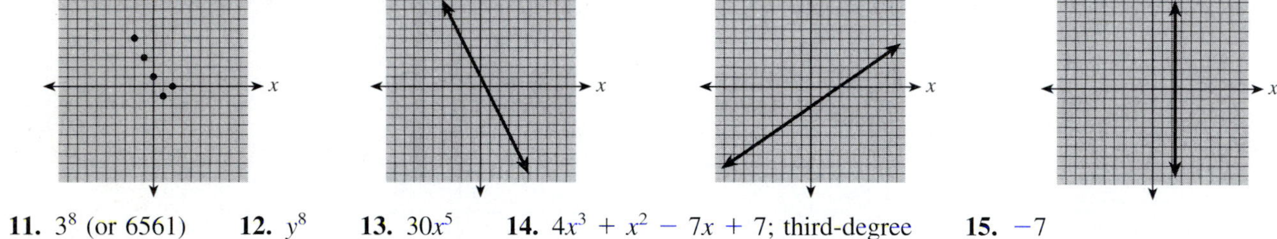

11. 3^8 (or 6561) **12.** y^8 **13.** $30x^5$ **14.** $4x^3 + x^2 - 7x + 7$; third-degree **15.** -7
16. $-y^3 + 5y^2 - 2y - 18$ **17.** $4x + 27$ **18.** $-2a^2 - 4a$ **19.** $8x^2 - 3x - 14$ **20.** $15x^3 + 10x^2 - 20.5x$
21. $8x^5 - 16x^4 + 10x^3$ **22.** $a^2 + 12a + 36$ **23.** $8y^2 - 10y - 33$ **24.** $5x^4 + 6x^3 - 23x^2 + 22x - 8$
25. $2x^4 - 9x^2 + 15x + 7$

Cumulative Review, p. T59

1. $x = 1\frac{13}{18}$ **2.** $y = 30\frac{13}{40}$ **3.** $t = 23.69$ **4.** $w = 9.068$ **5.** $\frac{7}{36}$ **6.** $1\frac{1}{2}$ **7.** 74.875 **8.** 4.47

9. $x = 72{,}000{,}000$ **10.** $y = \frac{1}{7}$ **11.** $q = 16$ **12.** $p = -1.9684$ **13.** $\frac{5}{6}$ **14.** $7\frac{1}{2}$ **15.** 0.608

16. undefined **17.** $2^3 \cdot 3^2 \cdot 11$ **18.** Yes. The factors of 45 are in the product. **19.** 5990 **20.** 723.1

21. 22 **22.** $\frac{5}{36}$ **23.** $3\sqrt{5} + 20\sqrt{3}$ **24.** $-|-7| < -(-7)$ **25.** $2(n + 6) - 5$ **26.** $x = -5$

27. $-12 = x$ **28.** $x = 18$ **29.** $\frac{1}{4} > x \ge -\frac{1}{2}$

30. $5\sqrt{5}$ in. (or 11.18 in.) **31.** $\{(-1, 0), (0, 2), (1, 4), (2, 2), (3, 0)\}$

32.

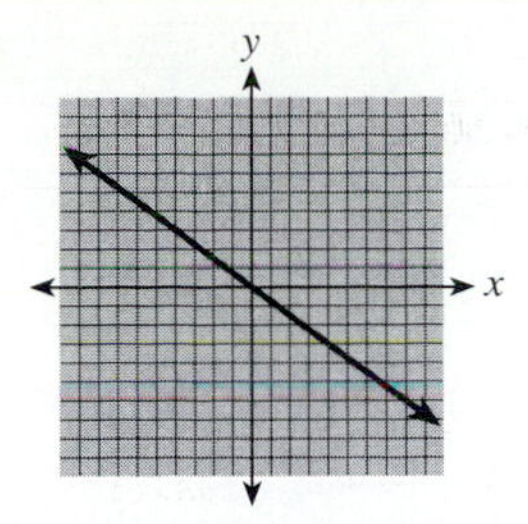

33.

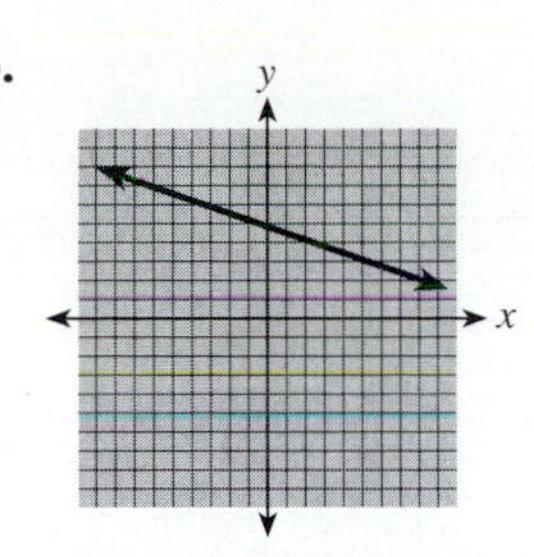

34. 36 **35.** 210% **36.** (a) 17.8 ft (b) 15.6 ft^2 **37.** (a) 125.6 in. (b) 1256 in.^2 **38.** 225 mi
39. (a) $1969 (b) $419 **40.** $41 **41.** -113 **42.** 3.45×10^6 **43.** (a) 45 (b) 36 (c) 72
44. 12 hr 2 min 10 sec **45.** 4 yd 8 in. **46.** $42x^{12}$ **47.** $-x^3 + 9x^2 - 11x + 23$ **48.** $5x^2 - 4x + 18$
49. $10x^2 - 9x - 9$ **50.** $x^4 - 2x^3 - 5x^2 - 10x - 8$

Index